from micro to macro
quantum systems

a unified formalism
with superselection rules
and its applications

from micro to macro quantum systems

a unified formalism with superselection rules and its applications

K. Kong Wan
*School of Physics and Astronomy,
University of St Andrews,
St Andrews, Scotland, UK*

Imperial College Press

Published by

Imperial College Press
57 Shelton Street
Covent Garden
London WC2H 9HE

Distributed by

World Scientific Publishing Co. Pte. Ltd.
5 Toh Tuck Link, Singapore 596224
USA office: 27 Warren Street, Suite 401-402, Hackensack, NJ 07601
UK office: 57 Shelton Street, Covent Garden, London WC2H 9HE

British Library Cataloguing-in-Publication Data
A catalogue record for this book is available from the British Library.

FROM MICRO TO MACRO QUANTUM SYSTEMS:
A Unified Formalism with Superselection Rules and Its Applications

Copyright © 2006 by Imperial College Press

All rights reserved. This book, or parts thereof, may not be reproduced in any form or by any means, electronic or mechanical, including photocopying, recording or any information storage and retrieval system now known or to be invented, without written permission from the Publisher.

For photocopying of material in this volume, please pay a copying fee through the Copyright Clearance Center, Inc., 222 Rosewood Drive, Danvers, MA 01923, USA. In this case permission to photocopy is not required from the publisher.

ISBN 1-86094-625-9

Printed in Singapore by World Scientific Printers (S) Pte Ltd

This book is dedicated to
my wife Chong and my daughters, Kay and Ray,
for making my life the way life should be.

Preface

Orthodox quantum mechanics is an extremely successful physical theory. The conceptual foundations and the mathematical formalism of the theory are so rich that after 80 years there are still many fundamental issues to be explored. The rapid development of technology leads to the construction and discovery of exciting new physical systems with quantum properties such as free atom Bose-Einstein condensate at very low temperatures and various low dimensional systems, nanostructures and quantum circuits. The advance of nanotechnology opens up the possibility of designing and assembling structures atom by atom. These systems differ from the traditional microscopic systems described by orthodox quantum theory in a number of ways. They could be macroscopic in dimensions, spatially confined in a circuit geometry, and more importantly they may possess classical properties as well as quantum properties. Orthodox quantum mechanics has a very rigid structure making it very difficult to accommodate new and novel properties. A generalization of orthodox quantum mechanics is proposed. We shall present a flexible quantum formalism to provide a unified theory of physical systems, from microscopic and macroscopic quantum to classical. Our aim is not to produce an all embracing general theory in a highly abstract form; the objective is to generalize orthodox quantum theory in a concrete form and to an extent that it can be directly applied to describe a wide range of physical systems.

The basic mathematical language used here is that of Hilbert space and operators. The relationship between classical and quantum quantities is made transparent by adopting a geometric method for quantizing basic physical quantities. The book is divided into four parts.

Part one presents a study of mathematical preliminaries. We shall concentrate on topics which are seldom discussed in conventional exposition of quantum theory. Firstly, there is the geometric language used for quantization. The central ideas involve the definition of vector fields in a manifold as operators, the concept of completeness of vector fields, and the geometric formulation of classical dynamical systems. Secondly, there are the less familiar aspects of operator theory essential for this book. These include a systematic discussion of symmetric operators and their maximal symmetric extensions, local operators, and a study of direct integrals of Hilbert spaces and operators. Spectral functions and spectral measures are presented via their direct link to classical probability functions and classical probability measures. To make this book self-contained, we shall devote two chapters to these discussions. We have kept the mathematics to a minimum, summarizing only those of immediate relevance to physical discussions later on. For a better understanding of the mathematics and its applications, we have included a large number of comments and explicit examples which are numbered and are often referred to in later chapters. This enables us to concentrate on the discussions of physical ideas in subsequent chapters avoiding many interruptions and digressions into mathematical technicalities. The presentation takes the form of brief summaries of definitions and theorems together with comments and examples to demonstrate their relevance. Readers familiar with these mathematical preliminaries can skip most of these discussions and go straight to part two of the book.

To facilitate cross reference in later chapters, we have numbered these definitions, theorems, comments and examples according to the sections they appear in. For instance, we have:

Definition 2.12.5(1) indicates the first definition in §2.12.5, i.e., in Chapter 2, Section 12.5.

§1.6.3E(1) E3 indicates the third example in the first set of examples presented in §1.6.3.

§2.5.1C(2) C4 indicates the fourth comment in the second set of comments presented in §2.5.1.

We shall also follow the convention used in most physics texts of putting a "hat" over a symbol, e.g., $\widehat{A}$, to denote an operator. A new symbol, $\widehat{i}$, to be referred to as *ibar*, is introduced to denote the ratio $i/\hbar$ which appears all too frequently.

Part two presents the main mathematical and theoretical framework for orthodox and generalized theories, in three chapters. Chapter 3 is on orthodox quantum mechanics. After a brief presentation of the postulate on quantum statics, this chapter launches into the quantization problem in order to establish basic quantum observables. The quantization methods introduced here will also be used in later chapters. A postulate on quantum dynamics based on the conventional unitary time evolution is then introduced. This is followed by a discussion on the asymptotic behaviour of quantum dynamics which leads to the concepts of asymptotic localization and separation. A general theory for state preparation is then presented. The chapter ends with a measurement theory. Chapter 4 sets out to generalize orthodox quantum mechanics. The

PREFACE

generalization is based mathematically on direct integrals of Hilbert spaces and operators and physically on the notion of superselection rules. This leads to a unified and flexible theory which reduces to orthodox quantum mechanics as a special case, and is capable of describing non-orthodox systems from macroscopic quantum to classical. Chapter 5 begins a further generalization by incorporating strictly maximal symmetric operators as quantum observables.

Part three investigates the implications and applications of our generalized theory to demonstrate its relevance, in two chapters. Chapter 6 lays the mathematical foundations by reviewing the theory of selfadjoint extensions of symmetric operators. This is followed by a detailed discussion of point interactions essential in the construction of any model theory of quantum systems in a circuit geometry. Chapter 7 applies our generalized theory and point interactions to describe superconducting systems in various circuit geometry, especially those having a Josephson junction. The Josephson equation in superconductivity is seen to be derivable in a rigorous manner within our theory. Strictly maximal symmetric operators are seen to be necessary for certain circuit configurations.

The final part of the book is devoted to some topical issues arising from previous discussions. Chapter 8 investigates Schrödinger's cat states, dynamic and asymptotic decoherence, entanglement, chronological disordering and the formulation of an asymptotically separable quantum mechanics. Chapter 9 presents a path space formulation of quantum mechanics which lends further support for the emergence of superselection rules.

This book is not a comprehensive review of various theories and formulations existing in the vast literature on quantum mechanics. The materials presented in this book reflect the author's view and interest over a number of years working in the foundations of quantum mechanics. These materials have not previously been fully and systematically discussed in an accessible form. The author aims to demonstrate that quantum theory together with its mathematical structure and physical interpretation is capable of restrictions as well as generalizations. It is this flexibility which enables the theory to be so applicable. The richness of the formalism is likely to allow the theory to adapt and to cope with future demands arising from the discoveries of new physical phenomena for years to come.

No attempt is made to present a comprehensive review of other more familiar theories and formalism, since these are fully discussed by many existing monographs. We have not endeavoured to set out a grand scheme to encompass everything such as various interpretations, environmental influences and gravity. Our aim is modest. We shall keep close to the orthodox formalism and we do not claim that the theories presented are universal; they are designed to be applicable to specific types of physical systems. We do believe that the fundamental structure of quantum theory should not be rigid and set in stone. It must be allowed to evolve in order to keep abreast with technological development and the discovery of new physical phenomena. We have therefore devoted much space in this monograph to developing a flexible quantum theory and to the treatment of some typical non-orthodox quantum systems. This may help to demonstrate the relevance of fundamental studies of quantum theory to the understanding and the exploration of a rapidly expanding set of novel quantum systems.

There are monographs on fundamental issues of quantum theory motivated by mathematical or conceptual and philosophical considerations. This book is motivated mainly by physical considerations with an eye on possible applications to non-orthodox quantum systems. Although we have formulated our theories and various concepts in rigorous mathematics terms our primary interest is in understanding and developing applicable physical theories, not mathematics. Our analysis is motivated by simple and intuitive physical ideas. One can appreciate the physical ideas involved, e.g., on state preparation, measurement, asymptotic superselection rules and asymptotic notion of decoherence, without having to delve too deeply into mathematics. However, it is pleasing and reassuring to know that physical ideas can be formulated axiomatically and treated in a mathematically vigorous manner. There is a bibliographical list at the end of each chapter which is for immediate reference and not meant to be complete and exhaustive. The author apologizes for inevitable omissions.

This monograph aims at a readership of theoretical physicists, mathematical physicists, mathematicians and philosophers of science with an interest in the foundations of quantum mechanics and its applications. Hopefully the self-contained nature of the presentation will render this book useful to a wide range of readers.

I am deeply indebted to R. H. Fountain, F. E. Harrison, P. Guest and C. Trueman for their comments on the manuscript which have led to many improvements.

K. Kong Wan
St Andrews

Contents

Preface   vii

I   Aspects of Geometric and Operator Theories

1 Manifolds and Dynamical Systems   3
   1.1 Topological Spaces and Topological Equivalence 4
       1.1.1 Basic concepts and definitions 4
       1.1.2 Topological equivalence 9
   1.2 Euclidean Spaces 11
       1.2.1 Basic concepts and definitions 11
       1.2.2 Coordinate systems and coordinate transformations . . 13
       1.2.3 Contravariant and covariant vectors in $I\!\!E^n$ 15
       1.2.4 Contravariant, covariant and mixed tensors 17
   1.3 Differential Operators, Vectors and Fields 20
       1.3.1 Differential operators and derivations 21
       1.3.2 Tangent vectors, tangent vector fields and their integral curves 26
       1.3.3 Transformation groups and complete vector fields 35
   1.4 Cotangent Vectors and Differential Forms 40
       1.4.1 Cotangent vectors, differentials and one-forms 41
       1.4.2 Tensor fields and two-forms 47
       1.4.3 Exterior differentiation 51
       1.4.4 Interior products, closed and exact forms 53
   1.5 Differentiable Manifolds 56
       1.5.1 Definition and examples 56
       1.5.2 Riemannian manifolds 60
       1.5.3 Hamiltonian manifolds 63
   1.6 Classical Dynamical Systems 67
       1.6.1 Classical systems of finite order 67
       1.6.2 First-order systems 68

	1.6.3 Second-order Hamiltonian systems	69
	1.6.4 Momentum observables, vector fields and operators	73
	1.6.5 Concluding remarks	77
References		78

2 Operators and their Direct Integrals — 81

2.1 Hilbert Spaces — 81
2.2 Operators: Basic Definitions — 87
    2.2.1 Boundedness, adjoints, extensions and restrictions, continuity and closure — 87
    2.2.2 Convergence of a family of bounded operators — 92
    2.2.3 Tensor products of Hilbert spaces and operators — 94
2.3 Types of Operators and their Reductions — 97
2.4 Unitary Operators and Unitary Transforms — 107
2.5 Extensions of Symmetric Operators — 113
    2.5.1 Selfadjoint and maximal symmetric extensions — 113
    2.5.2 Von Neumann's formula for selfadjoint extensions — 128
2.6 Probability and Expectation Values — 130
    2.6.1 Borel sets, measures and measurable functions — 132
    2.6.2 Probability measures and probability functions — 137
    2.6.3 Expectation values, variances and uncertainties — 140
2.7 Spectral Measures and Probability — 142
2.8 Selfadjointness and Spectral Decomposition — 148
    2.8.1 Spectral theorem — 148
    2.8.2 Functions of a selfadjoint operator — 154
    2.8.3 Spectra of selfadjoint operators — 157
    2.8.4 Spectral representation spaces and spectral representations of selfadjoint operators — 161
2.9 Generalized Spectral Measures and Probability — 167
2.10 Spectral Functions of Symmetric Operators — 170
    2.10.1 Symmetric operators and their spectral functions — 170
    2.10.2 Strictly maximal symmetric operators and their spectral functions — 173
    2.10.3 The square of maximal symmetric operators — 175
    2.10.4 Spectra of symmetric operators — 178
2.11 Probability and Operators — 180
    2.11.1 Probability measures, spectral measures and selfadjoint operators — 180
    2.11.2 Probability measures, generalized spectral measures and strictly maximal symmetric operators — 183

CONTENTS   xiii

 2.12 Local Operators in Coordinate Space 185
   2.12.1 Definitions 185
   2.12.2 Localization of bounded operators 187
   2.12.3 Local operator algebras 188
   2.12.4 Localization of unbounded operators 1 191
   2.12.5 Localization of unbounded operators 2 192
   2.12.6 Local momentum and local Hamiltonian 195
 2.13 Direct Integrals of Hilbert Spaces 195
   2.13.1 Discrete composition of Hilbert spaces 196
   2.13.2 Continuous composition of Hilbert spaces 198
 2.14 Direct Integrals of Operators 209
   2.14.1 Direct sums of operators 209
   2.14.2 Direct integrals of operators 213
   2.14.3 Density operators 218
   2.14.4 Statistical operators 221
 2.15 Direct Integrals of Tensor Products 224
   2.15.1 Direct integrals of tensor product Hilbert spaces 224
   2.15.2 Direct integrals and tensor product of operators 225
 References 226

II   Orthodox and Generalized Quantum Mechanics

3 Orthodox Quantum Mechanics   231
 3.1 Introduction 231
   3.1.1 Structure of physical theories 231
   3.1.2 Mathematical framework of quantum mechanics 234
 3.2 Orthodox Quantum Statics 236
   3.2.1 Postulate on orthodox quantum statics 236
   3.2.2 Pure and mixed states 239
   3.2.3 Correlation between states 246
   3.2.4 Discretization of bounded and unbounded observables . 248
   3.2.5 Approximate nature of measurements 250
 3.3 Quantization in $I\!\!E^n$ 252
   3.3.1 Preliminaries on quantization 252
   3.3.2 Failure of general schemes 256
   3.3.3 Complete momentum observables 259
   3.3.4 Observables linear in momenta 269
   3.3.5 Incomplete momentum observables 272
   3.3.6 Kinetic energy and the Hamiltonian 275
   3.3.7 Constraint and quantization in circuit geometry 282

3.4	Orthodox Quantum Dynamics		286
	3.4.1	Postulate on orthodox quantum dynamics	286
	3.4.2	Asymptotic localization and separation: Free systems	290
	3.4.3	Asymptotic localization and separation: Scattering systems	294
3.5	Quantum State Preparation		300
	3.5.1	The problem	300
	3.5.2	Mathematical preliminaries	302
	3.5.3	Ideal particle source	303
	3.5.4	Random particle source	305
	3.5.5	Extension to spin-$\frac{1}{2}$ particles	307
3.6	Quantum Measurement		310
	3.6.1	Local position observables and their measurability	310
	3.6.2	Reduction to local position measurements	313
	3.6.3	Spectral separation for spinless particles	314
	3.6.4	Spectral separation for spin-$\frac{1}{2}$ particles	319
	3.6.5	Local position measurement as an ionization process	320
	3.6.6	A model ionization propagator	324
	3.6.7	Projection postulate, local position measurements and uncertainty relations	328
	3.6.8	Concluding remarks	329
References			331

4 Physical Theory in Hilbert Space 337
4.1	Introduction		337
4.2	Unified Statics in Direct Integral Space		338
	4.2.1	A unified postulate on quantum statics	339
	4.2.2	Discrete and continuous direct integral decompositions	339
4.3	States and Superposition Principle		341
	4.3.1	Regular and singular states, pure and mixed states	341
	4.3.2	Coherence and superposition principle	344
	4.3.3	Superselection rules, their origins and classical observables	345
4.4	Unified Dynamics in Direct Integral Space		351
	4.4.1	Preliminaries	351
	4.4.2	Preserving dynamics	352
	4.4.3	Non-preserving dynamics 1: Motivation	356
	4.4.4	Linear functionals for state description	358
	4.4.5	Extensions and restrictions of linear functionals	361
	4.4.6	Non-preserving dynamics 2: A general scheme	364
	4.4.7	Non-preserving evolution and environments	367

CONTENTS

- 4.5 Classical Systems of Finite Order 368
  - 4.5.1 First-order systems in Hilbert space 368
  - 4.5.2 Second-order Hamiltonian systems in Hilbert space ... 373
- 4.6 Mixed Quantum Systems 378
  - 4.6.1 A model system 378
  - 4.6.2 Classification of physical systems 379
  - 4.6.3 Quantum/Classical divide 1 381
  - 4.6.4 Equilibrium and mixed quantum systems 383
- 4.7 Coupling of Systems of Different Types 384
  - 4.7.1 Measuring devices 384
  - 4.7.2 Coupling of orthodox quantum and classical systems .. 385
  - 4.7.3 Coupling of orthodox and mixed quantum systems ... 388
  - 4.7.4 Coupling of classical and mixed quantum systems ... 390
- 4.8 Concluding Remarks 390
- References 392

5 Generalized Quantum Mechanics — 395
- 5.1 Introduction 395
- 5.2 Maximal Symmetric Operators and Observables 400
  - 5.2.1 Observables: Concept and description 400
  - 5.2.2 Measurement of intrinsically unsharp observables 406
- 5.3 Approximate and Related Observables 407
  - 5.3.1 Approximate observables 407
  - 5.3.2 Related family of observables 408
- 5.4 Implications on Quantization 409
- 5.5 Time Operators and Uncertainty Relation 409
- 5.6 Local Values in Coordinate and in Phase Spaces 413
  - 5.6.1 Expectation values in terms of local values 413
  - 5.6.2 Local values and semi-local observables 415
  - 5.6.3 Local values in generalized phase space 418
- 5.7 Appendix on Maximal Probability Families 420
- 5.8 Appendix on Time Operators 423
- 5.9 Concluding Remarks 425
- References 426

III Point Interactions, Macroscopic Quantum Systems and Superselection Rules

6 Point Interactions — 431
- 6.1 Introduction 431

6.2	Extensions of Symmetric Operators	433
6.3	Extensions of Direct Sum Operators	435
	6.3.1 Direct sums and their selfadjoint extensions	435
	6.3.2 Selfadjoint extensions in terms of boundary conditions	439
6.4	Quantization by Parts and Point Interactions	443
6.5	Classification of Point Interactions in $I\!\!E$	446
	6.5.1 Type 1 (BC1): The step potential	450
	6.5.2 Type 2 (BC2): δ-interaction as high-pass filters	451
	6.5.3 Type 3 (BC3): δ'-interaction as low-pass filters	455
	6.5.4 Type 4 (BC4): Perfect reflector	463
	6.5.5 Type 5 (BC5): Elastic reflectors	464
	6.5.6 Type 6 (BC6): Open end	464
	6.5.7 Type 7 (BC7): Ideal π-phase shifters	465
	6.5.8 Type 8 (BC8): High-pass π-phase shifters	467
	6.5.9 Type 9 (BC9): Low-pass π-phase shifters	469
	6.5.10 Type 10 (BC10): Ideal mid-pass $\frac{1}{2}\pi$-phase shifters	471
	6.5.11 Type 11 (BC11): Partial mid-pass filter	473
	6.5.12 Type 12 (BC12): Ideal tunable phase shifters	476
6.6	Remarks on Quantization by Parts	478
6.7	Charged Particles in Circular Motion	480
	6.7.1 Charged particles constrained to move in a circle	480
	6.7.2 Charged particles in 3-dimensions	486
6.8	Point Interactions in a Circle	489
	6.8.1 Momentum operators	490
	6.8.2 Hamiltonians with reflection symmetry	491
6.9	Classification of Point Interactions in $\mathcal{C}$	495
	6.9.1 Type 1 (BCC1): Free motion	495
	6.9.2 Type 2 (BCC2): δ-interaction	495
	6.9.3 Type 3 (BCC3): δ'-interaction	497
	6.9.4 Type 4 (BCC4): Perfect reflector	498
	6.9.5 Type 5 (BCC5): Elastic reflector	498
	6.9.6 Type 6 (BCC6): Open end	499
	6.9.7 Type 7 (BCC7): Ideal dynamic π-phase shifter	500
	6.9.8 Type 8 (BCC8): Static π-phase shifter	500
	6.9.9 Type 9 (BCC9): Gradient π-phase shifter	501
	6.9.10 Type 10 (BCC10): Ideal $\frac{1}{2}\pi$-phase shifter	502
	6.9.11 Type 11 (BCC11): Static junction correlator	503
	6.9.12 Type 12 (BCC12): Ideal tunable phase shifters	504
6.10	Current and Stationary States in a Circle	505
References		506

7 Macroscopic Quantum Systems — 509

- 7.1 Single-Particle Representation 509
- 7.2 Macroscopic Wave Function Hypothesis 512
- 7.3 Uniformly Thick Superconducting Rings 513
  - 7.3.1 Physical properties 513
  - 7.3.2 Superconducting rings: Preliminaries 514
  - 7.3.3 Superconducting rings as equilibrium mixed quantum systems 520
- 7.4 Superconducting Rings with a Junction 522
  - 7.4.1 Josephson junction and dc Josephson effect 522
  - 7.4.2 Supercurrent and magnetic flux operators 524
  - 7.4.3 The Hamiltonian: Preliminary results 525
  - 7.4.4 Superconducting ring with a Josephson junction as an equilibrium mixed quantum system 528
  - 7.4.5 Superconducting ring with a π-junction 530
  - 7.4.6 Superconducting ring with a $\frac{1}{2}\pi$-junction 530
  - 7.4.7 Superconducting ring with a Josephson junction in an external magnetic field 531
- 7.5 Feynman's Derivation of Josephson's Equation 533
- 7.6 Superconducting Wire with a Junction 535
  - 7.6.1 Point interactions 535
  - 7.6.2 Momentum and supercurrent operators 535
  - 7.6.3 Hamiltonian operator 1: π-junction 536
  - 7.6.4 Hamiltonian operator 2: $\frac{1}{2}\pi$-junction 536
  - 7.6.5 Hamiltonian operator 3: Josephson junction 537
  - 7.6.6 Superconducting wire with a Josephson junction as a mixed equilibrium quantum system 539
- 7.7 Y-Shape Circuits 542
  - 7.7.1 Momentum and supercurrent operators: Special cases . 542
  - 7.7.2 Hamiltonian operators: Special cases 544
  - 7.7.3 Physics of strictly maximal symmetric operators 544
  - 7.7.4 Momentum and supercurrent operators: General cases 545
  - 7.7.5 Hamiltonian operators: General cases 547
  - 7.7.6 Correlation 547
  - 7.7.7 Superselection rules 548
  - 7.7.8 Condensate in a pure or in a mixed state 549
- 7.8 Continuous Y-Shape Circuit 551
- 7.9 Superconducting Quantum Interference Devices 552
- 7.10 Non-Equilibrium Mixed Quantum System 554
- 7.11 BCS Theory and Superselection Rules 558
- 7.12 Conceptual Analyses 562

7.12.1 Non-uniqueness of quantization 562
7.12.2 Y-shape circuits, equilibrium mixed quantum systems and non-locality 562
7.12.3 Equilibrium states, globalization and non-locality 566
7.12.4 Quantum/Classical divide 2 568
7.13 Orthodox Quantum Systems 569
7.14 Prospects and Other Approaches 573
References 575

IV Asymptotic Disjointness, Asymptotic Separability, Quantum Mechanics on Path Space and Superselection Rules

8 Separability and Decoherence 581
8.1 Introduction 581
8.2 Scattering Systems and de Broglie Paradox 586
    8.2.1 Scattering systems 586
    8.2.2 de Broglie paradox 587
8.3 Schrödinger's Cat States 589
    8.3.1 Classical-like states 589
    8.3.2 Classical cats and their states 592
    8.3.3 Quantum cats and their states 593
    8.3.4 Disjointness and Schrödinger's cat states 594
    8.3.5 Scattering systems and Schrödinger's cat states 595
    8.3.6 Quantized oscillator and Schrödinger's cat states 595
    8.3.7 Weak Schrödinger's cat states 597
    8.3.8 Periodic Schrödinger's cat states 599
    8.3.9 Double-well potentials and chiral molecules 601
    8.3.10 Dynamic and asymptotic decoherence 607
8.4 Superconducting Schrödinger's Cat States 608
    8.4.1 Breakdown of superselection rules and capacitive junction 608
    8.4.2 Schrödinger's cat states in superconducting systems . . 618
8.5 Asymptotically Separable Quantum Theory 620
    8.5.1 Motivation 620
    8.5.2 Asymptotically separable quantum mechanics 620
8.6 Entanglement and Decoherence 622
    8.6.1 Distinguishable particles 623
    8.6.2 Identical Fermions and Pauli exclusion principle 626
8.7 Chronological Disordering 629

		8.7.1	The concept of chronological disordering	629
		8.7.2	Two-particle correlation and conservation laws	630
	References			633

9 Quantum Mechanics on Path Space   637

    9.1   Introduction 637
    9.2   Physical Space and Path Space 638
    9.3   Functions on Path Space 644
    9.4   Quantum Mechanics on Path Space 649
          9.4.1   Hilbert spaces $\mathcal{H}_\gamma\big(\mathbf{\Pi}(\mathcal{C})\big)$ on path space $\mathbf{\Pi}(\mathcal{C})$ 649
          9.4.2   Comparing $\mathcal{H}_\gamma\big(\mathbf{\Pi}(\mathcal{C})\big)$ and $L^2(\mathcal{C}_c)$ 651
          9.4.3   Position operators in $\mathcal{H}_\gamma\big(\mathbf{\Pi}(\mathcal{C})\big)$ 654
          9.4.4   Momentum operators in $\mathcal{H}_\gamma\big(\mathbf{\Pi}(\mathcal{C})\big)$ 654
    9.5   Josephson Effect and Superselection Rules 655
    9.6   Concluding Remarks 657
    References 657

    Bibliography 659

    Index 675

Part I
Aspects of Geometric and Operator Theories

Chapter 1

Manifolds and Dynamical Systems

In this chapter we shall introduce some basic topological and geometric concepts used in the studies of differentiable manifolds. In particular we shall present an intrinsic definition of vectors and tensors. These are the basic quantities for the formulation of classical mechanics and classical field theory. They also play an essential role in a geometric approach to quantization, providing a clear link between classical quantities and their quantized counterparts. The notations and materials, including many examples, are selected for their relevance to later physical applications. Mathematical technicality is kept to a minimum. More details on manifold theory are available in the references provided at the end of this chapter.

We shall denote the set of all real numbers by $I\!R$, i.e., $I\!R = (-\infty, \infty)$. The set of all ordered n-tuples of real numbers will be denoted by $I\!R^n$, i.e.,

$$I\!R^n = I\!R \times I\!R \times \cdots \times I\!R \tag{1.1}$$

with elements

$$\alpha = (\alpha^1, \alpha^2, \ldots, \alpha^n), \tag{1.2}$$

where α^j, $j = 1, 2, \ldots, n$, are real numbers. For $I\!R$ which corresponds to $n = 1$ we shall simply write $\alpha = \alpha^1$. Generally we call a set endowed with some kind of geometric or algebraic structure a *space*. There are two well-known structures which can be built on $I\!R^n$ making it respectively into a topological space and a Euclidean space.

1.1 Topological Spaces and Topological Equivalence

1.1.1 Basic concepts and definitions

We shall start by looking into structures arising from within $I\!R^n$. A simple structure is the family of subsets of $I\!R^n$.[1] For simplicity let us take the set $I\!R$ to begin with. Not all subsets of $I\!R$ are the same in nature. First we have the open intervals (a, b), $a, b \in I\!R$ defined by[2]

$$(a, b) = \{\alpha \in I\!R : a < \alpha < b\}. \tag{1.3}$$

Every element in (a, b) is contained in an open interval which is itself contained in (a, b). This enables us to generalize the notion of open intervals to that of open sets in $I\!R$.

Definition 1.1.1(1)   **Open sets in $I\!R$**

- A subset Λ of $I\!R$ is an *open set* in $I\!R$ if every element α in Λ is contained in an open interval Λ_α inside Λ, i.e., $\alpha \in \Lambda \Rightarrow \alpha \in \Lambda_\alpha \subset \Lambda$.

- The complement of a subset Λ of $I\!R$ is the subset which contains all the elements of $I\!R$ that are not in Λ.

- A subset of $I\!R$ is a *closed set* in $I\!R$ if its complement is open.

Comments 1.1.1(1)   **Properties of open and closed sets**

C1   The notation $\Lambda_1 \subset \Lambda_2$ means that $\alpha \in \Lambda_1 \Rightarrow \alpha \in \Lambda_2$, e.g., Λ_1 could be equal to Λ_2.

C2   Open intervals are open sets. Closed intervals are closed sets. A subset containing a single number $\{a\}$ is also closed. Open sets are generalizations of open intervals, e.g., $(1, 2) \cup (3, 4)$ is an open set but it is not an interval. Similarly closed sets are a generalization of closed intervals.

C3   There are sets which are neither open nor closed, e.g., a semi-open interval $(a, b]$.

---

[1] $I\!R^n$ itself is formally included as a subset, and so is the empty set $\emptyset$, which contains no elements.

[2] We adopt the standard notation that $[a, b]$ represents a closed interval and $(a, b]$ a semi-open interval, i.e.,

$$[a, b] = \{\alpha \in I\!R : a \leq \alpha \leq b\}, \quad (a, b] = \{\alpha \in I\!R : a < \alpha \leq b\}.$$

1.1. TOPOLOGICAL SPACES AND TOPOLOGICAL EQUIVALENCE 5

C4 Regarded as a subset $\mathbb{R}$ itself is clearly an open set. The empty set $\emptyset$ formally satisfies the requirement of an open set, since there is no element in $\emptyset$ which is not contained in an open interval in $\emptyset$. So, the empty set shall be regarded as an open set in $\mathbb{R}$. It then follows that $\mathbb{R}$ and $\emptyset$ also formally satisfy the requirement of a closed set, since their respective complements are open. Therefore, $\mathbb{R}$ and $\emptyset$ shall also be regarded as closed sets. These two are the only subsets of $\mathbb{R}$ which are both open and closed.

C5 The set of positive real numbers $(0, \infty)$, denoted by $\mathbb{R}^+$ hereafter, is an open set, and so is the set of negative real numbers $(-\infty, 0)$, denoted hereafter by $\mathbb{R}^-$.

C6 A fundamental property of closed sets is that its points cannot get arbitrarily close to any point outside it, e.g., points in $[a, b]$ cannot get arbitrarily close any point outside $[a, b]$. This is in sharp contrast to the situation for open sets. For example points in (a, b) can get arbitrarily close to a and b which lie outside (a, b).

C7 The fundamental structural differences between open sets and closed sets in $\mathbb{R}$ are:

1. The union of any, possibly *infinite*, number of open sets is an open set, and the intersection of any *finite* number of open sets is also open.

2. The union of any *finite* number of closed sets is closed, and the intersection of any, possibly *infinite*, number of closed sets is also closed.

The fact that the intersection of an infinite number of open sets is not necessarily open is exemplified as follows:

$$(-1, 1) \cap \left(-\frac{1}{2}, \frac{1}{2}\right) \cap \left(-\frac{1}{3}, \frac{1}{3}\right) \cap \left(-\frac{1}{4}, \frac{1}{4}\right) \cap \cdots = \{0\} \quad \text{which is closed.}$$

The fact that the union of an infinite number of closed sets is not always closed is seen in the following example:

$$\{1\} \cup \left\{\frac{1}{2}\right\} \cup \left\{\frac{1}{3}\right\} \cup \left\{\frac{1}{4}\right\} \cup \cdots.$$

This union is not closed because points in the union can get arbitrarily close to 0, which lies outside the union.

C8 Every open set in $\mathbb{R}$ can be shown to be a union of open intervals.

Definition 1.1.1(2)   **Topological structure and topological spaces**[3]

- Let $\mathcal{T}$ be a non-empty set. A collection $\mathbf{C}$ of subsets of $\mathcal{T}$ is called a *topological structure* or a *topology* on $\mathcal{T}$ if:

  1. Both $\mathcal{T}$ and the empty set $\emptyset$ belong to $\mathbf{C}$.
  2. The union of any, possibly infinite, number of sets in $\mathbf{C}$ again belongs to $\mathbf{C}$.
  3. The intersection of any finite number of sets in $\mathbf{C}$ again belongs to $\mathbf{C}$.

- Members of $\mathbf{C}$ are called *open sets* and the set $\mathcal{T}$ together with a topological structure is called a *topological space*.

- A subset of $\mathcal{T}$ is said to be *closed* if its complement is open.

Comments 1.1.1(2)   **Neighbourhoods, closure and dense sets**

C1   The open sets in $I\!R$ introduced earlier in terms of open intervals constitute a topological structure, known as the *standard topology* on $I\!R$. With this topology $I\!R$ becomes a topological space. The structure of open sets in a topological space defined above is a generalization of that of open sets in $I\!R$. From now on we shall always adopt the standard topology on $I\!R$.

C2   Closed sets in a topological space possess the properties of closed sets in $I\!R$ stated in §1.1.1C(1) C7. Also $\mathcal{T}$ and the empty set $\emptyset$ are both closed.

C3   An arbitrary open interval $(a,b) \subset I\!R$ is a topological space in its own right with the standard topology formed by the open sets of $I\!R$ that are subsets of (a,b).

C4   Generally it is possible to single out a different collection of subsets to form a different topological structure; this will result in a different topological space. A trivial example is to take the class $\mathbf{C}$ to consist of just the empty set $\emptyset$ and the entire set $\mathcal{T}$.

C5   An element of a topological space is often referred to as a *point* in the space. A useful concept is that of a *neighbourhood* of a point which is defined to be an open set containing the point.

C6   The terms *closure* and *denseness* are topological concepts useful also in the context of Hilbert spaces:

---

[3]See Lipschutz (1965), Simmons (1963) and Sutherland (1995) for an introduction to topology as a subject.

1.1. TOPOLOGICAL SPACES AND TOPOLOGICAL EQUIVALENCE

- The *closure* of a subset Λ of a toplogical space $\mathcal{T}$, denoted by $\bar{\Lambda}$, is the smallest closed set in $\mathcal{T}$ containing Λ. In other words $\bar{\Lambda}$ is the intersection of all closed sets in $\mathcal{T}$ containing Λ.

- A set Λ is *dense* in $\mathcal{T}$ if its closure $\bar{\Lambda}$ coincides with $\mathcal{T}$.

As an example we can see that the closure of an open interval (a,b) in $I\!\!R$ is the closed interval $[a,b]$. The set of all rational numbers is a dense set in $I\!\!R$. We know that rational numbers permeate every part of $I\!\!R$. Intuitively this is precisely the property of a dense set, i.e., a dense subset Λ of $\mathcal{T}$ is a subset which permeates every part of $\mathcal{T}$ so that the only closed set containing Λ is the space $\mathcal{T}$ itself.

C7 If a and b are two distinct points in $I\!\!R$ then there exists a neighbourhood of a and a neighbourhood of b such that these neighbourhoods are disjoint.[4] This property is not shared by all topological spaces. We can formalize this property as follows:

- A topological space is called a *Hausdorff space* if any two distinct points have disjoint neighbourhoods.

All the topological spaces we shall encounter in this book are Hausdorff.

Definition 1.1.1(3)   **Open sets in $I\!\!R^n$**

- Let $\Lambda_1, \Lambda_2, \ldots, \Lambda_n$ be n open intervals in $I\!\!R$. Then the set $\Lambda_{rec} = \Lambda_1 \times \Lambda_2 \times \cdots \times \Lambda_n$ is called an *open rectangle* in $I\!\!R^n$.

- A subset Λ of $I\!\!R^n$ is called an open set in $I\!\!R^n$ if each point $\alpha \in \Lambda$ is contained in an open rectangle Λ_{rec} inside Λ, i.e., $\alpha \in \Lambda \Rightarrow \alpha \in \Lambda_{rec} \subset \Lambda$.

Comments 1.1.1(3)   **Standard topology on $I\!\!R^n$**

C1   For $I\!\!R^3$, the simplest example of an open rectangle is an *open cube* of width $w \in I\!\!R^+$, e.g.,

$$\Lambda_{cu}(w) = \left\{ \alpha \in I\!\!R^3 : -\frac{w}{2} < \alpha^j < \frac{w}{2}, j = 1, 2, 3 \right\}. \tag{1.4}$$

Its closure is simply the closed cube

$$\bar{\Lambda}_{cu}(w) = \left\{ \alpha \in I\!\!R^3 : -\frac{w}{2} \leq \alpha^j \leq \frac{w}{2}, j = 1, 2, 3 \right\}. \tag{1.5}$$

---

[4]In this chapter the term *disjointness* means having no common elements.

C2 The family of open sets defined above constitutes a topological structure, known as the *standard topology* on $\mathbb{R}^3$. A similar topology can be constructed in $\mathbb{R}^n$ which renders $\mathbb{R}^n$ a topological space.

C3 An open cube is a topological space in its own right, with the topology formed by the open sets of $\mathbb{R}^n$ that are subsets of the cube.

C4 We can also define closed rectangles and closed sets in $\mathbb{R}^n$ in terms of the openness of their complements as we did in $\mathbb{R}$.

Definition 1.1.1(4) **Functions on $\mathbb{R}^n$**

- A mapping $f : \mathbb{R}^n \mapsto \mathbb{R}$, assigning a real number $f(\alpha)$ to each point $\alpha \in \mathbb{R}^n$, is called a *function* on $\mathbb{R}^n$. The set of values $\mathcal{R} = \{f(\alpha) : \alpha \in \mathbb{R}^n\}$ is called the *range* of the function.

Comments 1.1.1(4) **Domain, support and smoothness of functions**

C1 We can also define *local functions*. Let $\mathcal{D}$ be an open subset of $\mathbb{R}^n$. A map $f : \mathcal{D} \mapsto \mathbb{R}$ is called a *local function* and $\mathcal{D}$ is called the *domain* of the function. By allowing $\mathcal{D} = \mathbb{R}^n$ all functions defined so far may be called local. Conversely a local function f can be extended to the entire space $\mathbb{R}^n$ by setting, for example $f(\alpha) = 0$ for every $\alpha \notin \mathcal{D}$. But from now on, a function means a function defined on the domain $\mathcal{D} = \mathbb{R}^n$, unless stated otherwise.

C2 The closure of the set of all points at which a function f is not zero is called the *support* of the function, to be denoted by $supp(f)$, i.e.,

$supp(f)$ is the closure of the set $\{\alpha \in \mathbb{R}^n : f(\alpha) \neq 0\}$.

C3 A function f on $\mathbb{R}^n$ is a function of n real variables α^j. We often write
$$f(\alpha) = f(\alpha^1, \alpha^2, \ldots, \alpha^n). \tag{1.6}$$
A function f is said to be of *class* C^ℓ if for all integers
$$\ell_k \geq 0, \; k = 1, 2, \ldots, n \tag{1.7}$$
such that
$$\ell_1 + \ell_2 + \cdots + \ell_n = \ell \tag{1.8}$$
the partial derivatives
$$\frac{\partial^\ell f}{(\partial \alpha^1)^{\ell_1} (\partial \alpha^2)^{\ell_2} \cdots (\partial \alpha^n)^{\ell_n}} \tag{1.9}$$

1.1. TOPOLOGICAL SPACES AND TOPOLOGICAL EQUIVALENCE

exist and are continuous. If f is of class C^ℓ for all positive integers ℓ we call f a C^∞ function. Such a function is also called *infinitely differentiable* or *smooth* for short.

C4 The following four groups of smooth functions prove to be particularly useful later on:

- $C^\infty(I\!R^n)$: The collection of all smooth functions on $I\!R^n$.

- $C_0^\infty(I\!R^n)$: The set of all smooth functions defined on $I\!R^n$ with bounded support, i.e., each of these functions vanishes outside some bounded rectangle in $I\!R^n$. These functions are generally known as smooth *functions of compact support*.[5]

- $C^\infty(\mathcal{D})$: Smooth local functions with domain $\mathcal{D}$.

- $C^\infty(\alpha)$: The set of all smooth local functions on $I\!R^n$ whose domains contain the point α in $I\!R^n$.

1.1.2 Topological equivalence

It is important to be able to compare and relate different topological spaces. Let $\mathcal{T}$ and $\mathcal{T}'$ be two topological spaces. To relate them we would need to map $\mathcal{T}$ to $\mathcal{T}'$, and it is the nature of this mapping which enables us to compare the two spaces. We shall confine ourselves to one-to-one mappings F of $\mathcal{T}$ onto $\mathcal{T}'$ unless otherwise is stated.[6] Clearly for such a mapping the inverse F^{-1}, which maps $\mathcal{T}'$ one-to-one onto $\mathcal{T}$, exists. When the mapping F is not one-to-one, the inverse mapping does not exist.[7]

Given a function F the *image* $F(\Lambda)$ of any subset Λ of $\mathcal{T}$ consists of all $\tau' \in \mathcal{T}'$ such that $\tau' = F(\tau)$ for some $\tau \in \Lambda$. We can also define the *inverse image* $F^{-1}(\Lambda')$ of any subset Λ' of $\mathcal{T}'$ to be the subset $\Lambda \in \mathcal{T}$ consisting of all $\tau \in \mathcal{T}$ such that $F(\tau) \in \Lambda'$, i.e.,

$$F(\Lambda) = \{\tau' = F(\tau) : \tau \in \Lambda\}, \quad (1.10)$$
$$F^{-1}(\Lambda') = \{\tau : \tau \in \mathcal{T}, F(\tau) \in \Lambda'\}. \quad (1.11)$$

---

[5] A rectangle $\Lambda_{rec} = \Lambda_1 \times \Lambda_2 \times \cdots \times \Lambda_n$ is bounded if all the intervals $\Lambda_1, \Lambda_2, \ldots, \Lambda_n$ are bounded. Generally a subset in $I\!R^n$ is bounded if it is contained in a bounded rectangle Λ_{rec}. We have avoided introducing the notion of *compact sets* in the context of a general topological space. In $I\!R^n$ with the standard topology a compact set is just a closed and bounded set.

[6] Simmons (1963).

[7] A mapping $F : A \mapsto B$ is one-to-one if distinct elements a_1 and a_2 in A are mapped to distinct images in B, i.e., $F(a_1,) \neq F(a_2)$ if $a_1 \neq a_2$. It is an *onto mapping* if every element in B is the image of an element in A, i.e., given any $b \in B$ there is $a \in A$ such that $b = F(a)$, otherwise it is called an *into* mapping.

CHAPTER 1. MANIFOLDS AND DYNAMICAL SYSTEMS

Definition 1.1.2(1) Open functions and continuous functions

- F is said to be an *open function* if the image of every open set Λ of $\mathcal{T}$ is an open set Λ' in $\mathcal{T}'$.

- F is said to be a *continuous function* if the inverse image of every open set Λ' of $\mathcal{T}'$ is an open set Λ in $\mathcal{T}$.

Comments 1.1.2(1) Agreement with usual notion of continuity

C1 The concepts of images and inverse images are defined whether the function is one-to-one or not. So, the above definitions of open and continuous functions apply to functions which are not necessarily one-to-one.

C2 Let $\mathcal{T} = \mathcal{T}' = I\!R$. Then, a mapping F of $\mathcal{T}$ onto $\mathcal{T}'$ is identifiable with a real-valued function of a real variable. The above definition of continuity can be shown to agree with the usual concept of continuity of real-valued functions used in elementary calculus. At first sight it may appear natural to define continuity in terms of open functions. However, this turns out to be wrong. Consider a constant function on $\mathcal{T} = I\!R$ which maps every point $\alpha \in \mathcal{T}$ to the value 1. This is not an open function since it maps every open set of $\mathcal{T}$ to the closed set $\{1\}$, but it is continuous in the usual sense. Open functions thus cannot serve to define the usual notion of continuity.

Definition 1.1.2(2) Homeomorphism and topological equivalence

- Two topological spaces $\mathcal{T}$ and $\mathcal{T}'$ are said to be *topologically equivalent* or *homeomorphic* if there is a one-to-one mapping F of $\mathcal{T}$ onto $\mathcal{T}'$ such that both F and F^{-1} are continuous. Such a mapping F is said to be a *homeomorphism* or to be *bicontinuous*.

Comments 1.1.2(2) Topological equivalence in $I\!R$

C1 Let $\mathcal{T} = (-\frac{1}{2}\pi, \frac{1}{2}\pi) \subset I\!R$ and $\mathcal{T}'$ be the real line $I\!R$. Then

$$F(x) = \tan x$$

regarded as a function from $\mathcal{T}$ to $\mathcal{T}'$ is one-to-one, onto and continuous. Its inverse is also continuous, making F a homeomorphism. Hence $(-\frac{1}{2}\pi, \frac{1}{2}\pi)$ is topologically equivalent to $I\!R$. Conversely $I\!R$ is also topologically equivalent to the open interval $(-\frac{1}{2}\pi, -\frac{1}{2}\pi)$. In fact the real line is topologically equivalent to any finite open interval.

C2 Topologically equivalent spaces have common topological properties.

1.2. EUCLIDEAN SPACES

Definition 1.1.2(3)   **Connectivity**

- A topological space $\mathcal{T}$ is said to be *disconnected* if it is the union of two open, non-empty and disjoint subsets, i.e., there are non-empty open subsets $\mathcal{M}_1$ and $\mathcal{M}_2$ of $\mathcal{T}$ such that

$$\mathcal{M}_1 \cup \mathcal{M}_2 = \mathcal{T} \quad \text{and} \quad \mathcal{M}_1 \cap \mathcal{M}_2 = \emptyset. \tag{1.12}$$

- A topological space $\mathcal{T}$ is said to be *connected* if it is not disconnected.

Comments 1.1.2(3)   **Connectivity**

C1   The symbol $\emptyset$ shall always denote the *empty set*.

C2   Consider the spaces $\mathcal{T}_1 = I\!R$ and $\mathcal{T}_2 = I\!R^- \cup I\!R^+$. Then $\mathcal{T}_1$ is connected and $\mathcal{T}_2$ is disconnected as we would intuitively expect. We shall have occasions to employ disconnected spaces in later chapters. However, unless it is stated otherwise all the topological spaces we shall consider are assumed to be connected.

1.2 Euclidean Spaces

1.2.1 Basic concepts and definitions

We can set up a real vector space structure in $I\!R^n$ by introducing scalar multiplication and addition.[8] Let a be a real number then $a\alpha$ is an element in $I\!R^n$ defined by

$$a\alpha = (a\alpha^1, a\alpha^2, \ldots, a\alpha^n), \tag{1.13}$$

and if $\beta = (\beta^1, \beta^2, \ldots, \beta^n)$ is another element in $I\!R^n$, then the sum $\alpha + \beta$ is defined to be the element in $I\!R^n$ given by

$$\alpha + \beta = (\alpha^1 + \beta^1, \alpha^2 + \beta^2, \ldots, \alpha^n + \beta^n). \tag{1.14}$$

The resulting vector space is clearly of dimension n with the zero element $(0, 0, \ldots, 0)$ which will simply be denoted by 0. We can go further to introduce a *norm* $||\alpha||$ to each element α by

$$||\alpha|| = \left(\sum_{j=1}^{n} \alpha^j \alpha^j \right)^{1/2}. \tag{1.15}$$

This norm possesses the following characteristic properties:

---
[8]Simmons (1963) §14 pp. 80-81.

1. $||a\alpha|| = |a|\,||\alpha||$.
2. $||\alpha|| \geq 0$, and $||\alpha|| = 0$ if and only if $\alpha = 0$.
3. $||\alpha + \beta|| \leq ||\alpha|| + ||\beta||$.

Another useful property is

$$\Big| ||\alpha|| - ||\beta|| \Big| \leq ||\alpha - \beta||.$$

Definition 1.2.1(1)   **Euclidean spaces**

- The set $I\!R^n$ endowed with a real vector space structure and a norm, defined by Eqs. (1.13), (1.14) and (1.15), is called an n-dimensional *Euclidean space*.

Comments 1.2.1(1)   **Distance, metric and scalar product**

C1   To emphasize the vector space structure we shall adopt the notation $I\!E^n$ to denote the above Euclidean space. The zero element is also referred to as the *origin* of the space $I\!E^n$. It is important to distinguish the space $I\!E^n$ from $I\!R^n$ which has no vector space structure. The one-dimensional Euclidean space is denoted simply by $I\!E$.

C2   The norm in $I\!E^n$ induces a *distance function* or a *metric* between any two elements α and β given by

$$||\alpha - \beta|| = \left(\sum_{j=1}^n (\alpha^j - \beta^j)^2 \right)^{1/2}. \tag{1.16}$$

C3   The Euclidean space $I\!E^n$ endowed with this metric becomes a *metric space*. With a concept of distance we can introduce the concept of *boundedness* of a set in $I\!E^n$. A subset Λ of $I\!E^n$ is *bounded* if there is a point $\alpha_0 \in \Lambda$ and a number $a \in I\!R$ such that $||\alpha_0 - \alpha|| < a$ for all $\alpha \in \Lambda$. This agrees with the boundedness definition in footnote 5 to §1.1.1C(4) C4.

C4   Another useful structure in $I\!E^n$ is the *scalar product* which assigns a real number $\langle \alpha \mid \beta \rangle$ to any two elements α and β in $I\!E^n$ by

$$\langle \alpha \mid \beta \rangle = \sum_{j=1}^n \alpha^j \beta^j. \tag{1.17}$$

This scalar product satisfies the following Schwarz's inequality:

$$|\langle \alpha \mid \beta \rangle| \leq ||\alpha||\, ||\beta||. \tag{1.18}$$

1.2. EUCLIDEAN SPACES

C5 By inheriting the standard topology of $\mathbb{R}^n$ the Euclidean space $\mathbb{E}^n$ is also a topological space. An alternative way to introduce a topological structure is to make use of the metric to define open sets. First, we define an *open sphere* $\mathcal{S}_r(\alpha_0)$ with centre α_0 and radius r in $\mathbb{E}^n$ to be the subset

$$\mathcal{S}_r(\alpha_0) = \left\{ \alpha \in \mathbb{E}^n : ||\alpha - \alpha_0|| < r \right\}. \tag{1.19}$$

One can then define an open set Λ to be a subset such that every element α in Λ is contained in an open sphere $\mathcal{S}_r(\alpha_0)$ inside Λ, i.e.,

$$\alpha \in \Lambda \;\;\Rightarrow\;\; \alpha \in \mathcal{S}_r(\alpha_0) \subset \Lambda. \tag{1.20}$$

This results in the same standard topology inherited from $\mathbb{R}^n$. We shall adopt this topology for $\mathbb{E}^n$ from now on.

C6 Subsets of a topological space can be given a topology, known as a *relative topology*, in a natural manner. An example is that of a circle $\mathcal{C}$ in $\mathbb{E}^2$. Such a circle is a topological space with a topology consisting of open sets defined by the intersections of $\mathcal{C}$ with all the open sets in $\mathbb{E}^2$. Similarly a sphere in $\mathbb{E}^3$ is also a topological space.

1.2.2 Coordinate systems and coordinate transformations

A *coordinate system* on the Euclidean space $\mathbb{E}^n$ is an assignment of n real numbers, referred to as *coordinates*, to specify points in the space. The original n-tuples of real numbers $(\alpha^1, \alpha^2, \ldots, \alpha^n)$ can serve as a coordinate system with $\alpha^1, \alpha^2, \ldots, \alpha^n$ as coordinates.

Let $f^1, f^2, \ldots, f^n$ be n smooth functions on $\mathbb{R}^n$ which induce a one-to-one map of $\mathbb{R}^n$ onto itself by[9]

$$(\alpha^1, \alpha^2, \ldots, \alpha^n) \mapsto \left(f^1(\alpha), f^2(\alpha), \ldots, f^n(\alpha) \right). \tag{1.21}$$

Let

$$x^1 = f^1(\alpha),\; x^2 = f^2(\alpha), \ldots,\; x^n = f^n(\alpha). \tag{1.22}$$

Then every point $\alpha \in \mathbb{E}^n$ has two n-tuples of numbers associated with it, $(\alpha^1, \alpha^2, \ldots, \alpha^n)$ and $(x^1, x^2, \ldots, x^n)$. To specify points in $\mathbb{E}^n$ we can use the new n-tuple $(x^1, x^2, \ldots, x^n)$, leading to a new coordinatization with coordinates $x^1, x^2, \ldots, x^n$. This shows that $\mathbb{E}^n$ admits many different coordinate systems. We shall denote both the coordinate system and the coordinates by x^j.

---

[9] Smooth functions are members of $C^\infty(\mathbb{R}^n)$ introduced in §1.1.1C(4) C4.

To avoid confusion we emphasize here that the mapping defined by Eq. (1.21) is not a mapping of $I\!\!E^n$ onto itself; it is a mapping of the n-tuples of real numbers $I\!\!R^n$ onto itself. For each point $\alpha \in I\!\!E^n$ specified by the original coordinates α^j this mapping leads to a new set of numbers x^j associated with the same point α in $I\!\!E^n$. We call such a mapping a *coordinate transformation*. We shall confine ourselves to coordinate systems which relate to the original system $(\alpha^1, \alpha^2, \ldots, \alpha^n)$ in a smooth manner as described by Eq. (1.21).

A simple example is a *linear coordinate transformation* of the form

$$x^j = f^j(\alpha) = \sum_{k=1}^{n} M^j_k \alpha^k, \quad M^j_k \in I\!\!R \tag{1.23}$$

where the constants M^j_k satisfy

$$\sum_{j=1}^{n} M^j_i M^j_k = \delta_{ik}, \tag{1.24}$$

where δ_{ik} is the *Kronecker delta* which takes the value 1 if $i = k$ and vanishes otherwise. The Kronecker delta is often denoted also by the symbol δ^i_k. This transformation, known as a *homogeneous orthogonal coordinate transformation*, is characterized by the preservation of the expression for the norm, i.e., we have

$$||\alpha||^2 = \sum_{j=1}^{n} \alpha^j \alpha^j = \sum_{j=1}^{n} x^j x^j. \tag{1.25}$$

The original coordinate system α^j and all other coordinate systems related to it in this way are called *rectangular Cartesian* coordinate systems.

We can define real-valued functions on $I\!\!E^n$ as mappings of $I\!\!E^n$ to $I\!\!R$, i.e.,

$$f : I\!\!E^n \mapsto I\!\!R \quad \text{by} \quad \alpha \mapsto f(\alpha) \in I\!\!R. \tag{1.26}$$

Such mappings are independent of any coordinate system. In a given coordinate system x^j a function f on $I\!\!E^n$ can be described as a function of the coordinates. The expression for the function will be different in different coordinate systems, but the function f itself has not changed in that the same point α in $I\!\!E^n$ is mapped to the same value $f(\alpha)$. A function on $I\!\!E^n$ is said to be *smooth* or *infinitely differentiable* if expressed as a function of the original coordinates α^j it is a smooth function on $I\!\!R^n$ in the sense described in §1.1.1C(4) C3. Since all other coordinates are smooth functions of α^j we can define smooth functions on $I\!\!E^n$ in terms of any coordinate system. We can also define local functions in $I\!\!E^n$ in the same way as we do in $I\!\!R^n$. As in §1.1.1C(4) C4 we can define four sets of functions:

1.2. EUCLIDEAN SPACES

- $C^\infty(I\!\!E^n)$, the set of all smooth functions on $I\!\!E^n$.
- $C_0^\infty(I\!\!E^n)$, the set of all smooth functions on $I\!\!E^n$ of compact support.
- $C_0^\infty(\mathcal{D})$, the set of all smooth local functions with domain $\mathcal{D}$.
- $C^\infty(\alpha)$, the set of all smooth local functions whose domains contain the point $\alpha \in I\!\!E^n$.

It is often useful to introduce coordinate systems which are not rectangular Cartesian or even global. A well-known example is the spherical coordinate system in the 3-dimensional Euclidean space $I\!\!E^3$. At the origin spherical coordinates are not defined. Generally let $\mathcal{D}$ be an arbitrary connected open subset of $I\!\!E^n$. Then the original coordinates α^j confined to $\mathcal{D}$ form a coordinate system on $\mathcal{D}$. We can introduce a new coordinate system on $\mathcal{D}$ in terms of n smooth functions f^j defined on $\mathcal{D}$ which maps $\mathcal{D}$ one-to-one onto a connected open set of $I\!\!R^n$. These functions then define new coordinates x^j on $\mathcal{D}$, forming what is known as a *local coordinate chart* on $\mathcal{D}$. Let $\mathcal{D}'$ be another connected open set of $I\!\!E^n$, and let x'^j be a local coordinate chart on $\mathcal{D}'$. On the intersection $\mathcal{D} \cap \mathcal{D}'$ the coordinates x^j and x'^j are smooth functions of each other due to their smooth relation with α^j. So, it does not matter which coordinate chart we use.[10]

Local coordinate charts are of fundamental importance in topological spaces which cannot be covered by a single global coordinate system. They have to be covered by overlapping local coordinate charts. We shall return to this problem of local coordinate charts later.

1.2.3 Contravariant and covariant vectors in $I\!\!E^n$

Coordinate systems and the way they transform are instrumental in establishing the notion of vectors in physics. The term *vector* has been used to refer to a wide variety of mathematical objects, most commonly to elements of vector spaces. What we are interested in here is the notion of vectors as geometric objects appearing in differential geometry, mechanics, Special and General Relativity. We come across the term *vector* in elementary mechanics describing quantities which possess a direction as well as a magnitude. Intuitively we can see that a direction has to be relative to something, e.g., relative to some coordinate axes. So, in more advanced texts, the rather qualitative notion of direction is sharpened in terms of a multi-component object in a given coordinate system, namely a set of n numbers X^j, $j = 1, 2, \ldots, n$, representing the components along some chosen coordinate axes. However, simply listing a set of n numbers in a given coordinate system is not enough. One has to know

---
[10] See Benn and Tucker (1987) p. 131 for a diagramatic illustration.

what would happen to these components when one goes to a new coordinate system. There are two kinds of n-component objects traditionally referred to as vectors, depending on how the components change on a transformation of coordinates.[11]

Definition 1.2.3(1)   **Contravariant and covariant vectors**

- An n-tuple of numbers X^j, $j = 1, 2, \ldots, n$, associated with a point α in $I\!\!E^n$ are said to be the components of a *contravariant vector* at the point α in a given coordinate system x^j if they transform, on a change of coordinate system to x'^j, according to

$$X'^k = \sum_{j=1}^n \left(\frac{\partial x'^k}{\partial x^j}\right)_\alpha X^j, \qquad (1.27)$$

where the subscript α indicates that the derivatives are evaluated at α. The components of the vector in the new coordinate system become X'^k.

- An n-tuple of numbers Y_j, $j = 1, 2, \ldots, n$, associated with a point α in $I\!\!E^n$ are said to be the components of a *covariant vector* at the point α in a given coordinate system x^j if they transform, on a change of coordinate system to x'^j, according to

$$Y'_k = \sum_{j=1}^n \left(\frac{\partial x^j}{\partial x'^k}\right)_\alpha Y_j. \qquad (1.28)$$

The components of the vector in the new coordinate system become Y'_k.

Comments 1.2.3(1)   **Coordinate independence and examples**

C1   The set of all contravariant vectors at a given point $\alpha \in I\!\!E^n$ form a vector space V_α under the usual component-wise addition and scalar multiplication rules for contravariant vectors. The same is true for covariant vectors.

C2   The above description of vectors appears explicitly coordinate dependent, in sharp contrast to the definition of functions. One would guess that such a description is merely a numerical representation of vectors in various coordinate systems, and that there should be an intrinsic definition which is not explicitly dependent on coordinates. This is indeed the case and we shall introduce such an intrinsic definition in the next section, where we will also

---
[11] Synge and Schild (1966).

1.2. EUCLIDEAN SPACES

see why the components of the vectors should transform in the way they do on a change of coordinates.

C3 In classical mechanics the state of a particle moving in $\mathbb{E}^3$ can be determined by its position specified by rectangular Cartesian coordinates x^j and its momentum p_j canonically conjugate to x^j. Associated with the particle's motion we have the following vectors:

1. Displacement vector The particle's coordinate displacement Δx^j form the components of a contravariant vector since a simple differentiation gives

$$\Delta x'^j = \sum_{k=1}^{n} \frac{\partial x'^j}{\partial x^k} \Delta x^k, \qquad (1.29)$$

which is the transformation law for contravariant vectors.

2. Momentum vector We know from mechanics that a particle's momentum is a covariant vector with components denoted by p_j. This knowledge enables us to calculate, for example, the components of the momentum in different coordinate systems. Consider the motion in $\mathbb{E}^3$ with the usual rectangular Cartesian coordinates (x, y, z) and the corresponding momenta (p_x, p_y, p_z). In the usual spherical coordinates (r, θ, φ) we shall denote the corresponding momenta conjugate to (r, θ, φ) by $(p_r, p_\theta, p_\varphi)$. Using the covariant vector transformation law we obtain:

$$\begin{aligned}
p_r &= \frac{\partial x}{\partial r} p_x + \frac{\partial y}{\partial r} p_y + \frac{\partial z}{\partial r} p_z \\
&= \frac{x}{\sqrt{x^2 + y^2 + z^2}} p_x + \frac{y}{\sqrt{x^2 + y^2 + z^2}} p_y + \frac{z}{\sqrt{x^2 + y^2 + z^2}} p_z. \quad (1.30) \\
p_\theta &= \frac{\partial x}{\partial \theta} p_x + \frac{\partial y}{\partial \theta} p_y + \frac{\partial z}{\partial \theta} p_z \quad (1.31) \\
&= \frac{xz}{\sqrt{x^2 + y^2}} p_x + \frac{yz}{\sqrt{x^2 + y^2}} p_y - \sqrt{x^2 + y^2}\, p_z. \quad (1.32) \\
p_\varphi &= \frac{\partial x}{\partial \varphi} p_x + \frac{\partial y}{\partial \varphi} p_y + \frac{\partial z}{\partial \varphi} p_z \\
&= -y p_x + x p_y. \quad (1.33)
\end{aligned}$$

1.2.4 Contravariant, covariant and mixed tensors

Definition 1.2.4(1) Different types of tensors

- A set of n^2 numbers C^{jk}, $j, k = 1, 2, \ldots, n$, associated with a point $\alpha \in \mathbb{E}^n$ are said to be the components of a *contravariant tensor* of the second order at the point α in a given coordinate system x^j if they transform, on a change of coordinate system to x'^j, according to

$$C'^{rs} = \sum_{j,k=1}^{n} \left(\frac{\partial x'^r}{\partial x^j}\right)_\alpha \left(\frac{\partial x'^s}{\partial x^k}\right)_\alpha C^{jk}, \qquad (1.34)$$

where the derivatives are evaluated at α. The components of the tensor in the new coordinate system become C'^{rs}.

- A set of n^2 numbers C_{jk}, $j,k = 1, 2, \ldots, n$, associated with a point α in $I\!\!E^n$ are said to be the components of a *covariant tensor* of the second order at the point α in a given coordinate system x^j if they transform, on a change of coordinate system to x'^j, according to

$$C'_{rs} = \sum_{j,k=1}^{n} \left(\frac{\partial x^j}{\partial x'^r}\right)_\alpha \left(\frac{\partial x^k}{\partial x'^s}\right)_\alpha C_{jk}. \qquad (1.35)$$

The components of the tensor in the new coordinate system become C'_{rs}.

- A set of n^2 numbers C^j_k, $j,k = 1, 2, \ldots, n$, associated with a point α in $I\!\!E^n$ are said to be the components of a *mixed tensor* of the second order at the point α in a given coordinate system x^j if they transform, on a change of coordinate system to x'^j, according to

$$C'^r_s = \sum_{j,k=1}^{n} \left(\frac{\partial x'^r}{\partial x^j}\right)_\alpha \left(\frac{\partial x^k}{\partial x'^s}\right)_\alpha C^j_k. \qquad (1.36)$$

The components of the tensor in the new coordinate system become C'^r_s.

Comments 1.2.4(1)   **Outer products, contraction and interior product, symmetric tensors**

C1   The above definitions are a generalization of contravariant and covariant vectors. They can be extended further in a straightforward manner to define tensors of higher orders. In view of the similarity of these definitions, we can refer to a vector as a tensor of first order. We can also formally call a coordinate independent number associated with a point α a tensor of zero order at α.

C2   A second order covariant tensor C_{jk} is called *symmetric* if $C_{jk} = C_{kj}$ and *anti-symmetric* if $C_{kj} = -C_{jk}$, and likewise for contravariant tensors. Anti-symmetric tensors will be seen to play a crucial role in formulating Hamiltonian mechanics.

1.2. EUCLIDEAN SPACES

C3 Vectors and tensors of different types may be combined together to produce tensors of higher orders. For example, a covariant vector Y_i and a contravariant vector X^j can be combined to form a mixed tensor with components $C_i^j = Y_i X^j$. Such a combination is known as an *outer product*. The outer product of a covariant vector of components Y_i and a contravariant tensor of the second order of components C^{jk} produces a mixed tensor of the third order of components $C_i^{jk} = Y_i C^{jk}$.

C4 The order of a mixed tensor can be reduced by two by summing over a superscript and a matching subscript. For example a third order mixed tensor C_i^{jk} is reduced to a first order tensor, i.e., a contravariant vector with components X^j, by

$$X^j = \sum_{k=1}^{n} C_k^{jk}. \tag{1.37}$$

Such a process is known as a *contraction*.

C5 Applying contraction to the outer product $C_j^k = Y_j X^k$ produces a tensor of zero order:

$$\sum_{k=1}^{n} C_k^k = \sum_{k=1}^{n} Y_k X^k. \tag{1.38}$$

This number is the same in different coordinate systems, i.e.,

$$\sum_{j=1}^{n} Y_j X^j = \sum_{r=1}^{n} Y_r' X'^r, \tag{1.39}$$

since

$$\begin{aligned}
\sum_{r=1}^{n} Y_r' X'^r &= \sum_{r=1}^{n} \left(\sum_{k=1}^{n} \frac{\partial x^k}{\partial x'^r} Y_k \sum_{j=1}^{n} \frac{\partial x'^r}{\partial x^j} X^j \right) \\
&= \sum_{j,k=1}^{n} Y_k \left(\sum_{r=1}^{n} \frac{\partial x^k}{\partial x'^r} \frac{\partial x'^r}{\partial x^j} \right) X^j \\
&= \sum_{j=1}^{n} Y_j X^j, \tag{1.40}
\end{aligned}$$

since

$$\sum_{r=1}^{n} \frac{\partial x^k}{\partial x'^r} \frac{\partial x'^r}{\partial x^j} = \delta_j^k. \tag{1.41}$$

We call such a number a *scalar* or an *invariant*.

C6 Another process which produces new tensors is a combination of outer product and contraction. For example, given a second rank covariant tensor C_{jk} and a contravariant vector X^r their outer product is a third rank mixed tensor $C_{jk}^r = C_{jk}X^r$. One can then carry out a contraction to produce a covariant vector

$$Y_j = \sum_{k=1}^{n} C_{jk}^k = \sum_{k=1}^{n} C_{jk}X^k. \tag{1.42}$$

This process enables us to relate a contravariant vector X^r to a covariant vector Y_j through a covariant tensor C_{jk}. We shall give a formal discussion of this procedure later in §1.4.4 under the heading of *interior product*.

C7 A covariant vector Y_j at a point α generates a mapping of the set of all contravariant vectors at α to the reals by contraction, i.e., the covariant vector Y_j maps a contravariant vector X^j to the number

$$\sum_{j=1}^{n} Y_j X^j \in \mathbb{R}. \tag{1.43}$$

We can put this mapping on a more formal footing. We know that the set of all contravariant vectors at α form a vector space V_α. A covariant vector Y_j at α then induces a mapping $\mathcal{Y}_\alpha$ of V_α to $\mathbb{R}$ by contraction:

$$\mathcal{Y}_\alpha : V_\alpha \mapsto \mathbb{R} \quad \text{by} \quad X^j \mapsto \mathcal{Y}_\alpha(X^j) = \sum_{j=1}^{n} Y_j X^j. \tag{1.44}$$

This mapping is linear. We shall see later that it is possible to turn this process around to define a covariant vector at α as a linear mapping of V_α to $\mathbb{R}$.

C8 We can introduce the concept of a *vector field* by assigning a vector of the same type at each point in $\mathbb{E}^n$, and similarly for *tensor fields*. Most of the operations introduced for vectors carry over for vector fields, e.g., contraction between contravariant and covariant vector fields.

1.3 Differential Operators, Vectors and Fields

In view of the non-uniqueness of coordinate systems it would be highly desirable to define geometric quantities without direct reference to coordinates. The idea is to start with functions defined on $\mathbb{E}^n$ as primary objects since they are coordinate-independent. We can then introduce other geometric objects in terms of their action on these functions. So, our starting point is the set $C^\infty(\mathbb{E}^n)$ of smooth functions on $\mathbb{E}^n$. Since all the functions are defined on the same domain, i.e., $\mathbb{E}^n$, we can carry out all usual algebraic operations,

1.3. DIFFERENTIAL OPERATORS, VECTORS AND FIELDS

e.g., we can add and multiply these functions, without having to impose any conditions. On the other extreme we have the set $C^\infty(\alpha)$ of smooth functions each defined on a neighbourhood of the point $\alpha \in I\!\!E^n$. Let f and g be members of $C^\infty(\alpha)$ with f defined on domain $\mathcal{D}(f)$ and g defined on domain $\mathcal{D}(g)$. Since both $\mathcal{D}(f)$ and $\mathcal{D}(g)$ contain the point α we would generally have

$$\mathcal{D}(f) \cap \mathcal{D}(g) \neq \emptyset, \quad \mathcal{D}(f) \neq \mathcal{D}(g). \tag{1.45}$$

We can define the sum $f + g$ and the product fg as functions on the domain $\mathcal{D}(f) \cap \mathcal{D}(g)$ according to:

$$f + g : \mathcal{D}(f) \cap \mathcal{D}(g) \mapsto I\!\!R \quad \text{by} \quad (f+g)(\alpha') = f(\alpha') + g(\alpha'), \tag{1.46}$$
$$fg : \mathcal{D}(f) \cap \mathcal{D}(g) \mapsto I\!\!R \quad \text{by} \quad (fg)(\alpha') = f(\alpha')g(\alpha'). \tag{1.47}$$

Here $\alpha' \in \mathcal{D}(f) \cap \mathcal{D}(g)$. With this introduction we can now define further operations on the functions in $C^\infty(\alpha)$ and in $C^\infty(I\!\!E^n)$.

1.3.1 Differential operators and derivations

Definition 1.3.1(1) Linear operators acting on $C^\infty(\alpha)$ and $C^\infty(I\!\!E^n)$

- A mapping $\widehat{A}_\alpha$ of $C^\infty(\alpha)$ into $I\!\!R$ with the property

$$\widehat{A}_\alpha(af + bg) = a\,\widehat{A}_\alpha(f) + b\,\widehat{A}_\alpha(g) \quad \forall f, g \in C^\infty(\alpha), \tag{1.48}$$

  where $a, b \in I\!\!R$, is called a *linear operator at a point* $\alpha \in I\!\!E^n$ on the set of smooth functions $C^\infty(\alpha)$.

- A mapping $\widehat{A}$ of $C^\infty(I\!\!E^n)$ into $C^\infty(I\!\!E^n)$ with the property

$$\widehat{A}(af + bg) = a\,\widehat{A}(f) + b\,\widehat{A}(g) \quad \forall f, g \in C^\infty(I\!\!E^n), \tag{1.49}$$

  where $a, b \in I\!\!R$, is called a *linear operator* on the set of smooth functions $C^\infty(I\!\!E^n)$.

Comments 1.3.1(1) Properties

C1 A linear operator $\widehat{A}_\alpha$ gives rise to a real number $\widehat{A}_\alpha(f)$ for any $f \in C^\infty(\alpha)$. If f is a zero function then $\widehat{A}_\alpha(f) = 0$. In contrast a linear operator $\widehat{A}$ on $C^\infty(I\!\!E^n)$ gives rise to a new smooth function $\widehat{A}(f)$ on $I\!\!E^n$.

C2 A linear operator $\widehat{A}_\alpha$ at a point α depends only on the behaviour of local functions in the neighbourhood of α. Consequently, if two local functions

f and g agree on some neighbourhood of α, then $\widehat{A}_\alpha(f) = \widehat{A}_\alpha(g)$. This is not true for a linear operator $\widehat{A}$ on $C^\infty(I\!\!E^n)$ in general.

Definition 1.3.1(2)   **Differential operators**

- A linear operator $\widehat{A}_\alpha$ on $C^\infty(\alpha)$ at a point α is called a *differential operator of the first order* at α if there is a coordinate system x^j in a neighbourhood of α such that

$$\widehat{A}_\alpha(f) = \sum_{j=1}^n A_\alpha^j \left(\frac{\partial f}{\partial x^j}\right)_\alpha \quad \forall f \in C^\infty(\alpha), \tag{1.50}$$

where $A_\alpha^j \in I\!\!R$ are independent of f but are dependent on the coordinate system used.

- A linear operator $\widehat{A}$ on $C^\infty(I\!\!E^n)$ is called a *smooth differential operator of the first order* on $C^\infty(I\!\!E^n)$ if there is a coordinate system x^j such that

$$\widehat{A}(f) = \sum_{j=1}^n A^j \left(\frac{\partial f}{\partial x^j}\right) \quad \forall f \in C^\infty(I\!\!E^n), \tag{1.51}$$

where $A^j \in C^\infty(I\!\!E^n)$ are independent of f, but are dependent on the coordinate system used.

Comments 1.3.1(2)   **Properties of differential operators**

C1   We often express $\widehat{A}_\alpha$ and $\widehat{A}$ respectively as

$$\widehat{A}_\alpha = \sum_{j=1}^n A_\alpha^j \left(\frac{\partial}{\partial x^j}\right)_\alpha \quad \text{and} \quad \widehat{A} = \sum_{j=1}^n A^j \left(\frac{\partial}{\partial x^j}\right). \tag{1.52}$$

A differential operator $\widehat{A}$ is not smooth if A^j is not smooth.

C2   These operators satisfy the following properties:

$$\widehat{A}_\alpha(fg) = \widehat{A}_\alpha(f)\,g(\alpha) + f(\alpha)\,\widehat{A}_\alpha(g) \quad \forall f, g \in C^\infty(\alpha), \tag{1.53}$$
$$\widehat{A}(fg) = \widehat{A}(f)\,g + f\,\widehat{A}(g) \quad \forall f, g \in C^\infty(I\!\!E^n). \tag{1.54}$$

These properties reflect the product rule of differentiation and are therefore characteristic of a differential operator. A linear operator which does not possess these properties is not a differential operator, e.g., a multiplication operator is a linear operator but not a differential operator. As we shall see

1.3. DIFFERENTIAL OPERATORS, VECTORS AND FIELDS

later Eqs. (1.53) and (1.54) can lead to an intrinsic definition of differential operators.

C3 Some authors would include an additive term in the definition of differential operators, i.e., they would regard operators of the form

$$\widehat{C}(f) = \sum_{j=1}^{n} A^j \left(\frac{\partial f}{\partial x^j} \right) + Bf, \tag{1.55}$$

where B is a function on $\mathbb{E}^n$, as differential operators. We shall not adopt this definition since operators of this form do not obey the product rule of differentiation shown in Eqs. (1.53) and (1.54) unless $B = 0$.

C4 Although defined explicitly through a specific coordinate system differential operators are in fact coordinate independent. Let us examine what would happen when we move to a new coordinate system x'^j. The function f is now a function of x'^j. An application of normal rules of differentiation yields:

$$\widehat{A}_\alpha(f) = \sum_{j=1}^{n} A^j_\alpha \left(\frac{\partial f}{\partial x^j} \right)_\alpha \tag{1.56}$$

$$= \sum_{j=1}^{n} A^j_\alpha \left\{ \sum_{k=1}^{n} \left(\frac{\partial f}{\partial x'^k} \right)_\alpha \left(\frac{\partial x'^k}{\partial x^j} \right)_\alpha \right\}. \tag{1.57}$$

It follows that we can express $\widehat{A}_\alpha(f)$ in coordinates x'^j as

$$\widehat{A}_\alpha(f) = \sum_{k=1}^{n} A'^k_\alpha \left(\frac{\partial f}{\partial x'^k} \right)_\alpha, \tag{1.58}$$

where

$$A'^k_\alpha = \sum_{j=1}^{n} \left(\frac{\partial x'^k}{\partial x^j} \right)_\alpha A^j_\alpha. \tag{1.59}$$

We conclude that $\widehat{A}_\alpha$ is a coordinate independent quantity in the sense that:

1. $\widehat{A}_\alpha$ maps every function $f \in C^\infty(\alpha)$ to a real number $\widehat{A}_\alpha(f)$, and this value $\widehat{A}_\alpha(f)$ is the same in all coordinate systems.

2. $\widehat{A}_\alpha$ can be written down in the same form in any coordinate system, i.e., we have

$$\widehat{A}_\alpha = \sum_{j=1}^{n} A^j_\alpha \left(\frac{\partial}{\partial x^j} \right)_\alpha = \sum_{k=1}^{n} A'^k_\alpha \left(\frac{\partial}{\partial x'^k} \right)_\alpha, \tag{1.60}$$

where A'^k_α to A^j_α are related by Eq. (1.59).

What has been said above applies to $\widehat{A}$. We have

$$\widehat{A}(f) = \sum_{j=1}^{n} A^j \left(\frac{\partial f}{\partial x^j}\right) = \sum_{k=1}^{n} A'^k \left(\frac{\partial f}{\partial x'^k}\right), \quad (1.61)$$

where

$$A'^k = \sum_{j=1}^{n} \left(\frac{\partial x'^k}{\partial x^j}\right) A^j. \quad (1.62)$$

In fact, differential operators can be introduced in a more abstract manner without explicit reference to any coordinate system. The idea is to employ the characteristic property of differentiation shown in Eqs. (1.53) and (1.54) to define differential operators.

C5 We shall proceed to use differential operators to establish an intrinsic definition of vectors and vector fields and other geometric objects.[12]

Definition 1.3.1(3) Derivations

- A *derivation on* $C^\infty(\alpha)$ at a point α is a linear operator $\widehat{X}_\alpha$ at the point α such that

$$\widehat{X}_\alpha(fg) = \widehat{X}_\alpha(f)g(\alpha) + f(\alpha)\widehat{X}_\alpha(g) \quad \forall f, g \in C^\infty(\alpha). \quad (1.63)$$

- A *derivation on* $C^\infty(I\!\!E^n)$ is a linear operator $\widehat{X}$ on $C^\infty(I\!\!E)$ such that

$$\widehat{X}(fg) = \widehat{X}(f)g + f\widehat{X}(g) \quad \forall f, g \in C^\infty(I\!\!E^n). \quad (1.64)$$

Comments 1.3.1(3) Differential operators, derivations and contravariant vectors

C1 It can be shown that given a derivation $\widehat{X}_\alpha$ at α there exists a coordinate system x^j and n real numbers X_α^j such that

$$\widehat{X}_\alpha(f) = \sum_{j=1}^{n} X_\alpha^j \left(\frac{\partial f}{\partial x^j}\right)_\alpha \quad \forall f \in C^\infty(\alpha), \quad (1.65)$$

where X_α^j are independent of f. It follows that we can express $\widehat{X}_\alpha$ as a differential operator, i.e., we can simply write

$$\widehat{X}_\alpha = \sum_{j=1}^{n} X_\alpha^j \left(\frac{\partial}{\partial x^j}\right)_\alpha. \quad (1.66)$$

---

[12] Brickell and Clark (1970), Isham (1989), Darling (1994).

1.3. DIFFERENTIAL OPERATORS, VECTORS AND FIELDS

It should not be so surprising that derivations are related to differential operators since they satisfy the distintive product rule of differentiation.

C2 Given a non-zero derivation $\widehat{X}_\alpha$, i.e., one with $X_\alpha^j \neq 0$ for some j, it is possible to transform to new coordinates x'^j in which $\widehat{X}_\alpha$ takes the simple form, i.e.,
$$\widehat{X}_\alpha = \frac{\partial}{\partial x'^1}. \tag{1.67}$$

C3 A derivation $\widehat{X}$ on $C^\infty(I\!\!E^n)$ can also be shown to be expressible as a differential operator on $C^\infty(I\!\!E^n)$, i.e., we have[13]
$$\widehat{X} = \sum_{j=1}^n X^j \left(\frac{\partial}{\partial x^j}\right), \quad X^j \in C^\infty(I\!\!E^n). \tag{1.68}$$

In other words smooth differential operators of the first order are identifiable with derivations.

C4 Physicists are more used to defining things explicitly or constructively. However, it is often more desirable to define a quantity in terms of its characteristic features. Such a definition enables us to appreciate the concept better, often leading to extension of the concept to new and more general situations. In the present case one can appreciate the coordinate independent nature of differential operators in terms of the concept of derivations which are manifestly coordinate independent.

C5 In a given coordinate system x^j the operator $\widehat{X}_\alpha$ is characterized by n numbers, X_α^j. On a coordinate transformation from x^j to x'^j we have, following Eqs. (1.59) and (1.60),
$$\widehat{X}_\alpha = \sum_{j=1}^n X_\alpha^j \left(\frac{\partial}{\partial x^j}\right)_\alpha \tag{1.69}$$
$$= \sum_{k=1}^n X_\alpha'^k \left(\frac{\partial}{\partial x'^k}\right)_\alpha, \tag{1.70}$$

where
$$X_\alpha'^k = \sum_{j=1}^n \left(\frac{\partial x'^k}{\partial x^j}\right)_\alpha X_\alpha^j. \tag{1.71}$$

Compared with Eq. (1.27) we can see that X_α^j change like the components of a contravariant vector on a coordinate transformation. Further studies reveal

---
[13] Matsushima (1972) p. 73.

that differential operators are in fact the intrinsic and coordinate independent definition of contravariant vectors. This being the case we can now appreciate why the numerical components of a contravariant vector transform the way they do.

C6 We shall see later that covariant vectors are definable in a coordinate independent manner in terms of differentials of functions.

C7 While physicists are more used to the term contravariant vectors, differential geometers prefer to use the term *tangent vectors*. So, we have in effect the same quantity being referred to by four different names, i.e., **contravariant vectors, tangent vectors, derivations and differential operators**. When we want to concentrate on its numerical components we would employ the terms contravariant vectors. In other situations we may prefer the terms tangent vectors or differential operators.

1.3.2 Tangent vectors, tangent vector fields and their integral curves

Definition 1.3.2(1)   **Tangent vectors, tangent spaces and tangent vector fields**

- A *tangent vector* at a point α is a derivation $\widehat{X}_\alpha$ on $C^\infty(\alpha)$.

- The set $\widehat{T}_\alpha(\mathbb{E}^n)$ of all tangent vectors at a point α, endowed with a natural vector space structure under addition and scalar multiplication defined by

$$\left(\widehat{X}_\alpha + \widehat{Y}_\alpha\right)(f) = \widehat{X}_\alpha(f) + \widehat{Y}_\alpha(f), \quad \left(a\widehat{X}_\alpha\right)(f) = a\widehat{X}_\alpha(f), \qquad (1.72)$$

  where $a \in \mathbb{R}$ and $\widehat{X}_\alpha, \widehat{Y}_\alpha \in \widehat{T}_\alpha(\mathbb{E}^n)$, is called the *tangent space* at α.

- A derivation $\widehat{X}$ on $C^\infty(\mathbb{E}^n)$ is called a *tangent vector field* on $\mathbb{E}^n$, or simply a *vector field*. A point at which $\widehat{X}$ vanishes is called a *critical point* of the vector field.

Comments 1.3.2(1)   **Coordinate independence and terminology**

C1 Tangent vectors and vector fields have been defined in a way which is manifestly independent of coordinates.

C2 From §1.3.1C(3) C1 we can see that:

1. Tangent vectors are expressible in the form of differential operators by Eq. (1.66).

1.3. DIFFERENTIAL OPERATORS, VECTORS AND FIELDS

2. Vector fields can be written in the form of smooth differential operators using Eq. (1.68). It follows from §1.3.1C(3) C2 that if a given vector field $\widehat{X}$ does not have a critical point in the neighbourhood of a given point α, then new coordinates x'^j exist in which the vector field takes the simple form

$$\widehat{X} = \frac{\partial}{\partial x'^1} \qquad (1.73)$$

in that neighbourhood. Generally $\widehat{X}$ may not be expressible in such a simple form globally over the entire $I\!\!E^n$.

C3   A vector field $\widehat{X}$ gives rise to a tangent vector $\widehat{X}_\alpha$ at each point $\alpha \in I\!\!E^n$ by

$$\widehat{X}_\alpha(f) = \Big(\widehat{X}(f)\Big)(\alpha). \qquad (1.74)$$

Explicitly we have

$$\widehat{X} = \sum_{j=1}^n X^j \left(\frac{\partial}{\partial x^j}\right) \quad \Rightarrow \quad \widehat{X}_\alpha = \sum_{j=1}^n X^j_\alpha \left(\frac{\partial}{\partial x^j}\right)_\alpha. \qquad (1.75)$$

Conversely a vector field $\widehat{X}$ may be regarded as a *smooth* assignment of tangent vectors throughout $I\!\!E^n$, with one tangent vector $\widehat{X}_\alpha$ at each point α. Here, *smoothness* means that the function defined by $\alpha \mapsto \widehat{X}_\alpha(f)$ is smooth for every $f \in C^\infty(I\!\!E^n)$.

C4   Differential operators at α are closely related to the tangents at α to curves passing through α. This is why $\widehat{X}_\alpha$ is called a tangent vector. Before discussing this in detail we have to introduce the concept of differentiable curves first.

Definition 1.3.2(2)   **Curves in $I\!\!E^n$**

- Let J be an open interval of $I\!\!R$ containing the origin 0. A mapping

$$\sigma : J \mapsto I\!\!E^n, \qquad (1.76)$$

which associates each real number $\tau \in J$ to a point $\alpha(\tau)$ in $I\!\!E^n$, is called a *curve* in $I\!\!E^n$ starting from the point $\alpha(0)$. The interval J is called the *domain of the curve*.

Comments 1.3.2(2)   **Curves, their tangents and operators**

C1   As the parameter τ varies, the map σ traces out a set of points $\alpha(\tau)$ in $I\!\!E^n$ with coordinates $x^j(\alpha(\tau))$. The curve is said to be *smooth* or *differentiable* if $x^j(\alpha(\tau))$ are smooth functions of τ. From now on a curve means a smooth curve, unless stated otherwise.

C2 Traditionally the *tangent to a curve* at a point $\alpha(\tau)$ on the curve is defined to be a contravariant vector with components

$$X_\alpha^j = \frac{dx^j(\alpha(\tau))}{d\tau} \tag{1.77}$$

in a given coordinate system x^j. In our present set-up we can construct a differential operator $\widehat{X}_\alpha$ at $\alpha(\tau)$ on the curve by

$$\widehat{X}_\alpha = \sum_{j=1}^{n} X_\alpha^j \left(\frac{\partial}{\partial x^j}\right)_\alpha, \quad \text{with} \quad X_\alpha^j = \frac{dx^j(\alpha(\tau))}{d\tau} \tag{1.78}$$

and call this differential operator the *tangent vector to the curve* σ at $\alpha(\tau)$. A curve generates a family of tangent vectors along itself. A family of curves covering the entire space $I\!\!E^n$ will generate a vector field on $I\!\!E^n$. More interesting is the converse, i.e., whether a given vector field can generate curves in $I\!\!E^n$.

Definition 1.3.2(3)   **Maximal integral curves of a vector field**

- The operator $\widehat{X}_\alpha$ at $\alpha(\tau)$ on a given curve σ defined by

$$\widehat{X}_\alpha = \sum_{j=1}^{n} X_\alpha^j \left(\frac{\partial}{\partial x^j}\right)_\alpha, \quad \text{with} \quad X_\alpha^j = \frac{dx^j(\alpha(\tau))}{d\tau} \tag{1.79}$$

  is called the *tangent vector to the curve* σ at $\alpha(\tau)$.

- Let σ be a curve. If at every point $\alpha(\tau)$ on the curve the tangent vector to the curve coincides with the tangent vector at $\alpha(\tau)$ given rise by a vector field $\widehat{X}$ according to Eqs. (1.74) and (1.75) the curve σ is called an *integral curve* of the vector field $\widehat{X}$ starting from $\alpha(0)$.

- The integral curve starting from $\alpha(0)$ defined on the union of the domains of all the integral curves of $\widehat{X}$ starting from $\alpha(0)$ is called the *maximal integral curve* of the vector field $\widehat{X}$ starting from $\alpha(0)$.

- A vector field is said to be *complete* if the maximal integral curve starting from every point in $I\!\!E^n$ has the entire real line $I\!\!R$ as its domain. A vector field which is not complete is said to be *incomplete*.

1.3. DIFFERENTIAL OPERATORS, VECTORS AND FIELDS

Comments 1.3.2(3)   Completeness of vector fields

C1   Intuitively the relationship between a vector field and its integral curves resembles that of a velocity field of a liquid in flow and its flow lines or an electric field and its field lines. The maximal integral curve starting from $\alpha(0)$ is the integral curve starting from $\alpha(0)$ defined on the largest domain. This is to contrast with other "shorter" integral curves starting from $\alpha(0)$. From now on an integral curve means a maximal integral curve.

C2   Based on the properties of differential equations we can show that there exists one and only one (maximal) integral curve of a vector field starting from any given point. Let σ be the integral curve of the vector field

$$\widehat{X} = \sum_{j=1}^{n} X^j \frac{\partial}{\partial x^j}, \tag{1.80}$$

starting from the point $\alpha(0)$. Let $x^j(\alpha(\tau))$ be the coordinates of the point $\alpha(\tau)$ on the curve. Then $x^j(\alpha(\tau))$ and $X^j(\alpha(\tau))$ are smooth functions of τ satisfying the following first order differential equations:

$$\frac{dx^j(\alpha(\tau))}{d\tau} = X^j(\alpha(\tau)). \tag{1.81}$$

Given $X^j(\alpha(\tau))$ one can solve for $x^j(\alpha(\tau))$ as functions of τ. If all these functions $x^j(\alpha(\tau))$ corresponding to integral curves starting from every point in $I\!\!E^n$ are defined for all values of $\tau \in (-\infty, \infty)$, then the vector field is complete. In §1.3.2E(1) below we shall examine a number of vector fields and their integral curves explicitly.

C3   Sometimes we want to confine ourselves to an open subset Λ of $I\!\!E^n$, i.e., we desire to define vector fields and examine their integral curves within a subset Λ. Clearly we can do this and everything introduced on $I\!\!E^n$ can be carried over on Λ.

C4   A Euclidean space structure is not necessary for the introduction of vector fields and their integral curves. A similar statement applies to the introduction of covariant vectors.

Examples 1.3.2(1)   Vector fields, integral curves and completeness

E1   A rectangular Cartesian coordinate system in $I\!\!E^3$ is often denoted by (x, y, z). A corresponding Cartesian coordinates in $I\!\!E^2$ is denoted by (x, y). A vector field on $I\!\!E^3$ is of the form

$$\widehat{X} = X^1 \frac{\partial}{\partial x} + X^2 \frac{\partial}{\partial y} + X^3 \frac{\partial}{\partial z}, \tag{1.82}$$

and its integral curves satisfy equations of the form

$$\frac{dx}{d\tau} = X^1, \quad \frac{dy}{d\tau} = X^2, \quad \frac{dz}{d\tau} = X^3. \tag{1.83}$$

We shall adopt this notation in the ensuing examples many of which will have direct physical applications in later chapters.

E2   Consider the vector field

$$\widehat{X} = \frac{\partial}{\partial x} \tag{1.84}$$

in $\mathbb{E}^3$. The integral curve starting from any point (x_0, y_0, z_0) satisfies the following equations:

$$\frac{dx}{d\tau} = 1, \quad \frac{dy}{d\tau} = 0, \quad \frac{dz}{d\tau} = 0. \tag{1.85}$$

The solution is

$$x(\tau) = \tau + x_0, \quad y(\tau) = y_0, \quad z(\tau) = z_0 \tag{1.86}$$

valid for $\tau \in (-\infty, \infty)$. This is true for any initial point (x_0, y_0, z_0) in $\mathbb{E}^3$. This vector field is therefore complete.

E3   Consider the vector field

$$\widehat{X} = x\frac{\partial}{\partial x} + y\frac{\partial}{\partial y} \tag{1.87}$$

in $\mathbb{E}^3$. An integral curve starting from a point (x_0, y_0, z_0) is a solution of

$$\frac{dx}{d\tau} = x, \quad \frac{dy}{d\tau} = y, \quad \frac{dz}{d\tau} = 0. \tag{1.88}$$

The solution satisfying initial conditions $x(0) = x_0$, $y(0) = y_0$, $z(0) = z_0$ is

$$x(\tau) = x_0\, e^\tau, \quad y(\tau) = y_0\, e^\tau, \quad z(\tau) = z_0, \quad \text{for} \quad \tau \in (-\infty, \infty). \tag{1.89}$$

This vector field vanishes along the z-axis so that the integral curve starting from any point $(0, 0, z_0)$ on the z-axis is simply the point itself, i.e.,

$$x(\tau) = 0, \quad y(\tau) = 0, \quad z(\tau) = z_0, \quad \text{for} \quad \tau \in (-\infty, \infty). \tag{1.90}$$

This vector field is complete.

E4   Consider the vector field

$$\widehat{X} = y\frac{\partial}{\partial x} - x\frac{\partial}{\partial y} \tag{1.91}$$

1.3. DIFFERENTIAL OPERATORS, VECTORS AND FIELDS

in $I\!\!E^3$. An integral curve starting from a point (x_0, y_0, z_0) is a solution of

$$\frac{dx}{d\tau} = y, \quad \frac{dy}{d\tau} = -x, \quad \frac{dz}{d\tau} = 0 \tag{1.92}$$

satisfying initial conditions $x(0) = x_0, y(0) = y_0, z(0) = z_0$. The solution can be written down conveniently in terms of the usual cylindrical coordinates (r, θ, z) where

$$r = \sqrt{x^2 + y^2}, \quad \theta = \tan^{-1}(y/x). \tag{1.93}$$

The solution is

$$x(\tau) = r\sin(\tau + \theta), \quad y(\tau) = r\cos(\tau + \theta), \quad z(\tau) = z_0, \tag{1.94}$$

where

$$x_0 = r\sin\theta, \quad y_0 = r\cos\theta \quad \text{and} \quad \tau \in (-\infty, \infty). \tag{1.95}$$

The integral curves defined by Eq. (1.94) are circles lying on a plane parallel the x-y planes and centered at a point on the z-axis. This vector field vanishes along the z-axis so that the integral curve starting from any point $(0, 0, z_0)$ on the z-axis is simply the point itself. This vector field is again complete.

This example can be compared with a related vector field defined on a circle given in §1.5.1E(1) E5 below.

E5 A vector field is said to have a *compact support* if it vanishes outside a closed and bounded cube in $I\!\!E^n$. As an illustration, consider a vector field of the form

$$\widehat{X} = \xi \frac{d}{dx} \tag{1.96}$$

in $I\!\!E$, where $\xi = \xi(x)$ is a smooth function of compact support $[a, b]$ on $I\!\!E$ and $\xi(x) > 0$ for every $x \in (a, b)$. An integral curve σ starting from $x_0 \in (a, b)$ satisfies

$$\frac{dx}{d\tau} = \xi(x), \tag{1.97}$$

or

$$\int_{x_0}^{x} \frac{dx}{\xi(x)} = \tau. \tag{1.98}$$

We can appreciate that the integral tends to $-\infty$ as x approaches a and to ∞ as x approaches b since $\xi(x)$ tends to zero. It follows that τ can range from $-\infty$ to ∞. In other words such a vector field is complete. This result is also true in $I\!\!E^n$, i.e., generally a vector field of compact support in $I\!\!E^n$ is complete.[14]

---
[14] Abraham and Marsden (1978) Corollary 2.1.19 on p. 70.

E6 Consider vector fields of the form

$$\widehat{X}^{(k)} = x^k \frac{d}{dx}, \quad k = 1, 2, \ldots \tag{1.99}$$

in $I\!\!E$. All these vector fields have a single critical point at $x = 0$. There are two distinct cases:

1. When $k = 1$ we have a vector field $\widehat{X}^{(1)}$ whose integral curves satisfy equation $dx/d\tau = x$. The integral curve starting from x_0 is given by

$$x(\tau) = x_0 e^\tau, \quad \tau \in (-\infty, \infty). \tag{1.100}$$

The integral curve from $x = 0$ is the point itself. The vector field is therefore complete.

2. When $k > 1$ we have vector fields $\widehat{X}^{(2)}, \widehat{X}^{(3)}, \ldots$ which are all incomplete. As an illustration consider the case with $k = 2$, i.e., the vector field is

$$\widehat{X}^{(2)} = x^2 \frac{d}{dx}. \tag{1.101}$$

We have the equation $dx/d\tau = x^2$ for integral curves. The integral curve starting from a point $x_0 > 0$ is given by

$$x(\tau) = \frac{x_0}{1 - x_0 \tau}. \tag{1.102}$$

This solution is not valid for $\tau = 1/x_0$, i.e., the domain of the curve is not the entire real line $I\!\!R$. The vector field is therefore incomplete.

E7 The sum of two complete vector fields is not necessarily complete.[15] Let us illustrate this by considering the following three vector fields on $I\!\!E^2$:

1. The vector field

$$\widehat{X}_1 = y \frac{\partial}{\partial x} \tag{1.103}$$

is complete with its integral curve starting from the point (x_0, y_0) being

$$x(\tau) = y_0 \tau + x_0, \quad y(\tau) = y_0. \tag{1.104}$$

2. The vector field

$$\widehat{X}_2 = x^2 \frac{\partial}{\partial y} \tag{1.105}$$

is complete with its integral curves starting from a point (x_0, y_0) given by

$$x(\tau) = x_0, \quad y(\tau) = x_0^2 \tau + y_0. \tag{1.106}$$

---
[15] For another example see Abraham and Marsden (1978) Exercise 2.2H (i) on p. 99.

1.3. DIFFERENTIAL OPERATORS, VECTORS AND FIELDS

3. Now consider
$$\widehat{X}_3 = \widehat{X}_1 + \widehat{X}_2 = y\frac{\partial}{\partial x} + x^2\frac{\partial}{\partial y}. \tag{1.107}$$

Its integral curves are given by solutions of
$$\frac{dx}{d\tau} = y, \quad \frac{dy}{d\tau} = x^2. \tag{1.108}$$

The integral curve starting from
$$x_0 = 6/a^2, \quad y_0 = 12/a^3, \quad a \in \mathbb{R}^+ \tag{1.109}$$

is given by
$$x(\tau) = \frac{6}{(a-\tau)^2}, \quad y(\tau) = \frac{12}{(a-\tau)^3}. \tag{1.110}$$

This curve is not defined at $\tau = a$, rendering $\widehat{X}_3$ incomplete.

E8 We may desire to restrict ourselves to a subset Λ of $\mathbb{E}^n$. Then whether a vector field is complete or not would depend also on Λ. As examples let us consider two cases:

1. For the open interval $\Lambda = (0, \pi) \subset \mathbb{E}$ with coordinate $x \in (0, \pi)$ we can define the vector field $\widehat{X} = d/dx$. The integral curve starting from, say, the point $x = \pi/2$ is given by $x = \tau + \pi/2$. Note that the domain of the curve is $(-\pi/2, \pi/2)$ instead of the entire real line, and hence $\widehat{X}$ is incomplete in Λ.

2. For the half line $\mathbb{E}^+ = (0, \infty) \subset \mathbb{E}$ with coordinate $x \in (0, \infty)$ we can again define the vector field $\widehat{X} = d/dx$. The integral curve starting from, say, the point $x = \pi/2$ is given by $x = \tau + \pi/2$. The domain of the curve is $(-\pi/2, \infty)$ instead of the entire real line, and hence $\widehat{X}$ is incomplete. Similarly we can introduce such a vector field in the half line $\mathbb{E}^- = (-\infty, 0)$.

E9 The completeness of vector fields is a concept which plays a fundamental role in quantum mechanics. The completeness or otherwise of vector fields is directly relevant to the problem of quantizability. The fact that the sum of two complete vector fields is not necessarily complete will have an important consequence in quantization. We shall return to some of the examples presented here when we investigate quantization in §3.3 in Chapter 3. Some of the seemingly trivial examples presented above turn out to be physically very important when it comes to quantization.

Definition 1.3.2(4) Lie brackets

- The Lie bracket of two vector fields $\widehat{X}, \widehat{Y}$ in $I\!\!E^n$ is the vector field $\widehat{Z}$ on $I\!\!E^n$ defined by

$$\widehat{Z}(f) = \widehat{X}\left(\widehat{Y}(f)\right) - \widehat{Y}\left(\widehat{X}(f)\right), \quad f \in C^\infty(I\!\!E^n). \tag{1.111}$$

Comments 1.3.2(4) Explicit expressions and properties

C1 Let

$$\widehat{X} = \sum_{j=1}^n X^j \frac{\partial}{\partial x^j}, \quad \widehat{Y} = \sum_{j=1}^n Y^j \frac{\partial}{\partial x^j}, \tag{1.112}$$

then we have

$$\widehat{Z} = \sum_{k=1}^n Z^k \frac{\partial}{\partial x^k}, \quad Z^k = \sum_{j=1}^n \left(X^j \frac{\partial Y^k}{\partial x^j} - Y^j \frac{\partial X^k}{\partial x^j} \right). \tag{1.113}$$

The minus sign in the definition of the bracket eliminates terms with second-order derivatives so that the bracket contains only first-order derivatives. The Lie bracket is often written symbolically as

$$\widehat{Z} = [\widehat{X}, \widehat{Y}] = \widehat{X}\widehat{Y} - \widehat{Y}\widehat{X}. \tag{1.114}$$

C2 It is easy to verify that the Lie bracket is linear in $\widehat{X}$ and $\widehat{Y}$, and that

$$[\widehat{X}, \widehat{Y}] = -[\widehat{Y}, \widehat{X}], \tag{1.115}$$

$$\left[[\widehat{X}, \widehat{Y}], \widehat{Z}\right] + \left[[\widehat{Z}, \widehat{X}], \widehat{Y}\right] + \left[[\widehat{Y}, \widehat{Z}], \widehat{X}\right] = 0. \tag{1.116}$$

The set of all vector fields forms a Lie algebra under Lie bracket.[16]

C3 The Lie bracket of two complete vector fields is not necessarily complete.[17] An example is

$$\widehat{X} = y \frac{\partial}{\partial x}, \quad \widehat{Y} = x^2 \frac{\partial}{\partial y} \tag{1.117}$$

in $I\!\!E^2$. We have

$$\widehat{Z} = [\widehat{X}, \widehat{Y}] = -x^2 \frac{\partial}{\partial x} + 2xy \frac{\partial}{\partial y} \tag{1.118}$$

---

[16] Abraham and Marsden (1978), Martin (1991).
[17] For another example see Brickell and Clark (1970) Problems 8.2 on p. 139.

1.3. DIFFERENTIAL OPERATORS, VECTORS AND FIELDS

whose integral curve starting from $(x_0 > 0, y_0)$ is

$$x(\tau) = \frac{x_0}{1 + x_0 \tau}, \quad y(\tau) = (1 + x_0 \tau)^2 y_0 \qquad (1.119)$$

which is not defined for $\tau = -1/x_0$. So, $\widehat{Z}$ is incomplete.

1.3.3 Transformation groups and complete vector fields

Consider one-to-one mappings of $I\!\!E^n$ onto itself, i.e.,

$$T : I\!\!E^n \mapsto I\!\!E^n \quad \text{by} \quad \alpha \mapsto \bar{\alpha} = T\alpha. \qquad (1.120)$$

In view of the one-to-one and onto nature of T the inverse map T^{-1} exists and is again one-to-one and onto in nature. Let the coordinates of α and $\bar{\alpha}$ in $I\!\!E^n$ be x^j and $\bar{x}^j$ respectively. Each coordinate $\bar{x}^k$ is a function of the coordinates x^j.

Definition 1.3.3(1)   Differentiable mappings and diffeomorphisms

- The mapping T above is said to be *differentiable* if the coordinates $\bar{x}^k$ are smooth functions of x^j.

- The mapping T above is *diffeomorphic* and is a *diffeomorphism* if the inverse T^{-1} is also differentiable. A diffeomorphism is also known as a *transformation*.

Comments 1.3.3(1)   Mappings of spaces of different dimensions

C1   We can define differentiable mappings of spaces of different dimensions. Consider the following mapping of $I\!\!E^n$ into $I\!\!E^m$:

$$M : I\!\!E^n \mapsto I\!\!E^m \quad \text{by} \quad \alpha \mapsto \beta = M\alpha. \qquad (1.121)$$

Let x^j be the coordinates of $\alpha \in I\!\!E^n$ and y^k, $k = 1, 2, \ldots, m$, be the coordinates of $\beta = M\alpha \in I\!\!E^m$. The mapping is said to be *differentiable* if y^k are smooth functions of x^j. These mappings are not diffeomorphisms which map spaces of the same dimension.

C2   The notion of differentiable mappings can be extended to more general situations. Consider a mapping of $I\!\!R \times I\!\!E^n$ into $I\!\!E^n$:

$$T : I\!\!R \times I\!\!E^n \mapsto I\!\!E^n \quad \text{by} \quad (\tau, \alpha) \mapsto \bar{\alpha}(\tau) = T(\tau, \alpha), \qquad (1.122)$$

where $\tau \in \mathbb{R}$ and α, $\bar{\alpha}(\tau) \in \mathbb{E}^n$. For each τ we have a mapping of $\mathbb{E}^n$ onto itself:

$$T_\tau : \mathbb{E}^n \mapsto \mathbb{E}^n \quad \text{defined by} \quad \alpha \mapsto T_\tau \alpha = \bar{\alpha}(\tau) = T(\tau, \alpha). \tag{1.123}$$

In other words T corresponds to a family of such mappings. Conversely a family of such mappings can be assembled into a single mapping T from $\mathbb{R} \times \mathbb{E}^n$ to $\mathbb{E}^n$.

Let the coordinates of α and $\bar{\alpha}(\tau)$ be x^j and $\bar{x}^j(\tau)$ respectively. Mapping T is said to be differentiable if $\bar{x}^j(\tau)$ are smooth functions of x^j and τ.

Definition 1.3.3(2)   **One-parameter group of transformations**

- Let $T = \{T_\tau : \tau \in \mathbb{R}\}$ be a family of transformations of $\mathbb{E}^n$ such that[18]

  1. $T_{\tau_2}(T_{\tau_1}\alpha) = T_{\tau_2+\tau_1}\alpha$.
  2. The mapping from $\mathbb{R} \times \mathbb{E}^n$ to $\mathbb{E}^n$ defined by $(\tau, \alpha) \mapsto T_\tau \alpha$ is differentiable.

Then the family of transformations is called a *one-parameter group of transformations* of $\mathbb{E}^n$ or simply a *one-parameter transformation group of $\mathbb{E}^n$*.

Comments 1.3.3(2)   **Transformation groups and vector fields**

C1   The group structure of the family of transformations manifests itself with the identity element T_0 corresponding to $\tau = 0$ and with the inverse T_τ^{-1} of T_τ equal to $T_{-\tau}$. Note that

$$(T_{\tau_2} T_{\tau_1})^{-1} = T_{\tau_1}^{-1} T_{\tau_2}^{-1}. \tag{1.124}$$

C2   Consider an example in $\mathbb{E}$ with rectangular Cartesian coordinate x. Let T_τ be a family of mappings of $\mathbb{E}$ onto itself defined by mapping every $\alpha \in \mathbb{E}$ to another point $T_\tau \alpha$ according to the following coordinate expression:

$$x(\alpha) \mapsto x(T_\tau \alpha) = x(\alpha) + \tau, \quad \tau \in (-\infty, \infty). \tag{1.125}$$

One can see that T_τ form a one-parameter transformation group of $\mathbb{E}$, translating every point in $\mathbb{E}$ by an amount τ. This group generates a curve σ starting from each point $\alpha(0)$ by

$$x(\alpha(\tau)) = x(\alpha(0)) + \tau, \quad \tau \in (-\infty, \infty). \tag{1.126}$$

---

[18]Matsushima (1972) p. 79.

1.3. DIFFERENTIAL OPERATORS, VECTORS AND FIELDS

This curve in turn gives rise to a complete vector field

$$\widehat{X} = X \frac{d}{dx}, \quad \text{where} \quad X = \frac{dx(\alpha(\tau))}{d\tau} = 1 \quad \Rightarrow \quad \widehat{X} = \frac{d}{dx}. \tag{1.127}$$

C3 We can easily reverse the argument in C2 above. Given the complete vector field $\widehat{X} = d/dx$ in E we can obtain its integral curve σ starting from x. This enables us to define a transformation T_τ of E by

$$x \mapsto T_\tau x = \sigma(\tau) = x + \tau, \quad \text{for every } \tau \in (-\infty, \infty). \tag{1.128}$$

Then $T = \{T_\tau, \tau \in \mathbb{R}\}$ constitutes a one-parameter transformation group of E. We can illustrate the situation with two more examples:

1. The vector field $\widehat{X} = \xi(x)d/dx$ in Eq. (1.96) is complete in E. The transformation group is given by $x(\tau) = T_\tau x_0$ where $x(\tau)$ is related to x_0 and τ by Eq. (1.98).

2. Consider the vector field in Eq. (1.91). The integral curves are circles given by Eq. (1.94). The transformation group consists of rotations of E^3 about the z-axis.

C4 Generally let $T : \mathbb{R} \times E^n \mapsto E^n$ be a one-parameter transformation group of E^n. We can generate a curve σ starting from any $\alpha \in E^n$ by

$$\alpha(\tau) = T_\tau \alpha, \quad \tau \in \mathbb{R}. \tag{1.129}$$

All these curves are defined on the domain $\mathbb{R}$. We can also generate a vector field with these curves as integral curves by assigning the vector

$$\widehat{X}_\alpha = \sum_{j=1}^{n} X_\alpha^j \frac{\partial}{\partial x^j}\bigg|_\alpha, \quad X_\alpha^j = \frac{dx^j(\alpha(\tau))}{d\tau}\bigg|_{\tau=0} \tag{1.130}$$

to every $\alpha \in E^n$. Moreover, this is a smooth assignment in view of the differentiable nature of the mapping T. The resulting vector field is complete since its integral curves are defined on the domain $\mathbb{R}$. The converse statement is also true. Given a complete vector field $\widehat{X}$ we can construct a transformation group T of E^n by defining its element T_τ in terms of the integral curves σ of $\widehat{X}$ by $T_\tau \alpha = \sigma(\tau)$. This group T is called the *transformation group of the vector field* $\widehat{X}$.

C5 We can summarize our discussions above as follows:[19]

a one-parameter transformation group of E^n gives rise to a complete vector field on E^n, and the converse is also true.

---
[19] Choquet-Bruhat, de Witt-Morette with Dillard-Bleick (1989) p. 145.

CHAPTER 1. MANIFOLDS AND DYNAMICAL SYSTEMS

Definition 1.3.3(3)   **Flow of vector fields and transformation groups**

- The *flow* of a complete vector field is defined to be the transformation group of the vector field.

Comments 1.3.3(3)   **The concept of flow**

C1   For a fluid in motion we have a *velocity field* to describe its motion. The fluid particles move along the integral curves of the velocity field. We have an intuitive notion of fluid flow to signify how far and fast the fluid particles move in time. This intuitive notion is now sharpened by a precise definition in terms of the transformation group of the velocity field. This makes sense since this group contains information on how fast and far the fluid moves.

C2   For an incomplete vector field $\widehat{X}$ we still have an integral curve σ starting from every point α. Generally the domain $J(\alpha)$ of the integral curve starting from α is dependent on α and it may not be the entire real line.

C3   It is possible to restrict one's attention to a neighbourhood $\mathcal{N}$ of α and consider the mappings of $\mathcal{N}$ onto new regions $\mathcal{N}'$ by moving every point $\alpha \in \mathcal{N}$ along its integral curve by a common parameter τ. This can be used to establish a concept of local transformations.

Definition 1.3.3(4)   **Lie derivative and derivative along an integral curve**

- The *Lie derivative* of a function f with respect to a vector field $\widehat{X}$ is defined to be the function $\widehat{X}(f)$.

- Let σ be an integral curve of a given vector field $\widehat{X}$ starting from α. Then the *derivative of a function f* along this integral curve at α is defined to be

$$\left.\frac{df(\alpha(\tau))}{d\tau}\right|_{\tau=0} = \lim_{\tau \mapsto 0} \frac{f(\alpha(\tau)) - f(\alpha(0))}{\tau}, \quad \alpha(0) = \alpha. \quad (1.131)$$

Comments 1.3.3(4)   **Relationship between the two derivatives**

C1   Given a vector field $\widehat{X} = \sum_{j=1}^{n} X^j \, d/dx^j$ we have

$$\widehat{X}(f) = \sum_{j=1}^{n} X^j \frac{\partial f}{\partial x^j}. \quad (1.132)$$

1.3. DIFFERENTIAL OPERATORS, VECTORS AND FIELDS

Let σ be an integral curve of $\widehat{X}$ starting from α. Then

$$\left.\frac{dx^i(\alpha(\tau))}{d\tau}\right|_{\tau=0} = \left. X^j \right|_\alpha. \tag{1.133}$$

We have

$$\left.\frac{df(\alpha(\tau))}{d\tau}\right|_{\tau=0} = \sum_{j=1}^{n}\left[\frac{\partial f(\alpha(\tau))}{\partial x^j(\alpha(\tau))}\frac{dx^j(\alpha(\tau))}{d\tau}\right]_{\tau=0} \tag{1.134}$$

$$= \sum_{j=1}^{n}\left[\frac{\partial f}{\partial x^j}X^j\right]_\alpha. \tag{1.135}$$

It follows from Eqs. (1.132) and (1.135) that

$$\left.\frac{df(\alpha(\tau))}{d\tau}\right|_{\tau=0} = \widehat{X}_\alpha(f). \tag{1.136}$$

In other words the derivative of a function at a point along the integral curve is equal to the Lie derivative of the function with respect to the vector field at that point.

C2 We can also write down an explicit expression of a complete vector field $\widehat{X}$ in terms of the elements of its transformation group T as follows:

$$\widehat{X} = \sum_{j=1}^{n} X^j \frac{\partial}{\partial x^j}, \quad \text{where} \quad X^j(\alpha) = \left.\frac{dx^j(T_\tau \alpha)}{d\tau}\right|_{\tau=0}. \tag{1.137}$$

We also have

$$\left.\frac{df(T_\tau \alpha)}{d\tau}\right|_{\tau=0} = \widehat{X}_\alpha(f), \tag{1.138}$$

or

$$\left.\frac{df(T_\tau^{-1} \alpha)}{d\tau}\right|_{\tau=0} = -\widehat{X}_\alpha(f). \tag{1.139}$$

These expressions are useful for later applications.

Definition 1.3.3(5)   **Gradient, divergence and the Laplacian**

- Let f be a smooth function on $\mathbb{E}^n$ with rectangular Cartesian coordinates x^j. The *gradient* of f, denoted by ∇f, is the vector field

$$\nabla f = \sum_{j=1}^{n} \frac{\partial f}{\partial x^j}\frac{\partial}{\partial x^j}. \tag{1.140}$$

- Let $\widehat{X} = \sum_{j=1}^{n} X^j \partial/\partial x^j$ be a vector field on $I\!\!E^n$. The *divergence* of $\widehat{X}$, denoted by div $\widehat{X}$, is the function

$$\text{div}\,\widehat{X} = \sum_{j=1}^{n} \frac{\partial X^j}{\partial x^j}. \tag{1.141}$$

$\widehat{X}$ is said to be *divergence-free* if div $\widehat{X}$ vanishes everywhere in $I\!\!E^n$.

- The *Laplacian* of a function f, denoted by $\nabla^2 f$, is the function

$$\nabla^2 f = \text{div}\,\nabla f = \sum_{j=1}^{n} \frac{\partial^2 f}{\partial (x^j)^2}. \tag{1.142}$$

Comments 1.3.3(5)   **Coordinate independence**

C1   The above quantities are actually coordinate independent. We have chosen what appear to be coordinate dependent definitions for familiarity.

C2   As will be seen in Chapter 3 divergence-free vector fields have some simplifying properties in quantization.

1.4 Cotangent Vectors and Differential Forms

Given any vector space V we can define linear mappings of V to $I\!\!R$. Such mappings are known as *linear functionals* on the vector space. Under usual addition and scalar multiplication rules the set of all these linear functionals form a vector space called the *dual space* to V, denoted by V^*. As an example consider the set of contravariant vectors X^j at a given point $\alpha \in I\!\!E^n$. We know that:

1. This set forms a vector space V_α.[20]

2. A covariant vector Y_j at α induces a mapping $\mathcal{Y}_\alpha$ of V_α to $I\!\!R$ through a contraction operation.[21]

In other words a covariant vector gives rise to a linear functional on the space of contravariant vectors at α. It turns out that we can actually identify a covariant vector with such a linear functional, i.e., we can identify the mapping $\mathcal{Y}_\alpha$ with the covariant vector Y_j. This argument will now be used to formulate the notion of cotangent vectors as linear functionals on the space of tangent vectors. Cotangent vectors are the intrinsic definition of covariant vectors in the same way that tangent vectors are the intrinsic definition of contravariant vectors.

---

[20] See §1.2.3C(1) C1.
[21] See §1.2.4C(1) C7 and §1.2.4C(1) C4, C5.

1.4.1 Cotangent vectors, differentials and one-forms

Definition 1.4.1(1)   **Cotangent vectors**

- A *cotangent vector* $\widehat{Y}^*_\alpha$ at a point α in $I\!\!E^n$ is a linear mapping of the tangent space $\widehat{T}_\alpha(I\!\!E^n)$ at α to $I\!\!R$.

- The set of all cotangent vectors at α endowed with the usual vector space structure of a dual space is called the *cotangent space* at α and is denoted by $\widehat{T}^*_\alpha(I\!\!E^n)$.

Comments 1.4.1(1)   **Representation of cotangent vectors**

C1   We often denote cotangent vectors at α by a capital letter with an asterisk, e.g., by $\widehat{W}^*_\alpha$, $\widehat{Z}^*_\alpha$ to highlight the fact that cotangent vectors are linear functionals on the tangent space and so on. A cotangent vector $\widehat{W}^*_\alpha$ acts on a tangent vector $\widehat{X}_\alpha$ to produce a number $\widehat{W}^*_\alpha(\widehat{X}_\alpha)$.

C2   Similar to Eq. (1.72) the vector space structure of a cotangent space $\widehat{T}^*_\alpha(I\!\!E^n)$ is defined by

$$\left(\widehat{W}^*_\alpha + \widehat{Z}^*_\alpha\right)(\widehat{X}_\alpha) = \widehat{W}^*_\alpha(\widehat{X}_\alpha) + \widehat{Z}^*_\alpha(\widehat{X}_\alpha), \tag{1.143}$$

$$\left(a\widehat{W}^*_\alpha\right)(\widehat{X}_\alpha) = a\left(\widehat{W}^*_\alpha(\widehat{X}_\alpha)\right). \tag{1.144}$$

C3   A tangent vector $\widehat{X}_\alpha$ is a differential operator, i.e.,

$$\widehat{X}_\alpha = \sum_{j=1}^n X^j_\alpha \, (\partial/\partial x^j)_\alpha, \tag{1.145}$$

which acts on functions $f \in C^\infty(\alpha)$ to yield a value

$$\widehat{X}_\alpha(f) = \sum_{j=1}^n X^j_\alpha \left(\frac{\partial f}{\partial x^j}\right)_\alpha. \tag{1.146}$$

To appreciate the significance of this in our present context let us make the following observations:

1. We can define a contravariant coordinate displacement vector at a point $\alpha \in I\!\!E^n$ by a set of numerical components $(\Delta x)^1_\alpha, (\Delta x)^2_\alpha, \ldots, (\Delta x)^n_\alpha$ representing a displacement.[22] Let us denote such a displacement vector by $(\Delta x)^j_\alpha$.

---
[22] See §1.2.3C(1) C3.

2. In calculus the differential $(df)_\alpha$ of a function f at a point α along a displacement vector $(\Delta x)_\alpha^j$ is given by

$$(df)_\alpha = \sum_{j=1}^n \left(\frac{\partial f}{\partial x^j}\right)_\alpha (\Delta x)_\alpha^j. \tag{1.147}$$

We can see that while at each point α the derivative $(\partial f/\partial x^j)_\alpha$ of f has a definite value determined by the function, the numerical value of the differential $(df)_\alpha$ is not determined by f alone. The displacement vector $(\Delta x)_\alpha^j$ is also involved, i.e., the differential $(df)_\alpha$ takes different values for different displacement vectors.

3. We know from §1.3.1C(3) C5 that X_α^j form a contravariant vector at α. If we associate and equate X_α^j with a displacement vector $(\Delta x)_\alpha^j$ we can, by comparing Eqs. (1.146) and (1.147), identify $\widehat{X}_\alpha(f)$ with the differential $(df)_\alpha$ of f along the direction specified by the contravariant displacement vector $X_\alpha^j = (\Delta x)_\alpha^j$. Since the value of the differential $(df)_\alpha$ is generally different for different displacement vectors X_α^j we can generate a mapping of the space V_α of contravariant displacement vectors to the reals by $X_\alpha^j \mapsto \widehat{X}_\alpha(f)$. We call this mapping the *differential of f at α* and denote it simply by $(df)_\alpha$, i.e., we have, for a given function f,

$$(df)_\alpha : V_\alpha \mapsto \mathbb{R} \quad \text{by} \quad X_\alpha^j \mapsto \widehat{X}_\alpha(f). \tag{1.148}$$

4. Equations (1.146), (1.147) and (1.148) are linear in f.

5. Since V_α is identifiable with the tangent space $\widehat{T}_\alpha(\mathbb{E}^n)$ we finally arrive at a mapping of $\widehat{T}_\alpha(\mathbb{E}^n)$ to $\mathbb{R}$ defined by Eq. (1.148). We also call this map the the *differential* of f at α. A formal definition of this is given below.

Definition 1.4.1(2)   Differentials of functions on $\mathbb{E}^n$

- The *differential* $(df)_\alpha$ of a function $f \in C^\infty(\alpha)$ at a point α is a mapping of the tangent space $\widehat{T}_\alpha(\mathbb{E}^n)$ at α to the reals

$$(df)_\alpha : \widehat{T}_\alpha(\mathbb{E}^n) \mapsto \mathbb{R} \tag{1.149}$$

  defined by

$$\widehat{X}_\alpha \mapsto (df)_\alpha(\widehat{X}_\alpha) = \widehat{X}_\alpha(f) \in \mathbb{R}. \tag{1.150}$$

1.4. COTANGENT VECTORS AND DIFFERENTIAL FORMS

Comments 1.4.1(2) **Differentials, cotangent and covariant vectors**

C1 It should be emphasized that despite its familiar expression the differential of a function at a point α is not a number or a function. It must be considered as a mapping of $\widehat{T}_\alpha(I\!\!E^n)$ to $I\!\!R$. In other words it is a **cotangent vector**. In particular we have to distinguish a contravariant coordinate displacement vector $(\Delta x)_\alpha^j$ and the differential $(dx^j)_\alpha$ of a coordinate variable x^j regarded as a function in its own right on $I\!\!E^n$. The former consists of a set of numerical components while the latter is a cotangent vector mapping the tangent space $\widehat{T}_\alpha(I\!\!E^n)$ to the reals by

$$(dx^j)_\alpha : \widehat{T}_\alpha(I\!\!E^n) \mapsto I\!\!R \quad \text{by} \quad \widehat{X}_\alpha \mapsto (dx^j)_\alpha(\widehat{X}_\alpha) = \widehat{X}_\alpha(x^j) = X_\alpha^j. \quad (1.151)$$

As the dual space to $\widehat{T}_\alpha(I\!\!E^n)$ the cotangent space is also n-dimensional. Since there are n independent coordinate differentials $(dx^j)_\alpha$, $j = 1, 2, \ldots, n$, we can conclude that $(dx^j)_\alpha$ span the cotangent space $\widehat{T}_\alpha^*(I\!\!E^n)$. In other words a cotangent vector is generally of the form

$$\widehat{Y}_\alpha^* = \sum_{j=1}^n Y_{\alpha,j} (dx^j)_\alpha, \quad Y_{\alpha,j} \in I\!\!R. \quad (1.152)$$

As an immediate application we can write

$$(df)_\alpha = \sum_{j=1}^n \left(\frac{\partial f}{\partial x^j}\right)_\alpha (dx^j)_\alpha. \quad (1.153)$$

This is similar to a tangent space being spanned by derivations $(\partial/\partial x^j)_\alpha$.

C2 Let us examine how the differential $(df)_\alpha$ of a function f is related to a covariant vector. To do this we must examine how the expression for a differential changes under a coordinate transformation. On a transformation to a new coordinate system x'^j the expression for $(df)_\alpha$ in Eq. (1.153) can be written as

$$(df)_\alpha = \sum_{j=1}^n \left(\frac{\partial f}{\partial x^j}\right)_\alpha \left(\sum_{k=1}^n \left(\frac{\partial x^j}{\partial x'^k}\right)_\alpha (dx'^k)_\alpha\right). \quad (1.154)$$

It follows that

$$(df)_\alpha = \sum_{k=1}^n \left(\frac{\partial f}{\partial x'^k}\right)_\alpha (dx'^k)_\alpha, \quad (1.155)$$

where

$$\left(\frac{\partial f}{\partial x'^k}\right)_\alpha = \sum_{j=1}^{n} \left(\frac{\partial x^j}{\partial x'^k}\right)_\alpha \left(\frac{\partial f}{\partial x^j}\right)_\alpha. \tag{1.156}$$

We can now rewrite Eq. (1.153) as

$$(df)_\alpha = \sum_{j=1}^{n} Y_{\alpha,j} (dx^j)_\alpha \tag{1.157}$$

$$= \sum_{k=1}^{n} Y'_{\alpha,k} (dx'^k)_\alpha, \tag{1.158}$$

where

$$Y_{\alpha,j} = \left(\frac{\partial f}{\partial x^j}\right)_\alpha, \quad Y'_{\alpha,k} = \sum_{j=1}^{n} \left(\frac{\partial x^j}{\partial x'^k}\right)_\alpha Y_{\alpha,j}. \tag{1.159}$$

The coefficients $Y_{\alpha,j}$ of the cotangent vector $Y_\alpha^* = (df)_\alpha$ are seen to transform as the components of a covariant vector. Note that a coordinate differential $(dx^i)_\alpha$ corresponds to a covariant vector having components of the form of a Kronecker delta, i.e., $Y_{\alpha,j} = \delta_i^j$. The transformed components in the dashed coordinates are given, according to Eq. (1.159), by

$$Y'_{\alpha,k} = \sum_{i=1}^{n} \left(\frac{\partial x^i}{\partial x'^k}\right)_\alpha \delta_i^j = \left(\frac{\partial x^j}{\partial x'^k}\right)_\alpha. \tag{1.160}$$

So, generally the components of $\widehat{Y}_\alpha^*$ in Eq. (1.152) are seen to transform according to Eq. (1.159). This leads us to conclude that cotangent vectors are the intrinsic definition of covariant vectors. We can now appreciate why the numerical components of a covariant vector transform the way they do.

C3 To sum up, we have the following results:

1. A coordinate differential $(dx^k)_\alpha$ is a cotangent vector.

2. A partial derivative $(\partial/\partial x^j)_\alpha$ is a tangent vector.

3. When using $(dx^k)_\alpha$ to act on tangent vectors $\widehat{X}_\alpha$ we have two cases:

    (a) For $\widehat{X}_\alpha = (\partial/\partial x^j)_\alpha$ we get

$$(dx^k)_\alpha \left(\left(\frac{\partial}{\partial x^j}\right)_\alpha\right) = \left(\frac{\partial x^k}{\partial x^j}\right)_\alpha = \delta_j^k. \tag{1.161}$$

1.4. COTANGENT VECTORS AND DIFFERENTIAL FORMS

(b) For $\widehat{X}_\alpha = \sum_j X_\alpha^j (\partial/\partial x^j)_\alpha$ we get

$$(dx^k)_\alpha(\widehat{X}_\alpha) = X_\alpha^k. \tag{1.162}$$

4. When using a general cotangent vector

$$\widehat{Y}_\alpha^* = \sum_{k=1}^n Y_{\alpha,k}(dx^k)_\alpha \tag{1.163}$$

to act on a general tangent vector

$$\widehat{X}_\alpha = \sum_j X_\alpha^j (\partial/\partial x^j)_\alpha, \tag{1.164}$$

we get

$$\widehat{Y}_\alpha^* \left(\widehat{X}_\alpha\right) = \sum_{j=1}^n Y_{\alpha,j} X_\alpha^j. \tag{1.165}$$

This agrees with the previous contraction process.

C4 We can introduce cotangent vector fields by assigning a cotangent vector at each point in $I\!E^n$. One way to achieve this is to extend the concept of the differential of a function at a point α to the differential of a function on $I\!E^n$. The differential of a function f on $I\!E^n$, denoted by df, can be regarded as a cotangent vector field in that it assigns a cotangent vector $(df)_\alpha$ at each point α in $I\!E^n$. A cotangent vector field is commonly called a *differential form*.

Definition 1.4.1(3) One-forms

- A *one-form* $\Omega^{(1)}$ on $I\!E^n$ is a smooth assignment of cotangent vectors on $I\!E^n$ with a cotangent vector $\widehat{Y}_\alpha^*$ to each point $\alpha \in I\!E^n$. Here smoothness means that the function g on $I\!E^n$ defined by $g(\alpha) = \widehat{Y}_\alpha^*(\widehat{X}_\alpha)$ is smooth for any given (smooth) vector field $\widehat{X}$.

Comments 1.4.1(3) Explicit description of one-forms

C1 The cotangent vector assigned to a point α by a one-form is generally denoted by $\Omega_\alpha^{(1)}$.

C2 A simple example of one-forms is simply the differential of a function $f \in C^\infty(I\!E^n)$ on $I\!E^n$. We can write down an expression for df as

$$\Omega^{(1)} = df = \sum_{j=1}^{n} \frac{\partial f}{\partial x^j} dx^j. \tag{1.166}$$

This one-form assigns a cotangent vector

$$\Omega^{(1)}_\alpha = (df)_\alpha = \sum_{j=1}^{n} \left(\frac{\partial f}{\partial x^j}\right)_\alpha (dx^j)_\alpha \tag{1.167}$$

to every point α. Coordinate differentials dx^j are also examples of one-forms.[23]

C3 One-forms dx^j and $\Omega^{(1)} = df$ act on vector fields $\widehat{X} = \sum_j X^j \partial/\partial x^j$ to yield

$$dx^j(\widehat{X}) = X^j, \tag{1.168}$$

$$\Omega^{(1)}(\widehat{X}) = \sum_{j=1}^{n} \frac{\partial f}{\partial x^j} X^j. \tag{1.169}$$

The right-hand side of both of these equations is a smooth function on $I\!E^n$.

C4 Generally, a one-form is expressible as

$$\Omega^{(1)} = \sum_{j=1}^{n} u_j dx^j, \quad u_j \in C^\infty(I\!E^n), \tag{1.170}$$

with

$$\Omega^{(1)}(\widehat{X}) = \sum_{j=1}^{n} u_j X^j, \tag{1.171}$$

which is a smooth function on $I\!E^n$. But $\Omega^{(1)}$ is *not necessarily the differential of a function*, i.e., there may not exist a function f on $I\!E^n$ such that $u_j = \partial f/\partial x^j$ at every point in $I\!E^n$. For example, consider a one-form $\Omega^{(1)}$ in $I\!E^2$ given in rectangular Cartesian coordinates (x, y) by

$$\Omega^{(1)} = y\, dx + x^2\, dy. \tag{1.172}$$

There does not exist a function f such that

$$\frac{\partial f}{\partial x} = y, \quad \frac{\partial f}{\partial y} = x^2 \quad \text{since} \quad \frac{\partial^2 f}{\partial y \partial x} \neq \frac{\partial^2 f}{\partial x \partial y}. \tag{1.173}$$

---

[23] Loomis and Sternberg (1968) pp. 393-397.

1.4.2 Tensor fields and two-forms

At any point $\alpha \in I\!E^n$ we have two vector spaces, the tangent space $\widehat{T}_\alpha(I\!E^n)$ and the cotangent space $\widehat{T}^*_\alpha(I\!E^n)$. We can form their tensor product spaces

$$\widehat{T}_\alpha(I\!E^n) \otimes \widehat{T}_\alpha(I\!E^n), \quad \widehat{T}^*_\alpha(I\!E^n) \otimes \widehat{T}^*_\alpha(I\!E^n), \quad \widehat{T}_\alpha(I\!E^n) \otimes \widehat{T}^*_\alpha(I\!E^n) \quad (1.174)$$

which lead to the definition of various types of tensors.[24]

Definition 1.4.2(1)   **Different types of tensors**

- Elements of $\widehat{T}_\alpha(I\!E^n) \otimes \widehat{T}_\alpha(I\!E^n)$ are called contravariant tensors of the second order at the point α.

- Elements of $\widehat{T}^*_\alpha(I\!E^n) \otimes \widehat{T}^*_\alpha(I\!E^n)$ are called covariant tensors of the second order at the point α.

- Elements of $\widehat{T}_\alpha(I\!E^n) \otimes \widehat{T}^*_\alpha(I\!E^n)$ are called mixed tensors of the second order at the point α.

Comments 1.4.2(1)   **Anti-symmetric tensors and wedge products**

C1   We can write down the expression for a contravariant tensor at a point α as

$$\widehat{C}^\otimes_\alpha = \sum_{j,k=1}^n C^{jk}_\alpha \left(\frac{\partial}{\partial x^j}\right)_\alpha \otimes \left(\frac{\partial}{\partial x^k}\right)_\alpha, \quad (1.175)$$

where C^{jk}_α are the numerical components of the tensor in coordinate system x^j. Similar expressions for covariant and mixed tensors are given respectively by

$$\widehat{C}^{\otimes *}_\alpha = \sum_{j,k=1}^n C_{\alpha,jk} \left(dx^j\right)_\alpha \otimes \left(dx^k\right)_\alpha, \quad (1.176)$$

$$\widehat{M}^\otimes_\alpha = \sum_{j,k=1}^n M^j_{\alpha,k} \left(\frac{\partial}{\partial x^j}\right)_\alpha \otimes \left(dx^k\right)_\alpha. \quad (1.177)$$

These definitions can be extended to define tensors of higher orders.

It is easy to verify that upon a coordinate transformation, the numerical components of a tensor transform according to one of the Eqs. (1.34), (1.35) and (1.36) for tensor transformations. Indeed we can say that these transformation equations come from the above intrinsic definitions of tensors.

---
[24] For an intuitive definition of tensor product of vector spaces see Darling (1994). See also §2.2.3 later for a definition of tensor product of Hilbert spaces.

C2   The relation between cotangent vectors and tangent vectors carries over to contravariant and covariant tensors. A second order covariant tensor $\widehat{C}_\alpha^{\otimes *}$ may be defined as a *bilinear functional* on the Cartesian product $\widehat{T}_\alpha(I\!E^n) \times \widehat{T}_\alpha(I\!E^n)$. In other words we have

$$\widehat{C}_\alpha^{\otimes *} : \widehat{T}_\alpha(I\!E^n) \times \widehat{T}_\alpha(I\!E^n) \mapsto I\!R \tag{1.178}$$

by

$$(\widehat{X}_\alpha, \widehat{Y}_\alpha) \mapsto \widehat{C}_\alpha^{\otimes *}(\widehat{X}_\alpha, \widehat{Y}_\alpha) \in I\!R \tag{1.179}$$

with properties:

$$\widehat{C}_\alpha^{\otimes *}(a\widehat{X}_\alpha + b\widehat{Y}_\alpha, \widehat{Z}_\alpha) = a\widehat{C}_\alpha^{\otimes *}(\widehat{X}_\alpha, \widehat{Z}_\alpha) + b\widehat{C}_\alpha^{\otimes *}(\widehat{Y}_\alpha, \widehat{Z}_\alpha), \tag{1.180}$$

$$\widehat{C}_\alpha^{\otimes *}(\widehat{Z}_\alpha, a\widehat{X}_\alpha + b\widehat{Y}_\alpha) = a\widehat{C}_\alpha^{\otimes *}(\widehat{Z}_\alpha, \widehat{X}_\alpha) + b\widehat{C}_\alpha^{\otimes *}(\widehat{Z}_\alpha, \widehat{Y}_\alpha). \tag{1.181}$$

C3   For a general covariant tensor $\widehat{C}_\alpha^{\otimes *}$ in Eq. (1.176) we have the following explicit expression:

$$\widehat{C}_\alpha^{\otimes *}\left(\widehat{X}_\alpha, \widehat{Y}_\alpha\right) = \sum_{j,k=1}^n C_{\alpha,jk}\, X_\alpha^j Y_\alpha^k. \tag{1.182}$$

This reduces to

$$\left((dx^j)_\alpha \otimes (dx^k)_\alpha\right)\left((\partial/\partial x^r)_\alpha, (\partial/\partial x^s)_\alpha\right) = \delta_r^j\, \delta_s^k, \tag{1.183}$$

$$\left((dx^j)_\alpha \otimes (dx^k)_\alpha\right)\left(\widehat{X}_\alpha, \widehat{Y}_\alpha\right) = X_\alpha^j Y_\alpha^k \tag{1.184}$$

when $\widehat{C}_\alpha^{\otimes *} = (dx^j)_\alpha \otimes (dx^k)_\alpha$.

Likewise, we can regard a contravariant tensor as a bilinear functional on the Cartesian product $\widehat{T}_\alpha^*(I\!E^n) \times \widehat{T}_\alpha^*(I\!E^n)$ of cotangent vectors.

C4   A second order covariant tensor $\widehat{C}_\alpha^{\otimes *}$ is said to be *anti-symmetric* if

$$\widehat{C}_\alpha^{\otimes *}(\widehat{X}_\alpha, \widehat{Y}_\alpha) = -\widehat{C}_\alpha^{\otimes *}(\widehat{Y}_\alpha, \widehat{X}_\alpha). \tag{1.185}$$

This is consistent with the definition in §1.2.4C(1) C2, i.e., for $\widehat{C}_\alpha^{\otimes *}$ to be anti-symmetric it must satisfy the condition $C_{\alpha,jk} = -C_{\alpha,kj}$. Symmetric covariant tensors are similarly defined. Many tensors of physical significance, such as the electromagnetic tensor, are anti-symmetric. A simple example is

$$\widehat{C}_\alpha^{\otimes *} = (dx^j)_\alpha \otimes (dx^k)_\alpha - (dx^k)_\alpha \otimes (dx^j)_\alpha \tag{1.186}$$

with

$$\widehat{C}_\alpha^{\otimes *}\left(\widehat{X}_\alpha, \widehat{Y}_\alpha\right) = X_\alpha^j Y_\alpha^k - X_\alpha^k Y_\alpha^j \tag{1.187}$$

1.4. COTANGENT VECTORS AND DIFFERENTIAL FORMS

clearly showing anti-symmetry. We can build up other anti-symmetric covariant tensors in terms of these simple ones. To simplify the notation we introduce the concept of an *exterior product*, denoted symbolically by a wedge, hence the alternative name *wedge product*:

$$(dx^j)_\alpha \wedge (dx^k)_\alpha = (dx^j)_\alpha \otimes (dx^k)_\alpha - (dx^k)_\alpha \otimes (dx^j)_\alpha. \tag{1.188}$$

We have

$$\left((dx^j)_\alpha \wedge (dx^k)_\alpha\right)\left(\widehat{X}_\alpha, \widehat{Y}_\alpha\right) = X_\alpha^j Y_\alpha^k - X_\alpha^k Y_\alpha^j. \tag{1.189}$$

In $I\!\!E^3$ there are only 3 such wedge products at each point:

$$(dx^1)_\alpha \wedge (dx^2)_\alpha, \quad (dx^2)_\alpha \wedge (dx^3)_\alpha, \quad (dx^1)_\alpha \wedge (dx^3)_\alpha \tag{1.190}$$

or equivalently

$$(dx)_\alpha \wedge (dy)_\alpha, \quad (dy)_\alpha \wedge (dz)_\alpha, \quad (dx)_\alpha \wedge (dz)_\alpha. \tag{1.191}$$

These products form the basis spanning the space of second order anti-symmetric covariant tensors at α, i.e., an anti-symmetric covariant tensor $\widehat{C}_\alpha^{\otimes *}$ is expressible as

$$\begin{aligned}\widehat{C}_\alpha^{\otimes *} &= C_{\alpha,12}\,(dx^1)_\alpha \wedge (dx^2)_\alpha \\ &+ C_{\alpha,23}\,(dx^2)_\alpha \wedge (dx^3)_\alpha + C_{\alpha,13}\,(dx^1)_\alpha \wedge (dx^3)_\alpha,\end{aligned} \tag{1.192}$$

or more succinctly

$$\widehat{C}_\alpha^{\otimes *} = \sum_{1 \leq j < k}^{3} C_{\alpha,jk}\,(dx^j)_\alpha \wedge (dx^k)_\alpha. \tag{1.193}$$

C5 Let

$$\widehat{X} = x\frac{\partial}{\partial x} + \frac{\partial}{\partial y} + \frac{\partial}{\partial z}, \tag{1.194}$$

$$\widehat{Y} = \frac{\partial}{\partial x} + 2y\frac{\partial}{\partial y} + z\frac{\partial}{\partial z} \tag{1.195}$$

be two vector fields on $I\!\!E^3$ expressed in usual Cartesian coordinates (x, y, z). At a point $\alpha = (1, 1, 3)$ we have two tangent vectors

$$\widehat{X}_\alpha = \left(\frac{\partial}{\partial x}\right)_\alpha + \left(\frac{\partial}{\partial y}\right)_\alpha + \left(\frac{\partial}{\partial z}\right)_\alpha, \tag{1.196}$$

$$\widehat{Y}_\alpha = \left(\frac{\partial}{\partial x}\right)_\alpha + 2\left(\frac{\partial}{\partial y}\right)_\alpha + 3\left(\frac{\partial}{\partial z}\right)_\alpha. \tag{1.197}$$

Let
$$\widehat{C}_\alpha^{\otimes *} = (dy)_\alpha \wedge (dz)_\alpha. \tag{1.198}$$

Using Eq. (1.189) we get
$$\widehat{C}_\alpha^{\otimes *}(\widehat{X}_\alpha, \widehat{Y}_\alpha) = 3 - 2 = 1. \tag{1.199}$$

C6   The wedge product is linear so that
$$(dx^1)_\alpha \wedge \left(a\,(dx^2)_\alpha + b\,(dx^3)_\alpha\right) = a\,(dx^1)_\alpha \wedge (dx^2)_\alpha + b\,(dx^1)_\alpha \wedge (dx^3)_\alpha. \tag{1.200}$$

Definition 1.4.2(2)   **Covariant tensor fields of second order**

- A covariant tensor field of second order $\widehat{C}^{\otimes *}$ is a smooth assignment of a second order covariant tensor to each point $\alpha \in I\!E^n$, smoothness being defined in a comment below.

Comments 1.4.2(2)   **Smoothness and coordinate expressions**

C1   Given two vector fields $\widehat{X}$ and $\widehat{Y}$ any assignment of covariant tensors throughout $I\!E^n$, with a tensor $\widehat{C}_\alpha^{\otimes *}$ at each point α, gives rise to a function g on $I\!E^n$ defined by
$$g(\alpha) = \widehat{C}_\alpha^{\otimes *}(\widehat{X}_\alpha, \widehat{Y}_\alpha). \tag{1.201}$$

The assignment is said to be *smooth* if g is a smooth function for every pair of vector fields $\widehat{X}$ and $\widehat{Y}$. A covariant vector field is expressible as
$$\widehat{C}^{\otimes *} = \sum_{j,k=1}^{n} C_{jk}\, dx^j \otimes dx^k, \tag{1.202}$$

where the components $C_{jk} \in C^\infty(I\!E^n)$. We can see that the function g defined by
$$g = \widehat{C}^{\otimes *}(\widehat{X}, \widehat{Y}) = \sum_{j,k=1}^{n} C_{jk}\, X^j Y^k \tag{1.203}$$

is smooth.

C2   A covariant tensor field $\widehat{C}^{\otimes *}$ is said to be *symmetric* if
$$\widehat{C}^{\otimes *}(\widehat{X}, \widehat{Y}) = \widehat{C}^{\otimes *}(\widehat{Y}, \widehat{X}), \tag{1.204}$$

and *anti-symmetric* if
$$\widehat{C}^{\otimes *}(\widehat{X}, \widehat{Y}) = -\widehat{C}^{\otimes *}(\widehat{Y}, \widehat{X}). \tag{1.205}$$

1.4. COTANGENT VECTORS AND DIFFERENTIAL FORMS

Explicitly we have

$$\widehat{C}^{\otimes *} = \sum_{j,k=1}^{n} C_{jk}\, dx^j \otimes dx^k \tag{1.206}$$

which is symmetric if $C_{jk} = C_{kj}$ and is anti-symmetric if $C_{jk} = -C_{kj}$. Anti-symmetry implies $C_{jj} = -C_{jj} = 0$. We can also make use of wedge product to express an anti-symmetric tensor as

$$\Omega^{(2)} = \sum_{1 \le j < k}^{n} \omega_{jk}\, dx^j \wedge dx^k, \quad \omega_{jk} \in C^{\infty}(I\!E^n). \tag{1.207}$$

In $I\!E^3$ in the usual Cartesian coordinates (x, y, z) an example is

$$\Omega^{(2)} = x^2\, dx \wedge dy. \tag{1.208}$$

Acting on the pair of vector fields $\widehat{X} = \partial/\partial x$ and $\widehat{Y} = y^2\, \partial/\partial y$ gives

$$\Omega^{(2)}(\widehat{X}, \widehat{Y}) = x^2 y^2. \tag{1.209}$$

Anti-symmetric tensor fields of second order are of such fundamental importance in physics and mathematics that a special name is to be given to them. We shall also use a separate notation, i.e., $\Omega^{(2)}$, to denote a *second order anti-symmetric covariant tensor*.

Definition 1.4.2(3)   Two-forms

- A *two-form* $\Omega^{(2)}$ is an anti-symmetric covariant tensor field of second order.

1.4.3 Exterior differentiation

A differential operation on a function gives rise to a one-form, and a two-form is generated by the wedge product of two one-forms, and so on. Let us call a function a *zero-form*, then we can build up one-forms and two-forms from zero-forms. For uniformity of notation let $\Omega^{(0)}$ denote a zero-form, i.e., $\Omega^{(0)}$ is just a function f. We can set up a formal procedure for generating higher forms in terms of an *exterior differential operator*. An exterior differential operator $\widehat{d}_e$ is a mapping which maps a p-form $\Omega^{(p)}$ into a $p+1$ form $\Omega^{(p+1)}$ satisfying the following properties:[25]

---
[25] Frankel (1997) p. 73.

1. Given a zero-form $\Omega^{(0)} = f$ the one-form $\Omega^{(1)} = \widehat{d}_e f$ is given by

$$\Omega^{(1)} = \widehat{d}_e \Omega^{(0)} = df = \sum_{j=1}^{n} \frac{\partial f}{\partial x^j} \, dx^j. \qquad (1.210)$$

In particular we have, treating a coordinate variable x^j as a function,

$$\widehat{d}_e x^j = dx^j. \qquad (1.211)$$

In other words the exterior differential is the same as the differential when acting on functions. This is why we employed the term exterior differential rather than the more commonly used term of *exterior derivative*. It is when applied to a p-form that the exterior differential operator really comes into its own.

2. Given any p-form we have $\widehat{d}_e(\widehat{d}_e \Omega^{(p)}) = 0$. It follows that $\widehat{d}_e(df) = \widehat{d}_e(\widehat{d}_e \Omega^{(0)}) = 0$. In particular we have $\widehat{d}_e(dx^j) = 0$. This does not mean that the exterior differentials of all one-forms vanish since not every one-form is the differential of a zero form, as pointed out in §1.4.1C(3) C4.

3. The exterior differential operator $\widehat{d}_e$ is linear, i.e., given two p-forms $\Omega_1^{(p)}$ and $\Omega_2^{(p)}$ we have

$$\widehat{d}_e(\Omega^{(p)} + \Omega^{(p)}) = \widehat{d}_e \Omega_1^{(p)} + \widehat{d}_e \Omega_2^{(p)}. \qquad (1.212)$$

4. Given the wedge product of a p-form $\Omega^{(p)}$ and a q-form $\Omega^{(q)}$ we have

$$\widehat{d}_e \left(\Omega^{(p)} \wedge \Omega^{(q)} \right) = \left(\widehat{d}_e \Omega^{(p)} \right) \wedge \Omega^{(q)} + (-1)^p \, \Omega^{(p)} \wedge \left(\widehat{d}_e \Omega^{(q)} \right), \qquad (1.213)$$

with the understanding that, when $\Omega^{(p)}$ is a zero-form $\Omega^{(0)} = f$, the expression $\Omega^{(p)} \wedge \Omega^{(q)}$ is taken to be $f \Omega^{(q)}$ so that

$$\widehat{d}_e(f \, \Omega^{(q)}) = df \wedge \Omega^{(q)} + f \left(\widehat{d}_e \Omega^{(q)} \right). \qquad (1.214)$$

Examples 1.4.3(1)   **Exterior differentiation and forms**

E1   The above prescription is sufficient to define the operator $\widehat{d}_e$. To see how it works in practice let us consider the exterior differential of a general one-form

$$\Omega^{(1)} = \sum_{k=1}^{n} u_k \, dx^k, \quad u_k \in C^\infty(I\!\!E^n). \qquad (1.215)$$

1.4. COTANGENT VECTORS AND DIFFERENTIAL FORMS

We can work out $\widehat{d}_e \Omega^{(1)}$ as follows:

$$\Omega^{(2)} = \widehat{d}_e \Omega^{(1)} = \sum_{k=1}^{n} \widehat{d}_e(u_k dx^k)$$

$$= \sum_{k=1}^{n} \left((\widehat{d}_e u_k) \wedge dx^k + u_k \widehat{d}_e(dx^k) \right)$$

$$= \sum_{j,k}^{n} \frac{\partial u_k}{\partial x^j} dx^j \wedge dx^k, \tag{1.216}$$

since

$$\widehat{d}_e u_k = \sum_{1=j}^{n} \frac{\partial u_k}{\partial x^j} dx^j \quad \text{and} \quad \widehat{d}_e(dx^k) = 0. \tag{1.217}$$

E2 The highest form we can get in $\mathbb{E}^n$ is an n-form. For example, in $\mathbb{E}^3$ we can have a 3-form

$$\Omega^{(3)} = f(x, y, z)\, dx \wedge dy \wedge dz. \tag{1.218}$$

It is obvious that any attempt to create a higher form, for example by applying the operator $\widehat{d}_e$ to $\Omega^{(3)}$, would annihilate $\Omega^{(3)}$ and hence would fail to produce a 4-form.

We have discussed several ways of increasing the order of a tensor field, in particular by using the exterior product and exterior differentiation. On the other hand, we can reduce the order of tensors by contraction. Another important method is through a process known as an *interior product* which we shall discuss in the next section.

1.4.4 Interior products, closed and exact forms

An *interior product* is a product between a vector field and a p-form resulting in a $(p-1)$-form. The generally adopted notation is

$$\widehat{X} \lrcorner \Omega^{(p)} = \Omega^{(p-1)}. \tag{1.219}$$

However, we shall not go into a general definition or a general prescription for arbitrary forms. It is sufficient for our purposes to see how this interior product works for one-forms and two-forms. For these cases the interior product satisfies the following properties:

1. $\widehat{X} \lrcorner \Omega^{(0)} = 0.$

2. $\widehat{X} \lrcorner \Omega^{(1)} = \Omega^{(0)} = \Omega^{(1)}(\widehat{X})$.

3. $\widehat{X} \lrcorner \Omega^{(2)} = \Omega^{(1)}$, where $\Omega^{(1)}$ is determined by

$$\Omega^{(1)}(\widehat{Y}) = \Omega^{(2)}(\widehat{X}, \widehat{Y}) \quad \text{for all} \quad \widehat{Y}. \tag{1.220}$$

4. The interior product obeys the following rule:

$$\widehat{X} \lrcorner (\Omega^{(p)} \wedge \Omega^{(q)}) = (\widehat{X} \lrcorner \Omega^{(p)}) \wedge \Omega^{(q)} + (-1)^p \Omega^{(p)} \wedge (\widehat{X} \lrcorner \Omega^{(q)}). \tag{1.221}$$

Note that:

(a) If $\Omega^{(p)}$ is a 0-form, i.e., a function f, then

$$\widehat{X} \lrcorner \left(f \Omega^{(q)}\right) = f\left(\widehat{X} \lrcorner \Omega^{(q)}\right), \tag{1.222}$$

since $(\widehat{X} \lrcorner \Omega^{(p)}) = 0$ and $1^{-p} = 1$ when $p = 0$. In particular we have

$$\widehat{X} \lrcorner \left(\sum_{1=j<k}^{n} \omega_{jk}\, dx^j \wedge dx^k \right)$$

$$= \sum_{1=j<k}^{n} \omega_{jk} \left(\widehat{X} \lrcorner (dx^j \wedge dx^k) \right) \tag{1.223}$$

$$= \sum_{1=j<k}^{n} \omega_{jk} (X^j dx^k - X^k dx^j). \tag{1.224}$$

(b) If $\Omega^{(p)} = \Omega^{(1)}$ is a one-form then $(\widehat{X} \lrcorner \Omega^{(1)})$ is a zero-form, i.e., $(\widehat{X} \lrcorner \Omega^{(1)}) = f$ which is a function. We should take $(\widehat{X} \lrcorner \Omega^{(1)}) \wedge \Omega^{(q)}$ to be $f \Omega^{(q)}$.

Examples 1.4.4(1)   Interior products

E1   For a zero-form $\Omega^{(0)} = f$ we have $\widehat{X} \lrcorner f = 0$.

E2   For a one-form $\Omega^{(1)} = \sum_j u_j dx^j$ we have

$$\widehat{X} \lrcorner \left(\sum_{j=1}^{n} u_j\, dx^j \right) = \left(\sum_{j=1}^{n} u_j\, dx^j \right)(\widehat{X}) = \sum_{j=1}^{n} u_j X^j. \tag{1.225}$$

Here we have used a result from §1.4.1C(3) C4. In particular we have

$$\widehat{X} \lrcorner dx^j = dx^j(\widehat{X}) = X^j. \tag{1.226}$$

1.4. COTANGENT VECTORS AND DIFFERENTIAL FORMS

An interior product is similar to the contraction operation between a covariant vector field and a contravariant vector field discussed in §1.2.4C(1) C7.

E3 To appreciate property 3 embodied in Eq. (1.220) let us consider a two-form

$$\Omega^{(2)} = \sum_{1 \leq j < k} \omega_{jk}\, dx^j \wedge dx^k. \tag{1.227}$$

Then

$$\Omega^{(1)} = \widehat{X} \lrcorner \Omega^{(2)} = \widehat{X} \lrcorner \left(\sum_{1=j<k}^{n} \omega_{jk}\, dx^j \wedge dx^k \right) \tag{1.228}$$

is a one-form given by Eq. (1.224). When this one-form $\Omega^{(1)}$ acts on an arbitrary vector field $\widehat{Y}$ we get

$$\Omega^{(1)}(\widehat{Y}) = \sum_{1=j<k}^{n} \omega_{jk} \left(X^j dx^k - X^k dx^j \right)(\widehat{Y}) \tag{1.229}$$

$$= \sum_{1=j<k}^{n} \omega_{jk} \left(X^j Y^k - X^k Y^j \right). \tag{1.230}$$

Using Eq. (1.189) we also find that

$$\Omega^{(2)}(\widehat{X}, \widehat{Y}) = \sum_{1=j<k}^{n} \omega_{jk} (X^j Y^k - X^k Y^j). \tag{1.231}$$

It follows that

$$\Omega^{(1)}(\widehat{Y}) = \Omega^{(2)}(\widehat{X}, \widehat{Y}) \tag{1.232}$$

as stated in Eq. (1.220). This enables us to relate a one-form $\Omega^{(1)}$ to a vector field $\widehat{X}$ through a chosen two-form $\Omega^{(2)}$. We shall return to make use of this later when we formulate Hamiltonian mechanics.

Definition 1.4.4(1)   Closed, exact and non-degenerate two-forms

- A p-form $\Omega^{(p)}$ is said to be *closed* if $\widehat{d}_e \Omega^{(p)} = 0$.

- $\Omega^{(p)}$ is said to be *exact* if there exists a $(p-1)$-form $\Omega^{(p-1)}$ such that $\Omega^{(p)} = \widehat{d}_e \Omega^{(p-1)}$.

- A two-form $\Omega^{(2)}$ is said to be *non-degenerate* if $\Omega^{(2)}(\widehat{X}, \widehat{Y}) = 0$ for all $\widehat{Y}$ implies $\widehat{X} = 0$.

Examples 1.4.4(2) Degenerate and non-degenerate forms

E1 An exact form is obtained from a lower form by exterior differentiation. Generally a closed form is not necessarily exact, but an exact form is also closed, e.g., $\Omega^{(1)} = df$ is exact and closed. The one-form in Eq. (1.172) is neither closed nor exact.

E2 $\Omega^{(2)} = dx^1 \wedge dx^2$ in $I\!\!E^2$ is closed and non-degenerate. In $I\!\!E^3$ the two form $\Omega^{(2)} = dx^1 \wedge dx^2$ is closed but degenerate since

$$\Omega^{(2)}(\widehat{X}, \widehat{Y}) = X^1 Y^2 - Y^1 X^2 = 0 \quad \text{for all } \widehat{Y} \tag{1.233}$$

implies

$$X^1 = X^2 = 0 \quad \text{but leaving } X^3 \text{ unrestricted, i.e., } \widehat{X} \neq 0. \tag{1.234}$$

1.5 Differentiable Manifolds

1.5.1 Definition and examples

Although Euclidean spaces are most useful in physical applications there are many occasions when we have to deal with non-Euclidean spaces either in principle or for practical convenience. The simplest examples are the circle and the sphere in $I\!\!E^3$. These spaces, known as differentiable manifolds, are non-Euclidean because they are not topologically equivalent to $I\!\!E^n$.

Basically a differentiable manifold of dimension ℓ is a Hausdorff space which is locally topologically equivalent to $I\!\!R^\ell$, i.e., it locally resembles an open cube of $I\!\!R^\ell$. This intuitive notion helps us to appreciate the concept of differentiable manifolds. The set of points forming a circle in $I\!\!R^2$ is a manifold since any small section of the circle resembles an interval of $I\!\!R$; a circle is a one-dimensional manifold. Note that the resemblance is local only. The circle as a whole (globally) certainly does not look anything like $I\!\!R$. A sphere is not topologically equivalent to $I\!\!R^2$. However it is locally like an open rectangle of $I\!\!R^2$, making it a two-dimensional manifold. An electrical circuit containing branches joined up at some points, referred to as *branch points*, does not have the geometry of a manifold; branch points with several wires coming in and out do not resemble any open set of $I\!\!R^n$.

Definition 1.5.1(1) Differentiable manifolds

- An ℓ-dimensional differentiable manifold $I\!\!M^\ell$ is a Hausdorff space with the following properties:

1.5. DIFFERENTIABLE MANIFOLDS

1. Every point $m \in M^\ell$ has a neighbourhood $\mathcal{N}_m$ which is topologically equivalent to an open cube Λ of $\mathbb{R}^\ell$. The homeomorphism F_m of $\mathcal{N}_m$ onto Λ which maps $m \in \mathcal{N}_m$ to $\alpha \in \Lambda$ defines a local coordinate chart on $\mathcal{N}_m$ by assigning the coordinates x^j of α to m.

2. Let $m' \in M$ be another point which has a neighbourhood $\mathcal{N}'_{m'}$ topologically equivalent to an open cube Λ' of $\mathbb{R}^\ell$. Let x'^j be the local coordinates assigned to points in $\mathcal{N}'_{m'}$. Then on the overlapped region $\mathcal{N}_m \cap \mathcal{N}'_{m'}$ the coordinates x^j and x'^j are smooth functions of one another.

Examples 1.5.1(1) **Spheres and circles**

E1 The Euclidean space $\mathbb{E}^n$ is clearly a differentiable manifold. In fact the space $\mathbb{R}^n$ itself is a differentiable manifold and so is any open cube of $\mathbb{R}^n$. For brevity we often refer to a differentiable manifold simply as a manifold. According to the definition above we can have a single coordinate system covering the entire manifold only if the manifold is topologically equivalent to an open cube of $\mathbb{R}^\ell$. A coordinate system covering the entire manifold is referred to as a *global coordinate system*.

E2 In the usual rectangular Cartesian coordinates (x, y) in $\mathbb{E}^2$ a circle $\mathcal{C}$ of radius a and centered at the origin is specified by the set of points $x^2 + y^2 = a^2$. With the topology induced from that of $\mathbb{E}^2$ defined in §1.2.1C(1) C6 such a circle is a one-dimensional differentiable manifold. In the usual polar coordinates (r, θ) we have[26]

$$\mathcal{C} = \{(r, \theta) : r = a, \ \theta \in [0, 2\pi]\}, \tag{1.235}$$

where $\theta = 0$ and $\theta = 2\pi$ refer to the same point. Although the parameter θ does cover the entire circle it is not a global coordinate in $\mathcal{C}$. By Definition 1.5.1(1) a coordinate variable must be single-valued and takes values in an open cube of $\mathbb{E}^n$. In the case of the circle $\mathcal{C}$ a coordinate must take values in an open interval of $\mathbb{E}$. However, it is not possible to find a coordinate variable which takes values in an open interval of $\mathbb{E}$ to cover the entire circle. In other words $\mathcal{C}$ is not topologically equivalent to an open interval of $\mathbb{E}$. A coordinatization of $\mathcal{C}$ requires two overlapping local coordinate charts. Let

1. $\mathcal{C}_c = \{(r, \theta) : r = a, \ \theta \in (0, 2\pi)\}$, i.e., $\mathcal{C}_c$ is the circle with the point $\theta = 0$ cut away.

---

[26] We follow a standard practice to denote the polar angle in $\mathbb{E}^2$ by θ although the angle variable θ here is quite different from the angle variable also denoted by θ in spherical coordinates in $\mathbb{E}^3$.

2. $C_\pi = \{(r, \theta) : r = a, \theta \in [0, \pi) \cup (\pi, 2\pi]\}$, i.e., C_π is the circle with the point $\theta = \pi$ removed.

We can define global coordinates in C_c and C_π as follows:

1. Introduce a new variable ϑ defined on C_c by

$$\vartheta = \theta \quad \text{for} \quad \theta \in (0, 2\pi). \tag{1.236}$$

Then ϑ takes values in the open interval $(0, 2\pi)$ and hence constitutes a coordinate covering C_c.

2. Introduce a new variable ϑ_π defined on C_π by

$$\vartheta_\pi = \begin{cases} \theta & \text{if } \theta \in (0, \pi) \\ \theta - 2\pi & \text{if } \theta \in (\pi, 2\pi] \end{cases}. \tag{1.237}$$

This new variable ϑ_π, which takes values in the open interval $(-\pi, \pi)$ and agrees with θ over the range $(0, \pi)$, forms a coordinate covering C_π.

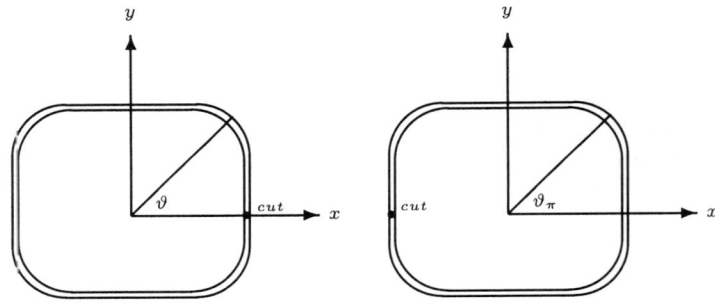

Fig. 1.5.1(1)   Circle C_c with point $\theta = 0$ removed (left) and circle C_π with point $\theta = \pi$ removed (right).

We have the following results:

1. $C_c \cup C_\pi = C$.

2. $C_c \cap C_\pi$ consists of two disjoint regions C_+ and C_-:

   (a) C_+ is the overlapping region specified by $\vartheta \in (0, \pi)$.
   (b) C_- is the region specified by $\vartheta_\pi \in (-\pi, 0)$.

1.5. DIFFERENTIABLE MANIFOLDS

3. In $\mathcal{C}_+$ we have $\vartheta = \vartheta_\pi$, and in $\mathcal{C}_-$ we have $\theta_\pi = \vartheta - 2\pi$.

4. The two coordinates ϑ and ϑ_π are smoothly related in the overlapping regions and together they cover the entire circle $\mathcal{C}$.

E3 A sphere $\mathcal{S}^2$ of radius a in $I\!\!E^3$ is a two-dimensional differentiable manifold. Similar to the circle, a sphere is not coverable by a single coordinate chart. This is related to the fact that $\mathcal{S}^2$ is not topologically equivalent to any open cube of $I\!\!R^2$.

E4 It is possible to introduce non-global coordinates in a Euclidean space. The usual spherical coordinates (r, θ, φ) in $I\!\!E^3$ are examples. These spherical coordinates do not form a global coordinate system in $I\!\!E^3$. In addition to the angle variables not being global they are also undefined at the origin $r = 0$.

E5 Vector fields can be established in a differentiable manifold $I\!\!M^\ell$, despite the possible absence of global coordinates. Clearly we can do this since vector fields are definable independent of coordinates. In fact the definitions of vector fields, tensor fields and differential forms introduced in $I\!\!E^\ell$ carry over into $I\!\!M^\ell$. There is just one complication: we have to match quantities expressed in terms of coordinates defined on overlapping local coordinate charts. Taking again the example of the circle $\mathcal{C}$ above, we can first define a vector field in region $\mathcal{C}_c$ and one in $\mathcal{C}_\pi$, e.g.,

$$\widehat{X}_c = \frac{d}{d\vartheta} \quad \text{on} \quad \mathcal{C}_c \tag{1.238}$$

$$\widehat{X}_\pi = \frac{d}{d\vartheta_\pi} \quad \text{on} \quad \mathcal{C}_\pi. \tag{1.239}$$

In the intersection $\mathcal{C}_c \cap \mathcal{C}_\pi$ we have $\widehat{X}_c = \widehat{X}_\pi$, a result derived from the relationship between ϑ and ϑ_π spelled out earlier. It follows that we have a vector field $\widehat{X}$ defined on the entire circle $\mathcal{C}$ with

$$\widehat{X} = \begin{cases} \widehat{X}_c & \text{on } \mathcal{C}_c \\ \widehat{X}_\pi & \text{on } \mathcal{C}_\pi \end{cases}. \tag{1.240}$$

We shall have many occasions to return to the circle and to this vector field on $\mathcal{C}$. It is clearly very tedious to have to explicitly spell out all these details every time. For notational convenience we shall formally regard $\theta \in [0, 2\pi]$ as a coordinate covering $\mathcal{C}$ and denote the above vector field symbolically by

$$\widehat{X} = \frac{d}{d\theta}. \tag{1.241}$$

To see if this vector field is complete or not we need to examine its integral curves. The integral curve of $\widehat{X}$ starting from, say, the point $\theta = 0$ in $\mathcal{C}$ is

given by
$$\theta(\tau) = \tau \quad \text{for} \quad \tau \in [0, 2\pi]. \tag{1.242}$$

We can extend the domain of the curve to the entire real line by writing down, for example,
$$\theta(\tau) = \tau - 2\pi \quad \text{for} \quad \tau \in [2\pi, 4\pi], \tag{1.243}$$

and so on. Intuitively we can see that this corresponds to translations along, and round and round, the circumference of $\mathcal{C}$. These translations form a one-parameter group of transformations of $\mathcal{C}$. The same applies to integral curves starting from all other points in $\mathcal{C}$. This vector field is therefore complete.[27] We can also consider the translations as rotations of $\mathcal{C}$ which are known to form a group.

1.5.2 Riemannian manifolds

Definition 1.5.2(1) **Pseudo-Riemannian and Riemannian metric**

- A *pseudo-Riemannian metric* on a manifold $I\!\!M^\ell$ is a symmetric and non-degenerate covariant tensor field $\widehat{G}^{\otimes *}$ of second order. A manifold with a pseudo-Riemannian metric is called a *pseudo-Riemannian manifold*.

- A pseudo-Riemannian metric $\widehat{G}^{\otimes *}$ on $I\!\!M^\ell$ is said to be *positive definite* if for all vector fields $\widehat{X}$ on $I\!\!M^\ell$ we have:

    1. $\widehat{G}_m^{\otimes *}(\widehat{X}_m, \widehat{X}_m) \geq 0 \quad \forall m \in \mathcal{M}^\ell$.
    2. $\widehat{G}_m^{\otimes *}(\widehat{X}_m, \widehat{X}_m) = 0 \quad \text{if and only if} \quad \widehat{X}_m = 0 \quad \forall m \in \mathcal{M}^\ell$.

- A positive definite pseudo-Riemannian metric on a manifold is called a *Riemannian metric* and a manifold with a Riemannian metric is called a *Riemannian manifold*.

Comments 1.5.2(1) **Volume elements and integration**

C1 A metric is an important geometric quantity. As a second order covariant tensor field it is able to act on vector fields, in accordance with §1.4.2C(2) C1, to give rise to smooth functions on the manifold. This opens the way to the introduction of a host of other useful quantities some of which will now be discussed.

---

[27] See also §1.3.2E(1) E4.

1.5. DIFFERENTIABLE MANIFOLDS

C2   Consider the Euclidean space $I\!E^n$ with rectangular Cartesian coordinates x^j. Introduce a covariant tensor by

$$\widehat{G}^{\otimes *} = \sum_{i,j=1}^{n} \delta_{ij}\, dx^i \otimes dx^j. \tag{1.244}$$

Then, we have

$$\widehat{G}^{\otimes *}(\widehat{X}, \widehat{Y}) = \sum_{i,j=1}^{n} \delta_{ij}\, X^i Y^j = \sum_{j=1}^{n} X^j Y^j, \tag{1.245}$$

showing that $\widehat{G}^{\otimes *}$ is symmetric, non-degenerate and positive definite. So, $\widehat{G}^{\otimes *}$ is a Riemannian metric on $I\!E^n$, making $I\!E^n$ a Riemannian manifold. On a transformation to a new coordinate chart x'^r, possibly defined only for an open subset $\mathcal{D}$ in $I\!E^n$, we have

$$dx^i = \sum_{r=1}^{n} \frac{\partial x^i}{\partial x'^r}\, dx'^r. \tag{1.246}$$

In the new coordinate chart the metric tensor on $\mathcal{D}$ will be of the form

$$\widehat{G}^{\otimes *} = \sum_{r,s=1}^{n} g'_{rs}\, dx'^r \otimes dx'^s, \tag{1.247}$$

$$g'_{rs} = \sum_{i,j=1}^{n} \frac{\partial x^i}{\partial x'^r} \frac{\partial x^j}{\partial x'^s} \delta_{ij} \tag{1.248}$$

with

$$\widehat{G}^{\otimes *}(\widehat{X}, \widehat{Y}) = \sum_{r,s=1}^{n} g'_{rs}\, X'^r Y'^s, \tag{1.249}$$

where X'^r and Y'^s are respectively the components of $\widehat{X}$ and $\widehat{Y}$ in the new coordinates.

Generally a Riemannian manifold $I\!M^\ell$ admits only local coordinate charts x'^j which may be non-rectangular, the sphere $I\!M^2 = \mathcal{S}^2$ in $I\!E^3$ being an example. In each coordinate chart the metric tensor is still of the form

$$\widehat{G}^{\otimes *} = \sum_{i,j=1}^{\ell} g'_{ij}\, dx'^i \otimes dx'^j. \tag{1.250}$$

On a coordinate transformation to $\bar{x}^j$ the expression changes to

$$\widehat{G}^{\otimes *} = \sum_{i,j=1}^{\ell} \bar{g}_{ij}\, d\bar{x}^i \otimes d\bar{x}^j, \tag{1.251}$$

with

$$\bar{g}_{ij} = \sum_{m,k=1}^{\ell} \frac{\partial x'^m}{\partial \bar{x}^i} \frac{\partial'^k}{\partial \bar{x}^j} g'_{mk}. \quad (1.252)$$

Comparing with Eq. (1.35) we can see that generally the coefficients g'_{ij} of the metric $\widehat{G}^{\otimes *}$ transform like the components of a covariant tensor. Hence g'_{ij} are often referred to as the components of the metric tensor in coordinate chart x'^j.

C3 A Riemannian metric generates a scalar product on each tangent space. In $I\!\!E^n$ the scalar product between two tangent vectors $\widehat{X}_\alpha$ and $\widehat{Y}_\alpha$ in $\widehat{T}_\alpha I\!\!E^n$ is given by $\widehat{G}^{\otimes *}(\widehat{X}, \widehat{Y})$ evaluated at α, i.e., by evaluating Eq. (1.245) at α. This agrees with our previous discussion in §1.2.1C(1) C4.

C4 A Riemannian manifold admits a *volume element* which enables us to set up integrations on $I\!\!M^\ell$. Let

$$\widehat{G}^{\otimes *} = \sum_{i,j=1}^{\ell} g'_{ij} \, dx'^i \otimes dx'^j \quad (1.253)$$

be the metric tensor in $I\!\!M^\ell$. Then in coordinate system x'^j the volume $\delta\mu'$ of an infinitesimal rectangle $(x'^j, x'^j + \Delta x'^j)$ is defined to be

$$\delta\mu' = \sqrt{g'}\, \Delta x'^1 \Delta x'^2 \cdots \Delta x'^\ell, \quad (1.254)$$

where g' is the determinant formed by the elements g'_{ij}. The volume of any bounded region Λ is given by the limit of the sum of these infinitesimal rectangles which make up Λ.[28] Such a volume is written in the form of the following integral:

$$\int_\Lambda d\mu' = \int_\Lambda \sqrt{g'}\, dx'^1 dx'^2 \cdots dx'^\ell. \quad (1.255)$$

Given any function f on $I\!\!M^\ell$ we also have an integral

$$\int_\Lambda f\, d\mu = \int_\Lambda f\sqrt{g'}\, dx'^1 dx'^2 \cdots dx'^\ell. \quad (1.256)$$

Most relevant is the set of functions whose square is integrable in $I\!\!M^\ell$, since such a set of functions will form the basis for the construction of a Hilbert space in quantum mechanics.

---

[28]Schutz (1980) p. 121.

1.5. DIFFERENTIABLE MANIFOLDS

The integral in Eq. (1.256)) is an invariant independent of any choice of coordinate system. On a transformation to new coordinates $\bar{x}^j$ the integral becomes

$$\int_\Lambda f d\bar{\mu} = \int_\Lambda f \sqrt{\bar{g}}\, d\bar{x}^1 d\bar{x}^2 \cdots d\bar{x}^\ell, \qquad (1.257)$$

where $\bar{g}$ is the determinant formed by the transformed $\bar{g}_{ij}$ which are related to g'_{ij} by Eq. (1.252). We have

$$\int_\Lambda f \sqrt{g'}\, dx'^1 dx'^2 \cdots dx'^\ell = \int_\Lambda f \sqrt{\bar{g}}\, d\bar{x}^1 d\bar{x}^2 \cdots d\bar{x}^\ell. \qquad (1.258)$$

Finally we note that the volume element is identifiable with the ℓ-form $dx'^1 \wedge dx'^2 \wedge \cdots \wedge dx'^\ell$ although we shall not delve into this more abstract concept of volume.[29]

C5 The 4-dimensional space-time in Special Relativity is an example of a pseudo-Riemannian space. We shall not pursue pseudo-Riemannian spaces further in this book.

1.5.3 Hamiltonian manifolds

Definition 1.5.3(1)   **Hamiltonian structures and Hamiltonian manifolds**

- A *Hamiltonian structure* $\Omega_H^{(2)}$ on a $2n$-dimensional manifold is a closed and non-degenerate two-form on the manifold.

- A $2n$-dimensional manifold together with a Hamiltonian structure is called a *Hamiltonian manifold*, to be denoted by $I\!\!M_H^{2n}$.

Theorem 1.5.3(1)   **Darboux's theorem on Hamiltonian manifolds**[30]

- In a $2n$-dimensional Hamiltonian manifold $I\!\!M_H^{2n}$ there exist coordinates $(\mathsf{q}^j, \mathsf{p}_j), j = 1, 2, \ldots, n$, in the neighbourhood of every point $m \in I\!\!M_H^{2n}$ such that the Hamiltonian structure $\Omega_H^{(2)}$ is given by

$$\Omega_H^{(2)} = \sum_{j=1}^n d\mathsf{q}^j \wedge d\mathsf{p}_j. \qquad (1.259)$$

These are called *canonical coordinates*.

---

[29] Nash and Sen (1983). We have avoided the definition of a volume element in $I\!\!M^\ell$ in terms of an ℓ-form. Instead we introduce the notion of a volume element in line with the usual and intuitive concept.

[30] Abraham and Marsden (1978).

Comments 1.5.3(1)   **Canonical transformations, functions and vector fields on M_H^{2n}**

C1   A Hamiltonian structure is also known as a *symplectic structure* and a Hamiltonian manifold as a *symplectic manifold*. The name Hamiltonian arises from the fact that such manifolds form the basis of Hamilton's formulation of classical mechanics. To see how a Hamiltonian structure comes about in the traditional Hamiltonian formulation of classical mechanics let us consider a particle moving in $I\!\!E^n$. Let the particle's position be given by rectangular Cartesian coordinates x^j in $I\!\!E^n$ and its motion specified by their conjugate momenta p_j. These momenta p_j form the components of a covariant vector[31] and this is why we use a subscript rather than a superscript. We can combine position and momentum variables to form a $2n$-dimensional space Γ known as the *phase space* which is coordinated by $2n$ variables x^j and p_j. We can construct a two-form

$$\Omega_H^{(2)} = \sum_{j=1}^n dx^j \wedge dp_j \qquad (1.260)$$

which is seen to be closed and non-degenerate, i.e., $\Omega_H^{(2)}$ is a Hamiltonian structure on Γ, rendering Γ a Hamiltonian manifold. This phase space in classical mechanics is a paradigm of Hamiltonian manifolds, and this also explains why one set of canonical variables is labelled by a superscript and the other set by a subscript in Darbourx's theorem.

C2   In a Hamiltonian manifold we may also call q^j coordinate variables and p_j momentum variables. Moreover the momentum variable p_j is also said to be *canonically conjugate* to q^j. We may consider that the momentum variables formally span a space, called the *momentum space*, in a similar way to that in which the coordinate variables q^j span a *coordinate space*.

C3   Canonical coordinates are not unique, i.e., we can have other canonical coordinates $(\bar{\mathsf{q}}^j, \bar{\mathsf{p}}_j), j = 1, 2, \ldots, n$, in the neighbourhood of any point satisfying Eq. (1.259) of Darboux's theorem. The transformation from one set of canonical coordinates $(\mathsf{q}^j, \mathsf{p}_j)$ to another set $(\bar{\mathsf{q}}^j, \bar{\mathsf{p}}_j)$ is called a *canonical transformation*. A simple example is the transformation

$$\bar{\mathsf{q}}^j = -\mathsf{p}_j, \quad \bar{\mathsf{p}}_j = \mathsf{q}^j. \qquad (1.261)$$

This shows that a coordinate variable $\bar{\mathsf{q}}^j$ need not necessarily refer to position and a momentum variable $\bar{\mathsf{p}}_j$ need not refer to motion.

C4   A function f on M_H^{2n} gives rise to a one-form $\Omega^{(1)} = df$. We can relate such a one-form to a vector field. To achieve this relationship we must

---
[31] See §1.1.3C(1) C3.

1.5. DIFFERENTIABLE MANIFOLDS

first choose a two-form, and then form the interior product between the vector field and the two-form.[32] In a Hamiltonian manifold $I\!M_H^{2n}$ the Hamiltonian structure $\Omega_H^{(2)}$ can serve as the chosen two-form. Then a one-form $\Omega^{(1)} = df$ together with $\Omega_H^{(2)}$ defines a vector field $\widehat{\mathcal{X}}_f$ on $I\!M_H^{2n}$ by

$$\widehat{\mathcal{X}}_f \,\lrcorner\, \Omega_H^{(2)} = df. \tag{1.262}$$

Explicitly we have

$$\widehat{\mathcal{X}}_f = \sum_{j=1}^{n} \left(\frac{\partial f}{\partial \mathsf{p}^j} \frac{\partial}{\partial \mathsf{q}^j} - \frac{\partial f}{\partial \mathsf{q}^j} \frac{\partial}{\partial \mathsf{p}_j} \right). \tag{1.263}$$

To see how such an expression is obtained we first note that following Eq. (1.75) a vector field $\widehat{\mathcal{X}}$ on $I\!M_H^{2n}$ is expressible as

$$\widehat{\mathcal{X}} = \sum_{j=1}^{n} \left(\mathcal{X}^j \frac{\partial}{\partial \mathsf{q}^j} + \mathcal{X}^{n+j} \frac{\partial}{\partial \mathsf{p}_j} \right). \tag{1.264}$$

Making use of Eqs. (1.221) and (1.224) we get

$$\widehat{\mathcal{X}} \,\lrcorner\, \Omega_H^{(2)} = \widehat{\mathcal{X}} \,\lrcorner\, \sum_{j=1}^{n} \left(d\mathsf{q}^j \wedge d\mathsf{p}_j \right) \tag{1.265}$$

$$= \sum_{j=1}^{n} \left\{ \left(\widehat{\mathcal{X}} \,\lrcorner\, d\mathsf{q}^j \right) \wedge d\mathsf{p}_j - d\mathsf{q}^j \wedge \left(\widehat{\mathcal{X}} \,\lrcorner\, d\mathsf{p}_j \right) \right\} \tag{1.266}$$

$$= \sum_{j=1}^{n} \left(-\mathcal{X}^{n+j} d\mathsf{q}^j + \mathcal{X}^j d\mathsf{p}_j \right), \tag{1.267}$$

on account of the fact that

$$\widehat{\mathcal{X}} \,\lrcorner\, d\mathsf{q}^j = \mathcal{X}^j, \quad \widehat{\mathcal{X}} \,\lrcorner\, d\mathsf{p}_j = \mathcal{X}^{n+j}. \tag{1.268}$$

Since

$$df = \sum_j^n \left(\frac{\partial f}{\partial \mathsf{q}^j} d\mathsf{q}^j + \frac{\partial f}{\partial \mathsf{p}_j} d\mathsf{p}_j \right), \tag{1.269}$$

we obtain from Eqs. (1.262) and (1.267) the following results:

$$\mathcal{X}_f^{n+j} = -\frac{\partial f}{\partial \mathsf{q}^j}, \quad \mathcal{X}_f^j = \frac{\partial f}{\partial \mathsf{p}_j}. \tag{1.270}$$

---

[32] See §1.4.4E(1) E3.

We conclude that a function f generates a vector field $\widehat{\mathcal{X}}_f$ through the Hamiltonian structure $\Omega_H^{(2)}$. This result leads to a number of concepts which are indispensable in formulating Hamiltonian mechanics geometrically. These are set out in Definition 1.5.3(2).

Definition 1.5.3(2)   **Hamiltonian vector fields, Poisson bracket**

- In a Hamiltonian manifold $I\!M_H^{2n}$ a vector field is called the *Hamiltonian vector field* generated by a function f on $I\!M_H^{2n}$ if Eq. (1.262) is satisfied. Such a Hamiltonian vector field is denoted by $\widehat{\mathcal{X}}_f$. Explicitly, we have

$$f \quad \to \quad \widehat{\mathcal{X}}_f = \sum_{j=1}^{n} \left(\frac{\partial f}{\partial \mathsf{p}_j} \frac{\partial}{\partial \mathsf{q}^j} - \frac{\partial f}{\partial \mathsf{q}^j} \frac{\partial}{\partial \mathsf{p}_j} \right). \qquad (1.271)$$

- The Poisson bracket $\{f, g\}$ between two functions f and g on $I\!M_H^{2n}$ is defined to be the function

$$\{f, g\} = -\widehat{\mathcal{X}}_f(g). \qquad (1.272)$$

Comments 1.5.3(2)   **Coordinate expressions and Lie algebras**

C1   The coordinate expression for the Poisson bracket $\{f, g\}$ is

$$\{f, g\} = \sum_{j=1}^{n} \left(\frac{\partial f}{\partial \mathsf{q}^j} \frac{\partial g}{\partial \mathsf{p}_j} - \frac{\partial f}{\partial \mathsf{p}_j} \frac{\partial g}{\partial \mathsf{q}^j} \right). \qquad (1.273)$$

This agrees with the usual definition of Poisson bracket in classical mechanics. Care should be taken if the coordinates employed are not global. Canonical variables $\mathsf{q}^i, \mathsf{p}_j$ satisfy the following characteristic Poisson bracket relations:

$$\{\mathsf{q}^i, \mathsf{p}_j\} = \delta_j^i. \qquad (1.274)$$

C2   One can verify that the Lie bracket between two Hamiltonian vector fields $\widehat{\mathcal{X}}_{f_1}$ and $\widehat{\mathcal{X}}_{f_2}$ is proportional to the Hamiltonian vector field $\widehat{\mathcal{X}}_f$ generated by the function f which is the Poisson bracket of f_1 and f_2, i.e.,[33]

$$[\widehat{\mathcal{X}}_{f_1}, \widehat{\mathcal{X}}_{f_2}] = -\widehat{\mathcal{X}}_f, \quad \text{where} \quad f = \{f_1, f_2\}. \qquad (1.275)$$

C3   The set of all (smooth) functions on $I\!M_H^{2n}$ constitutes a *Lie algebra* with Lie algebra multiplication defined by the Poisson bracket. The set of Hamiltonian vector fields under Lie bracket also forms a Lie algebra.

---
[33] Abraham and Marsden (1978) p. 194.

1.6 Classical Dynamical Systems

1.6.1 Classical systems of finite order

A *classical system of order* $\ell < \infty$ is traditionally defined by three properties:[34]

1. States of the system are specifiable by ℓ parameters $\gamma = \{\gamma^1, \gamma^2, \ldots, \gamma^\ell\}$. Each parameter γ^j assumes a continuous set of real values. To start with we shall assume that each γ^j ranges over the entire $\mathbb{R}$.

2. Physical observables are represented by real-valued functions $A(\gamma)$ of the above parameters.

3. The time evolution of the system is represented by the time-dependence of the parameters which satisfy certain first-order differential equations of the form
$$\frac{d\gamma^j}{dt} = \mathcal{X}^j(\gamma, t) \qquad (1.276)$$
for some smooth functions $\mathcal{X}^j(\gamma, t)$ of γ and t. These are referred to as *equations of motion*. Given any state γ_0 at time $t = 0$ the state at a later time is determined by the solution $\gamma(t)$ of these equations satisfying the initial conditions $\gamma_0 = \gamma(0)$, i.e., $\gamma_0^j = \gamma^j(0)$ for all j.

The collection of all states can be conveniently identified with the set Γ of ℓ-tuples $\mathbb{R} \times \mathbb{R} \times \cdots \times \mathbb{R}$. To formalize this we call

$$\Gamma = \mathbb{R} \times \mathbb{R} \times \cdots \times \mathbb{R} \qquad (1.277)$$

the *phase space*, or the *state space*, of the system. While it may be possible formally to impose a vector space structure together with a norm to convert Γ to a Euclidean space, the resulting structure, e.g., the norm of an element of Γ, may not have any particular physical meaning generally. So, when referring to Γ as a Euclidean space we should bear this in mind; it is often sufficient to treat the phase space as an ℓ-dimensional manifold without reference to any metric structure. We shall formally regard each set of parameter values as a *point* in Γ and the parameters as *coordinates* in Γ. If the range of variables γ^j is restricted the phase space becomes a subspace of $\mathbb{R} \times \mathbb{R} \times \cdots \times \mathbb{R}$.

In geometric terms a state corresponds to a point γ in the phase space Γ, and an observable corresponds to a function A on Γ. The dynamics of the system can then be described by a vector field

$$\widehat{\mathcal{X}} = \sum_{j=1}^{\ell} \mathcal{X}^j \frac{\partial}{\partial \gamma^j} \qquad (1.278)$$

---

[34] Percival and Richards (1982) §3.1, Perko (1991).

in Γ in that the motion of the system can be identified with the integral curves of $\widehat{\mathcal{X}}$ on the phase space. This vector field is called the *velocity field* of the system. Each point $\gamma_0 \in \Gamma$ can serve as an initial state and the starting point of an integral curve σ_{γ_0} of $\widehat{\mathcal{X}}$ with $\gamma(t) = \sigma_{\gamma_0}(t)$ representing the state at time t. Integral curves are obtained from the solutions of Eq. (1.276).

These dynamics may also be recast into the form of the flow of vector field $\widehat{\mathcal{X}}$, i.e., a one-parameter group of transformations of Γ

$$T_t : \Gamma \mapsto \Gamma, \quad t \in \mathbb{R} \quad \text{or} \quad T : \mathbb{R} \times \Gamma \to \Gamma \qquad (1.279)$$

with differentiable and group properties as stated in Definition 1.3.3(2). In other words the dynamics are expressible as a one-parameter group of transformations of the phase space onto itself. The situation is simpler if the vector field is not explicitly time-dependent, i.e., $\mathcal{X}^j = \mathcal{X}^j(\gamma)$.

Definition 1.6.1(1)   **Stationary states and phase flows**

- The map T in Eq. (1.279) is called the *phase flow* of the dynamics.

- A state $\gamma(0)$ is said to be *stationary* if $\gamma(t) = \gamma(0)$ for all t, or equivalently if $A(\gamma(t)) = A(\gamma(0))$ for all observables A.

Clearly a state $\gamma(0)$ is stationary if and only if the velocity field $\widehat{\mathcal{X}}$ vanishes at $\gamma(0)$, i.e., $\mathcal{X}^j(\gamma(0)) = 0$ for every j and for all times. In other words stationary states correspond to critical points of the velocity field. Stationary states can be further classified into various types, e.g., stable and unstable.[35]

1.6.2   First-order systems

The state of a *first-order classical system* is specified by a single real parameter $\gamma = \gamma^1$ lying in a certain range $\mathcal{R} \subset \mathbb{R}$. The phase space Γ of such a system is given by $\Gamma = \mathcal{R}$ with equation of motion[36]

$$\frac{d\gamma}{dt} = \mathcal{X}(\gamma), \quad \gamma \in \Gamma. \qquad (1.280)$$

The corresponding velocity field is

$$\widehat{\mathcal{X}} = \mathcal{X}\frac{d}{d\gamma}, \quad \mathcal{X} = \frac{d\gamma}{dt}. \qquad (1.281)$$

---

[35] Percival and Richards (1982) §3.3 to §3.5. Stationary states are often referred to as *fixed points* of the vector field $\widehat{\mathcal{X}}$.

[36] For many physical systems $\mathcal{X}$ can also contain the time variable t explicitly.

1.6. CLASSICAL DYNAMICAL SYSTEMS

Examples 1.6.2(1)   **Electrical circuits as first-order systems**

E1   Consider an idealized classical resistanceless circuit containing an inductor with an inductance L in a time-dependent external magnetic field. The applied magnetic flux Φ_{ex} threading through the inductor varies as the external magnetic field changes, and a current I is generated according to Faraday's and Lenz's laws:[37]

$$L\frac{dI}{dt} = -\frac{d\Phi_{ex}}{dt}. \tag{1.282}$$

Integrating from 0 to t we get

$$LI(t) - LI(0) = -\Big(\Phi_{ex}(t) - \Phi_{ex}(0)\Big), \tag{1.283}$$

or

$$LI(t) + \Phi_{ex}(t) = \Phi_T, \quad \Phi_T = LI(0) + \Phi_{ex}(0). \tag{1.284}$$

Physically we can interpret Φ_T as the total flux threading through the inductor, being the sum of the external flux and the induced flux due to the current. The fact that Φ_T is a constant means that the induced current $I(t)$ must vary with $\Phi_{ex}(t)$ in such a way as to keep the total flux Φ_T constant. The state of this system is specifiable by a single parameter, the current I. The energy of the system is known is to be given by

$$E = \frac{1}{2}LI^2 = \frac{1}{2L}(\Phi_T - \Phi_{ex})^2. \tag{1.285}$$

On the phase space $\Gamma = \{I \in \mathbb{R}\}$ the vector field corresponding to the dynamics of the circuit is

$$\widehat{\mathcal{X}} = \mathcal{X}\frac{d}{dI}, \quad \mathcal{X} = -\frac{1}{L}\frac{d\Phi_{ex}}{dt}. \tag{1.286}$$

We shall return to this system again in §4.5.1.

1.6.3 Second-order Hamiltonian systems

The phase space for second-order systems is parameterized by two coordinates, i.e., $\Gamma = \{(\gamma^1, \gamma^2)\}$. We are particularly interested in second-order Hamiltonian systems which are defined in geometric terms in Definition 1.6.3(1).

---

[37] Bleaney and Bleaney (1957) p. 151.

Definition 1.6.3(1)   Second-order Hamiltonian systems

- A second-order system is called a *Hamiltonian system* if the phase space Γ is a two-dimensional Hamiltonian manifold $I\!M_H^2$ with a Hamiltonian structure $\Omega_H^{(2)}$ and a chosen real-valued function H_g on $I\!M_H^2$, to be called the *Hamiltonian generator* of the system, such that the dynamics is generated by the Hamiltonian vector field $\widehat{\mathcal{X}}_{H_g}$ of H_g.

Comments 1.6.3(1)   Hamiltonians, Hamiltonian generators, phase flow and Liouville's theorem

C1   The dynamics of a Hamiltonian system is generated by the Hamiltonian vector field $\widehat{\mathcal{X}}_{H_g}$ acting as the velocity field. As a result the Hamiltonian generator H_g is often referred to as the *generator of time translations*.[38]

C2   The energy of a Hamiltonian system is often referred to as the *Hamiltonian of the system*, denoted by H. We must emphasize that in principle and in general H is not the same as H_g, despite the fact that in many practical cases the energy of the system turns out to be equal to the Hamiltonian generator.[39] We shall return to this important point later in quantum theory.

C3   According to Darboux's theorem there are canonical coordinates (q,p) in the neighbourhood of every point in Γ such that $\Omega_H^{(2)} = d\mathsf{q} \wedge d\mathsf{p}$. The Hamiltonian vector field $\widehat{\mathcal{X}}_{H_g}$ is given in accordance with Eq. (1.271), by

$$\widehat{\mathcal{X}}_{H_g} = \frac{\partial H_g}{\partial \mathsf{p}}\frac{\partial}{\partial \mathsf{q}} - \frac{\partial H_g}{\partial \mathsf{q}}\frac{\partial}{\partial \mathsf{p}}. \tag{1.287}$$

The dynamics is given by the integral curves of $\widehat{\mathcal{X}}_{H_g}$, i.e., solutions of the following equations:

$$\frac{d\mathsf{q}}{dt} = \frac{\partial H_g}{\partial \mathsf{p}}, \quad \frac{d\mathsf{p}}{dt} = -\frac{\partial H_g}{\partial \mathsf{q}}. \tag{1.288}$$

These are the well-known *Hamilton's equations of motion* in classical mechanics.[40]

C4   Stationary states occur at the critical points of the Hamiltonian vector field, i.e., where

$$\frac{\partial H_g}{\partial \mathsf{p}} = 0 \quad \text{and} \quad \frac{\partial H_g}{\partial \mathsf{q}} = 0. \tag{1.289}$$

---

[38] Goldstein (1950) p. 259.
[39] The fact that H_g may not be the energy of the system is clearly pointed out by Goldstein (1950) p. 221. For an example in which $H_g \neq H$ see Leech (1965) p. 48.
[40] Bishop and Goldberg (1968) Chapter 6.

1.6. CLASSICAL DYNAMICAL SYSTEMS

C5 The nature of the phase flow T in Hamiltonian dynamics is characterized by *Liouville's Theorem* which states that a phase flow in Hamiltonian dynamics preserves the *phase volume*.[41] For a second-order system an elementary phase volume is just $d\mathsf{q}d\mathsf{p}$. At time $t = \tau$ later the point (q,p) is mapped to the point $(\mathsf{q}_\tau, \mathsf{p}_\tau) = T_\tau(\mathsf{q},\mathsf{p})$. Liouville's Theorem implies

$$d\mathsf{q}_\tau d\mathsf{p}_\tau = d\mathsf{q}d\mathsf{p}. \tag{1.290}$$

Examples 1.6.3(1)   **Hamiltonian generators, Hamiltonians and Hamilton's equations**

E1 Consider a particle of mass m in one-dimensional motion in $I\!\!E$ under a potential V which is a differentiable function on $I\!\!E$. In classical mechanics the state of the particle is fixed by its position x and canonically conjugate momentum p. The dynamics are governed by Hamilton's Eqs. (1.288) with a Hamiltonian generator H_g given by

$$H_g = \frac{p^2}{2m} + V. \tag{1.291}$$

This is a typical example of a second-order Hamiltonian system. The phase space is

$$\Gamma = \{(\gamma^1 = x, \gamma^2 = p)\} \quad \text{with} \quad \Omega_H^{(2)} = dx \wedge dp. \tag{1.292}$$

A point $\gamma = (x, p)$ in Γ corresponds to a state of the particle. The Hamiltonian vector field for the dynamics is

$$\begin{aligned}\widehat{\mathcal{X}}_{H_g} &= \frac{\partial H_g}{\partial p}\frac{\partial}{\partial q} - \frac{\partial H_g}{\partial x}\frac{\partial}{\partial p} \\ &= \frac{p}{m}\frac{\partial}{\partial x} - \frac{\partial V}{\partial x}\frac{\partial}{\partial p}.\end{aligned} \tag{1.293}$$

The integral curves are then solutions of Hamilton's equations

$$\frac{dx}{dt} = \frac{p}{m}, \quad \frac{dp}{dt} = -\frac{\partial V}{\partial x}. \tag{1.294}$$

For such a system the Hamiltonian generator H_g is equal to the Hamiltonian (energy) H of the system. For a free particle the Hamiltonian vector field reduces to

$$\widehat{\mathcal{X}}_{H_g} = \frac{p}{m}\frac{\partial}{\partial x}. \tag{1.295}$$

---

[41] Arnold (1978) pp. 68-70, Abraham and Marsden (1978) Proposition 3.3.4. For a simple proof see Percival and Richards (1982) §4.9, §6.2 who use the term *phase area* rather than phase volume.

E2 A canonical momentum conjugate to a coordinate variable is not unique. In the case of a particle in motion in $I\!\!E$ the variable

$$p' = p + f(x) \tag{1.296}$$

is also canonically conjugate to x. This is seen in the Poisson brackets

$$\{x, p'\} = \{x, p\} = 1, \tag{1.297}$$

which agree with the Poisson bracket relations given by Eq. (1.274) of canonical variables. One can also confirm the canonical nature of x and p' by going back to Theorem 1.5.3(1) to check that $dx \wedge dp'$ and $dx \wedge dp$ give rise to the same Hamiltonian structure. Indeed (x, p') may even be regarded as a canonical transformation of (x, p).

E3 Another example is a classical resistanceless LC circuit consisting of a capacitor with capacitance C and an inductor with inductance L. When a current I flows there will be a charge Q on the capacitor linked to the current by $I = dQ/dt$ and a magnetic flux $\Phi = LI$ through the inductor. A voltage $V = Q/C$ will develop across the capacity. The circuit equation is

$$L\frac{dI}{dt} + \frac{Q}{C} = 0, \tag{1.298}$$

or

$$\frac{d^2Q}{dt^2} = -\omega^2 Q, \quad \omega^2 = 1/LC. \tag{1.299}$$

We can also express the circuit equation in terms of Φ, i.e., we have

$$\frac{d^2\Phi}{dt^2} = -\omega^2 \Phi. \tag{1.300}$$

This is a Hamiltonian system. We have a phase space

$$\Gamma = \{(\mathsf{q} = \Phi, \mathsf{p} = Q)\} \quad \text{with} \quad \Omega_H^{(2)} = d\Phi \wedge dQ. \tag{1.301}$$

Here Φ and Q are treated as a pair of canonical variables. One can easily verified that an appropriate Hamiltonian generator for this system is

$$H_g = \frac{Q^2}{2C} + \frac{\Phi^2}{2L}. \tag{1.302}$$

This circuit system is essentially the same as a harmonic oscillator in mechanics. The only stationary state is at $(\Phi = 0, Q = 0)$.

E4 Consider a particle confined to move in a circle $\mathcal{C}$ which is coordinated by a variable $\theta \in [0, 2\pi]$ as discussed in §1.5.1E(1) E2 and E5. The phase space

1.6. CLASSICAL DYNAMICAL SYSTEMS

Γ is parameterized by θ and its conjugate momentum[42] $P_\theta(\mathcal{C}) \in (-\infty, \infty)$. Here Γ can be visualized as a 2-dimensional cylinder, i.e., $\Gamma = \mathcal{C} \times I\!R$, showing the non-Euclidean nature of the Hamiltonian manifold.

E5 The definition of Hamiltonian systems can be extended to higher order, i.e., to order $2n$ with a $2n$-dimensional manifold $I\!M_H^{2n}$ and canonical coordinates $(\mathsf{q}^j, \mathsf{p}_j)$, $j = 1, 2, \ldots, n$.

1.6.4 Momentum observables, vector fields and operators

Many of the geometric methods developed in recent years are very powerful and are directly applicable to the formulation of many physical theories. These methods are capable of producing a unified treatment of seemingly diverse subjects such as differential equations, differential geometry and dynamical systems such as classical mechanical systems. Moreover, they lead to a deeper understanding of classical mechanics and its transition to quantum mechanics. This is why we have devoted so much space to their discussion. As illustrations we shall consider a class of observables, to be referred to as *momentum observables*, which play a fundamental role in both classical mechanics and quantum mechanics. This section is devoted to reviewing these observables within classical mechanics as preparation to their quantization which will be discussed in Chapter 3.

Consider a particle moving in $I\!E^n$ with rectangular Cartesian coordinates x^j. Let us call the space $I\!E^n$ in which the particle moves the *physical coordinate space*, or **physical space** or **coordinate space** for short. Let the particle's momenta conjugate to x^j be denoted by p_j. Then the phase space Γ is a Hamiltonian manifold coordinated by x^j, p_j with a Hamiltonian structure $\Omega_H^2 = \sum_j dx^j \wedge dp_j$. An observable is by definition a function A on the $2n$-dimensional phase space. We start by examining functions linear in p_j, i.e., functions of the form

$$P = \sum_{k=1}^{n} \xi^k(x) p_k, \qquad (1.303)$$

where $\xi^k(x)$ are functions of coordinates x^j only. According to Eq. (1.271) such a momentum observable generates a Hamiltonian vector field on Γ, i.e.,

$$\widehat{\mathcal{X}}_P = \sum_{j=1}^{n} \left(\xi^j(x) \frac{\partial}{\partial x^j} - \sum_k \left(\frac{\partial \xi^k}{\partial x^j} p_k \right) \frac{\partial}{\partial p_j} \right). \qquad (1.304)$$

---
[42] See §1.6.4E(1) E5.

The fact that ξ^j are functions of coordinates only enables us to project this vector field down from the phase space Γ to the coordinate space $I\!E^n$, i.e., we can generate a vector field $\widehat{X}_P$ on $I\!E^n$ given by

$$\widehat{X}_P = \sum_{j=1}^n \xi^j(x) \frac{\partial}{\partial x^j}. \tag{1.305}$$

We shall call $\widehat{X}_P$ the *Hamiltonian vector field* of P on $I\!E^n$.

Definition 1.6.4(1)   **Complete and incomplete momenta**

- For a Hamiltonian system an observable of the form

$$P = \sum_{k=1}^n \xi^k(x) p_k, \tag{1.306}$$

where ξ^k are functions of coordinates x^j only, is called a *momentum observable based on the physical space* $I\!E^n$ or simply a momentum.

- A momentum $P = \sum_{k=1}^n \xi^k(x) p_k$ is said to be

    - *complete* if its Hamiltonian vector $\widehat{X}_P$ given by Eq. (1.305) on the coordinate space $I\!E^n$ is complete;
    - *incomplete* if $\widehat{X}_P$ is incomplete in $I\!E^n$.

Comments 1.6.4(1)   **Generators of transformation groups**

C1   The simplest momenta are canonical variables p_j conjugate to coordinates x^j. It is possible to effect a canonical transformation from (x^j, p_j) to another canonical variables $(\bar{x}^j, \bar{p}_j)$ such that $\bar{x}^j$ do not relate solely to position in the coordinate space. We could have, according to Eq. (1.261),

$$\bar{x}^j = -p_j, \quad \bar{p}_j = x^j. \tag{1.307}$$

The canonical momentum $\bar{p}_j$ conjugate to canonical coordinate $\bar{x}^j$ is no longer a momentum observable based on the physical space $I\!E^n$. This example serves to demonstrate the distinction between a momentum observable based on the coordinate space and a canonical momentum

When there is no risk of confusion we shall simply call a momentum based on the coordinate space a momentum.

C2   A momentum generates a Hamiltonian vector field $\widehat{\mathcal{X}}_P$ on Γ and a separate Hamiltonian vector field $\widehat{X}_P$ on $I\!E^n$.

1.6. CLASSICAL DYNAMICAL SYSTEMS 75

C3   A *complete momentum* is often referred to as the *generator of a one-parameter group of transformations* of the coordinate space $I\!\!E^n$ in the form of translations along the integral curves of its Hamiltonian vector field $\widehat{X}_P$ in the coordinate space $I\!\!E^n$. An incomplete momentum does not generate a one-parameter group of transformations. The distinction between complete and incomplete momenta will have a dramatic effect in their quantization. In the examples presented below we shall consider the motion of a particle in $I\!\!E^3$ with the usual notation (x, y, z) for the rectangular Cartesian coordinates with (p_x, p_y, p_z) for their respective conjugate momenta. We also present an example of a particle constrained to move in a circle $\mathcal{C}$; we shall return to their quantization in §6.7, §6.8 and §6.9 in Chapter 6.

Examples 1.6.4(1)   Linear and angular momenta

E1   **Linear momenta in $I\!\!E^3$**   The momentum p_x conjugate to x is referred to as a *linear momentum*. Its Hamiltonian vector field on $I\!\!E^3$ is

$$\widehat{X}_{p_x} = \frac{\partial}{\partial x}. \tag{1.308}$$

This vector field is clearly complete. As pointed out in §1.3.3C(2) C2 such a vector field also generates a one-parameter group of transformations of $I\!\!E^3$ in the form of translations along the coordinate variable x by

$$x \mapsto x(\tau) = T_\tau x = \tau + x. \tag{1.309}$$

We shall call this a *group of translations* along x.

E2   **Angular momenta in $I\!\!E^3$**   The momentum $L_z = y p_x - x p_y$ gives rise to a Hamiltonian vector field

$$\widehat{X}_{L_z} = y \partial/\partial x - x \partial/\partial y \tag{1.310}$$

on $I\!\!E^3$ which generates rotations about z-axis. This vector field is complete, as shown in §1.3.2E(1) E4, and its integral curves are circles lying on planes parallel to the x-y plane and centred at z-axis. We identify L_z with the angular momentum component along z-axis. This momentum is complete and may be interpreted as the generator for rotations about z-axis. Both L_z and $\widehat{X}_{L_z}$ are zero along z-axis.

E3   **Momentum p_θ in $I\!\!E^3$**   In §1.2.3C(1) C3 we introduce the canonical momentum p_θ conjugate to the angle variable $\theta \in (0, \pi)$ in spherical coordinates. From Eq. (1.31) we can write down its Hamiltonian vector field on $I\!\!E^3$ as

$$\widehat{X}_{p_\theta} = \frac{xz}{\sqrt{x^2+y^2}} \frac{\partial}{\partial x} + \frac{yz}{\sqrt{x^2+y^2}} \frac{\partial}{\partial y} - \sqrt{x^2+y^2} \frac{\partial}{\partial z}. \tag{1.311}$$

In spherical coordinates this expression becomes

$$X_{p_\theta} = \frac{\partial}{\partial \theta}. \tag{1.312}$$

We can work out the integral curve starting from a point $(r_0, \theta_0, \varphi_0)$ explicitly. It is a solution of

$$\frac{dr}{d\tau} = 0, \quad \frac{d\theta}{d\tau} = 1, \quad \frac{d\varphi}{d\tau} = 0, \tag{1.313}$$

namely

$$r = r_0, \quad \theta = \tau + \theta_0, \quad \varphi = \varphi_0. \tag{1.314}$$

Since τ is confined to the range $(-\theta_0, \pi - \theta_0)$ to keep θ in the range $(0, \pi)$ the vector field, and hence the momentum, are both incomplete.

E4 Radial momentum in $\mathbb{E}^3$ The momentum p_r in §1.2.3C(1) C3 is called the *radial momentum*. Its Hamiltonian vector field $\widehat{X}_{p_r}$ on $\mathbb{E}^3$ is

$$\widehat{X}_{p_r} = \frac{x}{r}\frac{\partial}{\partial x} + \frac{y}{r}\frac{\partial}{\partial y} + \frac{z}{r}\frac{\partial}{\partial z}, \quad r > 0. \tag{1.315}$$

In the usual spherical coordinates (r, θ, φ) this expression becomes

$$\widehat{X}_{p_r} = \frac{\partial}{\partial r}, \quad r > 0. \tag{1.316}$$

The integral curve starting from a point $(r_0, \theta_0, \varphi_0)$ is given by

$$r = \tau + r_0, \quad \theta = \theta_0, \quad \varphi = \varphi_0. \tag{1.317}$$

Since τ is confined to the range $(-r_0, \infty)$ to keep r in the range $(0, \infty)$ the vector field is incomplete. To illustrate the link between the group nature of translations and completeness stated in §1.3.3C(2) C2 we can see here that the constraint on the values of r to the range $(0, \infty)$ also implies that translations along coordinate r do not form a group, e.g., there is no inverse translation since such a translation will render some r negative. Similarly we can see that the incompleteness of p_θ is related to the restricted range of the conjugate coordinate θ which makes a translation along θ undefinable for every value of θ.

In passing we should point out that p_r and $\widehat{X}_{p_r}$ are not defined along the z-axis where $x = y = 0$.

E5 Momentum $P_\theta(\mathcal{C})$ conjugate to θ in $\mathcal{C}$ For a particle constrained to move in a circle $\mathcal{C}$ of radius a in $\mathbb{E}^2$ and centered at the coordinate origin its position can be specified by the polar angle θ in §1.5.1E(1) E2. Let us denote the momentum conjugate to θ by $P_\theta(\mathcal{C})$. In accordance

1.6. CLASSICAL DYNAMICAL SYSTEMS

with Eq. (1.271) its associated Hamiltonian vector field $\widehat{X}_{P_\theta(\mathcal{C})}$ on $\mathcal{C}$ is given by Eq. (1.241). This vector field is complete. Hence $P_\theta(\mathcal{C})$ is a complete momentum for a particle constrained to move in a circle $\mathcal{C}$. Momentum $P_\theta(\mathcal{C})$ is identifiable with the traditional angular momentum. Note that this is quite different from p_θ in $I\!\!E^3$ defined by Eq. (1.31) and discussed in E3 above.

1.6.5 Concluding remarks

In this chapter we have presented an intrinsic approach to establish a number of geometric quantities on Euclidean spaces and manifolds. The starting point is the set of smooth functions which serve as the basic coordinate independent constructs on a manifold on which to build up other geometric quantities. Derivations are then introduced as linear operators satisfying the product rule of differentiation when acting on the set of functions. Naturally derivations are identifiable with differential operators. All these quantities are defined in a coordinate independent way. Fundamentally contravariant vectors are seen to be differential operators. In other words differential operators provide an intrinsic definition of contravariant vectors. Covariant vectors can then be identified with differentials of functions. Smooth assignments of vectors to every point in the manifold lead to the establishment of vector fields.

We formulate classical mechanics in a geometric language.[43] The basic geometric space is the phase space Γ as a Hamiltonian manifold. In the phase space every function f generates a vector field $\widehat{\mathcal{X}}_f$ known as its Hamiltonian vector field on the phase space. The dynamics can be described in terms of the Hamiltonian vector field $\widehat{\mathcal{X}}_{H_g}$ of a function H_g on Γ, known as the Hamiltonian generator. An important class of observables is the momentum observables P. A momentum P can generate a vector field $\widehat{X}_P$ on the coordinate space, referred to as its Hamiltonian vector field on the coordinate space. This property leads us to the following conclusions:

- We can interpret a momentum as a generator for spatial translations.

- We can classify momenta into complete ones and incomplete ones.

- Bearing in mind that vector fields are identifiable with differential operators we see that momentum observables are intrinsically linked to differential operators. This indicates a link between classical mechanics and quantum mechanics where observables are described directly in

---

[43] We have limited the discussion to what we need later for quantization purposes, and refrained from going deeper into the mathematical and geometric foundations of classical mechanics, e.g., we have not mentioned momentum mapping on sympletic geometry (Abraham and Marsden (1978)) and Poisson manifolds (Choquet-Bruhat and de Witt-Morette (1989)).

terms of operators, i.e., a classical momentum P may be quantized in terms of its Hamiltonian vector field $\widehat{X}_P$ in the coordinate space. This we shall do later in Chapter 3. This method of quantization, referred to as geometric quantization, will be seen to ease the conceptual transition from classical mechanics to quantum mechanics.

References

1. Abraham, R. and Marsden, J. E. (1978). *Foundations of Mechanics*, Benjamin/Cummings, Reading, Mass.
2. Arnold, V. I. (1978). *Mathematical Methods of Classical Mechanics*, Springer-Verlag, New York.
3. Benn, I. M. and Tucker, R. W. (1987). *An Introduction to Spinors and Geometry*, Adam Hilger, Bristol.
4. Bishop, R. L. and Goldberg, S. I. (1968). *Tensor Analysis on Manifolds*, Macmillan, New York.
5. Bleaney, B. I. and Bleaney, B. (1957). *Electricity and Magnetism*, Clarendon Press, Oxford.
6. Brickell, F. and Clark, R. S. (1970). *Differentiable Manifolds*, Van Nostrand Reinhold, London.
7. Choquet-Bruhat, Y., de Witt-Morette, C., with Dillard-Bleick, M. (1982). *Analysis, Manifolds and Physics Part I: Basics*, North-Holland, Amsterdam.
8. Choquet-Bruhat, Y., de Witt-Morette, C. (1989). *Analysis, Manifolds and Physics Part II: 92 Applications*, North-Holland, Amsterdam.
9. Darling, R. W. R. (1994). *Differential Forms and Connections*, Cambridge University Press, Cambridge.
10. Frankel, T. (1997). *The Geometry of Physics*, Cambridge University Press, Cambridge.
11. Goldstein, H. (1950). *Classical Mechanics*, Addison-Wesley, Reading, Mass.
12. Isham, C. J. (1989). *Modern Differential Geometry for Physicists*, World Scientific, Singapore.
13. Leech, J. W. (1965). *Classical Mechanics*, Methuen & Co Ltd, London.
14. Lipschutz, S. (1965). *General Topoloogy*, Schaum's Outline Series, McGraw-Hill, New York.
15. Loomis, L. H. and Sternberg, S. (1968). *Advanced Calculus*, Addison-Wesley, Reading, Mass.

REFERENCES

16. Martin, D. (1991). *Manifold Theory*, Ellis Harwood, New York.
17. Matsushima, Y. (1972). *Differentiable Manifolds*, Marcel Dekker, New York.
18. Nash, C. and Sen, S. (1983). *Topology and Geometry for Physicists*, Academic Press, London.
19. Percival, I. and Richards, D. (1982). *Introduction to Dynamics*, Cambridge University Press, Cambridge.
20. Perko, L. (1991). *Differential Equations and Dynamical Systems*, Springer-Verlag, New York.
21. Schutz, B. (1980). *Geometrical Methods of Mathematical Physics*, Cambridge University Press, Cambridge.
22. Simmons, G. F. (1963). *Introduction to Topology and Modern Analysis*, McGraw-Hill International Edition, Tokyo.
23. Sutherland, W. A. (1995). *Introduction to Metric and Topological Spaces*, Clarendon Press, Oxford.
24. Synge, J. L. and Schild, A. (1966). *Tensor Calculus*, University of Toronto Press, Toronto.

Chapter 2

Operators and their Direct Integrals

2.1 Hilbert Spaces

This chapter is devoted to the discussion of three aspects of operator theory which are not widely used in orthodox quantum mechanics, and are therefore less familiar to physicists:

1. Symmetric operators, their maximal symmetric extensions and their spectral measures and spectral functions;

2. Local operators;

3. Direct integrals of Hilbert spaces and operators.

The presentation consists of summaries of definitions and results supplemented by comments and examples. This knowledge is neccesary for our formulation of a generalized quantum theory later. In this introductory section we shall briefly introduce Hilbert spaces, subsets of a Hilbert space and some relevant terminology, assuming a knowledge of complex vector spaces. Throughout this chapter many useful results will be stated for later reference without proofs. More detailed discussions are available in some of the books and papers listed in the References at the end of this chapter.

Let V be a complex scalar product space, i.e., V is a complex vector space endowed with a scalar product $\langle \cdot \mid \cdot \rangle$.[1] We define the norm $||\phi||$ of a vector

---

[1] A scalar product is often referred to as an inner product and a scalar product space is also referred to as an inner product space.

ϕ in V in terms of the scalar product by

$$||\phi|| = \sqrt{\langle \phi | \phi \rangle}. \tag{2.1}$$

A scalar product is also expressible in terms of norms, i.e.,

$$\langle \psi | \phi \rangle = \frac{1}{2} \left\{ ||\psi + \phi||^2 - i\,||\psi + i\phi||^2 + (i-1)\left(||\psi||^2 + ||\phi||^2\right) \right\} \tag{2.2}$$

$$= \frac{1}{4} \left\{ ||\psi + \phi||^2 - ||\psi - \phi||^2 + i\,||\psi + i\phi||^2 - i\,||\psi - i\phi||^2 \right\}. \tag{2.3}$$

A *unit* vector is one with a unit norm, and two vectors ψ and ϕ are said to be

1. *orthogonal* if $\langle \psi | \phi \rangle = 0$, and

2. *orthonormal* if

$$\langle \psi | \phi \rangle = 0, \quad ||\psi|| = ||\phi|| = 1. \tag{2.4}$$

Perhaps the simplest complex vector space is $\mathbb{C}$, the set of complex numbers $a, b, \ldots$, with scalar product given by

$$\langle a | b \rangle = a^* b, \qquad ||a|| = |a| = \sqrt{a^* a} \tag{2.5}$$

where a^* is the complex conjugate of a and $|a|$ is the absolute value of a. A sequence of complex numbers $a_n, n = 1, 2, \ldots$, is said to converge to another complex number a if

$$|a_n - a| \to 0 \quad \text{as} \quad n \to \infty. \tag{2.6}$$

In $\mathbb{C}$ we have the well-known *Cauchy convergence criterion* which says that a sequence of complex numbers $a_n, n = 1, 2, \ldots$, converges to another complex number if and only if the sequence of numbers satisfies

$$|a_n - a_m| \to 0 \quad \text{as} \quad n, m \to \infty. \tag{2.7}$$

Let us call a sequence satisfying this condition a *Cauchy sequence*. Then every Cauchy sequence converges in $\mathbb{C}$.

In a general complex vector space V we can retain the above definitions. A sequence of vectors φ_n in V is said

1. to be a Cauchy sequence in V if [2]

$$||\varphi_n - \varphi_m|| \to 0 \quad \text{as} \quad m, n \to \infty, \quad \text{and} \tag{2.8}$$

---

[2] Weidmann (1980) p. 15.

2.1. HILBERT SPACES

2. to converge to a vector φ in V if [3]

$$||\varphi_n - \varphi|| \to 0 \quad \text{as} \quad n \to \infty. \tag{2.9}$$

We shall denote the convergence of a sequence of vectors φ_n by

$$\varphi_n \to \varphi \quad \text{as} \quad n \to \infty. \tag{2.10}$$

The question arises as to the validity of Cauchy's convergence criterion in V. Although Cauchy's criterion fails generally there are many vector spaces in which Cauchy's criterion remains valid and it turns out that these spaces are of great importance to physics. These spaces are known as Hilbert spaces:

- A complex vector space endowed with a scalar product becomes a *Hilbert space* if every Cauchy sequence in the space converges to a vector in the space.

Let ϕ and ψ be two vectors in a Hilbert space. Then one can establish the following inequalities:[4]

$$|\langle \phi \mid \psi \rangle| \leq ||\phi|| \, ||\psi|| \quad \text{Schwarz inequality,} \tag{2.11}$$
$$||\phi + \psi|| \leq ||\phi|| + ||\psi|| \quad \text{Triangle inequality 1,} \tag{2.12}$$
$$\Big| \, ||\phi|| - ||\psi|| \, \Big| \leq ||\phi - \psi|| \quad \text{Triangle inequality 2,} \tag{2.13}$$

and the following *parallelogram equality*:[5]

$$||\psi + \phi||^2 + ||\psi - \phi||^2 = 2\,||\psi||^2 + 2\,||\phi||^2. \tag{2.14}$$

An important type of Hilbert space is one formed by complex-valued functions $\phi, \psi, \ldots$, defined on a given manifold $I\!\!M^\ell$ with scalar product defined by an integral with respect to the volume element $d\mu$ over the manifold, i.e.,

$$\langle \psi \mid \phi \rangle = \int_{I\!\!M^\ell} \psi^* \phi \, d\mu, \tag{2.15}$$

---

[3] As seen later a weaker condition of convergence is also useful. A sequence of vectors ϕ_n is said to *converge weakly* to φ (Weidmann (1980) p. 76) if for every $\psi \in V$ we have $\langle \phi_n \mid \psi \rangle \to \langle \phi \mid \psi \rangle$ as $n \to \infty$. However, unless otherwise is stated a convergence means a convergence in the sense of Eq. (2.9).

[4] Amrein (1981) pp. 3-4.

[5] This equality can distinguish scalar product spaces from normed spaces which are vector spaces endowed directly with a norm. While a scalar product space is a norm space with a norm defined by Eq. (2.1) a norm space is generally not a scalar product space. In a normed space Eqs. (2.2) or (2.3) do not produce a scalar product since the parallelogram equality is not automatically satisfied (Kreyszig (1978) p. 130).

84   CHAPTER 2. OPERATORS AND THEIR DIRECT INTEGRALS

where ψ^* denotes the complex conjugate of ψ. The norm of a function ϕ regarded as a vector is then given by

$$||\phi|| = \left(\int_{I\!\!M^\ell} \phi^* \phi \, d\mu \right)^{1/2}. \qquad (2.16)$$

This type of Hilbert spaces will be denoted by $L^2(I\!\!M^\ell, d\mu)$ and these functions are called *square-integrable*.[6] An example is the Hilbert space formed by functions on the Euclidean space $I\!\!E^n$ square-integrable with respect to $d\mu = dx^1 dx^2 \cdots dx^n$ which is just the usual volume element in $I\!\!E^n$ in terms of rectangular Cartesian coordinates x^j. We shall denote this Hilbert space simply by $L^2(I\!\!E^n)$.

Many important quantities such as operators are defined on a subset of a Hilbert space. So, we shall start by introducing some useful subsets and their properties:

1. **Notation**   A Hilbert space is generally denoted by $\mathcal{H}$.

2. **Linear subsets**   A *linear subset* is a subset $\mathcal{L}$ of $\mathcal{H}$ such that if ϕ and ψ are elements of $\mathcal{L}$ then their *linear combination*, i.e., $a\phi + b\psi$, with any complex coefficients a, b is again an element of $\mathcal{L}$.

3. **Closed linear subsets and subspaces**   A vector $\phi \in \mathcal{H}$ is called a *limit vector* of a linear subset $\mathcal{L}$ if there exists a sequence of vectors ϕ_n within $\mathcal{L}$ converging to ϕ. We call $\mathcal{L}$ a *closed linear subset* if $\mathcal{L}$ contains all its limit vectors. It follows that a closed linear subset is a Hilbert space in its own right.[7] Hence, a closed linear subset is also called a *subspace*.[8]

4. **Closure**   If a linear subset $\mathcal{L}$ is not closed we can construct a bigger linear subset $\bar{\mathcal{L}}$ by adding to $\mathcal{L}$ all its limit vectors. We call $\bar{\mathcal{L}}$ the *closure* of $\mathcal{L}$.

5. **Dense linear subsets**   A linear subset $\mathcal{L}$ of $\mathcal{H}$ is said to be *dense* in $\mathcal{H}$ if its closure is $\mathcal{H}$. We also call $\mathcal{L}$ a dense subset of $\mathcal{H}$.

6. **Orthogonal subspaces**   A vector ϕ is said to be *orthogonal to a subspace* $\mathcal{L}$ if ϕ is orthogonal to every vector in $\mathcal{L}$, i.e.,

$$\langle \phi \mid \varphi \rangle = 0 \quad \forall \varphi \in \mathcal{L}. \qquad (2.17)$$

---

[6]To form a Hilbert space this way the integrals are meant to be Lebesgue integrals, as opposed to the familiar Riemann integrals (Williamson (1962), Burrill (1972)). However, a detailed knowledge of Lebesgue integration is not essential here or in the rest of the book.

[7]The condition that $\mathcal{L}$ contains all its limit vectors implies that every Cauchy sequence in $\mathcal{L}$ converges to a vector in $\mathcal{L}$. It follows that $\mathcal{L}$ is a Hilbert space.

[8]Some authors use the term *subspace* to mean a linear subset which is not necessarily closed and hence not neccesarily a Hilbert space in its own right.

2.1. HILBERT SPACES

Two subspaces $\mathcal{L}_1$ and $\mathcal{L}_2$ are said to be *orthogonal* if

$$\langle \phi_1 \mid \phi_2 \rangle = 0 \quad \forall \phi_1 \in \mathcal{L}_1, \ \phi_2 \in \mathcal{L}_2. \tag{2.18}$$

In other words every vector in $\mathcal{L}_1$ is orthogonal to $\mathcal{L}_2$ and *vice versa*.

7. **Orthogonal complements** The orthogonal complement $\mathcal{L}^\perp$ to a given subspace $\mathcal{L}$ is the set of all the vectors in $\mathcal{H}$ which are orthogonal to $\mathcal{L}$. One can verify that $\mathcal{L}^\perp$ is again a subspace. Such a pair of subspaces is said to be complementary.

8. **Orthonormal sequences and countable orthonormal basis** A sequence of vectors $\varphi_n, n = 1, 2, \ldots$, in $\mathcal{H}$ is said to be an *orthonormal sequence* if

$$\langle \varphi_n \mid \varphi_m \rangle = \delta_{n,m}, \quad \forall m, n. \tag{2.19}$$

For many Hilbert spaces important in physics we can find an orthonormal sequence $\varphi_n, n = 1, 2, \ldots$, in $\mathcal{H}$ to express every vector ϕ in $\mathcal{H}$ as a linear combination of φ_n. In other words we can write

$$\phi = \sum_{n=1}^\infty c_n \varphi_n, \quad c_n \in \mathbb{C} \tag{2.20}$$

in the sense that the sequence of vectors ϕ_N defined by

$$\phi_N = \sum_{n=1}^N c_n \varphi_n \tag{2.21}$$

converges to ϕ, i.e.,

$$\|\phi - \phi_N\| = \left\|\phi - \sum_{n=1}^N c_n \phi_n\right\| \to 0 \quad \text{as} \quad N \to \infty. \tag{2.22}$$

Such a sequence $\varphi_n, n = 1, 2, \ldots$, in $\mathcal{H}$ is called a *countable orthonormal basis* or a *complete orthonormal set* in $\mathcal{H}$. In view of the orthonormal nature of the sequence we can write down the coefficients of the linear combination in Eq. (2.20) as

$$c_n = \langle \varphi_n \mid \phi \rangle. \tag{2.23}$$

A set of vectors φ_n not necessarily orthonormal is called a *complete set* in $\mathcal{H}$ if every vector in $\mathcal{H}$ can be expressed as a linear combination of φ_n in the sense of Eqs. (2.20), (2.21) and (2.22). Then Eq. (2.23) is not valid and it is not so easy to write down the expansion coefficients c_n.

A Hilbert space $\mathcal{H}$ which admits a countable orthonormal basis

$$\{\varphi_n, n = 1, 2, \ldots\} \tag{2.24}$$

possesses the following properties:

(a) No non-zero vector can be orthogonal to all members of the orthonormal basis, i.e.,

$$\langle \varphi_n \mid \phi \rangle = 0 \quad \forall n \quad \Rightarrow \quad \phi = 0. \tag{2.25}$$

(b) A dense linear subset $\mathcal{D}$ of $\mathcal{H}$ necessarily contains an orthonormal basis so that

$$\langle \varphi \mid \phi \rangle = 0 \quad \forall \varphi \in \mathcal{D} \quad \Rightarrow \quad \phi = 0. \tag{2.26}$$

Hilbert spaces possessing these features are called *separable*. A formal statement is given in the next item.

9. **Separable Hilbert spaces**   A Hilbert space is said to be *separable* if it admits a countable orthonormal basis. In a separable Hilbert space a vector orthogonal to a countable orthonormal basis must be the zero vector; moreover a dense set neccesarily contains an orthonormal basis so that no non-zero vector can be orthogonal to a dense set. We shall call a countable orthonormal basis simply an *orthonormal basis* in a separable Hilbert space. From now on, unless it is stated otherwise, a Hilbert space $\mathcal{H}$ is meant to be complex and separable. If the orthonormal basis consists of only a finite number N of vectors the Hilbert space is called finite-dimensional, or N-dimensional. The simplest separable Hilbert space is the one-dimensional space $\mathbb{C}$ with scalar product defined by Eq. (2.5).

Compared with §1.1.1C(2) C6 we can see that concepts of denseness and closure are of topological origin. Intuitively a dense linear subset $\mathcal{L}$ has to be very large, large enough to permeate everywhere in $\mathcal{H}$ so that every vector in $\mathcal{H}$ can be approximated with arbitrary accuracy by vectors in $\mathcal{L}$. A well known dense linear subset in $L^2(I\!\!E^n)$, which we shall encounter frequently, is the set $C_0^\infty(I\!\!E^n)$ of smooth functions of compact support on $I\!\!E^n$ introduced in §1.2.2, i.e., the set of smooth functions each of which vanishes outside a closed and bounded set in $I\!\!E^n$.

In the next section we shall give a summary of some basic definitions on operators. A brief review of the theory concerning symmetric operators, including their spectral measures, their spectral functions and their extensions, will be given.

2.2 Operators: Basic Definitions

In §1.3 we introduce linear and differential operators acting on the set of smooth functions $C^\infty(I\!E^n)$ on $I\!E^n$. When considering operators acting in a Hilbert space a great deal more refinement to the definitions is required. For instance, the Hilbert space $L^2(I\!E^n)$, which contains discontinuous and non-differentiable functions, is bigger than the set $C^\infty(I\!E^n)$. Differential operators capable of acting on every function in $C^\infty(I\!E^n)$ are clearly unable to operate on every function in $L^2(I\!E^n)$. So, from the start we have to define operators on a linear subset of a Hilbert space rather than on the entire Hilbert space.

2.2.1 Boundedness, adjoints, extensions and restrictions, continuity and closure

A linear operator $\widehat{A}$ in a Hilbert space $\mathcal{H}$ is a linear mapping of a linear subset $\mathcal{D}(\widehat{A})$ of $\mathcal{H}$ onto another subset $\mathcal{R}(\widehat{A})$ of $\mathcal{H}$:

$$\widehat{A}: \mathcal{D}(\widehat{A}) \mapsto \mathcal{R}(\widehat{A}) \quad \text{by} \quad \phi \in \mathcal{D}(\widehat{A}) \mapsto \widehat{A}\phi \in \mathcal{R}(\widehat{A}). \tag{2.27}$$

The subsets $\mathcal{D}(\widehat{A})$ and $\mathcal{R}(\widehat{A})$ are respectively known as the *domain* and the *range* of $\widehat{A}$. Two operators $\widehat{A}_1$ and $\widehat{A}_2$ are deemed to be equal if

$$\mathcal{D}(\widehat{A}_1) = \mathcal{D}(\widehat{A}_2) = \mathcal{D} \quad \text{and} \quad \widehat{A}_1\phi = \widehat{A}_2\phi \quad \forall \phi \in \mathcal{D}. \tag{2.28}$$

Care has to be taken when we add and multiply two operators since the sum $\widehat{A}_1 + \widehat{A}_2$ and product $\widehat{A}_2\widehat{A}_1$ are defined on new domains:

$$\mathcal{D}(\widehat{A}_1 + \widehat{A}_2) = \mathcal{D}(\widehat{A}_1) \cap \mathcal{D}(\widehat{A}_2), \tag{2.29}$$
$$\mathcal{D}(\widehat{A}_2\widehat{A}_1) = \left\{\phi \in \mathcal{H} : \phi \in \mathcal{D}(\widehat{A}_1), \ \widehat{A}_1\phi \in \mathcal{D}(\widehat{A}_2)\right\}. \tag{2.30}$$

This means that it is possible, for example, to have $\mathcal{D}(\widehat{A}_1) \cap \mathcal{D}(\widehat{A}_2) = \emptyset$, the empty set in $\mathcal{H}$, so that we cannot have a well-defined sum.

It is often useful to introduce operators to relate two different Hilbert spaces. For example, if $\mathcal{H}_1$ and $\mathcal{H}_2$ are two Hilbert spaces we can define an operator which maps every vector $\phi_1 \in \mathcal{H}_1$ to a vector $\phi_2 \in \mathcal{H}_2$. We shall see an example of this later.

An operator in a Hilbert space $\mathcal{H}$ is often specified by an explicit operator expression together with the domain for the operator expression to act on. The domain of an operator is an integral part of the definition of the operator. An identical operator expression defined on different domains can lead to very different operators. Since different operators relate to different physical quantities in quantum mechanics we can see that the domain of an operator is of physical significance. Unless it is stated otherwise we shall assume from now on that every operator in a Hilbert space introduced in this book is a linear operator acting on a dense domain; such an operator is said to be *densely defined*. An operator $\widehat{B}$ is said to be *bounded* if there is a number $N < \infty$ such that

$$\frac{\|\widehat{B}\phi\|}{\|\phi\|} \leq N \qquad \forall \phi \neq 0 \in \mathcal{D}(\widehat{B}). \tag{2.31}$$

The smallest of all such N is called the *norm* of $\widehat{B}$, to be denoted by $\|\widehat{B}\|$. We have

$$\|\widehat{B}\phi\| \leq \|\widehat{B}\|\,\|\phi\|. \tag{2.32}$$

An operator is said to be *unbounded* if it is not bounded.

Let $\widehat{A}$ be an operator defined on a domain $\mathcal{D}(\widehat{A})$ in $\mathcal{H}$. Given a pair of vectors

$$\psi \in \mathcal{H} \quad \text{and} \quad \phi \in \mathcal{D}(\widehat{A}) \tag{2.33}$$

we can always find a vector $\varphi \in \mathcal{H}$ such that

$$\langle \psi \mid \widehat{A}\phi \rangle = \langle \varphi \mid \phi \rangle. \tag{2.34}$$

Such a vector φ is generally dependent on both ψ and ϕ and is clearly not unique since we can add to φ a vector orthogonal to ϕ without affecting the scalar product on the right hand side of the above equation. In an attempt to render φ unique let us impose a further condition. Suppose we require that for any arbitrarily chosen ψ we require one and the same φ for Eq. (2.34) to hold for all $\phi \in \mathcal{D}(\widehat{A})$. The question is whether such a vector φ exists. Unfortunately this condition is so stringent that φ may not exist. We can weaken the condition by abandoning the arbitrary nature of ψ. In other words we want to know whether such a φ exists for a suitably selected vector ψ. It turns out that for a densely defined operator $\widehat{A}$ the answer to this question is in the affirmative.[9] Indeed we can have a set of such suitably selected ψ. The one-to-one correspondence between ψ and φ enables us to introduce a new operator $\widehat{A}^\dagger$ defined on the set of suitably selected vectors ψ by

$$\widehat{A}^\dagger \psi = \varphi. \tag{2.35}$$

This new operator is called the *adjoint* of $\widehat{A}$. The selected set of vectors ψ serves as the domain of this new operator $\widehat{A}^\dagger$ and is naturally denoted by $\mathcal{D}(\widehat{A}^\dagger)$. We can now rewrite Eq. (2.34) as

$$\langle \psi \mid \widehat{A}\phi \rangle = \langle \widehat{A}^\dagger \psi \mid \phi \rangle \qquad \forall \phi \in \mathcal{D}(\widehat{A}),\ \forall \psi \in \mathcal{D}(\widehat{A}^\dagger). \tag{2.36}$$

---

[9] Prugovečki (1981) Theorem 2.5 on p. 187.

2.2. OPERATORS: BASIC DEFINITIONS

To sum up we conclude that every densely defined operator possesses a unique adjoint operator, although the adjoint operator may not be densely defined. Adjoint operators are very important since many important types of operators are defined in terms of their relationship with their adjoints.

Let $\widehat{A}$ be an operator defined on a dense domain $\mathcal{D}(\widehat{A})$ in a Hilbert space $\mathcal{H}$. We can introduce the concept of extensions and of restrictions of $\widehat{A}$. This is done in Definition 2.2.1(1).

Definition 2.2.1(1)   Extensions and restrictions; continuity, closedness and closure

- An operator $\widehat{A}_{ex}$ defined on a domain $\mathcal{D}(\widehat{A}_{ex})$ is called an *extension* of $\widehat{A}$ if

    1. $\mathcal{D}(\widehat{A}_{ex}) \supset \mathcal{D}(\widehat{A})$, i.e., $\mathcal{D}(\widehat{A})$ is a subset of $\mathcal{D}(\widehat{A}_{ex})$.
    2. $\widehat{A}_{ex}$ is the same as $\widehat{A}$ on $\mathcal{D}(\widehat{A})$, i.e., $\widehat{A}_{ex}\phi = \widehat{A}\phi \quad \forall \phi \in \mathcal{D}(\widehat{A})$.

    This relationship is denoted by $\widehat{A} \subset \widehat{A}_{ex}$, and $\widehat{A}$ is called a *restriction* of $\widehat{A}_{ex}$.

- $\widehat{A}$ is said to be *continuous* if whenever a sequence $\phi_1, \phi_2, \ldots$ of vectors in $\mathcal{D}(\widehat{A})$ converges to a vector $\phi \in \mathcal{D}(\widehat{A})$, then the sequence $\widehat{A}\phi_1, \widehat{A}\phi_2, \ldots$ of vectors converges to the vector $\widehat{A}\phi \in \mathcal{H}$.

- $\widehat{A}$ is said to be *closed* if whenever a sequence of vectors $\phi_1, \phi_2, \ldots$ in $\mathcal{D}(\widehat{A})$ converges to $\phi \in \mathcal{H}$ and the sequence $\widehat{A}\phi_1, \widehat{A}\phi_2, \ldots$ of vectors converges to some vector $\varphi \in \mathcal{H}$, then ϕ lies in $\mathcal{D}(\widehat{A})$ and $\widehat{A}\phi = \varphi$.

- $\widehat{A}$ is said to be *closable* if it has extensions which are closed. The smallest closed extension, assuming it exists, is called the *closure* of $\widehat{A}$, and is denoted by $\bar{A}$.[10]

Comments 2.2.1(1)   Continuity, closedness, expectation values and extensions

C1   The concept of continuity of an operator is really the same as that of continuity of a real-valued function f of a real variable τ, i.e., if a sequence of numbers τ_n converges to a number τ, then $f(\tau_n)$ converges to $f(\tau)$.

C2   Closedness is different from continuity; here we deduce that ϕ lies in $\mathcal{D}(\widehat{A})$ from the convergence of the sequences ϕ_n and $\widehat{A}\phi_n$. A continuous

---

[10] $\bar{A}$ is the smallest closed extension of $\widehat{A}$ in the sense that any other closed extension of $\widehat{A}$ is also an extension of $\bar{A}$.

operator can be shown to be closed but a closed operator is not necessarily continuous. Closedness is an important properties of operators, since we do want ϕ to be in $\mathcal{D}(\widehat{A})$ when the sequences ϕ_n and $\widehat{A}\phi_n$ converge. Without this closedness property many of the results on extensions of an operator to be introduced later will not hold.

C3 The expression $\langle \phi \mid \widehat{A}\phi \rangle$ for a unit vector $\phi \in \mathcal{D}(\widehat{A})$ leads to a value known as the *expectation value* of $\widehat{A}$ in ϕ. We shall return to discuss the physical relevance of expectation values later.

C4 Bounded operators possess the following properties:

1. A bounded operator is continuous, and the converse is true so that continuity and boundedness are equivalent for linear operators.

2. Bounded operators are closed, but the converse is generally false.

3. The adjoint $\widehat{B}^\dagger$ of a bounded operator $\widehat{B}$ is bounded. Moreover, we have
$$\widehat{B}^{\dagger\dagger} = \widehat{B}. \tag{2.37}$$

4. Let $\widehat{B}$ be a bounded operator defined on a dense domain $\mathcal{D}(\widehat{B})$. Then $\widehat{B}$ has a unique bounded extension $\widehat{B}_{ex}$, by continuity, to the entire $\mathcal{H}$ which preserves the norm, i.e., $||\widehat{B}|| = ||\widehat{B}_{ex}||$. To avoid unnecessary complications we shall always assume a bounded operator to be defined on the entire Hilbert space from now on.

5. Let V be a bounded function on $I\!E$, i.e., $V(x)$ is a bounded function of $x \in I\!E$.[11] Then V, acting on $L^2(I\!E)$ as a multiplication operator, is bounded. A very important class of bounded functions is the class of *characteristic functions* on $I\!E$. Let χ_Λ be the characteristic function of an interval Λ on $I\!E$, i.e.,

$$\chi_\Lambda : I\!E \mapsto \{0,1\} \quad \text{such that} \quad \chi_\Lambda(x) = \begin{cases} 1, & x \in \Lambda \\ 0, & x \notin \Lambda \end{cases}. \tag{2.38}$$

Acting as a multiplication operator on $L^2(I\!E)$ the characteristic function χ_Λ has the following effect:

$$\left(\chi_\Lambda \phi\right)(x) = \begin{cases} \phi(x), & x \in \Lambda \\ 0, & x \notin \Lambda \end{cases}. \tag{2.39}$$

As an operator χ_Λ is clearly bounded with norm $||\chi_\Lambda|| = 1$. It also has the striking property of being equal to its square, a property referred to as *idempotence*.

---
[11] A rectangular Cartesian coordinate in $I\!E$ is always denoted by x from now on.

2.2. OPERATORS: BASIC DEFINITIONS

6. Two bounded operators $\widehat{B}_1$ and $\widehat{B}_2$ are equal if they possess the same expectation values with respect to all vectors in $\mathcal{H}$, i.e.,

$$\langle \phi \mid \widehat{B}_1 \phi \rangle = \langle \phi \mid \widehat{B}_2 \phi \rangle \quad \forall \phi \in \mathcal{H} \quad \Longleftrightarrow \quad \widehat{B}_1 = \widehat{B}_2. \tag{2.40}$$

7. Many physically important operators such as unitary operators and projectors to be defined later are bounded.

8. A scalar product between ψ and $\widehat{B}\phi$ is expressible in terms of expectation values of $\widehat{B}$:

$$\begin{aligned}\langle \psi \mid \widehat{B}\phi \rangle &= \frac{1}{2}\Big\{ \langle \psi+\phi \mid \widehat{B}(\psi+\phi) \rangle - i\langle \psi+i\phi \mid \widehat{B}(\psi+i\phi) \rangle \\ &\quad + (i-1)\Big(\langle \psi \mid \widehat{B}\psi \rangle + \langle \phi \mid \widehat{B}\phi \rangle\Big)\Big\}. \end{aligned} \tag{2.41}$$

This is an extension of Eq. (2.2).

C5 Two possibly unbounded operators $\widehat{A}_1$ and $\widehat{A}_2$ defined on the same dense domain $\mathcal{D}$ in a Hilbert space $\mathcal{H}$ are equal if they possess the same expectation values with respect to all vectors in $\mathcal{D}$, i.e.,

$$\langle \phi \mid \widehat{A}_1 \phi \rangle = \langle \phi \mid \widehat{A}_2 \phi \rangle \quad \forall \phi \in \mathcal{D} \quad \Longleftrightarrow \quad \widehat{A}_1 = \widehat{A}_2. \tag{2.42}$$

To see this we first deduce from Eq. (2.41) that

$$\langle \psi \mid \widehat{A}_1 \phi \rangle = \langle \psi \mid \widehat{A}_2 \phi \rangle \quad \forall \psi, \phi \in \mathcal{D} \tag{2.43}$$

or

$$\langle \psi \mid (\widehat{A}_1 - \widehat{A}_2)\phi \rangle = 0 \quad \forall \psi, \phi \in \mathcal{D}. \tag{2.44}$$

From this we can conclude that $\widehat{A}_1 \phi = \widehat{A}_2 \phi$ on account of Eq. (2.26).

Some simple but important examples of unbounded operators are multiplication operators:

1. An unbounded function acts in $L^2(\mathbb{E})$ as an unbounded operator. A prime example of these is the *position operator*

$$\widehat{x} = x \quad \text{acting on} \quad \mathcal{D}(\widehat{x}) = \Big\{\phi \in L^2(\mathbb{E}) : x\phi \in L^2(\mathbb{E})\Big\} \tag{2.45}$$

and a simple harmonic oscillator potential $V = \widehat{x}^2$ defined on

$$\mathcal{D}(\widehat{x}^2) = \Big\{\phi \in L^2(\mathbb{E}) : x^2\phi \in L^2(\mathbb{E})\Big\}. \tag{2.46}$$

Note that $\mathcal{D}(\widehat{x}^2)$ is smaller than $\mathcal{D}(\widehat{x})$.

92   CHAPTER 2. OPERATORS AND THEIR DIRECT INTEGRALS

2. The Dirac delta function $\delta(x)$ does not define a multiplication operator in $L^2(I\!\!E)$ since $\delta(x)\phi(x)$ is clearly not square-integrable, and hence lies outside $L^2(I\!\!E)$.

Differential operators are also examples of unbounded operators. The momentum operator given by the differential expression

$$\widehat{p} = -i\hbar \frac{d}{dx} \qquad (2.47)$$

acting on the set of once differentiable functions on $I\!\!E$ is a well-known unbounded operator in $L^2(I\!\!E)$.[12]

C6   Let $\widehat{A}$ be a densely defined operator in $\mathcal{H}$. Then the following results can be established:[13]

1. The adjoint $\widehat{A}^\dagger$ exists and is closed, i.e., $\widehat{A}^\dagger = \overline{A^\dagger}$, where $\overline{A^\dagger}$ denotes the closure of $\widehat{A}^\dagger$. Note that $\widehat{A}^\dagger$ is generally not an extension of $\widehat{A}$ so that not every densely defined operator is closable.

2. $\widehat{A}$ is closable if and only if $\widehat{A}^\dagger$ is also densely defined.

3. If $\widehat{A}$ is closable, then its closure $\bar{A}$ is equal to the adjoint of its adjoint, i.e., we have
$$\bar{A} = \widehat{A}^{\dagger\dagger}. \qquad (2.48)$$

4. If $\widehat{A}$ is closed already then
$$\widehat{A} = \bar{A} = \widehat{A}^{\dagger\dagger}. \qquad (2.49)$$

It follows that if $\widehat{A}$ is closable then
$$\widehat{A}^\dagger = (\bar{A})^\dagger = \widehat{A}^{\dagger\dagger\dagger}. \qquad (2.50)$$

2.2.2   Convergence of a family of bounded operators

We often encounter convergence problems of sequences or of one-parameter families of operators. This is not a trivial matter. There are many different types of convergence. Let $\widehat{A}(\tau)$, $\tau \in \Lambda \subset I\!\!R$ be a one-parameter family of bounded operators. Three different types of convergence or limits are commonly used.[14]

---
[12] For a precise definition of the momentum operator see §2.5.1E(1) E3.
[13] Reed and Simon Vol. 1 (1972) p. 253.
[14] Prugovečki (1981) §5.3 p. 229.

2.2. OPERATORS: BASIC DEFINITIONS

Definition 2.2.2(1)   Weak, strong and uniform convergence

- An operator $\widehat{A}$ is the *weak limit* of $\widehat{A}(\tau)$ as $\tau \to \tau_0$, to be denoted by

$$\widehat{A} = \text{w-}\lim_{\tau \to \tau_0} \widehat{A}(\tau), \qquad (2.51)$$

if for all $\phi, \psi \in \mathcal{H}$ we have

$$\langle \psi \mid \widehat{A}\phi \rangle = \lim_{\tau \to \tau_0} \langle \psi \mid \widehat{A}(\tau)\phi \rangle. \qquad (2.52)$$

The family $\widehat{A}(\tau)$ is said to converge weakly to $\widehat{A}$ at τ_0.

- An operator $\widehat{A}$ is the *strong limit* of $\widehat{A}(\tau)$ as $\tau \to \tau_0$, to be denoted by

$$\widehat{A} = \text{s-}\lim_{\tau \to \tau_0} \widehat{A}(\tau), \qquad (2.53)$$

if for all $\phi \in \mathcal{H}$ we have

$$\lim_{\tau \to \tau_0} \left\| \left(\widehat{A} - \widehat{A}(\tau) \right) \phi \right\| = 0. \qquad (2.54)$$

The family $\widehat{A}(\tau)$ is said to converge strongly to $\widehat{A}$ at τ_0.

- An operator $\widehat{A}$ is the *uniform limit* of $\widehat{A}(\tau)$ as $\tau \to \tau_0$, to be denoted by

$$\widehat{A} = \text{u-}\lim_{\tau \to \tau_0} \widehat{A}(\tau), \qquad (2.55)$$

if

$$\lim_{\tau \to \tau_0} \left\| \left(\widehat{A} - \widehat{A}(\tau) \right) \right\| = 0. \qquad (2.56)$$

The family $\widehat{A}(\tau)$ is said to converge uniformly to $\widehat{A}$ at τ_0. A uniform convergence is also known as a *norm convergence*.

Comments 2.2.2(1)   Weak, strong and uniform convergence

C1   The limit operator $\widehat{A}$ is unique and bounded. If $\widehat{A}(\tau)$ converges uniformly to $\widehat{A}$, it would also converge strongly to $\widehat{A}$, and if $\widehat{A}(\tau)$ converges strongly to $\widehat{A}$, it would also converge weakly to $\widehat{A}$. The converse of these is not true.[15]

C2   What has been said about convergences of a family of operators applies equally well to sequences of bounded operators. Unless stated otherwise from

---
[15] For an example, see Reed and Simon Vol. 1 (1972) p. 184.

94   CHAPTER 2. OPERATORS AND THEIR DIRECT INTEGRALS

now on, convergence of a family or of a sequence of operators is meant in the strong sense.

C3   The conditions for convergence of operators is similar to that for the convergence of a sequence of vectors, i.e., we have the Cauchy convergence criterion. A sequence of bounded operators $\{\widehat{A}_n,\ n=1,2,\ldots\}$ is called

1. a *strong Cauchy sequence* if for every $\varepsilon > 0$ and every ϕ in the Hilbert space there is an integer n_0 such that[16]

$$\left\|\left(\widehat{A}_n - \widehat{A}_m\right)\phi\right\| < \varepsilon \quad \forall n, m \geq n_0, \tag{2.57}$$

and

2. a *uniform Cauchy sequence* if for every $\varepsilon > 0$ there is an integer n_0 such that[17]

$$\left\|\widehat{A}_n - \widehat{A}_m\right\| < \varepsilon \quad \forall n, m \geq n_0. \tag{2.58}$$

A strong Cauchy sequence can be shown to converge strongly to a bounded operator. The same applies to a uniform Cauchy sequence.

C4   The sequence of adjoint operators $\widehat{A}_n^\dagger$ may not converge strongly even if the sequence of operators $\widehat{A}_n$ converges strongly.[18]

C5   The situation becomes more complicated when dealing with sequences of unbounded operators because they are only densely defined.[19]

2.2.3   Tensor products of Hilbert spaces and operators

Given two Hilbert spaces $\mathcal{H}_1$ and $\mathcal{H}_2$ we can construct further Hilbert spaces. There are two physically important constructions, one leading to a tensor product space and the other giving rise to a direct sum space. Here we shall examine the tensor product space construction, leaving the discussion on direct sum spaces till the end of this chapter. We shall introduce the notion of tensor product in an intuitive manner.[20]

---

[16]Weidmann (1980) §4.5, p. 75.
[17]Smirnov Vol. 5 (1964) §104, p. 211, Prugovečki (1981) p. 230, p. 240.
[18]Weidmann (1980) p. 70.
[19]Reed and Simon Vol. 1 (1972) p. 283.
[20]Sewell (1986) pp. 41-42, Isham (1995) §8.4.1. For a more abstract definition see Reed and Simon Vol. 1 (1972) pp. 49-54, Blank, Exner and Havlíček (1994) §2.4.

2.2. OPERATORS: BASIC DEFINITIONS

Definition 2.2.3(1)   **Tensor products of Hilbert spaces**

- Let $\mathcal{H}^{(1)}$ and $\mathcal{H}^{(2)}$ be two Hilbert spaces. A Hilbert space is called the *tensor product* of $\mathcal{H}^{(1)}$ and $\mathcal{H}^{(2)}$, to be denoted by $\mathcal{H}^{\otimes}$ with its relation to $\mathcal{H}^{(1)}$ and $\mathcal{H}^{(2)}$ denoted by

$$\mathcal{H}^{\otimes} = \mathcal{H}^{(1)} \otimes \mathcal{H}^{(2)}, \tag{2.59}$$

if:

1. For each pair of vectors $\phi^{(1)} \in \mathcal{H}^{(1)}$ and $\phi^{(2)} \in \mathcal{H}^{(2)}$ there corresponds a unique vector in $\mathcal{H}^{\otimes}$, denoted by $\phi^{(1)} \otimes \phi^{(2)}$, such that

$$\langle \phi^{(1)} \otimes \phi^{(2)} \mid \phi^{(1)} \otimes \phi^{(2)} \rangle^{\otimes} = \langle \phi^{(1)} \mid \phi^{(1)} \rangle^{(1)} \langle \phi^{(2)} \mid \phi^{(2)} \rangle^{(2)}, \tag{2.60}$$

    where the superscripts denote scalar products in the Hilbert spaces $\mathcal{H}^{\otimes}$, $\mathcal{H}^{(1)}$ and $\mathcal{H}^{(2)}$ respectively.

2. Linear combinations of vectors of the form $\phi^{(1)} \otimes \phi^{(2)}$ belong to $\mathcal{H}^{\otimes}$ and the closure of the set of all such linear combinations can be identified with $\mathcal{H}^{\otimes}$.

The notation $\phi^{(1)} \otimes \phi^{(2)}$ is a symbolic one; the concept of a product lies in the scalar product operation which involves a multiplication of $\langle \phi^{(1)} \mid \phi^{(1)} \rangle^{(1)}$ and $\langle \phi^{(2)} \mid \phi^{(2)} \rangle^{(2)}$ in Eq. (2.60). Let $\{\phi_j^{(1)}\}$ be an orthonormal basis in $\mathcal{H}^{(1)}$ and $\{\phi_k^{(2)}\}$ be an orthonormal basis in $\mathcal{H}^{(2)}$. We have

$$\langle \phi_j^{(1)} \otimes \phi_k^{(2)} \mid \phi_m^{(1)} \otimes \phi_n^{(2)} \rangle^{\otimes} = \delta_{jm} \delta_{kn}. \tag{2.61}$$

A general element Φ in $\mathcal{H}^{\otimes}$ can be written in the form

$$\Phi = \sum_{j,k} c_{jk} \phi_j^{(1)} \otimes \phi_k^{(2)}, \qquad c_{jk} = \langle \phi_j^{(1)} \otimes \phi_k^{(2)} \mid \Phi \rangle^{\otimes}. \tag{2.62}$$

This shows the linear nature of tensor product operation as well as the fact that the set $\{\phi_j^{(1)} \otimes \phi_k^{(2)}\}$ forms an orthonormal basis in $\mathcal{H}^{\otimes}$.

Definition 2.2.3(2)   **Tensor products of subsets of Hilbert spaces**

- Let $\mathcal{L}^{(1)}$ be a subset of a Hilbert space $\mathcal{H}^{(1)}$, and $\mathcal{L}^{(2)}$ be a subset of another Hilbert space $\mathcal{H}^{(2)}$. Then the tensor product of $\mathcal{L}^{(1)}$ and $\mathcal{L}^{(2)}$, denoted by $\mathcal{L}^{(1)} \hat{\otimes} \mathcal{L}^{(2)}$, is defined as the subset of $\mathcal{H}^{(1)} \otimes \mathcal{H}^{(2)}$ composed of all finite linear combinations of vectors of the form $\phi^{(1)} \otimes \phi^{(2)}$, where $\phi^{(1)} \in \mathcal{L}^{(1)}$ and $\phi^{(2)} \in \mathcal{L}^{(2)}$.

If $\mathcal{L}^{(1)}$ is dense in Hilbert space $\mathcal{H}^{(1)}$ and $\mathcal{L}^{(2)}$ dense in Hilbert space $\mathcal{H}^{(2)}$ then $\mathcal{L}^{(1)} \hat{\otimes} \mathcal{L}^{(2)}$ is dense in $\mathcal{H}^{(1)} \otimes \mathcal{H}^{(2)}$. Note that generally $\mathcal{L}^{(1)} \hat{\otimes} \mathcal{L}^{(2)}$ is not closed, and is hence not a Hilbert space.

Definition 2.2.3(3) Tensor products of operators[21]

- Let $\widehat{A}^{(1)}$ be an operator in $\mathcal{H}^{(1)}$ with domain $\mathcal{D}(\widehat{A}^{(1)})$ and $\widehat{A}^{(2)}$ in $\mathcal{H}^{(2)}$ with domain $\mathcal{D}(\widehat{A}^{(2)})$, and let $\mathcal{D}(\widehat{A}^{(1)}) \hat{\otimes} \mathcal{D}(\widehat{A}^{(2)})$ denote the set of finite linear combinations Φ of vectors of the form

$$\Phi = \sum_{j,k} c_{jk}\, \phi_j^{(1)} \otimes \phi_k^{(2)}, \quad \phi_j^{(1)} \in \mathcal{D}(\widehat{A}^{(1)}),\quad \phi_k^{(2)} \in \mathcal{D}(\widehat{A}^{(2)}). \qquad (2.63)$$

Define an operator denoted by $\widehat{A}^{(1)} \hat{\otimes} \widehat{A}^{(2)}$ on $\mathcal{D}(\widehat{A}^{(1)}) \hat{\otimes} \mathcal{D}(\widehat{A}^{(2)})$ by

$$\left(\widehat{A}^{(1)} \hat{\otimes} \widehat{A}^{(2)}\right)\Phi = \sum_{j,k} c_{jk} \left(\widehat{A}^{(1)} \phi_j^{(1)}\right) \otimes \left(\widehat{A}^{(2)} \phi_k^{(2)}\right). \qquad (2.64)$$

- The *tensor product* of $\widehat{A}^{(1)}$ and $\widehat{A}^{(2)}$ is defined to be the closure of $\widehat{A}^{(1)} \hat{\otimes} \widehat{A}^{(2)}$ in $\mathcal{H}^{\otimes}$, i.e., $\overline{\widehat{A}^{(1)} \hat{\otimes} \widehat{A}^{(2)}}$. For notational brevity we shall denote the tensor product simply by $\widehat{A}^{(1)} \otimes \widehat{A}^{(2)}$.

The simplest operation of $\widehat{A}^{(1)} \otimes \widehat{A}^{(2)}$ is on a single product vector $\phi^{(1)} \otimes \phi^{(2)}$. We have

$$\left(\widehat{A}^{(1)} \otimes \widehat{A}^{(2)}\right)\left(\phi^{(1)} \otimes \phi^{(2)}\right) = \left(\widehat{A}^{(1)} \phi^{(1)}\right) \otimes \left(\widehat{A}^{(2)} \phi^{(2)}\right). \qquad (2.65)$$

We often encounter tensor product operators of the forms

$$\widehat{A}^{(1)} \otimes \widehat{I}^{(2)}, \quad \widehat{I}^{(1)} \otimes \widehat{A}^{(2)} \qquad (2.66)$$

and

$$\widehat{A}^{(1)} \otimes \widehat{I}^{(2)} + \widehat{I}^{(1)} \otimes \widehat{A}^{(2)}, \qquad (2.67)$$

where $\widehat{I}^{(1)}$ and $\widehat{I}^{(2)}$ are respectively the identity operators on $\mathcal{H}^{(1)}$ and $\mathcal{H}^{(2)}$. Tensor product operators may preserve some of the properties of their

---

[21] Reed and Simon Vol. 1 (1972) pp. 299-302, Amrein, Jauch and Sinha (1977) p. 85, Weidmann (1980) pp. 262-268, and Blank, Exner and Havlíček (1994) §2.4. Care has to be taken as different authors employ the notation $\widehat{A}^{(1)} \otimes \widehat{A}^{(2)}$ to mean different things, e.g., the notation $\widehat{A}^{(1)} \otimes \widehat{A}^{(2)}$ in Weidmann (1980) corresponds to $\widehat{A}^{(1)} \hat{\otimes} \widehat{A}^{(2)}$ in Definition 2.2.3(3) here. Reed and Simon Vol. 1 (1972) simply use the notation to mean both. We follow the usage of Amrein, Jauch and Sinha (1977).

2.3. TYPES OF OPERATORS AND THEIR REDUCTIONS

constituent operators, e.g., $\widehat{A}^{(1)} \otimes \widehat{A}^{(2)}$ is bounded if and only if $\widehat{A}^{(1)}$ and $\widehat{A}^{(2)}$ are both bounded. More examples will be presented at the end of next section in §2.3C(1) C12 and later in §2.8.1E(1) E3.

We can define the tensor products of a finite number of Hilbert spaces and operators in a similar manner.

2.3 Types of Operators and their Reductions

Operators in a Hilbert space $\mathcal{H}$ defined on a dense domain $\mathcal{D}(\widehat{A})$ can be classified into various types, as done in Definition 2.3(1).

Definition 2.3(1) **Important types of operators**

An operator $\widehat{A}$ in a Hilbert space $\mathcal{H}$ defined on a dense domain $\mathcal{D}(\widehat{A})$ is said to be

- *invertible* if $\widehat{A}\phi = 0 \Rightarrow \phi = 0$.

- *symmetric* if $\widehat{A} \subset \widehat{A}^\dagger$.

- *maximal symmetric* if it is symmetric and it possesses no further symmetric extensions.[22]

- *selfadjoint* if $\widehat{A} = \widehat{A}^\dagger$.

- *essentially selfadjoint* if it is symmetric and it has a unique selfadjoint extension.[23]

- *positive* if $\langle \phi \mid \widehat{A}\phi \rangle$ is real and $\langle \phi \mid \widehat{A}\phi \rangle \geq 0 \quad \forall \phi \in \mathcal{D}(\widehat{A})$.

- a *projection operator*, or a *projector* for short, if it is bounded, selfadjoint and idempotent, i.e., $\widehat{A}^2 = \widehat{A}$.

- An operator $\widehat{U}$ in a Hilbert space $\mathcal{H}$ is said to be a *unitary* if

    1. both its domain and its range coincide with the entire space $\mathcal{H}$, i.e., $\mathcal{D}(\widehat{U}) = \mathcal{R}(\widehat{U}) = \mathcal{H}$, and

    2. the norm of a vector is preserved under the action of $\widehat{U}$, i.e.,

$$\langle \widehat{U}\phi \mid \widehat{U}\phi \rangle = \langle \phi \mid \phi \rangle \quad \forall \phi \in \mathcal{H}. \qquad (2.68)$$

---

[22] An extension $\widehat{A}_{ex}$ of $\widehat{A}$ is a symmetric extension of $\widehat{A}$ if $\widehat{A}_{ex}$ is symmetric. $\widehat{A}$ is maximal symmetric if it possesses no symmetric extension apart from itself. See §2.3C(1) C4.

[23] An extension $\widehat{A}_{ex}$ of $\widehat{A}$ is a selfadjoint extension if $\widehat{A}_{ex}$ is selfadjoint. See §2.3C(1) C4.

Comments 2.3(1) Properties and examples

C1 An invertible operator maps vectors in its domain $\mathcal{D}(\widehat{A})$ to its range $\mathcal{R}(\widehat{A})$ in a one-to-one manner, i.e.,

$$\phi_1 \neq \phi_2 \Rightarrow \widehat{A}\phi_1 \neq \widehat{A}\phi_2. \tag{2.69}$$

This enables us to define an *inverse operator* $\widehat{A}^{-1}$ with domain $\mathcal{D}(\widehat{A}^{-1}) = \mathcal{R}(\widehat{A})$ and range $\mathcal{R}(\widehat{A}^{-1}) = \mathcal{D}(\widehat{A})$ by

$$\widehat{A}^{-1}\psi = \phi \quad \forall \psi \in \mathcal{R}(\widehat{A}), \quad \text{where } \psi = \widehat{A}\phi. \tag{2.70}$$

An invertible operator therefore admits an inverse.[24] Since the range $\mathcal{R}(\widehat{A})$ of an operator is not necessarily dense, the inverse $\widehat{A}^{-1}$ may not be densely defined.

C2 A symmetric operator $\widehat{A}$ is one whose domain $\mathcal{D}(\widehat{A})$ is less than or at most equal to that of its adjoint $\widehat{A}^\dagger$, i.e.,

$$\mathcal{D}(\widehat{A}) \subset \mathcal{D}(\widehat{A}^\dagger), \tag{2.71}$$

and on $\mathcal{D}(\widehat{A})$ the operator is equal to its adjoint, i.e.

$$\widehat{A}\phi = \widehat{A}^\dagger \phi, \quad \forall \phi \in \mathcal{D}(\widehat{A}). \tag{2.72}$$

In contrast, a selfadjoint operator is a symmetric operator whose domain equals that of its adjoint. In other words, if $\widehat{A}$ is symmetric but not selfadjoint, then there are vectors ψ lying in $\mathcal{D}(\widehat{A}^\dagger)$, but not in $\mathcal{D}(\widehat{A})$, such that

$$\langle \psi \mid \widehat{A}\phi \rangle = \langle \widehat{A}^\dagger \psi \mid \phi \rangle \quad \forall \phi \in \mathcal{D}(\widehat{A}). \tag{2.73}$$

A symmetric operator $\widehat{A}$ possesses the following properties:

1. From Eq. (2.36) we get

   $$\langle \psi \mid \widehat{A}\phi \rangle = \langle \widehat{A}\psi \mid \phi \rangle \quad \forall \psi, \phi \in \mathcal{D}(\widehat{A}). \tag{2.74}$$

   The converse is also true, i.e., an operator $\widehat{A}$ defined on a dense domain $\mathcal{D}(\widehat{A})$ is symmetric if the above equality is satisfied.

2. A symmetric operator possesses real expectation values, i.e.,

   $$\langle \phi \mid \widehat{A}\phi \rangle \in \mathbb{R} \quad \forall \phi \in \mathcal{D}(\widehat{A}). \tag{2.75}$$

---

[24]Some authors employ a different terminology, e.g., Roman Vol. 2 (1975). We follow the terminology of Kato (1966) here.

2.3. TYPES OF OPERATORS AND THEIR REDUCTIONS

This can also serve to define symmetric operators, i.e., an operator $\widehat{A}$ defined on a dense domain $\mathcal{D}(\widehat{A})$ is symmetric if the scalar product $\langle \phi \mid \widehat{A}\phi \rangle$ is real for every $\phi \in \mathcal{D}(\widehat{A})$.[25] This leads immediately to the results stated in the item below.

3. A symmetric operator defined on the entire Hilbert space is bounded and selfadjoint.[26]

4. Positive operators are symmetric, and bounded positive operators are selfadjoint.

5. Recall that a number a, possibly complex, is called an *eigenvalue* of an operator $\widehat{A}$ if a vector φ in the Hilbert space exists such that[27]

$$\widehat{A}\varphi = a\varphi, \quad \varphi \neq 0. \tag{2.76}$$

The vector φ is called an *eigenvector* of $\widehat{A}$ corresponding to, or belonging to, or associated with, the eigenvalue a. Eigenvectors corresponding to different eigenvalues of an operator are linearly independent. Generally there may be a number n of linearly independent eigenvectors belonging to the same eigenvalue; the eigenvalue is then said to be *degenerate* and the number n is called the *degeneracy* of the eigenvalue. The subspace spanned by the set of all linearly independent eigenvectors corresponding to an eigenvalue is called the *eigensubspace* of the eigenvalue.

If $\widehat{A}$ is a symmetric operator then

 (a) its eigenvalues are real, and

 (b) its eigenvectors belonging to different eigenvalues are orthogonal.

C3   Selfadjoint and essential selfadjoint operators are symmetric, but the converse is not necessarily true.

C4   An extension is called

- a *symmetric extension* if the extension operator is symmetric,

- a *selfadjoint extension* if the extension operator is selfadjoint.

---

[25] See Weidmann (1980) p. 72.
[26] Prugovečki (1981) Theorem 2.10 on p. 195.
[27] By definition, φ must be a member of the Hilbert space. Example §2.5.1E(1) E6 shows that in some specific cases it is useful to relax this condition. See also §2.8.3C(1) C4 for the notions of generalized and approximate eigenvalues and eigenvectors.

100 CHAPTER 2. OPERATORS AND THEIR DIRECT INTEGRALS

As shown in C6 below, a selfadjoint operator is also maximal symmetric. However, the converse is not true. To highlight this a maximal symmetric operator is called

- strictly maximal symmetric

if it is not selfadjoint.

C5 Generally a symmetric operator may have many symmetric or selfadjoint extensions. To constrast with essential selfadjointness we shall call a symmetric operator

- essentially strictly maximal symmetric

if it possesses a unique strictly maximal symmetric extension. In a later section we shall investigate the conditions for essential selfadjointness and essentially maximal symmetry.

C6 The adjoint operation satisfies the following relations:

1. Order relations:
$$\widehat{A} \supset \widehat{B} \Rightarrow \widehat{A}^\dagger \subset \widehat{B}^\dagger. \tag{2.77}$$

It follows that if $\widehat{A}$ is a symmetric then
$$\widehat{A} \subset \widehat{A}^{\dagger\dagger} \subset \widehat{A}^\dagger. \tag{2.78}$$

While the adjoint $\widehat{A}^\dagger$ of a symmetric operator $\widehat{A}$ is an extension of $\widehat{A}$, the adjoint is generally not a symmetric extension of $\widehat{A}$ since $\widehat{A}^\dagger$ is generally not a symmetric operator.

If $\widehat{A}_{ex}$ is a symmetric extension of $\widehat{A}$, i.e.,
$$\widehat{A} \subset \widehat{A}_{ex} \quad \text{and} \quad \widehat{A}_{ex} \subset \widehat{A}^\dagger_{ex}, \tag{2.79}$$

then we have
$$\widehat{A}^\dagger \supset \widehat{A}^\dagger_{ex} \quad \text{and} \quad \widehat{A} \subset \widehat{A}_{ex} \subset \widehat{A}^\dagger. \tag{2.80}$$

If $\widehat{A} = \widehat{A}^\dagger$ then $\widehat{A} = \widehat{A}_{ex}$, i.e., $\widehat{A}$ is maximal symmetric with no further symmetric extension. Consequently:

  (a) A selfadjoint operator is also maximal symmetric since it has no symmetric extensions.

  (b) Any selfadjoint extension $\widehat{A}_{ex}$ of a symmetric operator $\widehat{A}$ is related to the adjoint $\widehat{A}^\dagger$ by
$$\widehat{A} \subset \widehat{A}_{ex} \subset \widehat{A}^\dagger. \tag{2.81}$$

2.3. TYPES OF OPERATORS AND THEIR REDUCTIONS

2. Sum and product relations:

    (a) Generally for unbounded operators we have
    $$(\widehat{A}+\widehat{B})^\dagger \supset \widehat{A}^\dagger + \widehat{B}^\dagger, \quad (\widehat{A}\widehat{B})^\dagger \supset \widehat{B}^\dagger\widehat{A}^\dagger. \tag{2.82}$$

    We have assumed here that $\widehat{A}+\widehat{B}$ and $\widehat{A}\widehat{B}$ are densely defined so that their adjoints exist. It follows that

    > *the sum of two selfadjoint operators is symmetric but not necessarily selfadjoint, and the same is true for the following symmetrized product of two selfadjoint operators*

    $$\frac{1}{2}\left(\widehat{A}\widehat{B}+\widehat{A}\widehat{B}\right). \tag{2.83}$$

    These results will be seen to pose a severe problem for quantization.

    (b) If $\widehat{B}$ is a bounded operator, then[28]
    $$(\widehat{A}+\widehat{B})^\dagger = \widehat{A}^\dagger + \widehat{B}^\dagger, \quad (\widehat{B}\widehat{A})^\dagger = \widehat{A}^\dagger\widehat{B}^\dagger. \tag{2.84}$$

C7 The following statements on closedness and closure of operators are true:

1. Selfadjoint operators are closed and symmetric and essentially selfadjoint operators are closable.

2. If a symmetric operator $\widehat{A}$ is closed then §2.2.1C(1) C5 tells us that $\widehat{A}=\widehat{A}^{\dagger\dagger}$.

3. The closure of a symmetric operator is symmetric on account of Eqs. (2.77) and (2.78).

4. Every maximal symmetric operator is closed. For if it were not closed, it would have a symmetric extension, i.e., its closure.

5. If $\widehat{A}$ is closed, then $\widehat{A}^\dagger\widehat{A}$ is selfadjoint and positive.[29] It follows that:

    (a) $\widehat{A}\widehat{A}^\dagger$ is also selfadjoint and positive, since $\widehat{A}^\dagger$ is closed and that $\widehat{A}$ is closed implies $\widehat{A}^{\dagger\dagger} = \widehat{A}$ from §2.2.1C(1) C5.

    (b) The square $\widehat{A}^2$ of a selfadjoint operator $\widehat{A}$ is selfadjoint.

---
[28] See Weidmann (1980) p. 73.
[29] Akhiezer and Glazman Vol. 1 (1961) Theorem 2 on pp. 97-98. For explicit examples see Reed and Simon Vol. 2 (1975) pp. 180-181 and also later in §2.10.3 and §7.7.2.

C8 A symmetric operator $\widehat{A}$ is essentially selfadjoint if and only if its closure $\widehat{A}^{\dagger\dagger}$ is selfadjoint, i.e., $\widehat{A}^{\dagger\dagger\dagger} = \widehat{A}^{\dagger\dagger}$. Given that $\widehat{A}$ has a unique selfadjoint extension we conclude that its closure $\widehat{A}^{\dagger\dagger}$ is its unique selfadjoint extension. Since[30] $\widehat{A}^{\dagger\dagger\dagger} = \widehat{A}^{\dagger}$ we can conclude that the unique selfadjoint extension is equal to the adjoint $\widehat{A}^{\dagger}$. Essentially selfadjoint operators play a crucial role in quantization since the domain of an essentially selfadjoint operator is generally much easier to describe than the domain of its selfadjoint extension.

C9 Bounded selfadjoint operators:

1. Two bounded selfadjoint operators $\widehat{A}_1$ and $\widehat{A}_2$ are said to *commute* if their *commutator*

$$[\widehat{A}_1, \widehat{A}_2] = \widehat{A}_1\widehat{A}_2 - \widehat{A}_2\widehat{A}_1 \qquad (2.85)$$

vanishes. The situation becomes complicated if the two operators are unbounded since the commutator may not be well defined,[31] e.g., the product $\widehat{A}_1\widehat{A}_2$ may be defined only on the zero vector.[32] We shall return to this important point later in §2.8.1C(1) C7. A purely formal treatment of such a commutator can lead to paradoxes in physics. A well-known example involves attempts to write down a purely formal uncertainty relation between an angle variable and its conjugate momentum.[33]

2. Any bounded operator $\widehat{B}$ can be decomposed uniquely into a real part $\widehat{B}_{re}$ and an imaginary part $\widehat{B}_{im}$, where $\widehat{B}_{re}$ and $\widehat{B}_{im}$ are bounded selfadjoint operators, in the same way a complex number is so decomposable, i.e., we have[34]

$$\widehat{B} = \widehat{B}_{re} + i\,\widehat{B}_{im}. \qquad (2.86)$$

We can check that

$$\widehat{B}_{re} = \frac{1}{2}\left(\widehat{B} + \widehat{B}^{\dagger}\right), \quad \widehat{B}_{im} = \frac{1}{2i}\left(\widehat{B} - \widehat{B}^{\dagger}\right). \qquad (2.87)$$

C10 As pointed out earlier positive operators are also symmetric. If a positive operator is defined on the entire Hilbert space it becomes bounded and selfadjoint. Bounded positive operators play a fundamental role in a generalization of orthodox quantum mechanics. Some authors simply define

---

[30]Reed and Simon Vol. 1 (1972) p. 253, Prugovečki (1981) p. 374. $\widehat{A}^{\dagger\dagger\dagger} = \widehat{A}^{\dagger\dagger}$ together with $\widehat{A}^{\dagger\dagger\dagger} = \widehat{A}^{\dagger}$ means $\widehat{A}^{\dagger\dagger} = \widehat{A}^{\dagger}$, i.e., $\widehat{A}^{\dagger}$ is selfadjoint.
[31]Reed and Simon Vol. 1 (1972) p. 271.
[32]This is because of Eqs. (2.29) and (2.30). See also Weidmann (1980) p. 50.
[33]Fano (1971) pp. 407-408.
[34]Roman Vol. 2 (1975), Theorem 12.3b(5) on p. 537.

2.3. TYPES OF OPERATORS AND THEIR REDUCTIONS

positive operators on the entire Hilbert space so that they are automatically bounded and selfadjoint.[35] Of particular importance is the class of bounded positive operators whose norms are equal to or less than 1. This would include all the projectors which are also positive operators with norm equal 1.

C11 Projectors are a very important class of operators in quantum mechanics. They often appear to be rather simple. Take the Hilbert space $L^2(I\!\!E)$ as an example. Acting as a multiplication operator the characteristic function χ_Λ is obviously bounded, selfadjoint and idempotent. It follows that χ_Λ is a projector. Let $\phi \in L^2(I\!\!E)$, then $\phi_\Lambda = \chi_\Lambda \phi$ is a function in $L^2(I\!\!E)$ vanishing outside the interval Λ. Clearly the range of this projector forms a subspace $\mathcal{L}(\Lambda)$ consisting of functions on $I\!\!E$ vanishing outside the interval Λ. In other words χ_Λ projects every element of $L^2(I\!\!E)$ onto the subspace $\mathcal{L}(\Lambda)$. Note that χ_Λ has no effect on elements in the subspace, i.e., $\chi_\Lambda \phi_\Lambda = \phi_\Lambda$ for all $\phi_\Lambda \in \mathcal{L}(\Lambda)$. The characteristic function can be extended to $I\!\!E^n$ in a straightforward manner resulting in a corresponding projector on the Hilbert space $L^2(I\!\!E^n)$.

It turns out that what has just been said is generally true for all projectors. In other words every projector $\widehat{P}$ in a Hilbert space $\mathcal{H}$ is uniquely associated with a subspace $\mathcal{L}$ of $\mathcal{H}$ such that $\widehat{P}$ projects every element of $\mathcal{H}$ onto the subspace $\mathcal{L}$. In fact this property is often used to define projectors. We can illustrate this with a simple example. Given a one-dimensional subspace $\mathcal{L}_\varphi$ spanned by a unit vector φ we can define an operator $\widehat{P}_\varphi$ on $\mathcal{H}$ by

$$\widehat{P}_\varphi \phi = \langle \varphi \mid \phi \rangle \varphi \quad \forall \phi \in \mathcal{H}. \tag{2.88}$$

It is common practice in the physics literature to express $\widehat{P}_\varphi$ in *Dirac notation* as

$$\widehat{P}_\varphi = |\varphi\rangle\langle\varphi| \tag{2.89}$$

This operator clearly projects every vector $\phi \in \mathcal{H}$ onto $\mathcal{L}_\varphi$. It is easy to verify that $\widehat{P}_\varphi$ is bounded, selfadjoint and idempotent. Hence it is a projector. We shall call $\widehat{P}_\varphi$ the *projector generated by the vector* φ. The following definitions and properties on projectors are of particular interest:

1. The *dimension of a projector* is defined to be the dimension of the subspace onto which it projects, e.g., $\widehat{P}_\varphi$ is a *one-dimensional projector* and all one-dimensional projectors are of this form.

2. Being idempotent a projector $\widehat{P}$ possesses only two eigenvalues, i.e., 0 and 1.[36] Projectors may be regarded as a simple kind of selfadjoint operators, having only two eigenvalues. The expectation value of a projector

---
[35] Reed and Simon Vol. 1 (1972) p. 195.
[36] With the exception of the identity and the zero operators which are trivial examples of projectors.

with respect to any given unit vector ϕ in the Hilbert space lies in the range $[0, 1]$, i.e.,

$$0 \leq \langle \phi \mid \widehat{P}\phi \rangle \leq 1 \quad \forall \phi \in \mathcal{H}, \ \|\phi\| = 1. \tag{2.90}$$

This is a crucial property which enables us to relate projectors to probability in quantum mechanics. It is obvious that all projectors have a unit norm, i.e.,

$$\|\widehat{P}\| = 1 \quad \text{with} \quad \|\widehat{P}\phi\|^2 = \langle \phi \mid \widehat{P}\phi \rangle \leq 1. \tag{2.91}$$

3. Two projectors are said to be *orthogonal* if the subspaces onto which they project are orthogonal. One can show that two projectors $\widehat{P}_1$ and $\widehat{P}_2$ are orthogonal if and only if their products vanish, i.e., $\widehat{P}_1 \widehat{P}_2 = \widehat{P}_2 \widehat{P}_1 = 0$.

4. The sum of two orthogonal projectors is again a projector. The product of two projectors is a projector if and only if they commute.

5. Let $\mathcal{L}$ be a subspace and $\mathcal{L}^\perp$ be its orthogonal complement. Then the projectors $\widehat{P}_\mathcal{L}$ and $\widehat{P}_{\mathcal{L}^\perp}$ onto $\mathcal{L}$ and $\mathcal{L}^\perp$ respectively are orthogonal to each other. Moreover we have

$$\widehat{P}_\mathcal{L} + \widehat{P}_{\mathcal{L}^\perp} = \widehat{\mathbb{I}} \tag{2.92}$$

where $\widehat{\mathbb{I}}$ is the identity operator on $\mathcal{H}$. We also have

$$\widehat{P}_{\mathcal{L}^\perp} = \widehat{\mathbb{I}} - \widehat{P}_\mathcal{L}. \tag{2.93}$$

6. A *partial order relation*[37] can be introduced in the set of projectors. Let $\widehat{P}_1$ and $\widehat{P}_2$ be two projectors. Let $\mathcal{L}_1$ be the subspace onto which $\widehat{P}_1$ projects and $\mathcal{L}_2$ be the subspace onto which $\widehat{P}_2$ projects. Then $\widehat{P}_1$ is said to imply projector $\widehat{P}_2$, with the notation

$$\widehat{P}_1 \leq \widehat{P}_2, \tag{2.94}$$

if $\mathcal{L}_1$ is a subset of $\mathcal{L}_2$, i.e.,

$$\widehat{P}_1 \leq \widehat{P}_2 \quad \text{if} \quad \mathcal{L}_1 \subset \mathcal{L}_2. \tag{2.95}$$

We can also say that $\widehat{P}_1$ is smaller than $\widehat{P}_2$. With this relationship the set of projectors in a Hilbert space becomes a partially ordered set.[38]

---

[37] Lipschutz (1964).
[38] The partial order in the set of projectors is the basis for a logic formulation of quantum mechanics (Beltrametti and Cassinelli (1981) §10).

2.3. TYPES OF OPERATORS AND THEIR REDUCTIONS

The order is partial because there is no order relation between $\widehat{P}_1$ and $\widehat{P}_2$ if $\mathcal{L}_1$ and $\mathcal{L}_1$ are unrelated. The following property holds for related projectors:

$$\widehat{P}_1 \leq \widehat{P}_2 \quad \Leftrightarrow \quad \widehat{P}_1\widehat{P}_2 = \widehat{P}_2\widehat{P}_1 = \widehat{P}_1, \tag{2.96}$$
$$\Rightarrow \quad (\widehat{P}_2 - \widehat{P}_1)^2 = \widehat{P}_2 - \widehat{P}_1. \tag{2.97}$$

It follows that the difference $\widehat{P}_2 - \widehat{P}_1$ is again a projector.

7. Let $\{\varphi_j, j = 1, 2, \ldots\}$ be an orthonormal basis in a Hilbert space $\mathcal{H}$, and let $\left\{\widehat{P}_j = |\varphi_j\rangle\langle\varphi_j|\right\}$ be the corresponding set of projectors. One can verify that

$$\sum_{j=1}^{\infty} \widehat{P}_j = \sum_{j=1}^{\infty} |\varphi_j\rangle\langle\varphi_j| = \widehat{I}. \tag{2.98}$$

C12 The selfadjointness or otherwise of a tensor product operator relates closely to its constituent operators. In other words, if $\widehat{A}^{(1)}$ and $\widehat{A}^{(2)}$ are selfadjoint then generally we have:[39]

1. $\widehat{A}^{(1)} \hat{\otimes} \widehat{A}^{(2)}$ is essentially selfadjoint,

2. $\widehat{A}^{(1)} \otimes \widehat{A}^{(2)}$ is selfadjoint,

3. $\widehat{A}^{(1)} \otimes \widehat{I}^{(2)} + \widehat{I}^{(1)} \otimes \widehat{A}^{(2)}$ is essentially selfadjoint.

These statements are also true if $\widehat{A}^{(1)}$ and $\widehat{A}^{(2)}$ are only essentially selfadjoint.

C13 Unitary operators have many applications, as will be seen in §2.4.

Definition 2.3(2) **Invariant subspaces and reductions of operators**

- Let $\widehat{A}$ be an operator in a Hilbert space $\mathcal{H}$ with domain $\mathcal{D}(\widehat{A})$. A subspace $\mathcal{L}$ of $\mathcal{H}$ is called an *invariant subspace* of the operator $\widehat{A}$ if

$$\phi \in \mathcal{D}(\widehat{A}) \cap \mathcal{L} \Rightarrow \widehat{A}\phi \in \mathcal{L}. \tag{2.99}$$

- A subspace $\mathcal{L}$ is said to *reduce* the operator $\widehat{A}$ if

  1. $\mathcal{L}$ and its orthogonal complement $\mathcal{L}^\perp$ are invariant subspaces of the operator $\widehat{A}$, and

---
[39]Amrein, Jauch and Sinha (1977) p. 85, p. 98, p. 501, p. 599, Weidmann (1980) pp. 262-268, Reed and Simon Vol. 1 (1972) p. 299, p. 301, p. 316.

2. the projections of all $\phi \in \mathcal{D}(\widehat{A})$ onto $\mathcal{L}$ again lie in $\mathcal{D}(\widehat{A})$, i.e.,

$$\widehat{P}_{\mathcal{L}}\phi \in \mathcal{D}(\widehat{A}) \quad \forall \phi \in \mathcal{D}(\widehat{A}), \tag{2.100}$$

where $\widehat{P}_{\mathcal{L}}$ is the projector onto the subspace $\mathcal{L}$. We call $\mathcal{L}$ a *reducing subspace* of the operator $\widehat{A}$.

Comments 2.3(2)   Reduction of operators into parts

C1   If $\mathcal{L}$ is a reducing subspace of operator $\widehat{A}$, then so is its orthogonal complement $\mathcal{L}^{\perp}$. This is because $\mathcal{D}(\widehat{A})$ is a linear subset and hence

$$\phi, \widehat{P}_{\mathcal{L}}\phi \in \mathcal{D}(\widehat{A}) \quad \Rightarrow \quad \phi - \widehat{P}_{\mathcal{L}}\phi \in \mathcal{D}(\widehat{A}). \tag{2.101}$$

But

$$\phi - \widehat{P}_{\mathcal{L}}\phi = \left(\widehat{\mathbb{I}} - \widehat{P}_{\mathcal{L}}\right)\phi = \widehat{P}_{\mathcal{L}^{\perp}}\phi \in \mathcal{L}^{\perp}. \tag{2.102}$$

C2   If $\mathcal{L}$ reduces $\widehat{A}$ we can express $\widehat{A}$ as a sum of two operators. To see this let us recall that for every ϕ in $\mathcal{D}(\widehat{A})$ we have

$$\widehat{P}_{\mathcal{L}}\phi \in \mathcal{D}(\widehat{A}) \quad \text{and} \quad \widehat{P}_{\mathcal{L}^{\perp}}\phi \in \mathcal{D}(\widehat{A}). \tag{2.103}$$

So, it makes sense to define two operators, $\widehat{A}_{\mathcal{L}}$ and $\widehat{A}_{\mathcal{L}^{\perp}}$, on the domain $\mathcal{D}(\widehat{A})$ by[40]

$$\widehat{A}_{\mathcal{L}}\phi = \widehat{A}\widehat{P}_{\mathcal{L}}\phi, \quad \widehat{A}_{\mathcal{L}^{\perp}}\phi = \widehat{A}\widehat{P}_{\mathcal{L}^{\perp}}\phi \quad \forall \phi \in \mathcal{D}(\widehat{A}). \tag{2.104}$$

Then we have, for every ϕ on $\mathcal{D}(\widehat{A})$,

$$\left(\widehat{A}_{\mathcal{L}} + \widehat{A}_{\mathcal{L}^{\perp}}\right)\phi = \widehat{A}\widehat{P}_{\mathcal{L}}\phi + \widehat{A}\widehat{P}_{\mathcal{L}^{\perp}}\phi \tag{2.105}$$

$$= \widehat{A}\phi. \tag{2.106}$$

We conclude that $\widehat{A}$ is reducible to the sum of two parts. We call $\widehat{A}_{\mathcal{L}}$ the part of $\widehat{A}$ in $\mathcal{L}$ and $\widehat{A}_{\mathcal{L}^{\perp}}$ the part of $\widehat{A}$ in $\mathcal{L}^{\perp}$. Note that $\widehat{A}_{\mathcal{L}}$ has a range in $\mathcal{L}$ while $\widehat{A}_{\mathcal{L}^{\perp}}$ has a range in $\mathcal{L}^{\perp}$.

C3   A subspace $\mathcal{L}$ reduces a selfadjoint operator $\widehat{A}$ if and only if $\widehat{A}$ commutes with the projector $\widehat{P}_{\mathcal{L}}$ onto $\mathcal{L}$.[41]

---

[40] Roman Vol. 2 (1975) pp. 572-573.
[41] $\widehat{P}_{\mathcal{L}}$ commutes with $\widehat{A}$ if $\widehat{P}_{\mathcal{L}}\widehat{A} \subset \widehat{A}\widehat{P}_{\mathcal{L}}$. See Jauch (1968) p. 42, Amrein, Jauch and Sinha (1977) p. 187. See also §2.8.1C(1) C6, C7.

C4 If a selfadjoint operator $\widehat{A}$ possesses a reducing subspace $\mathcal{L}$ then the part of $\widehat{A}$ in $\mathcal{L}$ and the part of $\widehat{A}$ in $\mathcal{L}^\perp$ are both selfadjoint. We can make use of C3 above and Eq. (2.84) to show this, i.e., we have

$$\left(\widehat{A}\widehat{P}_\mathcal{L}\right)^\dagger = \left(\widehat{P}_\mathcal{L}\widehat{A}\right)^\dagger = \widehat{A}\widehat{P}_\mathcal{L}. \tag{2.107}$$

Hence the reduction $\widehat{A}_\mathcal{L}$ is selfadjoint, and so is $\widehat{A}_{\mathcal{L}^\perp}$.

2.4 Unitary Operators and Unitary Transforms

Definition 2.4(1)   **Unitary transforms; isometric operators**

- Let $\widehat{U}$ be a unitary operator on a Hilbert space $\mathcal{H}$. Then:

  1. A vector $\phi' = \widehat{U}\phi$ in $\mathcal{H}$ is called a *unitary transform* of the given vector $\phi \in \mathcal{H}$.
  2. An operator $\widehat{A}' = \widehat{U}\widehat{A}\widehat{U}^{-1}$ in $\mathcal{H}$ is called a *unitary transform* of the given operator $\widehat{A}$.[42]

- Let $\mathcal{H}_1$ and $\mathcal{H}_2$ be two Hilbert spaces, and let $\widehat{\mathcal{U}}$ be an operator which maps the whole of $\mathcal{H}_1$ onto the whole of $\mathcal{H}_2$. Then $\widehat{\mathcal{U}}$ is said to be a *unitary* operator if the norm of every vector is preserved under the action of $\widehat{\mathcal{U}}$, i.e.,

  $$\langle \widehat{\mathcal{U}}\phi \mid \widehat{\mathcal{U}}\phi \rangle_2 = \langle \phi \mid \phi \rangle_1 \quad \forall \phi \in \mathcal{H}_1 \tag{2.108}$$

  where $\langle \cdot \mid \cdot \rangle_1$ and $\langle \cdot \mid \cdot \rangle_2$ are respectively the scalar products in $\mathcal{H}_1$ and $\mathcal{H}_2$.

  The operator $\widehat{\mathcal{U}}$ is said to be an *isometric* operator, or an *isometry*, if $\widehat{\mathcal{U}}$ maps the whole of $\mathcal{H}_1$ to a subset of $\mathcal{H}_2$, i.e., if the range of $\widehat{\mathcal{U}}$ is a subset of $\mathcal{H}_2$.[43]

- Let $\widehat{A}_1$ be an operator in Hilbert space $\mathcal{H}_1$, and $\widehat{A}_2$ be an operator in $\mathcal{H}_2$. Then $\widehat{A}_2$ is called a *unitary transform* of $\widehat{A}_1$ if there is a unitary operator $\widehat{\mathcal{U}}$ from $\mathcal{H}_1$ to $\mathcal{H}_2$ such that

  $$\mathcal{D}(\widehat{A}_2) = \widehat{\mathcal{U}}\mathcal{D}(\widehat{A}_1) \quad \text{and} \quad \widehat{A}_2 = \widehat{\mathcal{U}}\widehat{A}_1\widehat{\mathcal{U}}^{-1}. \tag{2.109}$$

---

[42] Unitary operators are invertible on account of the following arguments:

$$\phi_2 \neq \phi_1 \Leftrightarrow \langle \widehat{U}(\phi_2 - \phi_1) \mid \widehat{U}(\phi_2 - \phi_1) \rangle = \langle \phi_2 - \phi_1 \mid \phi_2 - \phi_1 \rangle \neq 0 \Leftrightarrow \widehat{U}\phi_2 \neq \widehat{U}\phi_1.$$

[43] In other words $\widehat{\mathcal{U}}$ maps $\mathcal{H}_1$ into, rather than onto, $\mathcal{H}_2$.

- Two operators related by a unitary transformation are said to be *unitarily equivalent*.

Examples 2.4(1)   **Fourier transform and its inverse**

E1   As an example consider the familiar *Fourier transform* operation in $L^2(I\!E)$. Introduce the operator $\widehat{U}_F$ defined on $L^2(I\!E)$ formally by[44]

$$\phi(x) \mapsto \widetilde{\phi}(x) = \left(\widehat{U}_F \phi\right)(x) = \frac{1}{\sqrt{2\pi\hbar}} \int_{-\infty}^{\infty} e^{-\dot{i}xx'} \phi(x')\, dx'. \tag{2.110}$$

Here we have introduced the notation

$$\dot{i} = \frac{i}{\hbar}, \tag{2.111}$$

which will be used throughout the book. One can verify that

$$\int_{-\infty}^{\infty} \widetilde{\phi}^*(x)\widetilde{\phi}(x)\, dx = \int_{-\infty}^{\infty} \phi^*(x)\phi(x)\, dx. \tag{2.112}$$

It follows that $\widetilde{\phi}$ can be identified as a member of the Hilbert space $L^2(I\!E)$. In other words $\widehat{U}_F$ is a unitary operator on $L^2(I\!E)$, called the *Fourier transform operator*, and we call $\widetilde{\phi}$ the Fourier transform of ϕ. The inverse Fourier transform operator $\widehat{U}_F^{-1}$ is formally given by

$$\left(\widehat{U}_F^{-1}\widetilde{\phi}\right)(x) = \frac{1}{\sqrt{2\pi\hbar}} \int_{-\infty}^{\infty} e^{\dot{i}x'x} \widetilde{\phi}(x')\, dx' = \phi(x). \tag{2.113}$$

It is often convenient to relabel x' as p and rewrite Eqs. (2.110) and (2.113) as

$$\phi(x) \mapsto \widetilde{\phi}(p) = \left(\widehat{U}_F \phi\right)(p) = \frac{1}{\sqrt{2\pi\hbar}} \int_{-\infty}^{\infty} e^{-\dot{i}px} \phi(x)\, dx, \tag{2.114}$$

$$\widetilde{\phi}(p) \mapsto \phi(x) = \left(\widehat{U}_F^{-1}\widetilde{\phi}\right)(x) = \frac{1}{\sqrt{2\pi\hbar}} \int_{-\infty}^{\infty} e^{\dot{i}px} \widetilde{\phi}(p)\, dp. \tag{2.115}$$

As will become apparent later the various quantities introduced here have the following physical interpretation:

1. The coordinate variable x forms the coordinate space $I\!E$; in a similar way the dummy variable p also forms a geometric space. It turns out that in quantum mechanics when $\phi(x)$ is used to represent the wave

---

[44]For a discussion of the subtlety in Fourier transform operation in $L^2(I\!E)$ see Roman Vol. 2 (1975) §12.3c.

2.4. UNITARY OPERATORS AND UNITARY TRANSFORMS

function of a particle moving in $I\!\!E$ the variable p in the Fourier transform is related to the momentum of a particle. So, in accordance with §1.5.3C(1) C2 the geometric space formed by p is called the *momentum space* and is denoted by $\widetilde{I\!\!E}$.

2. The functions $\phi(x)$ form the Hilbert space $L^2(I\!\!E)$; in a similar way Fourier transforms $\widetilde{\phi}(p)$ may be taken to form a separate Hilbert space, to be denoted by $\widetilde{L}^2(\widetilde{I\!\!E})$, with scalar product defined by

$$\langle \widetilde{\phi} \mid \widetilde{\psi} \rangle_F = \int_{-\infty}^{\infty} \widetilde{\phi}^*(p)\widetilde{\psi}(p)\,dp. \qquad (2.116)$$

The Hilbert space $\widetilde{L}^2(\widetilde{I\!\!E})$ is often referred to as the *momentum representation space*. We can even introduce the following multiplication and differential operators[45]

$$\widetilde{p} = p, \quad \widetilde{x} = i\hbar\frac{d}{dp} \qquad (2.117)$$

to act on $\widetilde{\phi}(p) \in \widetilde{L}^2(\widetilde{I\!\!E})$. In contrast, we may call $L^2(I\!\!E)$ the *coordinate representation space*.

3. We can consider Fourier transform as a transformation between the coordinate representation space and the momentum representation space, i.e., we have

$$\widehat{U}_F : L^2(I\!\!E) \mapsto \widetilde{L}^2(\widetilde{I\!\!E}) \quad \text{by} \quad \phi(x) \mapsto \widetilde{\phi}(p). \qquad (2.118)$$

This is a unitary transformation since the scalar product is preserved:

$$\langle \widetilde{\phi} \mid \widetilde{\psi} \rangle_F = \int_{-\infty}^{\infty} \widetilde{\phi}^*(p)\widetilde{\psi}(p)\,dp \qquad (2.119)$$

$$= \int_{-\infty}^{\infty} \phi^*(x)\psi(x)\,dx \qquad (2.120)$$

$$= \langle \phi \mid \psi \rangle. \qquad (2.121)$$

E2 It is often useful to consider the unitary transforms of operators, in particular the Fourier transforms of operators. Examples of this can be found in §2.7E(1) E2, §2.8.1C(1) C9 and §2.8E(1) E2.

Comments 2.4(1)   Unitary group and Stone's theorem

C1   The following properties of a unitary operator $\widehat{U}$ are easily verified:

---
[45]These operators are related to the position and momentum operator $\widehat{x}$ and $\widehat{p}$ in $L^2(I\!\!E)$. See §2.4C(1) C7.

1. $\widehat{U}$ is bounded and possesses an inverse $\widehat{U}^{-1}$ which is again unitary. It follows that a unitary transformation is a mutual operation, i.e., we have

$$\phi = \widehat{U}\phi', \quad \widehat{A}' = \widehat{U}\widehat{A}\widehat{U}^{-1} \Leftrightarrow \phi' = \widehat{U}^{-1}\phi;, \quad \widehat{A} = \widehat{U}^{-1}\widehat{A}'\widehat{U}. \qquad (2.122)$$

2. $\widehat{U}$ can be characterized by the following property:

$$\widehat{U}^\dagger = \widehat{U}^{-1}, \quad \text{namely} \quad \widehat{U}^\dagger \widehat{U} = \widehat{U}\widehat{U}^\dagger = \widehat{\mathbb{I}}. \qquad (2.123)$$

3. The product of two unitary operators is again unitary.

C2 A unitary transformation of vectors preserves the norm, the scalar product and any orthonormal bases, i.e., if $\varphi_1, \varphi_2, \ldots$ form an orthonormal basis in $\mathcal{H}$, then their unitary transforms $\varphi_1', \varphi_2', \ldots$ also constitute an orthonormal basis in $\mathcal{H}$.[46] Moreover, any two orthonormal bases are unitarily related, i.e., if $\{\phi_1, \phi_2, \ldots\}$ is another orthonormal basis then there is a unitary operator $\widehat{U}$ such that

$$\phi_j = \widehat{U}\varphi_j, \quad j = 1, 2, \ldots. \qquad (2.124)$$

We can express this unitary operator in Dirac notation as

$$\widehat{U} = \sum_j |\phi_j\rangle\langle\varphi_j|. \qquad (2.125)$$

This notation means that

$$\widehat{U}\psi = \left(\sum_j |\phi_j\rangle\langle\varphi_j|\right)\psi = \sum_j \langle\varphi_j \mid \psi\rangle\, \phi_j, \quad \forall \phi \in \mathcal{H} \qquad (2.126)$$

and

$$\widehat{U}^\dagger = \sum_j |\varphi_j\rangle\langle\phi_j|. \qquad (2.127)$$

C3 Given any two unit vectors $\phi, \varphi \in \mathcal{H}$ there is a unitary operator $\widehat{U}$ in $\mathcal{H}$ such that $\phi = \widehat{U}\varphi$. To see this we can introduce two orthonormal bases φ_j and ϕ_j with $\varphi_1 = \varphi$ and $\phi_1 = \phi$. Then $\widehat{U}$ defined by Eq. (2.125) maps φ to ϕ.

C4 A unitary transformation of operators preserves closedness, symmetry and maximal symmetry, essential selfadjointness and selfadjointness properties

---
[46]Roman Vol. 2 (1975) Theorem 12.3c(3) on p. 559.

2.4. UNITARY OPERATORS AND UNITARY TRANSFORMS

of operators. A unitary transformation of operators from $\widehat{A}$ to $\widehat{A}' = \widehat{U}\widehat{A}\widehat{U}^{-1}$ preserves eigenvalues, i.e.,

$$\widehat{A}\varphi = a\,\varphi \quad \Leftrightarrow \quad \widehat{A}'\varphi' = a\,\varphi' \quad \text{with} \quad \varphi' = \widehat{U}\varphi. \tag{2.128}$$

The domain of the transformed operator is the transform of the domain of the original operator, i.e.,

$$\mathcal{D}(\widehat{A}') = \widehat{U}\,\mathcal{D}(\widehat{A}). \tag{2.129}$$

C5 A unitary transform of a projector is again a projector.

C6 A simultaneous unitary transformations of vectors and operators, i.e.,

$$\varphi \to \varphi' = \widehat{U}\varphi, \quad \widehat{A} \to \widehat{A}' = \widehat{U}\widehat{A}\widehat{U}^{-1}, \tag{2.130}$$

preserves expectation values, i.e.,

$$\langle \varphi \mid \widehat{A}\varphi \rangle = \langle \varphi' \mid \widehat{A}'\varphi' \rangle. \tag{2.131}$$

C7 The Fourier transform operator gives rise to the following familiar Fourier transforms of operators between $L^2(\mathbb{E})$ and $\widetilde{L}^2(\widetilde{\mathbb{E}})$:

1. From $L^2(\mathbb{E})$ to $\widetilde{L}^2(\widetilde{\mathbb{E}})$ we have

$$x \mapsto \widetilde{x} = \widehat{U}_F\, x\, \widehat{U}_F^{-1} = \left(i\hbar\frac{d}{dp}\right), \quad \widehat{p} \mapsto \widetilde{p} = \widehat{U}_F\,\widehat{p}\,\widehat{U}_F^{-1} = p. \tag{2.132}$$

2. From $\widetilde{L}^2(\widetilde{\mathbb{E}})$ to $L^2(\mathbb{E})$ we have

$$\widetilde{x} \mapsto \widehat{x} = \widehat{U}_F^{-1}\,\widetilde{x}\,\widehat{U}_F = x, \quad \widetilde{p} \mapsto \widehat{p} = \widehat{U}_F^{-1}\,\widetilde{p}\,\widehat{U}_F = -i\hbar\frac{d}{dx}. \tag{2.133}$$

C8 A selfadjoint operator $\widehat{A}$ generates a one-parameter family of operators $\widehat{U}_\tau$, parameterized by a real variable τ, by exponentiation:[47]

$$\widehat{U}_\tau = e^{-i\tau\widehat{A}}, \quad \tau \in \mathbb{R}. \tag{2.134}$$

This family of operators possesses the following properties:[48]

---

[47] See §2.8.2 for a definition of exponential functions, especially Eq. (2.367).

[48] Prugovečki (1981) Theorem 3.1 on p. 288 and Theorem 6.1 on pp. 334-335. The expression of $\widehat{U}_\tau$ gives it a rather intuitive meaning as an exponential function of $\widehat{A}$. As shown in §2.8.2 one can also introduce a precise definition $\widehat{U}_\tau$.

1. $\widehat{U}_\tau$ is unitary for all $\tau \in \mathbb{R}$ with $\widehat{U}_0$ equal to the identity operator and $\widehat{U}_{-\tau}$ equal to the inverse operator $\widehat{U}_\tau^{-1}$.

2. The family is weakly continuous in τ in the sense that $\langle \psi \mid \widehat{U}_\tau \phi \rangle$ is a continuous function of τ for all ψ and ϕ in $\mathcal{H}$.

3. $\widehat{U}_{\tau_2}\widehat{U}_{\tau_1} = \widehat{U}_{\tau_2+\tau_1}$ for any given $\tau_1, \tau_2 \in \mathbb{R}$. We shall therefore call the family a *weakly continuous one-parameter group of unitary operators*, or a *one-parameter group of unitary operators*, or simply a *one-parameter unitary group*.

4. The operator $\widehat{A}$ is expressible in terms of the unitary operators by[49]

$$\widehat{A}\phi = i\hbar \lim_{\tau \to 0} \frac{\widehat{U}_\tau - \widehat{\mathbb{I}}}{\tau} \phi, \quad \phi \in \mathcal{D}(\widehat{A}). \tag{2.135}$$

The correspondence between selfadjoint operators and groups of unitary operators is given by a theorem of Stone stated in Theorem 2.4(1).

C9 For unitary operators we have $\widehat{\mathcal{U}}$ mapping $\mathcal{H}_1$ onto $\mathcal{H}_2$ and its inverse $\widehat{\mathcal{U}}^{-1}$ mapping $\mathcal{H}_2$ to $\mathcal{H}_1$. These can be summarized into the following equations:[50]

$$\widehat{\mathcal{U}}^{-1}\widehat{\mathcal{U}} = \widehat{\mathbb{I}}_1, \quad \widehat{\mathcal{U}}\widehat{\mathcal{U}}^{-1} = \widehat{\mathbb{I}}_2, \tag{2.136}$$

where $\widehat{\mathbb{I}}_1$ and $\widehat{\mathbb{I}}_2$ are the identity operators on $\mathcal{H}_1$ and $\mathcal{H}_2$ respectively. More explicitly these mean that

$$\widehat{\mathcal{U}}^{-1}\widehat{\mathcal{U}}\phi_1 = \phi_1 \quad \forall \phi_1 \in \mathcal{H}_1 \quad \text{and} \quad \widehat{\mathcal{U}}\widehat{\mathcal{U}}^{-1}\phi_2 = \phi_2 \quad \forall \phi_2 \in \mathcal{H}_2. \tag{2.137}$$

Theorem 2.4(1)   **Stone's theorem**[51]

- Let $\widehat{U}_\tau$ be a weakly continuous one-parameter family of unitary operators on a Hilbert space $\mathcal{H}$ satisfying

$$\widehat{U}_{\tau_2}\widehat{U}_{\tau_1} = \widehat{U}_{\tau_2+\tau_1} \quad \forall \tau_1, \tau_2 \in \mathbb{R}. \tag{2.138}$$

Then there exists a unique selfadjoint operator $\widehat{A}$ in $\mathcal{H}$ such that Eq. (2.134) is satisfied.

---

[49] Vectors on the right-hand side converge to the vector $\widehat{A}\phi$ in the sense of Eq. (2.9).
[50] Weidmann (1980) p. 86.
[51] Jauch (1968) p. 57, Prugovečki (1981) p. 288, pp. 334-335, Reed and Simon Vol. 1 (1972) pp. 264-270, especially the discussion on weak and strong convergence on p. 268.

2.5. EXTENSIONS OF SYMMETRIC OPERATORS

Comments 2.4(2)   **One-parameter unitary groups and their generators**

C1   We call the selfadjoint operator $\widehat{A}$ the *generator* of the one-parameter group of unitary operators $\widehat{U}_\tau$.

C2   A well-known example in $L^2(I\!E)$ is the group

$$\widehat{T}_\tau = e^{-i\tau \widehat{p}}, \quad \tau \in I\!R, \tag{2.139}$$

where $\widehat{p}$ is the momentum operator defined in §2.5.1E(1) E3. The effect of $\widehat{U}_\tau$ on $\phi(x) \in L^2(I\!E)$ is well-known, i.e., it acts as a *translation operator* displacing $\phi(x)$ by a distance τ to the right, i.e.,[52]

$$\widehat{T}_\tau \phi(x) = \phi(x - \tau). \tag{2.140}$$

Unitary groups play a crucial role in quantization and in the formulation of quantum dynamics.

2.5 Extensions of Symmetric Operators

Generally a symmetric operator $\widehat{A}$ has many extensions, its adjoint being an obvious one. The interesting and physically relevant questions are

- *Under what conditions will a symmetric operator possess selfadjoint extensions and under what conditions will it possess only strictly maximal symmetric extensions?*

To answer this we have to introduce more definitions.

2.5.1 Selfadjoint and maximal symmetric extensions

Definition 2.5.1(1)   **Deficiency indices and deficiency subspaces**

- Let $\widehat{A}$ be a symmetric operator in a Hilbert space $\mathcal{H}$ and let $\widehat{A}^\dagger$ be its adjoint. Then the number n^+ of linearly independent eigenvectors $f_j^{(+i)} \in \mathcal{H}, j = 1, 2, \ldots, n^+$, of $\widehat{A}^\dagger$ corresponding to the eigenvalue i is called the *positive deficiency index* of $\widehat{A}$. Similarly the number n^- of linearly independent eigenvectors $f_k^{(-i)} \in \mathcal{H}, k = 1, 2, \ldots, n^-$, of $\widehat{A}^\dagger$ corresponding to the eigenvalue $-i$ is called the *negative deficiency index* of $\widehat{A}$. The subspace $\mathcal{L}^{(+i)}$ spanned by all the $f_j^{(+i)}$ is called the *positive*

---
[52] Fano (1971) pp. 286-287, Roman (1965) §5, Merzbacher (1998) pp. 69-70.

deficiency subspace of $\widehat{A}$ for the value i, and similarly the subspace $\mathcal{L}^{(-i)}$ spanned by all the $f_k^{(-i)}$ is called the *negative deficiency subspace* of $\widehat{A}$ for the value $-i$.

Comments 2.5.1(1)   Deficiency indices and selfadjointness

C1   The two deficiency indices are normally written as an ordered pair (n^+, n^-). To determine these indices we need to solve the eigenvalue equations

$$\widehat{A}^\dagger f_j^{(+i)} = +i\, f_j^{(+i)}, \quad \widehat{A}^\dagger f_k^{(-i)} = -i\, f_k^{(-i)} \qquad (2.141)$$

in order to find all the linearly independent eigenfunctions. Note that $f_j^{(+i)}$ and $f_k^{(-i)}$ must by definition be member of the Hilbert space.[53]

C2   A selfadjoint operator is known to admit no complex eigenvalues. If $\widehat{A}$ is selfadjoint, then its adjoint $\widehat{A}^\dagger$ is equal to $\widehat{A}$. Hence $\widehat{A}^\dagger$ admits no complex eigenvalues. In contrast, a symmetric operator is deficient in the sense that its adjoint $\widehat{A}^\dagger$ may admit complex eigenvalues. This deficiency is measured by a pair of non-zero deficiency indices. Clearly a selfadjoint operator has zero deficiency indices. Unfortunately the converse is not true. The adjoint of an essentially selfadjoint operator is selfadjoint, and hence admits no complex eigenvalues either. The difference here is that an essentially selfadjoint operator may not be closed.

Let $\widehat{A}$ be a closed symmetric operator with deficiency indices (n^+, n^-) in a Hilbert space $\mathcal{H}$. Then Theorem 2.5.1(1) applies, showing a close relationship between the values of the deficiency indices and the nature of the operator.

Theorem 2.5.1(1)   Deficiency indices and maximal symmetric extensions[54]

- $\widehat{A}$ is selfadjoint if $n^+ = n^- = 0$. If $\widehat{A}$ were not closed, then $n^+ = n^- = 0$ would imply only that $\widehat{A}$ is essentially selfadjoint.

- $\widehat{A}$ is strictly maximal symmetric if only one of the deficiency indices is zero, i.e., if

$$\text{either} \quad n^+ = 0,\, n^- \neq 0 \quad \text{or} \quad n^+ \neq 0,\, n^- = 0. \qquad (2.142)$$

---

[53] For some explicit examples see §2.5.1E(1) E1, E2 and E3.
[54] Weidmann (1980) p. 231, p. 239, Richtmyer (1978) p. 155, Akhiezer and Glazman Vol. 2 (1963) Theorems 3 and 4 on p. 97 (a symmetric operator is meant to be closed in this book).

2.5. EXTENSIONS OF SYMMETRIC OPERATORS 115

If $\widehat{A}$ were not closed this would only imply that $\widehat{A}$ is essentially strictly maximal symmetric, i.e., $\widehat{A}$ admits a unique strictly maximal symmetric extension.

- $\widehat{A}$ has an n^2-parameter family of selfadjoint extensions, i.e., $\widehat{A}$ possesses a family of selfadjoint extensions parameterizable by n^2 real variables, if the deficiency indices are equal, finite and non-vanishing, i.e., when $n^+ = n^- = n \neq 0$.

Comments 2.5.1(2)   Differential operators, absolutely continuous functions and Sobolev spaces

C1   Note that $\widehat{A}$ in Theorem 2.5.1(1) is a closed symmetric operator with deficiency indices (n^+, n^-). Generally, if a symmetric operator has two unequal and non-zero deficiency indices it would admit only strictly maximal symmetric extensions.

C2   We shall present a systematic procedure for working out the selfadjoint extensions of a symmetric operator later. For the moment let us gain some intuition by examining a number of physically important differential operators. A *differential operator* in $L^2(I\!\!E^n)$ is defined by its formal differential expression and the domain this differential expression operates on. The standard approach is to define a differential expression on $C_0^\infty(I\!\!E^n)$ first. This generally produces a symmetric operator which is not closed. As an example let $L^2(\Lambda)$ denote the Hilbert space of functions defined on a bounded open interval $\Lambda = (0, a)$ of $I\!\!E$ square-integrable with respect to dx in Λ. For any given positive integer ℓ we can introduce an operator $\widehat{p}_0^{(\ell)}(\Lambda)$ defined by the differential expression $(-i\hbar)^\ell \, d^\ell/dx^\ell$ on the domain $C_0^\infty(\Lambda)$, i.e., we have

$$\widehat{p}_0^{(\ell)}(\Lambda) = (-i\hbar)^\ell \frac{d^\ell}{dx^\ell} \quad \text{acting on} \quad \mathcal{D}\left(\widehat{p}_0^{(\ell)}(\Lambda)\right) = C_0^\infty(\Lambda). \tag{2.143}$$

This operator can be shown to be symmetric but not closed. One then proceeds to examine possible selfadjoint extensions. We shall present a number of examples which will serve to illustrate the procedure; these examples will also be directly relevant to many of the physical systems to be studied in later chapters. Before proceeding further we need to introduce two classes of functions.

C3   Absolutely continuous functions on $I\!\!E$ A function F on $I\!\!E$ is said to be *absolutely continuous* if there exists a function f on $I\!\!E$ integrable with respect to dx over every bounded interval such that

$$F(x) = F(0) + \int_0^x f(x) dx. \tag{2.144}$$

116 CHAPTER 2. OPERATORS AND THEIR DIRECT INTEGRALS

As a result the function F is differentiable almost everywhere with f as its derivative, i.e., $dF/dx = f$ everywhere except possibly at some isolated points, e.g., where f is discontinuous.[55] An absolutely continuous function is also continuous, but the converse is not necessarily true.[56] A continuous function may not be differentiable anywhere at all, while an absolutely continuous function is differentiable almost everywhere. This is the reason for introducing such a class of functions.

We shall denote the set of all absolutely continuous function on $I\!\!E$ by $AC(I\!\!E)$. Clearly we can similarly define absolutely continuous functions on an open interval Λ of $I\!\!E$. These functions will be denoted by $AC(\Lambda)$.

To appreciate the relevance of these discussions let us first introduce the largest set of absolutely continuous functions lying within the Hilbert space $L^2(I\!\!E)$, i.e.,

$$L^2(I\!\!E) \cap AC(I\!\!E). \tag{2.145}$$

Next we need to select functions in this set whose derivatives lie in the Hilbert space $L^2(I\!\!E)$. Let the largest set of this selection be denoted by $S_{2,1}(I\!\!E)$, i.e.,

$$S_{2,1}(I\!\!E) = \Big\{ F \in L^2(I\!\!E) \cap AC(I\!\!E) : dF/dx \in L^2(I\!\!E) \Big\}. \tag{2.146}$$

Here the first subscript 2 signifies square-integrability and the second subscript 1 signifies once differentiability almost everywhere on account of absolute continuity. The differential expression d/dx can act on any $F \in S_{2,1}(I\!\!E)$ to produce a derivative $f = dF/dx$ within $L^2(I\!\!E)$. This enables us to define first order differential operators in the Hilbert space $L^2(I\!\!E)$, e.g., using the operator expression d/dx to act on an appropriate subset of $S_{2,1}(I\!\!E)$. For example we can define the following operator:

$$\widehat{p}_0 = -i\hbar \frac{d}{dx}, \quad \text{with domain} \quad \mathcal{D}(\widehat{p}_0) = C_0^\infty(I\!\!E) \subset S_{2,1}(I\!\!E). \tag{2.147}$$

We can extend this concept to $I\!\!E^n$. Consider a differential expression in the form of a vector field

$$\widehat{X} = \sum_{j=1}^{n} X^j \partial/\partial x^j \tag{2.148}$$

in the Euclidean space $I\!\!E^n$. We know from C1.3.2(1) C2 that in the neighbourhood of any point in $I\!\!E^n$ in which $\widehat{X}$ does not vanish a new local coordinate system x'^j exists. Expressed in terms of x'^j the vector field takes the form

$$\widehat{X} = \partial/\partial x'^1 \tag{2.149}$$

---

[55] The phrase *almost everywhere* is a technical term in measure theory. A brief review of this term is given in §2.6.1C(1) C3.

[56] See Smirnov (1964) §76 for an example.

2.5. EXTENSIONS OF SYMMETRIC OPERATORS

in that neighbourhood. We shall call a function F on $I\!\!E^n$ *absolutely continuous with respect to* $\widehat{X}$ if it is absolutely continuous with respect to the coordinate x'^1 in every such neighbourhood. We denote the set of all these functions by $AC(\widehat{X}, I\!\!E^n)$. It follows that $\widehat{X}$ can act on functions in $AC(\widehat{X}, I\!\!E^n)$ in the same way that d/dx can act on $AC(I\!\!E)$. The idea here is to enable the differential expression for a vector field to act as an operator in the Hilbert space $L^2(I\!\!E^n)$ by acting on an appropriate subset of

$$L^2(I\!\!E^n) \cap AC(\widehat{X}, I\!\!E^n). \tag{2.150}$$

The set of smooth functions of compact support $C_0^\infty(I\!\!E^n)$ is clearly one such subset.

C4 Sobolev spaces[57]   Let Λ be an open interval in $I\!\!E$, and let

$$AC(d^\ell/dx^\ell, \Lambda) \tag{2.151}$$

denote the set of functions ϕ defined on Λ such that

1. ϕ, $d\phi/dx$, $d^2\phi/dx^2$, ..., $d^{\ell-2}\phi/dx^{\ell-2}$ are continuously differentiable on Λ, and

2. $d^{\ell-1}\phi/dx^{\ell-1}$ is absolutely continuous.

Note that the ℓth derivative exists almost everywhere on account of absolute continuity of the $(\ell-1)^{th}$ derivative. Two examples are:

1. For the case $\ell = 1$ we have $AC(d/dx, \Lambda) = AC(\Lambda)$ which is simply the set of absolutely continuous functions on Λ.

2. For the case $\ell = 2$ we have $AC(d^2/dx^2, \Lambda)$ which is the set of functions on Λ that is continuously differentiable and has an absolutely continuous derivative. Functions in $AC(d^2/dx^2, \Lambda)$ possess a second order derivative almost everywhere.

The *Sobolev space* $S_{2,\ell}(\Lambda)$ of order ℓ over the interval Λ is defined to be the following set of functions in $L^2(\Lambda)$:

$$S_{2,\ell}(\Lambda) = \{\phi \in L^2(\Lambda) \cap AC(d^\ell/dx^\ell, \Lambda) : d^\ell\phi/dx^\ell \in L^2(\Lambda)\}. \tag{2.152}$$

In other words $S_{2,\ell}(\Lambda)$ consists of functions belonging to both $L^2(\Lambda)$ and $AC(d^\ell/dx^\ell, \Lambda)$ whose ℓ^{th} derivatives remain in $L^2(\Lambda)$. We can similarly define

---
[57]See Weidmann (1980) pp. 157-159.

Sobolev spaces $S_{2,\ell}(I\!\!E)$. Examples are

$$S_{2,1}(\Lambda) = \left\{\phi \in L^2(\Lambda) \cap AC(d/dx, \Lambda) : d\phi/dx \in L^2(\Lambda)\right\}, \qquad (2.153)$$

$$S_{2,2}(\Lambda) = \left\{\phi \in L^2(\Lambda) \cap AC(d^2/dx^2, \Lambda) : d^2\phi/dx^2 \in L^2(\Lambda)\right\}, \qquad (2.154)$$

$$S_{2,1}(I\!\!E) = \left\{\phi \in L^2(I\!\!E) \cap AC(d/dx, I\!\!E) : d\phi/dx \in L^2(I\!\!E)\right\}, \qquad (2.155)$$

$$S_{2,2}(I\!\!E) = \left\{\phi \in L^2(I\!\!E) \cap AC(d^2/dx^2, I\!\!E) : d^2\phi/dx^2 \in L^2(I\!\!E)\right\}. \qquad (2.156)$$

The present expression for $S_{2,1}(I\!\!E)$ agrees with the expression in Eq. (2.146).

Intuitively we can see that the space $S_{2,\ell}(\Lambda)$ is the largest set of functions in $L^2(\Lambda)$ which can be meaningfully operated on by the differential expression d^ℓ/dx^ℓ serving as an operator in the Hilbert space $L^2(\Lambda)$. The relevance of Sobolev spaces is seen in the following examples:

1. Sobolev spaces can serve as the domain of the adjoint $\widehat{p}_0^{(\ell)\dagger}(\Lambda)$ of $\widehat{p}_0^{(\ell)}(\Lambda)$ defined by Eq. (2.143), i.e., one can show that the adjoint of $\widehat{p}_0^{(\ell)}(\Lambda)$ is given by[58]

$$\widehat{p}_0^{(\ell)\dagger} = (-i\hbar)^\ell \frac{d^\ell}{dx^\ell} \quad \text{acting on the Sobolev space } S_{2,\ell}(\Lambda). \qquad (2.157)$$

2. Sobolev spaces also help to specify the domains of the closure and maximal symmetric extensions of $\widehat{p}_0^{(\ell)}(\Lambda)$. According to Eq. (2.81), any symmetric extension to $\widehat{p}_0^{(\ell)}(\Lambda)$ is a restriction of $\widehat{p}_0^{(\ell)\dagger}(\Lambda)$, i.e., it is given by the same differential expression $(-i\hbar)^\ell d^\ell/dx^\ell$ operating on a domain smaller than $S_{2,\ell}(\Lambda)$. An example is the closure $\bar{p}_0^{(\ell)}(\Lambda)$ of $\widehat{p}_0^{(\ell)}(\Lambda)$ which is known to be given by $(-i\hbar)^\ell d^\ell/dx^\ell$ acting on a domain consisting of functions in $S_{2,\ell}(\Lambda)$ such that[59]

$$\phi = \frac{d\phi}{dx} = \frac{d^2\phi}{dx^2} = \cdots = \frac{d^{\ell-1}\phi}{dx^{\ell-1}} = 0 \quad \text{at } x = 0 \text{ and } x = a, \qquad (2.158)$$

where $x = 0$ and $x = a$ are the end points of the interval $\Lambda = (0, a)$. More explicit examples will be presented presently. Note that a function ϕ in $S_{2,\ell}(\Lambda)$ can be continuously extended to the end points $x = 0, a$; the same applies to its derivatives up to the order of $(\ell - 1)$.[60] This is why we can have the above equation for the values of $\phi, d\phi/dx, \ldots, d^{\ell-1}\phi/dx^{\ell-1}$ at $x = 0, a$.

---
[58] Weidmann (1980) Theorem 6.29 on p. 160.
[59] Weidmann (1980) Theorem 6.31 on p. 162.
[60] Weidmann (1980) Theorem 6.27 on p. 159.

2.5. EXTENSIONS OF SYMMETRIC OPERATORS

Examples 2.5.1(1)   **Deficiency indices, selfadjoint and maximal symmetric extensions**

E1   We shall begin by considering operators in the Hilbert space $L^2(\Lambda)$, where $\Lambda = (0, a) \subset I\!\!E$. Let $\widehat{p}_0(\Lambda)$ denote the operator $\widehat{p}_0^{(1)}(\Lambda)$, i.e., the operator $\widehat{p}_0^{(\ell)}(\Lambda)$ in $L^2(\Lambda)$ for the case $\ell = 1$. This operator is symmetric so that

$$\langle \psi \mid \widehat{p}_0(\Lambda)\phi \rangle = \langle \widehat{p}_0(\Lambda)\psi \mid \phi \rangle \quad \forall \psi, \phi \in C_0^\infty(\Lambda). \tag{2.159}$$

As shown in Eq. (2.157) the adjoint $\widehat{p}_0^\dagger(\Lambda)$ is defined by the same differential expression acting on the Sobolev space $S_{2,1}(\Lambda)$, i.e.,

$$\widehat{p}_0^\dagger(\Lambda) = -i\hbar d/dx \quad \text{acting on domain} \quad \mathcal{D}\left(\widehat{p}_0^\dagger(\Lambda)\right) = S_{2,1}(\Lambda). \tag{2.160}$$

To see if $\widehat{p}_0(\Lambda)$ admits selfadjoint extensions we need to work out its deficiency indices. This can be done by solving the following eigenvalue equations:

$$-i\hbar \frac{d}{dx} f^{(\pm i)}(x) = \pm i \, f^{(\pm i)}(x) \tag{2.161}$$

$$\Rightarrow \quad f^{(\pm i)} = c_\pm \, e^{\mp x/\hbar} \in S_{2,1}(\Lambda), \tag{2.162}$$

where the normalization constants are given by

$$c_+^2 = \frac{2}{\hbar}\left(1 - e^{-2a/\hbar}\right)^{-1}, \quad c_-^2 = \frac{2}{\hbar}\left(e^{2a/\hbar} - 1\right)^{-1} \tag{2.163}$$

$$\Rightarrow \quad c_+ = e^{a/\hbar} \, c_-. \tag{2.164}$$

The deficiency indices are therefore $(1,1)$. It follows from Theorem 2.5.1(1) that $\widehat{p}_0(\Lambda)$ possesses a one-parameter family of selfadjoint extensions, i.e., these extensions are parameterizable by a real variable λ. We shall denote these extensions by $\widehat{p}_\lambda(\Lambda)$, or $\widehat{p}_\lambda$ for short. They are defined by[61]

$$\widehat{p}_\lambda = -i\hbar \frac{d}{dx} \quad \text{acting on} \quad \mathcal{D}(\widehat{p}_\lambda) = \left\{\phi \in S_{2,1}(\Lambda) : \phi(0) = e^{i\lambda}\phi(a)\right\}, \tag{2.165}$$

where λ varies in the range $(-\pi, \pi]$. Each value of $\lambda \in (-\pi, \pi]$ gives rise to a selfadjoint extension $\widehat{p}_\lambda$. The condition

$$\phi(0) = e^{i\lambda} \, \phi(a) \tag{2.166}$$

---
[61] A systematic method to determine the domains of these selfadjoint extensions will be given in the next section. The quasi-periodic boundary condition will be derived in §2.5.2E(1) E1.

in Eq. (2.165) is known as a *quasi-periodic boundary condition*, and the usual *periodic boundary condition* corresponds to $\lambda = 0$. The differences of these extensions are obvious when we look at their eigenvalue equation

$$\widehat{p}_\lambda \varphi_{\lambda,n} = p_{\lambda,n} \varphi_{\lambda,n}. \qquad (2.167)$$

The eigenfunctions and eigenvalues are both dependent on λ. They are given by

$$\varphi_{\lambda,n} = \frac{1}{\sqrt{a}} \exp\left[i\left(\frac{2n\pi - \lambda}{a}\right)x\right], \quad n = 0, \pm 1, \pm 2, \ldots \qquad (2.168)$$

and

$$p_{\lambda,n} = \hbar \left(\frac{2n\pi - \lambda}{a}\right). \qquad (2.169)$$

So, a different extension, i.e., a different λ acting on a different domain means a different set of eigenvalues as well as eigenfunctions, even though the operator expression appears the same.

E2 Again consider the Hilbert space $L^2(\Lambda)$ with $\Lambda = (0, a)$. The second-order differential expression

$$-\frac{\hbar^2}{2m}\frac{d^2}{dx^2} \quad \text{acting on the domain} \quad C_0^\infty(\Lambda), \qquad (2.170)$$

where m is a real number, defines an operator[62]

$$\widehat{K}_0(\Lambda) = \frac{1}{2m}\widehat{p}_0^2(\Lambda) \qquad (2.171)$$

in the Hilbert space $L^2(\Lambda)$. This is a symmetric operator with many selfadjoint extensions. A selfadjoint extension, denoted by $\widehat{K}^\infty(\Lambda)$, is defined by[63]

$$\widehat{K}^\infty(\Lambda) = -\frac{\hbar^2}{2m}\frac{d^2}{dx^2} \qquad (2.172)$$

acting on

$$\mathcal{D}\left(\widehat{K}^\infty(\Lambda)\right) = \left\{\phi \in S_{2,2}(\Lambda) : \phi(0) = \phi(a) = 0\right\}. \qquad (2.173)$$

Normalized eigenfunctions of $\widehat{K}^\infty(\Lambda)$ are

$$\phi_\ell = \sqrt{\frac{2}{a}} \sin\left(\frac{\ell\pi}{a}x\right), \quad \ell = 1, 2, \ldots, \qquad (2.174)$$

---

[62] The notation is motivated by the fact that such an operator may represent the kinetic energy in quantum mechanics.

[63] The notation is motivated by the fact that this domain agrees with the domain of the Hamiltonian operator of a particle confined to the region Λ by an infinite square potential well.

2.5. EXTENSIONS OF SYMMETRIC OPERATORS

with eigenvalues
$$E_\ell = \frac{\pi^2 \hbar^2}{2ma^2} \ell^2. \tag{2.175}$$

The constant function $\phi_0(x) = 1/\sqrt{a}$ is not an eigenfunction since $\phi_0 = 1/\sqrt{a}$ does not satisfy the boundary condition in Eq. (2.173).

There are other selfadjoint extensions to $\widehat{K}_0(\Lambda)$, e.g., the one-parameter family of selfadjoint extensions
$$\widehat{K}_\lambda(\Lambda) = \frac{1}{2m} \widehat{p}_\lambda^2(\Lambda), \tag{2.176}$$

where $\widehat{p}_\lambda(\Lambda)$ is the operator defined by Eq. (2.165). We shall give a general discussion on the selfadjoint extensions of $\widehat{K}_0(\Lambda)$ and their physical applications later in Chapters 6 and 7.

In textbooks on introductory quantum mechanics, the selfadjoint extension $\widehat{K}^\infty(\Lambda)$ is taken as the Hamiltonian operator of a particle confined to the region Λ by an infinite square potential well. One should note that although $\widehat{K}^\infty(\Lambda)$ and $\widehat{K}_\lambda(\Lambda)$ have the same differential expression they are different operators, i.e.,

$$\text{the operator } \widehat{K}^\infty(\Lambda) \neq \text{the operator } \widehat{K}_\lambda(\Lambda). \tag{2.177}$$

This is because $\widehat{K}^\infty(\Lambda)$ and $\widehat{K}_\lambda(\Lambda)$ act on different domains. These two operators possess different eigenfunctions and different eigenvalues. Eigenfunctions $\varphi_{\lambda,n}$ of $\widehat{p}_\lambda(\Lambda)$ are also eigenfunctions of $\widehat{K}_\lambda(\Lambda)$. But $\varphi_{\lambda,n}$ do not satisfy the conditions of Eq. (2.173), hence they are not in the domain of $\widehat{K}^\infty(\Lambda)$. It follows that they are not the eigenfunctions of $\widehat{K}^\infty(\Lambda)$. Moreover, $\widehat{K}^\infty(\Lambda)\varphi_{\lambda,n}$ is not even defined, and neither is the expression $\langle \varphi_{\lambda,n} \mid \widehat{K}^\infty(\Lambda)\varphi_{\lambda,n} \rangle$ since $\varphi_{\lambda,n}$ lies outside the domain of $\widehat{K}^\infty(\Lambda)$. If one wants to calculate the expectation value by brute force, as it were, one may try to proceed by expanding $\varphi_{\lambda,n}$ in terms of the eigenfunctions ϕ_ℓ of $\widehat{K}^\infty(\Lambda)$, i.e.,

$$\varphi_{\lambda,n} = \sum_{\ell=1}^{\infty} c_\ell \phi_\ell, \quad c_\ell = \langle \phi_\ell \mid \varphi_{\lambda,n} \rangle. \tag{2.178}$$

This is permissible since ϕ_ℓ are known to constitute an orthonormal basis of $L^2(\Lambda)$. Then one may attempt to evaluate $\widehat{K}^\infty(\Lambda)\varphi_{\lambda,n}$ by linearity:

$$\widehat{K}^\infty(\Lambda)\varphi_{\lambda,n} = \sum_{\ell=1}^{\infty} c_\ell \widehat{K}^\infty(\Lambda)\phi_\ell = \sum_{\ell=1}^{\infty} c_\ell E_\ell \phi_\ell. \tag{2.179}$$

However, linearity fails here since the sum $\sum_{\ell=1}^{\infty} c_\ell E_\ell \phi_\ell$ cannot be a vector in $L^2(\Lambda)$ for the simple reason that the scalar product

$$\left\langle \sum_{\ell=1}^{\infty} c_\ell E_\ell \phi_\ell \middle| \sum_{\ell=1}^{\infty} c_\ell E_\ell \phi_\ell \right\rangle = \sum_{\ell=1}^{\infty} |c_\ell|^2 E_\ell^2 \qquad (2.180)$$

can be shown to diverge, i.e., a formal evaluation of $\langle \varphi_{\lambda,n} | \widehat{K}^\infty(\Lambda) \varphi_{\lambda,n} \rangle$ as the sum $\sum_\ell |c_\ell|^2 E_\ell$ fails to converge. All these serve to illustrate the importance of domain in the definition of an operator.

E3 We shall now consider operators in the Hilbert space $L^2(E)$. Let us examine the operator $\widehat{p}_0(E)$ defined by

$$\widehat{p}_0(E) = -i\hbar \frac{d}{dx} \quad \text{acting on} \quad C_0^\infty(E) \qquad (2.181)$$

in the Hilbert space $L^2(E)$. This operator is clearly symmetric, i.e.,

$$\langle \psi | \widehat{p}_0(E) \phi \rangle = \langle \widehat{p}_0(E) \psi | \phi \rangle \quad \forall \psi, \phi \in C_0^\infty(E). \qquad (2.182)$$

The adjoint $\widehat{p}_0^\dagger(E)$ can be shown to be defined by the same differential expression acting on the Sobolev space $S_{2,1}(E)$, i.e.,[64]

$$\widehat{p}_0^\dagger(E) = -i\hbar \frac{d}{dx} \quad \text{acting on} \quad \mathcal{D}\left(\widehat{p}_0^\dagger(E)\right) = S_{2,1}(E). \qquad (2.183)$$

In an attempt to find the deficiency indices we will again arrive at Eq. (2.162) with solutions of the form $f^{(\pm i)}$. However, these solutions are not square-integrable in E and are therefore not members of $L^2(E)$. In other words the eigenvalue equations of the form of Eq. (2.161) admit no solutions in $L^2(E)$. The deficiency indices are therefore $(0,0)$. Hence, according to Theorem 2.5.1(1), the operator $\widehat{p}_0(E)$ is essentially selfadjoint, and possesses a unique selfadjoint extension $\widehat{p}(E)$ which, as pointed out in §2.3C(1) C8, is equal to its adjoint $\widehat{p}_0^\dagger(E)$, i.e.,

$$\widehat{p}(E) = -i\hbar \, d/dx \quad \text{acting on} \quad \mathcal{D}\left(\widehat{p}(E)\right) = S_{2,1}(E). \qquad (2.184)$$

This is the usual *momentum operator* for a particle in one-dimensional motion in E in quantum mechanics. This operator is often denoted simply by $\widehat{p}$.

E4 The second-order differential expression $-(\hbar^2/2m)\, d^2/dx^2$ defines an operator

$$\widehat{K}_0(E) = \frac{1}{2m} \widehat{p}_0^2(E) \quad \text{acting on} \quad C_0^\infty(E) \qquad (2.185)$$

---

[64] Weidmann (1980) Theorem 6.30 on p. 162.

2.5. EXTENSIONS OF SYMMETRIC OPERATORS

in the Hilbert space $L^2(I\!E)$. It can be shown that $\widehat{K}_0(I\!E)$ admits a unique selfadjoint extension $\widehat{K}(I\!E)$ defined by the same differential expression acting on the domain $S_{2,2}(I\!E)$.[65] Indeed we can see that

$$\widehat{K}(I\!E) = \frac{1}{2m}\,\widehat{p}^{\,2}(I\!E). \tag{2.186}$$

We shall often denote this operator simply by $\widehat{K}$. The selfadjointness is obvious since $\widehat{K}$ is the square of selfadjoint operator $\widehat{p}(I\!E)$, apart from a multiplicative constant. It follows that $\widehat{K}_0(I\!E)$ is essentially selfadjoint.[66]

E5 Consider the operators obtained by

$$-i\hbar\, x^k\,\frac{d}{dx} \quad \text{acting on} \quad C_0^\infty(I\!E), \quad k = 1, 2, \ldots \tag{2.187}$$

in $L^2(I\!E)$. These operators are not even symmetric, let alone selfadjoint. They can be regarded as the product of two operators, i.e.,

$$x^k \quad \text{and} \quad -i\hbar\,\frac{d}{dx} \quad \text{both acting on } C_0^\infty(I\!E). \tag{2.188}$$

We can construct a symmetrized product in accordance with Eq. (2.83), i.e.,

$$\frac{1}{2}\left((x^k)\left(-i\hbar\frac{d}{dx}\right) + \left(-i\hbar\frac{d}{dx}\right)(x^k)\right)$$
$$= -i\hbar\left(x^k\frac{d}{dx} + \frac{k}{2}x^{k-1}\right) \tag{2.189}$$

to obtain a set of new operators $\widehat{(x^k p)}_0$ defined by

$$\widehat{(x^k p)}_0 = -i\hbar\left(x^k\frac{d}{dx} + \frac{k}{2}x^{k-1}\right) \quad \text{acting on} \quad C_0^\infty(I\!E). \tag{2.190}$$

These operators are symmetric in $L^2(I\!E)$. For differential operators of this kind the adjoint $\widehat{(x^k p)}_0^\dagger$ is known to be defined again by the same differential expression. This facilitates the solution of their deficiency indices. To find the deficiency indices of $\widehat{(x^k p)}_0$ we have to solve the eigenvalue equations

$$\widehat{(x^k p)}_0^\dagger f^{(\pm i)} = \pm i\, f^{(\pm i)}, \tag{2.191}$$

---

[65] Weidmann (1980) Theorem 6.30 on p. 162.
[66] Weidmann (1980) Theorem 6.30 on p. 162, Theorem 8.7 on p. 234 and Theorems 10.11 and 10.12 on pp. 299-301.

or explicitly
$$-i\hbar\left(x^k\frac{d}{dx}+\frac{k}{2}x^{k-1}\right)f^{(\pm i)}(x)=\pm i\,f^{\pm i}(x). \tag{2.192}$$

There are two distinct cases to consider here:

Case 1 When $k=1$ we have the operator $\widehat{(xp)}_0$. Equation (2.192) possesses the following formal solutions:

$$f^{(\pm i)}=x^{-(\hbar\pm 2)/(2\hbar)}. \tag{2.193}$$

These solutions are not square-integrable in $I\!\!E$. It follows that Eq. (2.192) with $k=1$ admits no solutions in $L^2(I\!\!E)$, i.e., $\widehat{(xp)}_0$ has deficiency indices $(0,0)$. According to Theorem 2.5.1(1) the operator

$$\widehat{xp}_0=-i\hbar\left(x\frac{d}{dx}+\frac{1}{2}\right)\quad\text{acting on}\quad C_0^\infty(I\!\!E) \tag{2.194}$$

is therefore essentially selfadjoint and possesses a unique selfadjoint extension.

Case 2 When $k\geq 2$ the situation becomes more complicated. The general solutions to Eq. (2.192) are

$$f^{(\pm i)}(x)=c\,x^{-k/2}\exp\left[\pm\frac{1}{(k-1)\hbar\,x^{k-1}}\right], \tag{2.195}$$

where c is a constant. These functions are ill-defined at $x=0$. Since $\widehat{(x^kp)}_0^\dagger$ vanishes at $x=0$ these solutions can be split up to produce two solutions for the eigenvalue i;

$$f_1^{(+i)}=\begin{cases}c\,x^{-k/2}\exp\left[\frac{1}{(k-1)\hbar\,x^{k-1}}\right] & \text{if }x\in I\!\!R^+\\ 0 & \text{if }x\notin I\!\!R^+\end{cases}, \tag{2.196}$$

$$f_2^{(+i)}=\begin{cases}0 & \text{if }x\notin I\!\!R^-\\ c\,x^{-k/2}\exp\left[\frac{1}{(k-1)\hbar\,x^{k-1}}\right] & \text{if }x\in I\!\!R^-\end{cases}, \tag{2.197}$$

and another two solutions for the eigenvalue $-i$;

$$f_1^{(-i)}=\begin{cases}c\,x^{-k/2}\exp\left[-\frac{1}{(k-1)\hbar\,x^{k-1}}\right] & \text{if }x\in I\!\!R^+\\ 0 & \text{if }x\notin I\!\!R^+\end{cases}, \tag{2.198}$$

$$f_2^{(-i)}=\begin{cases}0 & \text{if }x\notin I\!\!R^-\\ c\,x^{-k/2}\exp\left[-\frac{1}{(k-1)\hbar\,x^{k-1}}\right] & \text{if }x\in I\!\!R^-\end{cases}. \tag{2.199}$$

We can now identify two distinct cases:

2.5. EXTENSIONS OF SYMMETRIC OPERATORS 125

1. Suppose $k \geq 2$ is even. Then $k-1$ is odd and we have[67]

$$x^{-k/2} \exp\left[\frac{1}{(k-1)\hbar\, x^{k-1}}\right] \to \infty \quad \text{as} \quad x \to 0_+, \quad (2.200)$$

$$x^{-k/2} \exp\left[\frac{1}{(k-1)\hbar\, x^{k-1}}\right] \to 0 \quad \text{as} \quad x \to 0_-. \quad (2.201)$$

It follows that

$$f_1^{(+i)} \notin L^2(I\!\!E), \quad f_2^{(+i)} \in L^2(I\!\!E). \quad (2.202)$$

Similarly we have

$$f_1^{(-i)} \in L^2(I\!\!E), \quad f_2^{(-i)} \notin L^2(I\!\!E). \quad (2.203)$$

The corresponding deficiency indices are therefore (1,1). It follows from Theorem 2.5.1(1) that

$$\widehat{(x^k p)}_0 = -i\hbar \left(x^k \frac{d}{dx} + \frac{k}{2} x^{k-1} \right) \quad \text{acting on} \quad C_0^\infty(I\!\!E) \quad (2.204)$$

is not essentially selfadjoint, but it possesses a one-parameter family of selfadjoint extensions.

2. Suppose $k \geq 2$ is odd. Then $k-1$ is even and we have

$$f_1^{(+i)} \notin L^2(I\!\!E), \quad f_2^{(+i)} \notin L^2(I\!\!E), \quad (2.205)$$

and

$$f_1^{(-i)} \in L^2(I\!\!E), \quad f_2^{(-i)} \in L^2(I\!\!E). \quad (2.206)$$

The corresponding deficiency indices are therefore (0,2). It follows that $\widehat{(x^k p)}_0$ is not essentially selfadjoint. Note that $\widehat{(x^k p)}_0$ is not closed so that having a zero deficiency index does not imply that it is maximal symmetric. Instead it means that it is essentially strictly maximal symmetric and it possesses a unique strictly maximal symmetric extension.

E6 Finally let us investigate operators in the Hilbert space $L^2(I\!\!E^+)$. Here $I\!\!E^+$ corresponds to the half line first introduced E1.3.2(1) E8. In Cartesian coordinate x ranges only from 0 to ∞ in $I\!\!E^+$. Consider the operator

$$\widehat{p}_0(I\!\!E^+) = -i\hbar \frac{d}{dx} \quad \text{acting on} \quad C_0^\infty(I\!\!E^+) \quad (2.207)$$

---

[67] $x \to 0_+$ means limit from the right and $x \to 0_-$ means limit from the left.

in the Hilbert space $L^2(I\!\!E^+)$. This is again a symmetric operator. To find the deficiency indices we again formally obtain Eq. (2.162). This time $f^{(+i)}$ is square-integrable in $I\!\!E^+$ while $f^{(-i)}$ is not. This gives rise to deficiency indices $(1,0)$ and hence the operator has no selfadjoint extension. It has a unique strictly maximal symmetric extension, to be denoted by $\widehat{p}(I\!\!E^+)$, which can be shown to be defined by[68]

$$\widehat{p}(I\!\!E^+) = -i\hbar\frac{d}{dx} \qquad (2.208)$$

acting on

$$\mathcal{D}\Big(\widehat{p}(I\!\!E^+)\Big) = \Big\{\phi \in S_{2,1}(I\!\!E^+) : \phi(0) = 0\Big\}. \qquad (2.209)$$

Often we shall denote $\widehat{p}_0(I\!\!E^+)$ and $\widehat{p}(I\!\!E^+)$ by $\widehat{p}_0^+$ and $\widehat{p}^+$ respectively. The adjoint $(\widehat{p}^+)^\dagger$ of $\widehat{p}^+$ is defined by the same differential expression acting on the domain $S_{2,1}(I\!\!E^+)$.

We can see clearly that the same differential expression produces a selfadjoint operator $\widehat{p}$ in $L^2(I\!\!E)$ but only a strictly maximal symmetric operator $\widehat{p}^+$ in $L^2(I\!\!E^+)$. The difference between a selfadjoint and a strictly maximal symmetric operator can be seen in their formal eigenfunctions:

1. The operator $\widehat{p}$ possesses formal eigenvalues $p \in I\!\!R$ with corresponding formal eigenfunctions

$$\eta_p(x) = e^{ipx}, \quad x \in I\!\!E. \qquad (2.210)$$

These are formal eigenvalues and eigenfunctions in the sense that $\eta_p(x)$ are not square-integrable in $I\!\!E$, and are therefore not members of the Hilbert space $L^2(I\!\!E)$. However these functions, known as *plane waves*,[69] are very useful in many physical applications; they are complete and orthonormal in the Dirac δ-function sense, i.e.,

$$\int_{-\infty}^{\infty} \eta_{p'}^*(x)\eta_p(x)\,dx = 2\pi\hbar\,\delta(p-p'), \qquad (2.211)$$

$$\int_{-\infty}^{\infty} \eta_p^*(x')\eta_p(x)\,dp = 2\pi\hbar\,\delta(x-x'). \qquad (2.212)$$

We call p *generalized eigenvalues*, a concept which will be formally introduced later in §2.8.3C(1) C2, and η_p *generalized eigenfunctions* of $\widehat{p}$.

---

[68] Akhiezer and Glazman (1966) Vol. 1 §49.
[69] Functions of x and t of the form $\exp[i(px - \omega t)]$ are called *plane waves*. We also call functions of x of the form $\eta_p(x)$ plane waves.

2.5. EXTENSIONS OF SYMMETRIC OPERATORS

2. In contrast the operator $\hat{p}^+$ in $L^2(I\!\!E^+)$ does not even possess such a set of generalized eigenfunctions. Although the following functions

$$\eta_p^+(x) = e^{ipx}, \quad x \in I\!\!E^+ \tag{2.213}$$

formally satisfy the equation

$$-i\hbar \frac{d}{dx}\eta_p^+ = p\,\eta_p^+ \tag{2.214}$$

they do not satisfy the boundary condition at $x = 0$ for functions in $\mathcal{D}(\hat{p}^+)$.[70] Moreover, these functions are not orthonormal in the δ-function sense. Instead they satisfy:[71]

$$\int_0^\infty \eta_{p'}^{+*}(x)\eta_p^+(x)\,dx = 2\pi\hbar\left(\frac{1}{2}\delta(p-p') + \frac{1}{2\pi i\,(p-p')}\right). \tag{2.215}$$

Similarly we can introduce the Hilbert space $L^2(I\!\!E^-)$ of square-integrable functions on $I\!\!E^-$ which corresponds to the half line in which x ranges from $-\infty$ to 0. In $L^2(I\!\!E^-)$ we can introduce an operator

$$\hat{p}(I\!\!E^-) = -i\hbar d/dx \tag{2.216}$$

acting on

$$\mathcal{D}\big(\hat{p}(I\!\!E^-)\big) = \big\{\phi \in S_{2,1}(I\!\!E^-) : \phi(0) = 0\big\}. \tag{2.217}$$

This operator has deficiency indices $(0,1)$ and is hence strictly maximal symmetric. We shall often denote this operator by $\hat{p}^-$ for short.

E7 In analogy to §2.5.1E(1) E2 above let us consider the operator $\hat{K}_0(I\!\!E^+)$ defined by

$$\hat{K}_0(I\!\!E^+) = -\frac{\hbar^2}{2m}\frac{d^2}{dx^2} \quad \text{acting on} \quad C_0^\infty(I\!\!E^+) \tag{2.218}$$

in $L^2(I\!\!E^+)$. This operator is symmetric with deficiency indices $(1,1)$, and it is closable with closure $\bar{K}_0(I\!\!E^+)$ defined on the domain[72]

$$\mathcal{D}\big(\bar{K}_0(I\!\!E^+)\big) = \big\{\phi \in S_{2,2}(I\!\!E^+) : \phi'(0) = \phi(0) = 0\big\}, \tag{2.219}$$

where $\phi' = d\phi/dx$. According to Theorem 2.5.1(1) this operator admits a family of selfadjoint extensions parameterized by a real parameter λ. These

---
[70]The non-existence of generalized eigenfunctions satisfying boundary condition $\phi(0) = 0$ can have important physical consequences, as seen later in §7.7.
[71]Holevo (1982) p. 63.
[72]Weidmann (1980) p. 162.

128   CHAPTER 2. OPERATORS AND THEIR DIRECT INTEGRALS

extensions, to be denoted by $\widehat{K}_\lambda(I\!\!E^+)$, are known to be defined by [73]

$$\widehat{K}_\lambda(I\!\!E^+) = -\frac{\hbar^2}{2m}\frac{d^2}{dx^2} \qquad (2.220)$$

acting on

$$\mathcal{D}\left(\widehat{K}_\lambda(I\!\!E^+)\right) = \left\{\phi \in S_{2,2}(I\!\!E^+) : \phi'(0) + \lambda\phi(0) = 0\right\}, \qquad (2.221)$$

where $\lambda \in I\!\!R$. When $\lambda = \infty$ we have the selfadjoint extension $\widehat{K}_\infty(I\!\!E^+)$ defined by the same differential expression but on the domain

$$\mathcal{D}\left(\widehat{K}_\infty(I\!\!E^+)\right) = \left\{\phi \in S_{2,2}(I\!\!E^+) : \phi(0) = 0\right\}. \qquad (2.222)$$

The same is true for a similarly defined $\widehat{K}_0(I\!\!E^-)$ in $L^2(I\!\!E^-)$.

E8   Traditionally, strictly maximal symmetric operators are not employed in orthodox quantum mechanics to represent physical observables. We shall develop a generalized quantum mechanics where strictly maximal symmetric operators are utilized to represent physical observables in Chapter 4.

2.5.2   Von Neumann's formula for selfadjoint extensions

It is possible to construct all the selfadjoint extensions of a symmetric operator based on two formulae due to von Neumann. We shall state the formula directly relevant to our later applications.

Let $\mathcal{N}^{(+i)}$ and $\mathcal{N}^{(-i)}$ be the two deficiency subspaces of a closed symmetric operator $\widehat{A}_0$ defined on the domain $\mathcal{D}(\widehat{A}_0)$.

Theorem 2.5.2(1)   **A von Neumann formula on selfadjoint extensions**[74]

- An operator $\widehat{A}$ is a selfadjoint extension of $\widehat{A}_0$ if and only if there is a unitary operator $\widehat{\mathcal{U}}$ mapping $\mathcal{N}^{(+i)}$ onto $\mathcal{N}^{(-i)}$ such that

    1. the domain of $\widehat{A}$ is given by

$$\mathcal{D}(\widehat{A}) = \mathcal{D}(\widehat{A}_0) + \left\{\phi + \widehat{\mathcal{U}}\phi : \phi \in \mathcal{N}^{(+i)}\right\}, \qquad (2.223)$$

---

[73] Reed and Simon Vol. 2 (1975) pp. 144-145.
[74] Weidmann (19980) Theorem 8.12 on p. 238, Akhiezer and Glazman (1963) p. 98.

2.5. EXTENSIONS OF SYMMETRIC OPERATORS

2. and on $\mathcal{D}(\widehat{A})$, the operator $\widehat{A}$ is defined by

$$\widehat{A}(\phi_0 + \phi + \widehat{\mathcal{U}}\phi) = \widehat{A}^\dagger(\phi_0 + \phi + \widehat{\mathcal{U}}\phi) \qquad (2.224)$$
$$= \widehat{A}_0\phi_0 + i\phi - i\widehat{\mathcal{U}}\phi \qquad (2.225)$$

for every $\phi_0 \in \mathcal{D}(\widehat{A}_0)$ and $\phi \in \mathcal{N}^{(+i)}$.

Comments 2.5.2(1)   Deficiency spaces and unitary matrices

C1   The notation in Eq. (2.223) means that $\mathcal{D}(\widehat{A})$ is formed by linear combinations of vectors from $\mathcal{D}(\widehat{A}_0)$ and $\{\phi + \widehat{\mathcal{U}}\phi : \phi \in \mathcal{N}^{(+i)}\}$. Equation (2.225) follows from Eq. (2.224) since $\phi \in \mathcal{N}^{(+i)}$ is an eigenfunction of $\widehat{A}^\dagger$ for eigenvalue i while $\widehat{\mathcal{U}}\phi \in \mathcal{N}^{(-i)}$ is an eigenfunction of $\widehat{A}^\dagger$ for eigenvalue $-i$.

C2   For differential operators the deficiency subspaces $\mathcal{N}^{(+i)}$ and $\mathcal{N}^{(-i)}$ often have the same finite dimensions. It follows that the unitary mapping is specifiable by a finite-dimensional square unitary matrix.

Examples 2.5.2(1)   Selfadjoint extension of $\widehat{p}_0(\Lambda)$

E1   A simple situation is when $\mathcal{N}^{(+i)}$ and $\mathcal{N}^{(-i)}$ are both one-dimensional. This is the case for $\widehat{p}_0(\Lambda)$ defined on $C_0^\infty(\Lambda)$ in $L^2(\Lambda)$.[75] Let us see how we can arrive at a one-parameter family of selfadjoint extensions and how the quasi-periodic boundary condition in Eq. (2.166) comes about. By Theorem 2.5.2(1) a selfadjoint extension $\widehat{p}(\Lambda)$ of $\widehat{p}_0(\Lambda)$ has a domain of the form

$$\mathcal{D}(\widehat{p}(\Lambda)) = C_0^\infty(\Lambda) + \left\{\phi + \widehat{\mathcal{U}}\phi : \phi \in \mathcal{N}^{(+i)}\right\}. \qquad (2.226)$$

The deficiency subspaces $\mathcal{N}^{(\pm i)}$ are one-dimensional and spanned respectively by $f^{(\pm i)}$ of Eq. (2.161). So we have

$$\phi = \alpha f^{(+i)}, \quad \widehat{\mathcal{U}} f^{(+i)} = \beta f^{(-i)}, \quad \alpha, \beta \in \mathbb{C}. \qquad (2.227)$$

Since $\widehat{\mathcal{U}}$ is unitary and hence norm preserving, the constant β must have an absolute value of 1, i.e.,

$$\beta = e^{i\theta}, \quad \theta \in \mathbb{R}. \qquad (2.228)$$

So, we have

$$\mathcal{D}(\widehat{p}(\Lambda)) = \mathcal{D}(\widehat{p}_0) + \left\{\alpha f^{(+i)} + \alpha e^{i\theta} f^{(-i)}\right\}. \qquad (2.229)$$

---

[75] See §2.5.1E(1) E1.

From Eq. (2.164) we get

$$\alpha f^{(+i)}(x) + \alpha e^{i\theta} f^{(-i)}(x) = \alpha c_- \left(e^{a/\hbar} e^{-x/\hbar} + e^{i\theta} e^{x/\hbar} \right). \qquad (2.230)$$

Since functions ϕ_0 in $C_0^\infty(\Lambda)$ vanish at the boundaries of the interval Λ, i.e., $\phi_0(0) = \phi_0(a) = 0$, we have, for every $\Phi(x) \in \mathcal{D}(\widehat{p}(\Lambda))$, the following equation:

$$\Phi(0) = \alpha c_- \left(e^{a/\hbar} + e^{i\theta} \right) \qquad (2.231)$$

$$\Phi(a) = \alpha c_- \left(1 + e^{i\theta} e^{a/\hbar} \right) \qquad (2.232)$$

$$\Rightarrow \quad \frac{\Phi(0)}{\Phi(a)} = \frac{e^{a/\hbar} + e^{i\theta}}{1 + e^{i\theta} e^{a/\hbar}} = e^{i\theta} \left(\frac{1 + e^{-i\theta} e^{a/\hbar}}{1 + e^{i\theta} e^{a/\hbar}} \right) \qquad (2.233)$$

$$\Rightarrow \quad \left| \frac{\Phi(0)}{\Phi(a)} \right| = 1. \qquad (2.234)$$

It follows that we can express this result as

$$\Phi(0) = e^{i\lambda} \Phi(a), \quad \lambda \in \mathbb{R}. \qquad (2.235)$$

This is then the desired quasi-periodic boundary condition. Each value of λ specifies a selfadjoint extension, denoted by $\widehat{p}_\lambda(\Lambda)$ or simply $\widehat{p}_\lambda$, and as λ varies we obtain a one-parameter family of selfadjoint extensions.

E2 In Chapter 6 we shall see more explicit calculations to obtain selfadjoint extensions, especially when the deficiency subspaces are not one-dimensional and where a unitary matrix is required to specify the unitary mapping $\widehat{\mathcal{U}}$.

2.6 Probability and Expectation Values

Selfadjoint operators feature prominently in quantum mechanics because they possess real eigenvalues and they generate probability distributions. Before going into this let us recall some basic definitions of standard probability theory. This is written in such a way as to facilitate a parallel presentation of how symmetric operators, not just selfadjoint operators, in a Hilbert space can be related to probability. Note that we are interested in a mathematical formulation of a classical probability theory here rather than delving into the physical origins of probability.

Traditionally the notion of probability arises out of so-called *statistical experiments*.[76] A statistical experiment has many outcomes; the nature of a

---
[76]Penrose (1970) Chapter I §4. Probability can also arise from our ignorance or lack of complete knowledge of an otherwise deterministic physical system, e.g., a box of gas treated by classical kinetic theory and more generally by classical statistical mechanics.

2.6. PROBABILITY AND EXPECTATION VALUES

statitical experiment is such that we cannot predict exactly which outcome will occur. The experiment would lead to an outcome in an indeterministic manner, i.e., under identical conditions a repetition of the experiment may very well give rise to a different outcome. A collection of outcomes is called an *event*; for technical reasons we would also regard the empty set and the set of all outcomes as events. Fortunately many of these experiments also exhibit a certain regularity so that the results of many independent repetitions of such an experiment under identical conditions are describable in terms of a probability theory. A typical situation may involve a physical system with an observable A which possesses a discrete set of numerical values a_j such that

1. each individual experiment to find the value of A yields a value a_j in an unpredictable fashion, and

2. if the experiment is carried out n times under identical conditions and if the value a_j is obtained n_j times, then the ratio (n_j/n) appears to converge as n becomes larger and larger.

The regularity here is the convergence of (n_j/n) as n becomes larger and larger. This enables us to introduce the quantity

$$\wp_N(A, a_j) = \frac{n_j}{N}, \qquad (2.236)$$

for some large N, large enough for the convergence to become apparent. This quantity is then identified with the *probability* that the measured value of A is a_j. The average value $\mathcal{E}(A)$ is then given by

$$\mathcal{E}(A) = \frac{1}{N} \sum_j n_j\, a_j = \sum_j \wp_N(A, a_j)\, a_j. \qquad (2.237)$$

Ideally we would like N to tend to infinity. Of course in any practical experiment we can only achieve a large but finite N. This empirical understanding of probability is known as the *frequency interpretation* of probability. Although we shall adopt this frequency interpretation of probability we should also point out that there are many other interpretations of the notion of probability.[77] While one may proceed to establish a theory of probability based on this frequency interpretation,[78] it is mathematically more consistent to formulate a probability theory axiomatically. This we shall do in the following section.

---

[77] Penrose (1970) Chapter I §4. An example would be the statement saying there is a 50% probability of raining tomorrow afternoon. Clearly we cannot have a large ensemble of replicas of "tomorrow" to test the statement in the sense of a frequency interpretation. The statement is really a reflection of a subjective belief based on experience.

[78] Penrose (1970) Chapters I and II. It may be argued that a probability theory based on a frequency interpretation is not logically satisfactory since the definition of a probability cannot even be made precise on account of our failure to achieve the limiting value of $\wp_\infty(A, a_j)$ in a practical experiment.

132   CHAPTER 2.   OPERATORS AND THEIR DIRECT INTEGRALS

Since every outcome lies in some event an assignment of probability to all events will characterize the statistical nature of the experiment. For an experiment with only a finite number of outcomes, an equivalent characterization would be to assign a probability to each outcome. This seemingly simpler approach is not possible when the outcomes form a continuum. For many applications the outcomes are quantifiable by real numbers and the set of all outcomes identifiable with the set of real numbers $I\!R$. In other words each real number corresponds to an outcome. Then we would have difficulty assigning a probability to each individual outcome. In fact the probability of any individual outcome quantified by a single number would have to be zero, so that the total probability would not add up to infinity. So, the general approach should be to assign probability directly to events, not to individual outcomes.

Let us consider statistical experiments whose outcomes are quantified by the set of real numbers $I\!R$. We would have an enormous number of events if we were to regard every subset of $I\!R$ as an event. Since one expects that a run of the experiment will normally give a value in a certain interval there is no need to include all subsets of $I\!R$. It is sufficient to identify a structured family of subsets large enough to include all intervals and individual numbers. The commonly adopted family consists of *Borel sets* of the reals. Borel sets, which include all open and closed intervals, individual numbers as well the empty set and $I\!R$, are large enough to be able to quantify our experimental outcomes.

2.6.1   Borel sets, measures and measurable functions

Definition 2.6.1(1)[79] **Borel sets, measures and measurable functions**

- *Borel sets* of the reals, denoted by $I\!B(I\!R)$ or simply $I\!B$, comprise the smallest family of subsets of $I\!R$, which is closed under complements and countable unions, and contains all open intervals.

- A *measure* $\mathcal{M}$ is a set function[80] from the Borel sets to the extended reals $I\!R_{ex}$[81]

$$\mathcal{M} : I\!B \mapsto I\!R_{ex} \quad \text{by} \quad \Lambda \in I\!B \mapsto \mathcal{M}(\Lambda) \in I\!R_{ex} \qquad (2.238)$$

such that

---

[79] Reed and Simon Vol. 1 (1972) pp. 14-16.

[80] Pitt (1963) p. 4. A set function assigns a value to a set, as opposed to usual functions which assign a value to a point.

[81] Pitt (1963) p. 4, Prugovečki (1981) p. 67. The extended reals $I\!R_{ex}$ consists of the reals $I\!R = (-\infty, \infty)$ together with $-\infty$ and ∞, i.e., we have

$$I\!R_{ex} = -\infty \cup I\!R \cup \infty = [-\infty, \infty].$$

As a result we can have $\mathcal{M}(\Lambda) = \infty$ for some Λ.

2.6. PROBABILITY AND EXPECTATION VALUES

1. $\mathcal{M}(\emptyset) = 0$, where $\emptyset$ is the empty set,
2. $\mathcal{M}(\Lambda) \geq 0$, and
3. $\mathcal{M}(\Lambda_1 \cup \Lambda_2 \cup \cdots) = \mathcal{M}(\Lambda_1) + \mathcal{M}(\Lambda_2) + \cdots$, if $\Lambda_1, \Lambda_2, \ldots$ is any sequence of mutually disjoint Borel sets.

- A real-valued function f on $I\!R$ is said to be a *Borel function* or *Borel measurable* if the inverse image $f^{-1}(\Lambda)$ of every Borel set Λ defined by

$$f^{-1}(\Lambda) = \{\tau \in I\!R : f(\tau) \in \Lambda\} \tag{2.239}$$

is a Borel set. A complex-valued function is a Borel function if both its real and imaginary parts are Borel functions.

Comments 2.6.1(1) **Set of measure zero, Lebesgue-Stieltjes measures and integrals**

C1 All the subsets of $I\!R$ we shall encounter in this book, e.g., open intervals, semi-open intervals, closed intervals and sets containing a single number, are Borel sets. Borel sets are so pervasive that it is not an easy matter to write down a subset of real numbers which is not a Borel set.

C2 Intuitively the positive and the additive nature of a measure suggests that it is an assignment of a numerical value to some property associated with Borel sets. For example, we could have an assignment showing the size of Borel sets by requiring a measure to satisfy:

1. $\mathcal{M}\big((\tau_1, \tau_2)\big) = \tau_2 - \tau_1$ for any open interval (τ_1, τ_2),
2. $\mathcal{M}(\{\tau\}) = 0$ for any single point $\tau \in I\!R$.

It can be shown that these requirements define a unique measure, known as the *Lebesgue measure* on $I\!R$ used in Lebesgue integrals.[82] In other words there is a unique extension of the assignment of numerical values to all Borel sets. This Lebesgue measure is denoted simply by $d\tau$. In the Hilbert space $L^2(I\!E)$ we use just this measure, denoted by dx, to define the integrals for the scalar product.

C3 A set $\Lambda \in I\!B$ is called a *set of measure zero* if $\mathcal{M}(\Lambda) = 0$. A single point in $I\!R$ is an example of a *set of Lebesgue measure zero*. A property or equation which holds except on a set of measure zero is said to hold *almost everywhere*. The following two examples serve to illustrate this concept:

---
[82] Williamson (1962), Burrill (1972).

1. An absolutely continuous function on $I\!\!E$ is differentiable *almost everywhere*.

2. In $L^2(I\!\!E)$ a function $\varphi(x)$ of zero norm may not vanish everywhere, e.g.,
$$\int_{-\infty}^{\infty} |\varphi(x)|^2\, dx = 0 \;\not\Rightarrow\; \varphi(x) = 0 \;\forall x \in I\!\!R. \tag{2.240}$$

It would be sufficient for φ to vanish almost everywhere since the values of $\varphi(x)$ on a set of measure zero, i.e., some isolated points, does not contribute to the above integral. A function $\phi(x)$ in $L^2(I\!\!E)$ is therefore defined only up to a set of measure zero since we can add a function $\varphi(x)$ of zero norm to $\phi(x)$ without changing it.[83]

C4 Borel functions play an important role in the theory of Lebesgue integration since they are related to the integrability of functions. Borel functions are so pervasive that it is difficult to write down a function which is not Borel measurable. In practice all the functions we will encounter are Borel functions.

C5 A useful generalization of the Lebesgue measure is the *Lebesgue-Stieltjes* measure. Let g be a real-valued function on $I\!\!R$ which is right continuous and non-decreasing, i.e.,
$$g(x+0) = g(x) \text{ and } g(x_2) \geq g(x_1) \text{ if } x_2 > x_1. \tag{2.241}$$

We can define a measure $\mathcal{M}$ on the Borel sets by its values on any semi-open interval $(\tau_1, \tau_2]$ in terms of g, i.e.,
$$\mathcal{M}\big((\tau_1, \tau_2]\big) = g(\tau_2) - g(\tau_1), \tag{2.242}$$

in that there is a unique extension of this set function to all Borel sets of $I\!\!R$. The resulting measure is referred to as the **Lebesgue-Stieltjes** measure generated by the function g. Integrals with respect to such a measure are known as **Lebesgue-Stieltjes integrals** or simply *Stieltjes integrals*.[84] An integral of the form
$$\int_a^b f(\tau)\, d\tau, \tag{2.243}$$

is with respect to a continuous and independent variable τ. Stieltjes integrals are with respect to a function $g(\tau)$, not necessarily continuous. It is written as
$$\int_a^b f(\tau)\, d\mathcal{M}(\tau) \quad \text{or more explicitly} \quad \int_a^b f(\tau)\, dg(\tau), \tag{2.244}$$

---

[83] In a Hilbert space an element of zero norm is deemed to be the same as the zero element.
[84] Smirnov Vol. 5 (1964) §1.

2.6. PROBABILITY AND EXPECTATION VALUES

and is defined by

$$\int_a^b f(\tau)\,dg(\tau) = \lim_{\Delta\tau_j \to 0} \sum_{j=1}^n f(\tau'_j)\big(g(\tau_j) - g(\tau_{j-1})\big), \quad \tau'_j \in (\tau_{j-1}, \tau_j], \quad (2.245)$$

where

1. the points $a = \tau_0 < \tau_1 < \tau_2 < \cdots < \tau_n = b$ represent a partition of the interval $(a, b]$ with $\tau_0 = a$ and $\tau_n = b$, and

2. the limit is taken as every interval $(\tau_{j-1}, \tau_j]$ becomes arbitrarily small and n tends to infinity.

This new integral offers far more flexibility, hence is applicable to many bizarre situations, continuous or otherwise. It is especially useful in dealing with discontinuities, hence rendering it possible to have a more uniform treatment of both continuous and discrete cases. Let us consider a few examples of such integrals:

1. Let $g(\tau)$ be a differentiable function, then

$$\int_a^b f(\tau)\,dg(\tau) = \int_a^b f(\tau) \frac{dg(\tau)}{d\tau}\,d\tau, \qquad (2.246)$$

and the Stieltjes integral is reduced to an integral with respect to the variable τ.

2. Let $g(\tau)$ be a step function given by

$$g(\tau) = \begin{cases} g_0, & \text{if } \tau < \tau_1 \\ g_1, & \text{if } \tau_1 \leq \tau, \end{cases} \qquad (2.247)$$

where $g_0 < g_1$ are real constants. Then we have, for any given continuous function $f(\tau)$ and any interval (a, b) containing τ_1,

$$\int_a^b f(\tau)\,dg(\tau) = \lim_{\Delta\tau_j \to 0} \sum_{j=1}^n f(\tau'_j)\big(g(\tau_j) - g(\tau_{j-1})\big) \qquad (2.248)$$
$$= f(\tau_1)(g_1 - g_0),$$

since $g(\tau_j) - g(\tau_{j-1}) = 0$ for every subinterval $(\tau_{j-1}, \tau_j]$ except for the one containing τ_1.

C6 There is a simple link between the above Stieltjes integrals and the Dirac delta function. Consider a piecewise continuous function $g(\tau)$ which is discontinuous at some isolated point $\tau = \tau_1$. We can formally write down an expression for its derivative as follows:[85]

$$w(\tau) = \frac{dg(\tau)}{d\tau} = \Delta g(\tau_1)\, \delta(\tau - \tau_1) + \bar{g}'(\tau), \qquad (2.249)$$

where

1. $\delta(\tau - \tau_1)$ is the Dirac delta function centred at $\tau = \tau_1$,
2. $\Delta g(\tau_1) = g(\tau_1 + 0) - g(\tau_1 - 0)$,[86]
3. $\bar{g}'(\tau) = dg(\tau)/d\tau$ for $\tau \neq \tau_1$.[87]

Applying this prescription to $g(\tau)$ in Eq. (2.247) we get

$$w(\tau) = \frac{dg(\tau)}{d\tau} = (g_1 - g_0)\delta(\tau - \tau_1). \qquad (2.250)$$

This enables us to rewrite the integral in Eq. (2.248) in terms of the Dirac delta function as[88]

$$\int_a^b f(\tau)\, dg(\tau) = \int_a^b f(\tau)\, \frac{dg}{d\tau}\, d\tau \qquad (2.251)$$

$$= \int_a^b f(\tau)\,(g_1 - g_0)\delta(\tau - \tau_1)\, d\tau$$

$$= f(\tau_1)(g_1 - g_0). \qquad (2.252)$$

This result can be extended to the case where $g(\tau)$ is piecewise constant with many discontinuities. A simple example is given by the following *staircase function* $g(\tau)$:

$$g(\tau) = \begin{cases} g_0, & \tau < \tau_1 \\ g_1, & \tau_1 \leq \tau < \tau_2 \\ g_2, & \tau_2 \leq \tau < \tau_3 \\ g_3, & \tau_3 \leq \tau < \tau_4 \\ \vdots \end{cases}. \qquad (2.253)$$

---

[85] Friedman (1956) pp. 141-142.

[86] The notation $g(\tau_1 + 0)$ denotes the value of $g(\tau)$ as $\tau \to \tau_1$ from the right. The limit value as $\tau \to \tau_1$ from the left is denoted by $g(\tau_1 - 0)$.

[87] The value of $\bar{g}'(\tau)$ at τ_1 does not matter since this single value does not contribute to an integral with $w(\tau)$ as an integrand.

[88] Recall that

$$\int_a^b F(\tau)\delta(\tau - \tau_1)\, d\tau = F(\tau_1), \quad \tau_1 \in (a, b).$$

The δ-function picks out the value of the integrand F at the singular point $\tau = \tau_1$ of the δ-function.

2.6. PROBABILITY AND EXPECTATION VALUES

Then we have

$$\int_{-\infty}^{\infty} f(\tau)\, dg(\tau) = \sum_{j=1}^{\infty} f(\tau_j)(g_j - g_{j-1}) \qquad (2.254)$$

$$= f(\tau_1)(g_1 - g_0) + f(\tau_2)(g_2 - g_1) + \cdots. \qquad (2.255)$$

The integration process is seen to be equivalent to

1. picking up the value of the integrand $f(\tau_j)$ and the corresponding increment $\big(g(\tau_j) - g(\tau_{j-1})\big)$ at the discontinuity of $g(\tau)$ at τ_i,

2. multiplying them together to form the product $f(\tau_j)\big(g(\tau_j) - g(\tau_{j-1})\big)$, and finally

3. adding the resulting products $f(\tau_j)\big(g(\tau_j) - g(\tau_{j-1})\big)$ at all the discontinuities of $g(\tau)$.

C7 In most practical cases the condition $\mu(\emptyset) = 0$ in Definition 2.6.1(1) is redundant. Given any set Λ we have

$$\Lambda \cap \emptyset = \Lambda, \quad \Lambda \cup \emptyset = \emptyset, \text{ i.e., } \Lambda \text{ and } \emptyset \text{ are disjoint} \qquad (2.256)$$

so that

$$\mu(\Lambda) = \mu\big(\Lambda \cap \emptyset\big) = \mu(\Lambda) + \mu(\emptyset) \quad \Rightarrow \quad \mu(\emptyset) = 0, \qquad (2.257)$$

provided $\mu(\Lambda) \neq \pm\infty$.

C8 A useful expression is[89]

$$\mu(\Lambda_1 \cup \Lambda_2) = \mu(\Lambda_1) + \mu(\Lambda_2) - \mu(\Lambda_1 \cap \Lambda_2). \qquad (2.258)$$

Suppose $\mu(\Lambda_1) = \infty$ for some Λ_1. Then we cannot have $\mu(\Lambda_2) = -\infty$ for any Λ_2 within the same measure. If this were permitted then an attempt to evaluate the measure of the set $\Lambda_1 \cup \Lambda_2$ using the above expression would involve the addition of ∞ and $-\infty$ which is not defined.[90]

2.6.2 Probability measures and probability functions

Definition 2.6.2(1) Probability measures, probability distribution and density functions

- A measure $\mathcal{M}$ is called a *probability measure* if $\mathcal{M}(\mathbb{R}) = 1$.

---
[89]Lipschutz (1974) p. 41, Roman Vol. 1 (1975) Problems 7.2a-1 on p. 313.
[90]Pitt (1963) p. 4.

- A real-value function defined on $\mathbb{R}$ and having values in the interval $[0,1]$, i.e.,
$$\mathcal{F}: \mathbb{R} \to [0,1], \qquad (2.259)$$
is called a *probability distribution function*, or a *probability function* for short, on $\mathbb{R}$ if
  1. $\mathcal{F}(-\infty) = 0, \quad \mathcal{F}(\infty) = 1$;
  2. $\mathcal{F}(\tau + 0) = \mathcal{F}(\tau) \quad \forall \tau \in \mathbb{R}$;
  3. $\mathcal{F}(\tau_1) \leq \mathcal{F}(\tau_2)$ if $\tau_1 \leq \tau_2$.

- If a probability function $\mathcal{F}$ is differentiable, then its derivative $w = d\mathcal{F}/d\tau$ is called a *probability density function*.

Comments 2.6.2(1) Properties of probability measures and probability functions

C1 A probability measure $\mathcal{M}(\Lambda)$ takes a value in the interval $[0,1]$ for every Borel set Λ. Following §2.6.1C(1) C7 we have $\mathcal{M}(\emptyset) = 0$.

C2 A probability function is right continuous and non-decreasing.

C3 There is a one-to-one correspondence between probability functions and probability measures given by

$$\mathcal{F}(\tau) = \mathcal{M}\big((-\infty, \tau]\big), \qquad \mathcal{M}(\Lambda) = \int_\Lambda d\mathcal{F}(\tau). \qquad (2.260)$$

Clearly we have

$$\mathcal{M}(\mathbb{R}) = \int_\mathbb{R} d\mathcal{F}(\tau) = 1. \qquad (2.261)$$

When the probability density function w exists we have

$$\mathcal{F}(\tau) = \int_{-\infty}^\tau w(\tau)\, d\tau, \quad w(\tau) = \frac{d\mathcal{F}(\tau)}{d\tau}. \qquad (2.262)$$

The measure of a singleton set $\{\tau_0\}$ containing a single point τ_0 is given by

$$\mathcal{M}(\{\tau_0\}) = \mathcal{F}(\tau_0) - \lim_{\varepsilon \to 0} \mathcal{F}(\tau_0 - \varepsilon), \qquad (2.263)$$

which vanishes if $\mathcal{F}$ is also continuous from the left at τ_0.

C4 Probability measures, probability functions, and their corresponding probability density functions if they exist, are alternative and equivalent descriptions of the probabilistic nature of a statistical experiment with the following interpretations:

2.6. PROBABILITY AND EXPECTATION VALUES

1. A probability measure $\mathcal{M}$ maps each Borel set Λ to a number $\mathcal{M}(\Lambda) \in [0,1]$. This number is identified with the probability of occurrence of an event quantified by Λ. The probability of occurrence of an event corresponding to a singleton set $\{\tau_0\}$ is given by $\mathcal{M}(\{\tau_0\})$.

2. The value $\mathcal{F}(\tau)$ of a probability function at τ can be interpreted as the probability of an event quantified by the Borel set $(-\infty, \tau]$ on account of Eq. (2.260).

C5 The definition of Borel sets applies to subsets of the Euclidean space $\mathbb{E}$ since it is just $\mathbb{R}$ endowed with a real vector space structure, i.e., we can talk about Borel sets of $\mathbb{E}$ and probability functions defined on $\mathbb{E}$.

C6 Three types of probability functions are of particular interest:

1. A probability function $\mathcal{F}_r(\tau)$ is said to be *absolutely continuous* or *regular* if $\mathcal{F}_r(\tau)$ is an absolutely continuous function of τ over the entire range of values of τ. According to Eq. (2.263) any event corresponding to a singleton set $\{\tau_0\}$ has a zero probability of occurence.

2. A probability function $\mathcal{F}_s(\tau)$ is *purely discrete* or *singular* if it is a piecewise constant function of τ, i.e., if there is an increasing sequence of real numbers τ_j, $j = 1, 2, \ldots$, such that[91]

$$\mathcal{F}_s(\tau) = \begin{cases} 0 & \text{if } \tau < \tau_1 \\ \wp_1 & \text{if } \tau_1 \leq \tau < \tau_2 \\ \wp_1 + \wp_2 & \text{if } \tau_2 \leq \tau < \tau_3 \\ \cdots & \cdots \\ \sum_{j=1}^{k} \wp_j & \text{if } \tau_k \leq \tau < \tau_{k+1} \\ \cdots & \cdots \end{cases}, \quad (2.264)$$

where

$$\wp_j \geq 0 \quad \text{and} \quad \wp_1 + \wp_2 + \cdots = 1. \quad (2.265)$$

One can visualize $\mathcal{F}_s$ as a staircase function with discontinuous steps occurring at $\tau = \tau_j$ with height of the jth step equal to

$$\mathcal{F}_s(\tau_j) - \lim_{\varepsilon \to 0} \mathcal{F}_s(\tau_j - \varepsilon) = \wp_j. \quad (2.266)$$

It follows from Eq. (2.263) that an event corresponding to the singleton set $\{\tau_j\}$ will have a nonvanishing probability of occurrence given by

$$\mathcal{M}(\{\tau_j\}) = \wp_j. \quad (2.267)$$

---

[91] Lipschutz (1974) p. 86 for a graphical depiction.

So, $\wp_j$, the step height at $\tau = \tau_j$ of a staircase probability function $\mathcal{F}_s$ represents the probability of occurrence of an event corresponding to the singleton set $\{\tau_j\}$. Purely discrete probability functions apply to statistical experiments with a purely discrete set of outcomes, e.g., an experiment which admits only a finite number of outcomes quantified by $\tau_1, \tau_2, \ldots, \tau_n$ with a corresponding set of probabilities $\wp_1, \wp_2, \ldots, \wp_n$.

3. A probability function $\mathcal{F}_b(\tau)$ is said to be *blended* if it is absolutely continuous over certain range of values of τ and it is discrete over the remaining range of values of τ.

C7 We can introduce three types of probability density functions to correspond the above probability functions:

1. A probability density function is said to be *regular* if it is the derivative of a regular, i.e., an absolutely continuous, probability function. Any real-valued Borel function $w(\tau)$ on $I\!R$ can serve as a regular probability density function if it satisfies

$$w_r(\tau) \geq 0, \quad \int_{-\infty}^{\infty} w_r(\tau)\, d\tau = 1. \qquad (2.268)$$

2. A probability density function is said to be *singular* if it is the derivative of a singular, i.e., a purely discrete, probability function. Here the differentiability of a discrete probability function is taken in the Dirac δ-function sense as detailed in §2.6.1C(1) C6. So, we can write down the probability density function corresponding to the probability function (2.264) as

$$w_s(\tau) = \frac{d\mathcal{F}_s(\tau)}{d\tau} = \sum_j \wp_j\, \delta(\tau - \tau_j). \qquad (2.269)$$

We can integrate $w_s(\tau)$ to recover the probability function, i.e.,

$$\mathcal{F}_s(\tau) = \int_{-\infty}^{\tau} w_s(\tau)\, d\tau. \qquad (2.270)$$

3. A probability density function is said to be *blended* if it is the derivative of a blended probability function.

2.6.3 Expectation values, variances and uncertainties

Associated with a given probability function $\mathcal{F}$ we can introduce the concepts of expectation value, variance and uncertainty.

2.6. PROBABILITY AND EXPECTATION VALUES

Definition 2.6.3(1)   **Expectation value, variance and uncertainty**

- The *expectation value* $\mathcal{E}(\mathcal{F})$ is defined by

$$\mathcal{E}(\mathcal{F}) = \int_{-\infty}^{\infty} \tau \, d\mathcal{F}(\tau), \tag{2.271}$$

if the above integral with respect to the Lebesgue-Stieltjes measure $d\mathcal{F}(\tau)$ exists.

- The *variance* $\mathcal{V}(\mathcal{F})$ is defined by

$$\mathcal{V}(\mathcal{F}) = \int_{-\infty}^{\infty} \left(\tau - \mathcal{E}(\mathcal{F})\right)^2 d\mathcal{F}(\tau) \tag{2.272}$$

if the integral exists.

- The *uncertainty* $\Delta(\mathcal{F})$ is defined to be the square root of the variance, i.e.,

$$\Delta(\mathcal{F}) = \sqrt{\mathcal{V}(\mathcal{F})}. \tag{2.273}$$

Comments 2.6.3(1)   **Expectation values and average values**

C1   For computational purposes the following expression is useful:

$$\mathcal{V}(\mathcal{F}) = \int_{-\infty}^{\infty} \tau^2 \, d\mathcal{F}(\tau) - \mathcal{E}(\mathcal{F})^2. \tag{2.274}$$

In terms of the probability density function $w(\tau)$ the integrals for the expectation value and for the variance become

$$\mathcal{E}(\mathcal{F}) = \int_{-\infty}^{\infty} \tau \, w(\tau) \, d\tau. \tag{2.275}$$

$$\mathcal{V}(\mathcal{F}) = \int_{-\infty}^{\infty} \left(\tau - \mathcal{E}(\mathcal{F})\right)^2 w(\tau) \, d\tau \tag{2.276}$$

$$= \int_{-\infty}^{\infty} \tau^2 \, w(\tau) \, d\tau - \mathcal{E}(\mathcal{F})^2. \tag{2.277}$$

For the purely discrete probability function $\mathcal{F}_s$ given by Eq. (2.264) with a corresponding probability density function (2.269), the above integrals become

$$\mathcal{E}(\mathcal{F}_s) = \sum_j \wp_j \, \tau_j. \tag{2.278}$$

$$\mathcal{V}(\mathcal{F}_s) = \sum_j \wp_j \left(\tau_j - \mathcal{E}(\mathcal{F}_s)\right)^2 \tag{2.279}$$

$$= \sum_j \wp_j \, \tau_j^2 - \mathcal{E}(\mathcal{F}_s)^2. \tag{2.280}$$

In view of the frequency interpretation of probability these expressions clearly show that:

1. The expectation value is interpretable as the *mean*, or the *average value*, of experimental results in a statistical experiment.

2. The variance being the average of the square of the deviation from the mean, showing us how spread out the experimental results are from the average value. In practical calculations we can regard the variance as the mean of the square minus the square of the mean. If the mean happens to vanish then the variance is simply equal to the mean of the square.

For clarity it is desirable to call $\mathcal{E}(\mathcal{F})$ the expectation or expected value, reserving the terms the *mean*, or the *average value* for the empirical value in Eq. (2.237). Of course this empirical value should approach the expectation value as N becomes larger and larger.

C2 For an arbitrary probability function the expectation value may not exist, and if it exists the variance integral may not exist. For physical applications we would restrict ourselves to probability distribution functions with finite expectation values and finite variances. We shall return to this important restriction in our discussion for a generalization of orthodox quantum mechanics in §5.2.1 in Chapter 5 later.

C3 In the next four sections we shall detail how probability can be related to operators in Hilbert spaces, and in particular how selfadjoint and strictly maximal symmetric operators in a Hilbert space can generate probability measures and probability functions with finite expectation values and finite variances.

2.7 Spectral Measures and Probability

Probability measures and probability functions can be related to operators in a Hilbert space. This is done through operator-valued measures, known as *spectral measures*, and operator-valued functions known as *spectral functions*.

Definition 2.7(1)   **Spectral measures and spectral functions**[92]

- A *spectral measure* on a Hilbert space $\mathcal{H}$ is a *normalized projector-valued set function* $\widehat{M}$ on the Borel sets $\mathbb{B}$ of $\mathbb{R}$, namely

$$\widehat{M} : \mathbb{B} \mapsto \left\{ \text{projectors on } \mathcal{H} \right\} \quad (2.281)$$

---

[92] Prugovečki (1981) §5.4 on p. 231, §5.5 on p. 235, Akhiezer and Glazman Vol. 2 (1963) §61, Roman Vol. 2 (1975) §13.4b p. 633.

2.7. SPECTRAL MEASURES AND PROBABILITY

by
$$\Lambda \in \mathbb{B} \mapsto \widehat{M}(\Lambda) \in \left\{ \text{projectors on } \mathcal{H} \right\}, \tag{2.282}$$

such that

1. $\widehat{M}(\mathbb{R}) = \widehat{\mathbb{I}}$, the identity operator on $\mathcal{H}$,
2. $\widehat{M}(\Lambda_1 \cup \Lambda_2 \cup \cdots) = \widehat{M}(\Lambda_1) + \widehat{M}(\Lambda_2) + \cdots$, if $\Lambda_1, \Lambda_2, \ldots$ is any sequence of mutually disjoint Borel sets.

- A *spectral function* on a Hilbert space $\mathcal{H}$ is a *normalized projector-valued function* $\widehat{F}$ on $\mathbb{R}$, namely

$$\widehat{F} : \mathbb{R} \mapsto \left\{ \text{projectors on } \mathcal{H} \right\} \tag{2.283}$$

by
$$\tau \in \mathbb{R} \mapsto \widehat{F}(\tau) \in \left\{ \text{projectors on } \mathcal{H} \right\}, \tag{2.284}$$

such that

1. $\widehat{F}(-\infty) = \widehat{O}$, the zero operator on $\mathcal{H}$, and $\widehat{F}(\infty) = \widehat{\mathbb{I}}$,
2. $\widehat{F}(\tau + 0) = \widehat{F}(\tau) \ \forall \tau \in \mathbb{R}$,
3. $\widehat{F}(\tau_1) \leq \widehat{F}(\tau_2)$ if $\tau_1 \leq \tau_2$.

Comments 2.7(1)   PV measures, PV functions and probability

C1   Spectral measures are also known as *projector-valued measures* or *PV measures* or simply *PVMs* for short. Spectral functions are also referred to as *PV functions* for short. The order relation $\widehat{F}(\tau_1) \leq \widehat{F}(\tau_2)$ above is defined by Eq. (2.95) in §2.3C(1) C11. Following Eqs. (2.96) and (2.97) we deduce that $\widehat{F}(\tau_2) - \widehat{F}(\tau_1)$ is a projector, and hence positive and of norm 1.

Note that all limiting processes in Definition 2.7(1), e.g., the limit from the right $\widehat{F}(\tau + 0)$ and an infinite sum involved in a spectral measure, are meant to be operator limits in the strong sense.

C2   By comparing Definitions 2.6.1(1) and 2.6.2(1) with Definition 2.7(1) we can see the structural similarity between PV measures and probability measures, and between PV functions and probability functions. The difference is that $\widehat{M}$ associates each Borel set Λ with a projector $\widehat{M}(\Lambda)$. Similarly a spectral function $\widehat{F}$ assigns a projector $\widehat{F}(\tau)$ to every real number τ. Note that we do not impose the condition

$$\widehat{M}(\emptyset) = \widehat{O}. \tag{2.285}$$

144 CHAPTER 2. OPERATORS AND THEIR DIRECT INTEGRALS

By an analysis analogous to that in §2.6.1C(1) C7 we can show that this is a consequence of the conditions imposed in Definition 7.2(1).

Since $\widehat{M}(\Lambda)$ and $\widehat{F}(\tau)$ are projectors we have, for any unit vector ϕ in the Hilbert space,

$$\langle \phi \mid \widehat{M}(\Lambda)\phi \rangle \leq 1, \quad \langle \phi \mid \widehat{F}(\tau)\phi \rangle \leq 1. \tag{2.286}$$

These results enable us to

generate a probability measure from $\widehat{M}$ and a probability function from $\widehat{F}$ from any given unit vector ϕ in the Hilbert space by

$$\mathcal{M}_\phi(\Lambda) = \langle \phi \mid \widehat{M}(\Lambda)\phi \rangle, \quad \mathcal{F}_\phi(\tau) = \langle \phi \mid \widehat{F}(\tau)\phi \rangle. \tag{2.287}$$

A formal statement of this will be given in §2.11.1 later. Note that unless $\widehat{M}(\Lambda) = \widehat{O}$ there exist unit vectors such that probability $\mathcal{M}_\phi(\Lambda)$ is equal to 1; the same applies to $\widehat{F}(\tau)$. These results enables us to

conceptualize PV measures and PV functions as the projector equivalents of probability measures and probability functions.

C3 Probability measures are related to probability functions by Eq. (2.260). PV measures and PV functions are similarly related, i.e., there is a one-to-one relation between PV measures and PV functions given by

$$\widehat{F}(\tau) = \widehat{M}((-\infty, \tau]), \quad \widehat{M}(\Lambda) = \int_\Lambda d\widehat{F}(\tau). \tag{2.288}$$

For an interval $\Lambda = (a, b]$ we have[93]

$$\widehat{M}(\Lambda) = \widehat{F}(b) - \widehat{F}(a), \tag{2.289}$$

and for a singleton set $\Lambda = \{\tau_0\}$ containing a single number τ_0 we have

$$\widehat{M}(\{\tau_0\}) = \text{s-}\lim_{\varepsilon \to 0} \left(\widehat{F}(\tau_0) - \widehat{F}(\tau_0 - \varepsilon) \right). \tag{2.290}$$

Since $\widehat{M}(\mathbb{R}) = \widehat{I}$ we also have

$$\widehat{I} = \int_{-\infty}^{\infty} d\widehat{F}(\tau). \tag{2.291}$$

---

[93]See also footnotes to §2.8.3C(1) C1.

2.7. SPECTRAL MEASURES AND PROBABILITY

Hence a spectral function is also called a *decomposition of the identity* or a *resolution of the identity* in terms of projectors. Note that the above integral expressions should be understood in the sense of the following Stieltjes integrals:[94]

$$\left\|\widehat{M}(\Lambda)\phi\right\|^2 = \int_\Lambda d\langle\phi \mid \widehat{F}(\tau)\phi\rangle, \qquad \|\phi\|^2 = \int_{-\infty}^{\infty} d\langle\phi \mid \widehat{F}(\tau)\phi\rangle. \qquad (2.292)$$

C4 In view of Eq. (2.96) in §2.3C(1) C11 we can replace condition 3 in the definition for the spectral function by

$$\widehat{F}(\tau_2)\widehat{F}(\tau_1) = \widehat{F}(\tau_m), \quad \tau_m = \text{smaller of the two numbers } \tau_1 \text{ and } \tau_2. \quad (2.293)$$

This property, known as the *orthogonality property* of PV functions, is consistent with the projector-valued nature of the function, i.e., when $\tau_1 = \tau_2$ we recover the idempotent property of projectors.

C5 PV functions and PV measures possess the following properties:

$$[\widehat{F}(\tau_1), \widehat{F}(\tau_2)] = \widehat{O}, \qquad (2.294)$$

$$[\widehat{M}(\Lambda_1), \widehat{M}(\Lambda_2)] = \widehat{O}, \qquad (2.295)$$

$$\widehat{M}(\emptyset) = \widehat{O}, \qquad (2.296)$$

$$\widehat{M}(\Lambda^c) = \widehat{I} - \widehat{M}(\Lambda), \quad \Lambda^c = \mathbb{R} - \Lambda, \qquad (2.297)$$

$$\widehat{M}(\Lambda_1)\widehat{M}(\Lambda_2) = \widehat{O}, \quad \text{if } \Lambda_1 \cap \Lambda_2 = \emptyset, \qquad (2.298)$$

$$\widehat{M}(\Lambda_1 \cap \Lambda_2) = \widehat{M}(\Lambda_1)\widehat{M}(\Lambda_2) = \widehat{M}(\Lambda_2)\widehat{M}(\Lambda_1), \qquad (2.299)$$

$$\widehat{M}(\Lambda_1 \cup \Lambda_2) = \widehat{M}(\Lambda_1) + \widehat{M}(\Lambda_2) - \widehat{M}(\Lambda_1 \cap \Lambda_2). \qquad (2.300)$$

C6 It can be shown that the unitary transform of a PV measure $\widehat{M}$ is again a PV measure, and that the unitary transform of a PV function $\widehat{F}$ is again a PV function. Since the inverse $\widehat{U}^{-1}$ of a unitary operator $\widehat{U}$ is also unitary we can generate two new PV measures $\widehat{M}'$ and $\widehat{M}''$, and two new PV functions $\widetilde{F}'$ and $\widetilde{F}''$ given by

$$\widehat{M}' = \widehat{U}\,\widehat{M}\,\widehat{U}^{-1}, \quad \widehat{F}' = \widehat{U}\,\widehat{F}\,\widehat{U}^{-1}, \qquad (2.301)$$

and

$$\widehat{M}'' = \widehat{U}^{-1}\widehat{M}\,\widehat{U}, \quad \widehat{F}'' = \widehat{U}^{-1}\widehat{F}\,\widehat{U}. \qquad (2.302)$$

C7 The rather abstract definitions of PV measures and PV functions turn out to be quite simple in many practical applications as seen in the examples below.

---

[94] Naimark (1968) pp. 14-15, Roman Vol. 2 (1975) p. 653.

146 CHAPTER 2. OPERATORS AND THEIR DIRECT INTEGRALS

Examples 2.7(1)   Piecewise continuity and continuity

E1   Piecewise continuous spectral functions

1. Define a projector-valued function by

$$\widehat{F}^{(1)}(\tau) = \begin{cases} \widehat{O} & \tau < 0, \\ \widehat{I\!I} & \tau \geq 0. \end{cases} \quad (2.303)$$

This function takes the value $\widehat{O}$ for $\tau < 0$ and the value $\widehat{I\!I}$ for $\tau \geq 0$. It is discontinuous in τ at $\tau = 0$. One can verify that this is a PV function.

2. Define a projector-valued function by

$$\widehat{F}^{(2)}(\tau) = \begin{cases} \widehat{O} & \tau < 1, \\ \widehat{I\!I} & \tau \geq 1. \end{cases} \quad (2.304)$$

This function takes the value $\widehat{O}$ for $\tau < 1$ and the value $\widehat{I\!I}$ for $\tau \geq 1$. It is discontinuous in τ at $\tau = 1$. One can verify that this is a PV function.

3. We can construct piecewise constant PV functions with many discontinuities. Let $\{\varphi_j, j = 1, 2, \ldots\}$ be an orthonormal basis in a Hilbert space $\mathcal{H}$, and let $\{\widehat{P}_j = |\varphi_j\rangle\langle\varphi_j|\}$ be the corresponding set of projectors. Then we know from Eq. (2.98) in §2.3C(1) C11 that

$$\sum_j \widehat{P}_j = \sum_j |\varphi_j\rangle\langle\varphi_j| = \widehat{I\!I}. \quad (2.305)$$

Let $\tau_j, j = 1, 2, \ldots$, be a bounded set of real numbers arranged in an increasing order, i.e., $\tau_j < \tau_k$ if $j < k$, with the lowest upper bound denoted by τ_{ub}. In analogy to a purely discrete probability function given by Eq. (2.264) we can introduce a projector-valued function[95]

$$\widehat{F}^{(3)}(\tau) = \begin{cases} \widehat{O} & \text{if } \tau < \tau_1 \\ \widehat{P}_1 & \text{if } \tau_1 \leq \tau < \tau_2 \\ \widehat{P}_1 + \widehat{P}_2 & \text{if } \tau_2 \leq \tau < \tau_3 \\ \cdots & \cdots \\ \sum_{j=1}^k \widehat{P}_j & \text{if } \tau_k \leq \tau < \tau_{k+1} \\ \cdots & \cdots \\ \widehat{I\!I} & \text{if } \tau_{ub} \leq \tau \end{cases} . \quad (2.306)$$

---

[95] For a pictorial representation see Roman Vol. 2 (1975) p. 631.

2.7. SPECTRAL MEASURES AND PROBABILITY

One can verify that this is a PV function. This function is discontinuous in τ with discontinuities at $\tau = \tau_j$ as demonstrated by the following result:[96]

$$\operatorname*{s-lim}_{\varepsilon \to 0}\left(\widehat{F}^{(3)}(\tau_1) - \widehat{F}^{(3)}(\tau_1 - \varepsilon)\right) = \widehat{P}_1, \quad (2.307)$$

$$\operatorname*{s-lim}_{\varepsilon \to 0}\left(\widehat{F}^{(3)}(\tau_2) - \widehat{F}^{(3)}(\tau_2 - \varepsilon)\right) = \widehat{P}_2, \quad (2.308)$$

$$\cdots = \cdots$$

$$\operatorname*{s-lim}_{\varepsilon \to 0}\left(\widehat{F}^{(3)}(\tau_j) - \widehat{F}^{(3)}(\tau_j - \varepsilon)\right) = \widehat{P}_j. \quad (2.309)$$

E2 Continuous spectral functions

A simple example involves characteristic functions as multiplication operators in the Hilbert space $L^2(I\!\!E)$ introduced in §2.2.1C(1) C4. We can define a characteristic function χ_Λ for each Borel set Λ of $I\!\!E$ by Eq. (2.39). As pointed out in §2.3C(1) C11 every χ_Λ is a projector on $L^2(I\!\!E)$. Therefore we can define a projector-valued set function $\widehat{M}^{(4)}$ on $I\!\!B$

$$\widehat{M}^{(4)} : I\!\!B \mapsto \left\{\text{projectors on } \mathcal{H}\right\} \quad (2.310)$$

by

$$\Lambda \to \widehat{M}^{(4)}(\Lambda) = \chi_\Lambda. \quad (2.311)$$

This set function is a PV measure on $L^2(I\!\!E)$, the properties required of a PV measure being easily verified. The corresponding PV function is

$$\widehat{F}^{(4)}(\tau) = \chi_{(-\infty,\tau]}. \quad (2.312)$$

This PV function is continuous in τ, i.e.,

$$\widehat{F}^{(4)}(\tau + \varepsilon) - \widehat{F}^{(4)}(\tau - \varepsilon) \to \widehat{O} \quad \text{as} \quad \varepsilon \to 0. \quad (2.313)$$

This should not be confused with the discontinuous nature of $\chi_{(-\infty,\tau]}(x)$ as a function of x.

E3 Unitary transforms of spectral functions and measures

In the Hilbert space $L^2(I\!\!E)$ we have the Fourier transform operator $\widehat{U}_F$ as defined by Eq. (2.110). Now take the family of characteristic functions χ_Λ as a PV measure on $L^2(I\!\!E)$. For later applications we want to find out the inverse Fourier transform of χ_Λ. Let us denote the inverse Fourier transform

---

[96] We shall often omit the sign for a strong operator limit for notational brevity.

of χ_Λ by $\underset{\sim}{\chi}_\Lambda$.[97] We can write down the inverse Fourier transform of χ_Λ as

$$\underset{\sim}{\chi}_\Lambda = \left(\widehat{U}_F^{-1}\right)\chi_\Lambda\left(\widehat{U}_F^{-1}\right)^{-1} \qquad (2.314)$$

$$= \widehat{U}_F^{-1}\chi_\Lambda \widehat{U}_F. \qquad (2.315)$$

Then, according to §2.7C(1) C6, $\underset{\sim}{\chi}_\Lambda$ is a PV measure; the corresponding PV function $\underset{\sim}{F}(\tau)$ is equal to $\underset{\sim}{\chi}_{(-\infty,\tau]}$. Explicitly we have

$$\left(\underset{\sim}{F}(\tau)\phi\right)(x) = \left(\widehat{U}_F^{-1}\chi_{(-\infty,\tau]}\widehat{U}_F\phi\right)(x) \qquad (2.316)$$

$$= \frac{1}{\sqrt{2\pi\hbar}}\int_{-\infty}^{\tau} dx'\, e^{\pm x'x}\, \widetilde{\phi}(x'). \qquad (2.317)$$

So, Eq. (2.317) shows that the effect of $\underset{\sim}{F}(\tau)$ acting on ϕ amounts to

1. taking the Fourier transform $\widetilde{\phi}(x')$ of $\phi(x)$, and then

2. performing a "semi-inverse" Fourier transformation of $\widetilde{\phi}(x')$, on account of the fact that as a multiplication operator $\chi_{(-\infty,\tau]}$ acting on $\widetilde{\phi}(x')$ has the effect of truncating $\widetilde{\phi}(x')$ above $x' = \tau$ in accordance with Eq. (2.39).

2.8 Selfadjointness and Spectral Decomposition

A striking feature of selfadjoint operators is their one-to-one association with *spectral measures* and *spectral functions*. This feature helps us to study many properties of selfadjoint operators.

2.8.1 Spectral theorem

Theorem 2.8.1(1)   Spectral theorem on selfadjoint operators

- To each selfadjoint operator $\widehat{A}$ in a Hilbert space $\mathcal{H}$ there corresponds a unique spectral function $\widehat{F}^{\widehat{A}}$ such that the domain $\mathcal{D}(\widehat{A})$ of $\widehat{A}$ coincides with the set of vectors ϕ in $\mathcal{H}$ for which[98]

$$\int_{-\infty}^{\infty} \tau^2\, d\langle\phi\mid \widehat{F}^{\widehat{A}}(\tau)\phi\rangle < \infty, \qquad (2.318)$$

---

[97] As seen in Eq. (2.110) we denote the Fourier transform by a tilde over the quantity. We denote the inverse transform here by a tilde underlining the quantity.

[98] Prugovečki (1981) p. 250, Roman Vol. 2 (1975) §13.4b Theorem 13.4b(2) on p. 638.

2.8. SELFADJOINTNESS AND SPECTRAL DECOMPOSITION 149

and for any $\psi \in \mathcal{H}$ and $\phi \in \mathcal{D}(\widehat{A})$ we have

$$\langle \psi \mid \widehat{A}\phi \rangle = \int_{-\infty}^{\infty} \tau \, d\langle \psi \mid \widehat{F}^{\widehat{A}}(\tau)\phi \rangle, \qquad (2.319)$$

$$\left\| \widehat{A}\phi \right\|^2 = \int_{-\infty}^{\infty} \tau^2 \, d\langle \phi \mid \widehat{F}^{\widehat{A}}(\tau)\phi \rangle. \qquad (2.320)$$

- Every spectral function $\widehat{F}(\tau)$ defines a selfadjoint operator $\widehat{A}$ satisfying properties (2.318), (2.319) and (2.320) above.[99]

Comments 2.8.1(1) Spectral projectors and commuting selfadjoint operators

C1 The integrals appearing in the theorem are Stieltjes integrals.[100] In view of Eqs. (2.318), (2.319) and (2.320) we often express $\widehat{A}$ as[101]

$$\widehat{A} = \int_{-\infty}^{\infty} \tau \, d\widehat{F}^{\widehat{A}}(\tau). \qquad (2.321)$$

This integral expression is known as the *spectral decomposition* of $\widehat{A}$. On account of Eq. (2.291) we also have the integral expression

$$\widehat{\mathbb{I}} = \int_{-\infty}^{\infty} d\widehat{F}^{\widehat{A}}(\tau). \qquad (2.322)$$

The meaning of these integrals require some clarification. First, the integral expression for $\widehat{A}$ means that for every $\phi \in \mathcal{D}(\widehat{A})$ we have

$$\widehat{A}\phi = \int_{-\infty}^{\infty} \tau \, d\left(\widehat{F}^{\widehat{A}}(\tau)\phi\right). \qquad (2.323)$$

To understand the integral on the right hand side let

$$\cdots < \tau_{-2} < \tau_{-1} < \tau_0 < \tau_1 < \tau_2 < \cdots \qquad (2.324)$$

be a partition of the real line, and let N be a large positive integer. Introduce the sums

$$\Phi_N = \sum_{j=-N}^{N} \bar{\tau}_j \left(\widehat{F}^{\widehat{A}}(\tau_j) - \widehat{F}^{\widehat{A}}(\tau_{j-1})\right)\phi, \qquad (2.325)$$

$$\widehat{A}_N = \sum_{j=-N}^{N} \bar{\tau}_j \left(\widehat{F}^{\widehat{A}}(\tau_j) - \widehat{F}^{\widehat{A}}(\tau_{j-1})\right), \qquad (2.326)$$

---

[99] Akhiezer and Glazman Vol. 2 (1963), §66. Roman Vol. 2 (1975) §13.4b Theorem 13.4b(1) on p. 636 and Theorem 13.4b(3) on p. 641.
[100] Naimark (1968) Vol. 1 pp. 14-15.
[101] See also Eq. (2.375).

where $\phi \in \mathcal{D}(\widehat{A})$ and $\bar{\tau}_j \in (\tau_{j-1}, \tau_j]$. Then we have

$$\Phi_N = \widehat{A}_N \phi. \tag{2.327}$$

As $N \to \infty$ and as all the intervals vanish, i.e., $\tau_j - \tau_{j-1} \to 0$, the limits of Φ_N and $\widehat{A}_N$ can be shown to satisfy

$$\lim_{N \to \infty} ||\widehat{A}\phi - \Phi_N|| \to 0 \quad \text{and} \quad \lim_{N \to \infty} ||\widehat{A}\phi - \widehat{A}_N \phi|| \to 0. \tag{2.328}$$

In other words the sequence of operators $\widehat{A}_N$ in Eq. (2.326) converges to $\widehat{A}$ in a strong sense.[102]

C2 Each selfadjoint operator $\widehat{A}$ possesses a unique PV measure $\widehat{M}^{\widehat{A}}$ which is related to the PV function $\widehat{F}^{\widehat{A}}$ in accordance with Eq. (2.288), i.e.,

$$\widehat{M}^{\widehat{A}}(\Lambda) = \int_\Lambda d\widehat{F}^{\widehat{A}}(\tau), \quad \widehat{F}^{\widehat{A}}(\tau) = \widehat{M}^{\widehat{A}}\big((-\infty, \tau]\big). \tag{2.329}$$

We call $\widehat{M}^{\widehat{A}}$ and $\widehat{F}^{\widehat{A}}$ respectively the spectral measure and the spectral function of $\widehat{A}$. The projector $\widehat{M}^{\widehat{A}}(\Lambda)$ associated with a Borel set Λ is called the *spectral projector* of $\widehat{A}$ for Λ; $\widehat{F}^{\widehat{A}}(\tau)$ is the spectral projector for $\Lambda = (-\infty, \tau]$.

It turns out that the spectral measure and and the spectral function of $\widehat{A}$ can be regarded as functions of $\widehat{A}$. A discussion of this is given in §2.8.2 with the functions involved spelled out explicitly in Eq. (2.360).

C3 Each PV function, and hence each PV measure, generates a unique selfadjoint operator, symbolized by Eq. (2.321), having properties (2.318), (2.319) and (2.320).

C4 Since Spectral Theorem 2.8.1(1) ensures a one-to-one correspondence between the set of all selfadjoint operators and the set of all PV measures (PV functions) in a Hilbert space we can treat selfadjoint operators and PV measures (PV functions) as interchangeable terms.

C5 For a singleton set $\Lambda = \{\tau_0\}$ we have

$$\widehat{M}^{\widehat{A}}(\{\tau_0\}) = \lim_{\varepsilon \to 0} \big(\widehat{F}^{\widehat{A}}(\tau_0) - \widehat{F}^{\widehat{A}}(\tau_0 - \varepsilon)\big). \tag{2.330}$$

As will be seen presently, spectral projectors associated with singleton sets play an important role in the definition and classification of the *spectrum* of selfadjoint operators.

---

[102] Roman Vol. 2 (1975) pp. 640-641. Note that we are not saying that the sequence $\widehat{A}_N$ converges to $\widehat{A}$ strongly in the sense of Definition 2.2.2(1) since this would require Eq. (2.328) to apply to every ϕ in the Hilbert space. We say that $\widehat{A}_N$ converges to $\widehat{A}$ in a strong sense just to mean Eq. (2.328). In this context convergence in a weak sense means Eq. (2.319).

2.8. SELFADJOINTNESS AND SPECTRAL DECOMPOSITION

C6 Spectral projectors of $\widehat{A}$ mutually commute and commute with $\widehat{A}$, i.e.,

$$[\widehat{M}^{\hat{A}}(\Lambda_1), \widehat{M}^{\hat{A}}(\Lambda_2)] = \widehat{O}, \quad [\widehat{F}^{\hat{A}}(\tau_1), \widehat{F}^{\hat{A}}(\tau_2)] = \widehat{O}, \tag{2.331}$$

and for all $\phi \in \mathcal{D}(\widehat{A})$,

$$\widehat{A}\,\widehat{M}^{\hat{A}}(\Lambda)\phi = \widehat{M}^{\hat{A}}(\Lambda)\,\widehat{A}\,\phi, \quad \widehat{A}\,\widehat{F}^{\hat{A}}(\tau)\,\phi = \widehat{F}^{\hat{A}}(\tau)\,\widehat{A}\,\phi. \tag{2.332}$$

For a bounded interval Λ the product $\widehat{A}\,\widehat{M}^{\hat{A}}(\Lambda)$ is a bounded selfadjoint operator, and we have

$$\langle \widehat{A}\psi \mid \widehat{M}^{\hat{A}}(\Lambda)\,\phi \rangle = \langle \psi \mid \widehat{A}\,\widehat{M}^{\hat{A}}(\Lambda)\,\phi \rangle = \int_\Lambda \tau\, d\langle \psi \mid \widehat{F}^{\hat{A}}(\tau)\phi \rangle. \tag{2.333}$$

C7 Two selfadjoint operators are said to *commute* if their spectral projectors commute. While agreeing with the previous definition for bounded selfadjoint operators this definition also applies to unbounded selfadjoint operators.[103] Moreover, if $\widehat{A}$ is selfadjoint and unbounded, and $\widehat{B}$ is selfadjoint and bounded then $\widehat{A}$ commutes with $\widehat{B}$ if its spectral projectors $\widehat{M}^{\hat{A}}(\Lambda)$ commute with $\widehat{B}$.

C8 Given a selfadjoint operator $\widehat{A}$ and a unit vector ϕ in a Hilbert space $\mathcal{H}$ we can generate a probability measure $\mathcal{M}_\phi^{\hat{A}}$ and a probability function $\mathcal{F}_\phi^{\hat{A}}$ with the help of the spectral measure $\widehat{M}^{\hat{A}}$ and the spectral function $\widehat{F}^{\hat{A}}$ of $\widehat{A}$ by Eq. (2.287), i.e.,

$$\mathcal{M}_\phi^{\hat{A}}(\Lambda) = \langle \phi \mid \widehat{M}^{\hat{A}}(\Lambda)\phi \rangle, \quad \mathcal{F}_\phi^{\hat{A}}(\tau) = \langle \phi \mid \widehat{F}^{\hat{A}}(\tau)\phi \rangle. \tag{2.334}$$

We shall devote a section, i.e., §2.11, later to discuss this in greater detail since these results are of crucial importance in the formulation of quantum mechanics.

C9 If two selfadjoint operators are related by a unitary transformation then their spectral measures are also related by the same unitary transformation. This statement can be illustrated with the position and momentum operators in $L^2(I\!\!E)$. We know that the Fourier transform operator $\widehat{U}_F$ on $L^2(I\!\!E)$ defined by Eq. (2.110) and its inverse $\widehat{U}_F^{-1}$ are both unitary. One can verify that the position operator $\widehat{x}$ and the momentum operator $\widehat{p}$ in $L^2(I\!\!E)$ are unitarily related. Following from the result[104]

$$\widehat{x}\,\widehat{U}_F = \widehat{U}_F\,\widehat{p} \tag{2.335}$$

---

[103] Reed and Simon Vol. 1 (1972) p. 271, Prugovečki (1981) p. 261.
[104] Roman Vol. 2 (1975) §12.3c p. 561, Prugovečki (1981) p. 269, p. 331, p. 348. This result can be compared with the results presented in §2.4C(1) C7.

we get
$$\widehat{x} = \widehat{U}_F \, \widehat{p} \, \widehat{U}_F^{-1} \quad \text{and} \quad \widehat{p} = \widehat{U}_F^{-1} \, \widehat{x} \, \widehat{U}_F. \tag{2.336}$$

We may call the position operator the Fourier transform of the momentum operator, and the momentum operator the inverse Fourier transform of the position operator. Hence the momentum and position spectral measures and spectral functions are related by a Fourier transform, i.e., we have

$$\widehat{M^{\widehat{p}}}(\Lambda) = \widehat{U}_F^{-1} \widehat{M^{\widehat{x}}}(\Lambda) \widehat{U}_F \quad \text{and} \quad \widehat{F^{\widehat{p}}}(\tau) = \widehat{U}_F^{-1} \widehat{F^{\widehat{x}}}(\tau) \widehat{U}_F. \tag{2.337}$$

Examples 2.8.1(1)   **PV functions of selfadjoint operators**[105]

E1   **Selfadjoint operators with piecewise constant PV functions**

1. For the zero operator $\widehat{O}$ we have the PV function

$$\widehat{F^{\widehat{O}}}(\tau) = \begin{cases} \widehat{O} & \tau < 0, \\ \widehat{I} & \tau \geq 0. \end{cases} \tag{2.338}$$

This PV function is the one in §2.7E(1) E1.

2. For the identity operator $\widehat{I}$ we have the PV function

$$\widehat{F^{\widehat{I}}}(\tau) = \begin{cases} \widehat{O} & \tau < 1, \\ \widehat{I} & \tau \geq 1. \end{cases} \tag{2.339}$$

This PV function is the one in §2.7E(1) E2. By comparing with integrals (2.248) and (2.252) and Eq. (2.309) we can verify Eq. (2.321), i.e.,

$$\int_{-\infty}^{\infty} \tau \, d\widehat{F^{\widehat{I}}}(\tau) = \widehat{F^{\widehat{I}}}(1+0) - \widehat{F^{\widehat{I}}}(1-0) = \widehat{I}. \tag{2.340}$$

The same applies to the preceding and the following examples.

3. For a projector $\widehat{P}$ we have the PV function

$$\widehat{F^{\widehat{P}}}(\tau) = \begin{cases} \widehat{O} & \tau < 0, \\ \widehat{I} - \widehat{P} & 0 \leq \tau < 1, \\ \widehat{I} & \tau \geq 1. \end{cases} \tag{2.341}$$

From Eq. (2.330) we can verify the following properties of its associated probability measure:

$$\widehat{M^{\widehat{P}}}(\Lambda) = \begin{cases} \widehat{I} - \widehat{P} & \text{if } \Lambda = \{0\}, \\ \widehat{P} & \text{if } \Lambda = \{1\}, \\ \widehat{O} & \text{if } \Lambda \text{ does not contain 0 or 1}. \end{cases} \tag{2.342}$$

---

[105] Weidmann (1980) p. 195.

2.8. SELFADJOINTNESS AND SPECTRAL DECOMPOSITION 153

We shall return later to discuss some general properties of selfadjoint operators with a piecewise constant PV function.

E2   Selfadjoint operators with continuous PV functions

In $L^2(I\!E)$ there are physically two important selfadjoint operators, the position operator $\widehat{x}$ and the momentum operator $\widehat{p}$ defined by Eqs. (2.45) and (2.184) respectively. The PV measures and PV functions for these two operators are given as follows:

1. **The position operator, its spectral function and measure**

    The PV function and the PV measure of $\widehat{x}$ are known to be given in terms of the characteristic function, i.e., we have[106]

    $$\widehat{F}^{\widehat{x}}(\tau) = \chi_{(-\infty,\tau]}, \qquad (2.343)$$

    which is just $\widehat{F}^{(4)}$ in §2.7E(1) E2, and

    $$\widehat{M}^{\widehat{x}}(\Lambda) = \chi_\Lambda. \qquad (2.344)$$

    In other words we have, for any $\phi(x) \in L^2(I\!E)$,

    $$\left(\widehat{F}^{\widehat{x}}(\tau)\phi\right)(x) = \chi_{(-\infty,\tau]}(x)\phi(x), \quad \left(\widehat{M}^{\widehat{x}}(\Lambda)\right) = \chi_\Lambda(x)\phi(x). \qquad (2.345)$$

    The effect of $\widehat{F}^{\widehat{x}}(\tau)$ is to truncate the wave function for x bigger than τ, and the effect of $\widehat{M}^{\widehat{x}}(\Lambda)$ is to truncate the wave function for all x outside Λ.

2. **The momentum operator, its spectral function and measure**

    The PV function and the PV measure of $\widehat{p}$ are obtainable from the inverse Fourier transforms of $\widehat{F}^{\widehat{x}}(\tau)$ and $\widehat{M}^{\widehat{x}}(\tau)$ according to Eq. (2.337) in §2.8.1C(1) C9. From Eq. (2.317) in §2.7E(1) E2 we get[107]

    $$\left(\widehat{F}^{\widehat{p}}(\tau)\phi\right)(x) = \frac{1}{\sqrt{2\pi\hbar}} \int_{-\infty}^{\tau} dx'\, e^{\mp ix'x} \widetilde{\phi}(x'), \qquad (2.346)$$

    and from Eq. (2.315),

    $$\widehat{M}^{\widehat{p}}(\Lambda) = \chi_{\underset{\sim}{\Lambda}}. \qquad (2.347)$$

    In the same way $\widehat{F}^{\widehat{x}}(\tau)$ and $\widehat{M}^{\widehat{x}}(\Lambda)$ have a clear meaning in terms of a characteristic function of x in the coordinate space, we can give $\widehat{F}^{\widehat{p}}(\tau)$

---
[106] Prugovečki (1981) p. 263, p. 269.
[107] Prugovečki (1981) p. 263, p. 269, Byron and Fuller (1969) Vol. 1 §5 on p. 283.

and $\widehat{M^{\hat{p}}}(\Lambda)$ a similar meaning in terms of a characteristic function of p in the momentum space. We shall see how this comes about in §2.8.4 later.

Note that $\widehat{F^{\hat{p}}}(\tau)\phi$ is differentiable with respect to τ with

$$\frac{\partial}{\partial \tau}\left(\widehat{F^{\hat{p}}}(\tau)\phi\right)(x) = \frac{1}{\sqrt{2\pi\hbar}} e^{\frac{i}{\hbar}\tau x}\widetilde{\phi}(\tau). \tag{2.348}$$

It is more transparent in physical terms to rewrite the PV function as a function of the momentum variable p, i.e.,

$$\left(\widehat{F^{\hat{p}}}(p)\phi\right)(x) = \frac{1}{\sqrt{2\pi\hbar}} \int_{-\infty}^{p} dp'\, e^{\frac{i}{\hbar}p'x}\widetilde{\phi}(p'). \tag{2.349}$$

Then $\widehat{F^{\hat{p}}}(p)\phi$ becomes differentiable with respect to p with

$$\frac{\partial}{\partial p}\left(\widehat{F^{\hat{p}}}(p)\,\phi\right)(x) = \frac{1}{\sqrt{2\pi\hbar}} e^{\frac{i}{\hbar}px}\widetilde{\phi}(p). \tag{2.350}$$

In these two examples we have a *continuous spectral decomposition* of the two operators concerned, i.e.,

$$\widehat{x} = \int_{-\infty}^{\infty} \tau\, d\widehat{F^{\hat{x}}}(\tau), \qquad \widehat{p} = \int_{-\infty}^{\infty} \tau\, d\widehat{F^{\hat{p}}}(\tau). \tag{2.351}$$

E3   The spectral function and the spectral measure of a selfadjoint operator are expressible in terms of characteristic functions of that operator in the same way as in Eqs. (2.343) and (2.344). We shall spell this out explicitly after a discussion on functions of selfadjoint operators in the next section

E4   The spectral function of a tensor product operator of the form $\widehat{A}^{(1)} \otimes \widehat{I\!I}^{(2)}$ introduced in §2.2.3 is related to the spectral function of $\widehat{A}^{(1)}$ by

$$\widehat{F}^{\widehat{A}^{(1)}\otimes \widehat{I\!I}^{(2)}}(\tau) = \widehat{F}^{\widehat{A}^{(1)}}(\tau) \otimes \widehat{I\!I}^{(2)}. \tag{2.352}$$

2.8.2   Functions of a selfadjoint operator

The PV functions are instrumental in establishing a general definition of functions of a selfadjoint operator.[108] Let $\widehat{A}$ be a selfadjoint operator in a Hilbert space $\mathcal{H}$, and let $f(\tau)$ be a complex-valued function of a real variable τ. Then

---
[108] Roman Vol. 2 (1975) pp. 644-653, Prugovečki (1981) pp. 270-285.

2.8. SELFADJOINTNESS AND SPECTRAL DECOMPOSITION

there is a unique operator in $\mathcal{H}$, to be denoted by $f(\widehat{A})$ which acts on the domain

$$\mathcal{D}\big(f(\widehat{A})\big) = \left\{\phi : \int |f(\tau)|^2 \, d\langle \phi \mid \widehat{F}^{\widehat{A}}(\tau)\phi\rangle < \infty\right\} \quad (2.353)$$

and satisfies

$$\langle \psi \mid f(\widehat{A})\phi\rangle = \int f(\tau) \, d\langle \psi \mid \widehat{F}^{\widehat{A}}(\tau)\phi\rangle \quad \forall \psi \in \mathcal{H}, \ \phi \in \mathcal{D}\big(f(\widehat{A})\big). \quad (2.354)$$

One may express the above information by writing

$$f(\widehat{A}) = \int_{-\infty}^{\infty} f(\tau) \, d\widehat{F}^{\widehat{A}}(\tau). \quad (2.355)$$

The definition embodied in Eqs. (2.353) and (2.354) agrees with the more intuitive definition when dealing with, say, a polynomial, e.g., if $f(x) = \tau^2$, then $f(\widehat{A})$ defined above agrees with $\widehat{A}^2$, i.e.,

$$\langle \phi \mid \widehat{A}^2 \phi\rangle = \langle \widehat{A}\phi \mid \widehat{A}\phi\rangle = \left\|\widehat{A}\phi\right\|^2 \quad (2.356)$$

$$= \int_{-\infty}^{\infty} \tau^2 \, d\langle \phi \mid \widehat{F}^{\widehat{A}}(\tau)\phi\rangle \quad \forall \phi \in \mathcal{D}(\widehat{A}^2). \quad (2.357)$$

The last step follows from Spectral Theorem 2.8.1(1). The definition also applies to non-polynomial functions. An example is the square root of $\widehat{A}$ defined by

$$\widehat{A}^{1/2} = \int_{-\infty}^{\infty} \tau^{1/2} \, d\widehat{F}^{\widehat{A}}(\tau). \quad (2.358)$$

Moreover we can show that

$f(\widehat{A})$ *is selfadjoint if f is a real-valued function,*

e.g., the square of a selfadjoint operator is selfadjoint.

For an example in $L^2(I\!E)$ consider the characteristic function as a real-valued function $\chi_\Lambda(x)$ of the real variable x. We can define the characteristic function $\chi_\Lambda(\widehat{A})$ of a selfadjoint operator $\widehat{A}$. Let us investigate this simple function a little further:

1. When $\widehat{A} = \widehat{x} = x$, we have the operator

$$\chi_{(-\infty,\tau]}(\widehat{x}) = \chi_{(-\infty,\tau]}(x), \quad (2.359)$$

which is just the characteristic function $\chi_{(-\infty,\tau]}(x)$ acting as a multiplication operator on $L^2(I\!E)$. We know that $\chi_{(-\infty,\tau]}(x)$ is the spectral function of the position operator $\widehat{x}$. Generally the spectral function of a selfadjoint operator is similarly expressible in terms of characteristic functions of that operator. We shall state this more explicitly below.

156   CHAPTER 2. OPERATORS AND THEIR DIRECT INTEGRALS

2. The spectral function of a selfadjoint operator $\hat{A}$ is the characteristic function $\chi_{(-\infty,\tau]}(\hat{A})$ of $\hat{A}$. The spectral projector associated with a Borel set Λ is given by $\chi_\Lambda(\hat{A})$, i.e., we have[109]

$$\widehat{M}^{\hat{A}}(\Lambda) = \chi_\Lambda(\hat{A}), \quad \widehat{F}^{\hat{A}}(\tau) = \chi_{(-\infty,\tau]}(\hat{A}), \qquad (2.360)$$

since[110]

$$\chi_\Lambda(\hat{A}) = \int_{-\infty}^{\infty} \chi_\Lambda(\tau) d\widehat{F}^{\hat{A}}(\tau) = \int_\Lambda d\widehat{F}^{\hat{A}}(\tau) = \widehat{M}^{\hat{A}}(\Lambda). \qquad (2.361)$$

We can express the spectral function and spectral measure of a real-valued function $f(\hat{A})$ of a selfadjoint operator $\hat{A}$ in terms of the corresponding quantities of $\hat{A}$ as follows:[111]

$$\widehat{M}^{f(\hat{A})}(\Lambda) = \widehat{M}^{\hat{A}}\left(f^{-1}(\Lambda)\right), \qquad (2.362)$$

$$\widehat{F}^{f(\hat{A})}(\tau) = \widehat{M}^{f(\hat{A})}\left((-\infty,\tau]\right) = \widehat{M}^{\hat{A}}\left(f^{-1}((-\infty,\tau])\right), \qquad (2.363)$$

where

$$f^{-1}(\Lambda) = \left\{\tau \in I\!R : f(\tau) \in \Lambda\right\} \qquad (2.364)$$

is the inverse image of Λ under f. An example is when $f = x^2$. We have

$$f^{-1}\left((\tau_1,\tau_2]\right) = \left\{\tau \in I\!R : \tau^2 \in (\tau_1,\tau_2]\right\} \qquad (2.365)$$

and when $\tau_1 = -\infty$ we have

$$f^{-1}\left((-\infty,\tau_2]\right) = \begin{cases} \emptyset & \text{if } \tau_2 < 0 \\ [-\tau_2^{1/2}, \tau_2^{1/2}] & \text{if } \tau_2 \geq 0 \end{cases}. \qquad (2.366)$$

These enable us to work out $\widehat{M}^{\hat{A}^2}$ and $\widehat{F}^{\hat{A}^2}$ in terms of $\widehat{M}^{\hat{A}}$ and $\widehat{F}^{\hat{A}}$ from Eqs. (2.362) and (2.363).

When the function is complex, the situation is different. An example is the complex exponential function $f(\tau) = \exp ia\tau$, a being a real constant. The corresponding operator function is

$$e^{ia\hat{A}} = \int_{-\infty}^{\infty} e^{ia\tau} d\widehat{F}^{\hat{A}}(\tau). \qquad (2.367)$$

This gives rise to a unitary operator $\hat{U} = e^{ia\hat{A}}$ of a form directly related to Stone's Theorem 2.4(1).[112] Equation (2.367) is known as the spectral decomposition of the unitary operator $\hat{U}$.[113]

---
[109] Prugovečki (1981) Eq. (2.16) p. 277, Akhiezer and Glazman Vol. 2 (1963) p. 72.
[110] See also Eq. (2.329).
[111] Weidmann (1980) p. 197.
[112] See §2.4C(1) C8 and C9. Prugovečki (1981) pp. 286-287.
[113] Prugovečki (1981) pp. 242.

2.8. SELFADJOINTNESS AND SPECTRAL DECOMPOSITION

2.8.3 Spectra of selfadjoint operators

Definition 2.8.3(1)   **The spectrum of a selfadjoint operator**[114]

- The set of values $\tau \in I\!R$ is called the *spectrum* of a selfadjoint operator $\widehat{A}$, to be denoted by $sp(\widehat{A})$, if for each τ in $sp(\widehat{A})$ and for every open interval Λ_τ containing τ the spectral projector $\widehat{M}^{\widehat{A}}(\Lambda_\tau)$ does not vanish.

Comments 2.8.3(1)   **Point and continuous spectra, generalized and approximate eigenvectors**

C1   Consider a small open interval $\Lambda_\tau = (\tau - \varepsilon, \tau + \varepsilon)$ containing τ. If τ is an element of the spectrum of $\widehat{A}$ then we must have $\widehat{M}^{\widehat{A}}(\Lambda_\tau) \neq 0$.[115] This means that an element of the spectrum is a point of change of the spectral function $\widehat{F}^{\widehat{A}}$. There are two ways to change:

1. We call a point $\tau_0 \in sp(\widehat{A})$ a *point of continuous growth* if [116]

$$\left(\widehat{F}^{\widehat{A}}(\tau_0 + \varepsilon) - \widehat{F}^{\widehat{A}}(\tau_0 - \varepsilon)\right) \to \widehat{O} \quad \text{as } \varepsilon \to 0, \qquad (2.371)$$

2. We call a point $\tau_0 \in sp(\widehat{A})$ a *point of discontinuous growth*, or a *jump point* if

$$\left(\widehat{F}^{\widehat{A}}(\tau_0 + \varepsilon) - \widehat{F}^{\widehat{A}}(\tau_0 - \varepsilon)\right) \to \widehat{M}^{\widehat{A}}(\{\tau_0\}) \neq \widehat{O} \quad \text{as } \varepsilon \to 0. \qquad (2.372)$$

Note that $\widehat{M}^{\widehat{A}}(\{\tau_0\})$ is a projector, the spectral projector at the jump point τ_0, to be denoted conveniently by $\widehat{P}^{\widehat{A}}_{\tau_0}$.

The spectrum of $\widehat{A}$ can be divided into two parts: (1) the *point (discrete) spectrum* $sp(\widehat{A})_p$ which consists of all points of discontinuous growth in $sp(\widehat{A})$, and (2) the *continuous spectrum* $sp(\widehat{A})_c$ which consists of the rest of $sp(\widehat{A})$. We can single out two simple types of spectra:

---

[114]Prugovečki (1981) p. 253, Akhiezer and Glazman Vol. 2 (1963) §68.

[115]Following from Eqs. (2.289) and (2.290) and the continuity of $\widehat{F}^{\widehat{A}}(\tau)$ on the right, we have the following expressions (Weidmann (1980) p. 182, Naimark Vol. 1 (1968) p. 14):

$$\widehat{M}^{\widehat{A}}([a,b]) = \widehat{F}^{\widehat{A}}(b) - \widehat{F}^{\widehat{A}}(a-0), \qquad (2.368)$$
$$\widehat{M}^{\widehat{A}}([a,b)) = \widehat{F}^{\widehat{A}}(b-0) - \widehat{F}^{\widehat{A}}(a-0), \qquad (2.369)$$
$$\widehat{M}^{\widehat{A}}((a,b)) = \widehat{F}^{\widehat{A}}(b-0) - \widehat{F}^{\widehat{A}}(a). \qquad (2.370)$$

[116]We can replace $\widehat{F}^{\widehat{A}}(\tau + \varepsilon)$ by $\widehat{F}^{\widehat{A}}(\tau)$ in the definition since $\widehat{F}^{\widehat{A}}(\tau + \varepsilon) \to \widehat{F}^{\widehat{A}}(\tau)$ as $\varepsilon \to 0$.

1. **Purely point (discrete) spectrum**   A spectrum $sp(\widehat{A})$ is called a *purely point spectrum*, or a *purely discrete spectrum*, if it consists of only of points of discontinuous growth.

2. **Purely continuous spectrum**   A spectrum $sp(\widehat{A})$ is called a *purely continuous spectrum* if its point spectrum $sp(\widehat{A})_p$ is empty.

A spectrum could also be partly continuous and partly discrete.

C2   Conceptually the spectrum of a selfadjoint operator is a generalization of the notion of eigenvalues. Let us consider the following two cases:

1. Suppose $\widehat{A}$ has a purely point spectrum, i.e., $sp(\widehat{A}) = sp(\widehat{A})_p$. Then $\widehat{F}^{\widehat{A}}(\tau)$ remains unchanged except at jump points $\tau_j = a_j$, i.e., $\widehat{F}^{\widehat{A}}(\tau)$ is piecewise constant of the form of Eq. (2.306), i.e.,

$$\widehat{F}^{\widehat{A}}(\tau) = \begin{cases} \widehat{O} & \text{if } \tau < \tau_1 \\ \widehat{P}_1^{\widehat{A}} & \text{if } \tau_1 \leq \tau < \tau_2 \\ \widehat{P}_1^{\widehat{A}} + \widehat{P}_2^{\widehat{A}} & \text{if } \tau_2 \leq \tau < \tau_3 \\ \cdots \\ \sum_{j=1}^{k} \widehat{P}_j^{\widehat{A}} & \text{if } \tau_k \leq \tau < \tau_{k+1} \\ \cdots \\ \widehat{I} & \text{if } \tau_{ub} \leq \tau, \end{cases} \qquad (2.373)$$

where $\widehat{P}_j^{\widehat{A}}$ is the spectral projector at $\tau = a_j$. Note that the set of jump points need not be bounded, i.e., a_j can be arbitrarily large if $\widehat{A}$ is an unbounded operator. Eq. (2.319) in Theorem 2.8.1(1) reduces to

$$\langle \psi \mid \widehat{A}\phi \rangle = \sum_j a_j \langle \psi \mid \widehat{P}_j^{\widehat{A}} \phi \rangle. \qquad (2.374)$$

The spectral decomposition of $\widehat{A}$ expressed in Eq. (2.321) becomes a sum or a *discrete spectral decomposition*:

$$\widehat{A} = \sum_j a_j \widehat{P}_j^{\widehat{A}}. \qquad (2.375)$$

The spectral projectors $\widehat{P}_j^{\widehat{A}}$ are mutually orthogonal on account of Eq. (2.298). The jump points a_j coincide with the eigenvalues a_j of $\widehat{A}$, and the spectral projector at a jump point coincides with the projector onto the eigensubspace of the corresponding eigenvalue. To appreciate these results, let us consider the subspace $\mathcal{L}_k$ associated with the spectral projector $\widehat{P}_k^{\widehat{A}}$. Let $\varphi_{k\ell}$, $\ell = 1, 2, \ldots$, be an orthonormal basis in $\mathcal{L}_k$.

2.8. SELFADJOINTNESS AND SPECTRAL DECOMPOSITION

Since $\widehat{P}_k^{\widehat{A}}$ for different k are mutually orthogonal the corresponding subspaces $\mathcal{L}_k$ are also mutually orthogonal so that

$$\widehat{P}_j^{\widehat{A}} \varphi_{k\ell} = \delta_{jk} \varphi_{k\ell}. \tag{2.376}$$

We have

$$\widehat{A}\, \varphi_{k\ell} = \left(\sum_j a_j\, \widehat{P}_j^{\widehat{A}} \right) \varphi_{k\ell} = a_k\, \varphi_{k\ell}. \tag{2.377}$$

It follows that a_k is the eigenvalue of $\widehat{A}$ corresponding to eigenvectors $\varphi_{k\ell}$. The dimension of $\mathcal{L}_k$ corresponds to the *degeneracy* of the eigenvalue. The spectrum $sp(\widehat{A})$ is seen to consist of all the eigenvalues of $\widehat{A}$. We have already encountered several examples of selfadjoint operators with a purely point spectrum, e.g., $\widehat{p}_\lambda$ in §2.5.1E(1) E1 and $\widehat{K}^\infty(\Lambda)$ in §2.5.1E(1) E2. In both these cases the eigenvalues are non-degenerate so that every eigensubspace is one-dimensional spanned by a unit eigenvector. The spectral decomposition of these operators can be written explicitly as:

$$\widehat{p}_\lambda = \sum_j p_{\lambda,j}\, |\varphi_{\lambda,j}\rangle\langle\varphi_{\lambda,j}|, \tag{2.378}$$

$$\widehat{K}^\infty(\Lambda) = \sum_j E_j\, |\phi_j\rangle\langle\phi_j|. \tag{2.379}$$

2. Suppose $\widehat{A}$ has a purely continuous spectrum, i.e., $sp(\widehat{A}) = sp(\widehat{A})_c$. Then it can be shown that $\widehat{A}$ admits no eigenvectors in the Hilbert space. Nevertheless, we shall call

- a number τ in the continuous spectrum $sp(\widehat{A})_c$ a *generalized eigenvalue*, and

- a vector given in some formal sense[117] satisfying a formal eigenvalue equation for a generalized eigenvalue as a *generalized eigenvector*.

One can then say that the spectrum $sp(\widehat{A})_c$ consists of all the generalized eigenvalues of $\widehat{A}$. Operators $\widehat{x}$ and $\widehat{p}$ in the Hilbert space $L^2(I\!\!E)$ are prime examples of selfadjoint operators with a purely continuous spectrum. They possess generalized eigenvalues with generalized eigenfunctions presented earlier in §2.5.1E(1) E6.

---

[117] Generalized eigenvectors are not well-defined within the Hilbert space. One can recognize them in specific cases, e.g., the formal eigenfunction $\eta_p(x)$ of the momentum operator given in Eq. (2.210). Generalized eigenvectors, though not normalizable, are very useful in practical applications. They can be incorporated in a rigged Hilbert space and played a very useful role in quantum mechanics (Böhm (1978), (1979)).

C3   Let a_j be an eigenvalue of $\widehat{A}$ corresponding an eigenvector φ_j. Then

$$\left(\widehat{A} - a_j \, \widehat{I}\right) \varphi_j = 0, \quad \varphi_j \neq 0. \tag{2.380}$$

It follows from Definition 2.3(1) that the operator $(\widehat{A} - a_j \, \widehat{I})$ is not invertible. We can see that the non-invertibility of operators of the form $(\widehat{A} - a \, \widehat{I})$ can serve to define eigenvalues. We shall return to this later.

C4   Approximate eigenvectors[118]   This is an interesting concept. For any $a \in sp(\widehat{A})_c$ there does not exist a unit vector $\phi \in \mathcal{H}$ satisfying the following eigenvalue equation

$$\left(\widehat{A} - a \, \widehat{I}\right) \phi = 0. \tag{2.381}$$

But we can show that given any $\varepsilon > 0$, there is a unit vector $\xi_{a,\varepsilon} \in \mathcal{H}$ such that

$$\left\| \widehat{A} \, \xi_{a,\varepsilon} - a \, \xi_{a,\varepsilon} \right\| < \varepsilon. \tag{2.382}$$

We call such a vector an *approximate eigenvector* of $\widehat{A}$. Take the example of the momentum operator $\widehat{p}$ in $L^2(I\!\!E)$. In §2.5.1E(1) E6 we introduce generalized eigenfunctions $\eta_p(x)$ corresponding to generalized eigenvalues p. Let $f_\varepsilon(x)$ be a unit vector in $L^2(I\!\!E)$ such that its derivative $df_\varepsilon(x)/dx$ is also a vector in $L^2(I\!\!E)$. Introduce the function $\xi_{p,\varepsilon}(x) = f_\varepsilon(x)\eta_p(x)$. We have

$$|\eta_p(x)|^2 = 1 \quad \text{and} \quad |\xi_{p,\varepsilon}(x)|^2 = |f_\varepsilon(x)|^2. \tag{2.383}$$

As a result $\xi_{p,\varepsilon}(x)$ becomes a unit vector in $L^2(I\!\!E)$ since $|f_\varepsilon(x)|^2$ is integrable giving the value 1 when integrated from $-\infty$ to ∞, and we have

$$\|\widehat{p}\,\xi_{p,\varepsilon} - p\,\xi_{p,\varepsilon}\| = \hbar \left\| \frac{df_\varepsilon}{dx} \right\|. \tag{2.384}$$

As an example let

$$f_\varepsilon(x) = (2\varepsilon/\pi)^{1/4} \, e^{-\varepsilon x^2}. \tag{2.385}$$

Then $\|\widehat{p}\,\xi_{p,\varepsilon} - p\,\xi_{p,\varepsilon}\|$ becomes arbitrarily small as $\varepsilon \to 0$. It follows that $\xi_{p,\varepsilon}$ is an approximate eigenfunction of $\widehat{p}$ for a sufficiently small ε. The effect of $f_\varepsilon(x)$ is seen to produce a square-integrable function $\xi_{p,\varepsilon}(x)$.

Note that approximate eigenvectors can be used to characterize a generalized eigenvalue in that:

---

[118] Richtmyer Vol. 1 pp. 144-145, p. 190, Roman Vol. 2 (1975) Problem 13.1-8 on pp. 595-596.

2.8. SELFADJOINTNESS AND SPECTRAL DECOMPOSITION 161

- A value a is a *generalized eigenvalue*, i.e., $a \in sp(\widehat{A})_c$, if for every $\varepsilon > 0$ there is a unit vector $\xi_{a,\varepsilon}$ satisfying Eq. (2.382).

Since for $a \in sp(\widehat{A})_c$ there is no unit vector ϕ satisfying Eq. (2.381) the operator $(\widehat{A} - a\,\widehat{I})$ would admit an inverse. In view of Eq. (2.382) we can appreciate that this inverse, written as

$$\left(\widehat{A} - a\,\widehat{I}\right)^{-1} \quad \text{or} \quad \frac{1}{(\widehat{A} - a\,\widehat{I})} \tag{2.386}$$

is an unbounded operator. Moreover, it can be shown that this inverse operator is defined on a dense domain.[119] As will be seen later these properties of the inverse can serve to define the continuous spectrum of selfadjoint operators.

C5 Point and continuous spectra are disjoint by definition. An eigenvalue $\tau = a_j$ is a point of discontinuity and the associated spectral projector $\widehat{P}_j^{\widehat{A}} = \widehat{M}^{\widehat{A}}(\{a_j\})$ is not zero. We could have a situation that the intervals $(a_j - \varepsilon, a_j)$ and $(a_j, a_j + \varepsilon)$ are both in $sp(\widehat{A})_c$. Then the eigenvalue a_j appears to lodge in between intervals $(a_j - \varepsilon, a_j)$ and $(a_j, a_j + \varepsilon)$. We would say that the eigenvalue a_j is embedded in the continuous spectrum.[120]

2.8.4 Spectral representation spaces and spectral representations of selfadjoint operators

Following §2.4E(1) E1 we can regard the Fourier transform as a transform from the coordinate representation space $L^2(I\!\!E)$ to the momentum representation space $\widetilde{L}^2(\widetilde{I\!\!E})$. The momentum operator $\widehat{p} = -i\hbar\, d/dx$ in $L^2(I\!\!E)$ is seen in §2.4C(1) C7 to act as a multiplication operator in $\widetilde{L}^2(\widetilde{I\!\!E})$. We can highlight this situation with the following notation:

$$\phi(x) \quad \leftrightarrow \quad \widetilde{\phi}(p) = \left(\widehat{U}_F \phi\right)(p), \tag{2.387}$$

$$\widehat{p} = -i\hbar\, d/dx \quad \leftrightarrow \quad \widetilde{p} = \widehat{U}_F\, \widehat{p}\, \widehat{U}_F^{-1} = p, \tag{2.388}$$

$$\widehat{x} = x \quad \leftrightarrow \quad \widetilde{x} = \widehat{U}_F\, \widehat{x}\, \widehat{U}_F^{-1} = i\hbar\, d/dp, \tag{2.389}$$

so that

$$(\widehat{p}\phi)(x) \quad \leftrightarrow \quad \left(\widetilde{p\phi}\right)(p) = p\, \widetilde{\phi}(p). \tag{2.390}$$

$$(\widehat{x}\phi)(x) \quad \leftrightarrow \quad \left(\widetilde{x\phi}\right)(p) = i\hbar\, \frac{d}{dp}\, \widetilde{\phi}(p). \tag{2.391}$$

---
[119] Richtmyer Vol. 1 (1978) p. 148.
[120] Roman Vol. 2 (1975) p. 601, p. 651 and Fig. 13.3 on p. 652.

162   CHAPTER 2. OPERATORS AND THEIR DIRECT INTEGRALS

Here $\widetilde{p}$ is referred to as the *spectral representation of the momentum operator* in the momentum representation space $\widetilde{L}^2(\widetilde{I\!E})$. According to Eq. (2.360) the spectral projectors of the momentum operator $\widetilde{p}$ in $\widetilde{L}^2(\widetilde{I\!E})$ becomes characteristic functions $\chi_{\tilde{\Lambda}}(p)$ of the interval $\tilde{\Lambda}$ in the momentum space $\widetilde{I\!E}$. Here $\chi_{\tilde{\Lambda}}(p)$ is a function of the momentum variable p in the same way $\chi_\Lambda(x)$ is a function of the coordinate variable x.

It turns out that this feature is not restricted to the momentum operator. There is a mathematical theorem, known as the *Spectral Representation Theorem*, which assures us the existence of a unitary transformation capable of converting any selfadjoint operator to a multiplication operator on the transformed space, referred to as a *spectral representation space* of the operator.[121] This is why we call $\widetilde{L}^2(\widetilde{I\!E}) = \widehat{U}_F L^2(I\!E)$ the momentum representation space.

As another illustration let us consider the kinetic energy operator $\widehat{K}$ in $L^2(I\!E)$ defined to be

$$\widehat{K} = \frac{1}{2m}\widehat{p}^2, \tag{2.392}$$

where m is a numerical constant. We shall introduce a *kinetic energy representation space* in which the kinetic energy operator $\widehat{K}$ acts as a multiplication operator. There is a complication, i.e., $\widehat{K}$ possesses doubly degenerate generalized energy eigenvalues ϵ in a continuous spectrum $[0, +\infty)$. Let us denote the two degenerate generalized eigenfunctions corresponding to the eigenvalue ϵ by $f_+(\epsilon, x)$ and $f_-(\epsilon, x)$, i.e.,

$$f_\nu(\epsilon, x) = \left(\frac{m}{2\epsilon}\right)^{1/4} \frac{1}{\sqrt{2\pi\hbar}} \exp(\nu i \sqrt{2m\epsilon}\, x), \quad \nu = \pm. \tag{2.393}$$

These functions satisfy the following orthonormality conditions:

$$\int_{-\infty}^{\infty} f^*_{\nu'}(\epsilon', x) f_\nu(\epsilon, x) dx = \delta_{\nu'\nu}\, \delta(\epsilon' - \epsilon), \tag{2.394}$$

$$\sum_\nu \int_0^\infty f^*_\nu(\epsilon, x') f_\nu(\epsilon, x) d\epsilon = \delta(x' - x). \tag{2.395}$$

These conditions are similar to the orthonormality conditions in Eqs. (2.211) and (2.212) satisfied by the plane waves $\eta_p(x)$.[122]

The momentum representation space can be divided into two orthogonal and complementary subspaces corresponding to positive and negative values

---
[121] Weidmann (1981) p. 195, Reed and Simon Vol. 1 (1972) p. 227, Prugovečki (1981) IV §5.1 and §5.3. Spectral representation is generally not unique. For a selfadjoint operator with a discrete spectrum the spectral representation space would have a corresponding discrete measure of integration in the construction of the scalar product in Eq. (2.15).
[122] Gottfried (1966) pp. 52-55, Merzbacher (1998) pp. 63-64. One can verify these relations

2.8. SELFADJOINTNESS AND SPECTRAL DECOMPOSITION

of the momentum variable p, i.e.,

$$\widetilde{L}^2_+(\widetilde{I\!\!E}) = \{\widetilde{\phi}_+(p) \in \widetilde{L}^2(\widetilde{I\!\!E}) : \widetilde{\phi}_+(p) = 0 \text{ for } p < 0\}, \quad (2.396)$$
$$\widetilde{L}^2_-(\widetilde{I\!\!E}) = \{\widetilde{\phi}_-(p) \in \widetilde{L}^2(\widetilde{I\!\!E}) : \widetilde{\phi}_-(p) = 0 \text{ for } p > 0\}. \quad (2.397)$$

Since $\widetilde{\phi}_+(p)$ are defined non-trivially only on $p \in (0, \infty)$ and $\widetilde{\phi}_-(p)$ on $p \in (-\infty, 0)$ we have

$$\int_0^\infty |\widetilde{\phi}_+(p)|^2 \, dp = \int_{-\infty}^\infty |\phi_+(x)|^2 \, dx, \quad (2.398)$$

$$\int_{-\infty}^0 |\widetilde{\phi}_-(p)|^2 \, dp = \int_{-\infty}^\infty |\phi_-(x)|^2 \, dx. \quad (2.399)$$

Let $L^2_+(I\!\!E)$ and $L^2_-(I\!\!E)$ denote the two subspaces of $L^2(I\!\!E)$ corresponding to positive and negative momentum values, i.e.,

$$L^2_\pm(I\!\!E) = \widehat{U}_F^{-1} \widetilde{L}^2_\pm(\widetilde{I\!\!E}) \quad \text{and} \quad \widetilde{L}^2_\pm(x)(\widetilde{I\!\!E}) = \widehat{U}_F \, L^2_\pm(I\!\!E). \quad (2.400)$$

Clearly $L^2_+(I\!\!E)$ and $L^2_-(I\!\!E)$ are orthogonal and complementary in $L^2(I\!\!E)$. We shall label members of $L^2_\pm(I\!\!E)$ by $\phi_\pm(x)$ to correspond to their Fourier tranforms $\widetilde{\phi}_\pm(p)$, i.e.,

$$\phi_\pm(x) \mapsto \widetilde{\phi}_\pm(p) = \left(\widehat{U}_F \phi_\pm\right)(p) \quad (2.401)$$
$$\widetilde{\phi}_\pm(p) \mapsto \phi_\pm(x) = \left(\widehat{U}_F^{-1} \widetilde{\phi}_\pm\right)(x), \quad (2.402)$$

and label the kinetic energy operator acting on $L^2_+(I\!\!E)$ and $L^2_-(I\!\!E)$ by

$$\widehat{K}_+ \quad \text{and} \quad \widehat{K}_-. \quad (2.403)$$

The operator $\widehat{K}_+$ possesses non-degenerate generalized eigenvalues in $L^2_+(I\!\!E)$ and similarly for $\widehat{K}_-$ in $L^2_-(I\!\!E)$.

---

using Eqs. (2.211) and (2.212) for plane waves with

$$\epsilon = \frac{1}{2m} p^2, \quad \left(\frac{m}{2\epsilon}\right)^{1/2} d\epsilon = \pm dp,$$

and the following formula for manipulating δ-functions:

$$\delta\Big(g(x)\Big) = \frac{1}{|g'(x_0)|} \delta(x - x_0), \quad \text{if} \quad g'(x_0) = \left.\frac{dg}{dx}\right|_{x=x_0} \neq 0.$$

Here x_0 is the zero of $g(x)$, i.e., $g(x_0) = 0$. This formula is quoted in various texts, e.g., Friedman (1956) p. 136, Merzbacher (1998) p. 633, and it can be extended to $g(x)$ having more than one zero. See also Merzbacher (1998) p. 64.

Let us introduce a new function $\widetilde{\widetilde{\phi}}_+(\epsilon)$ to correspond to every $\phi_+(x)$ through the Fourier transform $\widetilde{\phi}_+(p)$ of $\phi_+(x)$ by

$$\widetilde{\widetilde{\phi}}_+(\epsilon) = \left(\frac{m}{2\epsilon}\right)^{1/4} \widetilde{\phi}_+(\sqrt{2m\epsilon}). \tag{2.404}$$

This function can be rewritten as

$$\widetilde{\widetilde{\phi}}_+(\epsilon) = \int_{-\infty}^{\infty} f_+^*(\epsilon, x)\, \phi_+(x)\, dx, \tag{2.405}$$

where

$$f_+(\epsilon, x) = \left(\frac{m}{2\epsilon}\right)^{1/4} \frac{1}{\sqrt{2\pi\hbar}} e^{\pm i\sqrt{2m\epsilon}\, x}. \tag{2.406}$$

Using the following change of variables:

$$p = \sqrt{2m\epsilon}, \quad dp = \left(\frac{m}{2\epsilon}\right)^{1/2} d\epsilon, \tag{2.407}$$

we can arrive at the inverse tranform as

$$\phi_+(x) = \int_0^{\infty} f_+(\epsilon, x)\, \widetilde{\widetilde{\phi}}_+(\epsilon)\, d\epsilon. \tag{2.408}$$

In analogy to the momentum space $\widetilde{\mathbb{E}}$ we can regard the set of energy values

$$\widetilde{\widetilde{\mathbb{E}}} = \left\{\epsilon : \epsilon \in (0, \infty)\right\} \tag{2.409}$$

as forming the *kinetic energy space*. Let $\widetilde{\widetilde{L}}_+^2(\widetilde{\mathbb{E}})$ denote the Hilbert space of functions $\widetilde{\widetilde{\phi}}(\epsilon)$ of ϵ, square-integrable over the kinetic energy space $\widetilde{\mathbb{E}}_+$. In analogy to the momentum representation space $\widetilde{L}^2(\widetilde{\mathbb{E}})$ we shall call $\widetilde{\widetilde{L}}^2(\widetilde{\mathbb{E}}_+)$ a *kinetic energy representation space*. We have a one-to-one corresponding between $\widetilde{L}_+^2(\widetilde{\mathbb{E}})$ and $\widetilde{\widetilde{L}}_+^2(\widetilde{\mathbb{E}})$ given by

$$\widetilde{\phi}_+(p) \mapsto \widetilde{\widetilde{\phi}}_+(\epsilon) = \left(\frac{m}{2\epsilon}\right)^{1/4} \widetilde{\phi}_+(\sqrt{2m\epsilon}), \tag{2.410}$$

$$\widetilde{\widetilde{\phi}}_+(\epsilon) \mapsto \widetilde{\phi}_+(p) = \left(\frac{m^2}{p^2}\right)^{-1/4} \widetilde{\widetilde{\phi}}_+(p^2/2m). \tag{2.411}$$

This mapping is unitary because of the scalar product in $\widetilde{\widetilde{L}}_+^2(\widetilde{\mathbb{E}})$ defined by

$$\langle \widetilde{\widetilde{\phi}}_+ | \widetilde{\widetilde{\psi}}_+ \rangle_{K_+} = \int_0^{\infty} \widetilde{\widetilde{\phi}}_+^*(\epsilon) \widetilde{\widetilde{\psi}}_+(\epsilon)\, d\epsilon \tag{2.412}$$

2.8. SELFADJOINTNESS AND SPECTRAL DECOMPOSITION

is preserved in the mapping, i.e.,

$$\int_0^\infty \tilde{\phi}_+^*(\epsilon)\tilde{\psi}_+(\epsilon)\, d\epsilon = \int_0^\infty \tilde{\phi}_+^*(p)\tilde{\psi}_+(p)\, dp. \tag{2.413}$$

It follows that there is a unitary transformatioin $\widehat{U}_{K_+}$ between $L_+^2(I\!\!E)$ and $\widetilde{\widetilde{L}}_+^2(\widetilde{I\!\!E})$, i.e., we have

$$\widehat{U}_{K_+} : L_+^2(I\!\!E) \mapsto \widetilde{\widetilde{L}}_+^2(\widetilde{I\!\!E}) \tag{2.414}$$

by

$$\phi_+(x) \mapsto \tilde{\phi}_+(\epsilon) = \left(\widehat{U}_{K_+}\phi_+\right)(\epsilon) = \left(\frac{m}{2\epsilon}\right)^{1/4} \tilde{\phi}_+(\sqrt{2m\epsilon}). \tag{2.415}$$

Moreover, one can check that the transformed kinetic energy operator $\widetilde{K}$ acts on $\tilde{\phi}_+(\epsilon)$ as a multiplication operator, i.e.,

$$\widetilde{K}_+ \tilde{\phi}_+(\varepsilon) = \left(\widehat{U}_{K_+} \widehat{K}_+ \widehat{U}_{K_+}^{-1}\right) \tilde{\phi}_+(\varepsilon) = \epsilon \tilde{\phi}_+(\varepsilon). \tag{2.416}$$

Similarly, we can introduce a Hilbert space of functions on the kinetic energy space $\widetilde{I\!\!E}$ composed of functions $\tilde{\phi}_-(\epsilon)$ defined in terms of functions of $\widetilde{L}_-^2(\widetilde{I\!\!E})$ and of $L_-^2(I\!\!E)$ by

$$\tilde{\phi}_-(\epsilon) = \left(\frac{m}{2\epsilon}\right)^{1/4} \tilde{\phi}_-(-\sqrt{2m\epsilon}) \tag{2.417}$$

$$= \int_{-\infty}^\infty f_-^*(\epsilon, x)\, \phi_-(x)\, dx, \tag{2.418}$$

with the inverse transform given by

$$\phi_-(x) = \int_0^\infty f_-(\epsilon, x) \tilde{\phi}_-(\epsilon)\, d\epsilon. \tag{2.419}$$

Here we have employed

$$f_-(\epsilon, x) = \left(\frac{m}{2\epsilon}\right)^{1/4} \frac{1}{\sqrt{2\pi\hbar}} e^{-i\sqrt{2m\epsilon}\, x} \tag{2.420}$$

with

$$p = -\sqrt{2m\epsilon} < 0 \quad \text{and} \quad dp = -\left(\frac{m}{2\epsilon}\right)^{1/2} d\epsilon. \tag{2.421}$$

A scalar product can be defined in terms of the following integral:

$$\langle \tilde{\phi}_- | \tilde{\psi}_- \rangle_{K_-} = \int_0^\infty \tilde{\phi}_-^*(\epsilon)\tilde{\psi}_-(\epsilon)\, d\epsilon. \tag{2.422}$$

CHAPTER 2. OPERATORS AND THEIR DIRECT INTEGRALS

The resulting Hilbert space will be denoted by $\tilde{\tilde{L}}_-^2(\tilde{\tilde{E}})$. There is a unitary mapping between $L_-^2(E)$ and $\tilde{\tilde{L}}_-^2(\tilde{\tilde{E}})$ by

$$\phi_-(x) \mapsto \tilde{\tilde{\phi}}_+(\epsilon) = \left(\hat{U}_{K_-} \phi_-\right)(\epsilon) = \left(\frac{m}{2\epsilon}\right)^{1/4} \tilde{\phi}_-(-\sqrt{2m\epsilon}), \tag{2.423}$$

on account of the preservation of scalar product

$$\int_0^\infty \tilde{\tilde{\phi}}_-^*(\epsilon)\tilde{\tilde{\psi}}_-(\epsilon)\,d\epsilon \;=\; \int_{-\infty}^0 \tilde{\phi}_-^*(p)\tilde{\psi}_-(p)\,dp \tag{2.424}$$

$$= \int_{-\infty}^\infty \phi_-^*(x)\psi_-(x)\,dx. \tag{2.425}$$

On $\tilde{\tilde{L}}_-^2(\tilde{\tilde{E}})$ the kinetic energy operator acts as a multiplication operator, i.e.,

$$\tilde{\tilde{K}}_- \tilde{\tilde{\phi}}_-(\varepsilon) = \left(\hat{U}_{K_-} \hat{K}_- \hat{U}_{K_-}^{-1}\right) \tilde{\tilde{\phi}}_-(\varepsilon) = \epsilon\,\tilde{\tilde{\phi}}_-(\varepsilon). \tag{2.426}$$

Finally we can combine $\tilde{\tilde{L}}_-^2(\tilde{\tilde{E}})$ and $\tilde{\tilde{L}}_+^2(\tilde{\tilde{E}})$ to construct a new Hilbert space $\tilde{\tilde{L}}^2(\tilde{\tilde{E}})$ composed of ordered pairs

$$\tilde{\tilde{\phi}} = \left\{\tilde{\tilde{\phi}}_-(\varepsilon),\,\tilde{\tilde{\phi}}_+(\varepsilon)\right\} \tag{2.427}$$

with scalar product defined by

$$\langle \tilde{\tilde{\phi}} \mid \tilde{\tilde{\phi}} \rangle_K \;=\; \langle \tilde{\tilde{\phi}}_- \mid \tilde{\tilde{\phi}}_- \rangle_{K_-} + \langle \tilde{\tilde{\phi}}_+ \mid \tilde{\tilde{\phi}}_+ \rangle_{K_+}. \tag{2.428}$$

Any $\phi \in L^2(E)$ can be written as a sum of ϕ_- and ϕ_+ so that we can define a new function $\tilde{\tilde{\phi}}(\varepsilon)$ on $\tilde{\tilde{E}}$ to correspond to ϕ by

$$\tilde{\tilde{\phi}}(\varepsilon) = \left\{\tilde{\tilde{\phi}}_-(\varepsilon),\,\tilde{\tilde{\phi}}_+(\varepsilon)\right\}. \tag{2.429}$$

The correspondence is one-to-one and unitary since scalar products are preserved, i.e.,

$$\langle \tilde{\tilde{\phi}} \mid \tilde{\tilde{\phi}} \rangle_K = \langle \phi \mid \phi \rangle. \tag{2.430}$$

The kinetic energy operator acts on $\tilde{\tilde{L}}^2(\tilde{\tilde{E}})$ as a multiplication operator:

$$\tilde{\tilde{K}} \left\{\tilde{\tilde{\phi}}_-(\varepsilon),\,\tilde{\tilde{\phi}}_+(\varepsilon)\right\} = \left\{\epsilon\,\tilde{\tilde{\phi}}_-(\varepsilon),\,\epsilon\,\tilde{\tilde{\phi}}_+(\varepsilon)\right\}. \tag{2.431}$$

The construction of a Hilbert space in terms of pairs of elements belonging to a pair of given Hilbert spaces is not restricted to our present special case. In §2.13.1 and §2.14.1 we shall introduce a formal theory of direct sum of Hilbert spaces and operators to describe such pairs of quantities. Our results here can then be rewritten neatly in direct sum notations; this is done in §2.13.1C(1) C3 and §2.14.1E(1) E3. The results presented here will be applied to discussions later in §3.6.6 and §5.8.

2.9 Generalized Spectral Measures and Probability

A symmetric operator which is not selfadjoint does not satisfy the Spectral Theorem 2.8.1(1). Still, we can have a generalized version of the spectral theorem based on generalized spectral measures and generalized spectral functions. While spectral measures map Borel sets of the reals to projectors, generalized spectral measures map Borel sets to positive operators whose norm is less than or equal to 1. These operators would also have expectation values in unit vectors not bigger than 1. Similar to projectors these positive operators can relate to probabilities.

Definition 2.9(1)   **Generalized spectral measures and functions**

- A *generalized spectral measure* on a Hilbert space $\mathcal{H}$ is a *normalized positive operator-valued set function* $\widehat{M}_+$ on the Borel sets $I\!B(I\!R)$ of $I\!R$, namely[123]

$$\widehat{M}_+ : I\!B(I\!R) \mapsto \left\{\text{positive operators of norm} \leq 1 \text{ on } \mathcal{H}\right\} \quad (2.432)$$

by

$$\Lambda \in I\!B(I\!R) \mapsto \widehat{M}_+(\Lambda) \in \left\{\text{positive operators of norm} \leq 1 \text{ on } \mathcal{H}\right\}$$

such that

1. $\widehat{M}_+(I\!R) = \widehat{I\!I}$, and
2. $\widehat{M}_+(\Lambda_1 \cup \Lambda_2 \cup \cdots) = \widehat{M}_+(\Lambda_1) + \widehat{M}_+(\Lambda_2) + \cdots$,   if $\Lambda_1, \Lambda_2, \ldots$ is any sequence of mutually disjoint Borel sets.

---

[123] Busch, Lahti and Mittelstaedt (1996), Schroeck (1996).

168 CHAPTER 2. OPERATORS AND THEIR DIRECT INTEGRALS

- A *generalized spectral function* on a Hilbert space $\mathcal{H}$ is a *normlized positive operator-valued function* $\widehat{F}_+$ on $\mathbb{R}$, namely[124]

$$\widehat{F}_+ : \mathbb{R} \mapsto \left\{\text{positive operators of norm } \leq 1 \text{ on } \mathcal{H}\right\} \tag{2.433}$$

by

$$\tau \in \mathbb{R} \mapsto \widehat{F}_+(\tau) \in \left\{\text{positive operators of norm } \leq 1 \text{ on } \mathcal{H}\right\}$$

such that

1. $\widehat{F}_+(-\infty) = \widehat{O}, \quad \widehat{F}_+(\infty) = \widehat{I}$;
2. $\widehat{F}_+(\tau + 0) = \widehat{F}_+(\tau) \; \forall \tau \in \mathbb{R}$; and
3. $\widehat{F}_+(\tau_2) - \widehat{F}_+(\tau_1)$ is a positive operator whenever $\tau_2 \geq \tau_1$.

Theorem 2.9(1)   Naimark's theorem on PV and POV functions[125]

- Let $\widehat{F}_+(\tau)$ be a POV function on a Hilbert space $\mathcal{H}$. Then there exists a PV function $\widehat{F}_{ex}(\tau)$ on an extended Hilbert space $\mathcal{H}_{ex}$ which contains $\mathcal{H}$ as a subspace such that

$$\widehat{F}_+(\tau)\phi = \widehat{P}_{\mathcal{H}}\left(\widehat{F}_{ex}(\tau)\phi\right) \quad \forall \phi \in \mathcal{H}, \tag{2.434}$$

where $\widehat{P}_{\mathcal{H}}$ is the projector on $\mathcal{H}_{ex}$ projecting onto the subspace $\mathcal{H}$.

Comments 2.9(1)   POV measures, POV functions and probability

C1   Generalized spectral measures are also known as *positive operator-valued measures* or *POV measures* or simply *POVM* for short. Generalized spectral functions are also referred to as *positive operator-valued functions* or *generalized resolutions of the identity*, or *POV functions* for short. POV functions $\widehat{F}_+$ correspond one-to-one to POV measures $\widehat{M}_+$ in the same way that PV functions are related to PV measures. As with PV measures one can picture POV measures as the positive operator equivalence of probability measures.

C2   POV measures and POV functions are defined in a similar way as PV measures and PV functions. Indeed PV measures and PV functions are special

---

[124] Akhiezer and Glazman Vol. 2 (1963) p. 121, Naimark (1968) Vol. 1 p. 45.
[125] Akhiezer and Glazman Vol. 2 (1963) p. 124, Naimark (1968) Vol. 1 pp. 45-46.

2.9. GENERALIZED SPECTRAL MEASURES AND PROBABILITY

cases of POV measures and POV functions. For example, the third property of POV functions suggests that a POV function is a non-decreasing function. A PV function $\widehat{F}$ clearly satisfies this condition since for $\tau_2 \geq \tau_1$, the difference $\widehat{F}(\tau_2) - \widehat{F}(\tau_1)$ is a projector.[126]

C3 POV functions reduce to PV functions if the orthogonality property in Eq. (2.293) for PV functions is satisfied. This condition implies

$$\widehat{F}_+(\tau)\widehat{F}_+(\tau) = \widehat{F}_+(\tau). \tag{2.435}$$

In other words $\widehat{F}_+(\tau)$ are projectors.

C4 Generally a POV measure and its corresponding POV function do not possess all the properties of PV measures and PV functions listed in §2.7C(1) C5. In particular, the positive operators involved in a given POV measure and its associated POV function may not mutually commute.

C5 The converse to the above Naimark's Theorem also holds. Let $\widehat{F}(\tau)$ be a PV function on $\mathcal{H}$. Let $\mathcal{L}$ be a subspace of $\mathcal{H}$ and let $\widehat{P}_\mathcal{L}$ be the projector onto the subspace $\mathcal{L}$. We can define an operator-valued function on $\mathcal{L}$, treated as a Hilbert space in its own right, by

$$\widehat{F}_\mathcal{L}(\tau)\varphi = \widehat{P}_\mathcal{L}\big(\widehat{F}(\tau)\varphi\big) \quad \forall \varphi \in \mathcal{L}. \tag{2.436}$$

Then $\widehat{F}_\mathcal{L}(\tau)$ can be shown to be a POV function on $\mathcal{L}$.
   One can use Naimark's Theorem to show that

$$\left\|\widehat{F}_+(\tau_2) - \widehat{F}_+(\tau_1)\right\| \leq 1. \tag{2.437}$$

To appreciate this result we can imagine $\widehat{F}_+(\tau)$ is obtained from a PV function $\widehat{F}_{ex}(\tau)$ on an extended Hilbert space $\mathcal{H}_{ex}$, i.e., $\widehat{F}_+(\tau) = \widehat{P}_\mathcal{H}\widehat{F}_{ex}(\tau)$, where $\widehat{P}_\mathcal{H}$ is the projector from $\mathcal{H}_{ex}$ onto $\mathcal{H}$. Using a well-known result $\|\widehat{A}_1\widehat{A}_2\| \leq \|\widehat{A}_1\| \|\widehat{A}_2\|$ in Eq. (2.525) we can deduce that

$$\left\|\widehat{F}_+(\tau_2) - \widehat{F}_+(\tau_1)\right\| = \left\|\widehat{P}_\mathcal{H}\big(\widehat{F}_{ex}(\tau_2) - \widehat{F}_{ex}(\tau_1)\big)\right\| \leq 1, \tag{2.438}$$

bearing in mind that $\widehat{F}_{ex}(\tau_2) - \widehat{F}_{ex}(\tau_1)$ is a projector on $\mathcal{H}_{ex}$ having unit norm.

C6 All the operators involved here are positive and bounded with a norm less than or equal to 1, and hence selfadjoint. We call $\widehat{M}_+(\Lambda)$ and $\widehat{F}_+(\tau)$ *generalized spectral operators*. Moreover, we have, for every unit vector $\phi \in \mathcal{H}$,

$$\langle \phi \mid \widehat{M}_+(\Lambda)\phi \rangle \leq 1, \quad \langle \phi \mid \widehat{F}_+(\tau)\phi \rangle \leq 1. \tag{2.439}$$

---
[126] See §2.3C(1) C11 and §2.7C(1) C1.

These results enable us to generate probability measures and probability functions. A formal statement of this is given in §2.11.2. However a new situation arises here. For some Λ and some τ we may well have

$$||\widehat{M}_+(\Lambda)|| < 1, \quad ||\widehat{F}_+(\tau)|| < 1. \tag{2.440}$$

It follows that for every unit vector $\phi \in \mathcal{H}$ we will have

$$\langle \phi \mid \widehat{M}_+(\Lambda)\phi \rangle < 1, \quad \langle \phi \mid \widehat{F}_+(\tau)\phi \rangle < 1. \tag{2.441}$$

If we interpret $\langle \phi \mid \widehat{M}_+(\Lambda)\phi \rangle$ and $\langle \phi \mid \widehat{F}_+(\tau)\phi \rangle$ as a probability then this probability is less than 1 for every unit vector ϕ, a situation quite different from the probability generated by a PV measure. This is a distinguishing feature of POV measures and POV functions.

Examples 2.9(1)   **POV functions and PV functions**[127]

E1   Let $\widehat{F}^{(1)}$ and $\widehat{F}^{(2)}$ be two arbitrary PV functions, and a_1, a_2 be two positive real numbers adding to one, i.e., $a_1 + a_2 = 1$. Then $a_1\widehat{F}^{(1)}(\tau) + a_2\widehat{F}^{(2)}(\tau)$ is a positive operator for any τ and the resulting operator-valued function

$$\widehat{F}_+(\tau) = a_1\widehat{F}^{(1)}(\tau) + a_2\widehat{F}^{(2)}(\tau) \tag{2.442}$$

is a POV function. One can check that $\widehat{F}_+(\tau)$ is generally not a projector.

E2   Let us illustrate the converse of Naimark's Theorem with an explicit example. The space $L^2(I\!\!E)$ contains $L^2(I\!\!E^+)$ as a subspace and the projector onto $L^2(I\!\!E^+)$ is simply the characteristic function $\chi_{(0,\infty)}(x)$ acting as a multiplication operator on $L^2(I\!\!E)$. We have on $L^2(I\!\!E)$ the PV function $\widehat{F}^{\hat{p}}$ of the momentum operator given in §2.8.1E(1) E2. This PV function generates a POV function $\widehat{F}_+^{\hat{p}^+}$ on $L^2(I\!\!E^+)$ by[128]

$$\widehat{F}_+^{\hat{p}^+}(\tau)\phi = \chi_{(0,\infty)} \widehat{F}^{\hat{p}}(\tau)\phi \quad \forall \phi \in L^2(I\!\!E^+). \tag{2.443}$$

2.10 Spectral Functions of Symmetric Operators

2.10.1 Symmetric operators and their spectral functions

Let $\widehat{S}$ be a symmetric operator in a Hilbert space $\mathcal{H}$. We can define generalized spectral functions to associated with $\widehat{S}$.

---
[127] Akhiezer and Glazman Vol. 2 (1963), Appendix I.
[128] The reason for the notation $\widehat{F}_+^{\hat{p}^+}$ with an additional superscript $+$ attached to the superscript $\hat{p}$ will become clear in §2.10.1C(1) C4 later.

2.10. SPECTRAL FUNCTIONS OF SYMMETRIC OPERATORS

Definition 2.10.1(1)   **Generalized spectral functions of symmetric operators**

- A generalized spectral function is called a *generalized spectral function* of $\widehat{S}$, or a *spectral function* of $\widehat{S}$ or a *POV function* of $\widehat{S}$, denoted by $\widehat{F}_+^{\widehat{S}}$, if for all $\psi \in \mathcal{H}$ and all $\phi \in \mathcal{D}(\widehat{S})$ we have

$$\langle \psi \mid \widehat{S}\phi \rangle = \int_{-\infty}^{\infty} \tau\, d\langle \psi \mid \widehat{F}_+^{\widehat{S}}(\tau)\phi \rangle, \qquad (2.444)$$

$$\left\| \widehat{S}\phi \right\|^2 = \int_{-\infty}^{\infty} \tau^2\, d\langle \phi \mid \widehat{F}_+^{\widehat{S}}(\tau)\phi \rangle. \qquad (2.445)$$

- We call the measure $\widehat{M}_+^{\widehat{S}}$ associated with a generalized spectral function $\widehat{F}_+^{\widehat{S}}$ of $\widehat{S}$ a *generalized spectral measure of* $\widehat{S}$, or a *POV measure of* $\widehat{S}$.

Comments 2.10.1(1)   **Naimark's construction of POV functions**

C1   There are important differences between PV measures and POV measures in their relation to operators:

1. PV measures and PV functions correspond one-to-one to selfadjoint operators in accordance with Spectral Theorem 2.8.1(1). The same is not true for POV measures and POV functions. A symmetric operator generally admits many different POV functions. One can gain an intuition as to why this the case in the discussion in C3 below.

2. An arbitrary POV function may not correspond to any symmetric operator since there are POV functions for which condition imposed by Eq. (2.445) is violated, i.e., the integral may not converge, for any ϕ in the Hilbert space.[129] In other words we cannot employ Eq. (2.318) in Spectral Theorem 2.8.1(1) for selfadjoint operators to construct a domain, and hence a symmetric operator for an arbitrary POV function.

C2   Following Eq. (2.321) one is tempted to symbolize the relations embodied in Eqs. (2.444) and (2.445) by writing

$$\widehat{S} = \int_{-\infty}^{\infty} \tau\, d\widehat{F}_+^{\widehat{S}}(\tau). \qquad (2.446)$$

However, unlike Eq. (2.321) for selfadjoint operators and PV functions which can be given a meaning as a limit in a strong sense, Equation (2.446) cannot be given such a meaning generally. So, Eq. (2.446) is more of a symbolic expression.[130] Equation (2.333) valid for selfadjoint operators again fails here.

---
[129] Akhiezer and Glazman Vol. 2 (1963) p. 132.
[130] Akhiezer and Glazman Vol. 2 (1963) pp. 132-133.

Instead we have[131]

$$\langle \widehat{S}\psi \mid \widehat{M}_+^{\hat{S}}(\Lambda)\,\phi \rangle = \langle \psi \mid \widehat{S}^\dagger\,\widehat{M}_+^{\hat{S}}(\Lambda)\,\phi \rangle = \int_\Lambda \tau\, d\langle \psi \mid \widehat{F}_+^{\hat{S}}(\tau)\phi \rangle. \qquad (2.447)$$

C3 We can make use of Naimark's Theorem 2.9(1) to obtain a POV function of a symmetric operator $\widehat{S}$ in a Hilbert space $\mathcal{H}$ in terms of the PV function of a selfadjoint operator in an extended Hilbert space. Let $\mathcal{H}_{ex}$ be a Hilbert space which contains $\mathcal{H}$ as a subspace. Let $\widehat{S}_{ex}$ be a selfadjoint operator in $\mathcal{H}_{ex}$ with domain $\mathcal{D}(\widehat{S}_{ex})$ containing the domain $\mathcal{D}(\widehat{S})$ of $\widehat{S}$ such that when acting in $\mathcal{D}(\widehat{S})$ the two operators are the same, i.e.,

$$\widehat{S}_{ex}\phi = \widehat{S}\phi \quad \forall \phi \in \mathcal{D}(\widehat{S}) \subset \mathcal{H} \subset \mathcal{H}_{ex}. \qquad (2.448)$$

We may then regard $\widehat{S}_{ex}$ as a selfadjoint extension of $\widehat{S}$ in the extended space $\mathcal{H}_{ex}$. Now, let $\widehat{P}_{ex}$ be the projector on $\mathcal{H}_{ex}$ projecting down to the subspace $\mathcal{H}$. Then Naimark's Theorem 2.9(1) enables us to obtain a POV functions $\widehat{F}_+^{\hat{S}}$ of $\widehat{S}$ in terms of the PV function $\widehat{F}^{\hat{S}_{ex}}$ of $\widehat{S}_{ex}$ by

$$\widehat{F}_+^{\hat{S}}\phi = \widehat{P}_{ex}\left(\widehat{F}^{\hat{S}_{ex}}\phi\right) \quad \forall \phi \in \mathcal{H}. \qquad (2.449)$$

Given that neither $\mathcal{H}_{ex}$ nor $\widehat{S}_{ex}$ is unique we can appreciate that a symmetric operator may admit many different POV functions.

C4 As an example of the procedure outline above let us consider $\widehat{p} = -i\hbar d/dx$ defined in $L^2(I\!\!E)$ introduced in §2.5.1E(1) E3 and $\widehat{p}^+ = -i\hbar d/dx$ defined in $L^2(I\!\!E^+)$ introduced in §2.5.1E(1) E6. We know that:

1. $\widehat{p}$ is selfadjoint in $L^2(I\!\!E)$ and $\widehat{p}^+$ is strictly maximal symmetric in $L^2(I\!\!E^+)$.

2. $L^2(I\!\!E)$ contains $L^2(I\!\!E^+)$ as a subspace. The projector on $L^2(I\!\!E)$ which projects onto $L^2(I\!\!E^+)$ is the characteristic function $\chi_{(0,\infty)}(x)$ on $I\!\!R$.

3. The domain of $\widehat{p}^+$ is

$$\mathcal{D}(\widehat{p}^+) = \{\phi_+(x) \in S_{2,1}(I\!\!E^+) : \phi_+(0) = 0\}. \qquad (2.450)$$

We can extend functions in $\mathcal{D}(\widehat{p}^+)$ to functions $\phi(x)$ on $I\!\!E$ by setting

$$\phi(x) = \begin{cases} 0 & \text{for } x \leq 0 \\ \phi_+(x) & \text{for } x \geq 0 \end{cases}. \qquad (2.451)$$

This renders $\mathcal{D}(\widehat{p}^+)$ a subset of $\mathcal{D}(\widehat{p})$ of $\widehat{p}$.

---
[131] Akhiezer and Glazman Vol. 2 (1963) p. 133.

2.10. SPECTRAL FUNCTIONS OF SYMMETRIC OPERATORS

4. Acting on $\mathcal{D}(\hat{p}^+)$ the operators $\hat{p}$ and $\hat{p}^+$ are the same.

It follows that a POV function $\widehat{F}_+^{\hat{p}^+}$ of $\hat{p}^+$ can be obtained in terms of the PV function $\widehat{F}^{\hat{p}}$ of $\hat{p}$. Using Eq. (2.317) for $\widehat{F}^{\hat{p}}$ we get, for any $\phi_+ \in L^2(I\!\!E^+) \subset L^2(I\!\!E)$,[132]

$$\left(\widehat{F}_+^{\hat{p}^+}(\tau)\phi_+\right)(x)$$

$$= \chi_{(0,\infty)}(x) \left(\widehat{F}^{\hat{p}}\phi_+\right)(x) \tag{2.452}$$

$$= \chi_{(0,\infty)}(x) \left[\frac{1}{\sqrt{2\pi\hbar}} \int_{-\infty}^{\tau} d\tau' \, e^{\frac{i}{\hbar}\tau' x} \widetilde{\phi}_+(\tau')\right] \tag{2.453}$$

$$= \chi_{(0,\infty)}(x) \left[\frac{1}{2\pi\hbar} \int_{-\infty}^{\tau} d\tau' \int_{-\infty}^{\infty} dx' \, e^{\frac{i}{\hbar}\tau'(x-x')} \phi_+(x')\right]$$

$$= \frac{1}{2\pi\hbar} \chi_{(0,\infty)}(x) \int_{-\infty}^{\tau} d\tau' \int_{0}^{\infty} dx' \, e^{\frac{i}{\hbar}\tau'(x-x')} \phi_+(x')$$

$$= \frac{1}{2\pi\hbar} \int_{-\infty}^{\tau} d\tau' \int_{0}^{\infty} dx' \, e^{\frac{i}{\hbar}\tau'(x-x')} \phi_+(x'), \quad x > 0. \tag{2.454}$$

This is the POV function obtained in §2.9E(1) E2.

The question now is whether there are other POV functions. As it turns out, $\hat{p}^+$ possesses no other POV functions. The reason is that $\hat{p}^+$ is strictly maximal symmetric, albeit not selfadjoint. As stated in Theorem 2.10.2(1) below, maximal symmetric operators are uniquely associated with their POV functions. Strictly maximal symmetric operators resemble selfadjoint operators in this respect and hence we expect these two types of operators to play a crucial role in quantum mechanics.

2.10.2 Strictly maximal symmetric operators and their spectral functions

The relationship between strictly maximal symmetric operators and their POV functions can be summarized in a theorem.

Theorem 2.10.2(1)   **Spectral theorem for maximal symmetric operators**[133]

- A closed symmetric operator possesses a unique POV function if and only if it is maximal.

---
[132] The function $\phi_+(x) \in L^2(I\!\!E^+)$ is extended to $L^2(I\!\!E)$ by setting $\phi_+(x) = 0$ for $x \leq 0$.
[133] Akhiezer and Glazman (1963) Vol. 2 p. 135, Naimark (1968) Vol. 1 pp. 45-47. Note that maximal symmetric operators include selfadjoint operators.

Examples 2.10.2(1)   Radial momentum operator and its generalized spectral function

E1   Since $\widehat{p}^+$ is maximal symmetric in $L^2(I\!\!E^+)$ Theorem 2.10.2(1) implies that the POV function $\widehat{F}_+^{\widehat{p}^+}$ given by Eq. (2.454) is unique to $\widehat{p}^+$.

E2   Using spherical polar coordinates (r, θ, φ) in $I\!\!E^3$ the Hilbert space $L^2(I\!\!E^3)$ can be decomposed into a radial part and an angular parts, i.e.,[134]

$$L^2(I\!\!E^3) = L^2(I\!\!R^+, r^2 dr) \otimes L^2(\mathcal{S}^2, \sin\theta d\theta d\varphi). \quad (2.455)$$

Here the variable r takes values in $I\!\!R^+$, i.e., $r \in (0, \infty)$, and $L^2(I\!\!R^+, r^2 dr)$ is the Hilbert space of functions on $I\!\!R^+$ square-integrable with respect to the volume element $r^2 dr$. The symbol $\mathcal{S}^2$ stands for the two-dimensional sphere of unit radius, centred at the origin in $I\!\!E^3$, with (θ, φ) as coordinates, and $L^2(\mathcal{S}^2, \sin\theta d\theta d\varphi)$ is the Hilbert space of functions on $\mathcal{S}^2$ square-integrable with respect to the volume element $\sin\theta d\theta d\varphi$.

Our object here is to define the *radial momentum operator* and obtain its generalized spectral function. Before we can do so we need to introduce another operator first. Let us define an operator $\widehat{p}_r$ in $L^2(I\!\!R^+, r^2 dr)$ by the differential expression[135]

$$\widehat{p}_r = -i\hbar \left(\frac{d}{dr} + \frac{1}{r} \right) \quad (2.456)$$

acting on the domain

$$\mathcal{D}(\widehat{p}_r) = \left\{ f \in AC(I\!\!R^+) : \widehat{p}_r f \in L^2(I\!\!R^+, r^2 dr), \lim_{r \to 0} r|f(r)| = 0 \right\}. \quad (2.457)$$

We can uncover the nature of this operator by a comparison with $\widehat{p}^+$ in $L^2(I\!\!E^+)$:

1. There is a unitary operator $\widehat{\mathcal{U}}_r$ linking $L^2(I\!\!R^+, r^2 dr)$ and $L^2(I\!\!E^+)$ by mapping every $f \in L^2(I\!\!R^+, r^2 dr)$ to $\phi_+ \in L^2(I\!\!E^+)$ and *vice versa* according to:

$$f(r) \to \phi_+(x) = \left(\widehat{\mathcal{U}}_r f \right)(x) = xf(x), \quad x \in (0, \infty), \quad (2.458)$$

$$\phi_+(x) \to f(r) = \left(\widehat{\mathcal{U}}_r^{-1} \phi_+ \right)(r) = \frac{1}{r} \phi_+(r), \quad r \in (0, \infty). \quad (2.459)$$

2. The domains $\mathcal{D}(\widehat{p}_r)$ and $\mathcal{D}(\widehat{p}^+)$ of the two operators are related unitarily, i.e., $\widehat{\mathcal{U}}_r \mathcal{D}(\widehat{p}_r) = \mathcal{D}(\widehat{p}^+)$. The boundary condition on ϕ in Eq. (2.450) is seen to be related to the condition on $f(x)$ in Eq. (2.457).

---
[134] Prugovečki (1981) p. 151.
[135] See Richtmyer Vol. 1 (1978) pp. 139-140, p. 157, Wan, Fountain and Tao (1995).

2.10. SPECTRAL FUNCTIONS OF SYMMETRIC OPERATORS

3. It is easy to verify that for every $f \in \mathcal{D}(\widehat{p}_r)$ and $\phi_+ = \widehat{\mathcal{U}}_r f$

$$\widehat{p}^+ \phi_+ = \widehat{\mathcal{U}}_r (\widehat{p}_r f), \qquad (2.460)$$

i.e., we have

$$\widehat{\mathcal{U}}_r \widehat{p}_r \widehat{\mathcal{U}}_r^{-1} = \widehat{p}^+ \quad \text{or} \quad \widehat{p}_r = \widehat{\mathcal{U}}_r^{-1} \widehat{p}^+ \widehat{\mathcal{U}}_r. \qquad (2.461)$$

It follows that $\widehat{p}_r$ and $\widehat{p}^+$ are unitarily equivalent, and that $\widehat{p}_r$ is strictly maximal symmetric.[136]

The *radial momentum operator* $\widehat{P}_r$ in $L^2(\mathbb{E}^3)$ is defined to be the closure of the tensor product operator[137]

$$\widehat{p}_r \otimes \widehat{\mathbb{I}}(\mathcal{S}^2), \quad \widehat{\mathbb{I}}(\mathcal{S}^2) = \text{the identity operator on } L^2(\mathcal{S}^2, \sin\theta d\theta d\varphi), \quad (2.462)$$

which is a strictly maximal symmetric operator in $L^2(\mathbb{E}^3)$.

The unitary equivalence of $\widehat{p}_r$ and $\widehat{p}^+$ enables us to write down the POV function of $\widehat{p}_r$ in terms of that of $\widehat{p}^+$, i.e., the POV function $\widehat{F}^{\widehat{p}_r}(\tau)$ of $\widehat{p}_r$ is equal to $\widehat{\mathcal{U}}_r^{-1} \widehat{F}^{\widehat{p}^+}(\tau) \widehat{\mathcal{U}}_r$. Explicitly we have, for $f \in L^2(\mathbb{R}^+, r^2 dr)$,

$$\begin{aligned}
\left(\widehat{F}^{\widehat{p}_r}(\tau) f\right)(r) \\
&= \widehat{\mathcal{U}}_r^{-1} \widehat{F}^{\widehat{p}^+}(\tau) \widehat{\mathcal{U}}_r f(r) \\
&= \widehat{\mathcal{U}}_r^{-1} \left(\widehat{F}_+^{\widehat{p}^+}(\tau) x f(x)\right) \\
&= \widehat{\mathcal{U}}_r^{-1} \left[\frac{1}{2\pi\hbar} \int_{-\infty}^{\tau} d\tau' \int_0^{\infty} dx' e^{i\tau'(x-x')} x' f(x')\right] \\
&= \frac{1}{2\pi\hbar r} \int_{-\infty}^{\tau} d\tau' \int_0^{\infty} dr' e^{i\tau'(r-r')} r' f(r'), \quad r > 0. \qquad (2.463)
\end{aligned}$$

2.10.3 The square of maximal symmetric operators

For selfadjoint operators we define their functions through their PV functions. In particular we know from §2.8.3 that a function $f(\widehat{A})$ of a selfadjoint operator $\widehat{A}$ is selfadjoint if f is a real-valued function of x. The situation becomes rather complicated for strictly maximal symmetric operators $\widehat{S}$. We cannot define functions of a strictly maximal symmetric operator in the same way.

---

[136]The operator expression in Eq. (2.456) is used in many books on quantum mechanics. Some authors even use the terms hermiticity and selfadjointness interchangeably, a practice which can be confusing (Flügge (1974) p. 151). Because of the boundary condition at the origin the radial momentum operator admits no eigenfunctions (Messiah Vol. 1 (1967) p. 346.

[137]We have taken the closure here since tensor product of two closed operators is not necessarily closed.

Let us illustrate the problem with an example which will also have direct physical applications later. Suppose we want to define the square of a strictly maximal symmetric operator $\widehat{S}$. We could simply take the product $\widehat{S}^2 = \widehat{S}\widehat{S}$ as the answer. However, $\widehat{S}^2$ is not maximal symmetric.[138] This presents a severe problem when it comes to its physical applications. In the next chapter we shall study quantization. There we need to consider the quantization of a classical observable A as well as its square A^2. If A is quantized as a selfadjoint operator $\widehat{A}$ then we can assume that A^2 is quantized as the operator $\widehat{A^2}$ which is automatically selfadjoint. The question arises as to what operator the classical observable A^2 should be quantized as, if A is quantized as a strictly maximal symmetric operator. Orthodox quantum theory would demand that A^2 be quantized as a selfadjoint operator. The best known example is the quantization of the square of the radial momentum p_r. As spelled out in §3.3.7 in the next chapter the quantity p_r^2 appears in the classical expression Eq. (3.200) for the kinetic energy in spherical coordinates. We can first quantize p_r as the strictly maximal symmetric radial momentum operator $\widehat{p}_r$ defined in §2.10.2E(1) E2. Next, we desire p_r^2 to be quantized as a maximal symmetric operator, or better still as a selfadjoint operator. The product operator $\widehat{p}_r^2$ does not satisfy this requirement. One way to proceed is to choose a selfadjoint extension to $\widehat{p}_r^2$ as the quantized operator for p_r^2 in order to produce a selfadjoint operator to relate to p_r^2.

Our discussion in §2.3C(1) C7 tells us that the product $\widehat{S}^2$ does admit selfadjoint extensions. An obvious selfadjoint extension is simply $\widehat{S}^\dagger \widehat{S}$, known as the *Friedrichs extension*[139] to $\widehat{S}^2$. We can go a step further to evaluate scalar products of the Friedrichs extension in terms of the POV function $\widehat{F}_+^{\widehat{S}}(\tau)$ of $\widehat{S}$. For $\psi \in \mathcal{D}(\widehat{S})$ and $\phi \in \mathcal{D}(\widehat{S}^\dagger \widehat{S})$, we can evaluate scalar products as follows:

$$\langle \psi \mid \widehat{S}^\dagger \widehat{S} \phi \rangle = \langle \widehat{S}\psi \mid \widehat{S}\phi \rangle = \int_{-\infty}^{\infty} \tau \, d\tau \langle \widehat{S}\psi \mid \widehat{F}_+^{\widehat{S}}(\tau)\phi \rangle. \tag{2.464}$$

From Eq. (2.447) we obtain, for all $\psi, \phi \in \mathcal{D}(\widehat{A})$:

$$\langle \widehat{S}\psi \mid \widehat{F}_+^{\widehat{S}}(\tau)\phi \rangle = \int_{-\infty}^{\tau} \tau' \, d\tau' \langle \psi \mid \widehat{F}_+^{\widehat{S}}(\tau')\phi \rangle. \tag{2.465}$$

So

$$\langle \psi \mid \widehat{S}^\dagger \widehat{S} \phi \rangle = \int_{-\infty}^{\infty} \tau \, d\tau \int_{-\infty}^{\tau} \tau' \, d\tau' \langle \psi \mid \widehat{F}_+^{\widehat{S}}(\tau')\phi \rangle \tag{2.466}$$

---

[138] To see this, let T be strictly maximal symmetric, $n = 2$ and $m = 1$ in Exercises 8.5 on p. 243 of Weidmann (1980) to arrive at a contradiction.

[139] Reed and Simon Vol. 2 (1975) p. 181.

2.10. SPECTRAL FUNCTIONS OF SYMMETRIC OPERATORS 177

$$= \int_{-\infty}^{\infty} \tau^2 \, d_\tau \langle \psi \mid \widehat{F}_+^{\hat{S}}(\tau)\phi \rangle. \qquad (2.467)$$

This is consistent with Eq. (2.445) in Definition 2.10.1(1). Compared this with Eq. (2.357) we see that scalar products of $\widehat{S}^\dagger \widehat{S}$ can be evaluated in a similar fashion as the square of a selfadjoint operator.

Again consider the strictly maximal symmetric operator $\widehat{p}^+ = -i\hbar d/dx$ in $L^2(I\!\!E^+)$. The product $(\widehat{p}^+)^2$ is defined by the differential expression $-\hbar^2 \, d^2/dx^2$ operating on the domain

$$\mathcal{D}\left((\widehat{p}^+)^2\right) = \left\{\phi \in \mathcal{D}(\widehat{p}^+) : \phi' \in \mathcal{D}(\widehat{p}^+)\right\} \qquad (2.468)$$

$$= \left\{\phi \in S_{2,2}(I\!\!E^+) : \phi(0) = 0 \text{ and } \phi'(0) = 0\right\}, \qquad (2.469)$$

where $\phi' = d\phi/dx$. This can be compared with the corresponding expression for the operator $\widehat{K}_0(I\!\!E^+)$ defined by Eqs. (2.218) and (2.219). As pointed out in §2.5.1E(1) E6 the domain of its adjoint $(\widehat{p}^+)^\dagger$ is $S_{2,1}(I\!\!E^+)$. It follows that the Friedrichs extension $(\widehat{p}^+)^\dagger \widehat{p}^+$ is defined by the same differential expression $-\hbar^2 \, d^2/dx^2$ acting on the domain

$$\mathcal{D}\left((\widehat{p}^+)^\dagger \widehat{p}^+\right) = \left\{\phi \in \mathcal{D}(\widehat{p}^+) : \phi' \in \mathcal{D}((\widehat{p}^+)^\dagger)\right\} \qquad (2.470)$$

$$= \left\{\phi \in S_{2,2}(I\!\!E^+) : \phi(0) = 0\right\}. \qquad (2.471)$$

This extension can be compared with the operator $\widehat{K}(I\!\!E^+)$ defined by Eqs. (2.220) and (2.222). There are other selfadjoint extensions with domains defined by[140]

$$\left\{\phi \in S_{2,2}(I\!\!E^+) : \phi'(0) + \lambda\phi(0) = 0\right\}, \quad \lambda \in I\!\!R. \qquad (2.472)$$

So, in the case of the radial momentum we can choose to quantize p_r^2 as the operator $\widehat{p}_r^\dagger \widehat{p}_r$; this turns out to be the correct choice. We shall return to this in more detail later in §3.3.6.

In view of the non-uniqueness of selfadjoint extensions of $\widehat{S}^2$ we are confronted with non-uniqueness in quantization. We shall study the physical implications of this non-uniqueness on quantization in great details in Chapters 6 and 7.

---

[140] Reed and Simon Vol. 2 (1975) p. 144, Akhiezer and Glazman (1963) p. 204. The fact that $(\widehat{p}^+)^2$ admits non-trivial selfadjoint extensions confirms that $(\widehat{p}^+)^2$ is not maximal symmetric. If it were maximal symmetric it would not have any further selfadjoint extension.

178   CHAPTER 2. OPERATORS AND THEIR DIRECT INTEGRALS

2.10.4   Spectra of symmetric operators

We have introduced the concept of spectra of selfadjoint operators. This notion can be generalized to symmetric and other operators. Given that a symmetric operator does not generally possess a unique POV function we would want to define its spectrum directly in terms of the operator rather than its POV function. Let $\widehat{S}$ be an arbitrary linear operator in a Hilbert space $\mathcal{H}$. Define a family of operators $\widehat{S}_z$, parameterized by a complex parameter z, associated with $\widehat{S}$ by

$$\widehat{S}_z = \widehat{S} - z\,\widehat{\mathit{I}}. \tag{2.473}$$

The interest here is in identifying the values of the parameter z for which the operators $\widehat{S}_z$ admit an inverse $\widehat{S}_z^{-1}$. The reason for this has been mentioned in §2.8.3C(1) C3 and will become more apparent as we go on.

Definition 2.10.4(1)   **The resolvent set and the spectrum**

- The *resolvent set* of an operator $\widehat{S}$ is the set of all complex numbers z for which the operators $\widehat{S}_z$ admit an inverse $\widehat{S}_z^{-1}$ which is bounded and defined on the entire Hilbert space $\mathcal{H}$.

- The *spectral set* or the *spectrum* of operator $\widehat{S}$ is the set $sp(\widehat{S})$ of all complex numbers not in the resolvent set. The spectrum can be divided into three parts:[141]

    - The *point or discrete spectrum* $sp(\widehat{S})_p$: This consists of all $z \in sp(\widehat{S})$ for which $\widehat{S}_z$ is not invertible, i.e., $\widehat{S}_z^{-1}$ does not exist.

    - The *continuous spectrum* $sp(\widehat{S})_c$: This consists of all $z \in sp(\widehat{S})$ for which $\widehat{S}_z$ is invertible, and its inverse $\widehat{S}_z^{-1}$ is densely defined and unbounded.

    - The *residual spectrum* $sp(\widehat{S})_r$: This consists of all $z \in sp(\widehat{S})$ for which $\widehat{S}_z$ is invertible, and its inverse $\widehat{S}_z^{-1}$ is not densely defined.

Comments 2.10.4(1)   **Point, continuous and residual spectra**

C1   Recall that an operator may or may not be invertible, and even when it is invertible the inverse operator may not be densely defined, let alone bounded. If z is to be in the resolvent set we require the corresponding operator $\widehat{S}_z$ to be invertible with an inverse $\widehat{S}_z^{-1}$ which is bounded and defined on the entire $\mathcal{H}$.

---

[141] Roman Vol. 2 (1975) p. 582.

2.10. SPECTRAL FUNCTIONS OF SYMMETRIC OPERATORS

If z_r is in the resolvent set, then $\widehat{S}_{z_r}$ admits an inverse. From Definition 2.3(1) we get

$$\left(\widehat{S} - z_r \widehat{I}\right)\phi = 0 \quad \Rightarrow \quad \phi = 0. \tag{2.474}$$

Therefore z_r is not an eigenvalue of $\widehat{S}$. In other words the resolvent set consists of non-eigenvalues of $\widehat{S}$, and this is why the spectrum of $\widehat{S}$ consists of values not in the resolvent set.

C2   The point spectrum $sp(\widehat{S})_p$ This consists of all eigenvalues of $\widehat{S}$. To appreciate this let z_p be an eigenvalue of $\widehat{S}$ and let φ_{z_p} be an eigenvector corresponding to this eigenvalue. Then

$$\widehat{S}_{z_p} \varphi_{z_p} = \left(\widehat{S} - z_p \widehat{I}\right) \varphi_{z_p} = 0, \quad \varphi_{z_p} \neq 0. \tag{2.475}$$

It follows that $\widehat{S}_{z_p}$ is not invertible, and z_p is an element of the point spectrum.

C3   The continuous spectrum $sp(\widehat{S})_c$ Suppose z_c is a value for which $\widehat{S}_{z_c}$ is invertible. Then z_c is not an eigenvalue of $\widehat{S}$. However, $\widehat{S}$ may admit approximate eigenvectors associated with z_c, i.e., for every $\epsilon > 0$ there is a unit vector $\xi_{\epsilon, z_c} \in \mathcal{H}$ such that Eq. (2.382) in §2.8.3C(1) C4 is satisfied. Then, as in the case of selfadjoint operator, we can consider z_c as a generalized eigenvalue, i.e., z_c is a member of the continuous spectrum. Since

$$\|(\widehat{S} - z_c)\xi_{\epsilon, z_c}\| \tag{2.476}$$

can become arbitrarily small we can intuitively appreciate that the inverse operator $\widehat{S}_{z_c}^{-1}$ written in the form

$$\widehat{S}_{z_c}^{-1} = \frac{1}{\widehat{S} - z_c} \tag{2.477}$$

can be unbounded. This is indeed the case. Hence we have the formal definition of $sp(\widehat{S})_c$ above.[142]

C4   The spectrum of a selfadjoint operator according to the present definition coincides with the previous definition in terms of spectral functions; the same remark applies to point spectrum and the continuous spectrum. For a selfadjoint operator the residual spectrum is empty. Point and the continuous spectra are a generalization of eigenvalues. The residual spectrum does not quite resemble eigenvalues and does not have direct physical applications in the context of this book (see C5 below).

---
[142] See Richtmyer Vol. 1 (1978) p. 145, p. 190 for an explicit example using approximation eigenvectors.

180   CHAPTER 2. OPERATORS AND THEIR DIRECT INTEGRALS

C5   The spectrum of a symmetric operator is real if and only if it is selfadjoint or essentially selfadjoint. If the operator is symmetric but neither selfadjoint nor essentially selfadjoint, then its spectrum neccesarily contains non-real values.[143] More specifically its point and continuous spectra are real but its residual spectrum is not entirely real. This is important if we try to use a symmetric operator to describe a physical quantity; we cannot identify its entire spectrum with the measured values of the physical quantity as we can when the operator is selfadjoint, since the measured values must be real numbers. For example, if we are to include maximal symmetric operators, e.g., the radial momentum operator $\widehat{p}_r$, into quantum mechanics to represent physical observables we cannot associate their entire spectra with the corresponding physical values which are taken to be real.

C6   The spectrum of a symmetric operator is generally not the same as the spectrum of its selfadjoint extensions.[144]

2.11   Probability and Operators

2.11.1   Probability measures, spectral measures and selfadjoint operators

We shall summarize the relationship between probability measures, PV measures and selfadjoint operators in this section in the form of a theorem.

Theorem 2.11.1(1)   Probability and selfadjoint operators

- Let ϕ be a unit vector in a Hilbert space $\mathcal{H}$, and let $\widehat{M}$ be a spectral measure in $\mathcal{H}$ and $\widehat{F}$ its associated spectral function. Then the set function $\mathcal{M}_\phi$ defined on the Borel sets $B(\mathbb{R})$ of $\mathbb{R}$

$$\mathcal{M}_\phi : B(\mathbb{R}) \mapsto \mathbb{R} \tag{2.478}$$

  by

$$\Lambda \in B(\mathbb{R}) \mapsto \mathcal{M}_\phi(\Lambda) = \langle \phi \mid \widehat{M}(\Lambda)\phi \rangle, \tag{2.479}$$

  is a probability measure on $B(\mathbb{R})$, and the real-valued function $\mathcal{F}_\phi$ defined on $\mathbb{R}$ by

$$\mathcal{F}_\phi(\tau) = \langle \phi \mid \widehat{F}(\tau)\phi \rangle, \quad \tau \in \mathbb{R} \tag{2.480}$$

  is a probability function.

---
[143] Roman Vol. 2 (1975) p. 606.
[144] Roman Vol. 2 (1975) p. 605.

2.11. PROBABILITY AND OPERATORS

- A selfadjoint operator $\widehat{A}$ together with a unit vector ϕ in $\mathcal{H}$ generates a unique probability measure $\mathcal{M}_\phi^{\widehat{A}}$ and a unique probability function $\mathcal{F}_\phi^{\widehat{A}}$ defined in terms of its spectral measure $\widehat{M}^{\widehat{A}}$ and spectral function $\widehat{F}^{\widehat{A}}$ in accordance with Eqs. (2.479) and (2.480).

Examples 2.11.1(1) Continuous and discrete probability functions

E1 The position operator Following Eqs. (2.343) and (2.344) in §2.8.1E(1) E2 we get[145]

$$\mathcal{M}_\phi^{\widehat{x}}(\Lambda) = \left\| \widehat{M}^{\widehat{x}}(\Lambda)\phi \right\|^2 = \int_\Lambda |\phi(x)|^2\, dx, \quad (2.481)$$

$$\mathcal{F}_\phi^{\widehat{x}}(x) = \left\| \widehat{F}^{\widehat{x}}(x)\phi \right\|^2 = \int_{-\infty}^x |\phi(x)|^2\, dx. \quad (2.482)$$

These results enable us to interpret $|\phi(x)|^2$ as a position probability density function and call a wave function $\phi(x)$ in quantum mechanics a *position probability amplitude function*.

E2 The momentum operator Following Eqs. (2.349) and (2.347) in §2.8.1E(1) E2

$$\mathcal{M}_\phi^{\widehat{p}}(\Lambda) = \left\| \widehat{M}^{\widehat{p}}(\Lambda)\phi \right\|^2 = \int_\Lambda |\widetilde{\phi}(p)|^2\, dp, \quad (2.483)$$

$$\mathcal{F}_\phi^{\widehat{p}}(p) = \left\| \widehat{F}^{\widehat{p}}(p)\phi \right\|^2 = \int_{-\infty}^p |\widetilde{\phi}(p)|^2\, dp. \quad (2.484)$$

These results enable us to interpret $|\widetilde{\phi}(p)|^2$ as a momentum probability density function and call $\widetilde{\phi}(p)$ a *momentum probability amplitude function* in quantum mechanics.

E3 Discrete spectra Suppose $\widehat{A}$ has a purely discrete spectrum. Then its spectral function $\widehat{F}^{\widehat{A}}(\tau)$ is piecewise constant of the form of Eq. (2.373). It follows that the corresponding probability function $\mathcal{F}_\phi^{\widehat{A}}(\tau)$ is also

---
[145] See §2.11.1C(1) C1.

piecewise constant of the form of Eq. (2.306) with $\wp_j = \langle \phi \mid \widehat{P}_j^{\hat{A}} \phi \rangle$,[146] i.e.,

$$\mathcal{F}^{\hat{A}}(\tau) = \begin{cases} 0 & \text{if } \tau < \tau_1 \\ \langle \phi \mid \widehat{P}_1^{\hat{A}} \phi \rangle & \text{if } \tau_1 \leq \tau < \tau_2 \\ \langle \phi \mid \widehat{P}_1^{\hat{A}} \phi \rangle + \langle \phi \mid \widehat{P}_2^{\hat{A}} \phi \rangle & \text{if } \tau_2 \leq \tau < \tau_3 \\ \cdots & \cdots \\ \cdots & \cdots \\ \sum_{j=1}^{k} \langle \phi \mid \widehat{P}_j^{\hat{A}} \phi \rangle & \text{if } \tau_k \leq \tau < \tau_{k+1} \\ \cdots & \cdots \\ \cdots & \cdots \end{cases} \quad (2.485)$$

Definition 2.11.1(1)   **Expectation values, variances and uncertainties**

- The *expectation value* $\mathcal{E}(\widehat{A}, \phi)$, the *variance* $\mathcal{V}(\widehat{A}, \phi)$ and the *uncertainty* $\Delta(\widehat{A}, \phi)$ of a selfadjoint operator $\widehat{A}$ in a unit vector ϕ in the domain $\mathcal{D}(\widehat{A})$ of $\widehat{A}$ are defined in terms of the probability distribution function

$$\mathcal{F}_\phi^{\hat{A}} = \langle \phi \mid \widehat{F}^{\hat{A}}(\tau) \phi \rangle \quad (2.486)$$

in accordance with Definition 2.6.3(1) by

$$\mathcal{E}(\widehat{A}, \phi) = \mathcal{E}(\mathcal{F}_\phi^{\hat{A}}) = \int_{-\infty}^{\infty} \tau \, d\mathcal{F}_\phi^{\hat{A}}(\tau), \quad (2.487)$$

$$\mathcal{V}(\widehat{A}, \phi) = \mathcal{V}(\mathcal{F}_\phi^{\hat{A}}) = \int_{-\infty}^{\infty} \left(\tau - \mathcal{E}(\widehat{A}, \phi) \right)^2 d\mathcal{F}_\phi^{\hat{A}}(\tau), \quad (2.488)$$

$$\Delta(\widehat{A}, \phi) = \mathcal{V}(\widehat{A}, \phi)^{1/2}. \quad (2.489)$$

Comments 2.11.1(1)   **Alternative expressions**

C1   In view of the projector nature of PV functions and PV measures Eqs. (2.479) and (2.480) can be rewritten as

$$\mathcal{M}_\phi(\Lambda) = \|\widehat{M}(\Lambda)\phi\|^2, \quad \mathcal{F}_\phi(\tau) = \|\widehat{F}(\tau)\phi\|^2. \quad (2.490)$$

C2   We can rewrite the probability measure and the probability function as

$$\mathcal{M}_\phi^{\hat{A}}(\Lambda) = \mathcal{E}(\widehat{M}^{\hat{A}}(\Lambda), \phi), \quad (2.491)$$

$$\mathcal{F}_\phi^{\hat{A}}(\tau) = \mathcal{E}(\widehat{F}^{\hat{A}}(\tau), \phi). \quad (2.492)$$

---

[146] For unbounded operators τ_{up} in Eq. (2.306) is infinite.

2.11. PROBABILITY AND OPERATORS

In other words we may regard $\mathcal{M}_\phi^{\hat{A}}(\Lambda)$ as the expectation value of the spectral projector $\widehat{M}^{\hat{A}}(\Lambda)$ in unit vector ϕ and $\mathcal{F}_\phi^{\hat{A}}(\tau)$ as the expectation value of the spectral projector $\widehat{F}^{\hat{A}}(\tau)$ in unit vector ϕ.

C3 Expectation values, variances and uncertainties are defined only if ϕ lies in the domain $\mathcal{D}(\hat{A})$ of $\hat{A}$. Then we can express these quantities in terms of $\hat{A}$ and $\phi \in \mathcal{D}(\hat{A})$ directly as

$$\mathcal{E}(\hat{A}, \phi) = \int \tau \, d\mathcal{F}_\phi^{\hat{A}}(\tau) \tag{2.493}$$

$$= \int \tau \, d\langle \phi \mid \widehat{F}_\phi^{\hat{A}}(\tau) \phi \rangle \tag{2.494}$$

$$= \langle \phi \mid \hat{A}\phi \rangle, \tag{2.495}$$

$$\mathcal{V}(\hat{A}, \phi) = \int_{-\infty}^{\infty} \tau^2 \, d\mathcal{F}_\phi^{\hat{A}}(\tau) - \mathcal{E}(\hat{A}, \phi)^2 \tag{2.496}$$

$$= \|\hat{A}\phi\|^2 - \langle \phi \mid \hat{A}\phi \rangle^2, \tag{2.497}$$

$$\Delta(\hat{A}, \phi) = \left(\|\hat{A}\phi\|^2 - \langle \phi \mid \hat{A}\phi \rangle^2 \right)^{1/2}. \tag{2.498}$$

Here we have used Eqs. (2.319) and (2.320) in Spectral Theorem 2.8.1(1). We have not defined the variance as

$$\left\langle \phi \mid \left(\hat{A} - \mathcal{E}(\mathcal{F}_\phi^{\hat{A}}) \right)^2 \phi \right\rangle \tag{2.499}$$

as is traditionally done since this would require ϕ to be in the domain of $\hat{A}^2$.

2.11.2 Probability measures, generalized spectral measures and strictly maximal symmetric operators

Theorem 2.11.2(1) Probability and strictly maximal symmetric operators

- Let ϕ be a unit vector in a Hilbert space $\mathcal{H}$, and let $\widehat{M}_+$ be a generalized spectral measure in $\mathcal{H}$ and $\widehat{F}_+$ its associated generalized spectral function. Then the real-valued set function $\mathcal{M}_\phi$ defined on the Borel sets $\mathbb{B}(\mathbb{R})$ of $\mathbb{R}$

$$\mathcal{M}_\phi : \mathbb{B}(\mathbb{R}) \mapsto \mathbb{R} \tag{2.500}$$

by

$$\Lambda \in \mathbb{B}(\mathbb{R}) \mapsto \mathcal{M}_\phi(\Lambda) = \langle \phi \mid \widehat{M}_+(\Lambda)\phi \rangle, \tag{2.501}$$

is a probability measure on $\mathbb{B}(\mathbb{R})$, and the real-valued function $\mathcal{F}_\phi$ defined on $\mathbb{R}$ by

$$\mathcal{F}_\phi(\tau) = \langle \phi \mid \widehat{F}_+(\tau)\, \phi \rangle, \quad \tau \in \mathbb{R} \tag{2.502}$$

is a probability function.

- A strictly maximal symmetric operator $\widehat{S}$ together with a unit vector ϕ in $\mathcal{H}$ generates a unique probability measure $\mathcal{M}_\phi^{\widehat{S}}$ and a unique probability function $\mathcal{F}_\phi^{\widehat{S}}$ defined in terms of its generalized spectral measure $\widehat{M}_+^{\widehat{S}}$ and its generalized spectral function $\widehat{F}_+^{\widehat{S}}$ by Eqs. (2.501) and (2.502).

Examples 2.11.2(1)   **Probability functions**

E1   The operator $\widehat{p}^+$ in $L^2(\mathbb{E}^+)$ in §2.10.1C(1) is strictly maximal symmetric with a corresponding POV function $\widehat{F}_+^{\widehat{p}^+}$ given by Eq. (2.454). Given any unit vector $\phi_+ \in L^2(\mathbb{E}^+)$ we can generate a probability function by

$$\mathcal{F}_{\phi_+}^{\widehat{p}^+}(\tau) = \langle \phi_+ \mid \widehat{F}_+^{\widehat{p}^+}(\tau)\, \phi_+ \rangle. \tag{2.503}$$

Using Eq. (2.453) we obtain the corresponding probability density function to be $|\widetilde{\phi}_+(\tau)|^2$. An important observation is that there is no normalized function $\phi_+(x)$ in $L^2(\mathbb{E}^+)$ which will lead to a probability of 1 for any finite interval $(\tau_1, \tau_2]$, i.e.,

$$\mathcal{F}_{\phi_+}^{\widehat{p}^+}(\tau_2) - \mathcal{F}_{\phi_+}^{\widehat{p}^+}(\tau_1) \neq 1 \quad \forall \phi_+ \in L^2(\mathbb{E}^+). \tag{2.504}$$

This follows from the properties of Fourier transforms, i.e., the Fourier transform $\widetilde{\phi}(\tau)$ of a function $\phi(x)$ in $L^2(\mathbb{E})$ which vanishes for $x \leq 0$ cannot be zero over any finite interval.[147] In other words $\widetilde{\phi}_+(\tau)$ will spread out over the entire range of values of τ. It follows that

$$\mathcal{F}_{\phi_+}^{\widehat{p}^+}(\tau_2) - \mathcal{F}_{\phi_+}^{\widehat{p}^+}(\tau_1) = \int_{\tau_1}^{\tau_2} |\widetilde{\phi}_+(\tau)|^2\, d\tau \neq 1. \tag{2.505}$$

This is in sharp contrast to the properties of PV measures and PV functions; as pointed out in §2.7C(1) C2 there exist unit vectors for a probability given rise by a PV function to take the value 1.

E2   The radial momentum operator $\widehat{p}_r$ defined by Eqs. (2.456) and (2.457) in §2.10.2E(1) E2 is strictly maximal symmetric in $L^2(\mathbb{R}^+, r^2 dr)$. Its spectral function $\widehat{F}^{\widehat{p}_r}(\tau)$ given by Eq. (2.463) generates a probability function

---

[147] Papoulis (1962) Problem 39 on p. 219.

$\mathcal{F}_\psi^{\hat{p}_r}(\tau) = \langle \psi \mid \widehat{F}^{\hat{p}_r}(\tau)\psi\rangle$ in any given unit vector ψ in $L^2(\mathbb{R}^+, r^2 dr)$. Again we can verify a property similar to those of $\mathcal{F}_{\phi_+}^{\hat{p}^+}(\tau)$ given by Eq. (2.504).

Comments 2.11.2(1)   Probability and symmetric operators

C1   For strictly maximal symmetric operators, Definition 2.11.1(1) for the expectation value, variance and uncertainty remains valid. On account of Eqs. (2.444) and (2.445) we can also use Eqs. (2.495) and (2.497) respectively for the expectation value and the variance.

C2   Generally a symmetric operator does not possess a unique POV measure. So, a symmetric operator does not generally give rise to a unique probability measure in a given unit vector ϕ.

C3   An arbitrary POV function $\widehat{F}_+$ not related to a strictly maximal symmetric operator also gives rise to a probability function $\mathcal{F}_\phi$ for each unit vector $\phi \in \mathcal{H}$. However such a probability function may not have a finite variance in a dense set of states. Indeed, following the comments in §2.10.1C(1) C1 we can appreciate that there are POV functions giving rise to probability functions which do not have a finite variance in any state at all because the following integral

$$\int_{-\infty}^{\infty} \tau^2 \, d\mathcal{F}_\phi(\tau) \tag{2.506}$$

may not converge for any ϕ.[148] This renders a finite expectation value as an average value not so meaningful. We shall return to discuss the physical implications of this in Chapter 4.

2.12 Local Operators in Coordinate Space

2.12.1 Definitions

The characteristic function χ_Λ of a bounded open interval Λ is an important quantity. It acts on $L^2(\mathbb{E})$ as a projector. Moreover, it is a spectral projector of the position operator $\widehat{x}$. We call χ_Λ a *local* operator since it vanishes outside a bounded region of coordinate space $\mathbb{E}$. In contrast, most of the operators one comes across in quantum mechanics, such as the position and the momentum operators, act everywhere and do not vanish outside any finite region in the coordinate space. We call them *global* operators. This section is devoted to a formal introduction of local operators and to the localization of

---
[148] Akhiezer and Glazman Vol. 2 (1963) p. 132.

global operators. This study enables us to introduce an algebraic approach to quantum mechanics based on local operators.[149]

Definition 2.12.1(1)   Global and local operators[150]

Let Λ be a bounded open subset of $I\!\!E^n$, and let χ_Λ be the characteristic function of Λ on $I\!\!E^n$. Then:

- An operator $\widehat{A}$ in $L^2(I\!\!E^n)$ is said to be a *local operator, localized* in the region Λ if

$$\widehat{A} = \chi_\Lambda \widehat{A} \chi_\Lambda. \tag{2.507}$$

- $\widehat{A}$ is said to be a *semi-local operator* if an operator $\widehat{A}'$ exists in $L^2(I\!\!E^n)$ such that $\widehat{A}$ can be written in the form

$$\widehat{A} = \frac{1}{2}\left(\chi_\Lambda \widehat{A}' + \widehat{A}' \chi_\Lambda\right). \tag{2.508}$$

- $\widehat{A}$ is *global* if it is not local or semi-local.

Comments 2.12.1(1)   Examples and characteristics

C1   The momentum operator $\widehat{p} = -i\hbar d/dx$ in $L^2(I\!\!E)$ is a global operator, and so is the usual position operator $\widehat{x} = x$.

C2   As a multiplication operator in $L^2(I\!\!E^n)$ the characteristic function χ_Λ for any bounded interval Λ is local. A local operator is also semi-local, but the conserve is not true. A semi-local operator $\widehat{A}$ becomes local if $\widehat{A}'$ in Eq. (2.508) commutes with χ_Λ, e.g.,

$$\frac{1}{2}(\chi_\Lambda \widehat{x} + \widehat{x}\chi_\Lambda) = \chi_\Lambda \widehat{x} = \chi_\Lambda \widehat{x} \chi_\Lambda. \tag{2.509}$$

On the other hand

$$\frac{1}{2}(\chi_\Lambda \widehat{p} + \widehat{p}\chi_\Lambda) \tag{2.510}$$

is semi-local but not local.[151]

---

[149] An algebraic approach to quantum field theory based on local observables was proposed by Haag and Kastler (1964). This has led to many applications to quantum statistical mechanics (Emch (1972), Bratteli and Robinson (1979), (1981), Haag (1992)).
[150] Wan and Jackson (1984).
[151] In §5.6.2 it is shown that semi-local observables play an important role in studying local values of quantum observables.

2.12. LOCAL OPERATORS IN COORDINATE SPACE

C3   Let $\widehat{A}$ be an operator localized in Λ. Let $\psi_{out} \in L^2(\mathbb{E}^n)$ be a function of support lying outside Λ and let $\phi_{in} \in \mathcal{D}(\widehat{A})$ be a function of support lying inside Λ. Following Eq. (2.507) we can see that

$$\langle \psi_{out} \mid \widehat{A}\psi_{out}\rangle = 0, \quad \langle \psi_{out} \mid \widehat{A}\phi_{in}\rangle = 0. \tag{2.511}$$

That is, an operator localized in Λ vanishes outside Λ and does not relate any function outside Λ to functions inside Λ.

2.12.2   Localization of bounded operators

Definition 2.12.2(1)   **Localization of operators**

Let Λ_0 be a closed set inside a bounded open set Λ in $\mathbb{E}^n$, and let $\widehat{A}$ be an operator in $L^2(\mathbb{E}^n)$, possibly global, with domain $\mathcal{D}(\widehat{A})$. Then:

- An operator $\widehat{A}_\Lambda$ in $L^2(\mathbb{E}^n)$ is said to be a *localization* of $\widehat{A}$ in Λ with *center of localization* Λ_0 if
  1. $\widehat{A}_\Lambda$ is a local operator localized in Λ, and
  2. for all functions $\phi_0 \in \mathcal{D}(\widehat{A})$ of support inside Λ_0, i.e., $supp(\phi_0) \subset \Lambda_0$, we have

$$\langle \phi_0 \mid \widehat{A}\phi_0\rangle = \langle \phi_0 \mid \widehat{A}_\Lambda \phi_0 \rangle. \tag{2.512}$$

Comments 2.12.2(1)   **Localization procedure**

C1   If $\widehat{A}$ is bounded, then a localization can be achieved by pre and post multiplication by χ_Λ, i.e.,

$$\widehat{A}_\Lambda = \chi_\Lambda \widehat{A} \chi_\Lambda. \tag{2.513}$$

Clearly $\widehat{A}_\Lambda$ acts like $\widehat{A}$ inside Λ in the sense of Eq. (2.512). Note that

$$\langle \phi_0 \mid \widehat{A}\phi_0\rangle = \langle \phi_0 \mid \widehat{A}_\Lambda \phi_0\rangle \not\Rightarrow \widehat{A}\phi_0 = \widehat{A}_\Lambda \phi_0. \tag{2.514}$$

The reason for this is that the function

$$(\widehat{A}_\Lambda \phi_0)(x) = \chi_\Lambda(x)\, (\widehat{A}\phi_0)(x) \tag{2.515}$$

must vanish outside Λ but the same need not be true for $(\widehat{A}\phi_0)(x)$. To see this let us consider a projector $\widehat{A} = |\varphi\rangle\langle\varphi|$, where φ is a function in $L^2(\mathbb{E}^n)$ which does not vanish outside Λ. Then

$$(\widehat{A}\phi_0)(x) = \big(|\varphi\rangle\langle\varphi|\big)\phi_0(x) = \big(\langle\varphi \mid \phi_0\rangle\big)\varphi(x) \tag{2.516}$$

does not vanish outside Λ.

C2 The localization procedure given by Eq. (2.513) preserves selfadjointness of bounded operators, i.e., for a bounded selfadjoint operator $\widehat{A}$ the localization $\widehat{A}_\Lambda$ is also selfadjoint, on account of Eq. (2.84). We call $\widehat{A}_\Lambda$ a *selfadjoint localization* of $\widehat{A}$.

C3 If $\widehat{A}$ is unbounded, then $\chi_\Lambda \widehat{A} \chi_\Lambda$ is generally not selfadjoint because of Eq. (2.82). For example, Eq. (2.513) does not produce a selfadjoint localization of the momentum operator. Intuitively one can see that $\widehat{p}$ will have difficulty operating on χ_Λ which is a discontinuous function. A simple way out is to replace χ_Λ by a smooth function.

C4 Bounded local operators have a distinctive mathematical structure which we shall introduce in the following section.

C5 The concept and procedure for localization in the coordinate representation space can be extended in an straight forward manner to other spectral representation space. To see this we can rewrite Eq. (2.513) as

$$\widehat{A}_\Lambda = \widehat{M^{\hat{x}}}(\Lambda)\, \widehat{A}\, \widehat{M^{\hat{x}}}(\Lambda). \tag{2.517}$$

We have replaced the characteristic function χ_Λ by the spectral projector $\widehat{M^{\hat{x}}}(\Lambda)$ of the position operator in accordance with Eq. (2.344). Equation (2.517) can be used to localize an observable in any chosen spectral representation space. For example we can localize operators in the momentum space $\widetilde{I\!\!E}$ in terms of its spectral projector $\widehat{M^{\hat{p}}}(\Lambda)$ by the following prescription:[152]

$$\widetilde{p}_{\widetilde{\Lambda}} = \widehat{M^{\hat{p}}}(\widetilde{\Lambda})\, \widehat{p}\, \widehat{M^{\hat{p}}}(\widetilde{\Lambda}), \tag{2.518}$$

where $\widetilde{\Lambda}$ is an interval in the momentum space.

2.12.3 Local operator algebras

Definition 2.12.3(1) **Local operator algebras**[153]

- The set $\mathcal{A}_\Lambda$ of all bounded operators on $L^2(I\!\!E^n)$ localized in a given region Λ is called the *local algebra* of operators in Λ.

- The union of local algebras on $L^2(I\!\!E^n)$ over all bounded Borel sets Λ, i.e.,

$$\mathcal{A}_L = \cup_\Lambda \mathcal{A}_\Lambda, \tag{2.519}$$

  is called the *local algebra of operators* on $L^2(I\!\!E^n)$.

---
[152] Wan and McLean (1985).
[153] Sewell (1986) pp. 20-25.

2.12. LOCAL OPERATORS IN COORDINATE SPACE

- Selfadjoint elements of $\mathcal{A}_L$ are called *local observables*.

Comments 2.12.3(1)   **Operator algebras**

C1   A set of bounded operators $\mathcal{A}$ is said to possess an *algebraic structure* and is called an *algebra of operators* if it is closed under addition and multiplication, i.e., for all $a_1, a_2 \in \mathbb{C}$

$$\widehat{A}_1, \ \widehat{A}_2 \in \mathcal{A} \ \Rightarrow \ a_1\widehat{A}_1 + a_2\widehat{A}_2 \in \mathcal{A} \quad \text{and} \quad \widehat{A}_1\widehat{A}_2 \in \mathcal{A}. \tag{2.520}$$

Clearly $\mathcal{A}_\Lambda$ and $\mathcal{A}_L$ possess such an algebraic structure. These algebras are called *non-commutative algebras* since multiplication is generally not commutative, i.e.,

$$\widehat{A}_1\widehat{A}_2 \neq \widehat{A}_2\widehat{A}_1. \tag{2.521}$$

C2   An operator algebra $\mathcal{A}$ is called a **-algebra* if it is closed under the adjoint operation, i.e.,

$$\widehat{A} \in \mathcal{A} \ \Rightarrow \ \widehat{A}^\dagger \in \mathcal{A}. \tag{2.522}$$

C3   The set $\mathcal{B}(\mathcal{H})$ of all bounded operators on a Hilbert space $\mathcal{H}$ form a *-algebra. Moreover, every operator in the set possesses a finite norm satisfying the following properties:[154]

$$\| a\widehat{A} \| = |a| \, \|\widehat{A}\|, \tag{2.523}$$
$$\| \widehat{A}_1 + \widehat{A}_2 \| \leq \| \widehat{A}_1 \| + \| \widehat{A}_2 \|, \tag{2.524}$$
$$\| \widehat{A}_1 \widehat{A}_2 \| \leq \| \widehat{A}_1 \| \, \| \widehat{A}_2 \|, \tag{2.525}$$
$$\| \widehat{A}^\dagger \widehat{A} \| = \| \widehat{A} \|^2. \tag{2.526}$$

These also imply:

$$\| \widehat{A}^\dagger \| = \| \widehat{A} \|. \tag{2.527}$$

C4   The set $\mathcal{B}(\mathcal{H})$ is closed with respect to uniform convergence, i.e., every uniform Cauchy sequence of bounded operators converges to a bounded operator, a fact pointed out in §2.2.2C(1) C3. This useful property will now be incorporated into a new class of operator algebras known as C*-*algebras*.

Definition 2.12.3(2)   **C*-algebra of operators on a Hilbert space**

- A set $\mathcal{A}$ of bounded operators on a Hilbert space is called a C*-algebra of operators if

---
[154] Roman Vol. 2 (1975) p. 500, Kadison and Ringrose Vol. 1 (1983) p. 101.

1. $\mathcal{A}$ constitutes a *-algebra, i.e., operators in $\mathcal{A}$ satisfy Eqs. (2.520) and (2.522).

2. $\mathcal{A}$ is closed with respect to uniform convergence, i.e., every uniform Cauchy sequence of operators in $\mathcal{A}$ converges uniformly to an operator in $\mathcal{A}$.

Comments 2.12.3(2) **Local and quasi-local algebras**

C1 In more mathematical literature the notion of C*-algebras is introduced as an abstract set of elements endowed with a norm. Then the *-algebraic properties in Eqs. (2.520) and (2.522) and the properties of the norm listed in Eqs. (2.524) and (2.526) have to be incorporated into the definition.[155] We shall not pursue such an abstract approach. In other words by a C*-algebra we shall always mean a C*-algebra of bounded operators in a given Hilbert space.

C2 Following §2.12.3C(1) C3 and C4 we can see that the set of all bounded operators $\mathcal{B}(L^2(I\!\!E^n))$ on the Hilbert space $L^2(I\!\!E^n)$ forms a C*-algebra. A local algebra $\mathcal{A}_\Lambda$ on $L^2(I\!\!E^n)$ also constitutes a C*-algebra since $\mathcal{A}_\Lambda$ on $L^2(I\!\!E^n)$ is identifiable with the set $\mathcal{B}(L^2(\Lambda))$ of bounded operators on the subspace $L^2(\Lambda)$.[156] A C*-algebra may or may not contain the identity operator of the Hilbert space. However, it is possible to annex the identity operator to the algebra.[157]

C3 The local algebra $\mathcal{A}_L = \cup_\Lambda \mathcal{A}_\Lambda$ on $L^2(I\!\!E^n)$ is a *-algebra but it is not a C*-algebra.[158] Consider the case of $L^2(I\!\!E)$. As an example consider a normalized element $\phi(x)$ of $L^2(I\!\!E)$ which does not vanish outside any bounded set in $I\!\!E$. Let $|\phi\rangle\langle\phi|$ be the projector generated by ϕ. Then $|\phi\rangle\langle\phi|$ is not a local operator. Now let Λ_n be an increasing sequence of bounded intervals converging to $I\!\!E$ and let $\phi_n = c_n \chi_{\Lambda_n} \phi$, where c_n are normalization constants. Then it can be shown that $\{|\phi_n\rangle\langle\phi_n|\}$ is a uniform Cauchy sequence of local operators which converges to $|\phi\rangle\langle\phi|$. Since $|\phi\rangle\langle\phi|$ is not an element of $\mathcal{A}_L$ we can see that $\mathcal{A}_L$ is not closed with respect to uniform convergence.[159] It is desirable to include projectors like $|\phi\rangle\langle\phi|$ in an operator algebra. We can achieve this by constructing a larger set which includes $\mathcal{A}_L$ and the limit operators of all uniform Cauchy sequences in $\mathcal{A}_L$. The resulting algebra is known as the *uniform closure* of $\mathcal{A}_L$ and it is denoted by $\bar{\mathcal{A}}_L$. Algebra $\bar{\mathcal{A}}_L$ then contains all projectors associated with finite dimensional subspaces.

---
[155] Bratteli and Robinson Vol. 1 (1979) p. 19-24, Dixmier (1977) Chapter 1.
[156] Wan and McLean (1984a), (1984b).
[157] Bratteli and Robinson Vol. 1 (1979) p. 23, Wan and McLean (1985).
[158] McLean (1983), Wan and McLean (1984a), Wan and Jackson (1984).
[159] Wan and McLean (1984a).

2.12. LOCAL OPERATORS IN COORDINATE SPACE

C4 The uniform closure $\bar{\mathcal{A}}_L$ of $\mathcal{A}_L$ is a C*-algebra called the *quasi-local algebra* of operators on $L^2(I\!\!E^n)$ and selfadjoint members of $\bar{\mathcal{A}}_L$ are called *quasi-local observables*. As will be seen in Chapter 8, this C*-algebra of quasi-local operators can be used to formulate a version of quantum mechanics which is asymptotically separable.

2.12.4 Localization of unbounded operators 1

For unbounded operators Eq. (2.513) fails to produce a selfadjoint localization. To illustrate the problem let us apply Eq. (2.513) to the momentum operator $\hat{p}$ in $L^2(I\!\!E)$. Introduce a local operator

$$\hat{p}_\Lambda = \chi_\Lambda \hat{p} \chi_\Lambda, \quad \Lambda = (a,b). \tag{2.528}$$

The domain of $\hat{p}_\Lambda$ is, according to Eqs. (2.30) and (2.184),

$$\mathcal{D}(\hat{p}_\Lambda) = \left\{ \phi \in L^2(I\!\!E) : \chi_\Lambda \phi \in S_{2,1}(I\!\!E) \right\}. \tag{2.529}$$

Clearly any function $\phi(x)$ in $\mathcal{D}(\hat{p}_\Lambda)$ must vanish at $x = a, b$ so that $\chi_\Lambda \phi$ can be absolutely continuous and be in the domain of $\hat{p}$. We can conveniently decompose $\hat{p}_\Lambda$ into a sum of two parts. Let Λ^c be the complement of Λ in $I\!\!E$ and let $L^2(\Lambda^c)$ be the Hilbert space of square-integrable functions on Λ^c. Then $L^2(I\!\!E)$ can be written as a sum of $L^2(\Lambda)$ and $L^2(\Lambda^c)$ and that $\hat{p}_\Lambda$ can be written as a sum of $\hat{p}(\Lambda)$, the restriction of $\hat{p}_\Lambda$ to the subspace $L^2(\Lambda)$ of $L^2(I\!\!E)$, and the zero operator on Λ^c.[160] The operator $\hat{p}(\Lambda)$ acts on domain

$$\mathcal{D}(\hat{p}(\Lambda)) = \{\phi \in L^2(\Lambda) : \phi \in S_{2,1}(\Lambda), \ \phi(a) = \phi(b) = 0\}. \tag{2.530}$$

Comparing with §2.5.1E(1) E1 we can see that $\hat{p}(\Lambda)$ admits a one-parameter family of selfadjoint extensions $\hat{p}_\lambda(\Lambda)$ acting on domains specified by Eq. (2.165). This shows that $\hat{p}_\Lambda$ is not selfadjoint, but it admits a one-parameter family of selfadjoint extensions. In other words the procedure based on Eq. (2.528) does not lead to a unique selfadjoint localization. Another obvious example is the localization of the kineitc energy operator $\hat{K}$ in $L^2(I\!\!E)$.[161] We can now see that unbounded operators generally admit no natural unique selfadjoint localizations.

---

[160] A discussion of such a sum, known as direct sums of Hilbert space and operators, is given in §2.13.1, espcially in §2.13.1C(1) C3.

[161] Bratteli and Robinson Vol. 1 (1979) p. 187, Bratteli and Robinson Vol. 2 (1981) pp. 50-51, p. 56, p. 75. A detailed discussions of selfadjoint extensions of local kinetic energy operators is presented in Chapter 6.

2.12.5 Localization of unbounded operators 2

Here we shall present an explicit selfadjoint localization scheme which also serves to illustrate the non-uniqueness of the procedure.[162] Consider the case where the Hilbert space is $L^2(I\!\!E)$. Let Λ be a bounded open interval (a, b) of $I\!\!E$, and let Λ_0 be a closed interval $[a_0, b_0]$ inside Λ.

Definition 2.12.5(1)   **Localizing functions and operators**

- A smooth real-valued function ξ on $I\!\!E$ is called a *localizing function with support* $\bar{\Lambda} = [a, b]$ and *centre of localization* Λ_0 if it takes the value 1 in Λ_0 and then decreases monotonically towards zero in the boundary regions $\Lambda - \Lambda_0$, and remains zero outside Λ.

- An operator $\widehat{L}_\xi$ on $L^2(I\!\!E)$ defined by

$$\left(\widehat{L}_\xi \phi\right)(x) = \begin{cases} \xi(x)^{-\frac{1}{2}} \phi(\sigma(x)) & \text{if } x \in \Lambda \\ 0 & \text{if } x \notin \Lambda \end{cases}, \qquad (2.531)$$

where σ is a function from Λ to $I\!\!R$ defined by

$$\sigma(x) = \int_{x_0}^{x} \frac{1}{\xi(x')} \, dx' + x_0, \quad \text{where } x_0 \in \Lambda_0 \text{ and } x \in \Lambda \qquad (2.532)$$

is called a *localizing operator* generated by the localizing function $\xi(x)$.

Comments 2.12.5(1)   **Localizing operators**

C1   A localizing function is of the following form, as shown in Fig. 2.12.5:

$$\xi(x) = \begin{cases} 1, & x \in \Lambda_0 \\ \xi(x) \in (0, 1), & x \in \Lambda - \Lambda_0 \\ \xi(x) = 0, & x \notin \Lambda \end{cases}. \qquad (2.533)$$

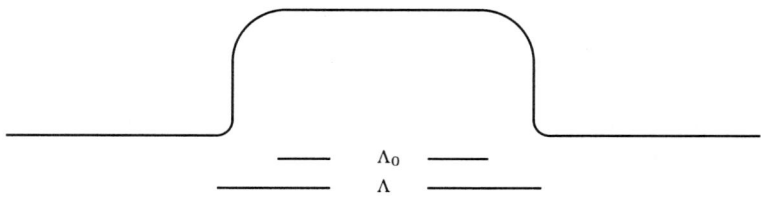

Fig. 2.12.5   Localizing function.

---
[162]Timson (1986), Wan (1988), Timson and Wan (1988).

2.12. LOCAL OPERATORS IN COORDINATE SPACE

C2 The function $\sigma(x)$ increases from $-\infty$ to ∞ monotonically as x varies from a to b. We also have

$$\sigma(x) = x \quad \forall x \in \Lambda_0. \tag{2.534}$$

This function clearly admits an inverse σ^{-1} which maps $\mathbb{R}$ to Λ. So, we have

$$\sigma : \Lambda \mapsto \mathbb{R}, \quad \sigma^{-1} : \mathbb{R} \mapsto \Lambda. \tag{2.535}$$

C3 $\widehat{L}_\xi$ is a *localizing operator* in the sense that:

1. It maps every function $\phi \in L^2(\mathbb{E})$ to a function $\widehat{L}_\xi \phi$ lying entirely in Λ, i.e.,

$$\left(\widehat{L}_\xi \phi\right)(x) = 0 \quad \text{if} \quad x \notin \Lambda. \tag{2.536}$$

2. It leaves the function ϕ in the center of localization unchanged, i.e.,

$$\left(\widehat{L}_\xi \phi\right)(x) = \phi(x) \quad \text{if} \quad x \in \Lambda_0. \tag{2.537}$$

3. It preserves the norm, i.e.,

$$\int_a^b |\widehat{L}_\xi \phi|^2(x)\, dx = \int_{-\infty}^{\infty} |\phi|^2(x)\, dx. \tag{2.538}$$

This is achieved by squeezing the parts of ϕ lying outside Λ_0 into the boundary regions $\Lambda - \Lambda_0$.[163]

Theorem 2.12.5(1)   **Localizing operators and their adjoints**

- The adjoint $\widehat{L}_\xi^\dagger$ of $\widehat{L}_\xi$ is defined on every $\phi \in L^2(\mathbb{E})$ by

$$\left(\widehat{L}_\xi^\dagger \phi\right)(x) = \xi\bigl(\sigma^{-1}(x)\bigr)^{1/2} \phi\bigl(\sigma^{-1}(x)\bigr), \quad x \in \mathbb{R}. \tag{2.539}$$

- The operators $\widehat{L}_\xi$ and $\widehat{L}_\xi^\dagger$ possesses the following properties:

$$\widehat{L}_\xi^\dagger \widehat{L}_\xi = \widehat{I}, \qquad \widehat{L}_\xi \widehat{L}_\xi^\dagger = \chi_\Lambda, \tag{2.540}$$

$$\chi_\Lambda \widehat{L}_\xi = \widehat{L}_\xi, \qquad \widehat{L}_\xi^\dagger \chi_\Lambda = \widehat{L}_\xi^\dagger, \tag{2.541}$$

and

$$\left(\widehat{L}_\xi^\dagger \phi\right)(x) = \left(\widehat{L}_\xi \phi\right)(x) = \phi(x) \quad \forall x \in \Lambda_0. \tag{2.542}$$

---
[163] Diagramatic illustrations are available in Wan, Jackson and McKenna (1984), Jackson (1985).

Theorem 2.12.5(2)   Selfadjoint localization

- Let $\widehat{A}$ be an arbitrary operator in $L^2(I\!\!E)$ and ξ be a localizing function with support $\bar{\Lambda} = [a,b]$. Then the operator $\widehat{A}_\xi$ defined by

$$\widehat{A}_\xi = \widehat{L}_\xi \, \widehat{A} \, \widehat{L}_\xi^\dagger \qquad (2.543)$$

possesses the following properties:

1. $\widehat{A}_\xi$ is localized in $\Lambda = (a,b)$, i.e.,

$$\widehat{A}_\xi = \chi_\Lambda \widehat{A}_\xi \chi_\Lambda. \qquad (2.544)$$

2. The domain of $\widehat{A}_\xi$ is

$$\mathcal{D}(\widehat{A}_\xi) = \widehat{L}_\xi\big(\mathcal{D}(\widehat{A})\big). \qquad (2.545)$$

3. $\widehat{A}_\xi$ is selfadjoint if and only if $\widehat{A}$ is selfadjoint.

4. Apart from an additional value 0 the spectrum of $\widehat{A}_\xi$ coincides with that of $\widehat{A}$.

Comments 2.12.5(2)   On selfadjoint localization

C1   Theorem 2.12.5(1) can be directly verified. Note that Eq. (2.540) is comparable with Eq. (2.136) in §2.4C(1) C1 on isometric operators. Theorem 2.12.5(2) follows from Theorem 2.12.5(1), e.g., the conclusion that $\widehat{A}_\xi$ is local is a direct result of Eq. (2.541).[164]

C2   We call $\widehat{A}_\xi$ the *localization of* $\widehat{A}$ *defined by the localizing function* ξ.

C3   If $\phi \in \mathcal{D}(\widehat{A})$ then $\phi_\xi = \widehat{L}_\xi \phi \in \mathcal{D}(\widehat{A}_\xi)$ and we have

$$\widehat{A}_\xi \phi_\xi = \widehat{L}_\xi \widehat{A} \phi. \qquad (2.546)$$

We can see from Eq. (2.542) that $\widehat{A}_\xi$ acts like $\widehat{A}$ in the center of localization, i.e.,

$$\big(\widehat{A}_\xi \phi_\xi\big)(x) = \big(\widehat{A}\phi\big)(x), \quad x \in \Lambda_0. \qquad (2.547)$$

C4   While the operator $\widehat{L}_\xi$ has the effect of localizing a widely spread-out wave function to the region Λ its adjoint $\widehat{L}_\xi^\dagger$ has the opposite effect. Let

---

[164] More detailed discussions on localizing operators and its adjoints, including the proofs of Theorem 2.12.5(1) and Theorem 2.12.5(2) are available in Wan (1988), Timson and Wan (1988), Timson (1986).

$\phi_\Lambda(x)$ be a function of support in Λ. Then Eq. (2.539) tells us that $\widehat{L}_\xi^\dagger \phi_\Lambda(x)$ may not vanish outside Λ, i.e. when acted on by $\widehat{L}_\xi^\dagger$ a wave function of support Λ may spread out all over E. One can say that while $\widehat{L}_\xi$ localizes, $\widehat{L}_\xi^\dagger$ globalizes.

2.12.6 Local momentum and local Hamiltonian

For the momentum operator $\widehat{p}$ in $L^2(E)$ the local momentum

$$\widehat{p}_\xi = \widehat{L}_\xi \widehat{p} \widehat{L}_\xi^\dagger \qquad (2.548)$$

defined by a given localizing function ξ can be shown to be equal to the symmetrized product of $\widehat{p}$ and ξ, i.e.,

$$\widehat{p}_\xi = \frac{1}{2}(\xi \widehat{p} + \widehat{p} \xi) = -i\hbar \left(\xi \frac{d}{dx} + \frac{1}{2} \frac{d\xi}{dx} \right). \qquad (2.549)$$

Clearly $\widehat{p}_\xi$ acts like $\widehat{p}$ in the center of localization. This local momentum operator admits the following generalized eigenfunctions

$$\varphi_p(x) = \begin{cases} (2\pi\hbar\xi(x))^{-1/2} e^{\frac{i}{\hbar}p \int_{x_0}^x \xi^{-1}(x)\, dx}, & x \in \Lambda \\ 0, & x \notin \Lambda \end{cases}. \qquad (2.550)$$

More details including diagramatic illustrations and applications are available in Wan, Jackson and McKenna (1984).

We can go on to localize the kinetic energy operator and the Hamiltonain. The global kinetic energy operator for a particle of mass m is $\widehat{K} = \widehat{p}^2/2m$, and it follows that the local kinetic energy associated with ξ is

$$\widehat{K}_\xi = \widehat{L}_\xi \widehat{K} \widehat{L}_\xi^\dagger = \frac{1}{2m} \widehat{p}_\xi^2. \qquad (2.551)$$

The global Hamiltonian operator for a particle of mass m is $\widehat{H} = \widehat{K} + V$, where $V = V(x)$ is a real-valued function on $\mathbb{R}$. The local Hamiltonian is then

$$\widehat{H}_\xi = \widehat{L}_\xi \widehat{H} \widehat{L}_\xi^\dagger = \frac{1}{2m} \widehat{p}_\xi^2 + V_\xi. \qquad (2.552)$$

2.13 Direct Integrals of Hilbert Spaces

Direct sums of a finite number of Hilbert spaces and operators are well-known and widely used in quantum mechanics. Less well-known and used are direct integrals of a continuous family of Hilbert spaces and operators. A systematic

196 *CHAPTER 2. OPERATORS AND THEIR DIRECT INTEGRALS*

and rigorous exposition of the relevant mathematics is available in Dixmier (1981, pp. 161-240). We shall summarize some relevant definitions and properties here to facilitate discussions in later chapters. Let us start with direct sums.

2.13.1   Discrete composition of Hilbert spaces

Definition 2.13.1(1)   **Direct sums**

- Let $\mathcal{H}_1$ and $\mathcal{H}_2$ be two arbitrary Hilbert spaces. The set of 2-tuples

$$\Big\{ \{\phi_1, \phi_2\} : \phi_1 \in \mathcal{H}_1,\ \phi_2 \in \mathcal{H}_2 \Big\} \quad (2.553)$$

can be given a natural Hilbert space structure with algebraic operations and scalar product of any two elements $\{\phi_1, \phi_2\}$ and $\{\psi_1, \psi_2\}$ of the set defined by[165]

$$\begin{aligned}
\alpha\{\phi_1, \phi_2\} &= \{\alpha\phi_1, \alpha\phi_2\}, \quad \alpha \in \mathbb{C}, & (2.554) \\
\{\phi_1, \phi_2\} + \{\psi_1, \psi_2\} &= \{\phi_1 + \psi_1, \phi_2 + \psi_2\}, & (2.555) \\
\langle \{\phi_1, \phi_2\} \mid \{\psi_1, \psi_2\} \rangle^{\oplus} &= \langle \phi_1 \mid \psi_1 \rangle_1 + \langle \phi_2 \mid \psi_2 \rangle_2. & (2.556)
\end{aligned}$$

The resulting Hilbert space $\mathcal{H}^{\oplus}$ is called the *direct sum* of $\mathcal{H}_1$ and $\mathcal{H}_2$ and is denoted by

$$\mathcal{H}^{\oplus} = \mathcal{H}_1 \oplus \mathcal{H}_2. \quad (2.557)$$

Elements $\{\phi_1, \phi_2\}$ of the direct sum are denoted by

$$\phi_1 \oplus \phi_2. \quad (2.558)$$

Comments 2.13.1(1)   **Internal and external direct sums**

C1   An example is the direct sum of the one-dimensional Hilbert space $\mathbb{C}$ with itself, denoted by $\mathbb{C}^2$, i.e.,

$$\mathbb{C}^2 = \mathbb{C} \oplus \mathbb{C}. \quad (2.559)$$

Clearly $\mathbb{C}^2$ is a two-dimensional Hilbert space.

Note that a direct sum $\phi_1 \oplus \phi_2$ is an ordered pair; only the scalar products in Eq. (2.556) involves an actual numerical sum, i.e.,

$$\langle \phi_1 \oplus \phi_2 \mid \psi_1 \oplus \psi_2 \rangle^{\oplus} = \langle \phi_1 \mid \psi_1 \rangle_1 + \langle \phi_2 \mid \psi_2 \rangle_2 \quad (2.560)$$

---

[165] The superscript $\oplus$ indicates the scalar product in the direct sum space $\mathcal{H}^{\oplus}$ while subscripts 1 and 2 denote the scalar products in $\mathcal{H}_1$ and $\mathcal{H}_2$ respectively.

2.13. DIRECT INTEGRALS OF HILBERT SPACES

is a numerical sum of scalar products $\langle \phi_1 \mid \psi_1 \rangle_1$ and $\langle \phi_2 \mid \psi_2 \rangle_2$, while

$$\phi_1 \oplus \phi_2 + \psi_1 \oplus \psi_2 = (\phi_1 + \psi_1) \oplus (\phi_2 + \psi_2) \qquad (2.561)$$

does not involve any numerical sum.

Let $\{\phi_j, j = 1, 2, \ldots\}$ be an orthonormal basis of $\mathcal{H}_1$. This gives rise to a set of elements $\phi_j \oplus 0$ in the direct sum space $\mathcal{H}^\oplus$. We shall denote these elements by $\phi_j^\oplus$, i.e., $\phi_j^\oplus = \phi_j \oplus 0$. In the same way an orthonormal basis $\{\psi_k, k = 1, 2, \ldots\}$ of $\mathcal{H}_2$ gives rise to a set of elements $\psi_k^\oplus = 0 \oplus \psi_k$ in $\mathcal{H}^\oplus$. Since $\phi \oplus \psi = \phi \oplus 0 + 0 \oplus \psi$ the following elements

$$\phi_j^\oplus, \psi_k^\oplus, j = 1, 2, \ldots, k = 1, 2, \ldots \qquad (2.562)$$

clearly form a complete orthonormal set in $\mathcal{H}_1 \oplus \mathcal{H}_2$. This shows that the direct sum space has the dimension of the sum of dimensions of the constituent spaces.

An extension of the definition to a direct sum of any finite number of Hilbert spaces is straightforward.

C2 We can form the direct sum of two subspaces of a Hilbert space. Let $\mathcal{L}$ be a subspace of $\mathcal{H}$ and let $\mathcal{L}^\perp$ be its orthogonal complement. Then $\mathcal{L}$ and $\mathcal{L}^\perp$ are Hilbert spaces in their own right and we can proceed to construct their direct sum $\mathcal{L}^\oplus = \mathcal{L} \oplus \mathcal{L}^\perp$. It is easy to check that we can identify $\mathcal{L}^\oplus$ with $\mathcal{H}$. Some authors call $\mathcal{L} \oplus \mathcal{L}^\perp$ an *internal direct sum* and $\mathcal{H}_1 \oplus \mathcal{H}_2$ in the definition above an *external direct sum*. We shall not make this distinction.

C3 The following example has important physical applications in Chapters 6 and 7. Consider the Hilbert spaces $\mathcal{H}_1 = L^2(I\!\!E^-)$ and $\mathcal{H}_2 = L^2(I\!\!E^+)$ introduced in §2.5.1E(1) E6. One can identify their direct sum with $L^2(I\!\!E)$ and write

$$L^2(I\!\!E) = L^2(I\!\!E^-) \oplus L^2(I\!\!E^+). \qquad (2.563)$$

To appreciate this let $\phi_1 \in \mathcal{H}_1$ and $\phi_2 \in \mathcal{H}_2$. Then $\phi_1 \oplus \phi_2$ defines a function ϕ in $L^2(I\!\!E)$ by

$$\phi(x) = \begin{cases} \phi_1(x), & x < 0 \\ \phi(0), & x = 0 \\ \phi_2(x), & x > 0 \end{cases}. \qquad (2.564)$$

The value of ϕ at the origin $x = 0$, shown as $\phi(0)$ above, can be chosen by a continuation of ϕ_1 or ϕ_2, or by any other choice without any consequence, since a function in $L^2(I\!\!E)$ is defined only up to a set of Lebesgue measure zero in the sense discussed in §2.6.1C(1) C3.

Similar arguments enable us to identify $L^2(I\!\!E)$ with the direct sum of the subspaces of positive and negative momentum values $L_-^2(I\!\!E)$ and $L_+^2(I\!\!E)$

introduced in Eq. (2.400), i.e., we can write

$$L^2(I\!\!E) = L^2_-(I\!\!E) \oplus L^2_+(I\!\!E). \qquad (2.565)$$

The corresponding momentum representation space $\widetilde{L}^2(\widetilde{I\!\!E})$ is expressible as

$$\widetilde{L}^2(\widetilde{I\!\!E}) = \widetilde{L}^2_-(\widetilde{I\!\!E}) \oplus \widetilde{L}^2_+(\widetilde{I\!\!E}), \qquad (2.566)$$

where $\widetilde{L}^2_-(\widetilde{I\!\!E})$ and $\widetilde{L}^2_+(\widetilde{I\!\!E})$ are given in Eqs. (2.396) and (2.397) in §2.8.4.

C4 Another example also comes from §2.8.4. The space $\widetilde{\widetilde{L}}^2(\widetilde{\widetilde{I\!\!E}})$ composed of ordered pairs in Eq. (2.427) with scalar product given by Eq. (2.428) is now seen to be the direct sum of $\widetilde{\widetilde{L}}^2_-(\widetilde{\widetilde{I\!\!E}})$ and $\widetilde{\widetilde{L}}^2_+(\widetilde{\widetilde{I\!\!E}})$, i.e., we can write

$$\widetilde{\widetilde{\phi}} = \widetilde{\widetilde{\phi}}_- \oplus \widetilde{\widetilde{\phi}}_+, \qquad \widetilde{\widetilde{L}}^2(\widetilde{\widetilde{I\!\!E}}) = \widetilde{\widetilde{L}}^2_-(\widetilde{\widetilde{I\!\!E}}) \oplus \widetilde{\widetilde{L}}^2_+(\widetilde{\widetilde{I\!\!E}}). \qquad (2.567)$$

2.13.2 Continuous composition of Hilbert spaces

When we have an arbitrary family of Hilbert spaces to compose the situation becomes technically far more complicated, although the idea involved is the same.

Let $\{\mathcal{H}(\tau) : \tau \in I\!\!R\}$ be a one-parameter family of Hilbert spaces $\mathcal{H}(\tau)$ and let $\boldsymbol{\Upsilon}$ be the set of all vector-valued functions $\boldsymbol{v}$ on $I\!\!R$ which map every $\tau \in I\!\!R$ to a vector $\boldsymbol{v}(\tau)$ in $\mathcal{H}(\tau)$, i.e.,

$$\boldsymbol{v} : I\!\!R \mapsto \{\mathcal{H}(\tau) : \tau \in I\!\!R\} \qquad \text{by} \qquad \tau \mapsto \boldsymbol{v}(\tau) \in \mathcal{H}(\tau). \qquad (2.568)$$

We have employed bold face symbols to emphasize the vector-valued nature of the function in this section to avoid confusion with the usual complex-valued functions. Let $\boldsymbol{v}$ and $\boldsymbol{w}$ be two elements of $\boldsymbol{\Upsilon}$. For each value τ there corresponds two vectors $\boldsymbol{v}(\tau)$ and $\boldsymbol{w}(\tau)$ in the Hilbert space $\mathcal{H}(\tau)$. Employing the operations of addition and scalar multiplication

$$\boldsymbol{v}(\tau) + \boldsymbol{w}(\tau) \quad \text{and} \quad \alpha\,\boldsymbol{v}(\tau) \qquad \forall \alpha \in \mathbb{C} \qquad (2.569)$$

within $\mathcal{H}(\tau)$ we can endow the set $\boldsymbol{\Upsilon}$ with the following algebraic operations

$$\begin{aligned}(\alpha \boldsymbol{v})(\tau) &= \alpha\,\boldsymbol{v}(\tau) \qquad \forall \alpha \in \mathbb{C}, & (2.570)\\ (\boldsymbol{v}+\boldsymbol{w})(\tau) &= \boldsymbol{v}(\tau)+\boldsymbol{w}(\tau) \qquad \forall \boldsymbol{v},\boldsymbol{w} \in \boldsymbol{\Upsilon}, & (2.571)\end{aligned}$$

which render $\boldsymbol{\Upsilon}$ a vector space.

Next, we can form the scalar product of two vectors of $\boldsymbol{v}(\tau)$ and $\boldsymbol{w}(\tau)$ in $\mathcal{H}(\tau)$, i.e.,

$$\langle \boldsymbol{v}(\tau) \mid \boldsymbol{w}(\tau) \rangle_\tau, \qquad (2.572)$$

2.13. DIRECT INTEGRALS OF HILBERT SPACES

where the subscript τ denotes a quantity defined in $\mathcal{H}(\tau)$. This results in a complex-valued function on $I\!R$ defined by the following mapping:

$$\tau \to \langle v(\tau) \mid w(\tau) \rangle_\tau. \qquad (2.573)$$

When $v = w$ we obtain a real-valued function

$$\tau \to \lVert v(\tau) \rVert_\tau, \qquad (2.574)$$

where

$$\lVert v(\tau) \rVert_\tau = \left(\langle v(\tau) \mid v(\tau) \rangle_\tau \right)^{1/2} \qquad (2.575)$$

is the norm of the vector $v(\tau)$ in $\mathcal{H}(\tau)$.

Definition 2.13.2(1)   Borel families of Hilbert spaces[166]

- A family $\{\mathcal{H}(\tau) : \tau \in I\!R\}$ of Hilbert spaces is called a *Borel family* of Hilbert spaces over $I\!R$ if there exists a linear subset Υ_0 of the set Υ of all vector-valued functions with the following properties:

  1. For every $v \in \Upsilon_0$, the real-valued function $\tau \to \lVert v(\tau) \rVert_\tau$ is a Borel function on $I\!R$.
  2. Any vector-valued function $v(\tau)$ in Υ is necessarily a member of Υ_0 if the complex-valued function

  $$\tau \to \langle v(\tau) \mid w(\tau) \rangle_\tau \qquad (2.576)$$

  arising from every $w \in \Upsilon_0$ is a Borel function.
  3. There exists a sequence $v_1(\tau), v_2(\tau), \ldots$ of members of Υ_0 such that for each τ the set $\{v_1(\tau), v_2(\tau), \ldots\}$ of vectors forms a complete set in the Hilbert space $\mathcal{H}(\tau)$.

Comments 2.13.2(1)   Clarification

C1   Note that $v(\tau)$ is a vector in the Hilbert space $\mathcal{H}(\tau)$ for each value of τ, hence v is called a vector-valued function on $I\!R$. These vector-valued functions form a vector space. There is as yet no definition of a scalar product in this new vector space.

C2   We do desire to introduce a scalar product in this new vector space, i.e., a scalar product between these vector-valued functions.[167] It is for this

---

[166] Dixmier (1981) Part II Chapter 1, Bratteli and Robinson Vol. 1 (1979) §4.4.1.
[167] For each value of τ we do have a scalar product between any two vectors $v(\tau)$ and $w(\tau)$ in the Hilbert space $\mathcal{H}(\tau)$, i.e., we have $\langle v(\tau) \mid w(\tau) \rangle_\tau$. This should not be confused with the need for a scalar product between v and w as two vector-valued functions.

reason that the three properties are included in Definition 2.13.2(1) for a Borel family of Hilbert spaces. First we can appreciate that an arbitrary vector-valued function w is not going to be an element of any useful Hilbert space, just like an arbitrary function on $I\!\!E$ cannot be an element of the Hilbert space $L^2(I\!\!E)$. We must restrict ourselves to a linear subset of $\boldsymbol{\Upsilon}$, e.g., $\boldsymbol{\Upsilon_0}$. The first defining property of this subset is to do with the need for square integrability of vector-valued functions $\boldsymbol{v}(\tau)$, i.e., we need integrability of $\|\boldsymbol{v}(\tau)\|_\tau^2$ with respect to τ. As already pointed out in §2.6.1C(1) C4 a function needs to be a Borel function in order for it to be integrable. The second and the third properties ensure that the set $\boldsymbol{\Upsilon_0}$ is big enough to be able to form a Hilbert space.

C3 The Hilbert spaces $\mathcal{H}(\tau)$ for different values of τ need not be related. They could, for example, have different dimensions. On the other hand we often have the simple situation that all $\mathcal{H}(\tau)$ are identical to a certain Hilbert space $\mathcal{H}(0)$. A family of such spaces $\mathcal{H}(\tau) = \mathcal{H}(0)$ is called a *constant family of Hilbert spaces*.[168]

C4 We can define a Borel family of Hilbert spaces over $I\!\!R^n$ in an identical manner. A further extension to a general measurable space is possible, although we shall not be needing such a generalization.

C5 In a given Borel family $\{\mathcal{H}(\tau) : \tau \in I\!\!R\}$ of Hilbert spaces, there is one and only one linear subset $\boldsymbol{\Upsilon_0}$ of vector-valued functions $\boldsymbol{v}$ satisfying the three properties set out in Definition 2.13.2(1).[169] We may regard a Borel family of Hilbert spaces, to be denoted by $\mathcal{H}_{\Upsilon_o}$, as a pair, i.e., we have

$$\mathcal{H}_{\Upsilon_o} = \Big\{ \{\mathcal{H}(\tau) : \tau \in I\!\!R\}, \boldsymbol{\Upsilon_0} \Big\}. \tag{2.577}$$

C6 Finally, we can have a Borel family of Hilbert spaces over an interval of $I\!\!R$. Everything we discuss remains valid if $I\!\!R$ above is replaced by an interval of $I\!\!R$.

Definition 2.13.2(2) **Square integrability and scalar product**

- Let $\mathcal{H}_{\Upsilon_o}$ be a Borel family of Hilbert spaces over $I\!\!R$ which is endowed with a Lebesgue-Stieltjes measure $dg(\tau)$ generated by a real-valued function $g(\tau)$. We call $\boldsymbol{v} \in \boldsymbol{\Upsilon_0}$ *square-integrable* with respect to $dg(\tau)$ if

$$\int_{-\infty}^{\infty} \|\boldsymbol{v}(\tau)\|_\tau^2 \, dg(\tau) < \infty. \tag{2.578}$$

---

[168] Reed and Simon Vol. 4 (1978) p. 280.
[169] Dixmier (1981) Proposition 4 on p. 167.

2.13. DIRECT INTEGRALS OF HILBERT SPACES

- A scalar product between v and w in Υ_0, to be denoted by $\langle v \mid w \rangle^\oplus$, is defined by

$$\langle v \mid w \rangle^\oplus = \int_{-\infty}^{\infty} \langle v(\tau) \mid w(\tau) \rangle_\tau \, dg(\tau) \tag{2.579}$$

whenever the integral exists.

Comments 2.13.2(2)   **Hilbert space structure**

C1   The flexibility of Lebesgue-Stieltjes measures, as discussed in §2.6.1C(1) C5 and C6, enables us to combine discrete and continuous composition of Hilbert spaces within one unified formalism. We shall return to demonstrate this explicitly.

C2   Consider the set of square-integrable members of Υ_0. With addition and scalar multiplication defined by Eqs. (2.570) and (2.571) this set forms a vector space. With the scalar product defined above this vector space of square-integrable vector-valued functions in Υ_0 forms a Hilbert space.[170]

Definition 2.13.2(3)[171]   **Direct integral space, pure and mixed vectors**

- We call the Hilbert space formed by the set of square-integrable members of Υ_0 the *direct integral* of the Borel family of Hilbert spaces $\mathcal{H}_{\Upsilon_0}$. This direct integral is denoted by

$$\mathcal{H}^\oplus = \int_{I\!R}^{\oplus} \mathcal{H}(\tau) \, dg(\tau), \tag{2.580}$$

with elements of $\mathcal{H}^\oplus$ denoted by

$$v^\oplus = \int_{I\!R}^{\oplus} v(\tau) \, dg(\tau), \tag{2.581}$$

and referred to as the *direct integral* of the family of vectors $\{v(\tau), \tau \in I\!R\}$ with respect to the measure $dg(\tau)$. For any given value τ_0 of τ the vector $v(\tau_0) \in \mathcal{H}(\tau_0)$ is called the *projection* of $v^\oplus$ on $\mathcal{H}(\tau_0)$.

- The scalar product between two elements $v^\oplus$ and $w^\oplus$ in $\mathcal{H}^\oplus$ is denoted

---

[170] Dixmier (1981) Proposition 5 on p. 169.
[171] Dixmier (1981) Part II Chapter 1.

by $\langle v^\oplus \mid w^\oplus \rangle^\oplus$ and given by[172]

$$\langle v^\oplus \mid w^\oplus \rangle^\oplus = \int_{-\infty}^{\infty} \langle v(\tau) \mid w(\tau) \rangle_\tau \, dg(\tau). \tag{2.582}$$

The norm of $v^\oplus$ is given by

$$\|v^\oplus\| = \left\{ \langle v^\oplus \mid v^\oplus \rangle^\oplus \right\}^{1/2} = \left\{ \int_{-\infty}^{\infty} \langle v(\tau) \mid v(\tau) \rangle_\tau \, dg(\tau) \right\}^{1/2}. \tag{2.583}$$

- $\mathcal{H}^\oplus$ is said to be *decomposable* as the direct integral of the Borel family of Hilbert spaces $\mathcal{H}_{\Upsilon_0}$.

- A non-zero vector $v^\oplus$ in the Hilbert space $\mathcal{H}^\oplus$ is called a *pure vector* in $\mathcal{H}^\oplus$ if its projections $v(\tau)$ as vectors in $\mathcal{H}(\tau)$ vanish except for a single value τ_0 of τ, i.e.,

$$v(\tau) = \begin{cases} v(\tau_0) \neq \mathbf{0}(\tau_0), & \text{for } \tau = \tau_0 \\ \mathbf{0}(\tau), & \text{for } \tau \neq \tau_0 \end{cases}, \tag{2.584}$$

where $\mathbf{0}(\tau)$ is the zero vector in $\mathcal{H}(\tau)$. Otherwise $v^\oplus$ is called a *mixed vector* in $\mathcal{H}^\oplus$.

Comments 2.13.2(3)   Choice of measures and singular vectors

C1   Recall that a direct sum $\phi_1 \oplus \phi_2$ is just a pair of vectors each belonging to a different Hilbert space. Conceptually a direct integral is similar to a direct sum, i.e., $v^\oplus$ is just a family of vectors $v(\tau)$ each generally belonging to a different Hilbert space $\mathcal{H}(\tau)$. The integral expression in Eq. (2.581) is just a symbolic expression and no integration of the vectors $v(\tau)$ is actually carried out. An integration in the usual sense occurs only in the calculation of scalar product in Eq. (2.582) and of norm in Eq. (2.583).

C2   We can incorporate discrete direct sums into the definition of direct integrals by using a piecewise constant real-valued function $g(\tau)$ with a finite or a countably infinite number of discontinuities to generate a discrete Lebesgue-Stieltjes measure for the integral as we do in §2.6.1C(1) C6 to arrive at Eq. (2.254). A simple example is given by the following function $g(\tau)$,

---

[172] The scalar product between two elements of the direct integral space must not be confused with the scalar product $\langle v(\tau) \mid w(\tau) \rangle_\tau$ between two vectors in the Hilbert space $\mathcal{H}(\tau)$.

2.13. DIRECT INTEGRALS OF HILBERT SPACES

known as a staircase function:

$$g(\tau) = \begin{cases} 0, & \tau < 1 \\ 1, & 1 \leq \tau < 2 \\ 2, & 2 \leq \tau < 3 \\ \vdots & \end{cases} \quad (2.585)$$

In accordance with Eq. (2.254) the scalar product in Eq. (2.582) reduces to a sum:

$$\langle v^\oplus \mid w^\oplus \rangle^\oplus = \int_{-\infty}^{\infty} \langle v(\tau) \mid w(\tau) \rangle_\tau \, dg(\tau) \quad (2.586)$$

$$= \sum_{\ell=1,2,\ldots} \langle v(\ell) \mid w(\ell) \rangle_\ell, \quad (2.587)$$

where

$$v(\ell) = v(\tau = \ell), \quad w(\ell) = w(\tau = \ell). \quad (2.588)$$

The direct integral reduces to a direct sum of a discrete set of Hilbert spaces $\mathcal{H}(\ell) = \mathcal{H}(\tau = \ell)$. Denoting this direct sum by $\mathcal{H}_d^\oplus$ we have

$$\mathcal{H}_d^\oplus = \int^\oplus \mathcal{H}(\tau) \, dg(\tau) = \oplus_\ell \mathcal{H}(\ell). \quad (2.589)$$

Elements of $\mathcal{H}_d^\oplus$ are of the form

$$v^\oplus = v(1) \oplus v(2) \oplus v(3) \oplus \cdots, \quad v(\ell) \in \mathcal{H}(\ell). \quad (2.590)$$

Let $v(1)$ and $w(1)$ be two non-zero vectors in $\mathcal{H}(1)$ and $v(2)$ and $w(2)$ be two non-zero vectors in $\mathcal{H}(2)$. Then vectors in $\mathcal{H}_d^\oplus$ of the form

$$v_1^\oplus = v(1) \oplus 0(2) \oplus 0(3) \oplus \cdots \quad (2.591)$$
$$v_2^\oplus = 0(1) \oplus v(2) \oplus 0(3) \oplus \cdots \quad (2.592)$$

are pure, while

$$w^\oplus = w(1) \oplus w(2) \oplus 0(3) \oplus \cdots \quad (2.593)$$

is mixed. In the case of such a direct sum we shall often adopt a more succinct notation with $\mathcal{H}_\ell$ denoting $\mathcal{H}(\ell)$ and v_ℓ denoting $v(\ell)$. It follows that

$$\mathcal{H}_d^\oplus = \oplus_\ell \mathcal{H}_\ell, \quad (2.594)$$
$$v^\oplus = \oplus_\ell v_\ell. \quad (2.595)$$

C3 In contrast to discrete measures we have absolutely continuous measures. These are generated by absolutely continuous real-valued functions g

on $\mathbb{R}$; the measure of any singleton set is zero. We can rewrite the direct integral as

$$\int^\oplus \mathcal{H}(\tau)\,dg(\tau) = \int^\oplus \mathcal{H}(\tau)\,f(\tau)\,d\tau, \quad f(\tau) = \frac{dg}{d\tau}. \tag{2.596}$$

The simplest continuous measure is the Lebesgue measure generated by $g(\tau) = \tau$. We shall denote the resulting direct integral by $\mathcal{H}_c^\oplus$, i.e.,

$$\mathcal{H}_c^\oplus = \int^\oplus \mathcal{H}(\tau)\,d\tau. \tag{2.597}$$

For such a continuous measure the direct integral space admits no pure vector, since Eq. (2.583) will give such a vector a zero norm in $\mathcal{H}_c^\oplus$. However, similar to generalized eigenfunctions and eigenvalues such vectors in $\mathcal{H}_c^\oplus$ have important physical applications. So we shall introduce a notion of *pure singular vectors* to describe them.

C4   **Pure singular vectors**   Consider an element of $\mathcal{H}_c^\oplus$ of the form

$$\boldsymbol{v}_{\tau_0}^\oplus = \begin{cases} \boldsymbol{v}(\tau_0) \neq \boldsymbol{0}(\tau_0), & \tau = \tau_0 \\ \boldsymbol{0}(\tau), & \tau \neq \tau_0 \end{cases}. \tag{2.598}$$

In other words $\boldsymbol{v}_{\tau_0}^\oplus$ has a zero projection on every $\mathcal{H}(\tau)$ except on $\mathcal{H}(\tau_0)$. It follows that $\boldsymbol{v}_{\tau_0}^\oplus$ has a vanishing norm. We call such an element of $\mathcal{H}_c^\oplus$ a *pure singular vector*. They may be symbolically written in the form of a direct integral with the help of a δ-function as[173]

$$\boldsymbol{v}_{\tau_0}^\oplus = \int^\oplus \delta(\tau - \tau_0)\boldsymbol{u}(\tau)\,d\tau. \tag{2.599}$$

There are no pure vectors of a non-zero norm in $\mathcal{H}_c^\oplus$.

C5   **Mixed singular vectors**   We call linear combinations of pure singular vectors of the form

$$a\,\boldsymbol{v}_{\tau_1}^\oplus + b\,\boldsymbol{v}_{\tau_2}^\oplus, \quad \tau_1 \neq \tau_2, \quad a, b \in \mathbb{C} \tag{2.600}$$

in $\mathcal{H}_c^\oplus$ *mixed singular vectors*. This term can be extended to apply to combinations of more than two singular pure vectors.

C6   We can also have a measure which is partly discrete and partly continuous. With any of these measures the resulting direct integral space $\mathcal{H}^\oplus$

---

[173] Written in the form of a direct integral involving a δ-function above does not help to give it a norm since a substitution into Eq. (2.579) will involve the square of a δ-function, rendering the integral undefined.

2.13. DIRECT INTEGRALS OF HILBERT SPACES

constitutes a separable Hilbert space.[174] Note that we have assumed separability in all the Hilbert spaces used in this book.

Examples 2.13.2(1) **Direct integral of one-dimensional spaces in coordinate and momentum representation spaces**

E1 In physical applications we often need only to employ direct integrals of a constant family of finite dimensional Hilbert spaces. Consider the direct integral with $\mathcal{H}(\tau) = \mathbb{C}$ for all τ over a range $\mathcal{R} \subset \mathbb{R}$ with respect to the Lebesgue measure $d\tau$. Let the resulting direct integral be denoted by $\mathcal{H}_\mathcal{R}^\oplus(\mathbb{C}, d\tau)$ or $\mathcal{H}_\mathcal{R}^\oplus(\mathbb{C})$ for short. We have

$$\mathcal{H}_\mathcal{R}^\oplus(\mathbb{C}, d\tau) = \int_\mathcal{R}^\oplus \mathcal{H}(\tau)\, d\tau, \quad \mathcal{H}(\tau) = \mathbb{C}. \tag{2.601}$$

An element of $\mathcal{H}_\mathcal{R}^\oplus(\mathbb{C}, d\tau)$ is of the form

$$\boldsymbol{f}^\oplus = \int_\mathcal{R}^\oplus \boldsymbol{f}(\tau)\, d\tau, \tag{2.602}$$

where, for each value of τ, $\boldsymbol{f}(\tau)$ is a member of $\mathcal{H}(\tau) = \mathbb{C}$. To avoid confusing the above direct integral with a usual Riemann or Lebesque integral we shall treat $\mathcal{H}(\tau) = \mathbb{C}$ as a Hilbert space seriously and formally introduce a unit vector $\boldsymbol{u}(\tau)$ in every $\mathcal{H}(\tau)$. An arbitrary element of $\mathcal{H}(\tau)$ is then of the form[175]

$$f(\tau)\, \boldsymbol{u}(\tau), \tag{2.603}$$

where $f(\tau)$ is a complex number. A vector-valued function in Υ_0 is then of the form

$$\boldsymbol{f}(\tau) = f(\tau)\, \boldsymbol{u}(\tau), \quad \forall \tau \in \mathcal{R}, \tag{2.604}$$

where $f(\tau)$ is a numerical function of τ, and we can rewrite $\boldsymbol{f}^\oplus$ in terms of $f(\tau)$ explicitly:

$$\boldsymbol{f}^\oplus = \int_\mathcal{R}^\oplus f(\tau)\, \boldsymbol{u}(\tau)\, d\tau. \tag{2.605}$$

Since

$$\langle \boldsymbol{f}^\oplus \mid \boldsymbol{f}^\oplus \rangle^\oplus = \int_{-\infty}^\infty \langle \boldsymbol{f}(\tau) \mid \boldsymbol{f}(\tau) \rangle_\tau\, d\tau \tag{2.606}$$

---
[174] Dixmier (1981) §1.5, Reed and Simon Vol. 1 (1972) p. 64.
[175] A unit vector $\boldsymbol{u}(\tau)$ in $\mathcal{H}(\tau) = \mathbb{C}$ is a complex number of magnitude 1. To avoid confusion and to achieve a uniform presentation applicable to $\mathcal{H}(\tau) \neq \mathbb{C}$ we have deliberately express an element in $\mathcal{H}(\tau) = \mathbb{C}$ in the form of a complex number multiplying into $\boldsymbol{u}(\tau)$. The usefulness of this presentation can be seen §2.13.2E(1) E2 and §2.13.2E(1) E3 below.

206   CHAPTER 2. OPERATORS AND THEIR DIRECT INTEGRALS

$$= \int_{-\infty}^{\infty} f^*(\tau) f(\tau) \langle u(\tau) \mid u(\tau) \rangle_\tau \, d\tau \qquad (2.607)$$

$$= \int_{\mathcal{R}} |f(\tau)|^2 \, d\tau, \qquad (2.608)$$

we require $f(\tau)$ to be square-integrable, i.e.,

$$\int_{\mathcal{R}} |f(\tau)|^2 \, d\tau < \infty. \qquad (2.609)$$

Note that the direct integral $f^\oplus$ in Eq. (2.602) is a symbol for a family of vectors $f(\tau)$. In contrast Eqs. (2.608) and (2.609) are ordinary Riemann or Lebesque integrals leading to a numerical value.

Finally we should point out that $\mathcal{H}_{\mathbb{R}}^\oplus(\mathbb{C}, d\tau)$ is identifiable with the usual Hilbert space $L^2(\mathbb{R})$ of square-integrable functions of a coordinate variable $x \in \mathbb{R}$, i.e., functions $f(x)$ satisfying Eq. (2.609).[176] By setting $\tau = x$ we have

$$\mathcal{H}_{\mathbb{R}}^\oplus(\mathbb{C}, dx) = \int_{\mathbb{R}}^{\oplus} \mathcal{H}(x) \, dx, \quad \mathcal{H}(x) = \mathbb{C} \qquad (2.610)$$

with

$$f^\oplus = \int_{\mathbb{R}}^{\oplus} f(x) \, u(x) \, dx. \qquad (2.611)$$

Clearly we can effect the following mappings:

$$f^\oplus \in \mathcal{H}_{\mathbb{R}}^\oplus(\mathbb{C}, dx) \quad \leftrightarrow \quad f \in L^2(\mathbb{R}). \qquad (2.612)$$

In other words $L^2(\mathbb{R})$, and hence $L^2(\mathbb{E})$, can be decomposed as a direct integral of one-dimensional Hilbert spaces $\mathcal{H}(x) = \mathbb{C}$. A pure singular vector $f_{x_0}^\oplus(x)$ corresponds to $f(x) = \delta(x - x_0)$.

E2 We can apply the results obtained above to the direct integrals of one-dimensional Hilbert spaces generally. Let $\mathcal{H}_{\mathcal{R}}^\oplus(\mathcal{H}, d\tau)$ denote the direct integral of a family of one-dimensional Hilbert spaces $\mathcal{H}(\tau)$ other than $\mathbb{C}$ with respect to the Lebesgue measure $d\tau$ over the range $\mathcal{R} \subset \mathbb{R}$. Then vectors in $\mathcal{H}_{\mathcal{R}}^\oplus(\mathcal{H}, d\tau)$ are of the form (2.605) with $u(\tau)$ denoting unit vectors in $\mathcal{H}(\tau)$. It is obvious that $\mathcal{H}_{\mathcal{R}}^\oplus(\mathcal{H}, d\tau)$ is unitarily equivalent to $\mathcal{H}_{\mathcal{R}}^\oplus(\mathbb{C}, d\tau)$.

E3 In §2.4E(1) E1 we introduce the notion of coordinate and momentum representation spaces. Equation (2.610) may be regarded as a decomposition of $L^2(\mathbb{R})$ as a direct integral of a constant family of Hilbert spaces $\mathbb{C}$ in the

---

[176] Roman Vol. 2 (1975) pp. 457-459.

2.13. DIRECT INTEGRALS OF HILBERT SPACES

coordinate (representation) space. We can signify this by

$$L^2(I\!E) = \mathcal{H}_{I\!E}^{\oplus}(\mathbb{C}, dx) \tag{2.613}$$

$$= \int_{I\!E}^{\oplus} \mathcal{H}(x)\,dx, \quad \mathcal{H}(x) = \mathbb{C}. \tag{2.614}$$

This is to contrast a direct integral decomposition in the momentum (representation) space. In §2.4E(1) E1 we introduce the Fourier transform operation and the idea that the Fourier transform $\widetilde{\phi}$ of an element $\phi(x)$ of $L^2(I\!E)$ may be regarded as a function on the momentum space $\widetilde{I\!E}$. We can write $\widetilde{\phi} = \widetilde{\phi}(p)$, where $\widetilde{\phi}(p)$ is square-integrable with respect to dp. It follows that the Fourier transform $\widetilde{L}^2(\widetilde{I\!E})$ of $L^2(I\!E)$ is decomposable into a direct integral of a constant family of Hilbert spaces $\widetilde{\mathcal{H}}(\tau) = \mathbb{C}$ in the same way as $L^2(I\!E)$ is. The analogy becomes transparent when we replace τ by the momentum variable p to get

$$\widetilde{L}^2(\widetilde{I\!E}) = \mathcal{H}_{\widetilde{I\!E}}^{\oplus}(\mathbb{C}, dp) \tag{2.615}$$

$$= \int_{\widetilde{I\!E}}^{\oplus} \widetilde{\mathcal{H}}(p)\,dp, \quad \widetilde{\mathcal{H}}(p) = \mathbb{C} \tag{2.616}$$

with

$$\widetilde{\boldsymbol{f}}^{\oplus} = \int_{\widetilde{I\!E}}^{\oplus} \widetilde{f}(p)\,\widetilde{\boldsymbol{u}}(p)\,dp, \tag{2.617}$$

where $\widetilde{\boldsymbol{u}}(p)$ is a unit vector in $\widetilde{\mathcal{H}}(p) = \mathbb{C}$. While mathematically this resembles Eq. (2.611) the physical interpretation in the context of quantum mechanics is quite different. In quantum mechanics each value p in the momentum space $\widetilde{I\!E}$ corresponds to a generalized eigenfunction η_p, given by Eq. (2.210) in §2.5.1E(1) E6, of the momentum operator $\widehat{p}$ in $L^2(I\!E)$. Consequently we may interpret Eq. (2.617) as a decomposition of $\widetilde{\boldsymbol{f}}^{\oplus}$ as a direct integral of η_p by identifying η_p with $\widetilde{\boldsymbol{u}}(p)$. Then a plane wave η_{p_0} for a particular momentum p_0 corresponds to a pure singular vector in $\mathcal{H}_{\widetilde{I\!E}}^{\oplus}(\mathbb{C}, dp)$, i.e.,

$$\widetilde{\boldsymbol{f}}_{p_0}^{\oplus} = \int^{\oplus} \delta(p - p_0)\,\widetilde{\boldsymbol{u}}(p)\,dp. \tag{2.618}$$

Definition 2.13.2(4)   **Trivial, discrete and continuous integral decompositions**

- A direct integral decomposition $\mathcal{H}^{\oplus} = \int^{\oplus} \mathcal{H}(\tau)\,dg(\tau)$ is called a *trivial decomposition* if the measure is a discrete one generated by the step

function
$$g(\tau) = \begin{cases} 0 & \tau < 0 \\ 1 & \tau \geq 0 \end{cases}. \tag{2.619}$$

In other words $\mathcal{H}^\oplus$ reduces to the constituent Hilbert space $\mathcal{H}(0)$ at $\tau = 0$. The decomposition is called *continuous* (*discrete*) if the measure is continuous (discrete).

Theorem 2.13.2(1)   Subspaces

- Let $\mathcal{H}_{\Upsilon_o} = \{\{\mathcal{H}(\tau) : \tau \in \mathbb{R}\}, \Upsilon_0\}$ be a Borel family of Hilbert spaces. Let $\mathcal{L}(\tau)$ be a subspace of $\mathcal{H}(\tau)$, and let $\Upsilon_{0\mathcal{L}}$ be the set of vector-valued functions $\boldsymbol{v} \in \Upsilon_0$ such that $\boldsymbol{v}(\tau)$ is in the subspace $\mathcal{L}(\tau)$ for each τ. If there exists a sequence of vector-valued functions $\{\boldsymbol{v}(\tau), \boldsymbol{w}(\tau), \ldots\}$ in $\Upsilon_{0\mathcal{L}}$ such that the sequence $\{\boldsymbol{v}(\tau), \boldsymbol{w}(\tau), \ldots\}$ forms a complete set in $\mathcal{L}(\tau)$ for each τ, then:

  1. The set $\{\{\mathcal{L}(\tau) : \tau \in \mathbb{R}\}, \Upsilon_{0\mathcal{L}}\}$ constitutes a Borel family of Hilbert spaces in its own right.

  2. The resulting direct integral space
     $$\mathcal{L}^\oplus = \int^\oplus \mathcal{L}(\tau) \, dg(\tau) \tag{2.620}$$
     is a subspace of
     $$\mathcal{H}^\oplus = \int^\oplus \mathcal{H}(\tau) \, dg(\tau). \tag{2.621}$$

Examples 2.13.2(2)   Subspaces and singular subspaces

E1   The family $\{\{\mathcal{L}(\tau) : \tau \in \mathbb{R}\}, \Upsilon_{0\mathcal{L}}\}$ is referred to as a *Borel family of subspaces*. A useful example is the family specified

$$\mathcal{L}_\Delta(\tau) = \begin{cases} \mathcal{H}(\tau) & \tau \in \Delta \\ 0 & \tau \notin \Delta \end{cases}, \tag{2.622}$$

where Δ is a bounded Borel set. Intuitively we can see that this family is a truncation of $\mathcal{H}_{\Upsilon_0}$. The resulting direct integral, to be denoted by $\mathcal{L}_\Delta^\oplus$ or $\mathcal{H}_\Delta^\oplus$, is given by

$$\mathcal{H}_\Delta^\oplus = \mathcal{L}_\Delta^\oplus = \int^\oplus \mathcal{L}_\Delta(\tau) \, dg(\tau), \tag{2.623}$$

is a subspace of $\mathcal{H}^\oplus$ by Theorem 2.13.2(1).

2.14. DIRECT INTEGRALS OF OPERATORS

E2 A physically interesting and useful situation arises when the Borel set Δ is of measure zero. For instance Δ consists of a discrete set of isolated values of τ, e.g., when Δ consisting of a single point τ_0 in a continuous direct integral $\mathcal{H}_c^\oplus$. Following the argument in §2.13.2E(2) E1 above we arrive at $\mathcal{H}_{\Delta=\{\tau_0\}}^\oplus$ as a subspace of $\mathcal{H}^\oplus$. This subspace is clearly singular in that elements of $\mathcal{H}_{\Delta=\{\tau_0\}}^\oplus$ are singular in the sense of §2.13.2C(3) C4. However, this subspace has important physical applications since it is identifiable with the constituent Hilbert space $\mathcal{H}(\tau_0)$. We can also introduce such a formal subspace in terms of singular vectors. Let $\mathcal{H}_{\tau_0}^\oplus$ denote the set of singular vectors $\boldsymbol{f}_{\tau_0}^\oplus$ at $\tau = \tau_0$ in $\mathcal{H}_c^\oplus$. Then this set is identifiable with the constituent Hilbert space $\mathcal{H}(\tau_0)$, and with $\mathcal{H}_{\Delta=\{\tau_0\}}^\oplus$. So, we call $\mathcal{H}_{\tau_0}^\oplus$, as well as the constituent space $\mathcal{H}(\tau_0)$, a *singular subspace* of $\mathcal{H}_c^\oplus$. As an example we may regard $\widetilde{\mathcal{H}}(p_0)$ in Eq. (2.616) as a singular subspace of $\widetilde{L}^2(\mathbb{E})$ spanned by a singular vector $\widetilde{\boldsymbol{f}}_{p_0}^\oplus$.

E3 In the case of a discrete measure given by Eqs. (2.585) and (2.589) we can have non-singular subspaces

$$\mathcal{H}_\ell^\oplus = \mathcal{H}_{\Delta=\{\ell\}}^\oplus \tag{2.624}$$

composed of vectors of the form

$$\boldsymbol{v}_\ell^\oplus = \boldsymbol{0}(1) \oplus \cdots \oplus \boldsymbol{0}(\ell-1) \oplus \boldsymbol{v}(\ell) \oplus \boldsymbol{0}(\ell+1) \oplus \cdots \tag{2.625}$$
$$= \boldsymbol{0}_1 \oplus \cdots \oplus \boldsymbol{0}_{\ell-1} \oplus \boldsymbol{v}_\ell \oplus \boldsymbol{0}_{\ell+1} \oplus \cdots. \tag{2.626}$$

We can also make the following identification:

$$\mathcal{H}_\ell^\oplus = \boldsymbol{0}(1) \oplus \cdots \oplus \boldsymbol{0}(\ell-1) \oplus \mathcal{H}(\ell) \oplus \boldsymbol{0}(\ell+1) \oplus \cdots \tag{2.627}$$
$$= \boldsymbol{0}_1 \oplus \cdots \oplus \boldsymbol{0}_{\ell-1} \oplus \mathcal{H}_\ell \oplus \boldsymbol{0}_{\ell+1} \oplus \cdots. \tag{2.628}$$

In our notation the vector $\boldsymbol{v}_\ell^\oplus$ is an element of $\mathcal{H}^\oplus$ while $\boldsymbol{v}_\ell$ is an element of $\mathcal{H}_\ell$.

2.14 Direct Integrals of Operators

2.14.1 Direct sums of operators

Let $\widehat{A}_1$ be an operator in a Hilbert space $\mathcal{H}_1$ with domain $\mathcal{D}(\widehat{A}_1)$ and $\widehat{A}_2$ be an operator in $\mathcal{H}_2$ with domain $\mathcal{D}(\widehat{A}_2)$. We can define the direct sum operator, denoted by $\widehat{A}_1 \oplus \widehat{A}_2$, or $\widehat{A}^\oplus$, acting in the direct sum space $\mathcal{H}^\oplus = \mathcal{H}_1 \oplus \mathcal{H}_2$ by

$$\left(\widehat{A}_1 \oplus \widehat{A}_2\right)\left(\phi_1 \oplus \phi_2\right) = \widehat{A}_1\phi_1 \oplus \widehat{A}_2\phi_2, \tag{2.629}$$

where $\phi_1 \in \mathcal{D}(\widehat{A}_1)$ and $\phi_2 \in \mathcal{D}(\widehat{A}_2)$. We have the following properties:[177]

1. The direct sum of two closed operators is closed.

2. If $\widehat{A}_1$ and $\widehat{A}_2$ are closed and symmetric operators, then, the deficiency indices (n^+, n^-) of their direct sum $\widehat{A}^\oplus = \widehat{A}_1 \oplus \widehat{A}_2$ are

$$n^+ = n_1^+ + n_2^+, \quad n^- = n_1^- + n_2^-. \qquad (2.630)$$

Here (n_1^+, n_1^-) and (n_2^+, n_2^-) are the deficiency indices of $\widehat{A}_1$ and $\widehat{A}_2$.

3. The direct sum of two selfadjoint operators is selfadjoint.

4. The direct sum of two strictly maximal symmetric operators is not necessarily strictly maximal symmetric. The example given below in §2.14.1E(1) E1 serves to demonstrate this statement. This result has important physical applications in Chapters 6 and 7.

Examples 2.14.1(1) **Direct sum and reduction of operators**

E1 We know from §2.5.1E(1) E6, especially from Eq. (2.209), that in $\mathcal{H}_2 = L^2(\mathbb{E}^+)$ the operator $\widehat{p}^+ = -i\hbar d/dx$ defined on functions $\phi_2(x)$ in the Sobolev space $S_{2,1}(\mathbb{E}^+)$ satisfying the boundary condition $\phi_2(0) = 0$ is strictly maximal symmetric. It has deficiency indices (1,0). Similarly in $\mathcal{H}_1 = L^2(\mathbb{E}^-)$ we have a strictly maximal symmetric operator $\widehat{p}^-$ defined by Eq. (2.217). This operator has deficiency indices (0,1). The direct sum operator

$$\widehat{p}^- \oplus \widehat{p}^+ \qquad (2.631)$$

defined in the direct sum space

$$\mathcal{H}^{(1,2)} = \mathcal{H}_1 \oplus \mathcal{H}_2 = L^2(\mathbb{E}^-) \oplus L^2(\mathbb{E}^+) \qquad (2.632)$$

is not strictly maximal symmetric, since its deficiency indices, according to Eq. (2.630), are (1,1). It follows that $\widehat{p}^- \oplus \widehat{p}^+$ possesses a one-parameter family of selfadjoint extensions. We shall present these selfadjoint extensions in Theorem 2.14.1(1).

E2 When $\widehat{A}$ admits a reducing subspace $\mathcal{L}$ we can reduce $\widehat{A}$ into two parts by Eq. (2.104):[178]

1. $\widehat{A}_\mathcal{L} = \widehat{A}\widehat{P}_\mathcal{L}$ as the part of $\widehat{A}$ in $\mathcal{L}$.

---

[177] Blank, Exner and Havlíček (1994) p. 145, p. 149
[178] See Definition 2.3(2).

2.14. DIRECT INTEGRALS OF OPERATORS   211

2. $\widehat{A}_{\mathcal{L}^\perp} = \widehat{A}\widehat{P}_{\mathcal{L}^\perp}$ as the part of $\widehat{A}$ in $\mathcal{L}^\perp$.

These operators act in $\mathcal{H}$ with $\widehat{A} = \widehat{A}_\mathcal{L} + \widehat{A}_{\mathcal{L}^\perp}$. Treating $\mathcal{L}$ and $\mathcal{L}^\perp$ as Hilbert spaces in their own right we can introduce:

1. An operator $\widehat{A}(\mathcal{L})$ acting in $\mathcal{L}$ with domain $\widehat{P}_\mathcal{L} \mathcal{D}(\widehat{A}) \in \mathcal{L}$ on which it agrees with $\widehat{A}$.

2. An operator $\widehat{A}(\mathcal{L}^\perp)$ acting in $\mathcal{L}^\perp$ with domain $\widehat{P}_{\mathcal{L}^\perp} \mathcal{D}(\widehat{A}) \in \mathcal{L}^\perp$ on which it agrees with $\widehat{A}$.

We call $\widehat{A}(\mathcal{L})$ the *reduction of* $\widehat{A}$ *in* $\mathcal{L}$ and $\widehat{A}(\mathcal{L}^\perp)$ the *reduction of* $\widehat{A}$ *in* $\mathcal{L}^\perp$. The Hilbert space $\mathcal{H}$ can be identified with the direct sum of $\mathcal{L}$ and $\mathcal{L}^\perp$, and $\widehat{A}$ as the direct sum of $\widehat{A}(\mathcal{L})$ and $\widehat{A}(\mathcal{L}^\perp)$, i.e.,

$$\mathcal{H} = \mathcal{L} \oplus \mathcal{L}^\perp \quad \text{and} \quad \widehat{A}(\mathcal{L}) \oplus \widehat{A}(\mathcal{L}^\perp). \tag{2.633}$$

We can also express $\widehat{A}_\mathcal{L}$ and $\widehat{A}_{\mathcal{L}^\perp}$ in terms of $\widehat{A}(\mathcal{L})$ and $\widehat{A}(\mathcal{L}^\perp)$:

$$\widehat{A}_\mathcal{L} = \widehat{A}(\mathcal{L}) \oplus \widehat{O}(\mathcal{L}^\perp), \quad \widehat{A}_{\mathcal{L}^\perp} = \widehat{O}(\mathcal{L}) \oplus \widehat{A}(\mathcal{L}^\perp), \tag{2.634}$$

where $\widehat{O}(\mathcal{L})$ and $\widehat{O}(\mathcal{L}^\perp)$ are the zero operators on $\mathcal{L}$ and $\mathcal{L}^\perp$ respectively. Following §2.3C(2) C4 we can conclude that if $\widehat{A}$ is selfadjoint in $\mathcal{H}$ then $\widehat{A}(\mathcal{L})$ is selfadjoint in $\mathcal{L}$ and $\widehat{A}(\mathcal{L}^\perp)$ is selfadjoint in $\mathcal{L}^\perp$.

It is often useful to keep $\widehat{A}_\mathcal{L}$ and $\widehat{A}(\mathcal{L})$ apart as seperate quantities to avoid confusion, even though we can often use them interchangeably.

E3   Another example comes from §2.8.4. We know from §2.13.1C(1) C4 that for a spectral representation of the kinetic energy in $L^2(I\!\!E)$ the representation space $\widetilde{\widetilde{L}}^2(\widetilde{I\!\!E})$ can be written as a direct sum as shown in Eq. (2.567). The kinetic energy operator itself can also be decomposed as a direct sum, i.e., we can write Eq. (2.431) as

$$\widetilde{\widetilde{K}}\widetilde{\widetilde{\phi}} = \widetilde{\widetilde{K}}_-\widetilde{\widetilde{\phi}}_- \oplus \widetilde{\widetilde{K}}_+\widetilde{\widetilde{\phi}}_+ \tag{2.635}$$

$$= \epsilon\widetilde{\widetilde{\phi}}_-(\epsilon) \oplus \epsilon\widetilde{\widetilde{\phi}}_+(\epsilon), \tag{2.636}$$

or

$$\widetilde{\widetilde{K}} = \widetilde{\widetilde{K}}_- \oplus \widetilde{\widetilde{K}}_+. \tag{2.637}$$

Theorem 2.14.1(1)   **Selfadjoint extensions of direct sum of symmetric operators**[179]

- The family of selfadjoint extensions in $\mathcal{H}^{(1,2)}$ of $\widehat{p}^- \oplus \widehat{p}^+$ in Eq. (2.631) is parameterizable by a real variable λ in the range $(-\pi, \pi]$, and for each $\lambda \in (-\pi, \pi]$ there corresponds a selfadjoint extension $\widehat{P}_\lambda^{(1,2)}$ defined on a domain $\mathcal{D}(\widehat{P}_\lambda^{(1,2)})$ consisting of the vectors

$$\phi^{(1,2)} = \phi_1 \oplus \phi_2 \in \mathcal{H}^{(1,2)} \qquad (2.638)$$

where

$$\phi_1 \in S_{2,1}(I\!\!E^-), \quad \phi_2 \in S_{2,1}(I\!\!E^+), \quad \phi_1(0) = e^{-i\lambda}\phi_2(0), \qquad (2.639)$$

and acting on $\phi^{(1,2)} \in \mathcal{D}(\widehat{P}_\lambda^{(1,2)})$ we have

$$\widehat{P}_\lambda^{(1,2)} \phi^{(1,2)} = -i\hbar \left(\frac{d\phi_1}{dx} \oplus \frac{d\phi_2}{dx} \right) \quad \forall \phi^{(1,2)} \in \mathcal{D}(\widehat{P}_\lambda^{(1,2)}). \qquad (2.640)$$

Here the superscripts $(1, 2)$ signify a linkage or coupling between $\phi_1 \in \mathcal{H}_1$ and $\phi_2 \in \mathcal{H}_2$ by the boundary condition in Eq. (2.639). Generalized eigenfunctions of $\widehat{P}_\lambda^{(1,2)}$ are of the form

$$\eta_{\lambda,p}^{(1,2)}(x) = e^{ipx_1} \oplus e^{i\lambda} e^{ipx_2} = \begin{cases} e^{ipx_1}, & x_1 \in I\!\!E^- \\ e^{i\lambda} e^{ipx_2}, & x_2 \in I\!\!E^+ \end{cases}. \qquad (2.641)$$

For $\lambda = 0$ the extension is just the usual momentum operator $\widehat{p}$ in $L^2(I\!\!E)$.

Theorem 2.14.1(2)   **The spectrum of a direct sum operator**

- The spectrum $sp(\widehat{A}^\oplus)$ of a direct sum operator $\widehat{A}^\oplus = \widehat{A}_1 \oplus \widehat{A}_2$ in $\mathcal{H}_1 \oplus \mathcal{H}_2$ is related to the spectra $sp(\widehat{A}_1)$ and $sp(\widehat{A}_2)$ of its constituent operators.[180] The main results are that

  1. $sp(\widehat{A}^\oplus) = sp(\widehat{A}_1) \cup sp(\widehat{A}_2)$.
  2. $sp(\widehat{A}^\oplus)_p = sp(\widehat{A}_1)_p \cup sp(\widehat{A}_2)_p$.
  3. $sp(\widehat{A}^\oplus)_c = sp(\widehat{A}_1)_c \cup sp(\widehat{A}_2)_c$.

---
[179] For a proof see Wan and Fountain (1996).
[180] Roman Vol. 2 (1975) p. 592.

2.14.2 Direct integrals of operators

Let $\mathcal{H}_{\Upsilon_0}$ be a Borel family of Hilbert spaces and let $\mathcal{B}(\mathcal{H}(\tau))$ denote the set of all bounded operators on the Hilbert space $\mathcal{H}(\tau)$.

Definition 2.14.2(1)   **Direct integrals of operators**[181]

- A mapping $\tau \to \widehat{A}(\tau) \in \mathcal{B}(\mathcal{H}(\tau))$ is called a *Borel family of operators* if the vector-valued function given by $\tau \to \widehat{A}(\tau)v(\tau)$ is a member of Υ_0 whenever v is.

- A Borel family of operators is *essentially bounded* if there exists a real number M_{ess} such that $\|\widehat{A}(\tau)\| < M_{ess}$ for almost all τ, i.e., for all τ except on a set of Lebesgue measure zero on $\mathbb{R}$. The smallest M_{ess} is referred to as the *essential supremum*.[182]

- An essentially bounded Borel family of operators $\{\widehat{A}(\tau) : \tau \in \mathbb{R}\}$ defines a bounded operator on the direct integral space $\mathcal{H}^\oplus$ in Eq. (2.580) by the following mapping:

$$v^\oplus = \int_{\mathbb{R}}^\oplus v(\tau)\,dg(\tau) \mapsto v'^\oplus = \int_{\mathbb{R}}^\oplus \widehat{A}(\tau)v(\tau)\,dg(\tau). \qquad (2.642)$$

This operator is called the *direct integral* of the essentially bounded Borel family of operators $\{\widehat{A}(\tau) : \tau \in \mathbb{R}\}$, and is denoted by

$$\widehat{A}^\oplus = \int_{\mathbb{R}}^\oplus \widehat{A}(\tau)\,dg(\tau) \qquad (2.643)$$

so that

$$\widehat{A}^\oplus v^\oplus = \int_{\mathbb{R}}^\oplus \widehat{A}(\tau)v(\tau)\,dg(\tau). \qquad (2.644)$$

Comments 2.14.2(1)   **Bounded and unbounded operators**

C1   In physical applications we sometimes need only to employ direct integrals of a constant family of finite dimensional Hilbert spaces. Then all the operators $\widehat{A}(\tau)$ are bounded. For an essentially bounded Borel family of such operators the existence of M_{ess} ensures that the vector-valued function defined by $\widehat{A}(\tau)v(\tau)$ is square-integrable if $v(\tau)$ is. This enables the family $\{\widehat{A}(\tau) : \tau \in \mathbb{R}\}$ of operators to map the direct integral space $\mathcal{H}^\oplus$ into itself.

---

[181] Dixmier (1981) Part II Chapter 2.
[182] Williamson (1962) p. 32.

This mapping defines an operator in $\mathcal{H}^\oplus$ which is the direct integral operator $\widehat{A}^\oplus$. The range of integration is often omitted to simplify the notation.

C2 One can consider direct integrals of unbounded family of bounded operators, i.e., we can form the direct integral of a Borel family of bounded operators whose corresponding family of norms $\|\widehat{A}(\tau)\|$ is unbounded. The resulting operator would be unbounded in $\mathcal{H}^\oplus$. We can go one step further to form the direct integral of a family of unbounded operators. To do so we have to take the domains of various operators involved carefully into account.[183] Unless otherwise is stated all direct integrals in what follows are meant to be direct integrals of an essentially bounded Borel family of bounded operators.

C3 When the measure $dg(\tau)$ is purely discrete a direct integral reduces to a direct sum, and a direct integral of operators reduces to a direct sum of operators.

Definition 2.14.2(2)   **Decomposable and diagonalizable operators**

- An operator in $\mathcal{H}^\oplus$ of the form of a direct integral $\widehat{A}^\oplus$ is said to be *decomposable*. The operator $\widehat{A}(\tau)$ is called the *reduction* of $\widehat{A}^\oplus$ in the constituent Hilbert space $\mathcal{H}(\tau)$.

- A *diagonalizable* operator in $\mathcal{H}^\oplus$ is one of the form

$$\widehat{C}^\oplus = \int_{\mathbb{R}}^{\oplus} c(\tau)\widehat{\mathbb{I}}(\tau)\, dg(\tau), \qquad (2.645)$$

where $\widehat{\mathbb{I}}(\tau)$ is the identity operator on $\mathcal{H}(\tau)$ and $c(\tau)$ is any essentially bounded complex-valued Borel function on $\mathbb{R}$.[184]

Comments 2.14.2(2)   **Direct integrals and reduction of operators**

C1 A decomposable operator in a direct sum space $\mathcal{H}^\oplus = \mathcal{H}_1 \oplus \mathcal{H}_2$ is an operator of the form $\widehat{A}_1 \oplus \widehat{A}_2$, where $\widehat{A}_1$ acts in $\mathcal{H}_1$ and $\widehat{A}_2$ in $\mathcal{H}_2$. A diagonalisable operator is an operator of the form $a_1\widehat{\mathbb{I}}_1 \oplus a_2\widehat{\mathbb{I}}_2$, $a_1, a_2 \in \mathbb{C}$.

More generally in a direct sum space $\mathcal{H}^\oplus = \oplus_\ell \mathcal{H}_\ell$ a decomposable operator is an operator of the form

$$\widehat{A}^\oplus = \oplus_\ell \widehat{A}_\ell, \qquad (2.646)$$

where $\widehat{A}_\ell$ is an operator in $\mathcal{H}_\ell$, and diagonalisable operators are of the form

$$\widehat{A}^\oplus = \oplus_\ell a_\ell \widehat{\mathbb{I}}_\ell. \qquad (2.647)$$

---

[183] Reed and Simon Vol. 4 (1978) pp. 280-286.

[184] A complex-valued Borel function $f(x)$ on $\mathbb{R}$ is essentially bounded if there exists a $M_{ess} \in \mathbb{R}$ such that $|f(x)| < M_{ess}$ for almost all x.

2.14. DIRECT INTEGRALS OF OPERATORS

In keeping with the notation used in Eqs. (2.594), (2.595), (2.626) and (2.628) we shall write

$$\widehat{A}_\ell^\oplus = \widehat{O}_1 \oplus \cdots \oplus \widehat{O}_{\ell-1} \oplus \widehat{A}_\ell \oplus \widehat{O}_{\ell+1} \oplus \cdots, \qquad (2.648)$$

where $\widehat{O}_\ell$ is the zero operator on $\mathcal{H}_\ell$.

An important feature of decomposable operators is that they cannot correlate vectors lying in different constituent Hilbert spaces, i.e.,[185]

$$\langle \boldsymbol{v}_j^\oplus \mid \widehat{A}^\oplus \boldsymbol{v}_k^\oplus \rangle^\oplus = 0, \quad j \neq k, \qquad (2.649)$$

where $\boldsymbol{v}_j^\oplus$ and $\boldsymbol{v}_k^\oplus$ are given by Eq. (2.626).

C2 Not all operators in $\mathcal{H}^\oplus$ are decomposable. For example, in $\mathcal{H}_1 \oplus \mathcal{H}_2$ we could have operators $\widehat{A}$ capable of correlating vectors in $\mathcal{H}_1$ and in $\mathcal{H}_2$ in the sense that:

$$\langle \boldsymbol{v}_1^\oplus \mid \widehat{A} \boldsymbol{v}_2^\oplus \rangle^\oplus \neq 0. \qquad (2.650)$$

As a simple illustration consider the case when both $\mathcal{H}_1$ and $\mathcal{H}_2$ are one-dimensional spanned by unit vectors $\boldsymbol{u}_1$ and $\boldsymbol{u}_2$ respectively. Then $\mathcal{H}_1 \oplus \mathcal{H}_2$ is two-dimensional spanned by unit vectors $\boldsymbol{u}_1^\oplus$ and $\boldsymbol{u}_2^\oplus$. We can define an operator $\widehat{A}$ by its action on these basis vectors, e.g.,

$$\widehat{A} \boldsymbol{u}_1^\oplus = 0, \quad \widehat{A} \boldsymbol{u}_2^\oplus = \boldsymbol{u}_1^\oplus. \qquad (2.651)$$

In Dirac notation as used in Eq. (2.125) this operator can be expressed as

$$\widehat{A} = |\boldsymbol{u}_1^\oplus\rangle\langle\boldsymbol{u}_2^\oplus|. \qquad (2.652)$$

Eq. (2.651) is clearly satisfied.

C3 We can extend the concept of operator reduction formally to singular subspaces. As an example following §2.13.2E(1) E3 we can introduce a formal reduction of the momentum operator $\widehat{p}$ on a singular subspace $\widetilde{\mathcal{H}}(p_0)$ as a multiplication operator in the form of a multiplicative constant p_0. These formal results will be seen to have important physical applications later.

C4 A decomposable operator $\widehat{A}^\oplus$ leaves any subspace $\mathcal{H}_\Delta^\oplus$ invariant:

$$\boldsymbol{v}^\oplus \in \mathcal{H}_\Delta^\oplus \quad \Rightarrow \quad \widehat{A}^\oplus \boldsymbol{v}^\oplus \in \mathcal{H}_\Delta^\oplus. \qquad (2.653)$$

This is why it is possible to define a reduction $\widehat{A}(\tau)$ of $\widehat{A}^\oplus$ in $\mathcal{H}(\tau)$. In general, a non-decomposable operator $\widehat{A}$ in $\mathcal{H}^\oplus$ does not have any reduction

---
[185] This is fundamental to later applications in the formulation of superselection rules in quantum mechanics.

in $\mathcal{H}(\tau)$ since $\mathcal{H}(\tau)$ is not invariant under $\widehat{A}$. This is especially obvious in the case of a direct sum.

C5 A diagonalizable operator is decomposable but the converse is not generally true. A simple but useful example of diagonalizable operators is

$$\widehat{\chi}_\Delta^\oplus = \int^\oplus \chi_\Delta(\tau) \widehat{\mathbb{I}}(\tau) \, dg(\tau). \tag{2.654}$$

It is obvious that $\widehat{\chi}_\Delta^\oplus$ will leave subspace $\mathcal{H}_\Delta^\oplus$ invariant.

Theorem 2.14.2(1) Useful properties of decomposable operators[186]

- Let $\widehat{A}^\oplus = \int^\oplus \widehat{A}(\tau) \, dg(\tau)$ and $\widehat{B}^\oplus = \int^\oplus \widehat{B}(\tau) \, dg(\tau)$ be two decomposable operators on $\mathcal{H}^\oplus$. Then:

$$\widehat{A}^\oplus + \widehat{B}^\oplus = \int^\oplus \left(\widehat{A}(\tau) + \widehat{B}(\tau)\right) dg(\tau),$$

$$\widehat{A}^\oplus \widehat{B}^\oplus = \int^\oplus \widehat{A}(\tau) \widehat{B}(\tau) \, dg(\tau),$$

$$\alpha \widehat{A}^\oplus = \int^\oplus \alpha \widehat{A}(\tau) \, dg(\tau),$$

$$\widehat{A}^{\oplus \dagger} = \int^\oplus \widehat{A}(\tau)^\dagger \, dg(\tau),$$

$$\|\widehat{A}^\oplus\| = \operatorname{ess\,sup}\left\{\|\widehat{A}(\tau)\| : \tau \in \mathbb{R}\right\}.$$

- A decomposable operator $\widehat{A}^\oplus = \int^\oplus \widehat{A}(\tau) \, dg(\tau)$ on $\mathcal{H}^\oplus$ is[187]

  1. selfadjoint if $\widehat{A}(\tau)$ is selfadjoint on $\mathcal{H}(\tau)$ for every τ,
  2. unitary if $\widehat{A}(\tau)$ is unitary on $\mathcal{H}(\tau)$ for every τ,
  3. a projector on $\mathcal{H}^\oplus$ if and only if $\widehat{A}(\tau)$ is a projector on $\mathcal{H}(\tau)$ for all τ.

Theorem 2.14.2(2) Decomposability conditions[188]

- A bounded operator $\widehat{A}$ on $\mathcal{H}^\oplus$ is decomposable if and only if one of the following conditions is satisfied:

---
[186] Dixmier (1981) Part II §2.4, Bratteli and Robinson Vol. 1 (1979) §4.4.1.
[187] The condition "for every τ" can be relaxed to "for almost every τ".
[188] Dixmier (1981) Part II §2.5, Wan, Bradshaw, Trueman and Harrison (1998) Appendix B.

2.14. DIRECT INTEGRALS OF OPERATORS

1. $\widehat{A}$ commutes with all selfadjoint diagonalizable operators on $\mathcal{H}^\oplus$.
2. $\widehat{A}$ commutes with $\widehat{\chi}_\Delta^\oplus$ for all Borel sets Δ of $\mathbb{R}$.
3. $\widehat{A}$ leaves the subspace $\mathcal{H}_\Delta^\oplus$ invariant for all Borel sets Δ of $\mathbb{R}$.

Corollary 2.14.2(1)   Decomposable selfadjoint operators

- A bounded selfadjoint operator $\widehat{A}$ on $\mathcal{H}^\oplus$ is decomposable if and only if all its spectral projectors $\widehat{M}^{\hat{A}}(\Lambda)$ are decomposable.

Comments 2.14.2(3)   Commutators and unbounded decomposable operators

C1   We can appreciate the intuitive contents of Theorem 2.14.2(1) without too much trouble.

C2   Theorem 2.14.2(2) is also intuitively reasonable since a diagonalizable operator clearly commutes with any diagonalizable and decomposable operators, even though two decomposable operators need not commute with each other. According to §2.8.1C(1) C7 two selfadjoint operators are said to commute if their spectral projectors commute. So, $\widehat{A}$ commutes with a selfadjoint diagonalizable operator $\widehat{C}^\oplus$ if and only if all $\widehat{M}^{\hat{A}}(\Lambda)$ commute with $\widehat{C}^\oplus$.

C3   The set of all bounded decomposable operators on $\mathcal{H}^\oplus$ will be denoted by $\mathcal{B}^\oplus(\mathcal{H}^\oplus)$. Unbounded decomposable operators can also be defined; their domains have to be specified carefully.[189] We can utilise Theorem 2.14.2(2) and Corollary 2.14.2(1) to define unbounded decomposable selfadjoint operators without having to deal with domains directly. Let $\widehat{A}$ be an arbitrary selfadjoint operator in $\mathcal{H}^\oplus$, not necessarily bounded, and let $\widehat{C}^\oplus$ be a selfadjoint diagonalizable operator in $\mathcal{B}^\oplus(\mathcal{H}^\oplus)$. From §2.8.1C(1) C7 we know that $\widehat{A}$ commutes with $\widehat{C}^\oplus$ if and only if its spectral projectors $\widehat{M}^{\hat{A}}(\Lambda)$ commute with $\widehat{C}^\oplus$. If $\widehat{A}$ commutes with all $\widehat{C}^\oplus$ in $\mathcal{B}^\oplus(\mathcal{H}^\oplus)$ then its spectral projectors $\widehat{M}^{\hat{A}}(\Lambda)$ also commutes with all $\widehat{C}^\oplus$. It follows from Corollary 2.14.1(1) that $\widehat{M}^{\hat{A}}(\Lambda)$ are decomposable. It is natural then to define a selfadjoint operator $\widehat{A}$, bounded or unbounded, to be decomposable if it commutes with all $\widehat{C}^\oplus$ in $\mathcal{B}^\oplus(\mathcal{H}^\oplus)$.

C4   Since decomposable operators leave subspaces $\mathcal{H}_\Delta^\oplus$ invariant they cannot correlate vectors belonging to subspaces $\mathcal{H}_{\Delta_1}^\oplus$ and $\mathcal{H}_{\Delta_2}^\oplus$ if Δ_1 and Δ_2 are disjoint, i.e., if

$$v_1^\oplus \in \mathcal{H}_{\Delta_1}^\oplus, \quad v_2^\oplus \in \mathcal{H}_{\Delta_2}^\oplus, \quad \Delta_1 \cap \Delta_2 = \emptyset, \qquad (2.655)$$

---
[189] Reed and Simon Vol. 4 (1975) pp. 280-286.

then
$$\langle v_1^\oplus \mid \widehat{A}^\oplus v_2^\oplus \rangle^\oplus = 0. \tag{2.656}$$

This reduces to Eq. (2.649) when the direct integral reduces to a direct sum.

2.14.3 Density operators

Let $\widehat{B}$ be a bounded positive operator in a Hilbert space $\mathcal{H}$. According to §2.3C(1) C2 operator $\widehat{B}$ is selfadjoint. Let φ_j be an orthonormal basis in $\mathcal{H}$ then the sum $\sum_j \langle \varphi_j \mid \widehat{B}\varphi_j \rangle$ can be shown to be finite and independent of any particular choice of an orthonormal basis, i.e., let φ'_j be another orthonormal basis in $\mathcal{H}$ then we have

$$\sum_j \langle \varphi_j \mid \widehat{B}\varphi_j \rangle = \sum_j \langle \varphi'_j \mid \widehat{B}\varphi'_j \rangle. \tag{2.657}$$

This sum turns out to be a very important property of the operator. We shall denote this sum by $\operatorname{tr}(\widehat{B})$, i.e.,

$$\operatorname{tr}(\widehat{B}) = \sum_j \langle \varphi_j \mid \widehat{B}\varphi_j \rangle, \tag{2.658}$$

and call it the *trace* of the operator. A bounded positive operator $\widehat{D}$ of unit trace, i.e., $\operatorname{tr}(\widehat{D}) = 1$, is called a *density operator*. Density operators possess the following properties:[190]

1. The trace of the sum and of the product of two density operators $\widehat{D}_1$ and $\widehat{D}_2$ are well defined and satisfies the following equations:

$$\operatorname{tr}(\widehat{D}_1 + \widehat{D}_2) = \operatorname{tr}(\widehat{D}_1) + \operatorname{tr}(\widehat{D}_2), \tag{2.659}$$
$$\operatorname{tr}(\widehat{D}_1 \widehat{D}_2) = \operatorname{tr}(\widehat{D}_2 \widehat{D}_1). \tag{2.660}$$

   Moreover we have, for any given positive real number a,

$$\operatorname{tr}(a\widehat{D}) = a \operatorname{tr}(\widehat{D}). \tag{2.661}$$

2. The trace of a density operator is invariant under a unitary transformation, i.e., given any unitary operator $\widehat{U}$ we have

$$\operatorname{tr}(\widehat{U}\widehat{D}\widehat{U}^\dagger) = \operatorname{tr}(\widehat{D}). \tag{2.662}$$

---

[190] Reed and Simon Vol. 1 (1972) p. 206, Prugovečki (1981) pp. 374-392, Jordan (1969).

2.14. DIRECT INTEGRALS OF OPERATORS

3. Let $\widehat{B}$ be a bounded operator. Then the trace of $\widehat{B}\widehat{D}$ is well defined and equal to that of $\widehat{D}\widehat{B}$, i.e.,

$$\operatorname{tr}(\widehat{B}\widehat{D}) = \operatorname{tr}(\widehat{D}\widehat{B}). \tag{2.663}$$

4. If a density operator $\widehat{D}$ is a projector then it is a one-dimensional projector.[191] Since one-dimensional projectors are generated by unit vectors we have $\widehat{D} = \widehat{P}_\phi$ for some unit vector ϕ. Let $\widehat{B}$ be a bounded operator. We can write down the trace of $\widehat{B}\widehat{D}$ directly in terms of ϕ, i.e., we have

$$\widehat{D} = \widehat{P}_\phi \quad \Rightarrow \quad \operatorname{tr}(\widehat{B}\widehat{D}) = \operatorname{tr}(\widehat{B}\widehat{P}_\phi) = \langle \phi \mid \widehat{B}\phi \rangle. \tag{2.664}$$

5. A density operator $\widehat{D}$ assigns a unique set of real numbers to the set $\mathcal{B}(\mathcal{H})$ of bounded selfadjoint operators $\widehat{B}$ on $\mathcal{H}$, i.e., we have

$$\mathcal{B}(\mathcal{H}) \mapsto \mathbb{R} \quad \text{by} \quad \widehat{B} \mapsto \operatorname{tr}(\widehat{B}\widehat{D}), \tag{2.665}$$

and conversely such a set of real numbers determines a unique density operator, i.e.,

$$\operatorname{tr}(\widehat{B}\widehat{D}) = \operatorname{tr}(\widehat{B}\widehat{D}') \quad \forall \widehat{B} \in \mathcal{B}(\mathcal{H}) \quad \Leftrightarrow \quad \widehat{D} = \widehat{D}'. \tag{2.666}$$

6. A density operator has a purely point spectrum $\{\omega_j\}$ so that we can express it in terms of its spectral decomposition as a combination of orthogonal spectral projectors, i.e.,

$$\widehat{D} = \sum_j \omega_j \widehat{P}_j, \tag{2.667}$$

where $\widehat{P}_j$ is the projector onto the eigensubspace associated with the eigenvalue ω_j.

In the case of all eigenvalues being non-degenerate the spectral projectors are all one-dimensional and of the form $\widehat{P}_j = |\psi_j\rangle\langle\psi_j|$, where ψ_j is the normalized eigenvector corresponding to the eigenvalue ω_j. We can rewrite Eq. (2.667) as[192]

$$\widehat{D} = \sum_j \omega_j \widehat{P}_j = \sum_j \omega_j |\psi_j\rangle\langle\psi_j|. \tag{2.668}$$

---

[191] It is easy to check that the trace of a one-dimensional projector is 1 and the trace of a two-dimensional projector is 2 and so on.
[192] We can still express $\widehat{D}$ in the form of Eq. (2.668) even if there are degeneracies. All we need to do is to allow some ω_j to be the same.

Since ψ_j form an orthonormal basis we can see that

$$\text{tr}(\widehat{D}) = \sum_\ell \langle \psi_\ell \mid \widehat{D}\psi_\ell \rangle = \sum_\ell \omega_\ell. \tag{2.669}$$

Moreover we have

$$\omega_\ell \geq 0, \quad \sum_\ell \omega_\ell = 1, \tag{2.670}$$

on account of $\widehat{D}$ being positive and of unit trace. It follows that given a set of real numbers w_j satisfying Eq. (2.670) and an orthonormal basis ψ_j in a Hilbert space we can construct a density operator using Eq. (2.668).

7. Let $\widehat{D}_\ell$, $\ell = 1, 2 \ldots, n$ be a finite set of density operators, and let w_ℓ, $\ell = 1, 2 \ldots, n$ be a set of positive real numbers adding to 1, i.e., these numbers satisfy Eq. (2.670). We call the following sum

$$\widehat{D} = \sum_\ell w_\ell \widehat{D}_\ell \tag{2.671}$$

a *convex combination* of density operators $\widehat{D}_\ell$ with weights w_ℓ. Clearly $\widehat{D}$ is positive, bounded and has a unit trace. In other words a convex combination of density operators is again a density operator.

8. A density operator may be written as a convex combination of non-orthogonal projectors. Let $|\xi_\ell\rangle\langle\xi_\ell|$ be a set of one-dimensional projectors associated with a set of non-orthogonal unit vectors ξ_ℓ and let w_ℓ be a set of real numbers satisfy Eq. (2.670). Then we can construct a density operator by a convex combination of $\widehat{P}_{\xi_\ell} = |\xi_\ell\rangle\langle\xi_\ell|$ with w_ℓ as weights. In other words,

$$\widehat{D}' = \sum_\ell w_\ell |\xi_\ell\rangle\langle\xi_\ell| = \sum_\ell w_\ell \widehat{P}_{\xi_\ell} \tag{2.672}$$

is a density operator. This expression is not a spectral decomposition of $\widehat{D}'$. The above density operator possesses a spectral decomposition in terms of orthogonal projectors of the form of Eq. (2.668), i.e., we have

$$\widehat{D}' = \sum_j \omega'_j |\psi'_j\rangle\langle\psi'_j| = \sum_j \omega'_j \widehat{P}_{\psi'_j}. \tag{2.673}$$

The fact that generally a density operator may be expressed as different convex combinations of distinct sets of other density operators, e.g., as in Eq. (2.672) and in Eq. (2.673), will be seen in §3.2.2 to have important physical consequences.

2.14. DIRECT INTEGRALS OF OPERATORS

9. Given a density operator of the form of Eq. (2.671) we have

$$\mathrm{tr}\,(\widehat{A}\widehat{D}) = \sum_\ell w_\ell \,\mathrm{tr}\,(\widehat{A}\widehat{D}_\ell). \tag{2.674}$$

In particular, if $\widehat{D}_\ell = \widehat{P}_{\phi_\ell} = |\phi_\ell\rangle\langle\phi_\ell|$ are one-dimensional projectors, not necessarily orthogonal, we have

$$\widehat{D} = \sum_\ell w_\ell \widehat{P}_{\phi_\ell}, \tag{2.675}$$

$$\mathrm{tr}\,(\widehat{A}\widehat{D}) = \sum_\ell w_\ell \,\mathrm{tr}\,(\widehat{A}\widehat{P}_{\phi_\ell}) = \sum_\ell w_\ell \,\langle \phi_\ell \mid \widehat{A}\phi_\ell \rangle, \tag{2.676}$$

showing that $\mathrm{tr}\,(\widehat{A}\widehat{D})$ is equal to a weighted sum of expectation values of $\widehat{A}$ in ϕ_ℓ.

2.14.4 Statistical operators

The state of an orthodox quantum system is known to be describable in terms of a density operator in an appropriate Hilbert space. We shall introduce the notion of statistical operators as a generalization to density operators in direct integral Hilbert spaces which can then be used to represent states of a wider class of physical systems.

To start with, consider a direct sum Hilbert space $\mathcal{H}_d^\oplus = \oplus_\ell \mathcal{H}(\ell)$. As in §2.14.2C(3) C2 we use the subscript d to indicate a direct sum as a special case of a direct integral with a discrete Lebesgue-Stieltjes measure given by Eq. (2.585). Let $\widehat{D}(\ell)$ be a density operator on $\mathcal{H}(\ell)$, and let $\{w_\ell\}$ be a set of real numbers satisfying Eq. (2.670).

Definition 2.14.4(1) **Statistical operators in direct sum spaces**

- A direct sum operator of the form

$$\widehat{S}^\oplus = \oplus_\ell w_\ell \widehat{D}(\ell) \tag{2.677}$$

is called a *statistical operator* on the direct sum space $\mathcal{H}_d^\oplus$.

- Let $\widehat{A}^\oplus = \oplus_\ell \widehat{A}(\ell)$ be a bounded decomposable operator on $\mathcal{H}_d^\oplus$. The *trace* in the direct sum space $\mathcal{H}_d^\oplus$ of the product operator $\widehat{A}^\oplus \widehat{S}^\oplus$ is defined to be

$$\mathrm{tr}^\oplus(\widehat{A}^\oplus \widehat{S}^\oplus) = \sum_\ell w_\ell \,\mathrm{tr}_\ell\,(\widehat{A}(\ell)\widehat{D}(\ell)), \tag{2.678}$$

where $\mathrm{tr}_\ell(\widehat{A}(\ell)\widehat{D}(\ell))$ is the trace of $\widehat{A}(\ell)\widehat{D}(\ell)$ in $\mathcal{H}(\ell)$.

Comments 2.14.4(1)   **Trace of operators in direct sum spaces**

C1   The trace in Eq. (2.678) is worked out in two stages. Noting that $\widehat{A}(\ell)\widehat{D}(\ell)$ is an operator on $\mathcal{H}(\ell)$ we evaluate of its trace $\operatorname{tr}_\ell(\widehat{A}(\ell)\widehat{D}(\ell))$ within the Hilbert space $\mathcal{H}(\ell)$. We then sum up these values in all the constituent Hilbert spaces to arrive at $\operatorname{tr}^\oplus(\widehat{A}^\oplus \widehat{S}^\oplus)$.

C2   Adopting the notation of Eq. (2.648) we can rewrite Eq. (2.677) in the form of a convex combination:

$$\widehat{S}^\oplus = \sum_\ell w_\ell \widehat{D}_\ell^\oplus. \tag{2.679}$$

C3   By definition a statistical operator is decomposable. It follows that while a statistical operator defined above is a density operator on $\mathcal{H}_d^\oplus$, the converse is generally false, e.g., a non-decomposable density operator on $\mathcal{H}_d^\oplus$ is not a statistical operator.

Definition 2.14.4(2)   **Statistical operators in direct integral spaces**

- Let $S_r(\tau)$ be a regular probability density function, and let $\{\widehat{D}(\tau)\}$ be a Borel family of density operators associated with a Borel family of Hilbert spaces $\mathcal{H}_{\Upsilon_0} = \{\{\mathcal{H}(\tau) : \tau \in \mathbb{R}\}, \Upsilon_0\}$. Then the direct integral operator

$$\widehat{S}_r^\oplus = \int^\oplus S_r(\tau)\widehat{D}(\tau)\, d\tau \tag{2.680}$$

  is called a *regular statistical operator* on the direct integral Hilbert space

$$\mathcal{H}_c^\oplus = \int^\oplus \mathcal{H}(\tau)\, d\tau. \tag{2.681}$$

- The direct integral operator above is called a *singular statistical*, or a *blended statistical*, operator on $\mathcal{H}_c^\oplus$ if the probability density function $S(\tau)$ is singular, or blended.

- Given any bounded decomposable operator $\widehat{A}^\oplus$ on $\mathcal{H}_c^\oplus$ the *trace* of the product operator $\widehat{A}^\oplus \widehat{S}^\oplus$ on $\mathcal{H}^\oplus$ is defined to be

$$\operatorname{tr}^\oplus(\widehat{A}^\oplus \widehat{S}^\oplus) = \int S(\tau)\operatorname{tr}_\tau(\widehat{A}(\tau)\widehat{D}(\tau))\, d\tau \tag{2.682}$$

  where $\operatorname{tr}_\tau(\widehat{A}(\tau)\widehat{D}(\tau))$ is the trace in $\mathcal{H}(\tau)$ of $\widehat{A}(\tau)\widehat{D}(\tau)$.

2.14. DIRECT INTEGRALS OF OPERATORS

Comments 2.14.4(2)   **Trace of operators in direct integral spaces**

C1   Note that $\text{tr}^\oplus(\widehat{A}^\oplus\widehat{S}^\oplus)$ is a real number given by a numerical Lebesgue integral.

C2   Some authors use the term statistical operators and density operators interchangeably.[193] We do not follow this practice for obvious reasons. We know from §2.14.4C(1) C3 that a density operator on $\mathcal{H}_d^\oplus$ is not neccesarily a statistical operator.[194]

C3   A *singular statistical operator* is typically one written in terms of a singular probability density function S_s as given by Eq. (2.269)

$$S_s(\tau) = \sum_\ell w_\ell\, \delta(\tau - \tau_\ell). \tag{2.683}$$

This effectively reduces the integral operator $\widehat{S}^\oplus$ into a formal sum

$$\widehat{S}_s^\oplus = \int^\oplus S_s(\tau)\,\widehat{D}(\tau)\,d\tau = \sum_\ell w_\ell\, \widehat{D}_{\tau_\ell}^\oplus. \tag{2.684}$$

In analogy with pure singular vectors in Eq. (2.599) we can have

$$\widehat{S}_{\tau_\ell}^\oplus = \int^\oplus \delta(\tau - \tau_\ell)\widehat{D}(\tau)\,d\tau. \tag{2.685}$$

Strictly speaking these singular operators, and their sum $\widehat{S}_s^\oplus$, do not define an operator in the direct integral Hilbert space since they act on singular subspaces $\mathcal{H}(\tau_\ell)$. However, these singular statistical operators, like pure singular vectors and singular subspaces, have important physical applications. We shall define the *trace* of $\widehat{S}_s^\oplus \widehat{A}^\oplus$ formally by

$$\text{tr}^\oplus\!\left(\widehat{S}_s^\oplus\widehat{A}^\oplus\right) = \int^\oplus S_s(\tau)\,\text{tr}_\tau\!\left(\widehat{D}(\tau)\widehat{A}(\tau)\right)d\tau \tag{2.686}$$

$$= \sum_{\ell=1} w_\ell\, \text{tr}_\ell\!\left(\widehat{D}(\tau_\ell)\widehat{A}(\tau_\ell)\right). \tag{2.687}$$

Statistical operators arising from blended probability functions are called *blended statistical operators*. We often attach subscripts r, s, b to highlight respectively the regular, the singular and the blended nature of the statistical operators involved. For uniformity of terminology, all statistical operators on $\mathcal{H}_d^\oplus$ are regarded as regular.

---
[193] Prugovečki (1981) p. 389, Blum (1981) p. 41.
[194] See an example in §4.4.5 in Chapter 4.

C4   We can combine Definitions 2.14.4(1) and 2.14.4(2) by considering direct integral spaces $\mathcal{H}^\oplus$ with a Lebesgue-Stieltjes measure $dg(\tau)$, i.e., a direct integral operator of the form

$$\widehat{S}^\oplus = \int^\oplus S(\tau)\widehat{D}(\tau)\,dg(\tau), \qquad (2.688)$$

where $S(\tau)$ is a probability density function, in the sense that:

1. When $g(\tau)$ is a staircase function as given in Eq. (2.585) and $S(\tau)$ satisfies $\sum_\ell S(\ell) = 1$ the direct integral $\mathcal{H}^\oplus$ reduces to $\mathcal{H}_d^\oplus$ and Eq. (2.688) becomes to Eq. (2.677).

2. When $g(\tau) = \tau$ the direct integral $\mathcal{H}^\oplus$ reduces to $\mathcal{H}_c^\oplus$ and Eq. (2.688) becomes to Eq. (2.680).

2.15   Direct Integrals of Tensor Products

2.15.1   Direct integrals of tensor product Hilbert spaces

Theorem 2.15.1(1)   **Tensor products of Borel families of Hilbert spaces**[195]

- Let $\{\{\mathcal{H}_1(\tau) : \tau \in \mathbb{R}\}, \boldsymbol{\Upsilon}_{10}\}$ and $\{\{\mathcal{H}_2(\tau) : \tau \in \mathbb{R}\}, \boldsymbol{\Upsilon}_{20}\}$ be two Borel families of Hilbert spaces over $\mathbb{R}$. Then there exists a unique Borel family of Hilbert spaces

$$\{\{\mathcal{H}(\tau) = \mathcal{H}_1(\tau) \otimes \mathcal{H}_2(\tau) : \tau \in \mathbb{R}\}, \boldsymbol{\Upsilon}_0\} \qquad (2.689)$$

such that given $\boldsymbol{v}_1 \in \boldsymbol{\Upsilon}_{10}$ and $\boldsymbol{v}_2 \in \boldsymbol{\Upsilon}_{20}$ the vector-valued function

$$\boldsymbol{v}: \mathbb{R} \mapsto \{\mathcal{H}(\tau) = \mathcal{H}_1(\tau) \otimes \mathcal{H}_2(\tau) : \tau \in \mathbb{R}\} \qquad (2.690)$$

defined by

$$\tau \to \boldsymbol{v}_1(\tau) \otimes \boldsymbol{v}_2(\tau) \in \mathcal{H}(\tau) \qquad (2.691)$$

is a member of $\boldsymbol{\Upsilon}_0$.

This theorem enables us to construct the following direct integral Hilbert space

$$\mathcal{H}^\oplus = \int^\oplus \mathcal{H}(\tau)\,dg(\tau) = \int^\oplus \bigl(\mathcal{H}_1(\tau) \otimes \mathcal{H}_2(\tau)\bigr)\,dg(\tau) \qquad (2.692)$$

---

[195] Dixmier (1981) Proposition 10 on p. 174.

2.15. DIRECT INTEGRALS OF TENSOR PRODUCTS

from any two Borel families of Hilbert spaces. A simple but useful case is when $\{\{\mathcal{H}_1(\tau) : \tau \in \mathbb{R}\}, \Upsilon_{10}\}$ is a constant family corresponding to Hilbert space $\mathcal{H}_1(0)$. We can identify the following spaces:

$$\mathcal{H}_1(0) \otimes \int^{\oplus} \mathcal{H}_2(\tau) \, dg(\tau) \quad \text{and} \quad \int^{\oplus} (\mathcal{H}_1(0) \otimes \mathcal{H}_2(\tau)) \, dg(\tau) \quad (2.693)$$

because of the theorem stated below.

Theorem 2.15.1(2)   **Tensor product with a constant family of Hilbert spaces**[196]

- There is a unique isomorphism between

$$\mathcal{H}_1(0) \otimes \int^{\oplus} \mathcal{H}_2(\tau) \, dg(\tau) \quad \text{and} \quad \int^{\oplus} (\mathcal{H}_1(0) \otimes \mathcal{H}_2(\tau)) \, dg(\tau) \quad (2.694)$$

achieved by mapping

$$\left(v_1(0) \otimes v_2^{\oplus}\right) \in \mathcal{H}_1(0) \otimes \int^{\oplus} \mathcal{H}_2(\tau) \, dg(\tau) \quad (2.695)$$

to

$$\int^{\oplus} \left(v_1(0) \otimes v_2(\tau)\right) dg(\tau) \in \int^{\oplus} (\mathcal{H}_1(0) \otimes \mathcal{H}_2(\tau)) \, dg(\tau). \quad (2.696)$$

Here we have

$$v_1(0) \in \mathcal{H}_1(0) \quad v_2(\tau) \in \mathcal{H}_2(\tau). \quad (2.697)$$

and

$$v_2^{\oplus} = \int^{\oplus} v_2(\tau) \, dg(\tau) \in \int^{\oplus} \mathcal{H}_2(\tau) \, dg(\tau). \quad (2.698)$$

2.15.2 Direct integrals and tensor product of operators

Theorem 2.15.2(1)   **Tensor product and direct integral of operators**[197]

- Let $\widehat{A}_1(0)$ be a bounded operator on $\mathcal{H}_1(0)$ and $\widehat{A}_2^{\oplus} = \int^{\oplus} \widehat{A}_2(\tau) \, dg(\tau)$ be a decomposable operator on the direct integral space $\int^{\oplus} \mathcal{H}_2(\tau) \, dg(\tau)$. Then we have

$$\widehat{A}_1(0) \otimes \widehat{A}_2^{\oplus} = \widehat{A}_1(0) \otimes \int^{\oplus} \widehat{A}_2(\tau) \, dg(\tau) \quad (2.699)$$

$$= \int^{\oplus} \left(\widehat{A}_1(0) \otimes \widehat{A}_2(\tau)\right) dg(\tau). \quad (2.700)$$

---
[196] Dixmier (1981) Proposition 11 on p. 174.
[197] Dixmier (1981) Proposition 8 on p. 188.

References

1. Akhiezer, N. I. and Glazman, I. M. (1961). *Theory of Linear Operators in Hilbert Space Vol. 1*, Frederick Ungar, New York.
2. Akhiezer, N. I. and Glazman, I. M. (1963). *Theory of Linear Operators in Hilbert Space Vol. 2*, Frederick Ungar, New York.
3. Amrein, W. O., Jauch, J. M. and Sinha, K. B. (1977). *Scattering Theory in Quantum Mechanics*, Benjamin, Reading, Mass.
4. Amrein, W. O. (1981). *Non-Relativistic Quantum Dynamics*, Reidel, Dordrecht.
5. Beltrametti, E. G. and Cassinelli, G. (1981). *The Logic of Quantum Mechanics*, Addison-Wesley, Reading, Mass.
6. Blank, J., Exner, P. and Havlíček, M. (1994). *Hilbert Space Operators in Quantum Physics*, American Institute of Physics Press, New York.
7. Blum, K. (1981). *Density Matrix Theory and Applications*, Plenum, New York.
8. Böhm, A. (1978). *The Rigged Hilbert Space and Quantum Mechanics* (*Lecture Notes in Physics 78*), Springer-Verlag, Berlin.
9. Böhm, A. (1979). *Quantum Mechanics*, Springer-Verlag, New York.
10. Bratteli, O. and Robinson, D. W. (1979). *Operator Algebras and Quantum Statistical Mechanics I*, Springer-Verlag, New York.
11. Bratteli, O. and Robinson, D. W. (1981). *Operator Algebras and Quantum Statistical Mechanics II*, Springer-Verlag, New York.
12. Burrill, G. W. (1972). *Measure, Integration and Probability*, McGraw-Hill, New York.
13. Busch, P., Lahti, P. and Mittelstaedt, P. (1996). *The Quantum Theory of Measurement*, 2nd edition, Springer-Verlag, Berlin.
14. Byron, F. W. and Fuller, R. W. (1969). *Mathematics of Classical and Quantum Physics Vol. 1*, Addison-Wesley, Reading, Mass.
15. Dixmier, J. (1977). C^*-*Algebras*, North-Holland, Amsterdam.
16. Dixmier, J. (1981). *Von Neumann Algebras*, North-Holland, Amsterdam.
17. Emch, G. G. (l972). *Algebraic Methods in Statistical Mechanics and Quantum Field Theory*, Wiley, New York.
18. Fano, G. (1971). *Mathematical Methods of Quantum Mechanics*, McGraw-Hill, New York.
19. Flügge, S. (1974). *Practical Quantum Mechanics*, Springer-Verlag, New York.
20. Friedman, B. (1956). *Principles and Techniques of Applied Mathematics*, Wiley, New York.
21. Gottfried, K. (1966). *Quantum Mechanics*, Benjamin, New York.
22. Haag, R. (1992). *Local Quantum Physics*, Springer-Verlag, Berlin.
23. Hagg, R. and Kastler, D. (1964). *J. Math. Phys.* **5** 848.
24. Holevo, A. S. (1982). *Probabilistic and Statistical Aspects of Quantum Theory*, North-Holland, Amsterdam.
25. Isham, C. (1995). *Lectures on Quantum Theory*, Imperial College Press, London.

REFERENCES

26. Jackson, T. D. (1985). *Quantum Mechanics, Locality and Asymptotic Separability*, St Andrews University PhD Thesis.
27. Jauch, J. M. (1968). *Foundations of Quantum Mechanics*, Addison-Wesley, Reading, Mass.
28. Jordan, T. F. (1969). *Linear Operators for Quantum Mechanics*, Wiley, Now York.
29. Kadison, R. V. and Ringrose, J. R. (1983). *Fundamentals of the Theory of Operator Algebras Vol. 1 Elementary Theory*, Academic Press, New York.
30. Kato, T. (1996). *Perturbation Theory for Linear Operators*, Springer-Verlag, Berlin.
31. Kreyszig, E. (1978). *Introductory Functional Analysis with Applications*, Wiley, New York.
32. Lipschutz, S. (1964). *Set Theory*, Schaum's Outline Series, McGraw-Hill, New York.
33. Lipschutz, S. (1974). *Theory and Problems of Probability*, Schaum's Outline Series, McGraw-Hill, New York.
34. McLean, R. G. D. (1983). *An Algebraic Formulation of Asymptotically Separable Quantum Mechanics*, St Andrews University PhD Thesis.
35. Merzbacher, E. (1998). *Quantum Mechanics*, 3rd edition, Wiley, New York.
36. Messiah, A. (l967). *Quantum Mechanics Vol. 1*, North-Holland, Amsterdam.
37. Naimark, M. A. (1968). *Linear Differential Operators Part II*, Harrap, London. Translated by E. R. Dawson.
38. Papoulis, A. (1962). *The Fourier Integral and its Applications*, McGraw-Hill, New York.
39. Penrose, O. (1970). *Foundations of Statistical Mechanics*, Pergamon Press, Oxford.
40. Pitt, H. R. (1963). *Integration, Measure and Probability*, Oliver & Boyd, Edinburgh.
41. Prugovečki, E. (1981). *Quantum Mechanics in Hilbert Spaces*, 2nd edition, Academic Press, New York.
42. Reed, M. and Simon, B. (1972). *Methods of Modern Mathematical Physics Vol. 1 Functional Analysis*, Academic Press, New York.
43. Reed, M. and Simon, B. (1975). *Methods of Modern Mathematical Physics Vol. 2 Fourier Analysis, Selfadjointness*, Academic Press, New York.
44. Reed, M. and Simon, B. (1978). *Methods of Modern Mathematical Physics Vol. 4 Analysis of Operators*, Academic Press, New York.
45. Richtmyer, R. D. (1978). *Principles of Advanced Mathematical Physics Vol. 1*, Springer-Verlag, New York.
46. Roman, P. (1965). *Advanced Qauntum Theory*, Addison-Wesley, Reading, Mass.
47. Roman, P. (1975). *Some Modern Mathematics for Physicists and Other Outsiders Vol. 1*, Pergamon, New York.
48. Roman, P. (1975). *Some Modern Mathematics for Physicists and Other Outsiders Vol. 2*, Pergamon, New York.
49. Schroeck, F. E. Jr. (1996). *Quantum Mechanics on Phase Space*, Kluwer, Dordrecht.
50. Sewell, G. L. (1986). *Quantum Theory of Collective Phenomena*, Clarendon Press, Oxford.
51. Smirnov, V. I. (1964). *A Course of Higher Mathematics Vol. 5*, Pergamon, London, translated by D. E. Brown and edited by I. N. Sneddon.

52. Timson, D. R. E. (1986). *Locality in Non-Relativistic and Relativistic Quantum Mechanics*, St Andrerws University PhD Thesis.
53. Timson, D. R. E. and Wan, K. K. (1988). 'Localizing Isometries, Local Comoving Evolution Operators and Observables in Quantum Mechanics' in L. Kostro, ed., *Problems in Quantum Physics; Gdansk 87*, World Scientific, Singapore.
54. Wan, K. K. (1988). *Found. Phys.* **18** 887.
55. Wan, K. K., Bradshaw, J., Trueman, C. and Harrison, F. (1998). *Found. Phys.* **28** 1739.
56. Wan, K. K. and Fountain, R. H. (1996). *Found. Phys.* **26** 1165.
57. Wan, K. K., Fountain R. H. and Tao, Z. Y. (1995). *J. Phys. A: Math. Gen.* **28** 2379.
58. Wan, K. K. and Jackson, T. D. (1984). *Phys. Lett. A* **106** 219.
59. Wan, K. K., Jackson, T. D. and McKenna, I. H. (1984). *Nuovo Cimento B* **81** 165.
60. Wan, K. K. and McLean, R. G. D. (1984a). *J. Phys. A: Math. Gen.* **17** 825. Corrigenda **17** 2363 (1984).
61. Wan, K. K. and McLean, R. G. D. (1984b). *J. Phys. A: Math. Gen.* **17** 837. Corrigenda **17** 2363 (1984).
62. Wan, K. K. and McLean, R. G. D. (1985). *J. Math. Phys.* **26** 2540.
63. Weidmann, J. (1980). *Linear Operators in Hilbert Spaces*, Springer-Verlag, New York.
64. Williamson, J. H. (1962). *Lebesgue Integration*, Holt, Rinehart and Winston, New York.

Part II

Orthodox and Generalized Quantum Mechanics

II

Orthodox Theology

Question of Theosis

Chapter 3

Orthodox Quantum Mechanics

3.1 Introduction

3.1.1 Structure of physical theories

Generally a physical theory should contain the following basic components:

1. A mathematical framework for the description of physical quantities and relations in the theory.

2. A description of possible states of the system.

3. A description of physically measurable quantities, referred to as *observables*, including their possible values.

4. A statement of the relationship between states and observables to arrive at a prescription for the probability distribution of the values of any observable of the system in a given state.

5. A description of dynamics, i.e., time evolution of the system.

Classical mechanics embraces these components in a very intuitive manner. The basic mathematical framework consists of finite-dimensional vector spaces, e.g., the theory of a single point particle is based on two 3-dimensional Euclidean spaces, the coordinate space and the momentum space, together with calculus and vector analysis. A state corresponds one-to-one to two vectors, one in the coordinate space for specifying the position and one in the momentum space for specifying the momentum of the particle. Observables are described by real-valued functions of the state, i.e., functions of position and momentum.[1] The relationship between observables and states is given explicitly by the very definition of observables. Given a state the value of any observable is uniquely determined by the function defining the observable. We shall call this the value *possessed* by the observable in the given state. A measurement serves only to reveal such a *possessed value*. Probability does not play an intrinsic role here. Time evolution of the system can be described in terms of the time dependence of the state which is governed by Hamilton's equations.

Quantum mechanics is again based on vector spaces. The difference is that we need a complex vector space of infinite dimensions, i.e., a Hilbert space, in general. A state is describable by a single unit vector in the vector space, although the correspondence is not one-to-one. We can see that a vector space structure plays a fundamental and unifying role in the description of states of physical systems, be it classical or quantum.[2] Structurally the departure from classical mechanics becomes obvious in the description of quantum observables. Quantum observables are no longer numerical functions of state. Instead they are described by a certain type of operators in an appropriate Hilbert space. This is not an entirely novel idea. As discussed in §1.6.3 and §1.6.4 observables in classical mechanics can also be related to operators in the form of the Hamiltonian vector fields they generate. These vector fields are differential operators acting on functions defined on the phase space.[3] The dynamics is also determinable by the vector field generated by the Hamiltonian generator in the sense discussed in C1.6.3(1). For quantum observables the difference is that their representative operators cannot be related to some numerical functions of the state. This breaks the transparent relationship between observables and states in classical mechanics. Operators by themselves do not generally have numerical values. They can generate numerical values, e.g., their eigenvalues. However, these eigenvalues are not directly related to an arbitrarily given state. So, an observable generally will not have a definite value in a given state. Therefore it is not meaningful to talk about possessed values in a given state in general. Of course we do have values obtained by physical measurements. A measurement may yield a range of possible values, to be called *measured values*. Probability will have to come in, and we have to give a prescription

---

[1] By combining the position and momentum variables one can construct a six-dimensional phase space. A single element of this phase space specifies a state. An observable is then a real-valued function on this phase space.

[2] The usual wave function description of a quantum state seems to be totally different from the way a state of a classical particle is described. The difference is more apparent than real since a wave function may be regarded as an element of a vector space.

[3] As discussed in C1.6.4(1) C3 and illustrated by the examples in E1.6.4(1) we can also gain an insight of the meaning of an observable through its associated vector field, e.g., a momentum as generator for translations.

3.1. INTRODUCTION

for the probability distribution of measured values of any observable of the system in a given state.[4]

The core theories of this book are set out in Chapters 3, 4 and 5. Chapter 3 starts the discussion on orthodox quantum theory. Given the mathematical structure of quantum mechanics based on a Hilbert space one is confronted with several preliminary questions:

1. Does every unit vector represent a state?

2. What kind of operators are eligible for the description of quantum observables? Do all eligible operators actually correspond to observables?

3. What is the prescription for the probability distribution of the values of an observable in a given state?

4. How to establish the quantum counterparts of familiar classical observables?

5. How to describe the dynamics of a quantum system?

To begin with it is natural to formulate a theory in which every unit vector does represent a state. One then establishes the mathematical conditions for an operator to be eligible for the description of quantum observables, and one would naturally begin with the assumption that every eligible operator does correspond to an observable. These basic ideas are set out in the form of Postulate OQS in §3.2. The way is then clear to consider a *geometric* approach to quantization of classical observables. A second postulate, Postulate OQD in §3.4 is introduced to describe time evolution. This enables us to establish a basic theory, referred to as an *orthodox quantum theory* or *orthodox quantum mechanics*. Quantum systems describable in terms of such an orthodox quantum theory are called *orthodox quantum systems*. The fact that observables and states are not directly related necessitates separate procedures for state preparation and measurement of observables. A general procedure for state preparation is presented in §3.5. Despite the fact that an operator representation of observables provides no obvious dynamic process for their measurement we are able to formulate a two-stage dynamic measurement theory in §3.6. Stage one involves the reduction of a quantum measurement to local position measurements through a process of spectral separation, and stage two models a local position measurement as an ionization process. Our model of quantum measurement emphasizes the separation and the identification of spectral components, but not the attribution of definite values of observables

---

[4] Of course one can turn this argument around to say that the reason for using operators for observables in quantum theory is because of empirical evidence of probability distribution of measured values of observables in a given state.

to individual systems. The richness of quantum formalism is demonstrated by the fact that the procedures and dynamic processes for state preparation and measurement can be formulated within orthodox quantum theory. There is no need to bring in external factors like the environment. This is as it should be for a satisfactory theory.

Having constructed the basic theory one can explore the effect of various restrictions and generalizations arising from the following questions:

> *What if not every unit vector can represent a state, and what if not every mathematically eligible operator can describe an observable?*

Attempts to tackle these questions lead us to some restrictions as well as some generalizations of the orthodox theory. These are discussed in Chapters 4 and 5 leading to the introduction of *superselection rules* and an extension of eligibility conditions of operators. Superselection rules effectively introduce classical properties into a quantum system. Many *macroscopic quantum systems* such as superconducting systems exhibit both quantum and classical properties. We call a physical system which possesses both quantum and classical properties *a mixed quantum system*. The introduction of superselection rules in Chapter 4 based on a direct integral Hilbert space and direct integrals of operators creates a highly flexible formalism capable of describing the behaviour of mixed quantum systems. Indeed the formalism is able to describe a large class of physical systems, ranging from orthodox quantum, mixed quantum to classical. In Chapter 5 we present a further generalization of orthodox quantum mechanics by relaxing the mathematical eligibility conditions so as to represent a larger class of observables, i.e., the extension of eligibility conditions of operators from selfadjointness to maximal symmetry.

3.1.2 Mathematical framework of quantum mechanics

In common with any physical theory, quantum mechanics begins with the notions of states and observables together with a mathematical framework for their description. An observable is related to some characteristic feature of the system which can be quantified in an empirical manner. The meaning of an observable has to be ascertained by physical means, e.g., by how the observable is affected by interactions and by how it is measured. This is especially true for observables which have no classical counterpart. An experimental procedure can be devised to obtain the numerical values of each observable, and we refer to such an experimental procedure as a *measurement process* or an *experiment* for short. In many traditional approaches states of a physical system are considered as the primary quantities. We must therefore establish the mathematical framework for state description first. A mathematical framework which proves to be successful, both for classical and quantum theories,

3.1. INTRODUCTION

consists of differentiable manifolds, vector spaces and operators. Generally vectors are related to states of a physical system while operators are related to physical observables and the dynamics of the system. The precise way vectors are related to states and operators to observables varies for different physical theories.

In a traditional formulation of quantum mechanics each quantum system has associated with it an appropriate Hilbert space $\mathcal{H}$ constructed from square-integrable functions on the physical space of the system. States of the system are describable in terms of unit vectors ϕ of $\mathcal{H}$, often referred to as *wave functions*.[5] Observables are assumed to be described by selfadjoint operators; indeed observables are assumed to correspond in a one-to-one manner to selfadjoint operators $\widehat{A}$ acting in $\mathcal{H}$.[6] The fact that the Hilbert space need to be complex can be understood from an analysis of the probabilistic behaviour of a quantum particle, as manifested in the probabilistic interpretation of the wave function and the superposition principle.[7]

There are other alternative mathematical formulations of physical theories.[8] For example, if one considers observables as the fundamental building blocks on account of their direct links to experiments, one would begin by establishing a mathematical structure for the set of observables first, and then set up states as a kind of superstructure on the set of observables. This is the thinking behind algebraic and quantum logic approaches to quantum theory. In the algebraic approach, the set of observables is given a certain algebraic structure, e.g., a C*-algebra structure, with states represented by linear functionals on the algebra of observables,[9] In the quantum logic approach, a subset of observables, known as propositions of the system, is given a lattice structure, with states regarded as probability measures on this set of propositions.[10] We shall not pursue these approaches here. Such theories would be too abstract and unwieldy as far as physical applications are concerned. The objective of this book is not to present abstract and general theories, and not to formulate theories for their own sake. Our aim is to establish theories in a form which are more familiar and hence more attractive and directly applicable to physics,

---

[5] Many authors simply refer to any element of $\mathcal{H}$ as a wave function. We shall use the term to mean normalized element.
[6] Beltrametti and Cassinelli (1981) p. 3.
[7] An explicit analysis is available in Ohanian (1995) p. 157.
[8] Gudder (1979).
[9] A brief discussion of the algebraic approach is given by Haag (1973), Bogolubov, Logunov and Todorov (1975), Müller-Herold (1978) and Primas (1983). Guenin (1966) presented a systematic discussion of the mathematics involved. More materials can be found in Emch (1972), Bratteli and Robinson (1979), Sewell (1986), Haag (1992), Landsman (1998) and Sewell (2002).
[10] MacKey (1963), Jauch (1968), Piron (1976), Beltrametti and Cassinelli (1981), Garden (1984), Pitowski (1989), Cohen (1989), Sudbery (1986). See §3.2.2 for a definition and a discussion on propositions.

e.g., theories which can directly model physical systems.

This explains why we have chosen to present a formulation of orthodox quantum mechanics based directly on Hilbert spaces and operators, rather than, say, on an abstract C*-algebra.[11]

3.2 Orthodox Quantum Statics

3.2.1 Postulate on orthodox quantum statics

A state provides a description of a physical system, so that when the system is in a known state we are able to predict the probability of any outcome of a measurement of a given observable. To be more specific, let $\mathcal{O}_{qm}$ and $\mathcal{S}_{qm}$ denote respectively the set of observables and states of a quantum system.[12] Let us assume that a measurement of any observable necessarily results in a value lying in a certain Borel set Λ of the reals. The fundamental question which any theory must provide an answer for is:

- What is the probability of a measurement of an observable of the system in a given state resulting in a value in a given Borel set Λ?

The answer should be in the form of a numerical probability measure $\mathcal{M}$ with the desired probability given by $\mathcal{M}(\Lambda)$. The need for such a probability measure lies the reason for the traditional choice of selfadjoint operators and unit vectors for the description of observables and states in quantum mechanics. A selfadjoint operator $\widehat{A}$ in a Hilbert space $\mathcal{H}$ possesses a unique spectral measure $\widehat{M}^{\hat{A}}$. Theorem 2.11.1(1) tells us that this spectral measure together with any unit vector ϕ generates a numerical probability measure $\mathcal{M}_\phi^{\hat{A}}$ by mapping each Borel set Λ to the value $\mathcal{M}_\phi^{\hat{A}}(\Lambda) = \langle \phi \mid \widehat{M}^{\hat{A}}(\Lambda)\phi \rangle$. The idea is then to identify this probability measure with the one required to describe the probability of measurement outcomes. The resulting expectation value is then given by $\mathcal{E}(\widehat{A}, \phi) = \langle \phi \mid \widehat{A}\phi \rangle$ in accordance with Eq. (2.495).

With the probability measure given by $\langle \phi \mid \widehat{M}^{\hat{A}}(\Lambda)\phi \rangle$ we can see that unit vectors different from ϕ only by a *phase factor*[13] will produce the same probability measure and are therefore physically indistinguishable. Since all unit

---

[11] Orthodox quantum mechanics is meant to be the theory presented in traditional texts on the subject, e.g., by Beltrammetti and Cassinelli (1981) and by Isham (1995). A mathematically rigorous treatment of quantum theory in Hilbert space goes back to von Neumann (1932). For a recent account of von Neumann's contributions to mathematics and physics see Rédei and Stöltzer (2001).

[12] Following common practice we shall denote both the observable and its representation by $\widehat{A}$. Similarly we denote both a state and its vector representation by ϕ.

[13] A phase factor is a complex multiplicative constant of magnitude 1.

3.2. ORTHODOX QUANTUM STATICS

vectors different from ϕ only by a phase factor span the same one-dimensional subspace $\mathcal{L}_\phi$ of $\mathcal{H}$, we can associate a state uniquely with a one-dimensional subspace of $\mathcal{H}$.[14] Since one-dimensional subspaces $\mathcal{L}_\phi$ are uniquely related to one-dimensional projectors $\widehat{P}_\phi = |\phi\rangle\langle\phi|$ we can describe states in terms of one-dimensional projectors in a unique manner.

So far we have considered states describable by a one-dimensional projector $\widehat{P}_\phi$. A state described by a given one-dimensional projector corresponds to a maximal knowledge of the system. We can generalize to include states corresponding to less than a maximal knowledge of the system. This may be due to our having only a partial specification of the state preparation process or to the sheer complexity of the system. As an example consider a less than ideal state preparation process which is unable to prepare the system in a desired state $\widehat{P}_\phi$. Instead, the state preparation process can only determine the system to within a set of possible states $\widehat{P}_{\phi_\ell}$, $\ell = 1, 2, \ldots$, namely the system may end up to be in state $\widehat{P}_{\phi_1}$, or in state $\widehat{P}_{\phi_2}$ and so on. Suppose the state preparation process can also tell us the probability w_ℓ of the system ending up in state $\widehat{P}_{\phi_\ell}$. Then we have a situation that we do not know for certain which state the system is actually in; we only know that the system has a probability w_ℓ to be in state $\widehat{P}_{\phi_\ell}$. The system is then said to be in a *classical mixture of states*.[15] The calculation of the expectation value of an observable $\widehat{A}$ in such a classical mixture of states should consist of two averaging processes:

1. Average of all measured values of $\widehat{A}$ in each possible state $\widehat{P}_{\phi_\ell}$, i.e.,

$$\mathcal{E}(\widehat{A}, \widehat{P}_{\phi_\ell}) = \langle \phi_\ell \mid \widehat{A}\phi_\ell \rangle. \tag{3.1}$$

2. Average of the expectation values $\mathcal{E}(\widehat{A}, \widehat{P}_{\phi_\ell})$ over the set of states $\{\widehat{P}_{\phi_\ell}\}$ with a matching set of probabilities $\{w_\ell\}$, i.e.,

$$\sum_\ell w_\ell \, \mathcal{E}(\widehat{A}, \widehat{P}_{\phi_\ell}) = \sum_\ell w_\ell \, \langle \phi_\ell \mid \widehat{A}\phi_\ell \rangle. \tag{3.2}$$

From the properties of density operators shown in Eqs. (2.675) and (2.676) we can see that the final average in Eq. (3.2) for a classical mixture of states may be obtained directly as the trace of the operator $\widehat{A}\widehat{D}$ formed by the product of $\widehat{A}$ and a density operator $\widehat{D}$ which is determined by the set of states $\widehat{P}_{\phi_\ell}$ and the set of probabilities w_ℓ in the form of Eq. (2.675). All this suggests that we can incorporate mixtures into our description of states if we represent states generally by density operators,[16] as summarized in the form of a postulate.

---
[14] A one-dimensional subspace is known as a *ray* so that some authors would describe states in terms of rays (Blank, Exner and Havlíček (1994) p. 253, Roman (1965) p. 18).
[15] This is to distinguish from a *quantum mixture of states* to be introduced later in §3.2.2.
[16] Recall that a one-dimension projector $|\phi\rangle\langle\phi|$ is a density operator.

Postulate OQS on orthodox quantum statics

- Each quantum system has an appropriate Hilbert space $\mathcal{H}$ associated with it such that:

  1. Density operators $\widehat{D}$ on $\mathcal{H}$ correspond one-to-one to the states of the system.
  2. Selfadjoint operators $\widehat{A}$ in $\mathcal{H}$ correspond one-to-one to observables of the system.
  3. When the system is in a state represented by a density operator $\widehat{D}$ the probability distribution for the measured values of an observable $\widehat{A}$ is determined by a probability measure $\mathcal{M}_{\widehat{D}}^{\widehat{A}}$ given by

$$\mathcal{M}_{\widehat{D}}^{\widehat{A}}(\Lambda) = \operatorname{tr}\left(\widehat{M}^{\widehat{A}}(\Lambda)\widehat{D}\right). \tag{3.3}$$

  In other words the probability of a measured value of $\widehat{A}$ lying in the Borel set Λ is given by $\mathcal{M}_{\widehat{D}}^{\widehat{A}}(\Lambda)$.

There are alternatives to some of the statements listed in the above postulate:

1. Spectral measures are directly linked to measurable quantities in the form of Eq. (3.3) in Postulate OQS. Since spectral measures corresponds one-to-one to selfadjoint operators we can replace statement 2 on observables in the postulate by a statement directly in terms of spectral measures:

   - *Spectral measures $\widehat{M}^{\widehat{A}}(\Lambda)$ correspond one-to-one to observables of the system.*

   By emphasizing probabilities and spectral measures rather than on self-adjoint operators we can broaden our outlook of quantum theory, leading to alternative as well as generalized theories and formulations.[17]

2. With statement 3 on probability measures we can work out the expectation values. As shown later in Eqs. (3.12) and (3.32) the expectation value $\mathcal{E}(\widehat{A}, \widehat{D})$ of an observable $\widehat{A}$ in state $\widehat{D}$ can be expressed in terms of $\operatorname{tr}(\widehat{A}\widehat{D})$, i.e.,

$$\mathcal{E}(\widehat{A}, \widehat{D}) = \operatorname{tr}(\widehat{A}\widehat{D}). \tag{3.4}$$

---

[17]The quantum logic approach is an example of an alternative formulation of orthodox quantum mechanics directly in terms of *propositions* and probability measures. Propositions are observables represented by projectors, e.g., spectral projectors. We shall look into the meaning of propositions in §3.2.2, especially in the discussion after Eqs. (3.18) and (3.19).

3.2. ORTHODOX QUANTUM STATICS

Comparing this with Eq. (3.3) we can identify the probability $\mathcal{M}_{\widehat{D}}^{\widehat{A}}(\Lambda)$ of a measured value of $\widehat{A}$ lying in the Borel set Λ with the expectation value of the spectral projector $\widehat{M}^{\widehat{A}}(\Lambda)$ in state $\widehat{D}$. In other words the expectation values of a spectral measure define a probability measure $\mathcal{M}_{\widehat{D}}^{\widehat{A}}$ by

$$\mathcal{M}_{\widehat{D}}^{\widehat{A}}(\Lambda) = \mathcal{E}\big(\widehat{M}^{\widehat{A}}(\Lambda), \widehat{D}\big). \tag{3.5}$$

This understanding enables us to rephrase statement 3 on probability in Postulate OQS by the following statement:

- The expectation value $\mathcal{E}(\widehat{B}, \widehat{D})$ of a bounded observable $\widehat{B}$ in state $\widehat{D}$ is given by $\operatorname{tr}(\widehat{B}\widehat{D})$.

Indeed we shall use such a statement in Chapter 4 to express two extensions of Postulate OQS. Note that this statement does not generally apply to unbounded observables since they do not have finite expectation values in every state. The expectation value of an unbounded observable $\widehat{A}$ in a suitable state $\widehat{D}$ can be calculated in terms of an associated probability measures $\mathcal{M}_{\widehat{D}}^{\widehat{A}}(\Lambda)$, i.e., we can obtain $\mathcal{M}_{\widehat{D}}^{\widehat{A}}(\Lambda)$ as an expectation value $\mathcal{E}\big(\widehat{M}^{\widehat{A}}(\Lambda), \widehat{D}\big)$ of the spectral projectors $\widehat{M}^{\widehat{A}}(\Lambda)$ which are bounded and hence included in the above statement.

The description of states in terms of density operators turns out to be far more significant than the mere inclusion of those classical mixtures of states. We shall classify and discuss the nature of states in the following section.

3.2.2 Pure and mixed states

States of a quantum system can be classified by the nature of their corresponding density operators.

Definition 3.2.2(1)   **On pure and mixed states**

- A state is said to be *pure* if its corresponding density operator is a projector and a state is called *mixed* if its corresponding density operator is not a projector.

A pure state represents a maximal knowledge of the system, while mixed states arise from our less than maximal knowledge of the system. We can appreciate this by an examination of the different structures of probability distributions and expectation values for pure and mixed states.

Case 1 Pure states For a pure state the associated density operator $\widehat{D}$ is of the form of a projector which is necessarily one-dimensional and hence generated by a unit vector ϕ, i.e., $\widehat{D} = \widehat{P}_\phi$. We can make use of the results in §2.11.1 and §2.14.3, especially Eq. (2.664), to recover the expressions for the probability measure $\mathcal{M}^{\widehat{A}}_{\widehat{P}_\phi}$ and the expectation value $\mathcal{E}(\widehat{A}, \widehat{P}_\phi)$ directly in terms of $\widehat{A}$ and the unit vector ϕ as follows:

1. The **probability measure** of observable $\widehat{A}$ in state $\widehat{P}_\phi$ is given by

$$\mathcal{M}^{\widehat{A}}_{\widehat{P}_\phi}(\Lambda) = \operatorname{tr}\left(\widehat{M}^{\widehat{A}}(\Lambda)\widehat{P}_\phi\right) = \langle \phi \mid \widehat{M}^{\widehat{A}}(\Lambda)\phi\rangle. \tag{3.6}$$

2. The **probability function** of observable $\widehat{A}$ in state $\widehat{P}_\phi$ is given by

$$\mathcal{F}^{\widehat{A}}_{\widehat{P}_\phi}(\tau) = \mathcal{M}^{\widehat{A}}_\phi((-\infty, \tau]) = \operatorname{tr}\left(\widehat{M}^{\widehat{A}}((-\infty, \tau])\widehat{P}_\phi\right) \tag{3.7}$$

$$= \operatorname{tr}\left(\widehat{F}^{\widehat{A}}(\tau)\widehat{P}_\phi\right) = \langle \phi \mid \widehat{F}^{\widehat{A}}(\tau)\phi\rangle. \tag{3.8}$$

Compared with Eq. (2.486) we see that

$$\mathcal{F}^{\widehat{A}}_{\widehat{P}_\phi}(\tau) = \mathcal{F}^{\widehat{A}}_\phi(\tau). \tag{3.9}$$

3. The **expectation value** of observable $\widehat{A}$ in state $\widehat{P}_\phi$ is given, according to Eq. (2.495), by

$$\mathcal{E}(\widehat{A}, \widehat{P}_\phi) = \int \tau \, d\mathcal{F}^{\widehat{A}}_{\widehat{P}_\phi}(\tau) = \int \tau \, d\mathcal{F}^{\widehat{A}}_\phi(\tau) \tag{3.10}$$

$$= \langle \phi \mid \widehat{A}\phi\rangle. \tag{3.11}$$

Alternatively we can express the expectation value directly in terms of $\widehat{P}_\phi$ as

$$\mathcal{E}(\widehat{A}, \widehat{P}_\phi) = \operatorname{tr}(\widehat{A}\widehat{P}_\phi). \tag{3.12}$$

4. As in Eq. (2.497) the **variance** of $\widehat{A}$ in state $\widehat{P}_\phi$ is expressible in terms of $\widehat{A}$ and ϕ as

$$\mathcal{V}(\widehat{A}, \widehat{P}_\phi) = \mathcal{V}(\widehat{A}, \phi) = \mathcal{V}(\mathcal{F}^{\widehat{A}}_\phi) = ||\widehat{A}\phi||^2 - \langle \phi \mid \widehat{A}\phi\rangle^2. \tag{3.13}$$

5. As pointed out in C2.11.1(1) C3 we require ϕ to lie in the domain of $\widehat{A}$ in order to obtain a finite expectation value and a finite variance.

3.2. ORTHODOX QUANTUM STATICS

In practice we can represent pure states in terms of unit vectors ϕ instead of one-dimensional projectors and rewrite various quantities in term of ϕ, i.e., we can rewrite

$$\mathcal{M}^{\hat{A}}_{\hat{P}_\phi}(\Lambda),\ \mathcal{F}^{\hat{A}}_{\hat{P}_\phi}(\tau),\ \mathcal{E}(\hat{A},\hat{P}_\phi) \quad \text{as} \quad \mathcal{M}^{\hat{A}}_{\phi}(\Lambda),\ \mathcal{F}^{\hat{A}}_{\phi}(\tau),\ \mathcal{E}(\hat{A},\phi). \quad (3.14)$$

As an illustration consider a particle in one-dimensional motion in $I\!\!E$. The associated Hilbert space is taken to be $L^2(I\!\!E)$. Let us examine some basic observables of such a particle, starting with the particle's position and momentum, known to be represented respectively by the operators $\hat{x}$ and $\hat{p}$ defined in Eqs. (2.45) and (2.184). We have:

1. **Position probability measure and function** in state ϕ are given by Eqs. (2.481) and (2.482), i.e., they are $\mathcal{M}^{\hat{x}}_{\phi}(\Lambda)$ and $\mathcal{F}^{\hat{x}}_{\phi}(x)$.

2. **Momentum probability measure and function** in state ϕ are given by Eqs. (2.483) and (2.484), i.e., they are $\mathcal{M}^{\hat{p}}_{\phi}(\Lambda)$ and $\mathcal{F}^{\hat{p}}_{\phi}(p)$. It follows that:

   (a) A particle with positive momentum values would correspond to a wave function ϕ_+ in $L^2(I\!\!E)$ whose Fourier transform $\tilde{\phi}_+(p)$ vanishes for $p < 0$.

   (b) A particle with negative momentum values would correspond to a wave function ϕ_- in $L^2(I\!\!E)$ whose Fourier transform $\tilde{\phi}_-(p)$ vanishes for $p > 0$.

3. **Kinetic energy probability measure and function** in state ϕ can be worked out explicitly. The kinetic energy operator $\hat{K}$ is a function of the momentum operator so that we can work out its spectral measure $\widehat{M}^{\hat{K}}(\Lambda)$ and spectral function $\widehat{F}^{\hat{K}}(\epsilon)$ from Eqs. (2.362) and (2.363). The operator has a doubly degenerate spectrum, making things a little messy. One can circumvent this by dealing with states corresponding to positive and negative momentum values separately to arrive at a more intuitive derivation of the kinetic energy probability measure and function. Let ϕ_+ be a state corresponding to positive momentum values. Based on the analysis in §2.8.4 we can identify the *energy probability amplitude function* in state ϕ_+ with the function $\tilde{\tilde{\phi}}_+(\epsilon)$ in Eq. (2.405) defined on the kinetic energy space $\tilde{I\!\!E}$. For a state ϕ_- of negative momentum values the associated energy probability amplitude function is the function $\tilde{\tilde{\phi}}_-(\epsilon)$ given in Eq. (2.418). We can now write down the probability measures $\mathcal{M}^{\hat{K}}_{\phi_\pm}(\Lambda)$ and the probability functions $\mathcal{F}^{\hat{K}}_{\phi_\pm}$ as

$$\mathcal{M}^{\hat{K}}_{\phi_\pm}(\Lambda) = \int_\Lambda |\tilde{\tilde{\phi}}_\pm(\epsilon)|^2 d\epsilon, \quad (3.15)$$

$$\mathcal{F}^{\hat{K}}_{\phi_\pm}(\epsilon) = \int_0^\epsilon |\tilde{\tilde{\phi}}_\pm(\epsilon))|^2 \, d\epsilon. \tag{3.16}$$

4. **Hamiltonian probability measure and function** in state ϕ has to be worked out individually. Suppose a quantum particle is in a potential such that its Hamiltonian $\hat{H}$ has a purely discrete and non-degenerate spectrum $\{a_j, j = 1, 2, \ldots\}$ with a corresponding set of orthonormal eigenfunctions $\{\varphi_j, j = 1, 2, \ldots\}$. Let $\hat{P}^{\hat{H}}_j = |\varphi_j\rangle\langle\varphi_j|$. Then the probability function $\mathcal{F}^{\hat{H}}_\phi(\tau)$ is given by Eq. (2.485). We have

$$\mathcal{M}^{\hat{H}}_\phi(\Lambda) = \begin{cases} \langle \phi \mid \hat{P}^{\hat{H}}_j \phi \rangle & \text{if } \Lambda \text{ is a singleton set containing only } a_j, \\ 0 & \text{if } \Lambda \text{ contains no eigenvalue.} \end{cases} \tag{3.17}$$

The interpretation is that a measurement always yields an eigenvalue and that the probability of a measured value equal to an eigenvalue a_j is equal to $\mathcal{M}^{\hat{H}}_\phi(\{a_j\}) = \langle \phi \mid \hat{P}^{\hat{H}}_j \phi \rangle$. These results apply to any observables with a purely discrete and non-degenerate spectrum. A generalization to include degeneracy is straightforward, i.e., we would identify $\hat{P}^{\hat{H}}_j$ with the projector onto the eigensubspace of a_j which will no longer be one-dimensional.

5. **Probability measure and function of a proposition** in state ϕ can also be calculated explicitly. Let us introduce the following definition:

   - a *proposition* is an observable represented by a projector $\hat{P}$.[18]

   From Eqs. (2.341) and (2.342) we obtain the following probability function and measure of a projector $\hat{P}$ in state ϕ:

$$\mathcal{F}^{\hat{P}}_\phi(\tau) = \begin{cases} 0 & \tau < 0 \\ 1 - \langle \phi \mid \hat{P}\phi \rangle & 0 \leq \tau < 1 \\ 1 & \tau \geq 1, \end{cases} \tag{3.18}$$

$$\mathcal{M}^{\hat{P}}_\phi(\Lambda) = \begin{cases} 1 - \langle \phi \mid \hat{P}\phi \rangle & \text{if } \Lambda = \{0\} \\ \langle \phi \mid \hat{P}\phi \rangle & \text{if } \Lambda = \{1\} \\ 0 & \text{if } \Lambda \text{ does not contain 0 or 1.} \end{cases} \tag{3.19}$$

To appreciate the meaning of propositions as observables let us consider the case of a projector $\hat{P}_\psi$ generated by a unit vector ψ. Such a projector

---

[18] Jauch (1968), Beltrametti and Cassinelli (1980). Mackey (1963) uses the term *questions*.

3.2. ORTHODOX QUANTUM STATICS

has two eigenvalues, 1 and 0. The value 1 is non-degenerate corresponding to the eigenvector ψ while the value 0 is degenerate corresponding to eigenvectors $\psi^\perp$ which are orthogonal to ψ. We interpret $\widehat{P}_\psi$ as

the proposition that the system is in state ψ

in the sense that when the system is in state ψ there is a probability of 1 of a measured value of $\widehat{P}_\psi$ being 1, i.e., $\mathcal{M}_\psi^{\widehat{P}_\psi}(\{1\}) = 1$. The point is that a proposition is an observable which corresponds to a definite statement about the system which is either true or false. An experiment to measure a proposition would have only two outcomes. One outcome would confirm the truth of the proposition and the other would ascertain the falsehood of the proposition. Such a measurement process which produces only two results is known as a *yes-no experiment.*[19] The outcome confirming the truth of the proposition is called the *yes* outcome (answer) while the other is known as the *no* outcome (answer).

For another example consider the Hamiltonian $\widehat{H}$ of a physical system. Let $\widehat{P}_j$ be the projector onto the eigensubspace of the Hamiltonian corresponding to eigenvalue E_j. If the system is in a corresponding eigenstate φ_j then $\mathcal{M}_{\varphi_j}^{\widehat{P}_j}(\{1\}) = 1$. We interpret $\widehat{P}_j$ as

the proposition that the system has energy E_j,

or

the proposition that the measured value of the Hamiltonian is E_j.

When the system is in state φ_j the answer to the proposition is a yes, and a measurement of $\widehat{P}_j$ would produce the outcome 1. More examples of propositions and their measurement are presented later in §3.6.1.

Case 2 Mixed states For mixed states the structure of the probability measures, functions and the expectation values are quite different. As an illustration consider the simple case where the density operator possesses a non-degenerate spectrum, i.e., all its eigenvalues ω_j are non-degenerate. Let the corresponding normalized eigenvectors be ψ_j. Following Eq. (2.668) we have the following spectral decomposition:

$$\widehat{D} = \sum_j \omega_j \widehat{P}_{\psi_j}, \quad \widehat{P}_{\psi_j} = |\psi_j\rangle\langle\psi_j|. \tag{3.20}$$

The following results are obvious:

---

[19] MacKey (1963), Jauch (1968), Beltrametti and Cassinelli (1981).

1. The probability measure of an observable $\widehat{A}$ in state $\widehat{D}$ is given by

$$\begin{aligned}
\mathcal{M}_{\widehat{D}}^{\widehat{A}}(\Lambda) &= \operatorname{tr}\left(\widehat{M}^{\widehat{A}}(\Lambda)\widehat{D}\right) & (3.21) \\
&= \sum_j \omega_j \operatorname{tr}\left(\widehat{M}^{\widehat{A}}(\Lambda)\widehat{P}_{\psi_j}\right) & (3.22) \\
&= \sum_j \omega_j \langle \psi_j \mid \widehat{M}^{\widehat{A}}(\Lambda)\psi_j \rangle. & (3.23)
\end{aligned}$$

2. The probability function of an observable $\widehat{A}$ in state $\widehat{D}$ is given by

$$\begin{aligned}
\mathcal{F}_{\widehat{D}}^{\widehat{A}}(\tau) &= \operatorname{tr}\left(\widehat{F}^{\widehat{A}}(\tau)\widehat{D}\right) & (3.24) \\
&= \sum_j \omega_j \operatorname{tr}\left(\widehat{F}^{\widehat{A}}(\tau)\widehat{P}_{\psi_j}\right) & (3.25) \\
&= \sum_j \omega_j \langle \psi_j \mid \widehat{F}^{\widehat{A}}(\tau)\psi_j \rangle. & (3.26)
\end{aligned}$$

3. The expectation value of observable $\widehat{A}$ in state $\widehat{D}$ is given by

$$\begin{aligned}
\mathcal{E}(\widehat{A}, \widehat{D}) &= \int \tau \, d\mathcal{F}_{\widehat{D}}^{\widehat{A}}(\tau) & (3.27) \\
&= \sum_j \omega_j \int \tau \, d\langle \psi_j \mid \widehat{F}^{\widehat{A}}(\tau)\psi_j \rangle & (3.28) \\
&= \sum_j \omega_j \left\langle \psi_j \,\Big|\, \int \tau \, d\widehat{F}^{\widehat{A}}(\tau)\psi_j \right\rangle & (3.29) \\
&= \sum_j \omega_j \langle \psi_j \mid \widehat{A}\psi_j \rangle & (3.30) \\
&= \sum_j \omega_j \operatorname{tr}\left(\widehat{A}\widehat{P}_{\psi_j}\right) & (3.31) \\
&= \operatorname{tr}\left(\widehat{A}\widehat{D}\right). & (3.32)
\end{aligned}$$

Compared with the corresponding results for pure states we can see that the results for mixed states are obtained from that of pure states by a further averaging process, i.e., we have:

$$\begin{aligned}
\mathcal{M}_{\widehat{D}}^{\widehat{A}}(\Lambda) &= \sum_j \omega_j \mathcal{M}_{\widehat{D}_j}^{\widehat{A}}(\Lambda) = \sum_j \omega_j \mathcal{M}_{\psi_j}^{\widehat{A}}(\Lambda), & (3.33) \\
\mathcal{F}_{\widehat{D}}^{\widehat{A}}(\tau) &= \sum_j \omega_j \mathcal{F}_{\widehat{D}_j}^{\widehat{A}}(\tau) = \sum_j \omega_j \mathcal{F}_{\psi_j}^{\widehat{A}}(\tau), & (3.34) \\
\mathcal{E}(\widehat{A}, \widehat{D}) &= \sum_j \omega_j \mathcal{E}(\widehat{A}, \widehat{P}_{\psi_j}) = \sum_j \omega_j \mathcal{E}(\widehat{A}, \psi_j). & (3.35)
\end{aligned}$$

3.2. ORTHODOX QUANTUM STATICS

Take the expectation as an illustration. Clearly $\mathcal{E}(\widehat{A}, \widehat{D})$ comprises two averaging processes set out in Eqs. (3.1) and (3.2):

1. First we obtain $\mathcal{E}(\widehat{A}, \widehat{P}_{\psi_j})$ which is the average over all possible values of $\widehat{A}$ in pure state $\widehat{P}_{\psi_j}$. This is an intrinsic quantum mechanical average in a pure state due to the inherent statistical nature of quantum systems.

2. Next we have to take the average of the expectation values $\mathcal{E}(\widehat{A}, \widehat{P}_{\psi_j})$ over all the pure states $\widehat{P}_{\psi_j}$ with ω_j serving as the probabilities for this averaging process. The need for this second averaging process is due to our lack of a maximal knowledge of the state of the system. For example, due to the complexity of the system, e.g., a statistical ensemble consisting of a large collection of quantum particles, we cannot know what state an individual particle is in. An individual particle in the collection may be in one of a set $\{\widehat{P}_{\psi_j}\}$ of pure states. Our knowledge may only tell us that an individual system has a probability ω_j to be in the pure state $\widehat{P}_{\psi_j}$, i.e., an individual is only known to be in a classical mixture of states $\widehat{P}_{\psi_j}$.[20]

3. The same analysis applies to the probability measure $\mathcal{M}_{\widehat{D}}^{\widehat{A}}(\Lambda)$ and the probability function $\mathcal{F}_{\widehat{D}}^{\widehat{A}}(\tau)$.

This lack of a full knowledge of the system represented by mixed states is more profound than the discussion presented above. While a classical mixture of states in the sense of the example which leads to the two averaging processes set out in Eqs. (3.1) and (3.2) can be incorparated in a description of states in terms of a densisty operator, the converse is not true. Our previous discussion in §2.14.3 tells us that the decomposition of a density operator in terms of a convex combination of projectors in not unique. Take the density operator $\widehat{D}'$ in Eq. (2.672) for example. The decomposition in Eq. (2.672) would suggest that the system is a classical mixture of states ξ_ℓ. But the decomposition of $\widehat{D}'$ in Eq. (2.673) suggests that the system is a classical mixture of states ψ'_ℓ. The point is that

> *a mixed state should be regarded as a state in its own right and we simply cannot say that a system in a mixed state must be a classical mixture of a certain set of pure states.*

---

[20] One can proceed to employ density operators to build up the state for the collection as a whole. This is one way to form the states for a quantum statistical ensemble (Penrose (1970) Chapters I and II. Capri (1985)). A simple example of a quantum statistical system is available in Isham (1995) p. 92. More examples and applications can be found in Blum (1981) and Balian (1991).

In the present context we really cannot assert that a mixed state described by a density operator $\widehat{D}$ in Eq. (3.20) necessarily implies that the system must be in one of the pure states $\widehat{P}_{\psi_j}$. The system may not really be in any of the pure states ψ_j or any pure state at all.[21] This is quite distinct from the corresponding situation in classical statistical mechanics where, despite our ignorance, the system is really in a certain definite pure state.[22] In other words such an interpretation of classical statistical states, known as an *ignorance interpretation*, does not generally apply to mixed states in quantum theory. To highlight this fundamental nature of mixed states in quantum theory we shall generally call a state represented by a density operator which is not a projector a *quantum mixture of states* or a *quantum mixture*. We shall often refer to a mixed state simply as a mixture without explicitly stating whether it is a quantum or a classical mixture, since the situation would be quite clear from the context in which the term appears.[23] All these discussions are crucial to the study of the fundamental problems on quantum state preparation and measurement to be presented in §3.5 and §3.6 at the end of this chapter.

3.2.3 Correlation between states

A mixed state is not a linear combination of pure states. For example, the mixed state $\widehat{D}$ given by Eq. (3.20) is intrinsically different from a pure state represented by the unit vector ϕ given by

$$\phi = \sum_j c_j \, \psi_j, \quad c_j = \sqrt{\omega_j}. \tag{3.36}$$

We can empirically distinguish the mixed state $\widehat{D}$ in Eq. (3.20) from the pure state ϕ since they lead to different expectation values for some observables:

$$\mathcal{E}(\widehat{A}, \psi) = \langle \phi \mid \widehat{A} \, \phi \rangle \tag{3.37}$$

$$= \sum_j |c_j|^2 \langle \psi_j \mid \widehat{A} \, \psi_j \rangle + \sum_{\ell \neq k} c_\ell^* \, c_k \langle \psi_\ell \mid \widehat{A} \, \psi_k \rangle \tag{3.38}$$

$$= \sum_j \omega_j \langle \psi_j \mid \widehat{A} \, \psi_j \rangle + \sum_{\ell \neq k} \sqrt{w_\ell \, w_k} \, \langle \psi_\ell \mid \widehat{A} \, \psi_k \rangle, \tag{3.39}$$

$$\mathcal{E}(\widehat{A}, \widehat{D}) = \text{Tr}(\widehat{A}\widehat{D}) = \sum_j w_j \, \langle \psi_j \mid \widehat{A} \, \psi_j \rangle. \tag{3.40}$$

---

[21] For an explicit example see Isham (1995) pp. 91-92. See also Beltrametti and Cassinelli (1981) p. 9. For a more detailed discussion see D'Espagnat (1989).

[22] Pure states here mean states embodying a maximal knowledge of the system. In the case of a classical particle a pure state corresponds to a unique position and momentum.

[23] It is more often that a classical mixture is meant in cases involving state preparation, measurement and superselection rules.

3.2. ORTHODOX QUANTUM STATICS

The pure state ϕ in Eq. (3.36) is said to be a *coherent superposition* of states ψ_j, while the mixed state $\widehat{D}$ in Eq. (3.20) is said to be a *quantum mixture* of states ψ_j.[24] Compared with $\mathcal{E}(\widehat{A}, \widehat{D})$ the expectation value $\mathcal{E}(\widehat{A}, \phi)$ contains additional terms equal to

$$I = \sum_{\ell \neq k} c_\ell^* c_k \langle \psi_\ell \mid \widehat{A} \psi_k \rangle. \tag{3.41}$$

This is known as the *interference term* or the *correlation term* due to operator $\widehat{A}$. We also call the individual terms

$$\langle \psi_\ell \mid \widehat{A} \psi_k \rangle \tag{3.42}$$

interference terms or *correlation terms* due to operator $\widehat{A}$. These terms represent the correlations between all the different constituent states in the superposition due to observable $\widehat{A}$. It is these correlation terms which distinguish a coherent superposition from a quantum mixture.

To generalize the concept of correlations further let ϕ and ψ be two arbitrary unit vectors and let $\widehat{P}_\phi$ and $\widehat{P}_\psi$ be the projectors generated by ϕ and ψ respectively. Following Eq. (3.42) we call $\langle \phi \mid \widehat{P}_\psi \psi \rangle$ the correlation term between ϕ and ψ due to $\widehat{P}_\psi$. Since

$$\langle \phi \mid \widehat{P}_\psi \psi \rangle = \langle \phi \mid \psi \rangle \tag{3.43}$$

we shall call $\langle \phi \mid \psi \rangle$, as well as its real part $\operatorname{Re} \langle \phi \mid \psi \rangle$, the correlation term or simply the *correlation* between ϕ and ψ. This terminology reflects the following interpretation. If the two vectors are not orthogonal we can express one in terms of the other plus a remainder, i.e., we have

$$\phi = c_1 \psi + c_2 \psi^\perp, \quad \text{where} \quad \langle \psi \mid \psi^\perp \rangle = 0 \tag{3.44}$$

and

$$\mathcal{M}_\phi^{\widehat{P}_\psi}(\{1\}) = \langle \phi \mid \widehat{P}_\psi \phi \rangle = |\langle \phi \mid \psi \rangle|^2. \tag{3.45}$$

In other words in state ϕ the probability of a measured value of $\widehat{P}_\psi$ being 1 is given by $|\langle \psi \mid \phi \rangle|^2$. Clearly the role of ϕ and ψ can be swopped to get the correlation between ψ and ϕ to be

$$\langle \psi \mid \widehat{P}_\phi \phi \rangle = \langle \psi \mid \phi \rangle \tag{3.46}$$

and

$$\mathcal{M}_\psi^{\widehat{P}_\phi}(\{1\}) = \langle \psi \mid \widehat{P}_\phi \psi \rangle = |\langle \psi \mid \phi \rangle|^2 = \mathcal{M}_\phi^{\widehat{P}_\psi}(\{1\}). \tag{3.47}$$

---

[24] We should stress again that a quantum mixture of ψ_j does not imply an ignorance interpretation which is applicable to classical mixtures.

Since $\widehat{P}_\psi$ represents the proposition that the system is in state ψ and similar for $\widehat{P}_\phi$ we can interpret

$$|\langle\phi\mid\psi\rangle|^2 = |\langle\psi\mid\phi\rangle|^2 \tag{3.48}$$

as the *transition probability* between states ϕ and ψ,[25] or the probability that the ψ occurs in state ϕ and vice versa.[26]

Physically one can imagine a scattering experiment in which an initial state ϕ is scattered into a new state ϕ_t at time t which is a superposition of various states ξ_j. Then $|\langle\xi_j\mid\phi_t\rangle|^2$ is the probability of finding the scattered system in state ξ_j at time t.[27]

3.2.4 Discretization of bounded and unbounded observables

The discussions in the preceding section demonstrate that probabilities arising from selfadjoint operators as presented in §2,11.1 play a crucial role in describing the probabilistic nature of quantum systems. In this section we shall look into some approximations to certain selfadjoint operators which will become relevant when discussing their physical measurement. In this context the simplest type of operators are bounded selfadjoint operators $\widehat{A}$ with a purely discrete spectrum $\{\tau_\ell\}$. Their spectral functions take the form of Eq. (2.306) and their spectral decompositions take the form of Eq. (2.375). In a given state ϕ, the probability function of $\mathcal{F}_\phi^{\widehat{A}}(\tau)$ of $\widehat{A}$ becomes a piecewise constant function of the form of Eq. (2.264) with

$$\wp_\ell = \langle\phi\mid\widehat{P}_\ell^{\widehat{A}}\phi\rangle. \tag{3.49}$$

Explicitly we have

$$\mathcal{F}_\phi^{\widehat{A}}(\tau) = \begin{cases} 0 & \text{if } \tau < \tau_1, \\ \wp_1 = \langle\phi\mid\widehat{P}_1^{\widehat{A}}\phi\rangle & \text{if } \tau_1 \leq \tau < \tau_2, \\ \cdots \\ \sum_{\ell=1}^{k}\wp_\ell = \sum_{\ell=1}^{k}\langle\phi\mid\widehat{P}_\ell^{\widehat{A}}\phi\rangle & \text{if } \tau_k \leq \tau < \tau_{k+1}, \\ \cdots \\ 1 & \text{if } \tau_{ub} \leq \tau, \end{cases} \tag{3.50}$$

where τ_{ub} is the lowest upper bound of the set of eigenvalues τ_ℓ. There is no probability of a measurement giving a value other than an eigenvalue, and the

---

[25] Beltrammetti and Cassinelli (1981) p. 12. We may call $\langle\phi\mid\psi\rangle$ and $\langle\psi\mid\phi\rangle$ *transition amplitudes*.
[26] Roman (1965) p. 26.
[27] Greiner (1989) p. 188 and Eq. (11.54) on p. 210, Merzbacher (1989) p. 316.

3.2. ORTHODOX QUANTUM STATICS

probability for a measured value being an eigenvalue τ_ℓ is equal to $\wp_\ell$. Using Eq. (2.278) we obtain the expectation value

$$\mathcal{E}(\mathcal{F}_\phi^{\widehat{A}}) = \sum_\ell \wp_\ell \tau_\ell = \sum_\ell \langle \phi \mid \widehat{P}_\ell \phi \rangle \tau_\ell \qquad (3.51)$$

$$= \langle \phi \mid \widehat{A}\phi \rangle, \qquad (3.52)$$

agreeing with $\mathcal{E}(\widehat{A}, \widehat{P}_\phi)$ in Eq. (3.11).

Unbounded observables represented by unbounded selfadjoint operators $\widehat{A}$ can be determined to an arbitrary degree of accuracy by their *bounded approximations* $\widehat{A}_N$ defined by[28]

$$\widehat{A}_N = \int_{-N}^{N} \tau \, d\widehat{F}^{\widehat{A}}(\tau) \qquad (3.53)$$

for sufficiently large N. Here $\widehat{A}_N$ tends to $\widehat{A}$ strongly as N tends to infinity, as pointed out in C2.8.1(1) C1. It follows that we have a similar approximation in expectation values:

$$\lim_{N \to \infty} \langle \phi \mid \widehat{A}_N \phi \rangle = \langle \phi \mid \widehat{A} \phi \rangle. \qquad (3.54)$$

In addition to bounded approximations we can also have what may be called *discretized approximations*. This applies to observables with a continuous spectrum. To be specific let $\widehat{A}$ be a bounded observable with a purely continuous spectrum. Then we have

$$\widehat{A} = \int_a^b \tau \, d\widehat{F}^{\widehat{A}}(\tau) \qquad (3.55)$$

for some finite numbers a and b. To *discretize* this observable we introduce a partition Π of $[a, b]$ by small intervals, i.e.,

$$\Pi = \{\tau_0\} \cup (\tau_0, \tau_1] \cup (\tau_1, \tau_2] \cup \cdots \cup (\tau_{n-1}, \tau_n] = [a, b]. \qquad (3.56)$$

Associated with this partition we introduce an operator[29]

$$\widehat{A}_\Pi = \sum_{\ell=1}^{n} \tau_\ell \, \widehat{M}^{\widehat{A}}\big((\tau_{\ell-1}, \tau_\ell]\big), \qquad (3.57)$$

---

[28] Sewell (1986) p. 13 and p. 41, Richtmyer (1978) p. 305, Naimark (1968) Part 2, p. 15.
[29] Roman Vol. 2 (1975) pp. 640-641, Kreyszig (1978) pp. 505-508. The sum in Eq. (3.57) should contain the term $\widehat{M}^{\widehat{A}}(\{\tau_0\})$. Since the spectrum is purely continuous we have $\widehat{M}^{\widehat{A}}(\{\tau_0\}) = \widehat{O}$. So the term was omitted.

where $\widehat{M}^{\hat{A}}((\tau_{\ell-1}, \tau_\ell])$ is the spectral projector of $\widehat{A}$ for the interval $(\tau_{\ell-1}, \tau_\ell]$. This is a selfadjoint operator with a purely discrete spectrum. Moreover, $\widehat{A}_\Pi$ converges strongly to $\widehat{A}$ as the partition becomes finer and finer. So, the expectation value of $\langle \phi \mid \widehat{A} \phi \rangle$ can be approximated by $\langle \phi \mid \widehat{A}_\Pi \phi \rangle$.

For a bounded observable whose spectrum consists of a discrete and a continuous part we can discretize the continuous part according to the above procedure. Since the spectrum of a selfadjoint operator is known to be a closed set the point τ_0 could be an eigenvalue so that we must amend the expression for $\widehat{A}_\Pi$ in Eq. (3.57) by an additional term $\tau_0 \widehat{M}^{\hat{A}}(\{\tau_0\})$.[30]

We can now conclude that an unbounded observable with a continuous spectrum can be approached by bounded observables with a discrete spectrum, i.e., we can carry out a bounded approximation $\widehat{A}_N$ followed by a discretization of $\widehat{A}_N$ to arrive at a bounded operator with a discrete spectrum. These results enable us to establish a general measurement theory in terms of a model for measuring bounded observables with a discrete spectrum.

3.2.5 Approximate nature of measurements

Generally a measurement process would involve inaccuracies, e.g., the average value in Eq. (2.237) obtained in an experiment only approximates the theoretical expectation value. We shall formalize this with the following general assumption on experiments:

Assumption 3.2.5   **On experiments**

- An experiment is performed within a finite time interval, and is capable of measuring only a finite set $\mathcal{O}_{qm}^n$ of n independent bounded observables of the system for the system in a limited or restricted set $\mathcal{S}_{qm}^{res}$ of states in order to achieve a finite upper bound of inaccuracy ε_{inc}. Here an inaccuracy for a measurement of $\widehat{A}$ in state ζ is defined to be

$$\varepsilon(\widehat{A}, \zeta) = |\mathcal{E}_{xp}(\widehat{A}, \zeta) - \mathcal{E}(\widehat{A}, \zeta)|, \quad (3.58)$$

where $\mathcal{E}_{xp}(\widehat{A}, \zeta)$ is the experimental expectation value of $\widehat{A}$ in state ζ, i.e., the measured value, and $\mathcal{E}(\widehat{A}, \zeta)$ is the theoretically calculated value. The desired upper bound of inaccuracy is then defined to be

$$\begin{aligned}\varepsilon_{inc} &= \varepsilon_{inc}(\mathcal{O}_{qm}^n, \mathcal{S}_{qm}^{res}) & (3.59) \\ &= \text{supremum}\left\{\varepsilon(\widehat{A}, \zeta) : \widehat{A} \in \mathcal{O}_{qm}^n \text{ and } \zeta \in \mathcal{S}_{qm}^{res}\right\}. & (3.60)\end{aligned}$$

---

[30]Akhiezer and Glazman (1961) Theorem 4 on p. 91, Roman Vol. 2 (1975) Theorem 13.2(6) on p. 605, Kreyszig (1978), Wan and McLean (1994).

3.2. ORTHODOX QUANTUM STATICS

Such an experiment is denoted by $\boldsymbol{E}_{xp}(\mathcal{O}_{qm}^n, \mathcal{S}_{qm}^{res}, \varepsilon_{inc})$.

The approximate nature of experiments can blur the distinction between a coherent superposition and a quantum mixture. To highlight this let us consider the case of two families of states $\varphi_{1\tau}$ and $\varphi_{2\tau}$ parameterized by a real parameter τ. For each value τ we have a pair of states $\varphi_{1\tau}, \varphi_{2\tau}$ from which we can form a coherent superposition

$$\zeta_\tau = c_1 \varphi_{1\tau} + c_2 \varphi_{2\tau} \tag{3.61}$$

to obtained a one-parameter family of superposed states ζ_τ. The contribution to the expectation value of any given observable $\widehat{A}$ from the correlation terms in the superposed state ζ_τ is, in accordance with Eq. (3.41), given by

$$I_\tau = 2\,\mathrm{Re}\left(c_1^* c_2 \langle \varphi_{1\tau} \mid \widehat{A}\,\varphi_{2\tau}\rangle\right), \tag{3.62}$$

where Re stands for the real part of a complex number. Now, suppose that the two families of states are such that for a given $\widehat{A}$ this correlation term vanishes asymptotically, i.e.,

$$I_\tau \to 0 \quad \text{as} \quad \tau \to \infty. \tag{3.63}$$

As a result a measurement of $\widehat{A}$ may fail to detect the existence of the correlation term. Then, observable $\widehat{A}$ will not be able to distinguish the superposed state ζ_τ from the mixture

$$\widehat{D}_\tau = |c_1|^2\,\widehat{P}_{\varphi_{1\tau}} + |c_2|^2\,\widehat{P}_{\varphi_{2\tau}} \tag{3.64}$$

at large τ, i.e., we have

$$|I_\tau| = \left|\mathcal{E}(\widehat{A}, \zeta_\tau) - \mathcal{E}(\widehat{A}, \widehat{D}_\tau)\right| \to 0 \quad \text{as } \tau \to \infty. \tag{3.65}$$

For reference later we shall formalize this situation as follows:

1. Let $\mathcal{O}_{qm}^n$ be a finite number of independent observables and ζ_τ be a one-parameter family of states given by Eq. (3.61).

2. Suppose for any small number $\delta > 0$ there is a number τ_0 such that for all $\widehat{A} \in \mathcal{O}_{qm}^n$ and $\tau \geq \tau_0$ we have

$$\varepsilon(\widehat{A}, \zeta_\tau) = \left|\mathcal{E}_{xp}(\widehat{A}, \zeta_\tau) - \mathcal{E}(\widehat{A}, \zeta_\tau)\right| < \frac{1}{2}\delta, \quad \text{and} \quad |I_\tau| < \frac{1}{2}\delta. \tag{3.66}$$

3. Then we have

$$\begin{aligned}
& |\mathcal{E}_{xp}(\widehat{A},\zeta_\tau) - \mathcal{E}(\widehat{A},\widehat{D}_\tau)| \\
&= |(\mathcal{E}_{xp}(\widehat{A},\zeta_\tau) - \mathcal{E}(\widehat{A},\zeta_\tau)) - (\mathcal{E}(\widehat{A},\widehat{D}_\tau) - \mathcal{E}(\widehat{A},\zeta_\tau))| \\
&= |\varepsilon(\widehat{A},\zeta_\tau) + I_\tau| \\
&\leq |\varepsilon(\widehat{A},\zeta_\tau)| + |I_\tau| < \delta.
\end{aligned} \quad (3.67)$$

4. Let $\mathcal{S}^{res}_{qm}(\tau_0) = \{\zeta_\tau : \tau \geq \tau_0\}$. We can now claim that experiment $\mathbf{E}_{xp}(\mathcal{O}^n_{qm}, \mathcal{S}^{res}_{qm}(\tau_0), \varepsilon_{inc} = \delta)$ is unable to distinguish a coherent superposition ζ_τ from a mixture $\widehat{D}_\tau$. In other words ζ_τ and $\widehat{D}_\tau$ are indistinguishable or equivalent respect to experiment $\mathbf{E}_{xp}(\mathcal{O}^n_{qm}, \mathcal{S}^{res}_{qm}(\tau_0), \varepsilon_{inc})$.

Since we shall have many occasions to return to this situation we shall adopt a usage proposed by John Bell,[31] by calling the coherent superposition ζ_τ and mixture $\widehat{D}_\tau$

equivalent or indistinguishable FAPP with respect to experiment $\mathbf{E}_{xp}(\mathcal{O}^n_{qm}, \mathcal{S}^{res}_{qm}(\tau_0), \varepsilon_{inc})$ (3.68)

if conditions (3.66) and (3.67) above are satisfied. We shall signify this situation by writing

$$\zeta_\tau \equiv \widehat{D}_\tau \quad \text{FAPP}(\mathcal{O}^n_{qm}, \mathcal{S}^{res}_{qm}(\tau_0), \varepsilon_{inc}). \quad (3.69)$$

In other words, as far as $\mathbf{E}_{xp}(\mathcal{O}^n_{qm}, \mathcal{S}^{res}_{qm}(\tau_0), \varepsilon_{inc})$ is concerned observables in $\mathcal{O}^n_{qm}$ are unable to lead to any detectable correlation between the two constituent states in ζ_τ for $\tau > \tau_0$. Arguments along this line are used to set up model theories for state preparation and quantum measurment in §3.5 and §3.6.[32] We shall also return to this later in Chapter 8 where we shall discuss the concept of decoherence.

3.3 Quantization in $I\!E^n$

3.3.1 Preliminaries on quantization

Having established a formal structure in Postulate OQS for the representation of states and physical observables we now want to know how to choose a Hilbert space and how to relate physical observables with selfadjoint operators in the chosen Hilbert space. There are two basic kinds of observables:

---
[31] Bell (1990). FAPP stands for *for all practical purposes*.
[32] The idea can be traced back to Daneri, Loinger and Prosperi (1962).

3.3. QUANTIZATION IN $I\!E^n$

1. Observables which have no classical counterpart: An example is the spin of a spin-$\frac{1}{2}$ particle. We have to find an appropriate Hilbert space and selfadjoint operators to reproduce the observed properties of spin, e.g., there are only two values of spin along any direction. This turns out to be relatively easy and the results for spin are well-known. So, we shall not delve into this type of observables here.

2. Observables which have a classical origin: Position, momentum and kinetic energy are typical examples. The question here is whether one could establish the selfadjoint operators for the description of these quantum observables from our knowledge of their classical counterparts by some kind of quantization scheme. This turns out to be a very difficult problem to solve. There are as yet no generally acceptable schemes. It will become clear that solutions to this quantization problem is unlikely to be soluble by purely mathematical and technical considerations. It must involve some physical principles. Moreover, solutions are generally not unique.

Quite apart from being of interest in its own right within orthodox quantum mechanics a study of quantization can serve to demonstrate the need for a generalization of orthodox quantum mechanics, something we shall do in Chapter 5, and the knowledge and techniques gained in such a study can also be applied to macroscopic quantum systems in a constrained environment, e.g., superconducting systems in circuit configurations. So, we shall devote several sections, from §3.3.2 to §3.3.7, within this chapter to a detailed study of the quantization problem. We shall return in Chapter 6 to extend our study to include point interactions; the results obtained are then applied to superconducting curcuit systems in Chapter 7 in the context of generalized quantum theories set out in Chapters 4 and 5.[33]

To avoid too much generality we shall consider the case of a point particle moving in Euclidean space $I\!E^n$ in classical mechanics. Let x^j be the chosen rectangular Cartesian coordinates which specify the particle's position, and let p_j be the chosen momenta canonically conjugate to x^j. The basic classical observables of the particle are position x^j and momentum p_j. All other observables A, excluding fixed quantities like the mass m of the particle, are just functions of position and momentum, i.e., $A = A(x,p)$. We shall denote by $\mathcal{O}_{cl}$ the set of all classical observables. To quantize we must first choose a Hilbert space to associate with the system. In common with standard practice, we associate the Hilbert space $\mathcal{H} = L^2(I\!E^n)$ of square-integrable functions $\phi(x)$ on $I\!E^n$ with the particle, assuming it to be spinless. Classically the

---
[33] The effort devoted to the quantization problem here is in keeping with what we set out to do in this book, i.e., to formulate an applicable form of quantum mechanics.

position of the particle is described by the coordinates x^j. We shall initially quantize these position variables as multiplication operators

$$\mathcal{Q}_0(x^j) = x^j \quad \text{on the domain} \quad C_0^\infty(I\!\!E^n) \subset L^2(I\!\!E^n). \tag{3.70}$$

It is easy to show that each $\mathcal{Q}_0(x^j)$ is essentially selfadjoint, and hence possesses a unique selfadjoint extension $\mathcal{Q}(x^j)$ acting as a multiplication operator on the domain

$$\mathcal{D}(\mathcal{Q}(x^j)) = \{\phi \in L^2(I\!\!E^n) : x^j \phi(x) \in L^2(I\!\!E^n)\}. \tag{3.71}$$

We identify $\mathcal{Q}(x^j)$ as the quantized position observables, denoted also by $\widehat{x}^j$ or simply by x^j.

The above choice of the Hilbert space based on the coordinate space $I\!\!E^n$ is referred to as the *coordinate representation* and we call normalized functions ϕ in $L^2(I\!\!E^n)$ *wave functions*. In this representation every classical observable which is a function V of position variables is quantized as a multiplication operator $\mathcal{Q}_0(V) = \widehat{V} = V(x)$ in $L^2(I\!\!E^n)$ defined on the domain

$$\mathcal{D}(\widehat{V}) = \{\phi \in L^2(I\!\!E^n) : V(x)\phi(x) \in L^2(I\!\!E^n)\}. \tag{3.72}$$

If the function is bounded then the operator $\widehat{V}$ is also bounded with the entire Hilbert space as its domain of operation. One such function is the characteristic function χ_Λ of an open rectangle Λ in $I\!\!E^n$ introduced in C2.2.1(1) C4 and C2.3(1) C10. As a multiplication operator, denoted by $\widehat{\chi}_\Lambda$ or simply by χ_Λ, it is a projector on $L^2(I\!\!E^n)$ projecting onto the subspace of wave functions vanishing outside Λ.

The stage is now set to quantize other observables as selfadjoint operators in $L^2(I\!\!E^n)$. Let $\mathcal{O}_{qm}$ denote the set of quantum observables in the form of selfadjoint operators in $L^2(I\!\!E^n)$. We seek a

quantization map $\quad \mathcal{Q} : \mathcal{O}_{cl} \to \mathcal{O}_{qm} \qquad (3.73)$

to associate classical observables $\mathcal{O}_{cl}$ to quantum observables $\mathcal{O}_{qm}$. Note that $\mathcal{O}_{cl}$ consists of functions of x and p while $\mathcal{O}_{qm}$ is identifiable with the set of selfadjoint operators in the chosen Hilbert space. Such a map cannot be arbitrary, i.e., we want such a quantization map to satisfy certain rules. In the next section we shall set out some of these rules explicitly. Before stating these rules let us set out step by step how we may proceed to establish a quantization map:

1. Let $C_0^\infty(I\!\!E^n)$ denote the set of smooth function of compact support on $I\!\!E^n$. These functions form a dense subset of $L^2(I\!\!E^n)$. Given a classical observable A we shall first establish an operator expression

3.3. QUANTIZATION IN $\mathbb{E}^n$

to act on $C_0^\infty(\mathbb{E}^n)$. The resulting operator, to be denoted by either $\mathcal{Q}_0(A)$ or $\widehat{A}_0$ where the subscript signifies the domain $C_0^\infty(\mathbb{E}^n)$, must be symmetric. In other words we first establish a mapping of the set $\mathcal{O}_{cl}$ of classical observables to the set $\mathcal{O}_{0qm}$ of symmetric operators in $L^2(\mathbb{E}^n)$, all defined on the domain $C_0^\infty(\mathbb{E}^n)$. We shall call this mapping

$$\mathcal{Q}_0 : \mathcal{O}_{cl} \to \mathcal{O}_{0qm} \qquad (3.74)$$

an **initial quantization map** and the operator $\mathcal{Q}_0(A)$ or $\widehat{A}_0$ as the **initially quantized operator**.

2. We then check if $\widehat{A}_0$ is essentially selfadjoint by examining its deficiency indices. If the deficiency indices is (0,0) the operator is essentially selfadjoint and possesses a unique selfadjoint extension $\widehat{A}$ acting on an appropriate domain $\mathcal{D}(\widehat{A})$. We can then identify $\widehat{A}$ as the quantized observable, i.e., we shall assume $\mathcal{Q}(A) = \widehat{A}$.

   Unfortunately no reasonable initial quantization map seems to exist capable of producing essentially selfadjoint operator $\widehat{A}_0$ for every classical observable A. Here the term "reasonable" means a quantization map satisfying certain reasonable rules, e.g., those set out in the next section.

3. If $\widehat{A}_0$ is not essentially selfadjoint, then the following possibilities open up:

   (a) $\widehat{A}_0$ may admit many selfadjoint extensions. This raises a fundamental question as to whether the quantization map $\mathcal{Q}$ in Eq. (3.73) should be single-valued or not. Our investigation shows that the quantization map is not single-valued. We shall demonstrate that this is a genuine nature of quantization and it is not a problem to be circumvented. Instead, non-uniqueness in quantization will be seen in Chapters 6 and 7 to have important physical applications.

   (b) $\widehat{A}_0$ may admit no selfadjoint extensions. However, $\widehat{A}_0$ would admit strictly maximal symmetric extensions. This raises an even more fundamental question as to whether we should include strictly maximal symmetric operators as quantum obervables in violation of Postulate OQS in §3.2. Orthodox quantum theory itself has to be generalized to accommodate strictly maximal symmetric operators. We shall devote Chapter 5 to formulate such a generalization. In this chapter we shall present a number of examples to highlight this particular case to demonstrate the need for a generalization of orthodox quantum theory.

3.3.2 Failure of general schemes

There has been a lot of interest in quantization problems over the years. A common approach is based on the idea that some mathematical structures, e.g., some algebraic structures, of classical observables are so characteristic of the physical system that such a structure should be preserved by their quantum counterparts, and that the requirement for the preservation of such a structure may be sufficient to define a quantization map.[34] A simple algebraic structure is that of linearity and squaring, i.e.,

$$a, b \in \mathbb{R}, \ A, B \in \mathcal{O}_{cl} \ \Rightarrow \ aA + bB \in \mathcal{O}_{cl} \ \text{and} \ A^2 \in \mathcal{O}_{cl}. \quad (3.75)$$

To preserve this property one is tempted to demand the quantization map to satisfy:

1. **The linearity rule**: $\mathcal{Q}(aA + bB) = a\mathcal{Q}(A) + b\mathcal{Q}(B)$.

2. **The squaring rule**: $\mathcal{Q}(A^2) = \mathcal{Q}(A)^2$.

Unfortunately these simple rules are applicable only in a limited number of cases such as in the construction of the Hamiltonian of a simple harmonic oscillator in $\mathbb{E}$, where we do have

$$H = \frac{1}{2m} p^2 + \frac{1}{2} k\, x^2 \ \rightarrow \ \mathcal{Q}(H) = \frac{1}{2m} \mathcal{Q}(p)^2 + \frac{1}{2} k\, \mathcal{Q}(x)^2. \quad (3.76)$$

As early as 1935 Temple discovered that these two rules, when applied generally, are inconsistent since these rules lead to the result that all operators representing physical observables commute.[35] This is known as *Temple's paradox*. The above rules are often dressed up into a more general form such as

1. **Linearity rule**: $\mathcal{Q}(aA + bB) = a\mathcal{Q}(A) + b\mathcal{Q}(B)$.

2. **Function preservation rule**: $\mathcal{Q}(f(A)) = f(\mathcal{Q}(A))$, where f is any real-valued function on $\mathbb{R}$.

Clearly Temple's paradox would prevent the generally application of this pair of rules. To circumvent Temple's paradox we can try to replace the function preservation rule by something else. An alternative is that of a Lie algebra structure. We know from C1.5.3(2) C3 that the set of smooth classical observables, i.e., smooth functions on the phase space, possesses a Lie algebra structure with Poisson bracket $\{A, B\}$ defining the Lie algebra multiplication of A and B. The Poisson bracket is an important construct in classical mechanics; it is to do with canonical variables as well as dynamics. In a Hilbert

---
[34] Isham (1995) §5.2.1.
[35] Temple (1935), Peierls (1935), Park and Margenau (1968), Wan and McKenna (1984).

3.3. QUANTIZATION IN $I\!\!E^n$

space there are selfadjoint operators having a Lie algebra structure with the commutator defining the Lie algebra multiplication. The question is whether we can find a corresponding Lie algebra of selfadjoint operators to correspond to the Lie algebra of smooth classical observables. Consider the simple case of a particle in one-dimensional motion in coordinate space $I\!\!E$. The associated Hilbert space is $L^2(I\!\!E)$. We may seek a quantization map $\mathcal{Q}$ from the set of smooth classical observables in $\mathcal{O}_{cl}$ into a Lie algebra of selfadjoint operators in $L^2(I\!\!E)$ satisfying the following rather natural requirements:

1. **Linearity rule:** $\mathcal{Q}(aA + bB) = a\,\mathcal{Q}(A) + b\,\mathcal{Q}(B)$, $a, b \in I\!\!R$.

2. **Lie algebra rule:** $\mathcal{Q}(\{A, B\}) = [\mathcal{Q}(A), \mathcal{Q}(B)]/i\hbar$.

3. **Canonical conjugation rule:** $\mathcal{Q}(x) = x$, $\mathcal{Q}(p) = -i\hbar\,d/dx$.

4. **Identity rule:** $\mathcal{Q}(I) = \widehat{I\!\!I}$, where I is a constant function of q, p of value 1.

A traditional rationale for identifying $-i\hbar\,d/dx$ as the momentum operator is based on:

1. The differential expression $-i\hbar\,d/dx$ acting on $C_0^\infty(I\!\!E)$ defines an essentially selfadjoint operator in $L^2(I\!\!E)$. We know this already in E2.5.1(1) E3. The operator $\mathcal{Q}(p)$ is meant to be the corresponding unique selfadjoint extension, i.e., $\mathcal{Q}(p) = \widehat{p}$.

2. The canonical pairs $\mathcal{Q}(x), \mathcal{Q}(p)$ satisfy the Lie algebra rule above.

Unfortunately this rather neat approach fails because such a quantization map does not exist.[36] In fact the situation is far worst than that. Each of the first three rules by itself is not generally applicable for lack of selfadjointness:

1. The linearity rule fails since C2.3(1) C6 tells us that the sum of two unbounded selfadjoint operators is not necessarily selfadjoint. A more concrete discussion of this will be in §3.3.4.

2. The Lie algebra rule fails since C2.3(1) C9 tells us that the commutator of two unbounded selfadjoint operators is not necessarily well-defined.

3. Canonical conjugation rule may also fail. An example is when the particle is confined to move in coordinate space $I\!\!E^+$. We know from E2.5.1(1) E6 that the operator $\widehat{p}_0(I\!\!E^+)$ defined by the differential expression $-i\hbar\,d/dx$ acting on $C_0^\infty(I\!\!E^+)$ is not essentially selfadjoint and that $\widehat{p}_0(I\!\!E^+)$ does not even admit any selfadjoint extensions. Another

---
[36] Abraham and Marsden (1978) §5.4, Wan and Viazminsky (1979).

example is when the particle is confined in an interval $\Lambda = (0, a)$. Then E2.5.1(1) E1 tells us that the operator $\hat{p}_0(\Lambda)$ defined by the differential expression $-i\hbar\, d/dx$ acting on $C_0^\infty(\Lambda)$ is again not essentially selfadjoint; $\hat{p}_0(\Lambda)$ admits a one-parameter group of selfadjoint extensions. We are then confronted with a non-uniqueness problem.

The squaring and functional preservation rule on their own do work well, since we know from §2.8.2 that a real function of a selfadjoint operator is again selfadjoint. However, even this rule encounters difficulties when we later extend orthodox quantum theory to include maximal symmetric operators to represent observables, since the square of a maximal symmetric operator is not necessarily maximal symmetric. The moral here is that mathematical structures by themselves are not really so characteristic of the system that they alone are sufficient to determine a satisfactory quantization. This is one of the reasons we have not adopted a purely algebraic approach to present an orthodox quantum theory.

Apart from various attempts based on structural considerations there are pragmatic rules to quantize certain types of observables. The best known ones are on the quantization of polynomials in position and momentum. Again let us confine ourselves to a point particle moving in Euclidean space $I\!\!E$ with position and momentum variables x and p. Consider a classical observable of the form of a polynomial

$$A = \sum_{k,\ell} a_{k\ell}\, x^k p^\ell, \quad a_{k\ell} \in I\!\!R. \tag{3.77}$$

In an attempt to quantize A one may first assume the linearity rule to obtain

$$\mathcal{Q}_0(A) = \sum_{k,\ell} a_{k\ell}\, \mathcal{Q}_0(x^k p^\ell). \tag{3.78}$$

The next step is to express $\mathcal{Q}_0(x^k p^\ell)$ in terms of $\mathcal{Q}_0(x)$ and $\mathcal{Q}_0(p)$. The problem here is that of factor ordering. While $x^k p^\ell$ is identical to $p^\ell x^k$ we can see that

$$\mathcal{Q}_0(x)^k \mathcal{Q}_0(p)^\ell \neq \mathcal{Q}_0(p)^\ell \mathcal{Q}_0(x)^k. \tag{3.79}$$

Moreoever, $\mathcal{Q}_0(x)^k \mathcal{Q}_0(p)^\ell$ is not even symmetric and neither is $\mathcal{Q}_0(p)^\ell \mathcal{Q}_0(x)^k$. By arranging the factors $\mathcal{Q}_0(x)^k$ and $\mathcal{Q}_0(p)^\ell$ appropriately it is possible to produce symmetric operators as a first step towards quantization. Three common schemes are:[37]

1. **The symmetrization rule**:

$$x^k p^\ell \to \mathcal{Q}_{0S}(x^k p^\ell) = \frac{1}{2}\left(\mathcal{Q}_0(x)^k \mathcal{Q}_0(p)^\ell + \mathcal{Q}_0(p)^\ell \mathcal{Q}_0(x)^k \right). \tag{3.80}$$

---

[37]Cohen (1966), Castellani (1978).

3.3. QUANTIZATION IN $I\!\!E^n$

This rule is often further generalized to quantize observables of the form a product $\xi(x)\eta(p)$ of a function of x and a function p as

$$\xi(x)\eta(p) \to \mathcal{Q}_{0S}(\xi(x)\eta(p)) \tag{3.81}$$

$$= \frac{1}{2}\left\{\mathcal{Q}_0(\xi(x))\,\mathcal{Q}_0(\eta(p)) + \mathcal{Q}_0(\eta(p))\,\mathcal{Q}_0(\xi(x))\right\}. \tag{3.82}$$

2. **The Weyl rule:**

$$x^k p^\ell \to \mathcal{Q}_{0W}(x^k p^\ell) = \frac{1}{2^k}\sum_{j=0}^{k} C_j^k \left(\mathcal{Q}_0(x)^{k-j}\,\mathcal{Q}_0(p)^\ell\,\mathcal{Q}_0(x)^j\right), \tag{3.83}$$

where C_j^k are the binomial coefficients.

3. **The Born-Jordan rule:**

$$x^k p^\ell \to \mathcal{Q}_{0BJ}(x^k p^\ell) = \frac{1}{(1+\ell)}\sum_{j=0}^{\ell} \mathcal{Q}_0(p)^{\ell-j}\,\mathcal{Q}_0(x)^k\,\mathcal{Q}_0(p)^j. \tag{3.84}$$

As seen in E2.5.1(1) E5, these rules are simply too crude to be able to achieve essential selfadjointness for $\mathcal{Q}_0(x^k p^\ell)$ on $C_0^\infty(I\!\!E)$ even in some very simple cases.

Apart from algebraic and pragmatic schemes there has been a lot of interest in some mathematically very sophisticated geometric and other methods of quantization, evident by the publication of a number of monographs on quantization with an aim of establishing some general quantization rules.[38] Unfortunately these schemes also fail to lead to an explicit and generally acceptable quantization map.

We shall be less ambitious here, i.e., we shall single out certain subsets of $\mathcal{O}_{cl}$ and quantize some of the observables in these subsets, on a need to know basis. We shall adopt an explicit geometric method of quantization. The linearity and the function preservation rules will be employed only in some specifc cases. Many of the results obtained will be applied to concrete physical systems in later chapters.

3.3.3 Complete momentum observables

For a classical point particle moving in $I\!\!E^n$ a momentum is an observable of the form

$$P = \sum_{j=1}^{n} \xi^j p_j, \tag{3.85}$$

---

[38] Woodhouse (1986), Śniatycki (1980), Hurt (1983), Namiki (1992), Landsman (1998).

where ξ^j are differentiable functions of x^j. The components of angular momentum along the three coordinate axes and the radial momentum are prime examples. Physically a momentum observable is to do with spatial movements. The discussions in §1.6.4 tell us that momenta can be divided into two types, complete ones and incomplete ones. A complete momentum is the generator of a one-parameter group of transformations of $I\!\!E^n$ corresponding to translations along the integral curves of its Hamiltonian vector field in $I\!\!E^n$ given by Eq. (1.305), while an incomplete momentum does not generate a one-parameter group of transformations. There is a fundamental difference between these two types of momenta when it comes to quantization.

In this section we shall consider complete momenta. The Hilbert space for the quantized particle is chosen to be $L^2(I\!\!E^n)$. Our task is to establish a unique selfadjoint operator in $L^2(I\!\!E^n)$ to represent the quantum counterpart of a given complete momentum P. The fact that P is a generator of a one-parameter group of transformations of $I\!\!E^n$ plays a crucial roles here since spatial transformations directly affect the wave function. The effect on the wave function due to a one-parameter group of spatial transformations can be realized by a one-parameter group of unitary operators acting on the wave function, a fact well-known in the studies of group representations.[39] The idea is then to relate the generator of the unitary group in $L^2(I\!\!E^n)$ to P.[40] There are two cases:

Case 1 Incompressible momentum observables   Intuitively one can imagine that a translation of points in the coordinate space $I\!\!E^n$ to the right carries a wave function in $L^2(I\!\!E^n)$ to the right as well.[41] For simplicity let us consider a one-dimensional case for which the coordinate space is $I\!\!E$. We know from E1.6.4(1) E1 and C1.3.3(2) C2 that momentum $P = p$ is a generator of a one-parameter group $T = \{T_\tau : \tau \in I\!\!R\}$ of translations of $I\!\!E$, with T_τ causing the following translation along Cartesian coordinate x:

$$T_\tau : I\!\!E \to I\!\!E \quad \text{by} \quad T_\tau x = x + \tau. \quad (3.86)$$

A wave function $\phi(x) \in L^2(I\!\!E)$ should be similarly translated, as shown in Figure 3.3.3.

---

[39] Cornwell (1984) §1.3.
[40] MacKey (1963) §2.6, Abraham and Marsden (1978) §5.4.
[41] See Greiner and Müller (1989) pp. 11-12 for a pictorial description.

3.3. QUANTIZATION IN $\mathbb{E}^n$

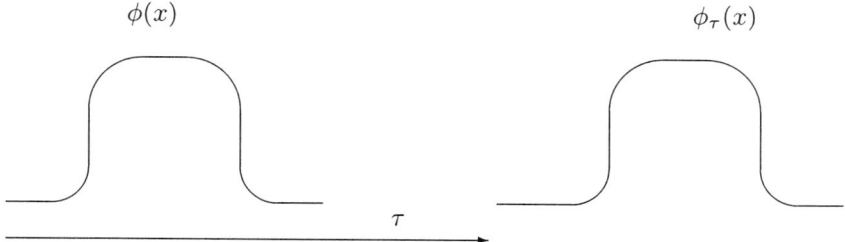

Fig. 3.3.3   Translations of the wave function.

In other words the translated wave function ϕ_τ is given by

$$\phi_\tau(x) = \phi(T_\tau^{-1}x) = \phi(x - \tau). \tag{3.87}$$

One can use these translations to define an operator, i.e., we define an operator $\widehat{T}_\tau$ to act on the wave function by

$$(\widehat{T}_\tau \phi)(x) = \phi_\tau(x) = \phi(T_\tau^{-1}x) \quad \forall \phi \in L^2(\mathbb{E}). \tag{3.88}$$

These translations are volume preserving transformations in that the volume element dx in $\mathbb{E}$ is translated unchanged, i.e., $dx = d(T_\tau x) = d(T_\tau^{-1}x)$ for every T_τ. As a result $\widehat{T}_\tau$ preserves the norm of the wave function since

$$\begin{aligned}
\langle \widehat{T}_\tau \phi \mid \widehat{T}_\tau \phi \rangle &= \int_{-\infty}^{\infty} \left|(\widehat{T}_\tau \phi)(x)\right|^2 dx = \int_{-\infty}^{\infty} \left|\phi(T_\tau^{-1}x)\right|^2 dx \\
&= \int_{-\infty}^{\infty} \left|\phi(T_\tau^{-1}x)\right|^2 d(T_\tau^{-1}x) = \langle \phi \mid \phi \rangle.
\end{aligned} \tag{3.89}$$

It follows that $\widehat{T}_\tau$ are unitary operators. These operators form a unitary representation of the group T.[42] We can check that

$$\begin{aligned}
((\widehat{T}_{\tau_2}\widehat{T}_{\tau_1})\phi)(x) &= \widehat{T}_{\tau_2}\phi(T_{\tau_1}^{-1}x) = \phi(T_{\tau_1}^{-1}T_{\tau_2}^{-1}x) \\
&= \phi((T_{\tau_2}T_{\tau_1})^{-1}x) \tag{3.90} \\
&= \phi((T_{\tau_2+\tau_1})^{-1}x), \tag{3.91}
\end{aligned}$$

---

[42] Cornwell (1984) §1.3, Merzbacher (1998) p. 69.

because $T_{\tau_2+\tau_1} = T_{\tau_2}T_{\tau_1}$. It follows that we can identify $\widehat{T}_{\tau_2}\widehat{T}_{\tau_1}$ with $\widehat{T}_{\tau_2+\tau_1}$, i.e., we have

$$(\widehat{T}_{\tau_2+\tau_1}\phi)(x) = ((\widehat{T}_{\tau_2}\widehat{T}_{\tau_1})\phi)(x). \tag{3.92}$$

Moreover Theorem 2.4(1) (Stone's Theorem) is satisfied so that this one-parameter unitary group is expressible in the form

$$\widehat{T}_\tau = e^{-i\tau\widehat{P}}, \quad i = \frac{i}{\hbar}, \tag{3.93}$$

where $\widehat{P}$ is the selfadjoint generator of the group. This is the group described in C2.4(2) C2 where it is pointed out that the generator $\widehat{P}$ is just the momentum operator $\widehat{p}$ given in E2.5.1(1) E3. Now, if we agree that fundamentally a momentum is a generator of translations along the Cartesian coordinate x in the coordinate space both classically and quantum mechanically, then the natural assumption is to take the quantized momentum $\mathcal{Q}(p)$ as the generator of the above unitary group. This is why for a particle in $\mathbb{E}$ we have the following quantization map:

$$p \quad \to \quad \mathcal{Q}(p) = \widehat{p} \quad \text{in } L^2(\mathbb{E}). \tag{3.94}$$

Generally transformations of $\mathbb{E}^n$ may not be volume preserving so that Eq. (3.88) will have to be modified if we are to produce unitary operators. To avoid this complication at this stage we can divide complete momenta into two different classes. Momentum p is typical of a class known as *incompressible* momenta. Generally a momentum $P = \sum_{j=1}^n \xi^j p_j$ is said to be *incompressible* if its Hamiltonian vector field $\widehat{X}_P$ in $\mathbb{E}^n$ given by Eq. (1.305) is divergence free, i.e., we have, from Eq. (1.141), that

$$\text{div } \widehat{X}_P = \sum_j \frac{\partial \xi_j}{\partial x^j} = 0. \tag{3.95}$$

Otherwise the momentum is said to be *compressible*. The terminology is borrowed from fluid dynamics where a fluid flow is *incompressible* if the divergence of its velocity field vanishes. An important characteristic of an incompressible momentum is that the transformations it generates is volume preserving.[43] We shall denote the transformations caused by translations along the integral curves of $\widehat{X}_P$ again by T_τ. These integral curves are given by the solutions of Eq. (1.81) rather then by Eq. (3.86). We can again define operators $\widehat{T}_\tau$ in terms of the corresponding transformations of $\phi(x) \in L^2(\mathbb{E}^n)$ by Eq. (3.88).

---

[43] Abraham and Marsden (1978) pp. 131-132, Arnold (1978) Theorem 2 on pp. 68-70.

3.3. QUANTIZATION IN $I\!E^n$

The volume-preserving nature of the transformations ensures that $\widehat{T}_\tau$ so defined are unitary, e.g., Eq. (3.89) is satisfied. We can then proceed to quantize P as the unique selfadjoint generator $\widehat{P}$ of the unitary group given by Eq. (2.135). To work out $\widehat{P}$ we shall first verify that $\widehat{P}$ possesses a restriction $\widehat{P}_0$ on the domain $C_0^\infty(I\!E^n)$. From Eq. (2.135) we have, for any $\phi(x) \in C_0^\infty(I\!E^n)$,

$$(\widehat{P}_0\phi)(x) = i\hbar \lim_{\tau \to 0} \frac{\widehat{T}_\tau - \widehat{I\!I}}{\tau}\phi = i\hbar \left.\frac{d\phi(T_\tau^{-1}x)}{d\tau}\right|_{\tau=0} \quad (3.96)$$

$$= -i\hbar(\widehat{X}_P\phi)(x). \quad (3.97)$$

We have used Eq. (1.139) in the final step. This shows that we have established an operator defined by the operator expression $-i\hbar\widehat{X}_P$ acting on $C_0^\infty(I\!E^n)$, i.e., we have

$$\widehat{P}_0 = -i\hbar\widehat{X}_P = -i\hbar \sum_{j=1}^n \xi^j \frac{\partial}{\partial x^j} \quad \text{acting on} \quad C_0^\infty(I\!E^n). \quad (3.98)$$

This operator is symmetric and essentially selfadjoint with a unique selfadjoint extension $\widehat{P}$ defined by the same differential expression acting on the domain[44]

$$\mathcal{D}(\widehat{P}) = \left\{\phi \in L^2(I\!E^n) : \phi \in AC(\widehat{X}_P, I\!E^n),\ \widehat{P}\phi \in L^2(I\!E^n)\right\}. \quad (3.99)$$

Here $AC(\widehat{X}_P, I\!E^n)$ is the set of functions introduced in C2.5.1(2) C3. Our conclusion is that the quantized operator for P is $\widehat{P}$.

For a particle in three dimensions we have the following familiar examples:

1. Canonical momenta p_j conjugate to Cartesian coordinates x^j: These canonical momenta are quantized as selfadjoint operators $\widehat{p}_j = -i\hbar\,\partial/\partial x^j$ in $L^2(I\!E^3)$ acting on a domain given by Eq. (3.99).

2. The z-component angular momentum $L_z = xp_y - yp_x$ in the usual notation in rectangular Cartesian coordinates (x, y, z): This momentum is complete and incompressible; its Hamiltonian vector field $\widehat{X}_{L_z}$ is given by Eq. (1.310). This vector field has a critical point at the origin. The quantized operator is selfadjoint and given by the differential expression:

$$\widehat{L}_z = -i\hbar\widehat{X}_{L_z} = -i\hbar\left(x\frac{\partial}{\partial y} - y\frac{\partial}{\partial x}\right). \quad (3.100)$$

---

[44]Mackey (1963), Abraham and Marsden (1978) Lemma 2.6.13, Proposition 2.6.14 and Theorem 2.6.15 on pp. 141-143. For an explicit proof see Wan and Viazminsky (1979) Appendix I. Knowing that $\widehat{P}_0$ is a restriction of a selfadjoint $\widehat{P}$ one can also use Lemma 2.6.13 of Abraham and Marsden (1978) to effect a proof.

This expression vanishes at the origin. As shown in C1.2.3(1) C3 the z-component angular momentum L_z is identifiable with the momentum p_φ conjugate to the angle variable φ in spherical coordinates. So, p_φ is quantized as $\widehat{p}_\varphi = \widehat{L}_z$ with the operator expression given in spherical coordinates away from the origin by

$$\widehat{p}_\varphi = -i\hbar \frac{\partial}{\partial \varphi}. \tag{3.101}$$

3. Not all the momenta discussed in E1.6.4(1) are included in the quantization scheme developed so far. For example the radial momentum p_r is incomplete. It does not generate a one-parameter group of transformations of the coordinate space. Our present procedure does not apply. We shall return to consider the quantization of incomplete momenta later.

Case 2 Compressible momentum observables For a compressible momentum the volume element is not preserved. Consider a momentum observable P and its associated vector field $\widehat{X}_P$ in E given by

$$P = \xi(x)\, p, \quad \widehat{X}_P = \xi(x)\, d/dx. \tag{3.102}$$

The transformations along the integral curves of the vector field $\widehat{X}_P$ will be denoted again by T_τ. Generally, we have:

$$x \;\to\; T_\tau x, \quad \text{with } T_\tau^{-1} = T_{-\tau}, \tag{3.103}$$

$$d_x(T_\tau x) = J_\tau\, dx, \quad \text{with } J_\tau(x) = \frac{\partial(T_\tau x)}{\partial x}, \tag{3.104}$$

$$d_x(T_{-\tau} x) = J_{-\tau}\, dx, \quad \text{with } J_{-\tau}(x) = \frac{\partial(T_{-\tau} x)}{\partial x}, \tag{3.105}$$

where d_x is a differentiation with respect to x keeping τ constant and $J_\tau(x)$ is commonly referred to as the Jacobian. Note that $T_\tau x$ and hence $J_\tau(x)$ may be regarded as functions of two independent variables x and τ. Since integral curves are smooth we have, for small values of τ,

$$T_\tau x = x + \xi(x)\,\tau + 0^2(\tau), \tag{3.106}$$

where $0^2(\tau)$ denotes terms of order equal to or higher than τ^2. This is consistent with Eqs. (1.129) and (1.130) relating a vector field to its integral curves, i.e., we have

$$\left.\frac{\partial(T_\tau x)}{\partial \tau}\right|_{\tau=0} = \xi(x). \tag{3.107}$$

3.3. QUANTIZATION IN $\mathbb{E}^n$

For the Jacobian we have[45]

$$J_\tau(x) = \frac{\partial}{\partial x}(T_\tau x) = 1 + \xi_{,x}(x)\tau + 0^2(\tau), \qquad (3.108)$$

where

$$\xi_{,x}(x) = \frac{d\xi(x)}{dx}. \qquad (3.109)$$

This leads to an integral expression for $J_\tau(x)$:[46]

$$J_\tau(x) = \exp\left\{\int_0^\tau \xi_{,x}(T_\tau x)\, d\tau\right\} \qquad (3.110)$$

which reduces to Eq. (3.108) as $\tau \to 0$. Comparing with Eq. (3.106) we can see that for small values of τ we have

$$J_{-\tau} = 1 - \xi_{,x}(x)\tau + 0^2(\tau) \qquad (3.111)$$

and

$$J_{-\tau}^{1/2} = 1 - \frac{1}{2}\xi_{,x}(x)\tau + 0^2(\tau). \qquad (3.112)$$

It follows that

$$\frac{\partial J_{-\tau}^{1/2}}{\partial \tau} = -\frac{1}{2}\xi_{,x}(x). \qquad (3.113)$$

There is one more result on the Jacobian we need:

$$J_{-\tau_2}(x) J_{-\tau_1}(T_{\tau_2}^{-1}x)$$

$$= \exp\left\{\int_0^{-\tau_2} \xi_{,x}(T_{\tau_2'}x)\, d\tau_2' + \int_0^{-\tau_1} \xi_{,x}(T_{\tau_1'} T_{\tau_2}^{-1}x)\, d\tau_1'\right\}$$

$$= \exp\left\{\int_0^{-\tau_2} \xi_{,x}(T_{\tau_2'}x)\, d\tau_2' + \int_0^{-\tau_1} \xi_{,x}(T_{\tau_1'-\tau_2}x)\, d\tau_1'\right\}$$

$$= \exp\left\{\int_0^{-\tau_2} \xi_{,x}(T_\tau x)\, d\tau + \int_{-\tau_2}^{-(\tau_1+\tau_2)} \xi_{,x}(T_\tau x)\, d\tau\right\}$$

$$= \exp\left\{\int_0^{-(\tau_2+\tau_1)} \xi_{,x}(T_\tau x)\, d\tau\right\}$$

$$= J_{-(\tau_2+\tau_1)}(x). \qquad (3.114)$$

---

[45] As in Eq. (3.106) the symbol $0^2(\tau)$ in Eq. (3.108) denotes terms of order equal to or higher than τ^2.

[46] Arnold (1978) pp. 68-70, Anosov and Arnold (1988) p. 22, Alanson (1992), Kowalski (1994) pp. 23-27.

266   CHAPTER 3. ORTHODOX QUANTUM MECHANICS

Now let us define an operator $\widehat{T}_\tau$ on $L^2(I\!\!E)$ to associate with each T_τ by[47]

$$\left(\widehat{T}_\tau \phi\right)(x) = \phi(T_\tau^{-1}x) J_{-\tau}^{1/2}(x) \tag{3.115}$$

$$= \phi(T_\tau^{-1}x) \exp\left\{\frac{1}{2}\int_0^{-\tau} \xi_{,x}(T_\tau x)\, d\tau\right\}. \tag{3.116}$$

Using the fact that $T_\tau^{-1} = T_{-\tau}$ pointed out in C1.3.3(2) C1 we finally obtain

$$\langle \widehat{T}_\tau \phi \mid \widehat{T}_\tau \phi \rangle = \int_{-\infty}^{\infty} \left|\phi(T_\tau^{-1}x)\right|^2 J_{-\tau}\, dx \tag{3.117}$$

$$= \int_{-\infty}^{\infty} \left|\phi(T_\tau^{-1}x)\right|^2 d(T_{-\tau}x) \tag{3.118}$$

$$= \int_{-\infty}^{\infty} \left|\phi(T_\tau^{-1}x)\right|^2 d(T_\tau^{-1}x) \tag{3.119}$$

$$= \langle \phi \mid \phi \rangle. \tag{3.120}$$

This result renders $\widehat{T}_\tau$ unitary. Using Eq. (3.114) above we can verify that $\widehat{T}_\tau$ form a group, giving a unitary representation of the transformation group $T = \{T_\tau : \tau \in I\!\!R\}$, e.g.,

$$(\widehat{T}_{\tau_2}\widehat{T}_{\tau_1}\phi)(x) = \widehat{T}_{\tau_2}\left(\phi(T_{\tau_1}^{-1}x) J_{-\tau_1}^{1/2}(x)\right) \tag{3.121}$$

$$= \phi(T_{\tau_1}^{-1}T_{\tau_2}^{-1}x) J_{-\tau_1}^{1/2}(T_{\tau_2}^{-1}x) J_{-\tau_2}^{1/2}(x) \tag{3.122}$$

$$= \phi(T_{\tau_2+\tau_1}^{-1}x) J_{-(\tau_2+\tau_1)}^{1/2}(x) \tag{3.123}$$

$$= (\widehat{T}_{\tau_2+\tau_1}\phi)(x). \tag{3.124}$$

The generator of this unitary group can be obtained as before, i.e., when acting on $\phi \in C_0^\infty(I\!\!E)$ we have

$$(\widehat{P}_0 \phi)(x) = i\hbar \lim_{\tau \to 0} \frac{\widehat{T}_\tau - \widehat{I\!\!I}}{\tau} \phi = i\hbar \left.\frac{\partial}{\partial \tau}\left(\phi(T_\tau^{-1}x) J_{-\tau}^{1/2}\right)\right|_{\tau=0} \tag{3.125}$$

$$= i\hbar \left.\left(\frac{\partial \phi(T_\tau^{-1}x)}{\partial \tau} J_{-\tau}^{1/2} + \phi(T_\tau^{-1}x)\frac{\partial J_{-\tau}^{1/2}}{\partial \tau}\right)\right|_{\tau=0} \tag{3.126}$$

$$= -i\hbar \left(\widehat{X}_P \phi + \frac{1}{2}\frac{d\xi}{dx}\phi\right)(x), \tag{3.127}$$

or

$$\widehat{P}_0 = -i\hbar \left(\widehat{X}_P + \frac{1}{2}\frac{d\xi}{dx}\right) \quad \text{acting on domain} \quad C_0^\infty(I\!\!E). \tag{3.128}$$

---
[47]Alanson (1992), Kowalski (1994) pp. 23-27.

3.3. QUANTIZATION IN $\mathbb{E}^n$

This is the initially quantized momentum operator. As in the incompressible case this operator is essentially selfadjoint. The quantized operator $\mathcal{Q}(P)$ is the unique selfadjoint extension of $\widehat{P}_0$. This result agrees formally with the symmetrization rule of quantization, i.e.,

$$\mathcal{Q}\left(\sum_{j=1}^{n} \xi^j p_j\right) = \frac{1}{2}\sum_{j=1}^{n}\Big(\mathcal{Q}(\xi^j)\mathcal{Q}(p_j) + \mathcal{Q}(p_j)\mathcal{Q}(\xi^j p_j)\Big). \tag{3.129}$$

Of course the symmetrization rule itself provides no information on the selfadjointness or otherwise of the resulting operator.

To highlight the compressible nature of P we can make use of Eq. (1.141) for the divergence of a vector field to rewrite $\widehat{P}_0$ as

$$\widehat{P}_0 = -i\hbar\left(\widehat{X}_P + \frac{1}{2}\operatorname{div}\widehat{X}_P\right). \tag{3.130}$$

It is straightforward to quantize a complete momentum

$$P = \sum_{j=1}^{n} \xi^j(x)\, p_j \tag{3.131}$$

of a particle in $\mathbb{E}^n$. The corresponding vector field is

$$\widehat{X}_P = \sum_{j=1}^{n} \xi^j\, \partial/\partial x^j. \tag{3.132}$$

The Jacobian determinant $J_{-\tau}(x)$ is composed of elements $\partial(T_\tau^{-1}x)^j/\partial x^k$. We have the following approximations for small values of τ:

$$\left(T_\tau^{-1}x\right)^j = x^j - \xi^j\,\tau + 0^2(\tau) \tag{3.133}$$

and

$$\frac{\partial\left(T_\tau^{-1}x\right)^j}{\partial x^k} = \delta_{jk} - \frac{\partial \xi^j}{\partial x^k}\tau + 0^2(\tau) \tag{3.134}$$

with the Jacobian determinant J_τ expressible as

$$J_{-\tau} = 1 - \sum_j \frac{\partial \xi^j}{\partial x^j}\tau + 0^2(\tau), \tag{3.135}$$

$$J_{-\tau}^{1/2} = 1 - \frac{1}{2}\sum_j \frac{\partial \xi^j}{\partial x^j}\tau + 0^2(\tau) \tag{3.136}$$

$$= 1 - \frac{1}{2}\left(\operatorname{div}\widehat{X}_P\right)\tau + 0^2(\tau). \tag{3.137}$$

We can define an operator $\widehat{T}_\tau$ on $L^2(I\!\!E^n)$ to associate with each T_τ by

$$\left(\widehat{T}_\tau \phi\right)(x) = \phi(T_\tau^{-1} x) \, J_{-\tau}^{1/2}(x) \tag{3.138}$$

$$= \phi(T_\tau^{-1} x) \exp\left\{\frac{1}{2} \int_0^{-\tau} (\operatorname{div} \widehat{X}_P)(T_\tau x) \, d\tau\right\}. \tag{3.139}$$

Following previous procedure we can obtain the expression for the generator of the unitary group, i.e., when acting on $C_0^\infty(I\!\!E^n)$ we get

$$\widehat{P}_0 = -i\hbar \left(\widehat{X}_P + \frac{1}{2} \sum_{j=1}^n \frac{\partial \xi^j}{\partial x^j}\right) \tag{3.140}$$

as the initially quantized momentum operator. Its unique selfadjoint extension $\widehat{P}$ has the same operator expression. In other words we have

$$\widehat{P} = -i\hbar \left(\widehat{X}_P + \frac{1}{2} \operatorname{div} \widehat{X}_P\right) \tag{3.141}$$

acting on the domain

$$\mathcal{D}(\widehat{P}) = \left\{\phi \in L^2(I\!\!E^n) : \phi \in AC(\widehat{X}_P, I\!\!E^n), \, \widehat{P}\phi \in L^2(I\!\!E^n)\right\}. \tag{3.142}$$

The expression for $\widehat{P}$ reduces to Eq. (3.98) if $\operatorname{div} \widehat{X}_P = 0$. Naturally we shall identify $\widehat{P}$ with the quantized momentum $\mathcal{Q}(P)$.

As examples let us consider:

1. Momentum $P = xp$ in $I\!\!E$: The associated vector field $\widehat{X}_P = x \, d/dx$ is complete.[48] This momentum is clearly compressible. Our quantization procedure produces an essentially selfadjoint operator

$$Q_0(xp) = -i\hbar \left(x \frac{d}{dx} + \frac{1}{2}\right) \tag{3.143}$$

   acting on $C_0^\infty(I\!\!E)$. The quantized operator $Q(xp)$ is the selfadjoint extension of $Q_0(xp)$.

2. Local momentum $P = \xi(x)p$: Here $\xi = \xi(x)$ is a smooth function of compact support in $I\!\!E$. This is called a *local momentum* since it vanishes outside the support of $\xi(x)$. The associated vector field $\widehat{X}_P = \xi \, d/dx$ is complete.[49] These momenta are therefore complete and also compressible. Their selfadjoint quantized counterparts are given by Eq. (3.128).

---

[48] See E1.3.2(1) E6.
[49] See E1.3.2(1) E5.

3.3. QUANTIZATION IN $\mathbb{E}^n$

These quantized operators agree with local momentum operators shown in Eq. (2.549). It is not too difficult to work out the spectral functions of these operators.[50]

The present quantization scheme is based on the intrinsic meaning of momenta as generators for transformations both classically and quantum mechanically, as evident by the link of both P and $\widehat{P}$ to the vector field $\widehat{X}_P$.

Finally we should point out that Eq. (3.141) is not explicitly coordinate dependent and is applicable in curvilinear coordinates x'^j in $\mathbb{E}^n$. In terms of x'^j there corresponds to a metric g'_{ij} given by Eq. (1.248). We have

$$\widehat{X}_P = \sum_j \xi'^j \frac{\partial}{\partial x'^j}, \quad \text{where } \xi'^j \text{ is related to } \xi^j \text{ by Eq. (1.71).} \quad (3.144)$$

The divergence in coordinates x'^j is known to be given by[51]

$$\operatorname{div} \widehat{X}_P = \frac{1}{\sqrt{g'}} \sum_j \frac{\partial(\sqrt{g'}\, \xi'^j)}{\partial x'^j}, \quad (3.145)$$

where g' is the determinant formed by g'_{ij}. Explicity we have the operator expression[52]

$$\widehat{P} = -i\hbar \left(\sum_j \xi'^j \frac{\partial}{\partial x'^j} + \frac{1}{2} \frac{1}{\sqrt{g'}} \sum_j \frac{\partial(\sqrt{g'}\, \xi'^j)}{\partial x'^j} \right). \quad (3.146)$$

3.3.4 Observables linear in momenta

Consider observables of the form

$$\mathcal{P} = P + f(x), \quad P = \sum_{j=1}^n \xi^j p_j \quad (3.147)$$

where P is a complete momentum and f is a real-valued function on $\mathbb{E}^n$. Then the following operator

$$\widehat{\mathcal{P}} = \widehat{P} + f(x) = -i\hbar \left(\widehat{X}_P + \frac{1}{2} \operatorname{div} \widehat{X}_P \right) + f(x) \quad (3.148)$$

acting on the domain

$$\mathcal{D}(\widehat{\mathcal{P}}) = \left\{ \phi \in AC(\widehat{X}_P, \mathbb{E}^n),\, \widehat{\mathcal{P}}\phi \in L^2(\mathbb{E}^n) \right\} \quad (3.149)$$

---

[50] Wan, Jackson and McKenna (1984).
[51] Martin (1991) p. 200.
[52] This expression can be applied to quantization in curved space. See Mackey (1963), Wan and Viazminsky (1977).

is selfadjoint in $L^2(I\!\!E^n)$.

We can summarize the results obtained so far in terms of a quantization axiom

Quantization axiom for complete momenta

- A complete momentum P is quantized as the selfadjoint generator $\widehat{P}$, given by Eqs. (3.141) and (3.142), of the one-parameter unitary group representing the one-parameter group of transformations of $I\!\!E^n$ generated by P.

- An observable $\mathcal{P} = P + f(x)$ linear in a complete momentum P is quantized as the operator $\widehat{\mathcal{P}}$ given by Eqs. (3.148) and (3.149).

Note that $\widehat{\mathcal{P}}$ and $\widehat{P}$ are unitarily related. To see this let $\mathcal{F}(x)$ be a solution of the equation
$$\widehat{X}_P \, \mathcal{F}(x) = f(x). \tag{3.150}$$
Then we can verify that[53]
$$\widehat{\mathcal{P}} = \widehat{U}_f \, \widehat{P} \, \widehat{U}_f^\dagger, \quad \widehat{U}_f = e^{-i\mathcal{F}(x)}. \tag{3.151}$$

Before proceeding further we would make two comments, one on the linearity rule and the second on the canonical quantization scheme:

1. The quantization scheme for $\mathcal{P}$ does not justify the general application of the linearity rule of quantization since
$$\widehat{\mathcal{P}} = \mathcal{Q}\big(P + f(x)\big) = \widehat{P} + f(x) = \mathcal{Q}(P) + \mathcal{Q}(f) \tag{3.152}$$
is just a fortuitous result. Let us test the validity or otherwise of the linearity rule in quantizing momenta. Let
$$P_1 = \sum_{j=1}^n \xi_1^j p_j \quad \text{and} \quad P_2 = \sum_{j=1}^n \xi_2^j p_j \tag{3.153}$$

---

[53] We have assumed the existence of a global solution $\mathcal{F}(x)$, i.e., a function $\mathcal{F}(x)$ defined for all $x \in I\!\!E^n$. For the case of $\mathcal{P} = p + f(x)$ in $I\!\!E$ we can see that
$$\mathcal{F}(x) = \int^x f(x)dx, \quad \widehat{U}_f = \exp\left(-i \int^x f(x)dx\right).$$
For more details see Wan and McFarlane (1983).

3.3. QUANTIZATION IN $\mathbb{E}^n$

be two complete momenta so that they can be separately quantized as $\mathcal{Q}(P_1)$ and $\mathcal{Q}(P_2)$. The question arises as to how one would quantize the sum

$$P = P_1 + P_2 = \sum_{j=1}^{n} \left(\xi_1^j + \xi_2^j \right) p_j \qquad (3.154)$$

which is again a momentum. The answer depends on whether the momentum P is complete or not. If so, we can quantize P as before as the unique selfadjoint operator $Q(P)$. When acting on $C_0^\infty(\mathbb{E}^n)$ its restriction $Q_0(P)$ is given by the following operator expression:

$$\mathcal{Q}_0(P) = -i\hbar \sum_{j=1}^{n} \left((\xi_1^j + \xi_2^j) \frac{\partial}{\partial x^j} + \frac{1}{2} \frac{\partial(\xi_1^j + \xi_2^j)}{\partial x^j} \right) \qquad (3.155)$$

$$= \mathcal{Q}_0(P_1) + \mathcal{Q}_0(P_2). \qquad (3.156)$$

Unfortunately as demonstrated in E1.3.2(1) E7 the sum of two complete vector fields, and hence their associated momenta, is not necessarily complete. It follows that $Q_0(P)$ is not necessarily essentially selfadjoint. This is reflected also in the fact that the sum of two selfadjoint operators is not necessarily selfadjoint. This clearly shows up the reason for the general failure of the linearity rule.

2. The widely adopted *canonical quantization scheme* is based on the idea that the quantized position variables $\mathcal{Q}(x^j)$ and their canonically conjugate momenta $\mathcal{Q}(p_k)$ satisfy the Lie algebra rule

$$[\mathcal{Q}(x^j), \mathcal{Q}(p_k)] = i\hbar\, \delta_{jk} \qquad (3.157)$$

so as to correspond to the classical Poisson brackets $\{x^j, p_k\} = \delta_{jk}$. Clearly

$$\mathcal{Q}(x^j) = x^j, \quad \mathcal{Q}(p_k) = -i\hbar\, \partial/\partial x^k \qquad (3.158)$$

formally obey the above commutation relations. However,

$$\mathcal{Q}(x^j) = x^j, \quad \mathcal{Q}(p_k) = -i\hbar\, \partial/\partial x^j + f(x) \qquad (3.159)$$

do also, showing that a method based just on canonical commutation relations does not lead to a unique quantization. However, in view of Eq. (3.151) we can say that this canonical scheme does lead to a quantization unique up to a unitary transformation.[54] Everything seems to

---

[54]This is an example of von Neumann's theorem on the representations of canonical commutation relations (Prugovečki (1981) p. 342, Jauch (1968) §12.3, Reed and Simon Vol. 1 (1972) pp. 274-275). Classically the canonical momentum conjugate to coordinate variable x is not unique. As pointed out in E1.6.3(1) E2 observable $p' = p + f(x)$ is also a possible momentum canonically conjugate to x. One would expect the corresponding quantized pairs to be related by a unitary transformation.

work well for global rectangular Cartesian coordinates and their conjugate momenta in $I\!E^n$. The question then arises as to whether the same would apply to the quantization of momenta conjugate to non-rectangular Cartesian coordinates. The answer turns out to be in the negative. One can appreciate that a formal operator representation of the commutation relations gives no information on the selfadjointness or otherwise of the operators. So, generally when we deal with (1) non-rectangular Cartesian coordinates and their conjugate momenta, and (2) coordinate spaces other than $I\!E^n$, e.g., $I\!E^+$, the commutation relations may not admit any representation in terms of selfadjoint operators because the canonical momenta involved may not be complete. The best known example is the radial momentum. Explicit examples to demonstrate this conclusion will be given in E3.3.5(1) later after a discussion of the quantization of incomplete momenta.

3.3.5 Incomplete momentum observables

An incomplete momentum does not generate a one-parameter group of transformations of $I\!E^n$. Consequently Eq. (3.115) does not lead to a unitary group of operators, and hence no unique selfadjoint generator can be singled out to associate with the momentum through Stone's theorem. It follows that the idea embodied in the quantization axiom for complete momenta stated in the preceding section is not applicable to incomplete momenta.

However, something can be salvaged. Using the incomplete Hamiltonian vector field asociated with the momentum we can still employ Eq. (3.140) to construct an operator $\widehat{P}_0$ to act on $C_0^\infty(I\!E^n)$. This operator is merely symmetric and not essentially selfadjoint. This being so $\widehat{P}_0$ would admit many selfadjoint or strictly maximal symmetric extensions. We have the rather awkward situation of having selfadjointness without uniqueness or no selfadjointness at all. As a result incomplete momenta are often regarded as unquantizable and ignored in orthodox quantum mechanics, although physically it is hard to see why they should be ignored in quantum theory.

In order to incorporate incomplete momenta such as the radial momentum into quantum theory we really need to generalize orthodox quantum mechanics embodied in Postulate OQS in §3.2. In Chapter 5 we shall present a generalized formulation of quantum mechanics which utilizes strictly maximal symmetric as well as selfadjoint operators to represent physical observables. We shall also demonstrate the adaptability and usefulness of such a theory by applying it to a number of physical systems. Incomplete momenta will be seen to play an important role in these quantum systems. Even the non-uniqueness of quantization turns out to be a blessing and of fundamental importance since a larger class of physical systems and observables can be accommodated within

3.3. QUANTIZATION IN $\mathbb{E}^n$

the theory, a situation which we shall demonstrate in Chapters 6 and 7.

Here we shall illustrate the non-uniqueness and non-selfadjointness in the quantization of incomplete momenta with a few examples.

Examples 3.3.5(1)   Non-selfadjointness and non-uniqueness in quantization

E1   In $\mathbb{E}$ consider classical momenta of the form $P = x^k p, k = 2, 3, \ldots$. For such simple cases the symmetrization rule, the Weyl rule, the Born-Jordan rule and Eq. (3.140) all produce the same formal differential expression for the initially quantized operator $\widehat{P}_0$, i.e.,

$$\widehat{P}_0 = \mathcal{Q}_0(x^k p) = -i\hbar \left(x^k \frac{d}{dx} + \frac{k}{2} x^{k-1} \right) \qquad (3.160)$$

acting on $C_0^\infty(\mathbb{E})$. There are two cases to consider:

1. **Selfadjointness without uniqueness**   Let k be an even integer, i.e., $k = 2, 4, \ldots$. The deficiency indices of $\widehat{P}_0$ have been shown to be (1,1) in E2.5.1(1) E5. It follows that $\widehat{P}_0$ is not essentially selfadjoint in $L^2(\mathbb{E})$. It has a one-parameter family of selfadjoint extensions.[55]

2. **Uniqueness without selfadjointness**   Let k be an odd integer, i.e., $k = 3, 5, \ldots$. Then the corresponding operator $\widehat{P}_0$ has deficiency indices (0,2). It follows that $\widehat{P}_0$ is symmetric with no selfadjoint extension in $L^2(\mathbb{E})$. However, it does have a unique strictly maximal symmetric extension, as pointed out in E2.5.1(1) E5.

E2   Recall the radial momentum of a particle in $\mathbb{E}^3$ discussed in C1.2.3(1) C3 and E1.6.4(1) E4. It is defined everywhere except at the origin by

$$p_r = \frac{x}{r} p_x + \frac{y}{r} p_y + \frac{z}{r} p_z, \quad r = \sqrt{x^2 + y^2 + z^2}. \qquad (3.161)$$

Its Hamiltonian vector field $\widehat{X}_{p_r}$ is given in Cartesian coordinates by Eq. (1.315) and in spherical coordinates by Eq. (1.316). We may attempt to quantize p_r by employing Eq. (3.141). We obtain the initially quantized operator

$$\begin{aligned}\widehat{p}_{r0} &= -i\hbar \left(\widehat{X}_{p_r} + \frac{1}{2} \operatorname{div} \widehat{X}_{p_r} \right) & (3.162) \\ &= -i\hbar \left(\frac{x}{r} \frac{\partial}{\partial x} + \frac{y}{r} \frac{\partial}{\partial y} + \frac{z}{r} \frac{\partial}{\partial z} + \frac{1}{r} \right), \quad r > 0 & (3.163)\end{aligned}$$

---

[55] Wan and Sumner (1991), Zhu and Klauder (1993a), (1993b).

acting on $C_0^\infty(I\!\!E^3)$. In the usual spherical coordinates (r, θ, φ) the metric g_{jk} consists of diagonal elements $1, r^2$ and $r^2 \sin^2 \theta$. We can employ Eq. (3.146) to obtain an expression for $\widehat{p}_{r0}$ in spherical coordinates:

$$\widehat{p}_{r0} = -i\hbar \left(\frac{\partial}{\partial r} + \frac{1}{r} \right), \quad r > 0. \tag{3.164}$$

Since p_r is not complete, $\widehat{p}_{r0}$ is not essentially selfadjoint in $L^2(I\!\!E^3)$. Discussions in E2.10.2(1) E2 tell us that $\widehat{p}_{r0}$ possesses no selfadjoint extension. Instead it has a unique strictly maximal symmetric extension $\widehat{p}_r$ defined by Eqs. (2.456) and (2.457). For this reason the radial momentum is regarded as not a quantum mechanical observable and is often simply ignored in orthodox quantum mechanics.[56]

E3 The situation above also applies when we consider a particle constrained to move along the half line $I\!\!E^+ = (0, \infty)$. The Hilbert space is $L^2(I\!\!E^+)$. The linear momentum p^+ conjugate to the position x of a particle confined to $I\!\!E^+ = (0, \infty)$ is incomplete since its Hamiltonian vector field is not complete, a fact pointed in E1.3.2(1) E8. As shown in E2.5.1(1) E6 the initially quantized operator $\widehat{p}_0^+(I\!\!E^+)$ acting on $C_0^\infty(I\!\!E^+)$ is not essentially selfadjoint and it has no selfadjoint extension. Instead it possesses a unique strictly maximal symmetric extension $\widehat{p}^+$. We shall see that this operator will have important physical applications in the context of a generalized quantum mechanics to be presented in Chapters 6 and 7.

E4 Another closely related quantity is the canonical momentum p_θ conjugate to the angle variable θ in spherical coordinates. This observable is introduced in C1.2.3(1) C3. Its Hamiltonian vector field $\widehat{X}_{p_\theta}$ is shown to be incomplete in E1.6.4(1) E3. In an attempt to quantize p_θ in Cartesian coordinates we may use Eq. (3.141) we get

$$\widehat{p}_{\theta 0} = -i\hbar \left(\widehat{X}_{p_\theta} + \frac{1}{2} \operatorname{div} \widehat{X}_{p_\theta} \right) \tag{3.165}$$

$$= -i\hbar \left(\widehat{X}_{p_\theta} + \frac{z}{2\sqrt{x^2 + y^2}} \right) \tag{3.166}$$

as the initially quantized operator. In spherical coordinates this expression becomes

$$\widehat{p}_{\theta 0} = -i\hbar \left(\frac{\partial}{\partial \theta} + \frac{1}{2} \cot \theta \right), \quad \theta \in (0, \pi). \tag{3.167}$$

---

[56]Messiah Vol. 1 (1967) p. 346, Dicke and Wittke (1960) p. 143, De Lange and Raab (1991) footnote on p. 359, Liboff (1980) pp. 366-368. It is possible to construct selfadjoint *local radial momentum operators* (Wan, Jackson and McKenna (1984)).

The same result is obtained using Eq. (3.146). This operator expression admits many different selfadjoint extensions in $L^2(I\!\!E^3)$.[57]

E5   A simple example discussed in many quantum mechanics textbooks involves a particle of mass m confined to a finite interval $\Lambda = (0, a)$ in $I\!\!E$ by an *infinite square potential well*. The Hilbert space for the system is $L^2(\Lambda)$ since the particle is permanently confined to Λ. A serious problem emerges when one enquires what the momentum operator for the particle is. The classical momentum p conjugate to $x \in \Lambda$ is incomplete since we cannot have translations along x regidly confined Λ, let alone any group structure for translations within the interval Λ. It follows that the initially quantized operator

$$\widehat{p}_0(\Lambda) = -i\hbar d/dx \quad \text{acting on} \quad C_0^\infty(\Lambda) \in L^2(\Lambda) \tag{3.168}$$

is not essentially selfadjoint. Discussions in E2.5.1(1) E1 show that this operator has a one-parameter family of selfadjoint extensions $\widehat{p}_\lambda(\Lambda)$ arising from a quasi-periodic boundary condition. We shall see later in Chapters 6 and 7 again that this kind of non-uniqueness has important physical applications.

E6   The operators $\widehat{p}_{0r}$ and $\widehat{p}_{0\theta}$ together with their respective coordinates r and θ do formally satisfy the canonical commutation relations on a restricted domain. The same is true for $\widehat{p}_\lambda(\Lambda)$ and x of a particle in an infinite square potential well. Once again this shows that as a quantization rule, formal canonical commutation relations do not provide sufficient information on selfadjointness or uniqueness.

3.3.6   Kinetic energy and the Hamiltonian

The above geometric quantization scheme, appealing as it may be, applies directly only to momentum observables since these observables generates a vector field in $I\!\!E^n$. It is tempting to generalize this to quantize an arbitrary classical observable $A(x, p)$, e.g., quadratic functions in p^j. There are two obvious ways to proceed:

1. Quantization in canonical coordinate space: The idea is to carry out a (classical) canonical transformation to new canonical variables $\bar{x}^j, \bar{p}^j$ such that $A(x, p)$ becomes a canonical momentum observable, i.e., $A = \sum_{j=1}^n \bar{\xi}^j(\bar{x})\bar{p}_j$, in the new canonical coordinate system in the phase space Γ. The idea is to use the new canonical variables $\bar{x}^j$ as "position variables" to construct a Hilbert space $\bar{\mathcal{H}}$ of square-integrable functions of $\bar{x}^j$. We then consider A as a momentum, and construct an operator for A in $\bar{\mathcal{H}}$ in a similar fashion as we have done before. Finally

---
[57] Wan and Viazminski (1977).

we require a procedure to establish a natural unitary map to relate $\bar{\mathcal{H}}$ to the usual Hilbert space $L^2(I\!\!E^n)$ in the coordinate representation.[58]

2. Quantization in phase space: The idea is to set up a Hilbert space $L^2(\Gamma)$ of functions on the phase space Γ square-integrable with respect to the volume element $dxdp$ of Γ. Any smooth function $A(x,p)$ would generate a Hamiltonian vector field in Γ which in turn will give rise to transformations of Γ. We can then construct operators in $L^2(\Gamma)$ to correspond to $A(x,p)$. This is known as *pre-quantization*. The final step would be to relate or to project $L^2(\Gamma)$ to $L^2(I\!\!E^n)$ in such a way as to obtain the corresponding operators in $L^2(I\!\!E^n)$.

Unfortunately both these schemes do not work well generally. There appears to be no universal scheme to obtain a desired mapping to relate $\bar{\mathcal{H}}$ or $L^2(\Gamma)$ to $L^2(I\!\!E^n)$ applicable to an arbitrary function $A(x,p)$.[59] Nevertheless these ideas prove to be successful in a limit number of physically important cases which we shall now discuss. First consider classical observables of the form

$$A(x,p) = \xi(p)x \tag{3.169}$$

for a particle in motion in $I\!\!E$. This is not in the form of a momentum so that our previous quantization procedure does not apply. Recall the canonical transformation in Eq. (1.307), i.e.,

$$x, p \quad \to \quad \bar{x} = -p, \quad \bar{p} = x. \tag{3.170}$$

As in C1.5.3(1) C2 and E2.4(1) E1 we can take the new coordinate variable $\bar{x}$ to form a geometric space. This space can be identified with the momentum space $\widetilde{I\!\!E}$ introduced in E2.4.1 E1. We can rewrite $A(x,p)$ as $\bar{A}(\bar{x},\bar{p})$ with

$$\bar{A}(\bar{x},\bar{p}) = \xi(-\bar{x})\bar{p}. \tag{3.171}$$

This new function function $\bar{A}(\bar{x},\bar{p})$ now appears in the form of a momentum observable in the new coordinate system. The Hamiltonian vector field generated by $\bar{A}(\bar{x},\bar{p})$ in phase space Γ is given, according to Eq. (1.271), by

$$\begin{aligned}\widehat{\mathcal{X}}_{\bar{A}} &= \frac{\partial \bar{A}(\bar{x},\bar{p})}{\partial \bar{p}}\frac{\partial}{\partial \bar{x}} - \frac{\partial \bar{A}(\bar{x},\bar{p})}{\partial \bar{x}}\frac{\partial}{\partial \bar{p}} \\ &= \xi(-\bar{x})\frac{\partial}{\partial \bar{x}} - \frac{\partial \xi(-\bar{x})}{\partial \bar{x}}\bar{p}\frac{\partial}{\partial \bar{p}}.\end{aligned}$$

---

[58] Wan, McKenna and Pinto (1984a), (1984b).
[59] Some mathematically very sophisticated procedure, known as pairing, constructed in geometric quantization scheme (Woodhouse (1986)) also fails to be generally applicable.

3.3. QUANTIZATION IN $I\!\!E^n$

Much the same as a momentum $P = \xi(x)p$ generates a vector field in the coordinate space $I\!\!E$ in accordance with Eq. (1.305) this vector field $\widehat{\mathcal{X}}_{\bar{A}}$ can be projected down onto the space $\widetilde{I\!\!E}$ formed by $\bar{x}$, i.e., $\bar{A}$ generates a vector field

$$\widetilde{X}_{\bar{A}} = \xi(-\bar{x})\frac{d}{d\bar{x}} \tag{3.172}$$

in $\widetilde{I\!\!E}$. We can also similarly introduce a definition of completeness of $\widetilde{X}_{\bar{A}}$

Let us now quantize $\bar{A}(\bar{x},\bar{p})$ as a momentum observable in the new canonical coordinate $\bar{x}$ in the same way as we do for a momentum observable in the original Cartesian coodinate system x:

1. We first set up a Hilbert space $\widetilde{L}^2(\widetilde{I\!\!E})$ consisting of square-integrable functions $\widetilde{\varphi}(\bar{x})$ on $\widetilde{I\!\!E}$ with respect to the volume element $d\bar{x}$.

2. Then quantize $\bar{A}$ as a differential operator in $\widetilde{L}^2(\widetilde{I\!\!E})$. Let us assume that $\bar{A}$ is a complete momentum in $\widetilde{I\!\!E}$. Then the quantized operator $\widetilde{A}$ is selfadjoint in $\widetilde{L}^2(\widetilde{I\!\!E})$ and we have

$$\widetilde{A} = -i\hbar\left(\widetilde{X}_{\bar{A}} + \frac{1}{2}\frac{d\xi(-\bar{x})}{d\bar{x}}\right) \tag{3.173}$$

$$= -i\hbar\left(\xi(-\bar{x})\frac{d}{d\bar{x}} + \frac{1}{2}\frac{d\xi(-\bar{x})}{d\bar{x}}\right). \tag{3.174}$$

Re-labelling variable $\bar{x}$ by $-p$ we can rewrite the operator as

$$\widetilde{A} = i\hbar\left(\xi(p)\frac{d}{dp} + \frac{1}{2}\frac{d\xi(p)}{dp}\right) \tag{3.175}$$

acting on $\widetilde{\varphi}(-p) \in C_0^\infty(\widetilde{I\!\!E}) \subset L^2(\widetilde{I\!\!E})$. This turns out to agree with the symmetrization rule, i.e.,

$$\widetilde{A} = \frac{1}{2}\left(\xi(p)\,i\hbar\frac{d}{dp} + i\hbar\frac{d}{dp}\,\xi(p)\right). \tag{3.176}$$

Note that we can regard $\widetilde{I\!\!E}$ as consisting of functions of $\bar{x}$ or p. In what follows we shall consider $\widetilde{I\!\!E}$ as consisting of functions p, i.e., we shall identify $\widetilde{I\!\!E}$ with the momentum space and $\widetilde{L}^2(\widetilde{I\!\!E})$ with the momentum representation space introduced in E2.4(1) E1.

3. The last step is to relate this operator to an operator in $L^2(I\!\!E)$ by Fourier transformation from $\widetilde{L}^2(\widetilde{I\!\!E})$ to $L^2(I\!\!E)$. In other words, using

the results in Eq. (2.133), we can quantize $F(x,p) = \xi(p)x$ in $I\!\!E$ as the following operator in $L^2(I\!\!E)$:

$$\begin{aligned}
\widehat{A} &= \widehat{U}_F^{-1} \widetilde{A} \widehat{U}_F \\
&= \frac{1}{2} \left\{ \widehat{U}_F^{-1} \left(\xi(p) i\hbar \frac{d}{dp} + i\hbar \frac{d}{dp} \xi(p) \right) \widehat{U}_F \right\} \\
&= \frac{1}{2} \left\{ \widehat{U}_F^{-1} \xi(p) \widehat{U}_F \widehat{U}_F^{-1} \left(i\hbar \frac{d}{dp} \right) \widehat{U}_F + \widehat{U}_F^{-1} \left(i\hbar \frac{d}{dp} \right) \widehat{U}_F \widehat{U}_F^{-1} \xi(p) \widehat{U}_F \right\} \\
&= \frac{1}{2} \left(\xi(\widehat{p}) x + x \xi(\widehat{p}) \right). \quad (3.177)
\end{aligned}$$

Since $\widetilde{A}$ is selfadjoint in $\widetilde{L}^2(\widetilde{I\!\!E})$ the transformed operator $\widehat{A}$ is selfadjoint in $L^2(I\!\!E)$.

We can go on to quantize observables the form

$$\mathcal{A} = \xi(p)x + \eta(p). \quad (3.178)$$

First we quantize this observable in terms of the operator expression

$$\widetilde{\mathcal{A}} = \frac{1}{2} \left(\xi(p) i\hbar \frac{d}{dp} + i\hbar \frac{d}{dp} \xi(p) \right) + \eta(p) \quad (3.179)$$

in $\widetilde{L}^2(\widetilde{I\!\!E})$. The corresponding quantized operator in $L^2(I\!\!E)$ is then obtained by a Fourier transform, i.e., we have

$$\widehat{\mathcal{A}} = \frac{1}{2} \left(\xi(\widehat{p}) x + x \xi(\widehat{p}) \right) + \eta(\widehat{p}). \quad (3.180)$$

The kinetic energy $K = p^2/2m$ of a particle of mass m in $I\!\!E$ is of the form of $\mathcal{A}$ with $\xi(p) = 0$ and $\eta = p^2/2m$. In the momentum representation space the kinetic energy is then quantized as the multiplication operator

$$\widetilde{K} = p^2/2m \quad (3.181)$$

in $\widetilde{L}^2(\widetilde{I\!\!E})$. The corresponding operator $\widehat{K}$ in the coordinate representation space $L^2(I\!\!E)$ is

$$\widehat{K} = \widehat{U}_F^{-1} \widetilde{K} \widehat{U}_F = \frac{1}{2m} \widehat{p}^2. \quad (3.182)$$

It may seem that we have used a huge hammer to crack a small nut here since we could have obtained the result using

1. the squaring rule of quantization, and

3.3. QUANTIZATION IN $\mathbb{E}^n$

2. the essentially selfadjoint operator $\widehat{K}_0(\mathbb{E})$ in Eq. (2.185) as the initially quantized kinetic operator which will then give the above operator $\widehat{K}$ as the quantized kinetic energy operator.

The point here is that we want to avoid treating the squaring rule as a general quantization axiom. It is not always correct to quantize the square A^2 of a classical observable A as the square $\widehat{A}^2$ of the quantized operator $\widehat{A}$ when A cannot be quantized to a unique selfadjoint operator. We will be able to appreciate this when we come to quantize K in spherical coordinates in what follows.

For a particle in three dimensions the selfadjoint kinetic energy operator can be constructed in terms of quantities in one dimension along x, y, z directions. Along the x-axis, denoted by $\mathbb{E}_x$, we have the Hilbert space $L^2(\mathbb{E}_x, dx)$ of square-integrable functions of x. Similarly along the y-axis we have the Hilbert space $L^2(\mathbb{E}_y, dy)$. A similar notation applies along the z-axis. The Hilbert space $L^2(\mathbb{E}^3)$ can be decomposed into a tensor product as

$$L^2(\mathbb{E}^3) = L^2(\mathbb{E}_x, dx) \otimes L^2(\mathbb{E}_y, dy) \otimes L^2(\mathbb{E}_z, dz). \tag{3.183}$$

The kinetic energy operator is then taken to be

$$\widehat{K} = \frac{1}{2m}\left(\widehat{p}_x^2 \otimes \widehat{\mathbb{I}}_y \otimes \widehat{\mathbb{I}}_z + \widehat{\mathbb{I}}_x \otimes \widehat{p}_y^2 \otimes \widehat{\mathbb{I}}_z + \widehat{\mathbb{I}}_x \otimes \widehat{\mathbb{I}}_y \otimes \widehat{p}_z^2\right), \tag{3.184}$$

where $\widehat{\mathbb{I}}_x, \widehat{\mathbb{I}}_y$ and $\widehat{\mathbb{I}}_z$ are the identity operators respectively on

$$L^2(\mathbb{E}_x, dx), \ L^2(\mathbb{E}_y, dy) \text{ and } L^2(\mathbb{E}_z, dz). \tag{3.185}$$

It is often convenient to write

$$\widehat{p}_x \otimes \widehat{\mathbb{I}}_y \otimes \widehat{\mathbb{I}}_z \quad \text{simply as} \quad \widehat{p}_x \tag{3.186}$$

with an explicit expression $\widehat{p}_x = -i\hbar\partial/\partial x$, and

$$\widehat{p}_x^2 \otimes \widehat{\mathbb{I}}_y \otimes \widehat{\mathbb{I}}_z \quad \text{simply as} \quad \widehat{p}_x^2 \tag{3.187}$$

and so on to enable us formally to write

$$\widehat{K} = \frac{1}{2m}\left(\widehat{p}_x^2 + \widehat{p}_y^2 + \widehat{p}_z^2\right). \tag{3.188}$$

What we want now is to exhibit various problems which could arise in quantizing K in spherical coordinates:

1. In quantum mechanics we often write down the expression of $\widehat{K}$ in Eq. (3.188) formally in spherical coordinates as

$$\widehat{K} = -\frac{\hbar^2}{2m}\left(\frac{\partial^2}{\partial r^2} + \frac{2}{r}\frac{\partial}{\partial r}\right) + \frac{1}{2mr^2}\widehat{L}^2, \qquad (3.189)$$

where

$$\widehat{L}^2 = -\hbar^2\left(\frac{1}{\sin\theta}\frac{\partial}{\partial\theta}\left(\sin\theta\frac{\partial}{\partial\theta}\right) + \frac{1}{\sin^2\theta}\frac{\partial^2}{\partial\varphi^2}\right) \qquad (3.190)$$

is the total angular momentum operator square.

2. $L^2(I\!\!E^3)$ can be decomposed into the tensor product of a radial part $L^2(I\!\!R^+, r^2 dr)$ and an angular part $L^2(\mathcal{S}^2, \sin\theta d\theta d\varphi)$, in accordance with Eq. (2.455). Similarly the kinetic energy operator $\widehat{K}$ is the sum of two terms each of which is decomposable as the tensor product of a radial part and an angular part:

$$\widehat{K} = -\frac{\hbar^2}{2m}\left(\frac{\partial^2}{\partial r^2} + \frac{2}{r}\frac{\partial}{\partial r}\right) \otimes \widehat{I\!\!I}(\mathcal{S}^2) + \frac{1}{2mr^2} \otimes \widehat{L}^2, \qquad (3.191)$$

where $\widehat{I\!\!I}(\mathcal{S}^2)$ is the identity operator on $L^2(\mathcal{S}^2, \sin\theta d\theta d\varphi)$. We also know from quantum theory of angular momentum that $\widehat{L}^2$, regarded as an operator in $L^2(\mathcal{S}^2, \sin\theta d\theta d\varphi)$, possesses a purely point spectrum with eigenvalues $\ell(\ell+1)\hbar^2$, $\ell = 0, 1, 2, \ldots$. Let us denote the corresponding eigensubspaces by $\mathcal{L}_\ell$. We can decompose $L^2(\mathcal{S}^2, \sin\theta d\theta d\varphi)$ into a direct sum of these eigensubspaces. This results in the following direct sum decompositions:

$$L^2(\mathcal{S}^2, \sin\theta d\theta d\varphi) = \oplus_\ell \mathcal{L}_\ell. \qquad (3.192)$$

It follows that
$$L^2(I\!\!R^+, r^2 dr) \otimes L^2(\mathcal{S}^2, \sin\theta d\theta d\varphi) \qquad (3.193)$$
$$= \oplus_\ell \left(L^2(I\!\!R^+, r^2 dr) \otimes \mathcal{L}_\ell\right). \qquad (3.194)$$

In other words $L^2(I\!\!E^2)$ can be decomposed as a direct sum of subspaces

$$\mathcal{H}_\ell = L^2(I\!\!R^+, r^2 dr) \otimes \mathcal{L}_\ell. \qquad (3.195)$$

These subspaces reduce the kinetic energy operator. Let the reduction of $\widehat{K}$ on $\mathcal{H}_\ell$ be $\widehat{K}_\ell$. Then

$$\widehat{K}_\ell = \left\{-\frac{\hbar^2}{2m}\left(\frac{\partial^2}{\partial r^2} + \frac{2}{r}\frac{\partial}{\partial r}\right) + \frac{\ell(\ell+1)\hbar^2}{2mr^2}\right\} \otimes \widehat{I\!\!I}(\mathcal{L}_\ell). \qquad (3.196)$$

3.3. QUANTIZATION IN $I\!\!E^n$

The radial part of this operator defined by

$$-\frac{\hbar^2}{2m}\left(\frac{\partial^2}{\partial r^2}+\frac{2}{r}\frac{\partial}{\partial r}\right)+\frac{\ell(\ell+1)\hbar^2}{2mr^2} \qquad (3.197)$$

acting on $C_0^\infty(I\!\!R^+)$ is not essentially selfadjoint.[60] This result is intuitively obvious when one realizes that the first term in Eq. (3.197) is the same as the restriction to $C_0^\infty(I\!\!R^+)$ of the square of the radial momentum operator $\widehat{p}_r$ defined by Eqs. (2.456) and (2.457). Using the Friedrichs extension $\widehat{p}_r^\dagger \widehat{p}_r$ in §2.10.3 we arrive at a selfadjoint extension[61]

$$\widehat{p}_r^\dagger \widehat{p}_r + \frac{\ell(\ell+1)\hbar^2}{2mr^2}. \qquad (3.198)$$

This choice is consistent with the boundary conditions imposed on the generalized eigenfunctions of the kinetic energy operator in quantum mechanics.[62] The preceding discussion enables us to rewrite the kinetic energy operator as

$$\widehat{K}=\frac{1}{2m}\widehat{p}_r^\dagger \widehat{p}_r \otimes \widehat{I\!\!I}(\mathcal{S}^2)+\frac{1}{2mr^2}\otimes \widehat{L}^2. \qquad (3.199)$$

The above expression for $\widehat{K}$ is manifestly selfadjoint, given that $\widehat{L}^2$ is well-known to be selfadjoint in quantum mechanics.

3. The kinetic energy has a classical expression in spherical coordinates:

$$K=\frac{1}{2m}\left(p_r^2+\frac{1}{r^2}\left(p_\theta^2+\frac{1}{\sin^2\theta}p_\varphi\right)\right). \qquad (3.200)$$

The question is whether we can obtain the selfadjoint operator $\widehat{K}$ in Eq. (3.199) by directly quantizing K in Eq. (3.200). This would entail quantizing the squares of the momenta conjugate to spherical coordinates:

(a) First let us quantize p_r^2. From our discussion in §2.10.3 we know that $\widehat{p}_r^2$ is not the answer and that we need to seek an appropriate selfadjoint extension to $\widehat{p}_r^2$. Equation (3.198) tells as that the Friedrichs extension $\widehat{p}_r^\dagger \widehat{p}_r$ is the answer.[63]

---

[60] Reed and Simon Vol. 2 (1975) p. 161. This should not be confused with the fact that the restriction of $\widehat{K}$ of Eq. (3.188) to $C_0^\infty(I\!\!E^3)$ is essentially sefladjoint (Reed and Simon Vol. 2 (1975) p. 166).

[61] There are other selfadjoint extensions, e.g., $\widehat{p}_r \widehat{p}_r^\dagger$.

[62] Zettili (2001) p. 326.

[63] Although $\widehat{p}_r^2$ and $\widehat{p}_r^\dagger \widehat{p}_r$ have the same differential expression they operate on different domains which render the resulting operators different.

(b) Next let us try to establish the operator $\widehat{L}^2$ in $L^2(\mathcal{S}^2, \sin\theta d\theta d\varphi)$. For p_φ we have the selfadjoint operator $\widehat{p}_\varphi$ in Eq. (3.101) so that we can quantize p_φ^2 unambiguously as $\widehat{p}_\varphi^2$. The situation is different for the incomplete momentum p_θ since $\widehat{p}_{\theta 0}$ in Eq. (3.167) has many selfadjoint extensions. Moreover $\widehat{p}_{\theta 0}^2$ does not even produce the correct differential expression for $\widehat{L}^2$.

(c) We can now see that on account of the problems in quantizing p_r and p_θ we cannot employ the squaring rules to quantize K in spherical coordinates.

For a particle of mass m moving in $I\!\!E^3$ under a potential V which is a function on $I\!\!E^3$ the kinetic energy K and the Hamiltonian H are expressed in terms of rectangular Cartesian coordinates x^j and their conjugate momenta p_j as

$$K = \frac{1}{2m}\left(p_x^2 + p_y^2 + p_z^2\right), \qquad H = K + V(x). \tag{3.201}$$

Assuming the linearity rule in this particular case the Hamiltonian can be successfully quantized as the following selfadjoint operator

$$\begin{aligned}\widehat{H} = \mathcal{Q}(H) &= \mathcal{Q}(K) + \mathcal{Q}(V) & (3.202)\\ &= \widehat{K} + V & (3.203)\\ &= \frac{1}{2m}\left(\widehat{p}_x^2 + \widehat{p}_y^2 + \widehat{p}_z^2\right) + V & (3.204)\end{aligned}$$

in $L^2(I\!\!E^3)$ without any complication for a large class of potential functions.[64]

3.3.7 Constraint and quantization in circuit geometry

The situation becomes rather more complex when we try to quantize the kinetic energy and the Hamiltonian of a system subject to some geometric constraint, e.g., a particle confined to an interval or a half line, or an electron going round an excluded region in a circular path in the Aharonov-Bohm experiment.[65] More interesting and general are systems confined to some circuit geometry. These could be electrons confined to some low dimensional nanostructures or superconducting condensate flowing in some circuits. A circuit may have several branches and junctions which give rise to a non-manifold geometry. We have to generalize present quantization schemes as well as orthodox quantum theory itself in order to be able to deal with such systems. The problems arising in quantizing these systems are not just technical ones. There are new principles involved as seen in the two examples below.

---

[64]Cycon, Froese, Kirsch and Simon (1987).
[65]Aharonov and Bohm (1959), Peshkin and Tonomura (1989).

3.3. QUANTIZATION IN $\mathbb{E}^n$

As a first example consider a particle of mass m confined to a finite interval $\Lambda = (0, a)$ in $\mathbb{E}$ and is otherwise free. We desire to investigate the quantized kinetic energy and the Hamiltonian of such a system. As shown in E3.3.5(1) E5 such a deceptively simple system has a very complex quantum counterpart because of the following considerations:

1. The momentum p conjugate to x is incomplete so that there is no unique selfadjoint operator to correspond to p. Instead there is a one-paramter group of possibile selfadjoint operators $\widehat{p}_\lambda$ given in E2.5.1(1) E1 to choose from. One could also have local momentum inside Λ.

2. Since the classical Hamiltonian H coincides with the kinetic energy K for the particle within $\Lambda = (0, a)$, i.e.,

$$H = K = \frac{1}{2m} p^2, \quad x \in \Lambda. \tag{3.205}$$

It is not obvious if the quantized Hamiltonian should be the same as the quantized kinetic energy. Indeed the question arises as to how K and H should be quantized.

Let us try to quantize K first. Following our general procedure we shall first establish the initially quantized operator $\widehat{K}_0$ acting on $C_0^\infty(\Lambda)$. An eminently suitable choice is

$$\widehat{K}_0(\Lambda) = -\frac{\hbar^2}{2m} \frac{d^2}{dx^2}. \tag{3.206}$$

We are immediately confronted with the problem of non-uniqueness. This operator has many selfadjoint extensions. For examples Eq. (2.176) lists a one-parameter family of selfadjoint extensions

$$\widehat{K}_\lambda(\Lambda) = \frac{1}{2m} \widehat{p}_\lambda^2. \tag{3.207}$$

The non-uniqueness problem is more severe than this. Taking $\widehat{K}_0(\Lambda)$ as the initially quantized kinetic energy operator we find ourselves confronted by a huge number of selfadjoint extensions.[66] Each extension can be specified in terms of some specific conditions on the functions ϕ in its domain of operation at the boundaries of Λ at $x = 0$ and a, known as *boundary conditions*. Three well-known boundary conditions are:[67]

---

[66] We shall spell out all the extensions in §6.2 in Chapter 6.
[67] Bratteli and Robinson (1981) p. 56.

1. **Quasi-periodic boundary conditions** The values of the wave function $\phi(x)$ at the well boundaries are related by

$$\phi(0) = e^{i\lambda} \phi(a). \tag{3.208}$$

The resulting operators are $\widehat{K}_\lambda(\Lambda)$ quoted above.

2. **Dirichlet conditions** The wave function $\phi(x)$ must vanish at the well boundaries:

$$\phi(0) = \phi(a) = 0. \tag{3.209}$$

The resulting operator is $\widehat{K}^\infty(\Lambda)$ introduced in Eqs. (2.172) and (2.173).

3. **Neumann conditions** The derivative of the wave function must vanish at the well boundaries:

$$\phi'(0) = \phi'(a) = 0, \quad \phi' = d\phi/dx. \tag{3.210}$$

Clearly not all the resulting selfadjoint operators can be interpreted as kinetic energy operators. The reason is quite fundamental. Take for example the selfadjoint operator $\widehat{K}^\infty(\Lambda)$ arising from the Dirichlet conditions. Physically the Dirichlet conditions signify the impossibility of the wave function to go into the boundary and beyond due to the presence of an infinite square potential rising up abruptly to infinity at the boundaries. Clearly the conditions imposed on the wave function at the boundaries represent the presence of a potential at the boundaries. The corresponding selfadjoint extention $\widehat{K}^\infty(\Lambda)$ embodies the boundary conditions and hence the potential at the boundaries. It follows that $\widehat{K}^\infty(\Lambda)$ cannot correspond to the kinetic operator since kinetic energy does not involve any potential. The natural assumption to make here is that such a selfadjoint extension should represent a quantized Hamiltonian. We therefore arrive at an interesting situation:

> *An attempt to achieve selfadjointness in the quantization of the kinetic energy operator of a constrained system can produce a family of possible Hamiltonians with corresponding potentials incorporated into the boundary conditions. It follows that we can interpret different selfadjoint extensions as corresponding to different potentials or geometric constraints used to confine the particle in the region Λ.*

For examples, we have:[68]

1. Dirichlet conditions represent a particle confined by a rigid square potential well rising abruptly to infinity at $x = 0$ and $x = a$.

---
[68] Bratteli and Robinson (1981) p. 56.

3.3. QUANTIZATION IN $I\!E^n$

2. The Neumann conditions represent a particle confined by a kind of *elastic potential well* which does not require the wave function to vanish at the well boundaries. Instead we have the derivative of the wave function vanishes at the well boundaries.

3. Another example involves the confinement of an otherwise free particle of mass m on the half line $I\!E^+$. To quantize the kinetic energy $K^+ = (p^+)^2/2m$ we may try to quantize p^+ first. We will obtain the strictly maximal symmetric operator $\widehat{p}^+$ in $L^2(I\!E^+)$. From the discussion in §2.10.3 we know that we cannot simply take the square $(\widehat{p}^+)^2$ as the quantized counterpart to $(p^+)^2$. We have to select a suitable selfadjoint extension of $(\widehat{p}^+)^2$. Again different selfadjoint extensions correspond to different ways the particle is physically confined to $I\!E^+$. For example we have the Friedrichs extension given by

$$\frac{1}{2m}(\widehat{p}^+)^\dagger(\widehat{p}^+), \tag{3.211}$$

which imposes the boundary condition $\phi(0) = 0$ in its domain of operation given by Eq. (2.470). This condition may be considered as representing a rigid square potential rising up to infinite abruptly at the coordinate origin. Intuitively we can appreciate that a rigid wall at the origin will cause any incoming quantum wave to be reflected back with a π phase change to produce a node at the origin, i.e., the boundary condition $\phi(0) = 0$. The resulting operator becomes a Hamiltonian. Selfadjoint extensions characterized by domains given in Eq. (2.472) may be considered as describing a boundary wall having a certain elasticity so that the reflected quantum wave does not suffer a π phase change. The corresponding boundary conditions are known as an *elastic boundary condition*. Note that each value of λ represents a different physical boundary, and hence a different physical system.

We have demonstrated that when a system is under certain geometric constraint the taking of a selfadjoint extension of $\widehat{K}_0$ could automatically incorporate a potential to arrive at a Hamiltonian operator. As pointed out by Reed and Simon[69] different selfadjoint extensions actually correspond to different physics. In this way we can regard the non-uniqueness of quantization as a blessing rather than a problem.

Conceptually it is interesting to see how powerful the selfadjointess requirement in Postulate OQS is and how a purely mathematical operation of taking selfadjoint extensions can create a Hamiltonian with an effective physical potential. Recall the standard situation of a quantum particle in one-dimension

---
[69] Reed and Simon Vol. 2 (1975) p. 145.

with the Hamiltonian given by the sum of the kinetic energy term and a potential energy term in $L^2(I\!\!E)$:

$$\widehat{H} = \widehat{K} + \widehat{V} \qquad (3.212)$$

defined on the domain

$$\mathcal{D}(\widehat{H}) = \mathcal{D}(\widehat{K}) \cap \mathcal{D}(V). \qquad (3.213)$$

The domain of the Hamiltonian is seen to be related to the potential $\widehat{V}$ which is usually expressible in the form a potential function $V(x)$ on the coordinate space $I\!\!E$. Given V the domain $\mathcal{D}(\widehat{H})$ is fixed. Conversely a knowledge of $\mathcal{D}(\widehat{H})$ enables us to say something about the potential. If $\mathcal{D}(\widehat{H}) = \mathcal{D}(\widehat{K})$ the potential is likely to be bounded, e.g., a finite potential well, and if $\mathcal{D}(\widehat{H})$ is smaller than $\mathcal{D}(\widehat{K})$ the potential is likely to be unbounded, e.g., a harmonic oscillator potential. Equations (3.212) and (3.213) fail when the potential V becomes singular in such a way that it cannot be represented by a selfadjoint multiplication operator. Point interactions fall into this category. For example we could have a potential which is zero over an extended region but rises to infinity abruptly at some isolated points in the coordinate space, a δ-function potential being a case in point. There are point interactions which simply cannot be written down in terms of a potential function, not even in terms of delta functions. This kind of potentials are not just of academic interest. Many quantum circuits have junctions whose effect is describable with great accuracy in terms of singular potentials of this kind; it turns out that we can handle these situations in terms of selfadjoint extensions of some initially quantized symmetric operators. So, what we describe here serves as a demonstration of the physical significance and usefulness of non-uniqueness of quantization, and as a precursor to the discussions in Chapters 6 and 7 on point interactions and quantum circuits.

Finally we should point out that a constrained system is likely to involve incomplete momenta, e.g., a particle confined to the half line. This would present a severe difficulty in its quantization within orthodox quantum mechanics on account of the appearance of strictly maximal symmetric operators. We shall return to examine how this problem can be accommodated within the context of a generalized quantum mehcanics in Chapter 5.

3.4 Orthodox Quantum Dynamics

3.4.1 Postulate on orthodox quantum dynamics

Dynamics is about time evolution of a physical system. We can formulate time evolution in terms of propagators and time evolution operators:

3.4. ORTHODOX QUANTUM DYNAMICS

Definition 3.4.1(1)   Propagators[70]

- A two-parameter family $\widehat{\mathcal{U}}$ of unitary operators $\widehat{U}_{r,s}$, $r, s \in I\!R$, is called a *unitary propagator*, or simply a *propagator*, acting on a given Hilbert space $\mathcal{H}$ if

  1. $\widehat{U}_{r,s}$ is strongly continuous in r and s.
  2. $\widehat{U}_{r,r} = \widehat{I\!\!I}$.
  3. $\widehat{U}_{r,s}\widehat{U}_{s,t} = \widehat{U}_{r,t}$.

We can now summarize orthodox quantum dynamics in a postulate.

Postulate OQD on orthodox quantum dynamics[71]

- Each quantum system has associated with it:

  1. A family of selfadjoint operators $\widehat{H}_g(t)$ with domains $\mathcal{D}(\widehat{H}_g(t))$, to be referred to as the *Hamiltonian generators* of the system.
  2. A propagator $\widehat{\mathcal{U}}$ related to the Hamiltonian generators of the system by

$$i\hbar \lim_{\epsilon \to 0} \frac{\widehat{U}_{t+\epsilon,t} - \widehat{I\!\!I}}{\epsilon} \varphi = \widehat{H}_g(t)\varphi, \quad \varphi \in \mathcal{D}(\widehat{H}_g(t)). \qquad (3.214)$$

- If the system is in a state represented by a density operator $\widehat{D}_{t_0}$ at time t_0, then its state at time t is given by the density operator[72]

$$\widehat{D}_t = \widehat{U}_{t,t_0} \widehat{D}_{t_0} \widehat{U}^\dagger_{t,t_0}. \qquad (3.215)$$

- Observables which are not explicitly time dependent in their definition remain unchanged in time.

This traditional postulate on time evolution is known as the *Schrödinger picture* with $\widehat{U}_{t,t_0}$ referred to as *time evolution operators* or simply as *evolution operators*. This postulate gives rise to the following results:

---

[70] Reed and Simon Vol. 2 (1975) p. 282.
[71] Prugovečki (1981) p. 292.
[72] Beltrametti and Cassinelli (1981) pp. 52-56. The differential equation for $\widehat{D}_t$ is

$$i\hbar \frac{d\widehat{D}_t}{dt} = [\widehat{H}_g, \widehat{D}_t].$$

1. If the system is in a pure state represented by a projector $\widehat{P}_{\phi_{t_0}}$ at time $t = t_0$, then its state at time t given by

$$\widehat{P}_{\phi_t} = \widehat{U}_{t,t_0} \widehat{P}_{\phi_{t_0}} \widehat{U}_{t,t_0}^\dagger \qquad (3.216)$$

is again a projector. It follows that a pure state always evolves into a pure state in any finite period of time. Alternatively, if a state is represented by a unit vector ϕ_{t_0} at time $t = t_0$ then the unit vector representing the state at time t is

$$\phi_t = \widehat{U}_{t,t_0} \phi_{t_0}. \qquad (3.217)$$

On account of Eq. (3.214) the evolved state ϕ_t satisfies the traditional Schrödinger equation:[73]

$$i\hbar \frac{d}{dt} \phi_t = \widehat{H}_g(t) \phi_t, \quad \text{provided} \quad \phi_t \in \mathcal{D}(\widehat{H}_g(t)). \qquad (3.218)$$

2. The density operator $\widehat{D}_t$ in Eq. (3.215) cannot be a projector if $\widehat{D}_{t_0}$ is not. It follows that a mixed state cannot evolve into a pure state in any finite period of time.

3. The fact that pure and mixed states cannot evolve into each other has a profound impact on the measurement and state preparation processes in quantum theory, a subject which we shall investigate later in this chapter and in later chapters.

4. In the simplest case the physical system may have a time independent Hamiltonian generator $\widehat{H}_g$ associated with it. Then we can write down an explicit relation between $\widehat{H}_g$ and the evolution operators as

$$\widehat{U}_{t,t_0} = e^{-\dot{i}(t-t_0)\widehat{H}_g}, \quad \dot{i} = \frac{i}{\hbar}. \qquad (3.219)$$

The operator $\widehat{U}_{t,t_0}$ depends only on the time difference $t - t_0$, not on the starting point t_0. It is more convenient to assume a starting time $t_0 = 0$. The propagator is then seen to compose of a *one-parameter group of unitary operators*, which can be conveniently denoted by

$$\widehat{U}_t = e^{-\dot{i}t\widehat{H}_g}, \qquad (3.220)$$

---

[73] To derive the result we observe that

$$\phi_{t+\epsilon} = \widehat{U}_{t+\epsilon,t_0} \phi_{t_0} = \widehat{U}_{t+\epsilon,t} \widehat{U}_{t,t_0} \phi_{t_0} = \widehat{U}_{t+\epsilon,t} \phi_t \quad \Rightarrow \quad \phi_{t+\epsilon} - \phi_t = \left(\widehat{U}_{t+\epsilon,t} - \widehat{I}\right) \phi_t.$$

See also Reed and Simon Vol. 2 (1975) p. 282.

3.4. ORTHODOX QUANTUM DYNAMICS

rather than by $\widehat{U}_{t,0}$, and the evolved state is given by

$$\phi_t = \widehat{U}_t \phi_0. \tag{3.221}$$

The following group properties show up clearly:

$$\widehat{U}_{t_2}\widehat{U}_{t_1} = \widehat{U}_{t_2+t_1}, \quad \widehat{U}_t^{-1} = \widehat{U}_{-t}, \tag{3.222}$$

or in the original notation:

$$\widehat{U}_{t_2,t_0}\widehat{U}_{t_1,t_0} = \widehat{U}_{t_2+t_1,t_0}, \quad \widehat{U}_{t,t_0}^{-1} = \widehat{U}_{-t,t_0}. \tag{3.223}$$

The Schrödinger equation becomes

$$i\hbar \frac{d}{dt}\phi_t = \widehat{H}_g \phi_t. \tag{3.224}$$

5. The description of dynamics directly in terms of a Schrödinger equation is less general than the description in terms of a propagator. For example, the Schrödinger equation above holds only if the initial state $\phi_{t=0}$ at $t = 0$ is in the domain $\mathcal{D}(\widehat{H}_g)$ of the Hamiltonian generator to start with. An initial state lying outside the domain of the Hamiltonian generator will evolve under the propagator in accordance with Eq. (3.221). It follows that any interpretation or properties which depend directly on the Schrödinger equation for evolution, e.g., the quantum potential interpretation, cannot be applied to $\phi_0 \notin \mathcal{D}(\widehat{H}_g)$.[74]

6. We can write down an expression for the evolution operators to correspond to a family of Hamiltonian generators $\widehat{H}_g(t)$ provided the Hamiltonian generators at different times commute, i.e., $\widehat{H}_g(t_1)$ and $\widehat{H}_g(t_2)$ commute for any t_1, t_2. Then we have

$$\widehat{U}_{t,t_0} = e^{-i \int_{t_0}^{t} \widehat{H}_g(t)dt}. \tag{3.225}$$

Generally, when the Hamiltonian generators at different times do not commute, there is no such simple relationship between $\widehat{U}_{t,t_0}$ and $\widehat{H}_g(t)$ and the evolution operators do not possess group properties.[75]

7. A simple but important expression can be obtained for the evolution of the expectation value of an observable. Let $\widehat{A}$ be an observable which is not explicitly time dependent and let ϕ_t be the state at time t. From

---
[74] Wan and Sumner (1988).
[75] Jauch (1968) p. 159, Baym (1969) pp. 140-145.

C2.11.1(1) C3 we know that the expectation value $\mathcal{E}(\widehat{A}, \phi_t)$ of $\widehat{A}$ in state ϕ_t at time t is equal to $\langle \phi_t \mid \widehat{A}\phi_t \rangle$. Differentiating with respect to t we get, assuming $\widehat{H}_g$ to be time-independent,

$$\frac{d\mathcal{E}(\widehat{A}, \phi_t)}{dt} = \frac{d\langle \phi_t \mid \widehat{A}\phi_t \rangle}{dt} = \frac{1}{i\hbar} \langle \phi_t \mid [\widehat{A}, \widehat{H}_g] \phi_t \rangle. \qquad (3.226)$$

It follows that the expectation value is *conserved*, i.e., time independent, if $\widehat{A}$ commutes with the Hamiltonian generator. Observable $\widehat{A}$ is also said to be *conserved*, e.g., the momentum and its expectation value of a free particle are conserved.

8. To avoid confusion we shall emphasize that, as for classical systems discussed in C1.6.3(1) C2, the Hamiltonian generator $\widehat{H}_g$ is responsible for time evolution and it is not generally assumed to be the Hamiltonian $\widehat{H}$ which represents the total energy of the system. This is contrary to a commonly adopted interpretation which simply identifies $\widehat{H}_g$ with $\widehat{H}$.[76] We will encounter systems for which $\widehat{H}_g \neq \widehat{H}$ in later chapters

Unless explicitly stated otherwise we shall only consider evolutions generated by a time independent Hamiltonian generator from now on. We shall also set $t_0 = 0$ so that the propagator is composed of a one-parameter group of unitary operators given by Eq. (3.220).

We call the Hamiltonian generator of a free particle a *free Hamiltonian generator*, to be denoted by $\widehat{H}_g^{(0)}$. The corresponding propagator and evolution operators are denoted by $\widehat{\mathcal{U}}^{(0)}$ and $\widehat{U}_t^{(0)}$ and are referred to as a *free propagator* and *free evolution operators* respectively. For example for motion in E we have

$$\widehat{H}_g^{(0)} = \frac{1}{2m}\widehat{p}^2. \qquad (3.227)$$

We shall refer to the Hamiltonian of a free particle as a free Hamiltonian which will be denoted by $\widehat{H}^{(0)}$. For a free particle in E we have $\widehat{H}_g^{(0)} = \widehat{H}^{(0)}$.

3.4.2 Asymptotic localization and separation: Free systems

Quantum dynamics is determined by the Hamiltonian generator. Despite the complexity of the dynamical equation involved it is possible to arrive at some simple and important conclusions on how the wave function would behave at large times. There are many excellent texts on quantum scattering dealing

---
[76] Prugovečki (1981) Axiom H1 on p. 296.

3.4. ORTHODOX QUANTUM DYNAMICS

with this kind of problems in great details.[77] What we are interested in here are situations immediately relevant to to our pending discussions on state preparation and measurement within orthodox quantum mechanics.

Consider the free motion of a classical particle of mass m in $I\!E$ with Cartesian coordinate x. If the particle is initially at position x_0 near the origin at time $t = 0$, and if the particle's velocity v is known to lie in a certain range $\Delta v = [v_1, v_2]$, then the particle will be found at a later time t to be in the region $[x_0 + v_1 t, x_0 + v_2 t]$. This region tends to $[v_1 t, v_2 t]$ at large times as x_0 becomes insignificant.

A similar result is also true for the free motion of a quantum particle of mass m in $I\!E$. To establish this result let us consider a free quantum particle whose time evolution arises from the free Hamiltonian generator $\widehat{H}_g^{(0)} = \widehat{H}^{(0)}$ in $L^2(I\!E)$. Let $\phi_0(x) \in L^2(I\!E)$ be the (normalized) wave function of the particle at $t = 0$. The evolved wave function is then given by $\phi_t = \widehat{U}_t^{(0)} \phi_0$.

The asymptotic behaviour of the wave function under free evolution is summarized in the following theorem.[78]

Theorem 3.4.2(1)   **Asymptotic localization for free evolution 1**

- Let $t\Delta v$ denote the interval $[v_1 t, v_2 t]$ and $m\Delta v$ denote the corresponding range of momentum values $[m v_1, m v_2]$. Then for free evolution given by $\phi_t = \widehat{U}_t^{(0)} \phi_0$ we have

$$\lim_{t \to \infty} ||\widehat{M}^{\hat{x}}(t\Delta v)\phi_t||^2 = ||\widehat{M}^{\hat{p}}(m\Delta v)\phi_0||^2. \tag{3.228}$$

To appreciate the physical content of this theorem we can, with the help of Eqs. (2.481) and (2.483), rewrite Eq. (3.228) in terms of position and momentum probability measures as

$$\lim_{t \to \infty} \mathcal{M}^{\hat{x}}_{\phi_t}(t\Delta v) = \mathcal{M}^{\hat{p}}_{\phi_0}(m\Delta v). \tag{3.229}$$

This tells us that at large times the probability of finding the particle in the region $t\Delta v$ is approximately equal to the probability of the particle's momentum being initially in the range $m\Delta v$. If the particle's momentum is known to be definitely in the range $m\Delta v$, i.e., if [79]

$$\mathcal{M}^{\hat{p}}_{\phi_0}(m\Delta v) = ||\widehat{M}^{\hat{p}}(m\Delta v)\phi_0||^2 = 1, \tag{3.230}$$

---

[77] Amrein, Jauch and Sinha (1977), Thirring (1979), Amrein (1981), Pearson (1988).
[78] Park and Margenau (1968), Pfeifer (1980), Enss (1983). For a proof see Wan and McLean (1983a), (1983b).
[79] Note that
$$\mathcal{M}^{\hat{p}}_{\phi_t}(m\Delta v) = \mathcal{M}^{\hat{p}}_{\phi_0}(m\Delta v).$$
This is due to the conservation of momentum for free particles.

then we have

$$\lim_{t\to\infty} \left|\left|\widehat{M^{\hat{x}}}(t\Delta v)\, \phi_t\right|\right|^2 = \lim_{t\to\infty} \left|\left|\chi_{t\Delta v}\, \phi_t\right|\right|^2 = 1. \tag{3.231}$$

It follows that the wave function ϕ_t becomes localized in the region $t\Delta v$ at large times. This is the quantum analogue to the classical situation mentioned earlier.

Let ϕ_0 and ψ_0 be two states corresponding to momentum in the ranges $m\Delta v$ and $m\Delta w$ respectively. If $m\Delta v$ and $m\Delta w$ are disjoint, then the two states will evolve into spatially disjoint regions $t\Delta v$ and $t\Delta w$ at large times. We can formalize this in terms of a definition.

Definition 3.4.2(1)   On asymptotic localization and separation

- A state ϕ_0 is said to be *asymptotically localizable* under a given propagator $\widehat{\mathcal{U}}$ if a bounded and closed interval Δv in E exists such that

$$\lim_{t\to\infty} \left|\left|\widehat{M^{\hat{x}}}(t\Delta v)\phi_t\right|\right|^2 = 1, \quad \text{where} \quad \phi_t = \widehat{U}_t \phi_0. \tag{3.232}$$

This state is also said to be asymptotically localizable into the region $t\Delta v$ in E.

- Two states ϕ_0 and ψ_0 are said to be *spatially separable*, or *asymptotically separable* under a given propagator $\widehat{\mathcal{U}}$ if two bounded, closed and disjoint intervals Δv and Δw in E exist such that ϕ_0 and ψ_0 are asymptotically localizable respectively into the regions $t\Delta v$ and $t\Delta w$ in E.

The criteria for asymptotic localization and separation for free evolution is given as follows.

Corollary 3.4.2(1)   Criteria for localization and separation

- A state ϕ_0 is asymptotically localizable into the region $t\Delta v$ under free evolution if

$$\left|\left|\widehat{M^{\hat{p}}}(m\Delta v)\phi_0\right|\right|^2 = 1. \tag{3.233}$$

- Two states ϕ_0 and ψ_0 are asymptotically separable into disjoint regions $t\Delta v$ and $t\Delta w$ under free evolution if

$$\left|\left|\widehat{M^{\hat{p}}}(m\Delta v)\phi_0\right|\right|^2 = 1 \quad \text{and} \quad \left|\left|\widehat{M^{\hat{p}}}(m\Delta w)\psi_0\right|\right|^2 = 1. \tag{3.234}$$

3.4. ORTHODOX QUANTUM DYNAMICS

We can interpret Corollary 3.4.2(1) in terms of Fourier transform of the wave function. We know from E2.11.1(1) E2 that $|\widetilde{\phi}_0(p)|^2$ is the momentum probability density function. The wave function will be asymptotically localized in the region $t\Delta v$ if $|\widetilde{\phi}_0(p)|^2$ has a support $m\Delta v$. Two states ϕ_0 and ψ_0 correspond to disjoint momentum values if and only if their Fourier transforms $\widetilde{\phi}_0$ and $\widetilde{\psi}_0$ have disjoint supports in the momentum space. Such states would be asymptotically separable. Note that the separation of these disjoint regions will become arbitrarily large as time tends to infinity. In practice the existence of various limits enables us to talk meaningfully about large but finite separations at large times.

These results can be extended to motion in three dimensions, using the following notation:

1. Let (x, y, z) be the usual Cartesian coordinates in $I\!E^3$, and $\Delta v_x = [v_{x_1}, v_{x_2}]$, $\Delta v_y = [v_{y_1}, v_{y_2}]$ and $\Delta v_z = [v_{z_1}, v_{z_2}]$.

2. Let $\Delta \mathbf{v}$ denote the Cartesian product $\Delta v_x \times \Delta v_y \times \Delta v_z$ which is a rectangle in $I\!E^3$, representing a chosen range of velocity of the particle.

3. Let $m\Delta \mathbf{v} = m\Delta v_x \times m\Delta v_y \times m\Delta v_z$ denote the corresponding range of momentum values.

4. Let $t\Delta \mathbf{v} = t\Delta v_x \times t\Delta v_y \times t\Delta v_z$.

A classical particle with momentum in the range $m\Delta \mathbf{v}$ will be most likely found to be in the region $t\Delta \mathbf{v}$ at large times. The Hilbert space for a quantum particle in $I\!E^3$ can be written in the form of a tensor product as in Eq. (3.183). As before $\widehat{M}^{\hat{x}}$ and $\widehat{M}^{\hat{p}_x}$ denote the spectral measures of $\widehat{x}$ and $\widehat{p}_x$ in $L^2(I\!E_x, dx)$. Along the y-axis we have the Hilbert space $L^2(I\!E_y, dy)$ with $\widehat{M}^{\hat{y}}$ and $\widehat{M}^{\hat{p}_y}$ denoting the spectral measures of $\widehat{y}$ and $\widehat{p}_y$. A similar notation applies in $L^2(I\!E_z, dz)$. Now, define two projectors on $L^2(I\!E^3)$ by

$$\widehat{M}^{\hat{\mathbf{x}}}(t\Delta \mathbf{v}) = \widehat{M}^{\hat{x}}(t\Delta v_x) \otimes \widehat{M}^{\hat{y}}(t\Delta v_y) \otimes \widehat{M}^{\hat{z}}(t\Delta v_z), \quad (3.235)$$

$$\widehat{M}^{\hat{\mathbf{p}}}(m\Delta \mathbf{v}) = \widehat{M}^{\hat{p}_x}(m\Delta v_x) \otimes \widehat{M}^{\hat{p}_y}(m\Delta v_y) \otimes \widehat{M}^{\hat{p}_z}(m\Delta v_z). \quad (3.236)$$

Since position spectral measures along x, y, z directions are equal to characteristic functions in x, y, z respectively we have

$$\widehat{M}^{\hat{\mathbf{x}}}(t\mathbf{v}) = \chi_{t\Delta v_x} \otimes \chi_{t\Delta v_y} \otimes \chi_{t\Delta v_z}. \quad (3.237)$$

The tensor product of characterisitc functions on the right hand side shall be denoted by $\chi_{t\Delta \mathbf{v}}$ which is clearly a projector on $L^2(I\!E^3)$. Acting on $\phi \in L^2(I\!E^3)$ the operator $\widehat{M}^{\hat{\mathbf{x}}}(t\Delta \mathbf{v})$ projects a wave function ϕ onto the subspace of wave functions of support lying in $t\Delta \mathbf{v} \subset I\!E^3$, while $\widehat{M}^{\hat{\mathbf{p}}}(m\Delta \mathbf{v})$

294   CHAPTER 3. ORTHODOX QUANTUM MECHANICS

projects a wave function onto the subspace of wave functions of momentum values lying in the range $m\Delta \mathbf{v}$. We can define asymptotic localization and separation as before, with three-dimensional rectangle $\Delta \mathbf{v}$ replacing interval Δv.

Free evolution in three dimensions is generated by the kinetic energy operator $\widehat{K}$ in Eq. (3.184) acting as the free Hamiltonian generator $\widehat{H}_g^{(0)} = \widehat{H}^{(0)} = \widehat{K}$. The evolved wave function behaves in the same way as in one dimension in large times, as seen in the following theorem.

Theorem 3.4.2(2)   Asymptotic localization for free evolution 2

- Let ϕ_0 be the initial state, and let $\phi_t = \widehat{U}_t^{(0)} \phi_0$ be the state at time t. Then we have

$$\lim_{t \to \infty} \left|\left| \widehat{M^{\hat{x}}}(t\Delta \mathbf{v}) \phi_t \right|\right|^2 = \left|\left| \widehat{M^{\hat{p}}}(m\Delta \mathbf{v}) \phi_0 \right|\right|^2. \qquad (3.238)$$

- A state ϕ_0 is asymptotically localizable into the region $t\Delta \mathbf{v}$ under free evolution if

$$\left|\left| \widehat{M^{\hat{p}}}(m\Delta \mathbf{v}) \phi_0 \right|\right|^2 = 1. \qquad (3.239)$$

- Two states ϕ_0 and ψ_0 are asymptotically separable into disjoint regions $t\Delta \mathbf{v}$ and $t\Delta \mathbf{w}$ under free evolution if

$$\left|\left| \widehat{M^{\hat{p}}}(m\Delta \mathbf{v}) \phi_0 \right|\right|^2 = 1 \quad \text{and} \quad \left|\left| \widehat{M^{\hat{p}}}(m\Delta \mathbf{w}) \psi_0 \right|\right|^2 = 1. \qquad (3.240)$$

3.4.3 Asymptotic localization and separation: Scattering systems

The evolution of the particle under a large class of Hamiltonian generators turns out to resemble free evolution at large times.[80] Before we discuss this result let us introduce the concepts of bound states, scattering states and simple scattering systems.

Definition 3.4.3(1)   Bound and scattering states

- A state ϕ_0 is called a *scattering state* of a propagator $\widehat{\mathcal{U}}$ at positive times if the evolved state $\phi_t = \widehat{U}_t \phi_0$ satisfies

$$\lim_{t \to \infty} \left|\left| \widehat{M^{\hat{x}}}(\Lambda_{cu}(w)) \phi_t \right|\right| = 0, \qquad (3.241)$$

---

[80] The Hamiltonian generator $\widehat{H}_g$ in many simple but important cases can be written in the form $\widehat{H}_g = \widehat{H}^{(0)} + V$, where $V(x)$ is a time independent potential representing the interaction. There is a class of potentials under which the wave function evolves in a way resembling free evolution at large times (Amrein (1981) p. 176).

3.4. ORTHODOX QUANTUM DYNAMICS

for every bounded cube[81] $\Lambda_{cu}(w)$ in $\mathbb{E}^3$. It is called a *scattering state at negative times* if the above limit holds for $t \to -\infty$.

- A state ϕ_0 is called a *bound state* of the propagator $\widehat{\mathcal{U}}$ if, for any (small) $\epsilon > 0$, there exists $w > 0$ dependent on ϵ such that the evolved state $\phi_t = \widehat{U}_t \phi_0$ satisfies[82]

$$||\widehat{M}^{\hat{x}}(\Lambda_{cu}^c(w))\phi_t|| < \epsilon \quad \text{for all times, i.e., } \forall t \in \mathbb{R}, \tag{3.242}$$

where $\Lambda_{cu}^c(w) = \mathbb{E}^3 - \Lambda_{cu}(w)$ is the complement of $\Lambda_{cu}(w)$ in $\mathbb{E}^3$.

A scattering state would move to spatial infinity, i.e., the probability of finding the particle inside any given bounded region is small at large times. This is the condition imposed by Eq. (3.241). A bound state would remain largely in a certain bounded region in space, i.e., the probability of finding the particle outside some large bounded region is small.

For simplicity we shall confine ourselves to evolution under which the set of scattering states at positive times coincide with the set of scattering states at negative times. This set will be denoted by $\mathcal{S}_{sc}(\widehat{H}_g)$. The set of bound states will be denoted by $\mathcal{S}_{bd}(\widehat{H}_g)$. It can be shown that $\mathcal{S}_{sc}(\widehat{H}_g)$ forms a subspace of $L^2(\mathbb{E}^3)$ invariant under $\widehat{U}_t$, and so does $\mathcal{S}_{bd}(\widehat{H}_g)$.[83] We also call $\mathcal{S}_{bd}(\widehat{H}_g)$ and $\mathcal{S}_{sc}(\widehat{H}_g)$ respectively the set of bound states and the set of scattering states of the Hamiltonian generator $\widehat{H}_g$.

In a scattering experiment the evolution of a scattering state describes the following situation:

> *Starting at large negative times a particle far away from the scattering center comes in, interacts with the scattering potential and then goes out to spatial infinity at large positive times, having been scattered. A typical interaction lasts for a nano second or less.*[84]

Generally we have some scattering states and some bounded states under a given evolution. There are two extreme cases:

---

[81] As in Eq. (1.4) the parameter w is the width of the cube.
[82] More intuitively we have

$$||\widehat{M}^{\hat{x}}(\Lambda_{cu}^c(w))\phi_t|| \to 0 \quad \text{as} \quad w \to \infty \quad \forall t \in \mathbb{R}.$$

[83] Amrein (1981) Proposition 5.2 on p. 128. We shall confine ourselves to evolutions generated by time-independent Hamiltonian generators.
[84] Amrein, Jauch and Sinha (1977) p. 140 for a pictorial presentation.

1. For a free evolution generated by a free Hamiltonian generator $\widehat{H}_g^{(0)} = \widehat{H}^{(0)}$ all states are scattering states, i.e., $\mathcal{S}_{sc}(\widehat{H}_g^{(0)}) = L^2(I\!E^3)$. There are no bound states.[85]

2. Suppose that the Hamiltonian generator $\widehat{H}_g$ possesses an eigenvalue E with a corresponding eigenfunction φ. Applying Eq. (3.220) to φ we get

$$\widehat{U}_t \varphi = e^{-iEt} \varphi. \tag{3.243}$$

Hence,

$$||\widehat{M^{\hat{x}}}(\Lambda_{cu}^c(w))\widehat{U}_t \varphi|| = ||\widehat{M^{\hat{x}}}(\Lambda_{cu}^c(w)) e^{-iEt} \varphi|| \tag{3.244}$$
$$= ||\widehat{M^{\hat{x}}}(\Lambda_{cu}^c(w)) \varphi|| \tag{3.245}$$

which tends to zero as $w \to \infty$ since $\varphi(x)$ tends to zero at spatial infinity on account of its square-integrability. It follows that

(a) every eigenvector of the Hamiltonian generator represents a bound state, which is also called a *stationary state*,[86] and

(b) a Hamiltonian generator possessing a purely point spectrum admits no scattering state.

Definition 3.4.3(2)   Wave operators

- Let $\widehat{\mathcal{U}}$ be the propagator given rise by a Hamiltonian generator $\widehat{H}_g$ in $L^2(I\!E^3)$ and let $\widehat{\mathcal{U}}^{(0)}$ be the free propagator corresponding to the free Hamiltonian generator $\widehat{H}_g^{(0)} = \widehat{K}$ in $L^2(I\!E^3)$. The two operators $\widehat{\Omega}_\pm(\widehat{H}_g, \widehat{H}_g^{(0)})$ defined by

$$\widehat{\Omega}_\pm(\widehat{H}_g, \widehat{H}_g^{(0)}) = \text{s-}\lim_{t \to \pm\infty} \widehat{U}_t^\dagger \widehat{U}_t^{(0)} \tag{3.246}$$

are called the *wave operators* associated with the evolution groups $\widehat{\mathcal{U}}$ and $\widehat{\mathcal{U}}^{(0)}$, when the limits exist. The ranges of these operators shall be denoted by $\mathcal{R}(\widehat{\Omega}_\pm)$.

---

[85] Amrein (1981) p. 132. Note that it is usual, as in Amrein (1981) pp. 125-129, to define bound and scattering states in terms of open balls in $I\!E^3$ defined by

$$\Lambda_b(r) = \{x = (x, y, z) \in I\!E^n : x^2 + y^2 + z^2 < r\}.$$

[86] See Definition 4.4.2(1).

3.4. ORTHODOX QUANTUM DYNAMICS

Definition 3.4.3(3) **Simple scattering systems**[87]

- A physical system with a propagator $\widehat{\mathcal{U}}$ such that
  1. the wave operators associated with $\widehat{\mathcal{U}}$ and $\widehat{\mathcal{U}}^{(0)}$ exist, and
  2. the ranges of the wave operators are both equal to the set of scattering states of $\widehat{\mathcal{U}}$, i.e.,

$$\mathcal{R}(\widehat{\Omega}_+) = \mathcal{R}(\widehat{\Omega}_-) = \mathcal{S}_{sc}(\widehat{H}_g) \tag{3.247}$$

is called a *simple scattering system*. The corresponding wave operators are said to be *complete*.

Examples of interaction potentials for the Hamiltonian generators for simple scattering systems are available in various texts.[88] These include finite potential wells.[89] A simple scattering system possesses the following features:[90]

1. Its scattering operators have the following properties:

   (a) Every $\varphi \in L^2(I\!\!E^3)$ gives rise to a scattering state ϕ_s of $\widehat{\mathcal{U}}$ by

   $$\phi_s = \widehat{\Omega}_+ \varphi, \tag{3.248}$$

   on account of Eq. (3.247).

   (b) Interwining relations:

   $$\widehat{U}_t \widehat{\Omega}_\pm = \widehat{\Omega}_\pm \widehat{U}_t^{(0)}, \quad \widehat{\Omega}_\pm^\dagger \widehat{U}_t = \widehat{U}_t^{(0)} \widehat{\Omega}_\pm^\dagger. \tag{3.249}$$

   The second expression is obtained from the first by taking the adjoint.[91]

---

[87] Amrein, Jauch and Sinha (1977) p. 141.
[88] Amrein, Jauch and Sinha (1977) Proposition 9.9 on p. 368, Amrein (1981) Proposition 5.34 on p. 176.
[89] Wan and McLean (1994) Lemma 2.
[90] Amrein, Jauch and Sinha (1977) Propositions 4.2 and 4.3 on pp. 142-143.
[91] As pointed out in C2.2.2(1) C4 the adjoint of a strong Cauchy sequence may not converge strongly. The adjoint $\widehat{\Omega}_\pm^\dagger$ is not equal to the straightforward taking of the adjoint of Eq. (3.246). Instead we have

$$\widehat{\Omega}_\pm^\dagger\left(\widehat{H}_g, \widehat{H}_g^{(0)}\right) = \lim_{t\to\pm\infty} \widehat{U}_t^{0\dagger} \widehat{U}_t \widehat{P}\left(\mathcal{S}_{sc}(\widehat{H}_g)\right),$$

where $\widehat{P}\left(\mathcal{S}_{sc}(\widehat{H}_g)\right)$ is the projector onto the subspace $\mathcal{S}_{sc}(\widehat{H}_g)$ of scattering states of $\widehat{H}_g$. To achieve a sense of uniformity we can rewrite the wave operators as

$$\widehat{\Omega}_\pm\left(\widehat{H}_g, \widehat{H}_g^{(0)}\right) = \lim_{t\to\pm\infty} \widehat{U}_t^\dagger \widehat{U}_t^{(0)} \widehat{P}\left(\mathcal{S}_{sc}(\widehat{H}_g^{(0)})\right),$$

where $\widehat{P}(\mathcal{S}_{sc}(\widehat{H}_g^{(0)}))$ is the projector onto the subspace of scattering states of $\widehat{H}_g^{(0)}$. Since $\widehat{P}(\mathcal{S}_{sc}(\widehat{H}_g^{(0)})) = \widehat{I\!\!I}$ we recover the definition in Eq. (3.246).

(c) We have, for every φ in $L^2(\mathbb{E}^3)$,

$$\widehat{\Omega}_+ \widehat{\Omega}_+^\dagger \phi_s = \phi_s \quad \forall \phi_s \in \mathcal{S}_{sc}(\widehat{H}_g), \quad \widehat{\Omega}_+^\dagger \widehat{\Omega}_+ \varphi = \varphi. \qquad (3.250)$$

We can see that $\widehat{\Omega}_+^\dagger$ is acting like an inverse to $\widehat{\Omega}_+$.

(d) Every scattering state ϕ_s of $\widehat{\mathcal{U}}$ gives rise to a vector $\varphi \in L^2(\mathbb{E}^3)$ satisfying Eq. (3.248), i.e.,

$$\varphi = \widehat{\Omega}_+^\dagger \phi_s. \qquad (3.251)$$

2. The evolution of a scattering state at large times becomes indistinguishable from a free evolution of an associated state. To make this statement precise, let ϕ_s be a scattering state under $\widehat{\mathcal{U}}$. Equation (3.251) then associates ϕ_s with another vector φ. At large times the evolution of ϕ_s under $\widehat{\mathcal{U}}$ becomes indistinguishable from the free evolution of φ under $\widehat{\mathcal{U}}^{(0)}$ in the following sense:[92]

$$||\widehat{U}_t \phi_s - \widehat{U}_t^{(0)} \varphi|| = ||\widehat{U}_t(\phi_s - \widehat{U}_t^\dagger \widehat{U}_t^{(0)} \varphi)|| \qquad (3.252)$$
$$= ||(\phi_s - \widehat{U}_t^\dagger \widehat{U}_t^{(0)} \varphi)||. \qquad (3.253)$$

Using Eq. (3.251) we arrive at the following result:

$$||\widehat{U}_t \phi_s - \widehat{U}_t^{(0)} \varphi|| = ||\phi_s - \widehat{U}_t^\dagger \widehat{U}_t^{(0)} \widehat{\Omega}_+^\dagger \phi_s||. \qquad (3.254)$$

Taking the limit as $t \to \infty$ we get, on account of Eq. (3.250),

$$\lim_{t \to \infty} ||\widehat{U}_t \phi_s - \widehat{U}_t^{(0)} \varphi|| = ||\phi_s - \widehat{\Omega}_+ \widehat{\Omega}_+^\dagger \phi_s|| = 0. \qquad (3.255)$$

3. Physically Eq. (3.255) means that at large times $\widehat{U}_t \phi_s$ can be approximated by $\widehat{U}_t^{(0)} \varphi$. In particular, if $\widehat{U}_t^{(0)} \varphi$ asymptotically localizes into a certain region, so does $\widehat{U}_t \phi_s$. This enables us to generalize previous results on asymptotic localization and separation for free systems to simple scattering systems. We can appreciate this generalization through the following analysis. On account of the Schwarz and Triangle inequalities in §2.1 and the fact that all projectors have a unit norm we get, for a pair of vectors ϕ_s and φ related by Eq. (3.251),

$$\left| ||\widehat{M^{\hat{x}}}(t\Delta\mathbf{v})\widehat{U}_t \phi_s|| - ||\widehat{M^{\hat{x}}}(t\Delta\mathbf{v})\widehat{U}_t^{(0)} \varphi|| \right| \qquad (3.256)$$
$$\leq ||\widehat{M^{\hat{x}}}(t\Delta\mathbf{v})\widehat{U}_t \phi_s - \widehat{M^{\hat{x}}}(t\Delta\mathbf{v})\widehat{U}_t^{(0)} \varphi||$$

---
[92] Amrein (1981) p. 140, p. 146.

3.4. ORTHODOX QUANTUM DYNAMICS

$$\begin{aligned}
&= ||\widehat{M}^{\hat{x}}(t\Delta\mathbf{v})\,(\widehat{U}_t\phi_s - \widehat{U}_t^{(0)}\varphi)|| \\
&\leq ||\widehat{M}^{\hat{x}}(t\Delta\mathbf{v})||\,||\widehat{U}_t\phi_s - \widehat{U}_t^{(0)}\varphi|| \\
&= ||\widehat{U}_t\phi_s - \widehat{U}_t^{(0)}\varphi|| \to 0 \quad \text{as } t \to \infty \quad (3.257)
\end{aligned}$$

by Eq. (3.255). It follows from Theorem 3.4.2(2) that

$$\lim_{t\to\infty} ||\widehat{M}^{\hat{x}}(t\Delta\mathbf{v})\widehat{U}_t\phi_s|| = \lim_{t\to\infty} ||\widehat{M}^{\hat{x}}(t\Delta\mathbf{v})\widehat{U}_t^{(0)}\varphi||$$

$$= ||\widehat{M}^{\hat{p}}(m\Delta\mathbf{v})\varphi|| \tag{3.258}$$

$$= ||\widehat{M}^{\hat{p}}(m\Delta\mathbf{v})\Omega^\dagger\phi_s|| \tag{3.259}$$

$$= \lim_{t\to\infty} ||\widehat{M}^{\hat{p}}(m\Delta\mathbf{v})\widehat{U}_t^{0\dagger}\widehat{U}_t\phi_s||$$

$$= \lim_{t\to\infty} ||\widehat{M}^{\hat{p}}(m\Delta\mathbf{v})\widehat{U}_t\phi_s||. \tag{3.260}$$

In the last step we make use of the fact that free evolution operators $\widehat{U}_t^{(0)}$ and their adjoints commute with the momentum operator. We can now summarize our results into a theorem.

Theorem 3.4.3(1)   Asymptotic localization

- A scattering state ϕ_s of a simple scattering system with a propagator $\widehat{\mathcal{U}}$ satisfies the following equation:[93]

$$\lim_{t\to\infty} ||\widehat{M}^{\hat{x}}(t\Delta\mathbf{v})\,\phi_{st}||^2 = ||\widehat{M}^{\hat{p}}(m\Delta\mathbf{v})\,\varphi|| \tag{3.261}$$

where $\phi_{st} = \widehat{U}_t\phi_s$ is the evolved state and $\varphi = \Omega_+^\dagger\phi_s$.

Corollary 3.4.3(1)   Asymptotic localization and separation

- A scattering state ϕ_s of a simple scattering system is asymptotically localizable into the region $t\Delta\mathbf{v}$ if

$$||\widehat{M}^{\hat{p}}(m\Delta\mathbf{v})\,\varphi|| = 1, \quad \text{where } \varphi = \Omega_+^\dagger\phi_s. \tag{3.262}$$

- Two scattering states ϕ_{s1} and ϕ_{s2} of a simple scattering system are asymptotically separable into disjoint regions $t\Delta\mathbf{v}_1$ and $t\Delta\mathbf{v}_2$ if

$$||\widehat{M}^{\hat{p}}(m\Delta\mathbf{v}_1)\,\varphi_1|| = 1, \quad \text{where } \varphi_1 = \Omega_+^\dagger\phi_{s1}, \tag{3.263}$$

$$||\widehat{M}^{\hat{p}}(m\Delta\mathbf{v}_2)\,\varphi_2|| = 1, \quad \text{where } \varphi_2 = \Omega_+^\dagger\phi_{s2}. \tag{3.264}$$

---
[93] Wan and McLean (1983a), (1983b).

3.5 Quantum State Preparation

3.5.1 The problem

An *a priori* notion of state preparation and a notion of quantum measurement are assumed at the outset of quantum theory to enable Postulate OQS in §3.2 on quantum statics to be expressed in terms of *known states* and *measured values* of observables. However, for these *a priori* notions to be physically meaningful we need a prescription to prepare the system in an arbitrary pure state, and a prescription for the measurement of an arbitrary observable. These prescriptions must lie within Postulates OQS and OQD, otherwise we would have to insert further structures or postulates into the theory.

There have been numerous discussions of these two problems through out the years. Traditionally the **projection postulate** of von Neumann is supposed to be able to provide both of these prescriptions at a stroke.[94] This projection postulate assumes the existence of an instantaneous and non-unitary process, i.e., a so-called "measurement process" or "state reduction process", which can evolve a pure state into a mixed state instantly. To illustrate this let the system be in a pure state ϕ at the instant of measurement of an observable $\widehat{A}$. Let us suppose that $\widehat{A}$ possesses a non-degenerate and purely point spectrum $\{a_j\}$ with a corresponding complete orthonormal set of eigenvectors φ_j. Postulate OQS tells us that the measured values of $\widehat{A}$ are the eigenvalues, i.e., each measurement act will result in a value a_j, with probability $\wp_j = \langle \phi \mid \widehat{P}_{\varphi_j} \phi \rangle$. The projection postulate then asserts that the state immediately after such a measurement act is given by the eigenvector φ_j, this being totally independent of the initial state ϕ. Now, consider a repetition of such a process, i.e., the execution of an identical measurement act performed on the same system in the same state ϕ again. This time the process may yield the value a_i. The state right afterward will then be φ_i. So, the situation is that the same measurement act could project the same initial state ϕ into different final states. In other words, the projection postulate assumes the transition of an initial pure state into a classical mixture of states according to:

$$\widehat{P}_\phi \quad \to \quad \widehat{D} = \sum_j \wp_j \widehat{P}_{\varphi_j}. \tag{3.265}$$

Clearly this projection process does not obey Postulate OQD in §3.4.1 on time evolution. This is the crux of the so-called quantum measurement problem which has attracted a great deal of attention in recent years, as seen in the discussions of the subject in text books as well as monographs.[95]

---

[94] Roman (1965) p.12 formally includes the projection postulate into the quantum theory. See also Isham (1995) §8.3, Beltrammetti and Cassinelli (1981) §8, and Jauch (1968) §11.

[95] Wheeler and Zurek (1983), Ludwig Vol. 2 (1985) Chapter XVIII, especially pp. 355-357,

3.5. QUANTUM STATE PREPARATION

The projection postulate also enables a well-chosen single measurement act to transform an unknown state abruptly into any desired pure state, albeit with less than 100% efficiency. For example, if we desire to prepare a state ψ, we could perform a measurement of an observable which admits ψ as an eigenvector with a corresponding non-degenerate eigenvalue. For instance we can perform a measurement of the proposition represented by the projector $\widehat{P}_\psi = |\psi\rangle\langle\psi|$. When the measurement yields the value 1 the state immediately after would be ψ.

The projection postulate, which effectively combines state preparation and measurement, is clearly an additional structure to Postulates OQS and OQD. There is no equation of motion to describe the projection process; there could not be any if the process is instantaneous. As such it is considered by many to be unsatisfactory.[96] We would argue that all state preparation and physical measurement processes performed in a laboratory involve an interaction between the quantum system and some physical devices and that any state preparation and measurement act must in principle take a finite duration to complete, contrary to the projection postulate. Therefore we shall not pursue such instantaneous projection processes any further.

There are other approaches to quantum measurement. An intuitively appealing approach is based on the introduction of *measuring devices* with classical properties.[97] This would again involve additional structures and assumptions going beyond Postulates OQS and OQD.

We now pose the following fundamental question:

Is it possible to construct a theory of state preparation and a measurement theory within the structure of orthodox quantum theory embodied in Postulates OQS and OQD?

A careful analysis reveals that the answer to this question is in the affirmative in the sense that we can formulate state preparation processes as well as measurement processes consistent with Assumption 3.2.5 in §3.2.5 on the approximate nature of experiments without imposing additional structures or postulates on the theory. To show that this is indeed the case we shall present a concrete and general theory of state preparation based on quantum dynamics stated in Postulate OQD. A similar theory for measurement will be discussed in the next section.

The physical situation is this. Suppose we have a *random particle source*, namely a source which produces (identical) particles in various pure states in

---

Greenberger (1986), van Fraassen (1991), Braginsky and Khalili (1992), Mensky (1993), Isham (1995) §8.5. Busch, Lahti and Mittelstaedt (1995), Busch, Grabowski and Lahti (1997).

[96] Cini and Levy-Leblond (1990).
[97] We shall return to discuss the concept of measuring devices in §4.7.

a random manner at the rate of one particle in every specified period of time. Here randomness means:

1. The set of states of the particles emitted by the particle source forms a dense set $\mathcal{S}^{(rd)}$ in the Hilbert space $\mathcal{H}$ associated with the particle.

2. Particles are produced by the source in various states in $\mathcal{S}^{(rd)}$ with a certain probability distribution pre-determined by the physical properties of the source. We do not require a random particle source to emit particles in various states in $\mathcal{S}^{(rd)}$ with an equal probability.

We want to know if we can *process* the particles emitted by the source to produce at least some particles in a desired state ϕ. By processing the particles it is meant the introduction of an interaction, characterizable by a Hamiltonian generator, to evolve the particles coming from the source in a unitary fashion in accordance with Postulate OQD to the desired state ϕ. The difficulty here is that the precise state of any particular particle emitted by the particle source is unknown. If this initial state is known the problem becomes trivial since we can always find a unitary operator to relate any two known unit vectors in $\mathcal{H}$, a fact pointed out earlier in C2.4(1) C3.

3.5.2 Mathematical preliminaries

Our analysis of the state preparation process is based on the relationship between scattering states and bounded states of a scattering system. Intuitively one expects the overlap between a bound state ϕ_b and a scattering state ϕ_s to tend to zero at large times as the scattering state moves away to spatial infinity, i.e., we have

$$\lim_{t\to\infty} \langle \phi_{bt} \mid \phi_{st} \rangle = 0. \tag{3.266}$$

We can go further to show that the correlations between a bound state and a scattering state due to observables, as manifested in the correlation terms in Eqs. (3.41) and (3.42), also tend to zero at large times. These results are stated in two theorems.

Theorem 3.5.2(1) Asymptotically vanishing correlations[98]

- Let $\phi_b \in \mathcal{S}_{bd}(\widehat{H}_g)$ be a bound state and $\phi_s \in \mathcal{S}_{sc}(\widehat{H}_g)$ be a scattering state of a simple scattering system with evolution operators $\widehat{U}_t$ generated by $\widehat{H}_g$ according to Eq. (3.221). Then the correlations between these two

---

[98] Wan and Harrison (1994) Appendix A.

3.5. QUANTUM STATE PREPARATION

states vanish at large times in the sense that for any bounded observable $\widehat{B}$ we have

$$\lim_{t \to \infty} \langle \widehat{U}_t \phi_b \mid \widehat{B} \widehat{U}_t \phi_s \rangle = 0. \tag{3.267}$$

Another useful result concerns the existence of well-behaved time evolution due to Hamiltonian generators with a pre-determined set of eigenvectors and eigenvalues.

Theorem 3.5.2(2) **Hamiltonian generator with pre-assigned eigenvectors and eigenvalues**[99]

- Let $\{\varphi_1, \ldots, \varphi_k\}$ be a finite orthonormal set in $L^2(I\!\!E^3)$ which spans a subspace $\mathcal{M}$, and let $a_1, \ldots, a_k$ be a set of real numbers. Then there is a selfadjoint operator $\widehat{A}$ in $L^2(I\!\!E^3)$ with the property

$$\widehat{A} \varphi_\ell = a_\ell \varphi_\ell, \qquad 1 \leq \ell \leq k \tag{3.268}$$

such that:

1. When serving as a Hamiltonian generator, i.e., taking $\widehat{H}_g = \widehat{A}$, the wave operators $\Omega_\pm(\widehat{H}_g, \widehat{H}_g^{(0)})$ exist and are complete.

2. The scattering subspace $\mathcal{S}_{sc}(\widehat{H}_g)$ of the propagator is equal to $\mathcal{M}^\perp$, the orthogonal complement to $\mathcal{M}$.

3.5.3 Ideal particle source

Suppose that we have an *ideal particle source* finely tuned to emit particles, one at a time, all in the same pure state ζ, albeit unknown. We want to process these particles to produce some particles in a desired state ϕ, despite our ignorance of the initial state ζ. First, we can relate ζ to ϕ by

$$\zeta = c\phi + c^\perp \phi^\perp, \tag{3.269}$$

where $\phi^\perp$ is a unit vector orthogonal to ϕ and c, $c^\perp$ are constants. Here ζ, ϕ and $\phi^\perp$ are all normalized members of $L^2(I\!\!E^3)$. According to the Theorem 3.5.2(2) we have the following results:

1. There exists a selfadjoint operator $\widehat{A}$ which admits ϕ as an eigenfunction corresponding to eigenvalue 0.

---

[99] Wan and McLean (1991), (1994).

2. The system becomes a simple scattering system under the evolution

$$\widehat{U}_t = e^{-itH_g}, \quad \widehat{H}_g = \widehat{A} \tag{3.270}$$

generated by $\widehat{A}$ serving as the Hamiltonian generator.

3. The state ϕ is a bound state under this evolution and every state orthogonal to ϕ, e.g., $\phi^\perp$, is a scattering state. To be specific we have

$$\phi_t = \widehat{U}_t \phi = \phi, \tag{3.271}$$
$$\zeta_t = \widehat{U}_t \zeta = c\widehat{U}_t \phi + c^\perp \widehat{U}_t \phi^\perp = c\phi + c^\perp \phi_t^\perp. \tag{3.272}$$

The fact that $\widehat{U}_t \phi = \phi$ is due to ϕ being an eigenvector of $\widehat{H}_g$ with eigenvalue 0.

With ϕ and $\phi_t^\perp$ being a bound and a scattering states respectively we have, for any bounded observable $\widehat{B}$,

$$\lim_{t \to \infty} \langle \phi_t \mid \widehat{B} \phi_t^\perp \rangle = 0, \tag{3.273}$$

on account of Theorem 3.5.2(1). It follows that at large times the state $\phi_t^\perp$ is not correlated to the bound state ϕ, for all practical purposes, by any observables. This is an important result which can lead to the blurring of the distinction between a coherent superposition and a mixture. Let us go into more details to see how the blurring can occur.

Consider the pure state ζ in Eq. (3.269). Regarded as a coherent superposition of ϕ and $\phi^\perp$ the state ζ is physically distinguishable from the mixed state represented by the density operator

$$\widehat{D} = |c|^2 \widehat{P}_\phi + |c^\perp|^2 \widehat{P}_{\phi^\perp} \tag{3.274}$$

on account of the interference term shown in Eq. (3.41). Next consider the evolution of state ζ. The evolved state ζ_t in Eq. (3.272) is also a pure state. Let us compare ζ_t with the mixed state represented by the density operator

$$\widehat{D}_t = |c|^2 \widehat{P}_\phi + |c^\perp|^2 \widehat{P}_{\phi_t^\perp}. \tag{3.275}$$

Because of Eq. (3.273) we have

$$\lim_{t \to \infty} \langle \zeta_t \mid \widehat{B} \zeta_t \rangle = \lim_{t \to \infty} \operatorname{tr}(\widehat{B}\widehat{D}_t). \tag{3.276}$$

It follows that given any finite set of bounded observables $\mathcal{O}_{qm}^n$ conditions (3.66) and (3.67) are satisfied at sufficiently large times, i.e., for t bigger

3.5. QUANTUM STATE PREPARATION

than some large time τ_0. We can conclude, in accordance with Eqs. (3.68) and (3.69), that

$$\zeta_\tau \equiv \widehat{D}_\tau \quad \text{FAPP}(\mathcal{O}_{qm}^n, \mathcal{S}_{qm}^{res}(\tau_0), \epsilon_{inc}). \tag{3.277}$$

Physically this means that for all practical purposes state ζ_t will evolve into a classical mixture of the desired state ϕ and the state $\phi_t^\perp$ at large times. Since $\phi^\perp$ is a scattering state, $\phi_t^\perp$ will move away to spatial infinity. So, given an ideal particle source tuned to a single unknown state ζ there exists a unitary evolution with a corresponding Hamiltonian generator to produce, with arbitrary accuracy, particles in the desired pure state ϕ. It takes time to achieve this of course. In other words we can conclude that

> a pure state in the form of a coherent superposition of ϕ and $\phi^\perp$ will evolve asymptotically into a state indistinguishable for all practical purposes from a classical mixture of ϕ and a state $\phi_t^\perp$ at a large distance away.

Unfortunately this process fails if the unknown state ζ happens to be orthogonal to the desired state ϕ. So, we would generally need a random particle source. By definition a random particle source will emit some particles in states not orthogonal to the desired state ϕ to enable the above process to operate successfully.

3.5.4 Random particle source

Let us assume that particles coming out of the random particle source is characterizable by a density operator $\widehat{D}^{(rd)}$ of the form[100]

$$\widehat{D}^{(rd)} = \sum_{j=1}^{\infty} w_j \widehat{P}_{\varphi_j}, \tag{3.278}$$

where $\{\varphi_j\}$ is an orthonormal basis in $\mathcal{H}$ contained within a dense set $\mathcal{S}^{(rd)}$ of states emitted by the random particle source. We shall subject these particles to the interaction represented by the Hamiltonian generator $\widehat{H}_g$ introduced in the preceding section with an aim to preparing the state ϕ.

Under the interaction the mixed state $\widehat{D}^{(rd)}$ will evolve in accordance with Eq. (3.215):

$$\widehat{D}_t^{(rd)} = \widehat{U}_t \widehat{D}^{(rd)} \widehat{U}_t^\dagger, \tag{3.279}$$

where $\widehat{U}_t$ is given by Eq. (3.270). Now let

$$\varphi_j = c_j \phi + c_j^\perp \phi_j^\perp, \tag{3.280}$$

---
[100] It does not matter whether the particle source produces a classical or a quantum mixture.

where $\phi_j^\perp$ is a unit vector orthogonal to ϕ and hence it is a scattering state of $\widehat{H}_g$. We have

$$\varphi_{jt} = \widehat{U}_t \varphi_j = c_j \phi + c_j^\perp \phi_{jt}^\perp, \quad \phi_{jt}^\perp = \widehat{U}_t \varphi_j^\perp. \tag{3.281}$$

Let us calculate the trace $\operatorname{tr}(\widehat{D}_t^{(rd)} \widehat{B})$ for a bounded observable $\widehat{B}$, using the orthonormal basis $\{\varphi_j\}$, i.e.,

$$\begin{aligned}
\operatorname{tr}(\widehat{D}_t^{(rd)} \widehat{B}) &= \operatorname{tr}(\widehat{U}_t \widehat{D}^{(rd)} \widehat{U}_t^\dagger \widehat{B}) = \operatorname{tr}(\widehat{B} \widehat{U}_t \widehat{D}^{(rd)} \widehat{U}_t^\dagger) \\
&= \operatorname{tr}(\widehat{U}_t^\dagger \widehat{B} \widehat{U}_t \widehat{D}^{(rd)}) = \sum_{j=1}^\infty \langle \varphi_j \mid \widehat{U}_t^\dagger \widehat{B} \widehat{U}_t \, \widehat{D}^{(rd)} \varphi_j \rangle \\
&= \sum_{j=1}^\infty \langle \widehat{U}_t \varphi_j \mid \widehat{B} \widehat{U}_t \, \omega_j \, \varphi_j \rangle \\
&= \sum_{j=1}^\infty \omega_j \, \langle (c_j \phi + c_j^\perp \phi_{jt}^\perp) \mid \widehat{B} (c_j \phi + c_j^\perp \phi_{jt}^\perp) \rangle. \tag{3.282}
\end{aligned}$$

Using Theorem 3.5.2(1) we obtain

$$\lim_{t \to \infty} \operatorname{tr}(\widehat{D}_t^{(rd)} \widehat{B}) = |c|^2 \langle \phi \mid \widehat{B} \phi \rangle + \sum_{j=1}^\infty \omega_j |c_j^\perp|^2 \left(\lim_{t \to \infty} \langle \phi_{jt}^\perp \mid \widehat{B} \phi_{jt}^\perp \rangle \right), \tag{3.283}$$

where

$$|c|^2 = \sum_{j=1}^\infty \omega_j |c_j|^2. \tag{3.284}$$

We can rewrite Eq. (3.283) as

$$\lim_{t \to \infty} \operatorname{tr}(\widehat{D}_t^{(rd)} \widehat{B}) = \lim_{t \to \infty} \operatorname{tr}(\widehat{D}_t \widehat{B}), \tag{3.285}$$

where

$$\widehat{D}_t = |c|^2 \widehat{P}_\phi + \sum_{j=1}^\infty \left(\omega_j |c_j^\perp|^2 \right) \widehat{P}_{\phi_{jt}^\perp}, \tag{3.286}$$

and

$$|c|^2 + \sum_{j=1}^\infty \omega_j |c_j^\perp|^2 = \sum_{j=1}^\infty \omega_j \left(|c_j|^2 + |c_j^\perp|^2 \right) = 1. \tag{3.287}$$

Physically this means that at large times we have a classical mixture of particles in the desired state ϕ and particles in various scattering states $\phi_{jt}^\perp$ at spatial infinity.

3.5. QUANTUM STATE PREPARATION

We have now established what may be called an *asymptotic state preparation procedure*. This is not just a purely theoretical model. Many practical state preparation processes in laboratories utilize a *waiting strategy* to prepare the ground state of a given system, e.g., given time a system will settle down into its ground state. Indeed there have been attempts at establishing the kind of potentials required to prepare certain states under this waiting strategy.[101] Our present analysis provides a general mathematical and theoretical foundation for such a strategy.

3.5.5 Extension to spin-$\frac{1}{2}$ particles

The Hilbert space for a spin-$\frac{1}{2}$ particle in $I\!\!E^3$ in non-relativistic quantum mechanics is assumed to be[102]

$$\mathcal{H}^{(s)} = L^2(I\!\!E^3) \otimes \mathbb{C}^2. \tag{3.288}$$

Let $\alpha^\uparrow$ and $\alpha^\downarrow$ be an orthonormal basis in $\mathbb{C}^2$ corresponding to spin-up and spin-down along a certain chosen direction. A spin-up state for the particle is given by $\varphi^\uparrow \otimes \alpha^\uparrow$ where $\varphi^\uparrow$, which represents the spatial part of the state, is a normalized element of $L^2(I\!\!E^3)$. We call $\varphi^\uparrow$ a *spatial state* and $\alpha^\uparrow$ a *spin state*. Similarly a spin-down state is $\varphi^\downarrow \otimes \alpha^\downarrow$, where $\varphi^\downarrow$ is a normalized element in $L^2(I\!\!E^3)$. We have employed a small tensor product sign to represent the tensor product of the spatial part and the spin part of the state. Generally a pure state for a spin-$\frac{1}{2}$ particle is a normalized element of $\mathcal{H}^{(s)}$ of the form

$$\Psi^{(s)} = a\,\psi^\uparrow \otimes \alpha^\uparrow + b\,\psi^\downarrow \otimes \alpha^\downarrow \tag{3.289}$$

for some $a, b \in \mathbb{C}$ such that $|a|^2 + |b|^2 = 1$. Definition 3.4.3(1) on bound and scattering states can be extended to cover spin in a straight forward manner, e.g., a state $\Psi^{(s)} \in \mathcal{H}^{(s)}$ is defined to be a *scattering state* of a given propagator $\widehat{\mathcal{U}}^{(s)}$ acting on $\mathcal{H}^{(s)}$ if

$$\lim_{t \to \infty} \left\| \left(\widehat{M^{\hat{x}}}(\Lambda_{cu}(w)) \otimes \widehat{I} \right) \widehat{U}_t^{(s)} \Psi^{(s)} \right\| = 0, \tag{3.290}$$

for every bounded cube $\Lambda_{cu}(w)$ in $I\!\!E^3$. Here $\widehat{I}$ is the identity operator on $\mathbb{C}^2$.

We now want to establish a unitary process for the preparation of spin states. As before we shall begin with an ideal spin-$\frac{1}{2}$ particle source. Let us assume that the particles emitted by the source are in a pure state $\Theta^{(s)} \in \mathcal{H}^{(s)}$,

---
[101] Ballentine (1990) §8.1.
[102] Isham (1995) p. 146. The 2-dimensional Hilbert space $\mathbb{C}^2$ is introduced in Eq. (2.559). The notation $L^2(I\!\!E^3) \otimes \mathbb{C}^2$ denotes a tensor product.

albeit unknown. Suppose we desire to have some particles prepared in the spin-up state
$$\Phi^{(s)} = \phi^\uparrow \otimes \alpha^\uparrow. \tag{3.291}$$
Our problem is to ascertain whether or not there is a propogator which can lead particles to this state. We know, from Theorem 3.5.2(1), that there exists a selfadjoint operator $\widehat{A}$ in $L^2(I\!\!E^3)$ having $\phi^\uparrow$ as an eigenvector with a corresponding eigenvalue 0. Using $\widehat{A}$ as the Hamiltonian generator we obtain an evolution for which every vector orthogonal to $\phi^\uparrow$ is a scattering state. Now let
$$\widehat{P}_{\alpha^\uparrow} = |\alpha^\uparrow\rangle\langle\alpha^\uparrow| \quad \text{and} \quad \widehat{P}_{\alpha^\downarrow} = |\alpha^\downarrow\rangle\langle\alpha^\downarrow| \tag{3.292}$$
be the projectors on $\mathbb{C}^2$ projecting onto the spin-up and the spin-down subspaces respectively. Introduce a one-parameter family of operators $\widehat{U}_t^{(s)}$ on $\mathcal{H}^{(s)}$ by
$$\begin{aligned}\widehat{U}_t^{(s)} &= \widehat{U}_t \otimes \widehat{P}_{\alpha^\uparrow} + \widehat{U}_t^{(0)} \otimes \widehat{P}_{\alpha^\downarrow} \\ &= e^{-it\widehat{H}_g} \otimes \widehat{P}_{\alpha^\uparrow} + e^{-it\widehat{H}_g^{(0)}} \otimes \widehat{P}_{\alpha^\downarrow},\end{aligned} \tag{3.293}$$
where $\widehat{H}_g = \widehat{A}$ and $\widehat{H}_g^{(0)}$ is the usual free Hamiltonian generator in $L^2(I\!\!E^3)$. One can verify that $\widehat{U}_t^{(s)}$ form a one-parameter group of unitary operators on $\mathcal{H}^{(s)}$, although $\widehat{U}_t \otimes \widehat{P}_{\alpha^\uparrow}$ and $\widehat{U}_t^{(0)} \otimes \widehat{P}_{\alpha^\downarrow}$ are not unitary operators on $\mathcal{H}^{(s)}$.[103] It follows that we can define a propagator $\widehat{\mathcal{U}}^{(s)}$ on $\mathcal{H}^{(s)}$ in terms of $\widehat{U}_t^{(s)}$. The spin-up state in Eq. (3.291) is a bound state of $\widehat{\mathcal{U}}^{(s)}$, a fact we shall emphasize by re-writing the state as $\Phi_b^{(s)}$. We shall now state without proof a theorem on the nature on bound and scattering states of a spin-$\frac{1}{2}$ particle.

Theorem 3.5.5(1)   **Bound and scattering states with spin**[104]

- Under an evolution arising from the evolution operators $\widehat{U}_t^{(s)}$ in Eq. (3.293) the spin-up state $\Phi_b^{(s)}$ evolves as
$$\widehat{U}_t^{(s)} \Phi_b^{(s)} = \Phi_b^{(s)} \quad \text{or} \quad \widehat{U}_t^{(s)}(\phi^\uparrow \otimes \alpha^\uparrow) = \phi^\uparrow \otimes \alpha^\uparrow \quad \forall t \in I\!\!R. \tag{3.294}$$

- Every vector in $\mathcal{H}^{(s)}$ orthogonal to $\Phi_b^{(s)} = \phi^\uparrow \otimes \alpha^\uparrow$ is a scattering state of $\widehat{\mathcal{U}}^{(s)}$.

---

[103] Operators of the form $\widehat{U}_t \otimes \widehat{P}_{\alpha^\uparrow}$ are not unitary since its adjoint is not equal to its inverse. To verify that $\widehat{U}_t^{(s)}$ form a unitary group one can employ the following results:
$$\widehat{P}_{\alpha^\uparrow} \widehat{P}_{\alpha^\downarrow} = 0, \quad \widehat{P}_{\alpha^\uparrow} + \widehat{P}_{\alpha^\downarrow} = \widehat{I} \quad \text{on } \mathbb{C}^2.$$

[104] Wan and McLean (1994) Theorem 4.

3.5. QUANTUM STATE PREPARATION

- For every bounded operator $\widehat{B}^{(s)}$ and every scattering state $\Psi_s^{(s)}$ of $\widehat{\mathcal{U}}^{(s)}$ we have

$$\lim_{t\to\infty} \langle \Phi_{bt}^{(s)} \mid \widehat{B}^{(s)} \Psi_{st}^{(s)} \rangle = \lim_{t\to\infty} \langle \Phi_b^{(s)} \mid \widehat{B}^{(s)} \Psi_{st}^{(s)} \rangle = 0, \quad (3.295)$$

$$\lim_{t\to\infty} \langle \Psi_{st}^{(s)} \mid \widehat{B}^{(s)} \Phi_{bt}^{(s)} \rangle = \lim_{t\to\infty} \langle \Psi_{st}^{(s)} \mid \widehat{B}^{(s)} \Phi_b^{(s)} \rangle = 0, \quad (3.296)$$

where

$$\Phi_{bt}^{(s)} = \widehat{U}_t^{(s)} \Phi_b^{(s)} \quad \text{and} \quad \Psi_{st}^{(s)} = \widehat{U}_t^{(s)} \Psi_s^{(s)}. \quad (3.297)$$

A somewhat counter intuitive consequence of the theorem is that any vector in $\mathcal{H}^{(s)}$ of the form $\phi \otimes \alpha^\downarrow$, $\phi \in L^2(\mathbb{E}^3)$, is a scattering state. The reason is that the evolution of this state is only due to the term $\widehat{U}_t^{(0)} \otimes \widehat{P}_{\alpha^\downarrow}$ in $\widehat{U}_t^{(s)}$.[105]

Let us now consider the evolution of state $\Theta^{(s)}$ of the particles emitted by the particle source. In line with the spinless case we shall express $\Theta^{(s)}$ as a sum of the desired state $\Phi^{(s)}$ and another one, i.e.,

$$\Theta^{(s)} = c\,\Phi^{(s)} + c^\perp \Phi^{(s)\perp} = c\,\phi^\uparrow \otimes \alpha^\uparrow + c^\perp \Phi^{(s)\perp}, \quad (3.298)$$

where $\Phi^{(s)\perp}$ is orthogonal to $\Phi^{(s)}$. For time evolution we have

$$\Theta_t^{(s)} = \widehat{U}_t^{(s)} \Theta^{(s)} = c\,\Phi^{(s)} + c^\perp \Phi_t^{(s)\perp}, \quad \Phi_t^{(s)\perp} = \widehat{U}_t^{(s)} \Phi^{(s)\perp}. \quad (3.299)$$

Since $\Phi^{(s)\perp}$ is orthogonal to $\Phi^{(s)}$ it is a scattering state of $\widehat{\mathcal{U}}^{(s)}$. Let

$$\widehat{D}_t^{(s)} = |c|^2 \widehat{P}_{\Phi^{(s)}} + |c^\perp|^2 \widehat{P}_{\Phi_t^{(s)\perp}}, \quad (3.300)$$

where

$$\widehat{P}_{\Phi^{(s)}} = |\Phi^{(s)}\rangle\langle\Phi^{(s)}| \quad \text{and} \quad \widehat{P}_{\Phi_t^{(s)\perp}} = |\Phi_t^{(s)\perp}\rangle\langle\Phi_t^{(s)\perp}|. \quad (3.301)$$

It follows from Theorem 3.5.5(1) that

$$\lim_{t\to\infty} \langle \Theta_t^{(s)} \mid \widehat{B}^{(s)} \Theta_t^{(s)} \rangle \quad (3.302)$$

$$= \lim_{t\to\infty} \left(|c|^2 \langle \Phi^{(s)} \mid \widehat{B}^{(s)} \Phi^{(s)} \rangle + |c^\perp|^2 \langle \Phi_t^{(s)\perp} \mid \widehat{B}^{(s)} \Phi_t^{(s)\perp} \rangle \right) \quad (3.303)$$

$$= \lim_{t\to\infty} \operatorname{tr}\left(\left[|c|^2 \widehat{P}_{\Phi^{(s)}} + |c^\perp|^2 \widehat{P}_{\Phi_t^{(s)\perp}} \right] \widehat{B}^{(s)} \right) \quad (3.304)$$

$$= \lim_{t\to\infty} \operatorname{tr}\left(\widehat{D}_t^{(s)} \widehat{B}^{(s)} \right). \quad (3.305)$$

---

[105] $\phi^\uparrow \otimes \alpha^\downarrow$ is orthogonal to $\Phi_b^{(s)}$. We have

$$\widehat{U}_t^{(s)}\left(\phi^\uparrow \otimes \alpha^\downarrow\right) = \left(e^{-it\widehat{H}_g^{(0)}} \phi^\uparrow\right) \otimes \alpha^\downarrow.$$

The fact that $\phi^\uparrow$ is a scattering state of the free Hamiltonian generator $\widehat{H}_g^{(0)}$ causes $\widehat{U}_t^{(s)}\left(\phi^\uparrow \otimes \alpha^\downarrow\right)$ to evolve to spatial infinity.

Equations (3.302) to (3.305) above hold for all bounded operators $\widehat{B}^{(s)}$ on $\mathcal{H}^{(s)}$ for which the limits exist. Note that $\widehat{D}_t^{(s)}$ is a density operator. At large times the state $\Theta_t^{(s)}$ becomes indistinguishable from the mixture described by the density operator $\widehat{D}_t^{(s)}$. In other words we have, at large times, a mixture of particles in the desired state $\Phi^{(s)} = \phi^\uparrow \otimes \alpha^\uparrow$ and particles in the scattering state $\Phi_t^{(s)\perp}$ at spatial infinity.

As in the spinless case an ideal spin-$\frac{1}{2}$ particle source fails to be useful in our state preparation procedure if the state $\Theta^{(s)}$ of the particles emitted by the source is orthogonal to the desired state $\phi^\uparrow \otimes \alpha^\uparrow$. A random source is then required. The particles emitted by a random source are describable by a density operator $\widehat{D}^{(s)rd}$ on $\mathcal{H}^{(s)}$ of the form

$$\widehat{D}^{(s)rd} = \sum_{j=1}^{\infty} w_j \widehat{P}_{\Theta_j^{(s)}}, \tag{3.306}$$

where $\{\Theta_j^{(s)}, j = 1, 2, \ldots\}$ is an orthonormal basis of $\mathcal{H}^{(s)}$. Now let

$$\Theta_j^{(s)} = c_j \Phi^{(s)} + c_j^\perp \Phi_j^{(s)\perp} \tag{3.307}$$

where $\Phi_j^{(s)\perp}$ are orthogonal to $\Phi^{(s)}$. Then for any bounded operator $\widehat{B}^{(s)}$ on $\mathcal{H}^{(s)}$ we have, following the analysis which leads to Eq. (3.283),

$$\begin{aligned}
\lim_{t \to \infty} \operatorname{tr}\left(\widehat{D}_t^{(s)rd} \widehat{B}^{(s)}\right) &= \lim_{t \to \infty} \sum_{j=1}^{\infty} \langle \Theta_j^{(s)} \mid \widehat{U}_t^{(s)} \widehat{D}^{(s)rd} \widehat{U}_t^{(s)\dagger} \widehat{B}^{(s)} \Theta_j^{(s)} \rangle \\
&= \left(\sum_{j=1}^{\infty} w_j |c|^2\right) \langle \Phi^{(s)} \mid \widehat{B}^{(s)} \Phi^{(s)} \rangle \\
&\quad + \sum_{j=1}^{\infty} \left(w_j |c_j^\perp|^2\right) \lim_{t \to \infty} \langle \Phi_{jt}^{(s)\perp} \mid \widehat{B}^{(s)} \Phi_{jt}^{(s)\perp} \rangle. \tag{3.308}
\end{aligned}$$

Consequently at large times the state $\widehat{D}_t^{(s)rd}$ is indistinguishable from a classical mixture of the desired state $\Phi^{(s)} = \phi^\uparrow \otimes \alpha^\uparrow$ and various scattering states at spatial infinity.

Once we succeeded in preparing state $\phi^\uparrow \otimes \alpha^\uparrow$, we can go on to prepare an arbitrary state of the form of $\Psi^{(s)}$ in Eq. (3.289), given that any two known states can be related by a unitary transformation.

3.6 Quantum Measurement

3.6.1 Local position observables and their measurability

Recall §2.12 which introduces the notion of local and global observables. Orthodox quantum mechanics employs global observables, like position and mo-

3.6. QUANTUM MEASUREMENT

mentum, which do not vanish outside any finite region in the physical space $I\!\!E^n$. The question is how any physical devices can possibly measure such global quantities. Surely any device used would have to encompass the entire physical space.

To be definite let us consider a quantum particle in one-dimensional motion. The position of the particle are traditionally represented by the operator $\widehat{x} = x$ in $L^2(I\!\!E)$. We want to know how we can physically measure the position observable represented by $\widehat{x}$ to obtain the expectation value $\langle \phi \mid \widehat{x}\phi \rangle$ in a given pure state ϕ. A traditional answer is that we can measure $\widehat{x}$ using a Geiger counter. Let us look into this answer carefully.

A Geiger counter is necessarily of finite size and is effective in detecting the particle only over a finite region Λ of $I\!\!E$. A Geiger counter can tell whether the particle is detected in the region Λ or not. For this reason a counter is also known as a *detector*. Let us call this region Λ the *size* of the counter. To obtain the expectation value $\langle \phi \mid \widehat{x}\phi \rangle$ for an extended wave function ϕ, e.g., a Gaussian, we require an infinite number of Geiger counters of finite size to cover the entire $I\!\!E$. Clearly this is impossible in principle since there are not enough material resources on earth to construct such an infinite number of Geiger counters and place them all over the universe! Consequently, both in principle and in practice we have to be content with an approximation to the expectation value, using a finite number of counters covering a sufficiently large area. This is in line with Assumption 3.2.5 in §3.2.5 on experiments. Once we realize that a single Geiger counter at a given location is not capable of measuring the position observable $\widehat{x}$ to yield the desired expectation value, we are then left with the following question:

What does a Geiger counter of size Λ measure, if not the position observable $\widehat{x}$?

To answer this question we observe that a counter (detector) is a device capable of producing only two results, namely the counter either does fire or does not. When a counter of size Λ fires we say that the counter has detected a particle in the region Λ. It is in this sense that a counter is used to ascertain whether a particle is in a certain region Λ or not. The counter enables us to carry out a yes-no experiment which can ascertain whether the measured value of position $\widehat{x}$ is contained in Λ.[106] In accordance with Postulate OQS on quantum statics in §3.2, an observable which is measured by such a yes-no experiment must correspond to a selfadjoint operator; moreover this operator must have only two eigenvalues, one of which may be identified with the yes answer and the other with the no answer. Intuitively we can see that the characteristic function $\chi_\Lambda(x)$ as a multiplication operator fits our requirement.

---
[106] See §3.2.2 for the notion of yes-no experiments.

312	CHAPTER 3. ORTHODOX QUANTUM MECHANICS

This is a projector with eigenvalues 1 and 0. Its eigenfunctions corresponding to the eigenvalue 1 are members of $L^2(I\!\!E)$ vanishing outside Λ while its eigenfunctions corresponding to the eigenvalue 0 are members of $L^2(I\!\!E)$ vanishing inside Λ. In other words, its eigenfunctions corresponding to the eigenvalue 1 describe a particle with probability 1 of being found in Λ. This enables us to use $\chi_\Lambda(x)$ to represent the proposition that

the particle is in the interval Λ

with eigenvalues 1 and 0 corresponding to the answers yes (the counter fires) and no (the counter does not fire) respectively. From E2.8.1(1) E2 we know that $\chi_\Lambda(x)$ is the spectral projector $\widehat{M^{\hat{x}}}(\Lambda)$ of the position operator $\hat{x}$ for the interval Λ. We shall call

$$\widehat{M^{\hat{x}}}(\Lambda) = \chi_\Lambda(x) \quad \text{a local position observable.} \tag{3.309}$$

In contrast the traditional position observable $\hat{x}$ may be called a *global position observable* in $L^2(I\!\!E)$.

Our conclusion here is that directly measurable position observables are the local ones represented by $\widehat{M^{\hat{x}}}(\Lambda)$. There is yet another problem arising from the unbounded nature of $\hat{x}$. Since no physical device can register data of arbitrarily large values we have to approach an unbounded observable by its bounded approximations. Hence, we shall proceed as follows, in order to obtain a global position expectation value $\langle \phi \mid \hat{x}\phi \rangle$ in state ϕ:

1. Following Eq. (3.53) we shall first introduce a bounded approximation of the global position observable $\hat{x}$:

$$\hat{x}_N = \int_{-N}^{N} \tau \, d\widehat{F^{\hat{x}}}(\tau) \, d\tau, \tag{3.310}$$

for some large N.

2. Next we shall follow Eq. (3.57) to bring in a discretized approximation to $\hat{x}_N$. This is done by introducing a partition Π of $(-N, N]$ in terms of small finite intervals

$$\Lambda_j = (\tau_{j-1}, \tau_j]. \tag{3.311}$$

In other words we have

$$[-N, N] = \{\tau_0\} \cup (\tau_0, \tau_1] \cup (\tau_1, \tau_2] \cup \cdots \cup (\tau_{n-1}, \tau_n]. \tag{3.312}$$

We then define a *discretized position operator* $\hat{x}_{N,\Pi}$ by[107]

$$\hat{x}_{N,\Pi} = \sum_{j=1}^{n} \tau_j \widehat{M^{\hat{x}}}(\Lambda_j). \tag{3.313}$$

---

[107] We have $\tau_0 = -N$ and $\widehat{M^{\hat{x}}}(\{\tau_0\}) = 0$ since $\hat{x}$ has a purely continuous spectrum.

3.6. QUANTUM MEASUREMENT

Each operator $\widehat{M^{\hat{x}}}(\Lambda_j)$ in the sum represents a local position observable.

3. For the expectation value we have the following approximation:

$$\langle \phi \mid \hat{x}\phi \rangle \approx \langle \phi \mid \hat{x}_{N,\Pi} \phi \rangle = \sum_{j=1}^{n} \tau_j \langle \phi \mid \widehat{M^{\hat{x}}}(\Lambda_j) \phi \rangle. \quad (3.314)$$

4. The term $\langle \phi \mid \widehat{M^{\hat{x}}}(\Lambda_j) \phi \rangle$ represents the expectation value of local position observable $\widehat{M^{\hat{x}}}(\Lambda_j)$ in state ϕ. According to Postulate OQS this expectation value is also equal to the probability of the particle being found in the region Λ_j. This probability can be physically measured by a Geiger counter of size Λ_j, i.e., by the *count rate* of the counter.[108]

3.6.2 Reduction to local position measurements

Intuitively we can appreciate that observables are often measured indirectly. Take the example of the spin of an electron. We do not really measure the "spin" directly. Consider the familiar Stern-Gerlach experiment for spin measurement. An electron is sent along a horizontal path through a vertically directed inhomogeneous magnetic field. As a result the electron will acquire a vertical velocity component, either up or down, depending on the spin orientation of the electron, i.e., as a result of the interaction between the magnetic field and the spin magnetic moment the spin-up and the spin-down components move apart with the spin-up component going up and the spin-down component going down. This is a quantum mechanical evolution process, i.e., the spatial separation of the state is a quantum unitary time evolution process generated by a Hamiltonian generator which incorporates the interaction between the magnetic field and the spin magnetic moment. Next, we need to determine the vertical deviation of the electron emerging from the magnetic field in order tell its spin orientation. This is done by a direct detection of the electron in a particular location. In other words we claim to have measured the spin and found the spin to be up if we detected the particle by a Geiger counter placed in an "up" location. We can regard the Stern-Gerlach experiment as a two-stage process:

1. Stage one involves a quantum mechanical evolution which results in the spin-up and spin-down components evolving into different locations.

2. Stage two involves a yes-no experiment to ascertain the location of the electron. In other words, the spin measurement is executed at the end by a local position measurement.

---
[108] The count rate is the fraction of particles detected by the counter in repeated measurements under identical conditions.

This leads to the following question:[109]

- Can we reduce all quantum measurements to local position measurements in a similar manner?

An affirmative answer would be significant. First, it would reduce quantum measurement problems in concrete terms to local position measurements. It then follows that in principle a satisfactory theory of local position measurement would provide the basis for a general solution of many fundamental problems relating to quantum measurement. To avoid confusion and extravagant claims we shall devote the rest of this chapter to spell things out in concrete terms.

3.6.3 Spectral separation for spinless particles

Let us consider the measurement of an observable $\widehat{A}$ having a purely point and non-degenerate spectrum $sp(\widehat{A}) = \{a_1, a_2, \ldots\}$ with a corresponding complete orthonormal set of eigenfunctions $\{\varphi_1, \varphi_2, \ldots\}$.[110] In other words we have the following spectral decomposition:

$$\widehat{A} = \sum_{j=1}^{\infty} a_j \widehat{P}_{\varphi_j}, \quad \widehat{P}_{\varphi_j} = |\varphi_j\rangle\langle\varphi_j|. \tag{3.315}$$

Our objective is to obtain the expectation value of $\widehat{A}$ in a given state ϕ. In analogy to the Stern-Gerlach experiment we want to establish a two-stage measurement process to achieve our objective:

Stage 1: Spectral separation Express the initial wave function ϕ as a linear combination of the eigenfunctions, i.e.,

$$\phi = \sum_{j=1}^{\infty} c_j \varphi_j, \quad c_j = \langle \varphi_j \mid \phi \rangle, \tag{3.316}$$

or

$$\phi = \sum_{j=1}^{\infty} \phi_j, \quad \phi_j = c_j \varphi_j. \tag{3.317}$$

The terms in the sum on the right, ϕ_j, are to be referred to as the *spectral components* of $\widehat{A}$ in ϕ. These components are associated with the eigenvalues of $\widehat{A}$, with the square of the norm of each component $||\phi_j||^2 = |c_j|^2$ being

---

[109] Wan and McLean (1991).
[110] If desired, degeneracy can be taken into account by allowing some of the eigenvalues to be the same.

3.6. QUANTUM MEASUREMENT

the probability of a measurement obtaining the value a_j. We would introduce a propagator to evolve the spectral components into spatially disjoint regions where counters are situated. We call this the *spectral separation stage*.[111] This would be an asymptotic process. We would like the originally overlapping spectral components to separate gradually so that their spatial overlaps become arbitrarily small at large times. This is a rather stringent requirement.[112]

Stage 2: Local position measurement   This involves the direct detection of the particle by counters at various locations. We call this the *local position measurement* stage.

A spectral separation is a quantum evolution process in accordance with Postulate OQD on quantum dynamics set out in §3.4.1. The question is whether there exists a propagator generated by a time independent Hamiltonian generator $\widehat{H}_g$ which can produce the desired spectral separation of the initial wave function. The fact that this is not a trivial problem can be seen in the controversy in spin measurement. Busch and Schreock (1989) has shown that the traditional Stern-Gerlach set-up does not lead to a spin measurement with arbitrary accuracy since the spatial overlap of the "spin-up" and the "spin-down" beams is not eliminated even asymptotically. In other words the spectral separation process is not achievable in a traditional Stern-Gerlach experiment. It is then tempting to view this as a fundamental result and go on to argue that as a matter of principle spin can never be accurately or sharply measured. This would necessitate the introduction of a new concept of observables, e.g., the concept of *unsharp observables*, to describe this new type of observables which can only be *unsharply measured* in principle.[113] It is not obvious how one can divide observables, all represented by selfadjoint operators in orthodox quantum mechanics, into those which can be sharply measured and those which cannot be sharply measured.[114] In order to see whether or not this development is inevitable we need to examine whether a spectral separation of different spin components is possible. For example we could try some modified and amended Stern-Gerlach set-up. Before delving into the problem of spin we shall show that for spinless particles it is in principle possible to reduce the measurement of a general observable $\widehat{A}$ to local position measurements by a spectral separation process.

---

[111] The present notion of spectral separation should not be confused with spectral decomposition or spectral resolution used in the context of the spectral theorem.

[112] Busch and Schreock (1989) has shown that the traditional Stern-Gerlach experiment is unable to separate the spin-up and spin-down components into spatially disjoint regions, i.e., a certain finite amount of overlap of the components remains even asymptotically. See also Scully, Lamb and Barut (1989).

[113] Busch, Lahti and Mittelstaedt (1995).

[114] We shall return to this topic in Chapter 5.

To achieve a spectral separation we want the spectral components to evolve into different spatial regions at large times. This idea can be formulated mathematically in the following way:

1. Let $\Delta \mathbf{v}_1, \Delta \mathbf{v}_2, \ldots$ be a collection of mutually disjoint regions in E^3. Then for each $t > 0$ the regions $t\Delta \mathbf{v}_1, t\Delta \mathbf{v}_2, \ldots$ in E^3 will remain mutually disjoint, and their separation will become arbitrarily large as $t \to \infty$.

2. Engineer a propagator to evolve the spectral components ϕ_j in Eq. (3.317) into these disjoint regions. In other words we want an evolution

$$\widehat{U}_t \phi = \sum_{j=1}^{\infty} \widehat{U}_t \phi_j \tag{3.318}$$

such that $\widehat{U}_t \phi_j$ would localize asymptotically into the region $t\Delta \mathbf{v}_j$, i.e., at some large time T the component $\widehat{U}_T \phi_j$ will localize, for all practical purposes, in $T\Delta \mathbf{v}_j$.[115] This spectral seperation process is illustrated by the diagram below.

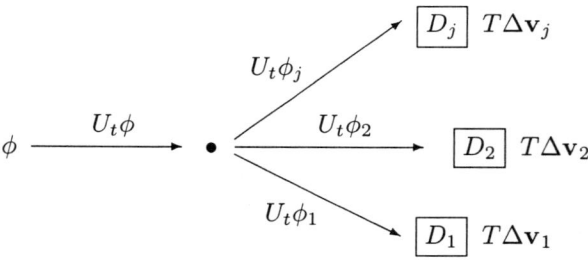

Fig. 3.6.3   Spectral separation.

3. Place detector (counter) D_j of size $T\Delta \mathbf{v}_j$ in each region $T\Delta \mathbf{v}_j$ to detect the arrival of the particle in that region.

4. Physically we can have only a finite number, say N, of counters in any experimental set-up. We can place these N counters in N disjoint regions,

---

[115]Strictly speaking it is $\widehat{U}_t \varphi_j = \widehat{U}_t \phi_j / c_j$ which localizes in the sense of Definition 3.4.2(1).

3.6. QUANTUM MEASUREMENT

$T\Delta \mathbf{v}_1, \ldots, T\Delta \mathbf{v}_N$ at time T. Such an array of detectors enables us to detect positively the evolved spectral components $\{\widehat{U}_T \phi_j,\ 1 \leq j \leq N\}$ at time T. This means that we can only measure a bounded approximation

$$\widehat{A}_N = \sum_{j=1}^{N} a_j \widehat{P}_{\varphi_j} \tag{3.319}$$

to $\widehat{A}$. The operator $\widehat{A}_N$ possesses the same eigenvalues a_j and eigenfunctions φ_j as $\widehat{A}$ for $j = 1, \ldots, N$, along with an additional infinitely degenerate eigenvalue 0. However, this can approximate $\widehat{A}$ with arbitrary accuracy, provided N is sufficiently large, on account of Eq. (3.54).

The following theorem ensures the existence of a propagator to effect the above spectral separation.

Theorem 3.6.3(1)   **Spectral separation**[116]

- Let $\{\psi_j, 1 \leq j \leq K\}$ be a finite orthonormal set in $L^2(I\!\!E^3)$ and let $\{\Delta \mathbf{v}_j, 1 \leq j \leq K\}$ be a set of mutually disjoint regions in $I\!\!E^3$, each of non-zero Lebesgue measure. Then there is a propagator $\widehat{\mathcal{U}}$ on $L^2(I\!\!E^3)$ generated by a time independent Hamiltonian generator $\widehat{H}_g$ such that:

  1. Every element of $L^2(I\!\!E^3)$ is a scattering state of $\widehat{\mathcal{U}}$.

  2. The wave operator $\widehat{\Omega}^+ = \text{s-lim}_{t \to \infty} \widehat{U}_t^{\dagger} \widehat{U}_t^{(0)}$ exists and is complete and unitary.

---

[116] See Wan and McLean (1991), (1994) for a proof of the first 3 statements of the theorem. For the last statement consider $j \neq k = \ell$ first. Since $\widehat{M}^{\hat{x}}(t\Delta \mathbf{v}_j) + \widehat{M}^{\hat{x}}(t\Delta \mathbf{v}_k)$ is equal to the projector $\widehat{M}^{\hat{x}}(t\Delta \mathbf{v}_j \cup t\Delta \mathbf{v}_k)$ we have

$$\langle \psi_{kt} \mid \widehat{M}^{\hat{x}}(t\Delta \mathbf{v}_j \cup t\Delta \mathbf{v}_k) \psi_{kt}\rangle \leq 1 \qquad \text{and}$$

$$\lim_{t \to \infty} \langle \psi_{kt} \mid \widehat{M}^{\hat{x}}(t\Delta \mathbf{v}_j \cup t\Delta \mathbf{v}_k) \psi_{kt}\rangle = \lim_{t \to \infty} \langle \psi_{kt} \mid \widehat{M}^{\hat{x}}(t\Delta \mathbf{v}_j) \psi_{kt}\rangle$$
$$+ \lim_{t \to \infty} \langle \psi_{kt} \mid \widehat{M}^{\hat{x}}(t\Delta \mathbf{v}_k) \psi_{kt}\rangle.$$

$\Rightarrow \qquad \lim_{t \to \infty} \langle \psi_{kt} \mid \widehat{M}^{\hat{x}}(t\Delta \mathbf{v}_j) \psi_{kt}\rangle = 0 \quad \text{since} \quad \lim_{t \to \infty} \langle \psi_{kt} \mid \widehat{M}^{\hat{x}}(t\Delta \mathbf{v}_k) \psi_{kt}\rangle = 1.$

When $k \neq \ell$ we get

$$|\langle \psi_{\ell t} \mid \widehat{M}^{\hat{x}}(t\Delta \mathbf{v}_j) \psi_{kt}\rangle| \leq \|\widehat{M}^{\hat{x}}(t\Delta \mathbf{v}_j) \psi_{kt}\| = \langle \psi_{kt} \mid \widehat{M}^{\hat{x}}(t\Delta \mathbf{v}_j) \psi_{kt}\rangle.$$

3. Under the propagator $\widehat{\mathcal{U}}$ members of the orthonormal set ψ_j will asymptotically localize into region $t\Delta\mathbf{v}_j$, i.e.,

$$\lim_{t\to\infty} \left\| \widehat{M}^{\hat{x}}(t\Delta\mathbf{v}_j)\,\widehat{U}_t\psi_j \right\|^2 = \lim_{t\to\infty} \langle \widehat{U}_t\psi_j \mid \widehat{M}^{\hat{x}}(t\Delta\mathbf{v}_j)\,\widehat{U}_t\psi_j\rangle \quad (3.320)$$
$$= 1. \quad (3.321)$$

4. We also have the following manifestation of asymptotic separation:

$$\lim_{t\to\infty} \langle \widehat{U}_t\psi_\ell \mid \widehat{M}^{\hat{x}}(t\Delta\mathbf{v}_j)\,\widehat{U}_t\psi_k\rangle = 0, \quad \text{unless } \ell = j = k. \quad (3.322)$$

Corollary 3.6.3(1)   **On asymptotic localization**

- Given any state ψ and a rectangular region Δ in $I\!\!E^3$ there exists a propagator $\widehat{\mathcal{U}}$ generated by a time independent Hamiltonian generator such that

$$\lim_{t\to\infty} \langle \widehat{U}_t\psi \mid \widehat{M}^{\hat{x}}(t\Delta)\,\widehat{U}_t\psi\rangle = 1, \quad (3.323)$$

i.e., $\widehat{U}_t\psi$ will localize asymptotically to $t\Delta$.

The aim of our experiment is to obtain the expectation value $\mathcal{E}(\widehat{A},\phi)$ of observable $\widehat{A}$ for the system in a given state ϕ. We have argued that $\mathcal{E}(\widehat{A},\phi)$ can be approximated by the expectation value of $\widehat{A}_N$ in ϕ. Using Eqs. (3.317) and (3.319) we can write down the expression for $\mathcal{E}(\widehat{A}_N,\phi)$ as[117]

$$\mathcal{E}(\widehat{A}_N,\phi) = \sum_j^N \langle \phi_j \mid \widehat{A}_N\phi_j\rangle = \sum_{j=1}^N |c_j|^2\,a_j. \quad (3.324)$$

Given $\widehat{A}$ we can calculate its eigenvalues a_j and eigenfunctions φ_j. If we also know the inital state ϕ we can calculate the coefficients c_j and obtain the right hand sum. Without the knowledge of ϕ this calculation cannot proceed. This is where a measurement process should come in. Our measurement procedure can obtain $|c_j|^2$, and hence the expectation value of $\widehat{A}$ in an unknown pure state. To see this we note that, on account of Theorem 3.6.3(1), there exists a

---

[117] Here we have assumed that the initial state ϕ is not orthogonal to any $\phi_j = c_j\varphi_j$, $j = 1, 2, \ldots, N$. If it is then we have to increase the value of N in order to obtain a non-trivial value. This is consistent with our discussion in §3.2.4 on the approximate nature of measurement, e.g., a particular measurement procedure is workable only on a restricted set of states.

3.6. QUANTUM MEASUREMENT

propagator under which the eigenvectors φ_j, $j = 1, \ldots, N$, will asymptotically localize into disjoint regions $t\Delta \mathbf{v}_j$ so that

$$\lim_{t \to \infty} \langle \widehat{U}_t \phi \mid \widehat{M}^{\widehat{x}}(t\Delta \mathbf{v}_j) \widehat{U}_t \phi \rangle$$

$$= \lim_{t \to \infty} \sum_{\ell,k=1}^{\infty} c_\ell^* c_k \langle \widehat{U}_t \varphi_\ell \mid \widehat{M}^{\widehat{x}}(t\Delta \mathbf{v}_j) \widehat{U}_t \varphi_k \rangle = |c_j|^2. \quad (3.325)$$

In other words at time t_{ss} large enough to effect a spectral separation to any desired accuracy, and hence the subscripts, we have

$$|c_j|^2 \approx \langle \widehat{U}_{t_{ss}} \phi \mid \widehat{M}^{\widehat{x}}(t_{ss}\Delta \mathbf{v}_j) \widehat{U}_{t_{ss}} \phi \rangle \quad \text{FAPP}. \quad (3.326)$$

This means that we can obtain each $|c_j|^2$ to any desired accuracy by an appropriate measurement of the local position observable $\widehat{M}^{\widehat{x}}(t_{ss}\Delta \mathbf{v}_j)$ at time t_{ss}, using a counter D_j sited in the region $t_{ss}\Delta \mathbf{v}_j$. As emphasized earlier counter D_j measures local position observable $\widehat{M}^{\widehat{x}}(t_{ss}\Delta \mathbf{v}_j)$, i.e., the count rate registered by D_j can be identified with the expectation value $\langle \widehat{U}_{t_{ss}} \phi \mid \widehat{M}^{\widehat{x}}(t\Delta \mathbf{v}_j) \widehat{U}_{t_{ss}} \phi \rangle$ of the local position observable $\widehat{M}^{\widehat{x}}(t_{ss}\Delta \mathbf{v}_j)$, and hence with $|c_j|^2$.

We can extend the theory to an observable $\widehat{A}$ whose spectrum is not purely discrete. For example, if $\widehat{A}$ has a purely continuous spectrum we can employ discretized approximations in Eq. (3.57) to enable the above measurement procedure to operate.

3.6.4 Spectral separation for spin-$\frac{1}{2}$ particles

We desire to measure the spin along a chosen direction. Any initial state $\Psi^{(s)}$ of a spin-$\frac{1}{2}$ particle can be written in the form of Eq. (3.289). We wish to find a propagator $\widehat{\mathcal{U}}^{(s)}$ on the Hilbert space $\mathcal{H}^{(s)} = L^2(I\!\!E^3)_\otimes \mathbb{C}^2$ of the spin-$\frac{1}{2}$ particle which will cause the spin-up and spin-down components in a given initial state to evolve asymptotically into disjoint spatial regions. The existence of such a propagator can be established as follows. Let $\widehat{U}_t^\uparrow$ be a one-parameter group of unitary operators on $L^2(I\!\!E^3)$, and $\widehat{U}_t^\downarrow$ be another group of unitary operators. Following Eq. (3.293) we introduce

$$\widehat{U}_t^{(s)} = \widehat{U}_t^\uparrow \otimes \widehat{P}_{\alpha^\uparrow} + \widehat{U}_t^\downarrow \otimes \widehat{P}_{\alpha^\downarrow}. \quad (3.327)$$

Then $\widehat{U}_t^{(s)}$ form a one-parameter group of unitary operators on $L^2(I\!\!E^3)_\otimes \mathbb{C}^2$. Note that $\widehat{U}_t^\uparrow \otimes \widehat{P}_{\alpha^\uparrow}$ will annihilate any spin-down state and $\widehat{U}_t^\downarrow \otimes \widehat{P}_{\alpha^\downarrow}$ will annihilate any spin-up state. So, employing $\widehat{U}_t^{(s)}$ as evolution operators we obtain

$$\widehat{U}_t^{(s)} \Psi^{(s)} = a \left(\widehat{U}_t^\uparrow \psi^\uparrow \right) \otimes \alpha^\uparrow + b \left(\widehat{U}_t^\downarrow \psi^\downarrow \right) \otimes \alpha^\downarrow. \quad (3.328)$$

Note that $\widehat{U}_t^\uparrow$ and $\widehat{U}_t^\downarrow$ are separate and independent unitary groups on $L^2(I\!\!E^3)$. On account of Corollary 3.6.3(1) we can choose $\widehat{U}_t^\uparrow$ and $\widehat{U}_t^\downarrow$ so that $\widehat{U}_t^\uparrow \phi^\uparrow$ and $\widehat{U}_t^\downarrow \phi^\downarrow$ evolve into disjoint regions. This result is summed up in the following theorem.

Theorem 3.6.4(1)   **Spectral separation with spin**

- Let $\Delta\mathbf{v}^\uparrow$, $\Delta\mathbf{v}^\downarrow$ be two disjoint Borel sets in $I\!\!E^3$, each of non-zero Lebesgue measure. Then there is a propagator $\widehat{\mathcal{U}}^{(s)}$ on $\mathcal{H}^{(s)} = L^2(I\!\!E^3) \otimes \mathbb{C}^2$ such that

$$\lim_{t\to\infty} \left\| \left(\widehat{M^{\hat{x}}}(t\Delta\mathbf{v}^\uparrow) \otimes \widehat{I}\right) \widehat{U}_t^{(s)} \left(\phi^\uparrow \otimes \alpha^\uparrow\right) \right\| = 1, \qquad (3.329)$$

$$\lim_{t\to\infty} \left\| \left(\widehat{M^{\hat{x}}}(t\Delta\mathbf{v}^\downarrow) \otimes \widehat{I}\right) \widehat{U}_t^{(s)} \left(\phi^\downarrow \otimes \alpha^\downarrow\right) \right\| = 1. \qquad (3.330)$$

Similar to our earlier result that a definite spin state can be prepared with an arbitrary degree of confidence we can now see that spin, along with other observables, can indeed be measured with an arbitrary accuracy. We can actually imagine how such a set-up can be achieved physically. The particles are sent through a Stern-Gerlach magnet first. On the screen a slit is made where the spin-up particles are most likely to land. Then, particles emerging from the slit would have a high spin-up probability. We can repeat this process by letting these particles though a further Stern-Gerlach magnet and a screen to filter out the spin-up beam. The particles emerging at the end, after many repeats of this filtering process, would be in a spin-up state with the desired accuracy.

To complete our measurement theory within orthodox quantum mechanics, we must know how in principle a local position measurement can be achieved. In particular we want to know to what extent classical devices have to be brought in for this purpose.

3.6.5   Local position measurement as an ionization process

Up to now we have taken for granted the properties of a detector as a physical device capable of detecting the presence of a particle in a given spatial region. We want to see how far orthodox quantum theory can go in establishing a theoretical model for a detector, and at what point classical physics comes into play.

We shall begin by an examination of some of the usual physical devices for detecting a particle. A cloud chamber is a good example. It contains supersaturated vapour which condenses on any ions in the chamber to form droplets

3.6. QUANTUM MEASUREMENT

that can grow rapidly to a visible size. An ionizing particle entering the cloud chamber will ionize some of the vapour molecules which will then form the nuclei for condensation. The presence of the particle can be ascertained by the subsequent emergence of visable droplets round the ions. The formation of droplets is an amplification process for the benefit of the human eye. Various studies suggest that this process can be satisfactorily understood in terms of classical physics.[118] The working of the Geiger counter follows a similar line, i.e., an ionization process followed by an amplification in the form an electrical discharge on the macroscopic scale.

Another familiar example is the photographic plate for the detection of a photon. A photographic film contains a huge number of individual grains each of which is composed of billions of molecules of a silver compound such as silver bromide. A single photon hitting a grain may be able to "ionize" a silver bromide molecule, i.e., to separate the silver atom from a compound molecule. Such a single neutral silver atom within a grain of billions of silver bromide molecules turns out to be unstable and will soon recombine to re-form a silver bromide molecule. However, if the photon having sufficient energy comes, or if a few photons come, and interact with a grain resulting in a cluster of several neutral silver atoms the situation becomes stable. Later, when the film is developed this small cluster of neutral silver atoms is able to cause all the silver ions in the grain to form a grain of visible metallic silver. Here we see that the formation of a stable situation of a few neutral silver atoms in a sea of silver bromide molecules in a grain, known as a *latent image*, corresponds to the ionization stage in the operation of the cloud chamber; the development of the film corresponds to the droplet formation, i.e., to an amplification process.[119] While the amplification process in a cloud chamber or a Geiger counter occurs soon after ionization,[120] the same need not be true in a photographic film which can be developed a long time after the latent images are formed, perhaps years after. We would therefore argue that the essential part of the quantum measurement process is at an end when the formation of the latent image is completed, not when the metallic silver dot becomes visible by developing the film years later. In other words a measurement is considered taken when the film is exposed, not when the exposed film has been developed. The development of the film is merely an amplification of already determined latent images. Of course one can go on to formulate separately a quantum mechanical model of the amplification process.[121] However there are cases where a classical treatment may also be satisfactory, cloud chambers

---

[118] Gupta and Ghosh (1946), Wilson (1951), Yuan (1961).
[119] Hey and Walters (1989), Mees (1963). It is of interest to note that manufacturers of photographic films such as Kodak are among the world's largest users of silver.
[120] In the order of milliseconds.
[121] Hepp (1972).

being an example.

We shall therefore take the view, as proposed by Jauch,[122] that the essential part of a local position measurement, i.e., ionization, is a purely quantum process describable in terms of quantum dynamics in accordance with Postulate OQD stated in §3.4.1 and involving a direct interaction of the incoming ionizing particle and another microscopic quantum system. This view differs from a widespread belief that in a quantum measurement process the system being measured necessarily has to interact directly with a macroscopic measuring device which is semi-classical in some sense.

For simplicity let us consider the ionization of an atom. An ionization process can be broken down into two stages:

1. *Initiation of the ionization process*   The electron, initially in a bound state of the atomic Hamiltonian generator, evolves into a scattering state through its interaction with the incoming ionizing particle.

2. *Completion of the ionization process*   The scattering state evolves and moves to spatial infinity. In the asymptotic limit the electron becomes dissociated from the remaining part of the atom in the sense of Theorems 3.5.2(1) and 3.5.5(1). The atom becomes an ion, and ionization is achieved.

Being an entirely quantum mechanical process we can produce various model propagators to provide a schematic model of ionization. Our reasoning on the role of ionization in quantum measurement does not depend on any particular model. The energy required for the ionization process is provided by the incoming particle whose position we want to measure. In practice the completion of the ionization is achieved, for all practical purposes, within finite time, and it is then followed by an amplification process by which the presence of the ion is manifested as a macroscopically observable event.

Let us illustrate the point of view presented so far with a simple and highly idealized model of ionization of a detector atom. A neutral detector atom is assumed to have an infinitely heavy nucleus serving as the centre of an attractive potential which vanishes at spatial infinity. Electrons are bound to the atom by the potential. For simplicity let us consider a detector atom with one single bound electron as a one-dimensional system, i.e., we have the following model:

1. The Hilbert space for the electron is taken as $\mathcal{H}_{(e)} = L^2(I\!\!E)$.

2. The Hamiltonian generator $\widehat{H}_{(e)g}$ for the evolution of the electron in the atom coincides with the Hamiltonian $\widehat{H}_{(e)}$ of the electron.

---

[122] Jauch (1968) p. 169.

3.6. QUANTUM MEASUREMENT

3. The Hamiltonian $\widehat{H}_{(e)}$ incorporates a potential which vanishes at spatial infinity. For simplicity we shall assume that:[123]

   (a) $\widehat{H}_{(e)}$ admits only one nondegenerate and negative eigenvalue $-\epsilon_b$, where $\epsilon_b > 0$, with eigenvector $\psi_b \in \mathcal{H}_{(e)}$. It follows that $\widehat{H}_{(e)}$ admits ψ_b as the only bound state.

   (b) All states orthogonal to ψ_b are scattering states with a positive energy expectation value, i.e., $\langle \psi_s \mid \widehat{H}_{(e)} \psi_s \rangle > 0$.

   (c) A neutral detector state corresponds to the electron being in the eigenstate ψ_b. This is a bound as well as a stationary state of the Hamiltonian generator $\widehat{H}_{(e)g}$.

   (d) An initial ionization process would cause the electron to evolve from ψ_b to a scattering state ψ_s.

For ease of reference and in order to avoid the complication of having to antisymmetrize states for identical Fermions, we shall take the ionizing particle to be a proton of mass m_p. The Hilbert space for the proton is taken to be $\mathcal{H}_{(p)} = L^2(\mathbb{E})$. For simplicity, we shall assume that the proton is evolving freely under a propagator generated by the usual free Hamiltonian $\widehat{H}_{(p)}^{(0)} = \widehat{K}_{(p)}$ in $\mathcal{H}_{(p)}$; then all states in $\mathcal{H}_{(p)}$ are scattering states. The proton Hilbert space can be decomposed into the direct sum of two subspaces $\mathcal{H}_{(p)+}$ and $\mathcal{H}_{(p)-}$, corresponding respectively to positive and negative momentum values, i.e.,

$$\xi_+(x) \in \mathcal{H}_{(p)+} \quad \Leftrightarrow \quad \text{Fourier transform } \widetilde{\xi}_+(p) \text{ vanishes for } p < 0, \quad (3.331)$$

$$\xi_-(x) \in \mathcal{H}_{(p)-} \quad \Leftrightarrow \quad \text{Fourier transform } \widetilde{\xi}_-(p) \text{ vanishes for } p > 0. \quad (3.332)$$

A proton in state $\xi_+ \in \mathcal{H}_{(p)+}$ has no probability of having a negative momentum value. An incoming proton and the electron form a composite system describable by the Hilbert space $\mathcal{H}^{(c)} = \mathcal{H}_{(p)} \otimes \mathcal{H}_{(e)}$.[124] The system is assumed to be initially in state $\xi_+ \otimes \psi_b$ which corresponds to an incoming proton on the far left and an electron bound to the atom near the coordinate origin. To tear the electron away from the atom the incoming proton must possess sufficient energy. We want an ionization propagator $\widehat{\mathcal{U}}^{(c)}$ on $\mathcal{H}^{(c)}$ capable of causing the initial ionization of the electron within a finite time interval δt, i.e.,

$$\xi_+ \otimes \psi_b \to \widehat{\mathcal{U}}_{\delta t}^{(c)}(\xi_+ \otimes \psi_b) = \xi'_+ \otimes \psi_s. \quad (3.333)$$

---

[123] Selfadjoint operators with the desired properties exist. For example, the depth and width of a finite square potential well may be chosen to produce a Hamiltonian for the well which would admit only a single nondegenerate eigenvalue (Merzbacher (1998) pp. 103-107).

[124] For composite systems see Jauch (1968) §11.8, or Peres (1995) Chapter 5.

Once the electron is in the scattering state ψ_s of $\widehat{H}_{(e)g}$ it will evolve to spatial infinity to complete the ionization process on its own.

For all practical purposes the completion stage of the ionization process may be considered effectively achieved at some large time t_c after the end of the initiation stage, and the whole ionization process is thus effectively completed in a time interval $\delta t + t_c \approx t_c$.

Both the ionization completion time t_c and the spectral separation time t_{ss} needed to produce an effective spectral separation in the first part of the measurement process are long on atomic scales. In laboratory terms, however, the spectral separation and ionization processes, which together constitute the *quantum mechanical* part of our quantum measurement process, are effectively achieved very quickly. Once this has happened, the classical amplification, by which the presence of the proton is revealed to the observer, can commence, although in some cases there may be a considerable lapse of time before it does so. The entire measurement procedure (comprising the quantum mechanical spectral separation and ionization stages and the classical amplification stage) and in particular the local position measurement process (ionization and amplification) are capable of repeated application. For example, in the latter case, it is possible to construct a model for the formation of a sequence of ionizations yielding, say, a track in a cloud chamber.

3.6.6 A model ionization propagator

Let us now fill in the technical details of the model ionization interaction presented in the preceding section.[125] We shall adhere to the same notation. Just before the ionization process begins the electron is in the bound state ψ_b and the proton is the state ξ_+. The bound state wave function ψ_b is assumed to spread mainly over a finite region Δ in space overlapping the proton wave function ξ_+. The initial ionization process causes the electron to move to a scattering state ψ_s. The Hilbert space for the composite system of the electron and the proton is $\mathcal{H}^{(c)} = \mathcal{H}_{(e)} \otimes \mathcal{H}_{(p)}$. Before the ionization interaction begins the proton's energy is described by its kinetic energy operator $\widehat{K}_{(p)}$.

Let $\{\psi_n\}$ be an orthonormal basis in $\mathcal{H}_{(e)}$ with $\psi_1 = \psi_b$. Then, all other members of this basis are scattering states of the atomic Hamiltonian $\widehat{H}_{(e)}$. Let us choose the basis so that $\psi_2 = \psi_s$. To lift the electron from the bound state $\psi_1 = \psi_b$ to the scattering state $\psi_2 = \psi_s$ we need a certain amount of energy to be provided by the proton. Since the electron in state ψ_b has an energy $-\epsilon_b$ the ionization energy required should be

$$\epsilon_{ion} = \epsilon_s + \epsilon_b, \quad \text{where } \epsilon_s = \langle \psi_s \mid \widehat{H}_{(e)s} \psi_s \rangle > 0. \quad (3.334)$$

---

[125] Wan and Harrison (1994).

3.6. QUANTUM MEASUREMENT

In other words if the proton energy is strictly bigger than ϵ_{ion}, i.e., it has no probability of having an energy less than ϵ_{ion}, the proton should be able to ionize the electron. In terms of the proton's kinetic energy probability amplitude function $\tilde{\xi}_+(\epsilon)$, which is given by Eq. (2.405) with interpretation given by Eqs. (3.15) and (3.16), we require that

$$\tilde{\xi}_+(\epsilon) = 0 \quad \forall \epsilon < \epsilon_{ion}. \tag{3.335}$$

We shall now introduce two new elements $\xi_{+d}(x)$ and $\xi_{+u}(x)$ in $\mathcal{H}_{(p)+}$ by the following prescription:

1. By shifting the energy probability amplitude function of the proton state down an amount ϵ_{ion} we get

$$\tilde{\xi}_{+d}(\epsilon) = \tilde{\xi}_+(\epsilon + \epsilon_{ion}). \tag{3.336}$$

By taking the inverse transform defined by Eq. (2.408) we obtain

$$\xi_{+d}(x) = \int_0^\infty f_+(\epsilon, x)\tilde{\xi}_{+d}(\epsilon)\, d\epsilon = \int_0^\infty f_+(\epsilon, x)\tilde{\xi}_+(\epsilon + \epsilon_{ion})\, d\epsilon. \tag{3.337}$$

2. By shifting the energy probability amplitude function of the proton state up an amount ϵ_{ion} we get

$$\tilde{\xi}_{+u}(\epsilon) = \tilde{\xi}_+(\epsilon - \epsilon_{ion}). \tag{3.338}$$

By taking the inverse transform we obtain

$$\xi_{+u}(x) = \int_0^\infty f_+(\epsilon, x)\tilde{\xi}_{+u}(\epsilon)\, d\epsilon = \int_0^\infty f_+(\epsilon, x)\tilde{\xi}_+(\epsilon - \epsilon_{ion})\, d\epsilon. \tag{3.339}$$

Next, we define a linear operator $\widehat{H}_I^{(c)}$ on $\mathcal{H}^{(c)}$ which has the effect of shifting the energy of the particles by ϵ_{ion}:

$$\begin{aligned}
\widehat{H}_I^{(c)}(\xi_+ \otimes \psi_b) &= \xi_{+d} \otimes \psi_s, & \forall \xi_+ \in \mathcal{H}_{(p)+}, \\
\widehat{H}_I^{(c)}(\xi_+ \otimes \psi_s) &= \xi_{+u} \otimes \psi_b, & \forall \xi_+ \in \mathcal{H}_{(p)+}, \\
\widehat{H}_I^{(c)}(\xi_+ \otimes \psi_n) &= 0, & \forall \xi_+ \in \mathcal{H}_{(p)+} \text{ and } n > 2, \\
\widehat{H}_I^{(c)}(\xi_- \otimes \psi_n) &= 0, & \forall \xi_- \in \mathcal{H}_{(p)-} \text{ and } \forall n.
\end{aligned} \tag{3.340}$$

This operator is bounded and selfadjoint. Since shifting the energy of $\xi_{+d}(x)$ up again in accordance with Eq. (3.339) by ϵ_{ion} recovers $\xi_+(x)$ we get

$$\widehat{H}_I^{(c)}(\xi_{+d} \otimes \psi_s) = \xi_+ \otimes \psi_b. \tag{3.341}$$

An explicit expression for $\widehat{H}_I^{(c)}$ can be written down in terms of the following symbolic operator expressions:

$$\widehat{P}_{sb}^{(c)}(\epsilon, x) = |f_+(\epsilon - \epsilon_{ion}, x) \otimes \psi_s\rangle \langle f_+(\epsilon, x) \otimes \psi_b|, \qquad (3.342)$$

$$\widehat{P}_{bs}^{(c)}(\epsilon, x) = |f_+(\epsilon, x) \otimes \psi_b\rangle \langle f_+(\epsilon - \epsilon_{ion}) \otimes \psi_s|. \qquad (3.343)$$

When acting on $\xi_+ \in \mathcal{H}_{(p)+}$ and $\xi_- \in \mathcal{H}_{(p)-}$ these operators have the following effects:[126]

$$\widehat{P}_{sb}^{(c)}(\epsilon, x)\left(\xi_+ \otimes \psi_b\right) = \tilde{\xi}_+(\epsilon) \, f_+(\epsilon - \epsilon_{ion}, x) \otimes \psi_s \qquad (3.344)$$

$$\widehat{P}_{bs}^{(c)}(\epsilon, x)\left(\xi_+ \otimes \psi_s\right) = \tilde{\xi}_+(\epsilon - \epsilon_{ion}) \, f_+(\epsilon, x) \otimes \psi_b, \qquad (3.345)$$

and

$$\widehat{P}_{sb}^{(c)}(\epsilon, x)\left(\xi_+ \otimes \psi_n\right) = \widehat{P}_{bs}^{(c)}(\epsilon, x)\left(\xi_+ \otimes \psi_n\right) = 0, \quad n > 2, \quad (3.346)$$

$$\widehat{P}_{bs}^{(c)}(\epsilon, x)\left(\xi_- \otimes \psi_n\right) = \widehat{P}_{sb}^{(c)}(\epsilon, x)\left(\xi_- \otimes \psi_n\right) = 0, \quad \forall n. \qquad (3.347)$$

It follows that we can express $\widehat{H}_I^{(c)}$ as

$$\widehat{H}_I^{(c)} = \int_0^\infty d\epsilon \left\{ \widehat{P}_{bs}^{(c)}(\epsilon, x) + \widehat{P}_{sb}^{(c)}(\epsilon, x) \right\}. \qquad (3.348)$$

Finally we shall introduce a time dependent Hamiltonian generator $\widehat{H}_g^{(c)}(t)$ in $\mathcal{H}^{(c)}$ by

$$\widehat{H}_g^{(c)}(t) = \begin{cases} \widehat{K}_{(p)} \otimes \widehat{\mathbb{I}}_{(e)} + \widehat{\mathbb{I}}_{(p)} \otimes \widehat{H}_{(e)}, & \text{for } t < 0 \\ \lambda \widehat{H}_I^{(c)}, \quad \lambda \in \mathbb{R} & \text{for } 0 \leq t \leq \delta t \\ \widehat{K}_{(p)} \otimes \widehat{\mathbb{I}}_{(e)} + \widehat{\mathbb{I}}_{(p)} \otimes \widehat{H}_{(e)}, & \text{for } \delta < t. \end{cases} \qquad (3.349)$$

Let us assume that at time $t = 0$ when the ionization interaction commences, the incoming proton is in state ξ_+ which corresponds to its kinetic

---

[126] Here the symbolic expression means

$$|f_+(\epsilon - \epsilon_{ion}, x) \otimes \psi_s\rangle \langle f_+(\epsilon, x) \otimes \psi_b| \left(\xi_+ \otimes \psi_b\right)$$

$$= \left\{\left(\int f_+^*(\epsilon, x)\xi_+(x)\, dx\right) \langle \psi_b \mid \psi_b\rangle\right\} f_+(\epsilon - \epsilon_{ion}, x) \otimes \psi_s$$

$$= \tilde{\xi}_+(\epsilon) \, f_+(\epsilon - \epsilon_{ion}, x) \otimes \psi_s.$$

We have used Eq. (2.405) to obtain $\tilde{\xi}_+(\epsilon)$.

3.6. QUANTUM MEASUREMENT

energy strictly bigger than ϵ_{ion}, i.e., its kinetic energy probability amplitude $\widetilde{\xi}_+(\epsilon)$ vanishes for $\epsilon < \epsilon_{ion}$. The state of the proton-electron composite system is

$$\Phi^{(c)}(0) = \xi_+ \otimes \psi_b. \tag{3.350}$$

During the interval $t \in [0, \delta]$ this state will evolve in accordance with the Schrödinger equation

$$i\hbar \frac{d}{dt} \Phi^{(c)}(t) = \widehat{H}_g^{(c)} \Phi^{(c)}(t) = \lambda \widehat{H}_I^{(c)} \Phi^{(c)}(t). \tag{3.351}$$

The solution satisfying the initial condition $\Phi^{(c)}(0) = \xi_+ \otimes \psi_b$ is

$$\Phi^{(c)}(t) = -i \left(\sin \frac{\lambda t}{\hbar} \right) \left(\xi_{+d} \otimes \psi_s \right) + \left(\cos \frac{\lambda t}{\hbar} \right) \left(\xi_+ \otimes \psi_b \right). \tag{3.352}$$

At the end of the interaction at $t = \delta t$ this state evolves to

$$\Phi^{(c)}(\delta t) = -i \left(\sin \frac{\lambda \delta t}{\hbar} \right) \left(\xi_{+d} \otimes \psi_s \right) + \left(\cos \frac{\lambda \delta t}{\hbar} \right) \left(\xi_+ \otimes \psi_b \right). \tag{3.353}$$

Assuming that λ and δt satisfy

$$\lambda \, \delta t = \hbar \pi / 2. \tag{3.354}$$

the state at time $t = \delta t$ becomes

$$\Phi^{(c)}(\delta t) = -i \, \xi_{+d} \otimes \psi_s, \tag{3.355}$$

which implies the completion of the initial ionization process.[127]

Once the electron is in the scattering state ψ_s the coupling with the proton is effectively switched off. The electron will evolve under the atomic propagator to spatial infinity to complete the ionization process on its own. In practice, the completion stage of the ionization process may be considered effectively over at some large time t_c after the end of the initiation stage.

We can now see that ideally a counter of size Λ can be designed to give a positive detection, i.e., a *yes* result, if the particle's wave function is localized in Λ. In our attempt to measure $\mathcal{E}(\widehat{A}, \phi)$ the wave function $\widehat{U}_{t_{ss}}\phi$ at the time of detection does not localize into the region of any single counter. Only a spectral component, say $\phi_j = c_j \varphi_j$, is localized into the region of detector D_j. It follows that detector D_j will not fire every time. Instead it will fire with probability $|c_j|^2$, i.e., the count rate is identifiable with probability $|c_j|^2$.

---

[127] A general situation when the ionizing particle's energy is not strictly bigger than the ionization energy can also be dealt with (Wan and Harrison (1994)).

3.6.7 Projection postulate, local position measurements and uncertainty relations

Quantum measurement and state preparation are now seen to be two separate processes. A measurement procedure is generally invasive, but not in the sense of the projection postulate. We regard a quantum measurement as having been completed for all practical purposes after the completion of a local position measurement, i.e., after the relevant ionization process. In our model an initial spectral component ϕ_j evolves during the spectral separation process to arrive at the detector at time T as a new vector $\widehat{U}_T \phi_j$. This vector is further altered by the detection process. So, generally the initial state of the system after a measurement act will not remain unchanged; it will change and moreover the changed state will generally not be an eigenvector of the observable concerned. Take the example of the cloud chamber for position measurement. When an ionization process is completed the incoming ionizing particle would have moved quite a distance away from the ion. In other words after a position measurement the particle is actually not in the vicinity of the measured position which is where the ion is. The particle would have travelled even further away by the time a magnification process is completed and ready for observation by a human. It follows that after a measurement the system is left in a state that does not necessarily correspond to an eigenvector of the observable. Indeed the system may not be in an eigenstate at any instant during the entire measurement process. It follows that a quantum measurement does not generally assign a definite eigenvalue of the observable to the system being measured, *either immediately before or after the measurement act*. It is in this sense that we say that the system generally does not **possess** a definite value of an observable or that the system does not have a **possessed value** of the observable. This is in sharp contrast to a classical system for which a measurement only reveals a value possessed by the system before, during and after the measurement. Our theory also contravenes the projection postulate. We would argue that a measurement is fundamentally and directly about detecting various spectral components of a given initial state, with numerical values being assigned to observables as a later consequence, e.g., the detection of a spin half particle in a Stern-Gerlach experiment in the "up" position leads to the assignment of spin $\frac{1}{2}\hbar$.

The detection of the spectral components yields a statistical distribution of triggerings of the particle detectors at various locations, which in turn provide the probabilities $|c_j|^2$. Together with eigenvalues these measured probabilities enable us to calculate the desired expectation value.

Since a measurement is about determining the expansion of the initial state in terms of a particular set of eigenvectors a measuring process designed to measure an observable $\widehat{A}$ could equally well be taken to be measuring any

3.6. QUANTUM MEASUREMENT

observable of the form $f(\widehat{A})$ or one sharing the same eigenvectors as $\widehat{A}$. By making an appropriate association of eigenvalues with the spectral components we can calculate the desired expectation value $\langle \Phi \mid f(\widehat{A})\Phi \rangle$.

By emphasizing local position measurements we may appear to be in conflict with the usual understanding of Heisenberg's uncertainty relation. It seems that in our model any observable can be measured simultaneously with position, irrespective of whether or not its operator commutes with the position operator. The impossibility of an experimental act simultaneously measuring two incompatible observables with an arbitrary accuracy is usually deduced from the projection postulate,[128] i.e., the final state of the system generally cannot simultaneously be an eigenvector of two non-commuting operators. This argument is obviously not applicable here, because the state vectors immediately before and after the local position measurement are not eigenvectors of the observable being measured. More importantly, the position is only ascertained to be within a region sufficiently large to be consistent with the usual uncertainty relation between position and momentum. For example a well-known method to determine the momentum of an elementary particle is by its deflection in an applied magnetic field. The determination of the deflection involves local position measurements, e.g., using a cloud chamber in a magnetic field or some kind of diaphragms. A simple analysis would reveal no conflict with the uncertainty relation between position and momentum.[129] That our model does not violate the usual uncertainty relation between position and momentum should be of no surprise since we have adopted a unitary time evolution in accordance with Postulate OQD.

3.6.8 Concluding remarks

We conclude this chapter by saying that orthodox quantum mechanics as embodied by Postulates OQS and OQD contains the mathematical framework for state preparation and measurement. There is no need to bring in macroscopic systems in the process, except when an amplification is required for the benefit of the human eye.[130] There are many different approaches and theories to tackle quantum measurement. These ranges from a fundamental change to the framework of quantum theory to fundamentally differently interpretations. Some often quoted examples of changes in the framework are:

1. Introduce measuring devices with some classical properties. This necessitates changing the very structure of the Hilbert space to bring in

---
[128]Roman (1965) p. 13.
[129]Messiah Vol. 1 (1967) pp. 142-147. The footnote on p. 142 on the cloud chamber is particularly relevant.
[130]The general idea that a quantum measurement does not necessarily require macroscopic systems has been pointed out by many authors, e.g., Percival (1991).

superselection rules in order to describe measuring devices.[131] This may include the introduction of *pointer observables* and the notion of *objectification*.[132] We shall present a systematic discussion of superselection rules and how they can help to formulate a measurement theory in Chapter 4. The origins of superselection rules will be discussed in §4.3.3.

2. Replace Postulate OQD to bring in nonlinear dynamical equations. Well-known examples are the nonlinear Schrödinger equation introduced by Ghirardi, Rimini and Weber (1986) to effect a spontaneous localization[133] and an approach based on the dynamics of quantum stochastic and diffusion processes.[134]

3. Introduce unsharp observables and the concept of unsharp measurement.[135] This would require the extension of Postulate OQS to include POV measures, in addition to the traditional PV measures, to accommodate unsharp observables and unsharp measurement. We shall return to present a systematic study of this in Chapter 5.

4. Adopt a phase space approach.[136] This involves formulating a quantum theory on the phase space rather than the coordinate space.

There are quite a number of well-known different interpretations of quantum theory.[137] For examples we have the consistent histories interpretation,[138] the causal interpretation,[139] the ensemble interpretations,[140] the many-worlds interpretation,[141] the modal interpretation,[142] and so on. All these different approaches and interpretations are well documented and widely available in various papers and monographs. Hence we shall not delve into them here.

---

[131] See §4.7 in Chapter 4.
[132] Busch, Lahti and Mittelstaedt (1996).
[133] Ghirardi, Rimini and Weber (1986), (1988), Bell (1987), Ghirardi and Rimini (1990).
[134] Lockhart and Misra (1986), Nelson (1985) Ch. IV. Grigolini (1993), Gisin and Percival (1992)
[135] Busch, Lahti and Mittelstaedt (1996). Busch, Grabowski and Lahti (1997).
[136] Kim and Noz (1991), Schroeck (1996).
[137] Bub (1997), Auletta (2000).
[138] Ommès (1994).
[139] Bohm (1952), Holland (1993).
[140] Ballentine (1970), Ballentine (1990), Home and Whitaker (1992).
[141] Everett (1957), Wheeler (1957).
[142] van Fraassen (1991).

References

1. Abraham, R. and Marsden, J. E. (1978). *Foundations of Mechanics*, Benjamin/Cummings, Reading, Mass.
2. Aharonov, Y. and Bohm, D. (1959). *Phys. Rev.* **115** 485.
3. Akhiezer, N. I. and Glazman I. M. (1961). *Theory of Linear Operators in Hilbert Space Vol. 1*, Frederick Ungar, New York.
4. Alanson, T. (1992). *Phys. Lett. A* **63** 41.
5. Amrein, W. O. (1981). *Non-Relativistic Quantum Dynamics*, Reidel, Dordrecht.
6. Amrein, W. O., Jauch, J. M. and Sinha, K. B. (1977). *Scattering Theory in Quantum Mechanics*, Benjamin, Reading, Mass.
7. Anosov, D. V. and Arnold, V. I. (1988). *Dynamical Systems I*, Springer-Verlag, Berlin.
8. Arnold, V. I. (1978). *Mathematical Methods of Classical Mechanics*, Springer-Verlag, New York.
9. Auletta, G. (2000). *Foundations and Interpretation of Quantum Mechanics*, World Scientific, Singapore. There are extensive bibliographic references.
10. Balian, R. (1991). *From Microphysics to Macrophysics*, Springer-Verlag, Berlin.
11. Ballentine, L. E. (1990). *Rev. Mod. Phys.* **42** 358.
12. Ballentine, L. E. (1990). *Quantum Mechanics*, Prentice-Hall, Eaglewood Cliffs, New Jersey.
13. Baym, G. (1969). *Lectures on Quantum Mechanics*, Benjamin, New York.
14. Bell, J. (1987). 'Are there quantum Jumps?' in J. Bell, *Speakable and Unspeakable in Quantum Mechanics*, Cambridge University Press, Cambridge.
15. Bell, J. (1990). 'Against "Measurement"' in A. I. Miller, ed., *Twenty-Six Years of Uncertainty*, a NATO ASI Series B Vol. 226, Plenum, New York.
16. Beltrametti, E. G. and Cassinelli, G. (1981). *The Logic of Quantum Mechanics*, Addison-Wesley, Reading, Mass.
17. Blank, J., Exner, P. and Havlíček, M. (1994). *Hilbert Space Operators in Quantum Physics*, American Institute of Physics Press, New York.
18. Blum, K. (1981). *Density Matrix Theory and Applications*, Plenum, New York.
19. Bogolubov, N., Logunov, A. and Todorov, I. (1975). *Introduction to Axiomatic Quantum Field Theory*, Benjamin, Reading Mass.
20. Bohm, D. (1952). *Phys. Rev.* **85** 166.
21. Braginsky, V. B. and Khalili, F. Y. (1992). *Quantum Measurement*, Cambridge University Press, Cambridge.
22. Bratteli, O. and Robinson, D. W. (1979). *Operator Algebras and Quantum Statistical Mechanics I*, Springer-Verlag, New York.
23. Bratteli, O. and Robinson, D. W. (1981). *Operator Algebras and Quantum Statistical Mechanics II*, Springer-Verlag, New York.
24. Bub, J. (1997). *Interpreting the Quantum World*, Cambridge University Press, Cambridge.

25. Busch, P., Grabowski, M. and Lahti, P. J. (1997). *Operational Quantum Physics*, Springer-Verlag, Berlin.
26. Busch, P., Lahti, P. J. and Mittelstaedt, P. (1996). *The Quantum Theory of Measurement*, 2nd edition, Springer-Verlag, Berlin.
27. Busch P. and Schroeck, F. E., Jr. (1989). *Found. Phys.* **19** 807.
28. Capri, A. Z. (1985). *Nonrelativistic Quantm Mechanics*, Benjamin/Cummings, Menlo Park, California.
29. Castellani, L. (1978). *Nuovo Cimento A* **48** 359.
30. Cini, M. and Levy-Leblond, J. M. (1990). *Quantum Theory without Reduction*, Adam Hilger, Bristol.
31. Cohen, D. W. (1989). *An Introduction to Hilbert Space and Quantum Logic*, Springer-Verlag, New York.
32. Cohen, N. (1966). *J. Math. Phys.* **7** 781.
33. Cornwell, J. F. (1984). *Group Theory in Physics Vol. 1*, Academic Press, London.
34. Cycon, H. L., Froese, R. G., Kirsch, W. and Simon, B. (1987). *Schrödinger Operators*, Springer-Verlag, Berlin.
35. Daneri, A., Loinger, A. and Prosperi, G. M. (1962). *Nucl. Phys.* **33** 297.
36. de Lange, O. L. and Raab, R. E. (1991). *Operator Methods in Quantum Mechanics*, Oxford University Press, Oxford.
37. D'Espagnat, B. (1989). *Conceptual Foundations of Quantum Mechanics*, Addison-Wesley, Reading, Mass.
38. Dicke, R. H. and Wittke, J. P. (1963). *Introduction to Quantum Mechanics*, Addison-Wesley, Reading Mass., World Student Series Edition.
39. Emch, G. G. (1972). *Algebraic Methods in Statistical Mechanics and Quantum Field Theory*, Wiley, New York.
40. Enss, V. (1983). *Com. Math. Phys.* **89** 245.
41. Everett, H. (1957). *Rev. Mod. Phys.* **29** 454.
42. Garden, R. W. (1984). *Modern Logic and Quantum Mechanics*, Adam Hilger, Bristol.
43. Ghirardi, G. C. and Rimini, A. (1990). 'Old and New Ideas in the Theory of Quantum Measurement' in A. I. Miller, ed, *Sixty-Two Years of Uncertainty and Physical Enquiries into the Foundations of Quantum Mechanics*, a NATO ASI Series B Vol. 226, Plenum, New York.
44. Ghirardi, G. C., Rimini, A. and Weber, T. (1986). *Phys. Rev. D* **34** 470.
45. Ghirardi, G. C., Rimini, A. and Weber, T. (1988). *Nuovo Cimento B* **102** 383.
46. Gisin, N. and Percival, I. (1992). *J. Phys. A: Math. Gen.* **25** 5677.
47. Greenberger, D. M. ed. (1986). *New Techniques and Ideas in Quantum Measurement Theory*, Annals of New York Academy of Sciences, New York.
48. Greiner, W. (1989). *Quantum Mechanics Vol. 1 - An Introduction*, Springer-Verlag, Berlin.
49. Greiner, W. and Müller, B. (1989). *Quantum Mechanics Vol. 2 - Symmetries*, Springer-Verlag, Berlin.
50. Grigolini, P. (1993). *Quantum Mechanical Irreversibility and Measurement*, World Scientific, Singapore.

REFERENCES

51. Gudder, S. P. (1979). *Stochastic Methods in Quantum Mechanics*, North-Holland, New York.
52. Guenin, M. (1966). 'Algebraic Methods in Quantum Field Theory' in W. E. Witten, A. O. Barut and M. Guenin, eds., *Boulder Lectures in Theoretical Physics Vol. IXA: Mathematical Methods of Theoretical Physics*, Gordan and Breach, New York.
53. Gupta, N. N. D. and Ghosh, S. K. (1946). *Rev. Mod. Phys.* **18** 225.
54. Haag, R. (1973). 'Infinite Quantum Systems' in W. E. Witten, ed., *Boulder Lectures in Theoretical Physics Vol. XIVB: Mathematical Methods of Theoretical Physics*, Gordan and Breach, New York.
55. Hagg, R. (1992). *Local Quantum Physics*, Springer-Verlag, Berlin.
56. Harrison, F. E. (1993). *Superselection rules, Quantum Measurement Problems and Macroscopic Quantum Systems*, St Andrews University PhD Thesis.
57. Hepp, K. (1972). *Helv. Phys. Acta.* **45** 237.
58. Hey, T. and Walters, P. (1987). *The Quantum Universe*, Cambridge University Press, Cambridge.
59. Holland, P. R. (1993). *The Quantum Theory of Motion*, Cambridge University Press, Cambridge.
60. Home, D. and Whitaker, A. (1992) *Physics Reports*, **210(4)** 223.
61. Hurt, N. E. (1983). *Geometric Quantization in Action*, Reidel, Dordrecht.
62. Isham, C. J. (1995). *Lectures on Quantum Theory*, Imperial College Press, London.
63. Jauch, J. M. (1968). *Foundations of Quantum Mechanics*, Addison-Wesley, Reading, Mass.
64. Kim, Y. S. and Noz, M. E. (1991). *Phase Space Picture of Quantum Mechanics: Group Theoretical Approach*, World Scientific, Singapore.
65. Kowalski, K. (1994). *Methods of Hilbert Spaces in the Theory of Nonlinear Dynamical Systems*, World Scientific, Singapore.
66. Kreyszig, E. (1978). *Introduction to Functional Analysis with Applications*, Wiley, New York.
67. Landsman, N. P. (1998). *Mathematical Topics between Classical and Quantum Mechanics*, Springer-Verlag, New York.
68. Liboff, R.L. (1980). *Introductory Quantum Mechanics*, Addison-Wesley, Reading, Mass.
69. Lockhart, C. M. and Misra, B. (1986). *Physica* **136A** 47.
70. Ludwig, G. (1985). *Foundations of Quantum Mechanics Vol. 2*, Springer-Verlag, New York.
71. MacKey, G. W. (1963). *The Mathematical Foundations of Quantum Mechanics*, Benjamin, Reading, Mass.
72. Martin, D. (1991). *Manifold Theory*, Ellis Harwood, New York.
73. Mees, C. R. K. (1963). *The Theory of Photographic Process*, Macmillan, New York.
74. Mensky, M. B. (1993). *Continuous Quantum Measurements and Path Integrals*, Institute of Physics Publishing, Bristol.
75. Merzbacher, E. (1998). *Quantum Mechanics*, 3rd edition, Wiley, New York.
76. Messiah, A. (l967). *Quantum Mechanics Vol. 1*, North-Holland, Amsterdam.

77. Naimark, M. A. (1968). *Linear Differential Operators Part II*, Harrap, London. Translated by E. R. Dawson.
78. Namiki, M. (1992). *Stochastic Quantization*, Springer-Verlag, Berlin.
79. Nelson, N. (1985). *Quantum Fluctuations*, Princeton University Press, Princeton.
80. Ohanian, H. C. (1995). *Modern Physics*, 2nd edition, Prentice-Hall, Englewood Cilffe, N. J.
81. Omnès, R. (1994). *The Interpretation of Quantum Mechanics*, Princeton University Press, Princeton, N. J.
82. Park, J. L. and Margenau, H. (1968). *Int. J. Theor. Phys.* **1** 211.
83. Pearson, D. B. (1988). *Quantum Scattering and Spectral Theory*, Academic Press, London.
84. Peierls, R. (1935). *Nature* **136** 395.
85. Penrose, O. (1970). *Foundations of Statistical Mechanics*, Pergamon Press, Oxford.
86. Percival, I. (1991). 'Quantum Measurement Theory and Experiment' in T. D. Clark, H. Prance, R. J. Prance and T. P. Spiller, eds., *Macroscopic Quantum Phenomena*, World Scientific, Singapore.
87. Peres, A. (1995). *Quantum Theory: Concepts and Methods*, Kluwer Academic Publishers, Dordrecht.
88. Peshkin, H. and Tonomura, A. (1989). *The Aharonov-Bohm Effect*, Springer-Verlag, Berlin.
89. Pfeifer, P. (1980). *Helv. Phys. Acta.* **53** 410.
90. Piron, C. (1976). *Foundations of Quantum Physics*, Benjamin, Reading, Mass.
91. Pitowski, I. (1989). *Quantum Probability - Quantum Logic*, Springer-Verlag, Berlin.
92. Primas, H. (1983). *Chemistry, Quantum Mechanics and Reductionism*, Springer-Verlag, Berlin.
93. Primas, H. and Müller-Herold, U. (1978). *Advances in Chemical Physics* **38** Chapters 1-6.
94. Prugovečki, E. (1981). *Quantum Mechanics in Hilbert Space*, 2nd edition, Academic Press, New York.
95. Rédei, M. and Stöltzer, M. (2001). *John von Neumann and the Foundations of Quantum Physics*, Kluwer, Dorchecht.
96. Reed, M. and Simon, B. (1972). *Methods of Modern Mathematical Physics Vol. 1 Functional Analysis*, Academic Press, New York.
97. Reed, M. and Simon, B. (1975). *Methods of Modern Mathematical Physics Vol. 2 Fourier Analysis, Selfadjointness*, Academic Press, New York.
98. Richtmyer, R. D. (1978). *Principles of Advanced Mathematical Physics Vol. 1*, Springer-Verlag, New York.
99. Roman, P. (1965). *Advanced Quantum Theory*, Addison-Wesley, Reading, Mass.
100. Roman, P. (1975). *Some Modern Mathematics for Physicists and Other Outsiders Vol. 2*, Pergamon, New York.
101. Schroeck, F. E., Jr. (1996). *Quantum Mechanics on Phase Space*, Kluwer, Dordrecht.
102. Scully, M. O., Lamb, W. E., Jr. and Barut, A. (1989). *Found. Phys.* **17** 575.

REFERENCES 335

103. Sewell, G. L. (1986). *Quantum Theory of Collective Phenomena*, Clarendon Press, Oxford.
104. Sewell, G. L. (2002). *Quantum Mechanics and its Emergent Macrophysics*, Princeton University Press, Princeton, N. J.
105. Śniatycki, J. (1980). *Geometric Quantization and Quantum Mechanics*, Springer-Verlag, New York.
106. Sudbery, A. (1986) *Quantum Mechanics and the Particles of Nature, An Outline for Mathematicians*, Cambridge University Press, Cambridge.
107. Temple, G. (1935). *Nature* **135** 957 (1935), **136** 179.
108. Thirring, W. (1979). *Quantum Mechanics of Atoms amd Molecules*, Springer-Verlag, New York.
109. van Fraassen, B. C. (1991). *Quantum Mechanics, An Empiricist View*, Clarendon Press, Oxford.
110. von Neumann, J. (1955). *Mathematical Foundations of Quantum Mechanics*, Princeton University Press, Princeton, N. J., translated from the original German book published in 1932 by R. T. Beyer.
111. Wan, K. K. and Harrison, F. E. (1994). *Found. Phys.* **24** 831.
112. Wan, K. K., Jackson, T. D. and McKenna, I. H. (1984). *Nuovo Cimento B* **81** 165.
113. Wan, K. K. and McFarlane, K. (1983). *Int. J. Theor. Phys.* **22** 55.
114. Wan, K. K. and McKenna, I. H. (1984). *Algebras, Groups and Geometries* **1** 154.
115. Wan, K. K., McKenna, I. H. and Pinto, J. (1984a). *Algebras, Groups and Geometries* **1** 344.
116. Wan, K. K., McKenna, I. H. and Pinto, J. (1984b). *Algebras, Groups and Geometries* **1** 372.
117. Wan, K. K. and McLean, R. D. G. (1983a). *Phys. Lett. A* **94** 198.
118. Wan, K. K. and McLean, R. D. G. (1983b). *Phys. Lett. A* **95** 76.
119. Wan, K. K. and McLean, R. G. D. (1984). *J. Phys. A: Math. Gen.* **17** 837. Corrigenda **17** 2363 (1984).
120. Wan, K. K. and McLean, R. D. G. (1991). *J. Phys. A: Math. Gen.* **24** L425.
121. Wan, K. K. and McLean, R. G. D. (1994). *Found. Phys.* **24** 715.
122. Wan, K. K. and Sumner, P. (1988). *Phys. Lett. A* **128** 458.
123. Wan, K. K. and Sumner, P. (1991). *Nuovo Cimento B* **106** 593.
124. Wan, K. K. and Viazminsky, C. (1977). *Prog. Theor. Phys.* **58** 1030.
125. Wan, K. K. and Viazminsky, C. (1979). *J. Phys. A: Math. Gen.* **12** 643.
126. Wheeler, J. A. (1957). *Rev. Mod. Phys.* **29** 463.
127. Wheeler, J. A. and Zurek, W. H. eds. (1983). *Quantum Theory of Measurement*, Princeton University Press, Princeton, N, J.
128. Wilson, J. L. (1951). *The Principles of Cloud-Chamber Technique*, Cambridge University Press, Cambridge.
129. Woodhouse, N. (1986). *Geometric Quantization*, Clarendon Press, Oxford.
130. Yuan, L. C. L. and Wu, C. S. (1961). *Methods of Experimental Physics Vol. 5A: Nuclear Physics*, Academic Press, New York.
131. Zettili, N. (2001) *Quantum Mechanics*, Wiley, Chichester.
132. Zhu, C. and Klauder, J. R. (1993a). *Am. J. Phys.* **61** 605.
133. Zhu, C. and Klauder, J. R. (1993b). *Found. Phys.* **23** 617.

Chapter 4

Physical Theory in Hilbert Space

4.1 Introduction

Traditionally classical systems and quantum systems are treated quite differently, especially in the mathematical language for their description. It would be desirable to have a unified mathematical formulation of the theories for these systems. This would facilitate any discussion on the interaction between these systems; it would also show up the similarity as well as the fundamental differences between them. Moreover, we believe that there are systems such as superconducting rings which possess both classical and quantum properties.[1] We call systems which possess both quantum and classical properties **mixed quantum systems**. A unified mathematical formulation capable of accommodating classical and quantum theories may allow us to describe mixed quantum systems as well.

In this chapter we shall present a unified treatment in Hilbert space of physical systems of finite degrees of freedom which can be categorized into **classical, orthodox quantum** and **mixed quantum**. Classical and orthodox quantum systems are describable by classical and orthodox quantum theory respectively. Mixed quantum systems are not described by either one of these two orthodox theories. Orthodox quantum mechanics based on Postulate OQS in §3.2.1 and Postulate OQD in §3.4.1 entails a one-to-one correspondence between observables and selfadjoint operators, and between pure states and one-dimensional projectors serving as density operators. These one-to-one correspondences allow the coherent superposition of any set of states, as seen in

---
[1] See Chapter 7 for details.

Eqs. (3.36) and (3.39). It is this possibility of a coherent superposition of an arbitrary set of states which prevents the emergence of classical properties. A convenient way to break up these one-to-one correspondences is by superselection rules, a concept introduced by Wick, Wightman and Wigner (1952) in particle physics. An incorporation of superselection rules into the mathematical structure can pave the way to a formulation of a unified theory in Hilbert space applicable to the three types of physical systems mentioned above. We shall do this in this chapter. To avoid any misunderstanding we would emphasize here that the aim in this chapter is to set out the mathematical formalism of a unified theory. A brief mention of various origins of superselection rules is given in §4.3.3. The applicability of the formalism presented in this chapter will be demonstrated in Chapter 7 where we apply the formalism to study concrete physical systems, e.g., superconducting circuit systems. So, this chapter should be read in conjunction with Chapter 7 in order to avoid the misconception that we are trying to tackle fundamental physical issues merely by *ad hoc* definitions and artificial mathematical structures.

There is a belief that all physical theories are derivable from a single universal theory.[2] Having such a belief should not prevent one from setting out a careful classification of various theories, e.g., believing that classical mechanics is derivable from quantum mechanics does not prevent us from classifying classical and quantum systems and having a well-developed classical mechanics which can be directly applied to classical systems.

4.2 Unified Statics in Direct Integral Space

Our approach[3] is based on the mathematical framework of a direct integral

$$\mathcal{H}^\oplus = \int^\oplus \mathcal{H}(\gamma)\, dg(\gamma) \qquad (4.1)$$

of a Borel family of separable Hilbert spaces $\mathcal{H}(\gamma)$, and the set of decomposable operators $\widehat{A}^\oplus$ in $\mathcal{H}^\oplus$, including statistical operators $\widehat{S}^\oplus$, introduced in §2.13, §2.14 and §2.15.

We shall, for simplicity and definiteness, restrict the measure $dg(\gamma)$ to either a purely discrete one generated by a staircase function of the form of Eq. (2.585) or a purely continuous one of the form of a Lebesgue measure $d\gamma$. The starting point of our unified theory is a postulate on statics.

---

[2] We shall return to comment on this belief further in the concluding section §4.8.
[3] Wan, Bradshaw, Trueman and Harrison (1998).

4.2. UNIFIED STATICS IN DIRECT INTEGRAL SPACE

4.2.1 A unified postulate on quantum statics

Postulate UQS on statics

- A physical system has associated with it a Hilbert space which has a **preferred** direct integral decomposition $\mathcal{H}^\oplus$, possibly trivial, into a Borel family of separable Hilbert spaces $\mathcal{H}(\gamma)$ such that:

  1. Physical observables correspond one-to-one to decomposable self-adjoint operators $\widehat{A}^\oplus$ in $\mathcal{H}^\oplus$.
  2. States correspond one-to-one to statistical operators $\widehat{S}^\oplus$ on $\mathcal{H}^\oplus$.
  3. The expectation value $\mathcal{E}(\widehat{A}^\oplus, \widehat{S}^\oplus)$ of a bounded observable $\widehat{A}^\oplus$ in state $\widehat{S}^\oplus$ is given by

$$\mathcal{E}(\widehat{A}^\oplus, \widehat{S}^\oplus) = \mathrm{tr}^\oplus(\widehat{A}^\oplus \widehat{S}^\oplus). \tag{4.2}$$

Mathematically a given Hilbert space admits many different direct integral decompositions, but the nature of the physical system will determine the preferred decomposition. For an orthodox quantum system the Hilbert space would have a preferred trivial decomposition with $\mathcal{H}^\oplus$ identified with an appropriate Hilbert space $\mathcal{H}(0)$, and Postulate UQS then reduces to Postulate OQS with $\mathcal{H}(0)$ as the associated Hilbert space.

If the preferred integral decomposition of the Hilbert space of the system is not a trivial one, then new physical features appear due to the emergence of superselection rules. It must be stressed that for the emergence of superselection rules a direct integral decomposition of the Hilbert space must be accompanied by a corresponding decomposable nature of all observables; just a direct integral decomposition of the Hilbert space is not enough.[4]

4.2.2 Discrete and continuous direct integral decompositions

Apart from the trivial decomposition with $\mathcal{H}^\oplus = \mathcal{H}(0)$ we need to consider two new classes of systems depending on whether the integral decomposition of the Hilbert space $\mathcal{H}^\oplus$ is purely discrete or purely continuous.

Class 1   Purely discrete decomposition   The Hilbert space is of the form

$$\mathcal{H}_d^\oplus = \oplus_n \mathcal{H}(n) \tag{4.3}$$

---

[4]See §8.4.1 and §8.4.2 for the breakdown of superselection rules.

as in Eq. (2.589). A statistical operator $\widehat{S}^\oplus$ is given by Eqs. (2.677) or (2.684) which is basically a convex combination of density operators $\widehat{D}(n)$ on the subspaces $\mathcal{H}(n)$, i.e., we have[5]

$$\widehat{S}^\oplus = \oplus_n w_n \widehat{D}(n) = \sum_n w_n \widehat{D}_n^\oplus. \tag{4.4}$$

Observables are direct sums of selfadjoint operators $\widehat{A}(n)$ on the subspaces $\mathcal{H}(n)$, i.e., we have

$$\widehat{A}^\oplus = \oplus_n w_n \widehat{A}(n) = \sum_n w_n \widehat{A}_n^\oplus. \tag{4.5}$$

The expectation value of observable $\widehat{A}^\oplus$ in state $\widehat{S}^\oplus$ is

$$\mathcal{E}(\widehat{A}^\oplus, \widehat{S}^\oplus) = \sum_n w_n \, \mathrm{tr}_n \left(\widehat{A}(n) \widehat{D}(n) \right). \tag{4.6}$$

Class 2 Purely continuous decomposition The Hilbert space is of the form given in Eq. (2.597), i.e.,

$$\mathcal{H}_c^\oplus = \int^\oplus \mathcal{H}(\tau) \, d\tau. \tag{4.7}$$

Regular statistical operators are given by Eq. (2.680), i.e.,

$$\widehat{S}_r^\oplus = \int^\oplus S_r(\tau) \widehat{D}(\tau) \, d\tau, \tag{4.8}$$

and observables are direct integrals of selfadjoint operators $\widehat{A}(\tau)$ on the subspaces $\mathcal{H}(\tau)$, i.e., we have

$$\widehat{A}^\oplus = \int^\oplus \widehat{A}(\tau) \, d\tau. \tag{4.9}$$

The expectation value of observable $\widehat{A}^\oplus$ in state $\widehat{S}^\oplus$ is

$$\mathcal{E}(\widehat{A}^\oplus, \widehat{S}_r^\oplus) = \int S_r(\tau) \, \mathrm{tr}_\tau \left(\widehat{A}(\tau) \widehat{D}(\tau) \right) d\tau. \tag{4.10}$$

---

[5] Operator $\widehat{D}_n^\oplus$ is the extension of $\widehat{D}(n)$ to $\mathcal{H}_d^\oplus$, as defined by Eq. (2.648).

4.3 States and Superposition Principle

4.3.1 Regular and singular states, pure and mixed states

We shall first introduce various kinds of states in terms of the corresponding properties of statistical operators defined in §2.14.4.

Definition 4.3.1(1) **Pure and mixed states**

- A state defined by a regular statistical operator $\widehat{S}_r^\oplus$ is said to be *regular*. A regular state is said to be *pure* if the statistical operator $\widehat{S}_r^\oplus$ is a projector onto a one-dimensional subspace spanned by a pure vector in $\mathcal{H}^\oplus$. Otherwise the state is said to be *mixed*.

- A state defined by a singular statistical operator $\widehat{S}_s^\oplus$ is said to be *singular*. A singular state is said to be *pure* if the associated statistical operator is generated by a density function $S_s(\tau)$ consisting of a single δ-function, i.e., $S_s(\tau) = \delta(\tau - \tau_0)$, and the corresponding density operator $\widehat{D}(\tau_0)$ is a projector in $\mathcal{H}(\tau_0)$. Otherwise the state is said to be *mixed*.

Let us spell things out more explicitly for the two classes of systems mentioned in §4.2.2.

Class 1 Purely discrete decomposition In Hilbert space $\mathcal{H}_d^\oplus$ all states are regular and there are no singular states. It is possible to represent some states in terms of pure and mixed vectors introduced in Definition 2.13.2(3). Pure regular states can be described by pure vectors and some mixed states can be described by mixed vectors:

1. **Pure States** As an example let $\boldsymbol{u}(1)$ be a unit vector in $\mathcal{H}(1)$, and following the notation in C2.13.2(3) C2 let its counterpart in $\mathcal{H}^\oplus$ defined by Eq. (2.591) be denoted by $\boldsymbol{u}_1^\oplus$. Then $\boldsymbol{u}_1^\oplus$ is a pure vector in $\mathcal{H}^\oplus$. Let

$$\widehat{D}(1) = |\boldsymbol{u}(1)\rangle\langle\boldsymbol{u}(1)| \qquad (4.11)$$

be the corresponding density opeator in $\mathcal{H}(1)$. Then

$$\widehat{S}_1^\oplus = \widehat{D}(1) \oplus \widehat{O}(2) \oplus \cdots \qquad (4.12)$$

is a projector onto the one-dimensional subspace spanned by the pure vector $\boldsymbol{u}_1^\oplus$. Hence $\widehat{S}_1^\oplus$ represents a pure state. Generally a pure state is represented by a statistical operator of the form

$$\widehat{S}_n^\oplus = \widehat{O}(1) \oplus \cdots \oplus \widehat{O}(n-1) \oplus \widehat{D}(n) \oplus \widehat{O}(n+1) \oplus \cdots, \qquad (4.13)$$

where
$$\widehat{D}(n) = |\boldsymbol{u}(n)\rangle\langle\boldsymbol{u}(n)| \qquad (4.14)$$
for some unit vector $\boldsymbol{u}(n) \in \mathcal{H}(n)$.

Clearly we can use pure vectors to represent pure states, e.g., $\boldsymbol{u}_n^\oplus$ given in Eq. (2.626) can be used to represent the state $\widehat{S}_n^\oplus$. The expectation value of an observable $\widehat{A}^\oplus$ is expressible in terms of either $\widehat{S}_1^\oplus$ or $\boldsymbol{u}_n^\oplus$:

$$\begin{aligned}\mathcal{E}(\widehat{A}^\oplus, \widehat{S}_n^\oplus) &= \operatorname{tr}^\oplus\left(\widehat{A}^\oplus \widehat{S}_n^\oplus\right) = \langle \boldsymbol{u}_n^\oplus \mid \widehat{A}^\oplus \boldsymbol{u}_n^\oplus\rangle^\oplus &(4.15)\\ &= \langle \boldsymbol{u}(n) \mid \widehat{A}(n)\,\boldsymbol{u}(n)\rangle_n. &(4.16)\end{aligned}$$

2. **Mixed States** If the density operator $\widehat{D}(n)$ in Eq. (4.13) is not a projector on $\mathcal{H}(n)$ then $\widehat{S}_n^\oplus$ would represent a mixed state. In addition there are other kind of mixed states. Of particular interest is a new type of mixed states which correspond to mixed vectors in $\mathcal{H}_d^\oplus$. For example, a normalized mixed vector

$$\boldsymbol{u}_{12}^\oplus = a_1\,\boldsymbol{u}_1^\oplus + a_2\,\boldsymbol{u}_2^\oplus = a_1\,\boldsymbol{u}(1) \oplus a_2\,\boldsymbol{u}(2) \qquad (4.17)$$

can represent a mixed state corresponding to the statistical operator

$$\begin{aligned}\widehat{S}_{12}^\oplus &= |a_1|^2\,\widehat{D}(1) \oplus |a_2|^2\,\widehat{D}(2) \oplus \widehat{O}(3) \oplus \cdots, &(4.18)\\ &= w_1\,\widehat{S}_1^\oplus + w_2\,\widehat{S}_2^\oplus, \quad w_1 = |a_1|^2,\ w_2 = |a_2|^2. &(4.19)\end{aligned}$$

The expectation value of any $\widehat{A}^\oplus$ in state $\widehat{S}_{12}^\oplus$ is expressible directly in terms of $\boldsymbol{u}_{12}^\oplus$:

$$\begin{aligned}\mathcal{E}(\widehat{A}^\oplus, \widehat{S}_{12}^\oplus) &= \operatorname{tr}^\oplus\left(\widehat{A}^\oplus \widehat{S}_{12}^\oplus\right) &(4.20)\\ &= w_1\,\langle \boldsymbol{u}(1) \mid \widehat{A}(1)\boldsymbol{u}(1)\rangle_1 + w_2\,\langle \boldsymbol{u}(2) \mid \widehat{A}(2)\boldsymbol{u}(2)\rangle_2 &(4.21)\\ &= \langle \boldsymbol{u}_{12}^\oplus \mid \widehat{A}^\oplus\,\boldsymbol{u}_{12}^\oplus\rangle^\oplus. &(4.22)\end{aligned}$$

Note that there is no correlation term in $\langle \boldsymbol{u}_{12}^\oplus \mid \widehat{A}^\oplus\,\boldsymbol{u}_{12}^\oplus\rangle^\oplus$ as $\widehat{A}^\oplus$ cannot correlate $\boldsymbol{u}_1^\oplus$ and $\boldsymbol{u}_2^\oplus$. The interpretation is that $\boldsymbol{u}_{12}^\oplus$ represents a classical mixture of $\boldsymbol{u}_1^\oplus$ and $\boldsymbol{u}_2^\oplus$.

Class 2 Purely continuous decomposition In $\mathcal{H}_c^\oplus$ we have singular states as well as regular states, and there are pure singular states but no pure regular states. Pure singular states play a very useful role for some physical systems.

4.3. STATES AND SUPERPOSITION PRINCIPLE

Examples 2.13.2(1) and 2.13.2(2) present several simple yet non-trivial direct integrals of one-dimensional constituent spaces. In view of their important physical applications we shall spell things out more explicitly here for later convenience. When the constituent Hilbert spaces $\mathcal{H}(\tau)$ are all one-dimensional we have the following results in $\mathcal{H}_c^\oplus$:

1. Every decomposable operator is unitarily equivalent to a diagonalizable operator, i.e.,

$$\widehat{A}^\oplus = \int^\oplus \widehat{A}(\tau)\, d\tau = \int^\oplus A(\tau)\, \widehat{I\!I}(\tau)\, d\tau, \tag{4.23}$$

for some numerical function $A(\tau)$ of τ. It follows that all observables commute; the same result also applies for the discrete case above.

2. There are no pure regular state. A mixed regular state is given by a regular statistical operator

$$\widehat{S}_r^\oplus = \int^\oplus S_r(\tau)\, \widehat{I\!I}(\tau)\, d\tau \tag{4.24}$$

with the expectation value of observable $\widehat{A}^\oplus$ in state $\widehat{S}_r^\oplus$ given by

$$\mathcal{E}(\widehat{A}^\oplus, \widehat{S}_r^\oplus) = \operatorname{tr}^\oplus\left(\widehat{S}_r^\oplus \widehat{A}^\oplus\right) = \int S_r(\tau)\, A(\tau)\, d\tau. \tag{4.25}$$

3. Let $\boldsymbol{u}(\tau)$ be a family of unit vectors, i.e., $\boldsymbol{u}(\tau)$ is a unit vector in $\mathcal{H}(\tau)$ for every τ. A normalized mixed vector in $\mathcal{H}^\oplus$ is then of the form

$$\boldsymbol{f}^\oplus = \int^\oplus f(\tau)\boldsymbol{u}(\tau)\, d\tau, \tag{4.26}$$

where $f(\tau)$ is a normalized complex-valued function of τ, i.e.,

$$\int |f(\tau)|^2\, d\tau = 1. \tag{4.27}$$

Such a mixed vector corresponds to a mixed state $\widehat{S}_r^\oplus$ with

$$S_r(\tau) = |f(\tau)|^2. \tag{4.28}$$

4. A pure singular state is given by a singular statistical operator

$$\widehat{S}_{\tau_0}^\oplus = \int^\oplus S_s(\tau)\, \widehat{I\!I}(\tau)\, d\tau = \int^\oplus \delta(\tau - \tau_0)\, \widehat{I\!I}(\tau)\, d\tau. \tag{4.29}$$

We shall call $\widehat{S}^{\oplus}_{\tau_0}$ a pure singular state at τ_0. Clearly pure singular states correspond one-to-one to singular subspaces $\mathcal{H}(\tau)$ or $\mathcal{H}^{\oplus}_\tau$. In a more traditional language a pure singular state corresponds to a unit vector $\boldsymbol{u}(\tau_0)$ lying in $\mathcal{H}(\tau_0)$, or to a pure singular vector $\boldsymbol{u}^{\oplus}_{\tau_0}$ of the form of Eq. (2.599) in $\mathcal{H}^{\oplus}_c$. The expectation value of $\widehat{A}^{\oplus}$ in a pure singular state $\widehat{S}^{\oplus}_{\tau_0}$ is given by

$$\mathcal{E}(\widehat{A}^{\oplus}, \widehat{S}^{\oplus}_{\tau_0}) = \int \delta(\tau - \tau_0) A(\tau)\, d\tau = A(\tau_0). \quad (4.30)$$

5. Mixed singular states are given by singular statistical operators of the form of Eq. (2.683) with expectation values given by Eq. (2.687).

A fundamental and distinguishing feature of the unified theory in both discrete and continuous cases is as follows:

- *A normalized mixed vector in the direct integral space represents a classical mixture of states while a pure vector represents a pure state.*

4.3.2 Coherence and superposition principle

We have introduced the concept of coherent superposition of states in orthodox quantum mechanics in §3.2.3. The coherence and superposition of states are fundamental concepts in any quantum theory. To avoid confusion we shall spell out their meaning in the context of our unified theory.

Definition 4.3.2(1)   **Coherence and the superposition principle**

- Let $\boldsymbol{v}^{\oplus}$, $\boldsymbol{w}^{\oplus}$,... be normalized pure vectors each representing a pure state. Then the state corresponding to a normalized linear combination $\boldsymbol{z}^{\oplus}$ of these pure vectors, i.e.,

$$\boldsymbol{z}^{\oplus} = a\,\boldsymbol{v}^{\oplus} + b\,\boldsymbol{w}^{\oplus} + \cdots, \quad ||\boldsymbol{z}^{\oplus}|| = 1, \quad (4.31)$$

is called a normalized linear combination of the pure states represented by pure vectors $\boldsymbol{v}^{\oplus}$, $\boldsymbol{w}^{\oplus}$,....

- Two or more pure states described by pure vectors are said to be *coherent* if any arbitrary normalized linear combination of these pure states again represents a pure state.

- A normalized linear combination of pure states which are coherent is referred to as a *coherent superposition* of the states.

4.3. STATES AND SUPERPOSITION PRINCIPLE

- A *superposition principle* is said to operate in a set of pure states if states in the set are coherent. Two or more states are said to satisfy the superposition principle if they are coherent.

Let us briefly recall the discussion in §3.2.3 for orthodox quantum mechanics. A coherent superposition of two states is a physical statement, meaning that a pure state ϕ is a weighted sum of other pure states φ_j such that the constituent pure states φ_j in the sum are correlatable by some observables. A coherent superposition is equivalent to a linear combination *plus* the possibility of observing correlations between the constituent states. These correlations are due the existence of physical observables represented by selfadjoint operators $\widehat{A}$ such that the correlation terms $\langle \varphi_j \mid \widehat{A} \varphi_k \rangle$ in Eq. (3.39) for the expectation value $\mathcal{E}(\widehat{A}, \phi)$ of $\widehat{A}$ are not all zero. Physically the correlations manifest themselves in the fact that $\mathcal{E}(\widehat{A}, \phi)$ is not simply a weighted sum of the expectation values $\mathcal{E}(\widehat{A}, \varphi_j)$ worked out in each constituent state separately. There are additional terms, the correlation terms $\langle \varphi_j \mid \widehat{A} \varphi_k \rangle$, which are dependent on both φ_j and φ_k. It follows that whether a given linear combination has the significance of a coherent superposition or not depends on the set of selfadjoint operators which are admissible for the representation of observables. In the case of orthodox quantum theory all selfadjoint operators correspond to observables. As a result we can always find observables for which some correlation terms do not vanish. Consequently all pure states are coherent and they satisfy the superposition principle.[6] The situation is quite different in the present unified theory. We shall formalize this in the next section.

4.3.3 Superselection rules, their origins and classical observables

As pointed out in C2.14.2(2) C1 and C2.14.2(3) C4 observables in our present unified theory, being decomposable, cannot correlate pure states belonging to different subspaces $\mathcal{H}_\tau^\oplus$. In the terminology of orthodox quantum mechanics this means a kind of selection rules forbidding such correlations.[7] Following Wick, Wightman and Wigner (1952) we shall call these superselection rules.[8]

---

[6]This statement is not inconsistent with the existence of states, like those detailed in §3.5 and §3.6 in relation to quantum messurements and state preparation, which are spatially so separated as to render any correlation terms negligible for all practical purposes. We may call this *asymptotic decoherence*, a topic which we shall return to investigate in Chapter 8.

[7]See Cisneros *et al.* (1998) for a list of limitations on the superposition principle.

[8]Bogolubov, Logunov and Todorov (1975), Wan (1980), Beltrammetti and Cassinelli (1981) §5, Van Fraassen (1991).

A formal definition is given in Definition 4.3.3(2) in the context of Postulate UQS, after Definition 4.3.3(1) which introduces some relevant terminology.

Definition 4.3.3(1)   Superselection operators, supersectors and disjoint states

- A selfadjoint diagonalizable operator $\widehat{C}^\oplus$ in $\mathcal{H}^\oplus$ defined by Eq. (2.645), other than the trivial case of the identity operator, is called a *superselection operator*.

- Subspaces $\mathcal{H}_\tau^\oplus$ and their corresponding constituent spaces $\mathcal{H}(\tau)$ are called *supersectors*.[9]

- A supersector is also referred to as a *coherent subspace*.

- Two pure states belonging to two different supersectors are called *disjoint*, as are the two supersectors.

Pure states corresponding to pure vectors within the same supersector are coherent and satisfy the superposition principle. This is why a supersector is called a coherent subspace. Two disjoint pure states are not coherent and do not satisfy the superposition principle. A linear combination of disjoint states is not a coherent superposition since there are no physical observables to correlate the constituent states in such a linear combination, i.e., no correlation term can be observed experimentally. Such a linear combination represents a classical mixture of states. This agrees with our earlier discussion leading to Eq. (4.21) for the expectation values.

Definition 4.3.3(2)   Classical observables and superselection rules

- Observables represented by superselection operators are called *classical observables*.[10]

- A physical system is said to possess *superselection rules* if it possesses classical observables, with each classical observable defining a superselection rule.

Generally observables do not commute. Being diagonalizable, superselection operators $\widehat{C}^\oplus$ commute among themselves and with all other observables of

---

[9] As discussed in E2.12.2(4) E2 these subspaces are singular if the preferred direct integral decomposition is continuous.

[10] Classical observables in the context of a quantum theory has been introduced by a number of authors, e.g., Primas and Müller-Herold (1978) §3.7.

4.3. STATES AND SUPERPOSITION PRINCIPLE

the system. Moreover, a pure state is an eigenvector of a superselection operator. Let us list some important properties:

1. For the case when the Hilbert space has a purely discrete direct integral decomposition $\mathcal{H}_d^\oplus$ a pure state corresponds to a pure vector $\boldsymbol{u}_n^\oplus$ and a diagonalizable operator reduces to the form

$$\widehat{C}^\oplus = \sum_j c_j \widehat{\mathbb{I}}_j^\oplus. \tag{4.32}$$

It follows that $\widehat{C}^\oplus$ can be considered to possess the value c_n in every pure state $\boldsymbol{u}_n^\oplus$ since

$$\widehat{C}^\oplus \boldsymbol{u}_n^\oplus = c_n \boldsymbol{u}_n^\oplus. \tag{4.33}$$

2. For the case of a purely continuous direct integral decompostion $\mathcal{H}_c^\oplus$ the situation is the same except that we will be dealing with singular quantities. We have

$$\widehat{C}^\oplus = \int^\oplus c(\tau) \widehat{\mathbb{I}}(\tau) \, d\tau, \tag{4.34}$$

and

$$\widehat{C}^\oplus \boldsymbol{u}_{\tau_0}^\oplus = c(\tau_0) \boldsymbol{u}_{\tau_0}^\oplus, \tag{4.35}$$

showing a possessed value $c(\tau_0)$ in a pure state $\boldsymbol{u}_{\tau_0}^\oplus$.

3. An observable of an orthodox quantum system characterized by Postulate OQS does not possess a definite value in an arbitrary pure state. As an observable a superselection operator does possess a definite value in an arbitrary pure state, e.g., c_n in the pure state $\boldsymbol{u}_n^\oplus$ in $\mathcal{H}_d^\oplus$ and $c(\tau_0)$ in the pure state $\boldsymbol{u}_{\tau_0}^\oplus$ in $\mathcal{H}_c^\oplus$; these observables are fundamentally different from any observable of orthodox quantum system. This is why $\widehat{C}^\oplus$ is called a classical observable.

In the present unified theory the one-to-one correspondence between observables and selfadjoint operators and between pure states and unit vectors in orthodox quantum theory breaks down. Classical observables and superselection rules emerge and normalized mixed vectors can be employed to represent a certain type of mixed states. Superselection rules can arise from a variety of reasons. We shall list a few recognized origins of superselection rules:

1. *Environmental origin*   The environmental origin of superselection rules proposed by Zurek[11] is well-established. The idea is that certain quantum systems can possess a special observable, often referred to as a

---
[11] Zurek (1981), (1982), (1991). Giulini, Kiefer and Zeh (1995) argued that even well-established superselection rules, e.g., those for charge, mass and particles with different spins in particle physics can be explained by environmental factors so that the *universality of superposition principle* can be maintained.

pointer observable, whose eigensubspaces can become disjoint in the sense of Definition 4.3.3(1), due to the interaction of the system with its environment. Naturally the term *environment* could mean external surroundings and influences, e.g., some kind of external background field.[12] However, an environment in the present context can arise entirely within the physical system. Let us pursue this argument a little futher. Consider a complex system composed of a large number of sub-systems corresponding to a large number of degrees of freedom. We may try to model such a complex system by a simpler system together with an auxiliary system. For example, the motion of a small object of a certain mass attached to the end of an elastic spring can be described in the first instance as a point particle in motion under a harmonic potential, i.e., we approximate the system by a simple harmonic oscillator. The physical system of an elastic spring with a small object attached to its end is clearly far more complex than a simple harmonic oscillator, e.g., the spring itself consists of a large number of interacting particles and these interactions may lead to energy dissipation and ultimately nonharmonic behaviour. To achieve a more realistic description we can try to incorporate the complexity of the physical system, e.g., all those additional degrees of freedom of the spring itself, into an auxiliary system. We would then describe the physical system, i.e., an object vibrating under the action of a spring, by a simple harmonic oscillator coupled to an appropriate auxiliary system. This auxiliary system is what we now call the *environment*, i.e., we have a simple harmonic oscillator coupled to an environment. It is possible then to set up some appropriate coupling interaction, e.g., an additional dissipative term to the oscillator Hamiltonian, to render the eigensubspaces of the uncoupled oscillator Hamiltonian disjoint. Another terminology used is *de-coherence*, e.g., we would say that the eigensubspaces of the uncoupled oscillator Hamiltonian become *de-cohered* when the coupling with its environment is taken into consideration. A systematic discussion with explicit examples is available in Omnès (1994) and Grigolini (1993).[13]

Physical systems with such a pointer observable can serve as measuring devices to effect a quantum measurement. We shall return to this idea of quantum measurement in terms of measuring devices in §4.7.

2. *Complexity and macroscopic origin*   Superselection rules can arise from complex and macroscopic systems in a way different from the environmental argument mentioned earlier. Using quasi-local observables

---

[12] Grigolini (1993) §3.5.
[13] Omnès (1994) Chapter 7, Grigolini (1993) Chapter 3. A series of articles on decoherence are available in Giulini *et al.* (1996).

4.3. STATES AND SUPERPOSITION PRINCIPLE

Hepp (1972) is able to construct several explicit examples of systems of a large number of degrees of freedom having disjoint states. Disjointness is achieved asymptotically as the number of degrees of freedom increases indefinitely. Mathematically this means the vanishing of those correlation terms in Eq. (3.39) as degrees of freedom increase indefinitely, i.e., we have an asymptotic decoherence. By considering the internal interaction of an infinite system of spins Bub (1988) demonstrates explicitly how a discrete superselection rule of Class 1 systems in §4.3.1 can arise. Model theories for continuous superselection rules are also available.[14] In Chapter 7 we discuss how superselection rules can arise from superconducting systems.

3. *Algebraic and quantum logic origin*   Traditionally states are regarded as the primary quantities and their Hilbert space structure is postulated first as in Postulate OQS. However, as pointed out in §3.1, if one takes the view that observables should be the primary quantities to build up a theory one may replace Postulate OQS by a postulate on the structure of the set of observables. In the algebraic approach the set of observables is assumed to have an abstract algebraic structure, e.g., an abstract C^*-algebra structure. For a given quantum system we can choose to represent the abstract C^*-algebra by operators on an appropriate Hilbert space. Operator representation of an abstract C^*-algebra is not unique, i.e., there are inequivalent representations which can lead to superselection rules with each representation space forming a superselection sector.[15] Algebraic approach is particularly useful in dealing with infinite systems, e.g., thermodynamical systems. For such systems one can introduce local observables and global macroscopic observables in a naturally manner.[16] In the quantum logic approach one can construct proposition systems which would admit superselection rules in a natural manner.[17]

4. *Dynamic origin*   Disjointness and the vanishing correlation terms for certain states can be achieved asymptotically by time evolution. We have several examples of this in Chapter 3 within the context of orthodox quantum mechanics. The simplest one is given by Theorem 3.5.2(1) which tells us that the correlations between a bound state $\widehat{U}_t \phi_b$ and a scattering state $\widehat{U}_t \phi_s$ vanish asymptotically as the scattering state move to spatial infinity at large times. In other words states $\widehat{U}_t \phi_b$ and

---

[14] Machida and Namiki (1980), Araki (1980), (1986), Namiki (1988), Namiki and Pascazio (1993).

[15] Haag and Kastler (1964), Roberts and Roepstorff (1969), Bogolubov, Logunov and Todorov (1975), Kastler (1990), Landsman (1991), Hagg (1992).

[16] Sewell (2002) pp. 42-44.

[17] Jauch (1968), Piron (1975).

$\widehat{U}_t \phi_s$ become disjoint asymptotically. We shall return to investigate this later in Chapter 8. It is also possible to go beyond orthodox quantum mechanics, i.e., to replace Postulate OQD by some non-linear evolution equation to achieve disjointness of states.[18] There are also theories based on stochastic and diffusion processes and thermodynamical irreversibility arguments to achieve disjointness of states.[19]

5. *Symmetry breaking, weak nuclear force, superselection rules in molecules*
It has been known for a long time that certain molecules exist in two types of structures which are mirror images of each other. In other words such a molecule can exist in one of two states reflecting two different molecular structures. Such states are known as *chiral states*. These two states are distinguishable. For a group of molecules known as *optical isomers* one structure can rotate the polarization of an incident polarized light clockwise while the other structure would rotate the polarization of an incident polarized light anti-clockwise.[20] The fact that these molecules are always found to be in a chiral state demonstrates the existence of a superselection rule forbidding any coherent superposition of these two chiral states.[21] Various reasons have been advanced to explain the origin of such superselection rules, including *symmetry breaking* and *electro-weak nuclear force* which is known to violate *parity conservation*.[22] In §8.3.9 we shall return to give a quantum mechanical treatment of these molecules.

6. *Topological origin* Quantum mechanics is usually set up on a Euclidean space topologically equivalent to $I\!\!E^n$. When the physical space is not topologically equivalent to $I\!\!E^n$ new features and new formulations can be established leading to a quantum theory with superselection rules. An example of such a quantum theory will be presented in Chapter 9.

Superselection rules are now seen to exist in microscopic[23] as well as in macroscopic quantum systems, e.g., from molecules to superconducting systems, and their origins can also range from microscopic, e.g., electro-weak nuclear force, to macroscopic, e.g., largeness and complexity. In §8.4.1 and §8.4.2 we shall also show that a superselection rule can be broken and a superposition principle restored. So, it is not appropriate to make sweeping and rigid statements.

---

[18] The theory of spontaneous localization of Ghirardi, Remini and Weber (1986) being a well-known example.

[19] Gisin and Percival (1992), Grigolini (1993) §3.6 and Chapter 4.

[20] Cram and Cram (1978) pp. 167-178, Williams (1982) p. 240.

[21] Wightman and Glance (1989), Müler-Herold (1985).

[22] Hegstrom and Kondepudi (1990).

[23] Some authors believe in superselection rules in very simple systems like the one-dimensional hydrogen atom (Núněz-Yépez, Vargas and Salas-Brito (1988) and (1998)).

Instead we have to study each physical system individually.

4.4 Unified Dynamics in Direct Integral Space

4.4.1 Preliminaries

We now consider the time evolution of our physical system and, in particular, the issue of whether the operator generating the evolution must necessarily be an observable of the system. To avoid possible complications we shall adopt the following statement regarding the time evolution of the system:

A preliminary and qualitative statement on dynamics

- The time evolution of a physical system is unitary, and in simple cases it is describable in terms of a group of unitary evolution operators $\widehat{U}_t$, $t \in I\!R$, acting on the direct integral space $\mathcal{H}^\oplus$ associated with the system.

In contrast to Postulate OQD in §3.4.1 we have deliberately left the above statement vague, i.e., we have not stated how initial and final states are related by unitary evolution operators. As will be seen later there are situations in which the precise relationship stated in Postulate OQD makes no sense. In other words different systems may require different expressions to relate the initial and final states. In our unified theory two types of evolutions may be identified. We shall present a number of explicit examples to demonstrate these evolutions. In what follows we shall confine ourselves to evolutions describable in terms of a unitary group. Theorem 2.3 (Stone's Theorem) then ensures that the unitary evolution operators are expressible as

$$\widehat{U}_t = e^{-i\widehat{H}_g t} \quad \text{acting on } \mathcal{H}^\oplus, \tag{4.36}$$

where $\widehat{H}_g$, again referred to as the Hamiltonian generator, is a time independent selfadjoint operator in $\mathcal{H}^\oplus$.

The Stone's Theorem does not require $\widehat{H}_g$ to be decomposable, i.e., while the Hamiltonian generator $\widehat{H}_g$ of the propagator $\widehat{\mathcal{U}} = \{\widehat{U}_t\}$ is selfadjoint in $\mathcal{H}^\oplus$ it is not neccessarily decomposable. It follows from Postulate UQS that the Hamiltonian generator may not be an observable, let alone the energy observable. To emphasize the distinction we shall denote the operator representing the energy by $\widehat{H}^\oplus$ and call $\widehat{H}^\oplus$ the Hamiltonian operator or simply the Hamiltonian.

Clearly the nature of the propagator would depend on the properties of $\widehat{H}_g$. We shall introduce two distinct classes of propagators and their associated

Hamiltonian generators based on whether the evolution operators preserves the set of observables or not.[24]

Definition 4.4.1(1) Preserving and non-preserving evolution

- A unitary operator $\widehat{V}$ on $\mathcal{H}^{\oplus}$ is said to preserve the set of observables in $\mathcal{H}^{\oplus}$ if given any observable $\widehat{A}^{\oplus}$ the operator $\widehat{V}^{\dagger}\widehat{A}^{\oplus}\widehat{V}$ is again an observable, i.e., if for every selfadjoint and decomposable operator $\widehat{A}^{\oplus}$ in $\mathcal{H}^{\oplus}$, the operator defined by $\widehat{V}^{\dagger}\widehat{A}^{\oplus}\widehat{V}$ is again decomposable, as well as selfadjoint.

- Let $\widehat{\mathcal{U}} = \{\widehat{U}_t\}$ be a given propagator. If $\widehat{U}_t$ preserves the set of observables for all times t then the propagator $\widehat{\mathcal{U}}$, the evolution operators $\widehat{U}_t$ and its Hamiltonian generator $\widehat{H}_g$ are all said to be *observables preserving* or *preserving* for short. Otherwise they are said to be *observables non-preserving* or *non-preserving*.

We shall discuss the physical meaning of these two fundamentally distinct kinds of propagators separately in the following two sections.

4.4.2 Preserving dynamics

Postulate UQD(P) on observable preserving dynamics

- For a preserving evolution in the Schrödinger picture a system has associated with it a preserving propagator $\widehat{\mathcal{U}}^{(p)}$ such that if the system is in a state represented by the statistical operator $\widehat{S}_0^{\oplus}$ at $t = 0$, its state at time t is given by a statistical operator $\widehat{S}_t^{\oplus}$ defined by

$$\widehat{S}_t^{\oplus} = \widehat{U}_t^{(p)}\,\widehat{S}_0^{\oplus}\,\widehat{U}_t^{(p)\dagger}. \tag{4.37}$$

This compares directly with the evolution Eq. (3.215) of Postulate OQD in orthodox quantum mechanics. Indeed orthodox dynamics is a special case of preserving evolution. Generally for Postulate UQD(P) to make sense we must require the operator $\widehat{S}_t^{\oplus}$ defined above to be decomposable in order to remain a statistical operator. A preserving evolution ensures this for all times.[25]

As in orthodox quantum mechanics we can also introduce the *Heisenberg picture* in which the state is time independent. A time evolution is given

---

[24] Bradshaw (1995).
[25] Note that $\widehat{U}_t^{(p)\dagger} = \widehat{U}_{-t}^{(p)}$ and $\widehat{U}_t^{(p)} = \widehat{U}_{-t}^{(p)\dagger}$.

4.4. UNIFIED DYNAMICS IN DIRECT INTEGRAL SPACE

directly in terms of the time dependence of observables, i.e., an observable $\widehat{A}_0^\oplus$ at time $t = 0$ will evolve into $\widehat{A}_t^\oplus$ given by

$$\widehat{A}_t^\oplus = \widehat{U}_t^{(p)\dagger} \widehat{A}_0^\oplus \widehat{U}_t^{(p)}. \tag{4.38}$$

The preserving condition set out in Definition 4.4.1(1) is neccesary to ensure that the evolved operator $\widehat{A}_t^\oplus$ remains an observable, i.e., $\widehat{A}_t^\oplus$ is selfadjoint and decomposable. The Heisenberg and Schrödinger pictures are equivalent on account of the equality of expectation values, i.e.,

$$\mathcal{E}(\widehat{A}_t^\oplus, \widehat{S}_0^\oplus) = \mathcal{E}(\widehat{A}_0^\oplus, \widehat{S}_t^\oplus) \quad \forall t \in (-\infty, \infty). \tag{4.39}$$

Whether an evolution is preserving or not depends on the nature of the Hamiltonian generator. Let us examine the effect of the Hamiltonian generator on evolution in the following two cases depending on whether it is an observable or not:

Case 1 The Hamiltonian generator being an observable If the Hamiltonian generator $\widehat{H}_g^{(p)}$ is an observable then the evolution will be preserving. To see this we first note that according to Postulate UQS the operator $\widehat{H}_g^{(p)}$, being an observable, must be decomposable. Hence, by Theorem 2.14.2(2), $\widehat{H}_g^{(p)}$ commutes with all selfadjoint diagonalizable operators. Being a bounded function of $\widehat{H}_g^{(p)}$ the evolution operators $\widehat{U}_t^{(p)}$ also commute with all selfadjoint diagonalizable operators. It follows that $\widehat{U}_t^{(p)}$ are decomposable, and so are their adjoints $\widehat{U}_t^{(p)\dagger}$. To highlight all this we can rewrite the above operators as

$$\widehat{H}_g^{(p)\oplus}, \quad \widehat{U}_t^{(p)\oplus} \quad \text{and} \quad \widehat{U}_t^{(p)\oplus\dagger}. \tag{4.40}$$

We can also conclude from Theorem 2.14.2(1) that the product

$$\widehat{U}_t^{(p)\oplus\dagger} \widehat{A}^\oplus \widehat{U}_t^{(p)\oplus} \tag{4.41}$$

is also decomposable. Being a unitary transform of $\widehat{A}^\oplus$ the above product operator is also selfadjoint whenever $\widehat{A}^\oplus$ is. In other words $\widehat{U}_t^{(p)\oplus}$ preserve the set of observables for all time, resulting in a preserving evolution.

A pure vector will evolve within a supersector and it will not be able to evolve from one supersector to another supersector. This is due to the decomposable nature of the unitary evolution operators $\widehat{U}_t^{(p)\oplus}$ and the fact that decomposable operators leave subspaces $\mathcal{H}_\Delta^\oplus$ invariant, a fact pointed out in C2.14.2(2) C3.

In the case of a purely discrete direct integral decomposition we have:

1. The Hilbert space is $\mathcal{H}_d^\oplus = \mathcal{H}(1) \oplus \mathcal{H}(2) \oplus \cdots$.

2. Suppose at time $t = 0$ the system is in a pure state $\widehat{S}_{10}^{\oplus}$ given by

$$\widehat{S}_{10}^{\oplus} = |\boldsymbol{u}_0(1)\rangle\langle \boldsymbol{u}_0(1)| \oplus \widehat{O}(2) \oplus \cdots. \tag{4.42}$$

This is equivalent to a representation in terms of the pure vector

$$\boldsymbol{u}_{10}^{\oplus} = \boldsymbol{u}_0(1) \oplus \widehat{O}(2) \oplus \cdots. \tag{4.43}$$

3. Following from Eq. (2.646) we know that the evolution operators $\widehat{U}_t^{(p)\oplus}$ are of the form

$$\widehat{U}_t^{(p)\oplus} = \widehat{U}_t(1) \oplus \widehat{U}_t(2) \oplus \cdots. \tag{4.44}$$

4. In accordance with Theorem 2.14.2(1) the state at a later time becomes

$$\begin{aligned}\widehat{S}_{1t}^{\oplus} &= \widehat{U}_t^{(p)\oplus} \widehat{S}_1^{\oplus} \widehat{U}_t^{(p)\oplus\dagger} \\ &= \widehat{U}_t(1)\big(|\boldsymbol{u}_0(1)\rangle\langle \boldsymbol{u}_0(1)|\big)\widehat{U}_t^{\dagger}(1) \oplus \widehat{O}(2) \oplus \cdots \\ &= |\boldsymbol{u}_t(1)\rangle\langle \boldsymbol{u}_t(1)| \oplus \widehat{O}(2) \oplus \cdots, \end{aligned} \tag{4.45}$$

where

$$\boldsymbol{u}_t(1) = \widehat{U}_t(1)\boldsymbol{u}_0(1), \quad \text{or} \quad \boldsymbol{u}_{1t}^{\oplus} = \boldsymbol{u}_t(1) \oplus \widehat{O}(2) \oplus \cdots. \tag{4.46}$$

This shows that a pure vector representing a pure state evolves within the same supersector.[26]

When the Hilbert space is $\mathcal{H}_c^{\oplus}$ the above arguments still go through except that a pure state is singular of the form $\widehat{S}_{\tau_0}^{\oplus}$. We have a continuous integral decomposition of the evolution operators so that

$$\widehat{S}_{\tau t}^{\oplus} = \widehat{U}_t^{(p)\oplus} \widehat{S}_{\tau_0}^{\oplus} \widehat{U}_t^{(p)\oplus\dagger} \tag{4.47}$$

$$= \int^{\oplus} \widehat{U}_t(\tau) \widehat{S}_{\tau_0}^{\oplus} \widehat{U}_t^{\dagger}(\tau)\, d\tau. \tag{4.48}$$

A preserving Hamiltonian generator need not be an observable. We shall see in §4.5 that in the Hilbert space formulation of classical dynamics the Hamiltonian generators are preserving but they are not observables, although by no means all systems with such preserving generators are classical.

Case 2 **The Hamiltonian generator not being an observable**   The corresponding unitary operators are not decomposable, and they will not preserve subspaces, e.g., they do not preserve supersectors. This means that there will be the possibility of evolution between different supersectors.

---
[26] An explicit example is given in §4.6.1.

4.4. UNIFIED DYNAMICS IN DIRECT INTEGRAL SPACE

We know that a statistical operator $\widehat{S}_0^\oplus$ corresponding to a pure state is a one-dimensional projector onto a supersector and that the unitary transform of a one-dimensional projector is again a one-dimensional projector, as pointed out in C2.4(1) C5. So, the evolved state $\widehat{S}_t^\oplus$ would also be a one-dimensional projector. Under a preserving unitary evolution $\widehat{S}_t^\oplus$ is also decomposable. It follows that $\widehat{S}_t^\oplus$ must be a one-dimensional projector onto a supersector, i.e., $\widehat{S}_t^\oplus$ would represent a pure state. Unlike Case 1 discussed earlier, the present statistical operator $\widehat{S}_t^\oplus$ can be a projector onto a new supersector. In other words, a pure state can evolve from one supersector to another supersector, the state remaining pure during the evolution.[27]

Let us summarize the features of preserving evolution in our present unified theory as follows:[28]

- *a pure state evolves into a pure state, not necessarily in the same supersector and a mixed state evolves into a mixed state, in any finite period of time.*

Before leaving this section we shall introduce the concept of stationary states, a concept which will be seen to help distinguish classical dynamics from quantum dynamics.

Definition 4.4.2(1)   **Stationary states**

- A pure state $\widehat{S}_0^\oplus$ is said to be *stationary* if $\widehat{S}_0^\oplus = \widehat{S}_t^\oplus$ for all times.

Physically a stationary state is equivalent to having time independent expectation values for all observables which are not explicitly time-dependent. As an example consider the simple yet physically important case of a direct integral space $\mathcal{H}_c^\oplus$ with a continuous decomposition into one-dimensional supersectors $\mathcal{H}(\tau)$. A pure singular state $\widehat{S}_0^\oplus$ at $t = 0$ evolving into $\widehat{S}_t^\oplus$ in accordance with Postulate UQD(P) is stationary if and only if

$$\mathcal{E}(\widehat{B}^\oplus, \widehat{S}_0^\oplus) - \mathcal{E}(\widehat{B}^\oplus, \widehat{S}_t^\oplus) = 0 \qquad (4.49)$$

for all bounded decomposable observables $\widehat{B}^\oplus$ on $\mathcal{H}_c^\oplus$ and for all $t \in (-\infty, \infty)$. An example of stationary state in orthodox quantum mechanics is seen in Eq. (3.243). Stationary states will be seen to play an important role in the study of quantum/classical divide in §4.6.2 and §4.6.3.

---
[27] Explicit examples are given in §4.5.1 and §4.5.2.
[28] Wan, Bradshaw, Trueman and Harrison (1998) Theorem 2.

356   CHAPTER 4. PHYSICAL THEORY IN HILBERT SPACE

4.4.3 Non-preserving dynamics 1: Motivation

Things are quite different when the evolution group is non-preserving. If the evolution operators do not preserve the decomposable nature of selfadjoint operators then time evolution in the Schrödinger picture and in the Heisenberg picture as defined respectively by Eqs. (4.37) and (4.38) do not make sense since the resulting operators are not decomposable. The operator $\widehat{U}_t \widehat{S}_0^\oplus \widehat{U}_t^\dagger$ cannot serve as a statistical operator and $\widehat{U}_t^\dagger \widehat{A}_0^\oplus \widehat{U}_t$ cannot serve as an observable. The question then arises as to whether there are situations when the evolution is non-preserving at all. The answer is a definite yes. Let us highlight the desirability of non-preserving evolution with a well-known example arising from the superselection rule approach to quantum measurement theory. In such an approach we shall come across model measurement interaction involving non-observable Hamiltonian generators giving rise to a non-preserving evolution.[29]

Consider a model physical system, referred to as Model 4.4.3 hereafter, described by a three-dimensional Hilbert space with the following preferred direct sum decomposition in accordance with Postulate UQS:

$$\mathcal{H}^\oplus = \mathcal{H}_- \oplus \mathcal{H}_0 \oplus \mathcal{H}_+, \qquad (4.50)$$

where $\mathcal{H}_-$, $\mathcal{H}_0$ and $\mathcal{H}_+$ are one-dimensional Hilbert spaces spanned by unit vectors $\boldsymbol{u}_-$, $\boldsymbol{u}_0$ and $\boldsymbol{u}_+$ respectively. We shall employ $\boldsymbol{0}_-$, $\boldsymbol{0}_0$, $\boldsymbol{0}_+$ and $\boldsymbol{0}^\oplus$ to denote the zero vectors on $\mathcal{H}_-$, $\mathcal{H}_0$, $\mathcal{H}_+$ and $\mathcal{H}^\oplus$ respectively. We have three distinct supersectors: $\mathcal{H}_-$, $\mathcal{H}_0$ and $\mathcal{H}_+$. A vector in $\mathcal{H}^\oplus$ is of the form

$$\boldsymbol{f}^\oplus = f_- \boldsymbol{u}_-^\oplus + f_0 \boldsymbol{u}_0^\oplus + f_+ \boldsymbol{u}_+^\oplus, \qquad (4.51)$$

where

$$\boldsymbol{u}_-^\oplus = \boldsymbol{u}_- \oplus \boldsymbol{0}_0 \oplus \boldsymbol{0}_+, \quad \boldsymbol{u}_0^\oplus = \boldsymbol{0}_- \oplus \boldsymbol{u}_0 \oplus \boldsymbol{0}_+, \quad \boldsymbol{u}_+^\oplus = \boldsymbol{0}_- \oplus \boldsymbol{0}_0 \oplus \boldsymbol{u}_+. \qquad (4.52)$$

Introduce two operators $\widehat{L}_+$ and $\widehat{L}_-$ on $\mathcal{H}^\oplus$ by:

$$\widehat{L}_+ \boldsymbol{u}_0^\oplus = \boldsymbol{u}_+^\oplus, \quad \widehat{L}_+ \boldsymbol{u}_+^\oplus = \boldsymbol{u}_0^\oplus, \quad \widehat{L}_+ \boldsymbol{u}_-^\oplus = \boldsymbol{0}^\oplus, \qquad (4.53)$$

$$\widehat{L}_- \boldsymbol{u}_0^\oplus = \boldsymbol{u}_-^\oplus, \quad \widehat{L}_- \boldsymbol{u}_-^\oplus = \boldsymbol{u}_0^\oplus, \quad \widehat{L}_- \boldsymbol{u}_+^\oplus = \boldsymbol{0}^\oplus. \qquad (4.54)$$

These two operators are not decomposable as they can relate vectors in different supersectors. Let

$$\widehat{H}_g^{(np)} = \frac{\lambda}{\sqrt{2}} (\widehat{L}_- + \widehat{L}_+), \quad \lambda \in \mathbb{R}. \qquad (4.55)$$

---

[29]Wan (1980), Beltrammetti and Cassinelli (1981), Bub (1988), Hughes (1989), Van Fraassen (1991).

4.4. UNIFIED DYNAMICS IN DIRECT INTEGRAL SPACE

This operator is selfadjoint but not decomposable. Consider an evolution with $\widehat{H}_g^{(np)}$ serving as the Hamiltonian generator, i.e., an evolution governed by the unitary group:[30]

$$\widehat{\mathcal{U}}^{(np)} = \left\{ \widehat{U}_t^{(np)} = e^{-i\widehat{H}_g^{(np)} t}, \; t \in \mathbb{R} \right\}. \tag{4.56}$$

Let $\boldsymbol{f}_0^\oplus$ be a unit vector defined by

$$\boldsymbol{f}_0^\oplus = i \, \boldsymbol{u}_0^\oplus. \tag{4.57}$$

If the system is initially in the pure state

$$\widehat{S}_0^\oplus = |\boldsymbol{f}_0^\oplus\rangle\langle \boldsymbol{f}_0^\oplus| \tag{4.58}$$

corresponding to the pure vector $\boldsymbol{f}_0^\oplus = i \, \boldsymbol{u}_0^\oplus$ we may be tempted to apply Eq. (4.37) in Postulate UQD(P) to work out the state at a later time. Unfortunately the operator

$$\widehat{U}_t^{(np)} \, \widehat{S}_0^\oplus \, \widehat{U}_t^{(np)\dagger} = \widehat{U}_t^{(np)} \left(|\boldsymbol{f}_0^\oplus\rangle\langle \boldsymbol{f}_0^\oplus| \right) \widehat{U}_t^{(np)\dagger} \tag{4.59}$$

is not decomposable since the projector $\widehat{U}_t^{(np)} \, \widehat{S}_0^\oplus \, \widehat{U}_t^{(np)\dagger}$ is able to correlate different supersectors, e.g.,

$$\langle \boldsymbol{u}_-^\oplus | \, \widehat{U}_t^{(np)} \left(|\boldsymbol{f}_0^\oplus\rangle\langle \boldsymbol{f}_0^\oplus| \right) \widehat{U}_t^{(np)\dagger} \, \boldsymbol{u}_+^\oplus \rangle \neq 0. \tag{4.60}$$

Hence it cannot serve as a statistical operator to describe the state of the system at time $t > 0$. This is an example of a non-preserving Hamiltonian generator with non-preserving evolution operators. Postulate UQD(P) with its evolution Eq. (4.37) cannot be applied here. A new prescription to relate initial and final states has to be brought in.

One natural possibility is to return to the Schrödinger equation, i.e., to use the solution of the Schrödinger equation to construct a statistical operator to serve as the state at time t in a way consistent with our general interpretation of mixed vectors discussed in §4.3.1, especially in Eqs. (4.17), (4.18) and (4.21). The following vector

$$\boldsymbol{f}_t^\oplus = \frac{\sin \lambda t/\hbar}{\sqrt{2}} \, \boldsymbol{u}_-^\oplus + i \, \cos \lambda t/\hbar \, \boldsymbol{u}_0^\oplus + \frac{\sin \lambda t/\hbar}{\sqrt{2}} \, \boldsymbol{u}_+^\oplus \tag{4.61}$$

satisfies the Schrödinger equation

$$i\hbar \frac{\partial}{\partial t} \boldsymbol{f}_t^\oplus = \widehat{H}_g^{(np)} \boldsymbol{f}_t^\oplus. \tag{4.62}$$

---

[30] The superscript (np) signifies a non-preserving evolution.

Moreover, this solution also satisfies the initial condition $\boldsymbol{f}^{\oplus}_{t=0} = i\,\boldsymbol{u}^{\oplus}_0$. Since $\boldsymbol{f}^{\oplus}_t$ is a mixed vector we expect the state to evolve from $\widehat{S}^{\oplus}_0$ to a mixed state. This gives rise to the idea that we should take the following operator[31]

$$\widehat{S}^{\oplus}_t = \frac{\sin^2 \lambda t/\hbar}{2}\;\widehat{I\!I}_- \;\oplus\; \cos^2 \lambda t/\hbar \;\widehat{I\!I}_0 \;\oplus\; \frac{\sin^2 \lambda t/\hbar}{2}\;\widehat{I\!I}_+, \qquad (4.63)$$

to serve as the statistical operator for the state at time t. As a consequence the evolution causes a transition from an initially pure state described by $\widehat{S}^{\oplus}_0$ or $\boldsymbol{f}^{\oplus}_0 = i\,\boldsymbol{u}^{\oplus}_0$ at $t=0$ to the mixed state $\widehat{S}^{\oplus}_t$ defined by Eq. (4.63) at time t. It is this type of evolution which paves the way to a superselection rule approach to quantum measurement theory.

While we may work out a link between $\widehat{S}^{\oplus}_0$ and $\widehat{S}^{\oplus}_t$ in individual cases directly through the Schrödinger equation we would ideally want an explicit and direct link in a general case. The root cause of the problem lies in the use of statistical operators for state description, since the set of statistical operators is not preserved under such an evolution. To achieve a direct link between initial and final states for a non-preserving evolution we really need an alternative representation of states. One simple way to achieve this is to employ linear functionals to bypass statistical operators. Since linear functionals are used for state description in the algebraic approach to quantum theory they can be easily adopted to suit our present situation.

4.4.4 Linear functionals for state description

A linear functional acting on a real vector space is a linear mapping of the vector space to the reals, a concept we first encountered in §1.4. This can be extended to linear functionals on an algebra. We know that operators on a Hilbert space possess algebraic structures.[32] Let $\mathcal{B}^{\oplus}(\mathcal{H}^{\oplus})$ denote the set of all bounded decomposable operators on $\mathcal{H}^{\oplus}$ and let $\mathcal{B}(\mathcal{H}^{\oplus})$ denote the set of all bounded operators on $\mathcal{H}^{\oplus}$, including non-decomposable ones. Clearly both $\mathcal{B}^{\oplus}(\mathcal{H}^{\oplus})$ and $\mathcal{B}(\mathcal{H}^{\oplus})$ form a *-algebra of bounded operators. The following definition[33] introduces some important types of linear functionals on $\mathcal{B}(\mathcal{H}^{\oplus})$ and on $\mathcal{B}^{\oplus}(\mathcal{H}^{\oplus})$.

Definition 4.4.4(1) **Normalized positive linear functionals on $\mathcal{B}(\mathcal{H}^{\oplus})$**

- A linear functional Ω on $\mathcal{B}(\mathcal{H}^{\oplus})$ is a mapping of $\mathcal{B}(\mathcal{H}^{\oplus})$ to the complex numbers, i.e.,

$$\Omega : \mathcal{B}(\mathcal{H}^{\oplus}) \mapsto \mathbb{C} \qquad (4.64)$$

---

[31] $\widehat{I\!I}_-$, $\widehat{I\!I}_0$ and $\widehat{I\!I}_+$ denote respectively the identity operators on $\mathcal{H}_-$, $\mathcal{H}_0$ and $\mathcal{H}_+$.
[32] See C2.12.3(1) C1 to C4 and Definition 2.12.3(2).
[33] The definition applies to the set of bounded operators on any Hilbert space $\mathcal{H}$, i.e., we can define NPLFs on $\mathcal{B}(\mathcal{H})$.

4.4. UNIFIED DYNAMICS IN DIRECT INTEGRAL SPACE

such that for any $\widehat{B}_1, \widehat{B}_2 \in \mathcal{B}(\mathcal{H}^\oplus)$ and $c_1, c_2 \in \mathbb{C}$ we have

$$\Omega(c_1 \widehat{B}_1 + c_2 \widehat{B}_2) = c_1 \Omega(\widehat{B}_1) + c_2 \Omega(\widehat{B}_2). \tag{4.65}$$

- A linear functional Ω on $\mathcal{B}(\mathcal{H}^\oplus)$ is

  1. normalized if acting on the identity operator $\widehat{I\!I}^\oplus$ on $\mathcal{H}^\oplus$ we get

  $$\Omega(\widehat{I\!I}^\oplus) = 1, \tag{4.66}$$

  2. positive if [34]

  $$\Omega(\widehat{B}^\dagger \widehat{B}) \in \mathbb{R} \quad \text{and} \quad \Omega(\widehat{B}^\dagger \widehat{B}) \geq 0 \quad \forall \widehat{B} \in \mathcal{B}(\mathcal{H}^\oplus). \tag{4.67}$$

An important feature is that an NPLF, short for *normalized positive linear functional*, Ω maps every selfadjoint operator $\widehat{B} \in \mathcal{B}(\mathcal{H}^\oplus)$ to a real number, i.e.,[35]

$$\widehat{B} = \widehat{B}^\dagger \;\Rightarrow\; \Omega(\widehat{B}) \in \mathbb{R}. \tag{4.68}$$

We can repeat Definition 4.4.4(1) to define NPLFs on $\mathcal{B}^\oplus(\mathcal{H}^\oplus)$ having the same properties in the same way. For clarity we shall employ $\Omega^\oplus$ to denote an NPLF on $\mathcal{B}^\oplus(\mathcal{H}^\oplus)$.

A simple way to obtain an NPLF is through a density operator. Let $\widehat{D}$ be a density operator on $\mathcal{H}^\oplus$, not necessarily decomposable. Then

$$\Omega(\widehat{B}) = \operatorname{tr}\left(\widehat{B}\widehat{D}\right) \tag{4.69}$$

is an NPLF on $\mathcal{B}(\mathcal{H}^\oplus)$.

Definition 4.4.4(2)   SNPLFs on $\mathcal{B}^\oplus(\mathcal{H}^\oplus)$

- A *statistically normalized positive linear functional*, abbreviated to SNPLF, $\Omega^\oplus$ acting on $\mathcal{B}^\oplus(\mathcal{H}^\oplus)$ is a linear functional on $\mathcal{B}^\oplus(\mathcal{H}^\oplus)$ defined in terms of a statistical operator $\widehat{S}^\oplus$ by

$$\Omega^\oplus(\widehat{B}^\oplus) = \operatorname{tr}^\oplus\left(\widehat{B}^\oplus \widehat{S}^\oplus\right) \quad \forall \widehat{B}^\oplus \in \mathcal{B}^\oplus(\mathcal{H}^\oplus). \tag{4.70}$$

---

[34]Condition $\Omega(\widehat{B}^\dagger \widehat{B}) \geq 0$ here is equivalent to the condition $\Omega(\widehat{B}\widehat{B}^\dagger) \geq 0$, since $\widehat{B}^{\dagger\dagger} = \widehat{B}$.
[35]This follows from Eq. (4.67) and $\widehat{B}^\dagger = \widehat{B}$, i.e.,

$$\Omega\big((\widehat{B} + \widehat{I\!I}^\oplus)^\dagger (\widehat{B} + \widehat{I\!I}^\oplus)\big) \in \mathbb{R} \;\Rightarrow\; \Omega(\widehat{B}^\dagger \widehat{B}) + 2\Omega(\widehat{B}) + \Omega(\widehat{I\!I}^\oplus) \in \mathbb{R} \;\Rightarrow\; \Omega(\widehat{B}) \in \mathbb{R}.$$

When the Hilbert space has a purely continuous decomposition we have regular statistical operators $\widehat{S}_r^\oplus$ of the form of Eq. (4.8) and bounded decomposable operators $\widehat{B}^\oplus$ of the form of Eq. (4.9) so that

$$\Omega^\oplus(\widehat{B}^\oplus) = \int S_r(\gamma) \operatorname{tr}_\gamma\big(\widehat{B}(\gamma)\widehat{D}(\gamma)\big)\, d\gamma. \tag{4.71}$$

When the Hilbert space has a purely discrete decomposition we have statistical operators $\widehat{S}^\oplus$ of the form of Eq. (4.4) and bounded decomposable operators $\widehat{B}^\oplus$ of the form of Eq. (4.5) so that

$$\Omega^\oplus(\widehat{B}^\oplus) = \sum_n w_n \operatorname{tr}_n\big(\widehat{B}(n)\widehat{D}(n)\big). \tag{4.72}$$

In both cases $\Omega^\oplus(\widehat{B}^\oplus)$ is equal to the expectation value $\mathcal{E}(\widehat{B}^\oplus, \widehat{S}^\oplus)$.

An SNPLF is clearly also an NPLF on $\mathcal{B}^\oplus(\mathcal{H}^\oplus)$. By definition we have a one-to-one correspondence between SNPLFs and statistical operators. Regular and singular SNPLFs can be introduced in terms of the regular or singular nature of their defining statistical operators. A similar terminology applies to pure and mixed SNPLFs.

As pointed out in C2.3(1) C9 and shown in Eq. (2.86) bounded operators are expressible in terms of bounded selfadjoint operators. Therefore we can determine an SNPLF $\Omega^\oplus$ on $\mathcal{B}^\oplus(\mathcal{H}^\oplus)$ by the its values $\Omega^\oplus(\widehat{A}^\oplus)$ for selfadjoint $\widehat{A}^\oplus$.

In view of what has been said we can employ SNPLFs to represent states. This is actually a standard way to represent states in the algebraic formulation of quantum mechanics. Let us now rephrase Postulate UQS in §4.2.1 by changing items 2 and 3 in terms of SNPLFs.

Postulate UQS(F) on statics

- A physical system has associated with it a Hilbert space which has a **preferred** direct integral decomposition $\mathcal{H}^\oplus$, possibly trivial, into a Borel family of separable Hilbert spaces $\mathcal{H}(\gamma)$ such that:

  1. Physical observables correspond one-to-one to decomposable selfadjoint operators $\widehat{A}^\oplus$ in $\mathcal{H}^\oplus$.

  2. States correspond one-to-one to statistically normalised positive linear functionals $\Omega^\oplus$ on the *-algebra of all bounded decomposable operators $\mathcal{B}^\oplus(\mathcal{H}^\oplus)$ on $\mathcal{H}^\oplus$.

  3. The expectation value $\mathcal{E}(\widehat{B}^\oplus, \Omega^\oplus)$ of a bounded observable $\widehat{B}^\oplus$ in state $\Omega^\oplus$ is given by

  $$\mathcal{E}(\widehat{B}^\oplus, \Omega^\oplus) = \Omega^\oplus(\widehat{B}^\oplus). \tag{4.73}$$

4.4. UNIFIED DYNAMICS IN DIRECT INTEGRAL SPACE

A state is a characterization of a physical system capable of fixing the expectation values of all observables. By its very definition a linear functional directly fixes these expectation values. So, it is actually natural to employ linear functionals to represent states. Such a description of states is in fact more flexible and general. The use of linear functionals often enables us to extend the class of states of a physical system, e.g., states at infinity.[36]

4.4.5 Extensions and restrictions of linear functionals

Generally the domain of operation of linear functionals can be extended or restricted.[37] In our present context we may extend the domain of operation of SNPLFs from $\mathcal{B}^\oplus(\mathcal{H}^\oplus)$ to a larger operator algebra. For example we can extend a given SNPLF $\Omega^\oplus$ on $\mathcal{B}^\oplus(\mathcal{H}^\oplus)$ to an NPLF Ω on $\mathcal{B}(\mathcal{H}^\oplus)$. A trivial extension is defined by:

$$\Omega^{(0)}(\widehat{B}) = \begin{cases} \Omega^\oplus(\widehat{B}) & \forall \widehat{B} \in \mathcal{B}^\oplus(\mathcal{H}^\oplus) \\ 0 & \forall \widehat{B} \in \mathcal{B}(\mathcal{H}^\oplus), \widehat{B} \notin \mathcal{B}^\oplus(\mathcal{H}^\oplus). \end{cases} \quad (4.74)$$

Another extension would be to use the statistical operator $\widehat{S}^\oplus$ associated with $\Omega^\oplus$ to define an NPLF by

$$\Omega^{(1)}(\widehat{B}) = \begin{cases} \Omega^\oplus(\widehat{B}) & \forall \widehat{B} \in \mathcal{B}^\oplus(\mathcal{H}^\oplus) \\ \operatorname{tr}\left(\widehat{B}\widehat{S}^\oplus\right) & \forall \widehat{B} \in \mathcal{B}(\mathcal{H}^\oplus), \widehat{B} \notin \mathcal{B}^\oplus(\mathcal{H}^\oplus). \end{cases} \quad (4.75)$$

Given an NPLF Ω on $\mathcal{B}(\mathcal{H}^\oplus)$ we may be able to find an SNPLF $\Omega^\oplus$ on $\mathcal{B}^\oplus(\mathcal{H}^\oplus)$ such that

$$\Omega^\oplus(\widehat{B}^\oplus) = \Omega(\widehat{B}^\oplus) \quad \forall \widehat{B}^\oplus \in \mathcal{B}^\oplus(\mathcal{H}^\oplus). \quad (4.76)$$

The extension from $\Omega^\oplus$ to Ω will not be unique. Also there will be many different NPLFs on $\mathcal{B}(\mathcal{H}^\oplus)$ giving rise to the same restriction $\Omega^\oplus$ to $\mathcal{B}^\oplus(\mathcal{H}^\oplus)$.

Definition 4.4.5(1)   **NPLFs on $\mathcal{B}(\mathcal{H}^\oplus)$ and SNPLFs on $\mathcal{B}^\oplus(\mathcal{H}^\oplus)$**

- An NPLF Ω on $\mathcal{B}(\mathcal{H}^\oplus)$ is said to be an extension of an SNPLF $\Omega^\oplus$ on $\mathcal{B}^\oplus(\mathcal{H}^\oplus)$ if

$$\Omega(\widehat{B}^\oplus) = \Omega^\oplus(\widehat{B}^\oplus) \quad \forall \widehat{B}^\oplus \in \mathcal{B}^\oplus(\mathcal{H}^\oplus). \quad (4.77)$$

Ω is also said to be generated by $\Omega^\oplus$.

---
[36] Wan and McLean (1984a), (1984b).
[37] Bratteli and Robinson (1979) Proposition 2.3.24 on p. 60.

- An SNPLF $\Omega^\oplus$ on $\mathcal{B}^\oplus(\mathcal{H}^\oplus)$ is said to be a restriction of an NPLF Ω on $\mathcal{B}(\mathcal{H}^\oplus)$ if Eq. (4.77) is satisfied. $\Omega^\oplus$ is also said to be generated by Ω.

Let us illustrate the non-uniqueness of the relationship by considering a simple system whose associated Hilbert space has the following preferred direct sum decomposition:
$$\mathcal{H}_{12}^\oplus = \mathcal{H}_1 \oplus \mathcal{H}_2, \tag{4.78}$$
where $\mathcal{H}_1$, $\mathcal{H}_2$ are one-dimensional subspaces spanned by unit vectors $\boldsymbol{u}_1$ and $\boldsymbol{u}_2$ respectively. The unit vectors in $\mathcal{H}^\oplus$ corresponding to $\boldsymbol{u}_1$ and $\boldsymbol{u}_2$ are
$$\boldsymbol{u}_1^\oplus = \boldsymbol{u}_1 \oplus \boldsymbol{0}_2, \quad \boldsymbol{u}_2^\oplus = \boldsymbol{0}_1 \oplus \boldsymbol{u}_2. \tag{4.79}$$

A general unit vector in $\mathcal{H}_{12}^\oplus$ is of the form:
$$\boldsymbol{f}^\oplus = f_1 \boldsymbol{u}_1^\oplus + f_2 \boldsymbol{u}_2^\oplus, \quad f_1, f_2 \in \mathbb{C} \text{ and } |f_1|^2 + |f_2|^2 = 1. \tag{4.80}$$

Let
$$\widehat{P}_{\boldsymbol{u}_1}^\oplus = |\boldsymbol{u}_1^\oplus\rangle\langle\boldsymbol{u}_1^\oplus|, \quad \widehat{P}_{\boldsymbol{u}_2}^\oplus = |\boldsymbol{u}_2^\oplus\rangle\langle\boldsymbol{u}_2^\oplus| \tag{4.81}$$
be the projectors onto the one-dimensional subspaces $\mathcal{H}_1$, $\mathcal{H}_2$ respectively. These operators are decomposable. Operators in $\mathcal{B}^\oplus(\mathcal{H}_{12}^\oplus)$ are of the form:
$$\widehat{B}^\oplus = b_1 \widehat{P}_{\boldsymbol{u}_1}^\oplus + b_2 \widehat{P}_{\boldsymbol{u}_2}^\oplus, \quad b_1, b_2 \in \mathbb{R}. \tag{4.82}$$

For later reference we shall call this system Model 4.4.5. We shall endow this system with different dynamics in §4.4.6 and §4.6.1 to study various types of evolution, both preserving and non-preserving.

The vector $\boldsymbol{f}^\oplus$ in Eq. (4.80) gives rise to two operators on $\mathcal{H}^\oplus$:

1. A statistical operator
$$\widehat{S}_{\boldsymbol{f}^\oplus}^\oplus = w_1 \widehat{P}_{\boldsymbol{u}_1}^\oplus + w_2 \widehat{P}_{\boldsymbol{u}_2}^\oplus, \quad w_1 = |f_1|^2, \ w_2 = |f_2|^2. \tag{4.83}$$

2. A density operator
$$\widehat{D}_{\boldsymbol{f}^\oplus} = |\boldsymbol{f}^\oplus\rangle\langle\boldsymbol{f}^\oplus|, \tag{4.84}$$
which is the projector onto the one-dimensional subspace of $\mathcal{H}_{12}^\oplus$ spanned by $\boldsymbol{f}^\oplus$. Since
$$\langle \boldsymbol{u}_1^\oplus | \left(|\boldsymbol{f}^\oplus\rangle\langle\boldsymbol{f}^\oplus| \right) \boldsymbol{u}_2^\oplus \rangle^\oplus \neq 0, \tag{4.85}$$
this density operator is not decomposable, and is hence not a statistical operator.

4.4. UNIFIED DYNAMICS IN DIRECT INTEGRAL SPACE

Now let us consider extensions of SNPLFs on $\mathcal{B}^\oplus(\mathcal{H}_{12}^\oplus)$ to $\mathcal{B}(\mathcal{H}_{12}^\oplus)$ and restrictions of NPLFs on $\mathcal{B}(\mathcal{H}_{12}^\oplus)$ to $\mathcal{B}^\oplus(\mathcal{H}_{12}^\oplus)$:

Extensions of SNPLFs   The statistical operator in Eq. (4.83) defines an SNPLF $\Omega_{f^\oplus}^\oplus$ on $\mathcal{B}^\oplus(\mathcal{H}_{12}^\oplus)$ by

$$\Omega_{f^\oplus}^\oplus(\widehat{B}^\oplus) = \mathrm{tr}^\oplus\left(\widehat{B}^\oplus \widehat{S}_{f^\oplus}^\oplus\right) = b_1\, w_1 + b_2\, w_2 \quad (4.86)$$

$$= \langle f^\oplus \mid \widehat{B}^\oplus f^\oplus\rangle^\oplus \quad \forall \widehat{B}^\oplus \in \mathcal{B}^\oplus(\mathcal{H}_{12}^\oplus). \quad (4.87)$$

We can construct two distinct types of extensions of $\Omega_{f^\oplus}^\oplus$ to $\mathcal{B}(\mathcal{H}_{12}^\oplus)$:

1. **Type 1**   We can use $\widehat{S}_{f^\oplus}^\oplus$ to generate an NPLF on $\mathcal{B}(\mathcal{H}_{12}^\oplus)$ in accordance with Eq. (4.75), i.e., for all $\widehat{B} \in \mathcal{B}(\mathcal{H}_{12}^\oplus)$, $\widehat{B} \notin \mathcal{B}^\oplus(\mathcal{H}_{12}^\oplus)$ we have[38]

$$\Omega_{f^\oplus}^{(1)}(\widehat{B}) = \mathrm{tr}\left(\widehat{B}\widehat{S}_{f^\oplus}^\oplus\right) \quad (4.88)$$

$$= \langle u_1^\oplus \mid \widehat{B}\widehat{S}_{f^\oplus}^\oplus u_1^\oplus\rangle^\oplus + \langle u_2^\oplus \mid \widehat{B}\widehat{S}_{f^\oplus}^\oplus u_2^\oplus\rangle^\oplus. \quad (4.89)$$

2. **Type 2**   We can define an NPLF on $\mathcal{B}(\mathcal{H}_{12}^\oplus)$ by

$$\Omega_{f^\oplus}^{(2)}(\widehat{B}) = \mathrm{tr}\left(\widehat{B}\widehat{D}_{f^\oplus}\right) = \langle f^\oplus \mid \widehat{B} f^\oplus\rangle^\oplus \quad \forall \widehat{B} \in \mathcal{B}(\mathcal{H}_{12}^\oplus). \quad (4.90)$$

This NPLF generates an SNPLF on $\mathcal{B}^\oplus(\mathcal{H}^\oplus)$ by

$$\Omega_{f^\oplus}^{(2)\oplus}(\widehat{B}^\oplus) = \Omega_{f^\oplus}^{(2)}(\widehat{B}^\oplus) = \langle f^\oplus \mid \widehat{B}^\oplus f^\oplus\rangle^\oplus \quad \forall \widehat{B}^\oplus \in \mathcal{B}^\oplus(\mathcal{H}_{12}^\oplus) \quad (4.91)$$

in accordance with Eq. (4.77). It follows that

$$\Omega_{f^\oplus}^\oplus(\widehat{B}^\oplus) = \Omega_{f^\oplus}^{(2)\oplus}(\widehat{B}^\oplus) \quad \forall\, \mathcal{B}^\oplus(\mathcal{H}_{12}^\oplus). \quad (4.92)$$

In other words we can consider $\Omega_{f^\oplus}^{(2)}$ as an extension of $\Omega_{f^\oplus}^\oplus$.

While equal on $\mathcal{B}^\oplus(\mathcal{H}^\oplus)$ extensions $\Omega_{f^\oplus}^{(1)}$ and $\Omega_{f^\oplus}^{(2)}$ are different since

$$\Omega_{f^\oplus}^{(1)}(\widehat{A}) \neq \Omega_{f^\oplus}^{(2)}(\widehat{A}) \quad \text{for some } \widehat{A} \in \mathcal{B}(\mathcal{H}_{12}^\oplus). \quad (4.93)$$

---

[38] The symbol 'tr' is the usual trace, as defined by Eq. (2.658), taken on the Hilbert space $\mathcal{H}_{12}^\oplus$. While $\langle u_1^\oplus \mid \widehat{B}\widehat{S}_{f^\oplus}^\oplus u_1^\oplus\rangle^\oplus$ is well-defined as a scalar product in $\mathcal{H}_{12}^\oplus$ the expression $\langle u_1 \mid \widehat{B} u_1\rangle_1$ is not defined within $\mathcal{H}_1$ if $\widehat{B}$ is not decomposable.

Restrictions of NPLFs Consider an NPLF on $\mathcal{B}(\mathcal{H}_{12}^\oplus)$ defined by

$$\Omega(\widehat{B}) = \operatorname{tr}\left(\widehat{B}\widehat{D}\right) \quad \forall \widehat{B} \in \mathcal{B}(\mathcal{H}_{12}^\oplus) \tag{4.94}$$

in terms of a density operator $\widehat{D}$, not necessarily decomposable, on $\mathcal{H}^\oplus$. We can generate an SNPLF $\Omega^\oplus$ on $\mathcal{H}_{12}^\oplus$ by

$$\Omega^\oplus(\widehat{B}^\oplus) = \Omega(\widehat{B}^\oplus) = \operatorname{tr}\left(\widehat{B}^\oplus \widehat{D}\right), \quad \widehat{B}^\oplus \in \mathcal{B}^\oplus(\mathcal{H}_{12}^\oplus). \tag{4.95}$$

To ascertain that $\Omega^\oplus$ so defined is an SNPLF let us introduce a decomposable operator $\widehat{D}^\oplus$ by

$$\widehat{D}^\oplus = d_1\, \widehat{P}_{\boldsymbol{u}_1}^\oplus + d_2\, \widehat{P}_{\boldsymbol{u}_2}^\oplus, \tag{4.96}$$

where

$$d_1 = \langle \boldsymbol{u}_1^\oplus \mid \widehat{D}\, \boldsymbol{u}_1^\oplus \rangle, \;\; d_2 = \langle \boldsymbol{u}_2^\oplus \mid \widehat{D}\, \boldsymbol{u}_2^\oplus \rangle. \tag{4.97}$$

Then we have

$$\operatorname{tr}\left(\widehat{D}^\oplus\right) = d_1 + d_2 = \operatorname{tr}\left(\widehat{D}\right) = 1. \tag{4.98}$$

It follows that $\widehat{D}^\oplus$ is a statistical operator. Using Eq. (4.82) we get, for all $\widehat{B}^\oplus \in \mathcal{B}^\oplus(\mathcal{H}_{12}^\oplus)$,

$$\Omega^\oplus(\widehat{B}^\oplus) = \operatorname{tr}^\oplus\left(\widehat{B}^\oplus \widehat{D}^\oplus\right) = d_1 b_1 + d_2 b_2 \tag{4.99}$$
$$= \operatorname{tr}\left(\widehat{B}^\oplus \widehat{D}\right). \tag{4.100}$$

4.4.6 Non-preserving dynamics 2: A general scheme

Returning to non-preserving evolution we can recall that the problems arising are due to the fact that $U_t^{(np)} \widehat{S}_0^\oplus U_t^{(np)\dagger}$ may not describe evolved states, and that $U_t^{(np)\dagger} \widehat{B}^\oplus U_t^{(np)}$ may not describe observables, both because of the lack of decomposability. An obvious way forward is to go outdside $\mathcal{B}^\oplus(\mathcal{H}^\oplus)$ to a bigger operator algebra.

For definiteness let us take $\mathcal{B}(\mathcal{H}^\oplus)$ as the bigger algebra. Let Ω_0 be an NPLF on $\mathcal{B}(\mathcal{H}^\oplus)$ generated by a given SNPLF $\Omega_0^\oplus$ on $\mathcal{B}^\oplus(\mathcal{H}^\oplus)$. Then the values

$$\Omega_0(U_t^{(np)\dagger} \widehat{B}^\oplus U_t^{(np)}) \tag{4.101}$$

are well-defined for all $\widehat{B}^\oplus \in \mathcal{B}^\oplus(\mathcal{H}^\oplus)$, since $U_t^{(np)\dagger} \widehat{B}^\oplus U_t^{(np)}$ belongs to $\mathcal{B}(\mathcal{H}^\oplus)$ on which Ω_0 acts. The idea now is to make use of Ω_0 to define an SNPLF $\Omega_t^\oplus$ on $\mathcal{B}^\oplus(\mathcal{H}^\oplus)$ by

$$\Omega_t^\oplus(\widehat{B}^\oplus) = \Omega_0(U_t^{(np)\dagger} \widehat{B}^\oplus U_t^{(np)}) \quad \forall \widehat{B}^\oplus \in \mathcal{B}^\oplus(\mathcal{H}^\oplus), \tag{4.102}$$

and then use $\Omega_t^\oplus$ to represent evolved states. We shall formalize this scheme in a postulate.

4.4. UNIFIED DYNAMICS IN DIRECT INTEGRAL SPACE

Postulate UQD(N) on non-preserving dynamics

- For a non-preserving evolution in the Schrödinger picture a system has associated with it a non-preserving propagator $\widehat{\mathcal{U}}^{(np)}$ such that if the system is in a state represented by an SNPLF $\Omega_0^\oplus$ at $t = 0$, its state at time t is given by an SNPLF $\Omega_t^\oplus$ defined by

$$\Omega_t^\oplus(\widehat{B}^\oplus) = \Omega_0(\widehat{U}_t^{(np)\dagger}\,\widehat{B}^\oplus\,\widehat{U}_t^{(np)}), \quad \forall \widehat{B}^\oplus \in \mathcal{B}^\oplus(\mathcal{H}^\oplus) \qquad (4.103)$$

where Ω_0 is a chosen NPLF on $\mathcal{B}(\mathcal{H}^\oplus)$ generated by $\Omega_0^\oplus$.

A striking feature here is that a given initial state $\Omega_0^\oplus$ may evolve in many inequivalent ways, depending on the extension Ω_0 chosen. Let us illustrate this with an explicit example.

Recall Model 4.4.5 which is described by a two-dimensional Hilbert space $\mathcal{H}_{12}^\oplus = \mathcal{H}_1 \oplus \mathcal{H}_2$. To endow a dynamics to this system let us assume a propagator generated by the following Hamiltonian generator:

$$\widehat{H}_g^{(np)} = E_0\,\widehat{I}^\oplus + \hbar\omega\,\widehat{P}, \quad \widehat{P} = |u_1^\oplus\rangle\langle u_2^\oplus| + |u_2^\oplus\rangle\langle u_1^\oplus|, \qquad (4.104)$$

where E_0, ω are real numbers, and $u_1^\oplus$, $u_2^\oplus$ are as in Eq. (4.79). We shall refer to the system endowed with a dynamics given rise by this Hamiltonian generator as Model 4.4.6. This Hamiltonian generator and the resulting evolution are non-preserving as seen in the action of various powers of $\widehat{P}$ on the unit vectors $u_1^\oplus$ and $u_2^\oplus$:

$$\begin{array}{llll}
\widehat{P}\,u_1^\oplus &= u_2^\oplus, & \widehat{P}^2\,u_1^\oplus &= u_1^\oplus, \\
\widehat{P}^3\,u_1^\oplus &= u_2^\oplus, & \widehat{P}^4\,u_1^\oplus &= u_1^\oplus, \\
\cdots &= \cdots, & \cdots &= \cdots.
\end{array} \qquad (4.105)$$

The evolution operators $\widehat{U}_t^{(np)}$ are given in the standard manner by Eq. (4.36). Since we are working in a finite dimensional space we can expand the evolution operators as a series:

$$\widehat{U}_t^{(np)} = e^{-i\widehat{H}_g^{(np)}t} = e^{-iE_0 t}\sum_{n=0}^{\infty}\frac{(-i\omega t)^n}{n!}\widehat{P}^n. \qquad (4.106)$$

When acting on $u_1^\oplus$ we get

$$\begin{aligned}
&\left(\sum_{n=0}^{\infty}\frac{(-i\omega t)^n}{n!}\widehat{P}^n\right)u_1^\oplus \\
&= \sum_{n=\text{even}}\frac{(-i\omega t)^n}{n!}u_1^\oplus + \sum_{n=\text{odd}}\frac{(-i\omega t)^n}{n!}u_2^\oplus \\
&= \cos\omega t\,u_1^\oplus - i\sin\omega t\,u_2^\oplus.
\end{aligned} \qquad (4.107)$$

It follows that

$$\widehat{U}_t^{(np)} \boldsymbol{u}_1^\oplus = e^{-iE_0 t} \left(\cos\omega t \, \boldsymbol{u}_1^\oplus - i\sin\omega t \, \boldsymbol{u}_2^\oplus \right), \tag{4.108}$$

$$\widehat{U}_t^{(np)} \boldsymbol{u}_2^\oplus = e^{-iE_0 t} \left(-i\sin\omega t \, \boldsymbol{u}_1^\oplus + \cos\omega t \, \boldsymbol{u}_2^\oplus \right). \tag{4.109}$$

For $\boldsymbol{f}^\oplus$ in Eq. (4.80) we have

$$\begin{aligned}\widehat{U}_t^{(np)} \boldsymbol{f}^\oplus &= e^{-iE_0 t} \left(f_1 \cos\omega t - i f_2 \sin\omega t \right) \boldsymbol{u}_1^\oplus \\ &+ e^{-iE_0 t} \left(f_2 \cos\omega t - i f_1 \sin\omega t \right) \boldsymbol{u}_2^\oplus.\end{aligned} \tag{4.110}$$

Returning to the problem of evolution, suppose the system is initially in state $\Omega_0^\oplus$ defined by the statistical operator $S_{\boldsymbol{f}^\oplus}^\oplus$ in Eq. (4.83). According to Postulate UQD(N) a non-preserving propagator $\{\widehat{U}_t^{(np)}\}$ does not determine the evolution of a system. We have to choose a particular extension Ω_0 of $\Omega_0^\oplus$ to an NPLF on $\mathcal{B}(\mathcal{H}^\oplus)$ first. Let us examine the evolution under the two types of extensions given by Eqs. (4.88) and (4.90) in §4.4.5.

Evolution under type 1 extension The initial state $S_{\boldsymbol{f}^\oplus}^\oplus$ defines an extension $\Omega_{\boldsymbol{f}^\oplus}^{(1)}$ by Eq. (4.88). Then the corresponding evolved state $\Omega_{1t}^\oplus$ is defined by its action on $\widehat{B}^\oplus$ according to:

$$\begin{aligned}\Omega_{1t}^\oplus(\widehat{B}^\oplus) &= \Omega_{\boldsymbol{f}^\oplus}^{(1)}(\widehat{U}_t^{(np)\dagger} \widehat{B}^\oplus \widehat{U}_t^{(np)}) \\ &= \operatorname{tr}(\widehat{U}_t^{(np)\dagger} \widehat{B}^\oplus \widehat{U}_t^{(np)} \widehat{S}_{\boldsymbol{f}}^\oplus) \\ &= w_1 \langle \boldsymbol{u}_1^\oplus \mid \widehat{U}_t^{(np)\dagger} \widehat{B}^\oplus \widehat{U}_t^{(np)} \boldsymbol{u}_1^\oplus \rangle + w_2 \langle \boldsymbol{u}_2^\oplus \mid \widehat{U}_t^{(np)\dagger} \widehat{B}^\oplus \widehat{U}_t^{(np)} \boldsymbol{u}_2^\oplus \rangle \\ &= w_1 \langle \widehat{U}_t^{(np)} \boldsymbol{u}_1^\oplus \mid \widehat{B}^\oplus \widehat{U}_t^{(np)} \boldsymbol{u}_1^\oplus \rangle + w_2 \langle \widehat{U}_t^{(np)} \boldsymbol{u}_2^\oplus \mid \widehat{B}^\oplus \widehat{U}_t^{(np)} \boldsymbol{u}_2^\oplus \rangle \\ &= w_1 \left(b_1 \cos^2 \omega t + b_2 \sin^2 \omega t \right) + w_2 \left(b_1 \sin^2 \omega t + b_2 \cos^2 \omega t \right).\end{aligned} \tag{4.111}$$

This dynamics is referred to as Model 4.4.6 E1.

Evolution under type 2 extension The initial state $S_{\boldsymbol{f}^\oplus}^\oplus$ defines an extension $\Omega_{\boldsymbol{f}^\oplus}^{(2)}$ by Eq. (4.90). Then the corresponding evolved state $\Omega_{2t}^\oplus$ is defined by its action on $\widehat{B}^\oplus$ according to:

$$\begin{aligned}\Omega_{2t}^\oplus(\widehat{B}^\oplus) &= \Omega_{\boldsymbol{f}^\oplus}^{(2)}(\widehat{U}_t^{(np)\dagger} \widehat{B}^\oplus \widehat{U}_t^{(np)}) \\ &= \operatorname{tr}(\widehat{U}_t^{(np)\dagger} \widehat{B}^\oplus \widehat{U}_t^{(np)} \widehat{D}_{\boldsymbol{f}^\oplus}) \\ &= \langle \widehat{U}_t^{(np)} \boldsymbol{f}^\oplus \mid \widehat{B}^\oplus \widehat{U}_t^{(np)} \boldsymbol{f}^\oplus \rangle^\oplus \\ &= w_1 \left(b_1 \cos^2 \omega t + b_2 \sin^2 \omega t \right) + w_2 \left(b_1 \sin^2 \omega t + b_2 \cos^2 \omega t \right) \\ &+ (b_1 - b_2) \operatorname{Re}\left(i f_1 f_2^* \right) \sin 2\omega t.\end{aligned} \tag{4.112}$$

4.4. UNIFIED DYNAMICS IN DIRECT INTEGRAL SPACE

We can see that

$$\Omega_{1t}^{\oplus}(\widehat{B}^{\oplus}) \neq \Omega_{2t}^{\oplus}(\widehat{B}^{\oplus}) \quad \text{for some } \widehat{B}^{\oplus} \in \mathcal{B}^{\oplus}(\mathcal{H}^{\oplus}). \tag{4.113}$$

This dynamics is referred to as Model 4.4.6 E2.

In the absence of further information about the physical system there is no *a priori* reason to prefer one type of extension over another. The choice of extension used to define the dynamics of the system has to be chosen to suit the particular physical situation under consideration. In the next section we shall look at the kind of physical situation relevant in the selection of dynamics.

4.4.7 Non-preserving evolution and environments

Generally a system is described by a Hilbert space $\mathcal{H}^{\oplus}$, with its bounded observables represented by the selfadjoint elements of $\mathcal{B}^{\oplus}(\mathcal{H}^{\oplus})$ in accordance with Postulate UQS or Postulate UQS(F). There is a superselection rule in operation.[39] When the system is isolated it would undergo a preserving evolution. No correlations between different supersectors occur and pure states will evolve within the same supersector. When the isolation is removed and the system then interacts with an external environment the system will evolve differently.[40] In analogy to the terminology introduced in §4.3.3 this external system shall be referred to as the environment the system is in. It would then be unavoidable that some important quantities which governs the dynamics of the system would involve the environment. These quantities cannot be constructed entirely from the observables of the system alone. A non-preserving Hamiltonian generator is an example. Such a Hamiltonian generator embodies the interaction between the system and its environment, e.g., an exchange of energy resulting in the energy of the system not being conserved in a non-preserving evolution process. The system may be forced to evolve from one supersector to another. Under such circumstances it clearly does not make sense to insist that the Hamiltonian generator must represent the energy of the system, since the Hamiltonian generator would involve the environment and hence may not be an observable of the system alone. A similar situation exists in classical physics. When isolated a thermodynamic system in equilibrium does not evolve away from that equilibrium state. When this system is brought in contact with an environment, say, a heat bath of a different temperature, an interaction will occur and the system will eventually evolve into another equilibrium state, with a possible change of energy. Obviously

---

[39] In the absence of any superselection rule the algebra of observables would simply be the set of all bounded operators on the Hilbert space associated with the system.

[40] Of course it is possible to treat the system plus the environment as a single isolated system. However, we do not want to do so here, since we are primarily interested in the system, its states and its physical observables.

the engine for change comes from an interaction with the environment and hence the evolution of the system cannot be described purely in terms of the observables of the system alone. In Chapter 7 we shall see more examples of this situation in superconducting systems.

When the evolution group is non-preserving, pure states will in general be able to evolve into mixed states and vice versa, as in the example presented in §4.4.3. In the context of §4.4.6 we can see why non-preserving evolution is not uniquely determined by the propagator. For example, the precise nature of the interaction with the environment will determine what extension to use for the evolution, e.g., type 1 extensions generated by mixed vectors of the system is often employed to formulate schematic models of quantum measurement.[41]

4.5 Classical Systems of Finite Order

Orthodox quantum theory as embodied in Postulates OQS and OQD is a special case of our unified theory when the preferred direct integral decomposition in Postulate UQS or Postulate UQS(F) is a trivial one. The same applied to the theory for a large class of classical systems. Let us consider how classical theory can be formulated in the Hilbert space language within our unified theory.

4.5.1 First-order systems in Hilbert space

Classical systems are defined in §1.6.1 and first-order systems are introduced in §1.6.2. A pure state of a first-order classical system is specified by a single real parameter γ in Eq. (1.280) lying in a certain range $\mathcal{R}$ which can be identified with the phase space Γ of such a system. The object here is to establish a Hilbert space theory for the description of such a classical system.

Consider an idealized classical resistanceless inductive circuit in the form of a perfect conducting ring in an external magnetic field descibed in E1.6.2(1) E1.[42] The state of this system is specifiable by a single parameter, the current I, and the phase space is the set of possible current values. Observables are real-valued functions $A(I)$ of I. The evolution of the system is governed by the time-dependence of the current $I = I(t)$ which is determined by the changes in the external flux according to Eq. (1.282) of classical dynamics.

To establish a theory in our unified framework we must first associate with the system a Hilbert space having a preferred direct integral decomposition. The idea is to build up a Hilbert space on the phase space Γ by

---

[41] Wan (1980), Bub (1988).
[42] Note that a classical resistanceless inductive ring is fundamentally different from a superconducting ring (Rose-Innes and Rhoderick (1980)).

4.5. CLASSICAL SYSTEMS OF FINITE ORDER

1. assigning the Hilbert space $\mathcal{H}(\gamma) = \mathbb{C}$ to each point γ in Γ, and then

2. constructing the direct integral $\mathcal{H}_\Gamma^\oplus(\mathbb{C}, d\gamma)$ of $\mathcal{H}(\gamma) = \mathbb{C}$ with respect to the continuous measure $d\gamma$ over the phase space Γ,

as done in Eq. (2.601) in E2.13.2(1) E1. For the specific example of the resistanceless circuit we just replace the parameter γ by the current I. The resulting theory for the circuit based on the direct integral space $\mathcal{H}_\Gamma^\oplus(\mathbb{C}, dI)$ possesses the following properties:

1. All observables commute, since all decomposable operators in $\mathcal{H}_\Gamma^\oplus(\mathbb{C}, dI)$ are diagonalizable. There is a continuous superselection rule. Every self-adjoint decomposable operator $\widehat{A}^\oplus$ corresponds to a real-valued numerical function $A(I)$ on Γ:

$$\widehat{A}^\oplus = \int^\oplus \widehat{A}(I)\, dI = \int^\oplus A(I) \widehat{\mathbb{I}}(I)\, dI. \tag{4.114}$$

This renders the operator representation of observables in full agreement with the classical description. For example, the Hamiltonian operator which represents the energy of the system shown in Eq. (1.285) is diagonalizable, i.e., we have

$$\widehat{H}^\oplus = \int^\oplus \frac{1}{2} L I^2\, \widehat{\mathbb{I}}(I)\, dI = \int^\oplus \frac{1}{2L}(\Phi_T - \Phi_{ex})^2\, \widehat{\mathbb{I}}(I)\, dI. \tag{4.115}$$

2. A statistical operator is of the form of Eq. (4.24). A regular statistical operator $\widehat{S}_r^\oplus$ represents a mixed state. For any bounded observable $\widehat{A}^\oplus$ the expectation value is given by Eq. (4.25).

3. A normalized mixed vector $\boldsymbol{f}^\oplus$ is of the form of Eq. (4.26) except with τ replaced by I, and it can be used to describe a mixed regular state since it corresponds to a regular statistical operator

$$\widehat{S}_r^\oplus = \int^\oplus S_r(I)\, \widehat{\mathbb{I}}(I)\, dI \quad \text{with} \quad S_r(I) = |f(I)|^2. \tag{4.116}$$

All regular statistical operators are of this form. Given any bounded observable $\widehat{A}^\oplus$ the expectation value $\mathcal{E}(\widehat{A}^\oplus; \widehat{S}_r^\oplus)$ can be calculated directly in terms of $f(I)$:

$$\mathcal{E}(\widehat{A}^\oplus; \widehat{S}_r^\oplus) = \operatorname{tr}^\oplus\left(\widehat{S}_r^\oplus \widehat{A}^\oplus\right) = \int |f(I)|^2 A(I)\, dI. \tag{4.117}$$

A regular mixed state may be identified with a classical statistical state defined by the probability density function $S_r(I)$.

4. There are no regular pure vectors in $\mathcal{H}_\Gamma^\oplus(\mathbb{C}, dI)$, and hence no regular pure states. There are pure singular vectors. Pure singular states correspond to singular statistical operators of the form of Eq. (4.29) with expectation values given by Eq. (4.30). Pure singular states correspond one-to-one to supersectors. The above analysis fits in with our intuition, that is, a δ-function $\delta(I - I_0)$ should correspond to the pure *singular state* specified by the current value I_0 in the classical theory.

Our next task is to express traditional classical dynamics of first-order systems in terms of unitary evolution in $\mathcal{H}_\Gamma^\oplus(\mathbb{C}, d\gamma)$ for a general first-order system. Classically the dynamics is given by Eq. (1.280), or equivalently by a phase flow T, i.e., a one-parameter group of transformations of the phase space onto itself:

$$T_t : \Gamma \to \Gamma, \quad t \in \mathbb{R}. \tag{4.118}$$

This transformation group is generated by a vector field $\widehat{\mathcal{X}} = \mathcal{X}(\gamma)\, d/d\gamma$ on Γ shown in Eq. (1.281). The idea now is to establish a unitary representation of this phase flow on $\mathcal{H}_\Gamma^\oplus(\mathbb{C}, d\gamma)$. Mathematically the problem is similar to the quantization of compressible complete momenta discussed in §3.3.3.

Recall that in coordinate space $\mathbb{E}^n$ the vector field $\widehat{X}_P$ associated with a complete momentum P generates a one-parameter group of transformations of $\mathbb{E}^n$. A unitary representation of this group in the Hilbert space $L^2(\mathbb{E}^n)$ of square-integrable functions defined on the coordinate space is given by Eq. (3.115) with its generator given by Eq. (3.140). We now have a vector field $\widehat{\mathcal{X}}$ generating a one-parameter group of transformations of the phase space Γ. Therefore we can set up a similar unitary representation of the group in the Hilbert space $L^2(\Gamma)$ of square-integrable functions defined on the phase space. Since $L^2(\Gamma)$ is unitarily equivalent to $\mathcal{H}_\Gamma^\oplus(\mathbb{C}, d\gamma)$ we can employ Eqs. (3.104) and (3.115) to obtain a desired unitary representation in $\mathcal{H}_\Gamma^\oplus(\mathbb{C}, d\gamma)$, i.e., we can define operators $\widehat{U}_t$ on $\mathcal{H}_\Gamma^\oplus(\mathbb{C}, d\gamma)$ by its action on normalized vectors $\boldsymbol{f}^\oplus \in \mathcal{H}_\Gamma^\oplus(\mathbb{C}, d\gamma)$,[43] i.e., for a given

$$\boldsymbol{f}^\oplus = \int^\oplus f(\gamma)\, \boldsymbol{u}(\gamma)\, d\gamma \tag{4.119}$$

we have

$$\widehat{U}_t \boldsymbol{f}^\oplus = \int^\oplus f(T_t^{-1}\gamma) \left(\frac{\partial(T_t^{-1}\gamma)}{\partial \gamma}\right)^{1/2} \boldsymbol{u}(\gamma)\, d\gamma. \tag{4.120}$$

---

[43] Reed and Simon Vol. 2 (1975) pp. 313-318, Alanson (1992), Kowalski (1994) §1.4.2. These authors do not utilize superselection rules. In using $\mathcal{H}_\Gamma^\oplus(\mathbb{C}, d\gamma)$ instead of $L^2(\Gamma)$ we have chosen a preferred non-trivial direct integral decomposition in the sense of Postulate UQS.

4.5. CLASSICAL SYSTEMS OF FINITE ORDER

The generator of this unitary group is the Hamiltonian generator as defined by Eq. (3.140). Following Eqs. (3.125) to (3.127) we get

$$\widehat{H}_{g0} \boldsymbol{f}^{\oplus} = -i\hbar \int^{\oplus} \left\{ \left(\widehat{\mathcal{X}} + \frac{1}{2} \frac{d\mathcal{X}}{d\gamma} \right) f(\gamma) \right\} \boldsymbol{u}(\gamma) \, d\gamma, \tag{4.121}$$

where $f(\gamma)$ are C_0^∞ functions on Γ in accordance with Eqs. (2.611) and (2.612). As in the quantization of complete momenta this operator is essentially selfadjoint, and its unique selfadjoint extension $\widehat{H}_g$ is the Hamiltonian generator. Despite its appearance $\widehat{H}_{g0}$ is not decomposable since $\widehat{\mathcal{X}}$ in the integrand is not an operator acting in any single supersector $\mathcal{H}(\gamma)$.

Following Eq. (3.214) we can also establish a differential equation for the time evolution of $\boldsymbol{f}^\oplus$. Let $\boldsymbol{f}_t^\oplus = \widehat{U}_t \boldsymbol{f}^\oplus$. Then we have

$$i\hbar \frac{d}{dt} \boldsymbol{f}_t^\oplus = \widehat{H}_g \boldsymbol{f}_t^\oplus. \tag{4.122}$$

Despite its appearance this is not the same as the Schrödinger equation of quantum mechanics. This equation does not involve any complex factor or the Planck's constant since the factor $i\hbar$ cancel out on both sides once we write down the expression for $\widehat{H}_g$. This becomes obvious when we introduce

$$\boldsymbol{f}_t^\oplus = \int^\oplus f_t(\gamma) \boldsymbol{u}(\gamma) \, d\gamma, \quad f_{t=0}(\gamma) = f(\gamma). \tag{4.123}$$

Then Eq. (4.122) reduces to

$$\frac{\partial}{\partial t} f_t(\gamma) = -\left(\widehat{\mathcal{X}} + \frac{1}{2} \frac{d\mathcal{X}}{d\gamma} \right) f(\gamma, t). \tag{4.124}$$

Comparing Eq. (4.123) with Eq. (4.120) we see that

$$f_t(\gamma) = f(T_t^{-1}\gamma) \left(\frac{\partial(T_t^{-1}\gamma)}{\partial \gamma} \right)^{1/2}. \tag{4.125}$$

The following features of the dynamics can be observed:

1. The Hamiltonian generator is a differential operator involving variations of $f(\gamma)$ between neighbouring supersectors, and hence it is not decomposable.[44] Consequently the evolution operators are not decomposable. It follows that quite unlike the Hamiltonian $\widehat{H}^\oplus$ in Eq. (4.115) the Hamiltonian generator $\widehat{H}_g$ is not an observable.

---

[44]Clearly $\widehat{H}_g$ does not commute with diagonalizable operators in $\mathcal{H}_\Gamma^\oplus(\mathbb{C}, d\gamma)$.

2. Despite the non-decomposable nature of the Hamiltonian generator and the evolution operators the resulting evolution is preserving, i.e., Postulate UQD(P) applies here. This is due to the fact that all observables are of the form of Eq. (4.114) and that

$$\left(\widehat{U}_t^\dagger \, \widehat{A}^\oplus \, \widehat{U}_t\right) f^\oplus$$

$$= \widehat{U}_t^\dagger \int^\oplus A(\gamma) \, f(T_t^{-1}\gamma) \left(\frac{\partial (T_t^{-1}\gamma)}{\partial \gamma}\right)^{1/2} u(\gamma) \, d\gamma \qquad (4.126)$$

$$= \int^\oplus A(T_t\gamma) \, f(\gamma) \, u(\gamma) \, d\gamma. \qquad (4.127)$$

It follows that

$$\widehat{U}_t^\dagger \, \widehat{A}^\oplus \, \widehat{U}_t = \int^\oplus A(T_t\gamma) \, \widehat{I}(\gamma) \, d\gamma \qquad (4.128)$$

which is again selfadjoint and decomposable.

3. Classically a pure state γ_0 evolves into another pure state γ_t in a different supersector. This corresponds to the statistical operator $\widehat{S}_{\gamma_0}^\oplus$ of the form Eq. (4.29) evolving into a pure singular state in a different supersector, i.e.,[45]

$$\widehat{S}_{\gamma_t}^\oplus = \widehat{U}_t \, \widehat{S}_{\gamma_0}^\oplus \, \widehat{U}_t^\dagger \qquad (4.129)$$

$$= \int^\oplus \delta(T_t^{-1}\gamma - \gamma_0) \, \widehat{I}(\gamma) \, d\gamma. \qquad (4.130)$$

$$= \int^\oplus \delta(\gamma - T_t\gamma_0) \, \widehat{I}(\gamma) \, d\gamma. \qquad (4.131)$$

4. We call a point γ_c in the phase space a *zero of the Hamiltonian generator* $\widehat{H}_g$ if

$$\left[\mathcal{X} \frac{d}{d\gamma} + \frac{1}{2} \frac{d\mathcal{X}}{d\gamma}\right]_{\gamma_c} = 0. \qquad (4.132)$$

A zero of $\widehat{H}_g$ corresponds to a critical point of the vector field $\mathcal{X} d/d\gamma + \frac{1}{2} d\mathcal{X}/d\gamma$.[46] A pure singular state is stationary only at a zero of $\widehat{H}_g$. In other words $\widehat{S}_{\gamma_c}^\oplus$ is stationary only if γ_c is a zero of $\widehat{H}_g$.

---

[45] Since $\widehat{U}_t^\dagger = \widehat{U}_{-t}$ and $\widehat{U}_t = \widehat{U}_{-t}^\dagger$, Equation (4.128) implies

$$\widehat{U}_t \, \widehat{A}^\oplus \, \widehat{U}_t^\dagger = \int^\oplus A(T_t^{-1}\gamma) \, \widehat{I}(\gamma) \, d\gamma.$$

Note that delta functions $\delta(T_t^{-1}\gamma - \gamma_0)$ and $\delta(\gamma - T_t\gamma_0)$ have the same singular point at $\gamma = T_t\gamma_0$.

[46] See comments following Definition 1.6.1(1).

4.5. CLASSICAL SYSTEMS OF FINITE ORDER

As an illustration consider the resistanceless circuit in E1.6.2(1) E1 again. Suppose the external flux varies linearly with time so that we can write

$$\frac{d\Phi_{ex}}{dt} = Lv, \quad v = \text{a constant and } L = \text{self inductance.} \qquad (4.133)$$

Comparing this with Eq. (1.286) we have

$$\mathcal{X} = -\frac{1}{L}\frac{d\Phi_{ex}}{dt} = -v, \quad \frac{d\mathcal{X}}{dI} = 0. \qquad (4.134)$$

The Hamiltonian generator $\widehat{H}_g$ acting on $\boldsymbol{f}^\oplus$ gives

$$\widehat{H}_g \boldsymbol{f}^\oplus = -i\hbar \int^\oplus \left(-\frac{1}{L}\frac{d\Phi_{ex}}{dt}\frac{df(I)}{dI}\right)\boldsymbol{u}(I)\,dI \qquad (4.135)$$

$$= i\hbar \int^\oplus \left(v\frac{df(I)}{dI}\right)\boldsymbol{u}(I)\,dI \qquad (4.136)$$

so that Eq. (4.124) reduces to

$$\frac{\partial}{\partial t}f(I,t) = v\frac{\partial}{\partial I}f(I,t). \qquad (4.137)$$

For an initial function $f(I,0) = \delta(I - I_0)$ the solution is

$$f(I,t) = \delta(I - I_t), \quad I_t = I_0 - vt. \qquad (4.138)$$

The corresponding statistical operator for the state at $t > 0$ is

$$\widehat{S}^\oplus_{I_t} = \int^\oplus \delta(I - I_t)\,\widehat{\boldsymbol{\mathit{I}}}(I)\,dI = \int^\oplus \delta\left(I - (I_0 - vt)\right)\widehat{\boldsymbol{\mathit{I}}}(I)\,dI. \qquad (4.139)$$

This is consistent with Eq. (1.284) for classical circuit theory. There are no stationary pure states.

4.5.2 Second-order Hamiltonian systems in Hilbert space

For a second-order Hamiltonian system introduced in §1.6.3 the phase space is parameterized by two coordinates. A particle in classical mechanics with one spatial degree of freedom is a typical example. The phase space is

$$\Gamma = \{(\gamma^1 = x, \gamma^2 = p)\}, \qquad (4.140)$$

and states are labelled by the phase space coordinates $\gamma = (\gamma^1, \gamma^2)$ with each pure state corresponding to a point γ in Γ.

CHAPTER 4. PHYSICAL THEORY IN HILBERT SPACE

Another example is the one introduced in E1.6.3(1) E3, i.e., a classical resistanceless LC circuit consisting of a capacitor with capacitance C and an inductor with inductance L. The state of the system is determined by two parameters. A choice which will render the circuit a standard Hamiltonian system is to take $\gamma^1 = \Phi$, the magnetic flux enclosed in the inductor, and $\gamma^2 = Q$, the charge on the capacity, with the phase space

$$\Gamma_{LC} = \{(\gamma^1 = \Phi, \gamma^2 = Q)\}. \tag{4.141}$$

In other words we can treat Φ and Q as a pair of canonical variables.

To apply our unified theory we first associate the Hilbert space $\mathcal{H}(\gamma) = \mathbb{C}$ with each point γ in Γ to obtain a family of Hilbert spaces $\mathcal{H}(\gamma)$ over the phase space Γ. We then form the direct integral Hilbert space. So, for a general second-order Hamiltonian system we associate with it the following direct integral Hilbert space:

$$\mathcal{H}_\Gamma^\oplus(\mathbb{C}, dg(\gamma)) = \int_\Gamma^\oplus \mathcal{H}(\gamma)\, dg(\gamma), \quad dg(\gamma) = d\gamma^1 d\gamma^2. \tag{4.142}$$

A vector in $\mathcal{H}_\Gamma^\oplus(\mathbb{C}, dg(\gamma))$ is of the form

$$\boldsymbol{f}^\oplus = \int^\oplus f(\gamma)\, \boldsymbol{u}(\gamma)\, dg(\gamma) = \int^\oplus f(\gamma^1, \gamma^2)\, \boldsymbol{u}(\gamma^1, \gamma^2)\, d\gamma^1 d\gamma^2. \tag{4.143}$$

Clearly $\mathcal{H}_\Gamma^\oplus(\mathbb{C}, dg(\gamma))$ is unitarily equivalent to $L^2(\Gamma)$ defined by

$$L^2(\Gamma) = \left\{ f(\gamma) : \int_\Gamma |f(\gamma)|^2\, dg(\gamma) < \infty \right\}. \tag{4.144}$$

But in $\mathcal{H}_\Gamma^\oplus(\mathbb{C}, dg(\gamma))$ we have chosen a preferred non-trivial direct integral decomposition of $L^2(\Gamma)$ in terms of $\mathcal{H}(\gamma) = \mathbb{C}$ in the sense of Postulate UQS.

All the four properties listed in §4.5.1 for first-order systems hold here. Physical observables correspond one-to-one to decomposable selfadjoint operators in $\mathcal{H}_\Gamma^\oplus(\mathbb{C}, dg(\gamma))$. For examples we have canonical variables[47]

$$\widehat{\gamma}^{1\oplus} = \int^\oplus \gamma^1\, \widehat{\boldsymbol{I}}(\gamma)\, dg(\gamma), \quad \text{and} \quad \widehat{\gamma}^{2\oplus} = \int^\oplus \gamma^2\, \widehat{\boldsymbol{I}}(\gamma)\, dg(\gamma) \tag{4.145}$$

---

[47]Similar to direct integrals of vectors pointed in E2.13.2(1) E1 this direct integral is a symbolic expression for a family of operators $\gamma^1\, \widehat{\boldsymbol{I}}(\gamma)$ and we do not actually integrate in the usual sense here. In contrast an indefinite Riemann integral

$$\int \gamma^1\, d\gamma^1 d\gamma^2$$

would have produced $\frac{1}{2}(\gamma^1)^2\, \gamma^2$ which becomes undefined when we try to substitute infinite limiting values to convert to a definite integral.

4.5. CLASSICAL SYSTEMS OF FINITE ORDER

and the Hamiltonian for the energy of the system

$$\widehat{H}^\oplus = \int^\oplus H(\gamma)\,\widehat{\mathbb{I}}(\gamma)\,dg(\gamma), \qquad (4.146)$$

where $H(\gamma)$ is the classical expression for the energy of the system. As in previous first-order examples, the set of decomposable selfadjoint operators in $\mathcal{H}_\Gamma^\oplus$ coincides with the set of real-valued functions $A(\gamma)$ on Γ and all observables commute.

There are no pure regular states, but there are pure singular states, e.g.,

$$\widehat{S}_{\gamma_0}^\oplus = \int^\oplus \delta(\gamma^1 - \gamma_0^1)\delta(\gamma^2 - \gamma_0^2)\,\widehat{\mathbb{I}}(\gamma)\,dg(\gamma). \qquad (4.147)$$

Expectation values of an observable in these singular states are given by

$$\mathcal{E}(\widehat{A}^\oplus, \widehat{S}_{\gamma_0}^\oplus) = \int^\oplus \left(\delta(\gamma^1 - \gamma_0^1)\delta(\gamma^2 - \gamma_0^2)\right) A(\gamma^1, \gamma^2)\,d\gamma^1 d\gamma^2 \qquad (4.148)$$

$$= A(\gamma_0^1, \gamma_0^2). \qquad (4.149)$$

For examples we have

$$\mathcal{E}(\widehat{\gamma}^{1\oplus}, \widehat{S}_{\gamma_0}^\oplus) = \gamma_0^1, \quad \mathcal{E}(\widehat{\gamma}^{2\oplus}, \widehat{S}_{\gamma_0}^\oplus) = \gamma_0^2, \quad \mathcal{E}(\widehat{H}^\oplus, \widehat{S}_{\gamma_0}^\oplus) = H(\gamma_0^1, \gamma_0^2). \qquad (4.150)$$

Regular mixed states are represented by regular statistical operators $\widehat{S}_r^\oplus$ or by normalized regular mixed vectors $\boldsymbol{f}^\oplus$. These correspond to classical statistical states.

As discussed in C1.6.3(1) C3 classical Hamiltonian dynamics is given by a phase flow T

$$T = \Gamma \mapsto \Gamma \quad \text{by} \quad T_t : \gamma \mapsto T_t\,\gamma \qquad (4.151)$$

generated by a Hamiltonian vector field $\widehat{\mathcal{X}}_{H_g}$ on the phase space Γ given explicitly by Eq. (1.287). Following the reasoning in the first-order case in §4.5.1, we can express this Hamiltonian dynamics in terms of unitary evolution in $\mathcal{H}_\Gamma^\oplus$. This time things are actually simpler. This is because the phase flow preserves the phase volume $d\gamma^1 d\gamma^2$ on account of Liouville's Theorem for Hamiltonian mechanics. The situation is similar to the quantization of complete incompressible momenta in §3.3.3. For a unitary representation of the phase flow T of a second order Hamiltonian system we have [48]

$$\widehat{U}_t \boldsymbol{f}^\oplus = \int^\oplus f(T_t^{-1}\gamma)\,\boldsymbol{u}(\gamma)\,d\gamma. \qquad (4.152)$$

---

[48] Koopman (1931), Alanson (1992), Kowalski (1994), Reed and Simon Vol. 2 (1975), Wan, Bradshaw, Trueman and Harrison (1998).

376 CHAPTER 4. PHYSICAL THEORY IN HILBERT SPACE

The Hamiltonian generator $\widehat{H}_g$ is given by Eq. (1.287) or equivalently by Eq. (3.98), i.e., we have[49]

$$\widehat{H}_g f^\oplus = -i\hbar \int^\oplus \left\{ \widehat{\mathcal{X}}_{H_g} f(\gamma) \right\} u(\gamma)\, dg(\gamma) \tag{4.153}$$

$$= -i\hbar \int^\oplus \left\{ \left(\frac{\partial H_g}{\partial \gamma^2} \frac{\partial}{\partial \gamma^1} - \frac{\partial H_g}{\partial \gamma^1} \frac{\partial}{\partial \gamma^2} \right) f(\gamma^1, \gamma^2) \right\} u(\gamma^1, \gamma^2)\, d\gamma^1 d\gamma^2. \tag{4.154}$$

As before a point γ_c in the phase space is called a *zero of the Hamiltonian generator* $\widehat{H}_g$ if $\widehat{\mathcal{X}}_{H_g}$ vanishes at γ_c, or

$$\left[\frac{\partial H_g}{\partial \gamma^1} \right]_{\gamma_c} = 0 \quad \text{and} \quad \left[\frac{\partial H_g}{\partial \gamma^2} \right]_{\gamma_c} = 0. \tag{4.155}$$

Let us rewrite $f(T_t^{-1}\gamma)$ as $f_t(\gamma)$. Following the arguments leading to Eqs. (4.122) and (4.124) we obtain an equation of motion in differential form:

$$\frac{\partial}{\partial t} f_t(\gamma) = -\left(\frac{\partial H_g}{\partial \gamma^2} \frac{\partial}{\partial \gamma^1} - \frac{\partial H_g}{\partial \gamma^1} \frac{\partial}{\partial \gamma^2} \right) f_t(\gamma). \tag{4.156}$$

For an initial function $f_0(\gamma) = \delta(\gamma^1 - \gamma_0^1)\delta(\gamma^2 - \gamma_0^2)$ the solution is

$$f_t(\gamma) = \delta(\gamma^1 - \gamma_t^1)\delta(\gamma^2 - \gamma_t^2), \tag{4.157}$$

where γ_t^1 and γ_t^1 are solutions of Hamilton's Eqs. (1.288) with initial values γ_0^1 and γ_0^2. It follows that pure singular states evolve as

$$\widehat{S}^\oplus_{\gamma_t} = \int^\oplus \delta(\gamma^1 - \gamma_t^1)\delta(\gamma^2 - \gamma_t^2)\, \widehat{\mathbb{I}}_\gamma\, d\gamma^1 d\gamma^2. \tag{4.158}$$

A pure singular state $\widehat{S}^\oplus_{\gamma_0}$ is stationary only if γ_0 is a zero of $\widehat{H}_g$. All these are consistent with classical motion, as expected.

As in the previous first-order case the evolution maps observables to observables. In other words the evolution is preserving despite the fact that $\widehat{U}_t$ and $\widehat{H}_g$ are not decomposable and $\widehat{H}_g$ is not an observable.

In classical Hamiltonian mechanics we have the Hamiltonian H which represents the energy of the system and the Hamiltonian generator H_g which generates the classical dynamics. For many systems we have $H = H_g$, i.e., the same quantity H can play the role of energy and of the generator for dynamics. In the Hilbert space theory these two roles have to be played by two

---

[49] Compared with the first-order case this Hamiltonian generator is notable by the absence of an additive function to the bracketed operator in the integrand; this is due to Liouville's theorem on the preservation of phase volume in Hamiltonian dynamics.

4.5. CLASSICAL SYSTEMS OF FINITE ORDER

different operators. More precisely, the classical Hamiltonian H, even when it is identical with H_g, gives rise to two distinct operators in $\mathcal{H}_\Gamma^\oplus$. The first is decomposable and is again called the Hamiltonion. This operator represents an observable in exactly the same way the original classical function H on Γ does and it is denoted by $\widehat{H}^\oplus$. The second is not decomposable and is referred to as the Hamiltonian generator $\widehat{H}_g$. This operator generates a preserving dynamics in $\mathcal{H}^\oplus$ which can cause pure states to evolve from one supersector to another. Here we can clearly see why Postulate UQD(P) does not demand the generator for dynamics to be an observable.

For the LC circuit we have:

1. The traditional classical formalism in the phase space

$$\gamma^1 = \Phi, \quad \gamma^2 = Q, \tag{4.159}$$

$$H = H_g = \frac{Q^2}{2C} + \frac{\Phi^2}{2L}. \tag{4.160}$$

2. The present unified formalism in a direct integral space

$$\widehat{H}^\oplus = \int^\oplus \left(\frac{Q^2}{2C} + \frac{\Phi^2}{2L}\right) \widehat{\mathbb{I}}(\Phi, Q) \, d\Phi dQ, \tag{4.161}$$

$$\widehat{H}_g = -i\hbar \int^\oplus \left(\frac{Q}{C}\frac{\partial}{\partial \Phi} - \frac{\Phi}{L}\frac{\partial}{\partial Q}\right) d\Phi dQ, \tag{4.162}$$

so that

$$\widehat{H}^\oplus \boldsymbol{f}^\oplus = \int^\oplus \left\{\left(\frac{Q^2}{2C} + \frac{\Phi^2}{2L}\right) f(\Phi, Q)\right\} \boldsymbol{u}(\Phi, Q) \, d\Phi dQ, \tag{4.163}$$

$$\widehat{H}_g \boldsymbol{f}^\oplus = -i\hbar \int^\oplus \left\{\left(\frac{Q}{C}\frac{\partial}{\partial \Phi} - \frac{\Phi}{L}\frac{\partial}{\partial Q}\right) f(\Phi, Q)\right\} \boldsymbol{u}(\Phi, Q) \, d\Phi dQ. \tag{4.164}$$

Clearly we have $\widehat{H}^\oplus \neq \widehat{H}_g$.

3. A pure state corresponding to a singular density function

$$f(\Phi, Q) = \delta(\Phi - \Phi_0)\delta(Q - Q_0). \tag{4.165}$$

There is just one stationary pure singular state corresponding to $\Phi_0 = 0$ and $Q_0 = 0$ which is the zero of the Hamiltonian generator $\widehat{H}_g$.

The present formulation of a classical LC circuit in the language of Hilbert space should be compared and clearly distinguished from a quantized LC circuit.[50]

---

[50] For a quantized LC circuit see §7.14 and Wan and Fountain (1996) Appendix D.

4.6 Mixed Quantum Systems

4.6.1 A model system

In addition to classical and orthodox quantum systems our unified theory embraces a third type of systems which we shall refer to as *mixed quantum systems*. These are systems whose associated Hilbert space has a preferred non-trivial decomposition. As a result these systems possess superselection rules and classical observables. In other words these systems possess both classical and quantum characteristics. A formal definition will be given in the next section. In §4.4.3, §4.4.4 and §4.4.5 we have already investigated some simple models of mixed quantum systems associated with a Hilbert space having a preferred discrete direct integral decomposition. We have also presented some explicit examples of non-preserving evolutions in these models. Here we shall present an illustration for preserving evolutions of a model mixed quantum system.

Let us start with the static system Model 4.4.5 which is a system whose associated Hilbert space has a preferred direct sum decomposition $\mathcal{H}_{12}^{\oplus} = \mathcal{H}_1 \oplus \mathcal{H}_2$ as given by Eq. (4.78), where $\mathcal{H}_1, \mathcal{H}_2$ are one-dimensional subspaces spanned by unit vectors u_1 and u_2. Generally observables of the system are of the form given by Eq. (4.82). It follows that all observables commute. They are therefore classical observables, having definite values in any pure state. Pure vectors can represent pure states and mixed vectors can represent mixed states.

Next we need to describe the dynamics for the system. It is possible to have different dynamics imposed on the Hilbert space $\mathcal{H}_{12}^{\oplus}$. In Model 4.4.6 the dynamics is given by the Hamiltonian generator in Eq. (4.104) which leads to a non-preserving dynamics. We shall consider a different dynamics here. Consider a Hamiltonian generator of the form

$$\widehat{H}_g^{(p)} = E_0 \, \widehat{I\!I}^{\oplus} + \hbar\,\omega_1 \, \widehat{P}_{u_1}^{\oplus} + \hbar\,\omega_2 \, \widehat{P}_{u_2}^{\oplus}, \qquad (4.166)$$

where E_0, ω_1 and ω_2 are real and positive constants, $\widehat{P}_{u_1}^{\oplus}$ and $\widehat{P}_{u_2}^{\oplus}$ are projectors onto subspaces $\mathcal{H}_1$ and $\mathcal{H}_2$ given by Eq. (4.81). This Hamiltonian generator is decomposable and it possesses two eigenvalues $E_0 + \hbar\,\omega_1$ and $E_0 + \hbar\,\omega_2$ corresponding to the eigenvectors $u_1^{\oplus}$ and $u_2^{\oplus}$ respectively. The resulting evolution with evolution operators

$$\widehat{U}_t^{(p)} = e^{\frac{i}{\hbar}\widehat{H}_g^{(p)} t} \qquad (4.167)$$

is preserving. There are two pure states:

$$\widehat{S}_{10}^{\oplus} = \widehat{P}_{u_1}^{\oplus}, \quad \widehat{S}_{20}^{\oplus} = \widehat{P}_{u_2}^{\oplus}. \qquad (4.168)$$

4.6. MIXED QUANTUM SYSTEMS

These states evolve in accordance of Eq. (4.37) of Postulate UQD(P). Since the evolution operators commute with $\widehat{S}_{10}^{\oplus}$ and $\widehat{S}_{20}^{\oplus}$ we have

$$\widehat{S}_{1t}^{\oplus} = \widehat{S}_{10}^{\oplus}, \quad \widehat{S}_{2t}^{\oplus} = \widehat{S}_{20}^{\oplus}. \tag{4.169}$$

These are stationary states. An explicit time dependence of these states can be expressed in terms of the evolution of the following pure vectors:

$$\boldsymbol{u}_{1t}^{\oplus} = e^{-i(\omega_0+\omega_1)t}\,\boldsymbol{u}_1^{\oplus}, \quad \boldsymbol{u}_{2t}^{\oplus} = e^{-i(\omega_0+\omega_2)t}\,\boldsymbol{u}_2^{\oplus}, \tag{4.170}$$

where $\omega_0 = E_0/\hbar$, with

$$\widehat{S}_{1t}^{\oplus} = |\boldsymbol{u}_{1t}^{\oplus}\rangle\langle\boldsymbol{u}_{1t}^{\oplus}| \quad \text{and} \quad \widehat{S}_{2t}^{\oplus} = |\boldsymbol{u}_{2t}^{\oplus}\rangle\langle\boldsymbol{u}_{2t}^{\oplus}|. \tag{4.171}$$

We conclude that the above model represents a mixed quantum system with a non-classical dynamics, highlighted by the fact that stationary states occur not at the zeros of the Hamiltonian generator. The state $\boldsymbol{u}_{1t}^{\oplus}$ is seen to evolve within the same supersector, and that this resembles a stationary state in Eq. (3.243) of an orthodox quantum system. We shall call the present model dynamics Model 4.6.1 for later reference.[51]

4.6.2 Classification of physical systems

In classifying physical systems we have to take both statics and dynamics into account. The involvement of dynamics can be traced back to the historical development of the theory of the atom. Rutherford produced a classical model which visualized an atom as having a heavy and positively charged nucleus with negatively charged electrons circling round under Coulomb attraction. This model fails since classical dynamics entails a continuous loss of energy by the electrons through radiation, leading to instability. Bohr was the first to abandon classical dynamics by introducing the idea of stationary states which circumvents the instability of the classical dynamics of the Rutherford atom, i.e., an electron in an atom can circle the nucleus in certain orbits without losing energy in the form of radiation. Classical dynamics based on Maxwell's theory of electromagnetism does not allow any such stationary state at all. This leads us to the notion that the fundamental structural differences between classical and quantum systems lies in their statics as well as their dynamics.

Definition 4.6.2(1)   On classical, orthodox quantum and mixed quantum systems

- A physical system described in accordance with Postulate UQS or Postulate UQS(F) is said to be:

---
[51] This model is an example of preserving dynamics Case 1 in §4.4.2.

1. **Classical** if the following conditions on statics and dynamics are satisfied:

   (a) **Statics** The Hilbert space $\mathcal{H}^\oplus$ associated with the system has a preferred direct integral decomposition in terms of a *family of trivial Hilbert spaces*, i.e.,
   $$\mathcal{H}^\oplus = \mathcal{H}_\Gamma^\oplus(\mathbb{C}, dg(\gamma)). \tag{4.172}$$

   (b) **Dynamics** Its dynamics is classical in the sense that it is derivable from classical dynamics in a manner as described in §4.5.[52]

2. **Orthodox quantum** if the following conditions on statics and dynamics are satisfied:

   (a) **Statics** The Hilbert space $\mathcal{H}^\oplus$ associated with it has a preferred *trivial decomposition*, i.e.,
   $$\mathcal{H}^\oplus = \mathcal{H}(0) \tag{4.173}$$
   for some Hilbert space $\mathcal{H}(0)$ so that Postulate OQS of §4.1 on orthodox quantum statics is satisfied.

   (b) **Dynamics** Its dynamics is given by Postulate OQD of §4.4.1 on orthodox quantum dynamics.

3. **Mixed quantum** if the following conditions on statics and dynamics are satisfied:

   (a) **Statics** The Hilbert space $\mathcal{H}^\oplus$ associated with it has a preferred *non-trivial decomposition*, e.g.,
   $$\mathcal{H}_d^\oplus = \oplus_n \mathcal{H}(n), \tag{4.174}$$
   or
   $$\mathcal{H}_c^\oplus = \int^\oplus \mathcal{H}(\gamma)\, dg(\gamma). \tag{4.175}$$

   (b) **Dynamics**

   i. The system can behave and evolve as an orthodox quantum system entirely within a single supersector.[53]

---

[52] Generally a classical system requires a continuous integral decomposition. However, it is possible to envisage a set of classical particles at fixed sites which would mean a degree of discreteness in the integral decomposition. For an example see §8.3.2 in Chapter 8.

[53] An example is the evolution in Eqs. (4.170) and (4.171).

4.6. MIXED QUANTUM SYSTEMS

ii. In a suitable environment the system can also evolve from one supersector to a different supersector, enabling a pure state to evolve to a pure state in another supersector or simply to a mixed state.[54]

There has been a great deal of controversy on whether mixed quantum systems exist or not, and if they do whether they are derivable from orthodox quantum systems. We shall devote Chapter 7 to demonstrate their existence by studying some concrete physical examples.

In §4.7.3 we will see that mixed quantum systems can serve as measuring devices to help solve various quantum measurement problems.

4.6.3 Quantum/Classical divide 1

The following comments may help to clarify the situation:

1. Classical systems as classified in Definition 4.6.2(1) coincide with those studied in §1.6 and §4.5.

2. For classical systems all observables mutually commute since the Hilbert space has a preferred direct integral decomposition in terms of a family of trivial Hilbert spaces. A distinguishing feature of classical dynamics is revealed in the condition for pure states to be stationary. For a classical system a supersector corresponds to a point in the phase space. Consequently there will not be any evolution within a supersector except when it corresponds to a stationary state at a zero of the Hamiltonian generator. In other words a pure state in the supersector $\mathcal{H}(\gamma_c)$ of a classical system is stationary if and only if γ_c is a zero of the Hamiltonian generator. This is generally not the case for orthodox and mixed quantum systems for which stationary states are identifiable with the eigenvectors of the Hamiltonian generator.

3. Orthodox quantum systems can be distinguished from classical and mixed quantum systems in having a preferred trivial direct integral decomposition. Physically there is the total absence of superselection rules and of classical observables. The superposition principle applies to any set of pure states.

4. Mixed quantum system can be separated from orthodox quantum systems by their Hilbert space having a preferred non-trivial direct integral decomposition and the consequent emergence of superselection rules.

---

[54] An example is the evolution in Eqs. (4.61) and (4.63).

The distinction between mixed quantum systems and classical systems is more subtle. Traditionally there is a belief that quantum systems and classical systems are distinguishable by their statics, i.e., commutivity of all observables can distinguish classical systems from quantum systems.[55] Such a belief is clearly contradicted by Model 4.6.1 where all observables commute but the system is clearly not classical. In fact, whenever the Hilbert space has a preferred direct integral decomposition in terms of a family of one-dimensional spaces all observables will mutually commute. This is why in Definition 4.6.2(1) dynamics is built in as an intrinsic requirement for the classification of physical systems. Commutivity is necessary but not sufficient for the definition of classical systems, i.e., if the dynamics is not classical as in Model 4.6.1 the system is not classical. The dynamics of a mixed system can lead to the following evolutions:

(a) A pure state lying in a given supersector can remain pure, evolving within the same supersector. Model 4.6.1 is an example of such evolution. In Chapter 7 we shall show that a thick superconducting ring in a superconducting state is a physical system having this type of evolution.

(b) An initial pure state can evolve into a mixed state in a non-preserving manner as in Model 4.4.6 E1 and Model 4.4.6 E2.

(c) A pure state lying in a given supersector can remain pure, but evolving from supersectors to supersectors in time. Such a dynamics can be visualized as due to environmental influences arising from a non-observable Hamiltonian generator. In other words a pure state, which would evolve within the same supersector if the system is left undisturbed, is forced to move out into another supersector by an interaction with the environment. For example a superconducting circuit with a Josephson junction behaves quantum mechanically with evolution within the same supersector. When an electric potential is applied across the junction of such a circuit its state will be forced to evolve from one supersector to another leading to a phenomenon known as an ac Josephson effect. We shall discuss this in more details in Chapter 7.

(d) We can further clarify the dynamics of mixed quantum systems in terms of a notion of equilibrium. This shall be done in the next section.

It should be pointed out that while many large systems composed of a macroscopic number of particles are classical this is not always the case. Superconducting systems are the familiar counter examples. This just illustrates

---

[55] Jauch (1964), Jauch (1968) p. 80, Busch, Lahti and Mittelstaedt (1996) p. 18.

4.6. MIXED QUANTUM SYSTEMS

the complex nature of classical/quantum devide which cannot be generally characterized just by some parameters becoming very large or small.

4.6.4 Equilibrium and mixed quantum systems

Definition 4.6.4(1)   **On equilibrium and non-equilibrium**

- A mixed quantum system is said to be in *equilibrium* if pure states evolve into pure states within the same supersector. Otherwise it is said to be in *non-equilibrium*. A mixed quantum system in equilibrium (non-equilibrium) is called an equilibrium (non-equilibrium) system.

For an equilibrium system two different supersectors are sufficiently divorced from each other that each supersector can represent a physical system in its own right. In other words an equilibrium mixed quantum system may be formally regarded as a collection of separate systems, each being an orthodox quantum system describable within a supersector and there are no correlations between different supersectors. However, such an equilibrium mixed quantum system is not a permanent collection of an arbitrary set of disparate quantum systems. The systems in the collection may evolve into each other in an appropriate environment, i.e., the system may become non-equilibrium. This is why the dynamics of mixed systems in Definition 4.6.2(1) is specified in two parts. Different supersectors may be correlated by a Hamiltonian generator to enable a pure state to evolve from one supersector to another. When the system turns non-equilibrium under the influence of its environment the nature of the Hamiltonian generator changes accordingly to incorparate the effect of the environment. The Hamiltonian generator will no longer be decomposable, and hence will not be an observable of the system itself. One can gain an insight into why the generator cannot be an observable of the system by looking at how one can break the equilibrium nature of a system. Take an ideal gas in a box as an example of thermodynamic systems mentioned in §4.4.7. When in equilibrium the gas is specified by the usual parameters of volume, pressure and temperature. The gas will not evolve out of such an equilibrium on its own. It will take an external agent to force a change, e.g., a change of pressure from outside the box. In other words the change has to come from outside the system and this is why the Hamiltonian generator which necessarily involves external influences cannot be an observable of the system alone.

Some important physical examples of equilibrium and non-equilibrium mixed quantum systems will be presented in Chapter 7.

4.7 Coupling of Systems of Different Types

4.7.1 Measuring devices

A selling point of a unified theory is that it can deal with the coupling of systems of different types within the same mathematical framework. The coupling of orthodox quantum systems has been well studied.[56] What is less familiar is the coupling of orthodox quantum systems with classical systems and with mixed quantum systems and the interaction of mixed quantum systems with classical systems. One application of such couplings would be to model a quantum measurement situation. Recall that a fundamental problem of quantum measurement is the lack of dynamics to effect a transition from a pure state to a mixture. In §3.6 we have discussed an asymptotic solution to this problem within orthodox quantum mechanics. With the inclusion of classical and mixed quantum systems we can introduce what may be called measuring devices to formulate some simple schematic measurement models, known as the

superselection rule approach to quantum measurement.

Quantum Measuring Devices:

- A *quantum measuring device* should have the following properties:[57]

    1. It possesses a superselection rule and is capable of coupling with an orthodox quantum system to form a *compound system*.

    2. The coupling should be such that the compound system would inherit the superselection rule from the measuring device. It is this superselection rule which decoheres certain linear combinations of pure states into a mixture to achieve a measurement.

In the following sections we shall see how measuring devices can help solve the measurement problem.

---

[56] Jauch (1968) §11-8, Beltrammetti and Cassinelli (1981) §7.

[57] Here we are concentrating on a device which can measure an observable of an orthodox quantum system. We can generalize this concept. For example, we can couple a mixed quantum system (serving as a measuring device) with a classical system to perform measurement of a classical observable of the classical system. A practical example of these is mentioned in §4.7.4.

4.7.2 Coupling of orthodox quantum and classical systems

We shall illustrate the possible interaction between orthodox quantum and classical systems by a simple model.[58] Let $\mathcal{H}^{(q)}$ be the Hilbert space associated with an orthodox quantum system, and let $\mathcal{H}^{(c)}$ be the Hilbert space for a second order classical Hamiltonian system, i.e., a classical system with canonical variables $\gamma = (\gamma^1, \gamma^2)$. We know that $\mathcal{H}^{(c)}$ has a preferred direct integral decomposition of the form of Eq. (4.142), i.e.,

$$\mathcal{H}^{(c)} = \mathcal{H}_\Gamma^\oplus(\mathbb{C}, dg(\gamma)) \tag{4.176}$$

$$= \int^\oplus \mathcal{H}(\gamma)\, dg(\gamma) = \int^\oplus \mathcal{H}(\gamma^1, \gamma^2)\, d\gamma^1 d\gamma^2. \tag{4.177}$$

A vector $\boldsymbol{f}^\oplus$ in $\mathcal{H}_\Gamma^\oplus(\mathbb{C}, dg(\gamma))$ is of the form of of the form of Eq. (4.143). The supersectors

$$\mathcal{H}(\gamma) = \mathcal{H}(\gamma^1, \gamma^2) = \mathbb{C} \tag{4.178}$$

are all one-dimensional. The Hilbert space associated with the compound system is assumed to be the tensor product space

$$\mathcal{H}^{(q,c)} = \mathcal{H}^{(q)} \otimes \mathcal{H}^{(c)} = \mathcal{H}^{(q)} \otimes \mathcal{H}_\Gamma^\oplus. \tag{4.179}$$

According to Theorems 2.15.2(1) and 2.15.2(2) this product space inherits a natural direct integral decomposition of the form:

$$\mathcal{H}^{(q,c)} = \int^\oplus \mathcal{H}^{(q)} \otimes \mathcal{H}(\gamma)\, dg(\gamma). \tag{4.180}$$

We shall therefore assume that the Hilbert space associated with the compound system has this particular preferred direct integral decomposition. In other words, the compound system inherits a superselection rule with supersectors $\mathcal{H}^{(q)} \otimes \mathcal{H}(\gamma)$. Consequently a pure state for the compound system corresponds to a pure state of the quantum system and a pure singular state of the classical system.

The interesting part comes when we consider the dynamics of the compound system. A simple model Hamiltonian generator for the coupled system is

$$\widehat{H}_g^{(q,c)} = \widehat{H}_g^{(q)} \otimes \widehat{\mathbb{I}}^{(c)} + \widehat{\mathbb{I}}^{(q)} \otimes \widehat{H}_g^{(c)} + \rho\,(\widehat{H}_I^{(q)} \otimes \widehat{H}_I^{(c)}), \tag{4.181}$$

where $\widehat{H}_g^{(q)}$ and $\widehat{H}_g^{(c)}$ are respectively the Hamiltonian generators of the quantum and of the classical systems separetely, ρ is a coupling constant, $\widehat{H}_I^{(q)}$

---
[58]There are many other models in the literature, e.g., Anderson (1995).

and $\widehat{H}_I^{(c)}$ are operators on $\mathcal{H}^{(q)}$ and $\mathcal{H}^{(c)}$ respectively, and ρ multiplied by $\widehat{H}_I^{(q)} \otimes \widehat{H}_I^{(c)}$ is the coupling interaction. By setting $\rho = 0$ we recover the evolution for each system separately. Using this particular coupling we can demonstrate how a classical system can serve as a measuring device to achieve a quantum measurement.

Consider the simple situation where the orthodox quantum system, on which we want to perform a measurement, is described by a two-dimensional Hilbert space $\mathcal{H}^{(q)}$. The observable we wish to measure $\widehat{A}^{(q)}$ has two orthonormal eigenvectors φ_1 and φ_2 corresponding to eigenvalues a_1 and a_2 respectively. These two eigenvectors form a basis in $\mathcal{H}^{(q)}$. The measuring device is a second order classical Hamiltonian system. For this measurement we shall assume an interaction Hamiltonian generator of the form of Eq. (4.181) with the coupling term formed by

1. $\widehat{H}_I^{(q)} = \widehat{A}^{(q)}$ and

2. $\widehat{H}_I^{(c)}$ defined by its action on $\boldsymbol{f}^\oplus \in \mathcal{H}_\Gamma^\oplus(\mathbb{C}, dg(\gamma))$ according to

$$\widehat{H}_I^{(c)} \boldsymbol{f}^\oplus = \int^\oplus \left(-i\hbar \frac{\partial}{\partial \gamma^1} f(\gamma^1, \gamma^2) \right) \boldsymbol{u}(\gamma)\, dg(\gamma). \tag{4.182}$$

The resulting evolution operators are

$$\widehat{U}_t^{(q,c)} = \exp\left\{ -i\left(\widehat{H}_g^{(q)} \otimes \widehat{\mathbb{I}}^{(c)} + \widehat{\mathbb{I}}^{(q)} \otimes \widehat{H}_g^{(c)} + \rho\left(\widehat{A}^{(q)} \otimes \widehat{H}_I^{(c)} \right) \right) t \right\}. \tag{4.183}$$

Suppose at $t = 0$ we have the following situation:

1. The quantum system is in a pure state $\widehat{S}^{(q)}$ corresponding to a coherent superposition of the eigenvectors of $\widehat{A}^{(q)}$, i.e.,

$$\widehat{S}^{(q)} = \widehat{P}_\varphi^{(q)} = |\varphi\rangle\langle\varphi|, \quad \varphi = c_1 \varphi_1 + c_2 \varphi_2 \in \mathcal{H}^{(q)} \tag{4.184}$$

with real coefficients c_1 and c_2 satisfying $c_1^2 + c_2^2 = 1$.

2. The classical system is in a state $\widehat{S}^{(c)}$ corresponding to a normalized vector $\boldsymbol{f}^\oplus$ in $\mathcal{H}_\Gamma^\oplus$, i.e.,

$$\widehat{S}^{(c)} = \int^\oplus S(\gamma)\, \widehat{\mathbb{I}}^{(c)}(\gamma)\, dg(\gamma), \quad S(\gamma) = |f(\gamma)|^2, \tag{4.185}$$

where $\widehat{\mathbb{I}}^{(c)}(\gamma)$ is the identity operator on $\mathcal{H}(\gamma)$.

4.7. COUPLING OF SYSTEMS OF DIFFERENT TYPES

It is more intuitive for our purposes here to describe the state for the compound system in terms of a vector in $\mathcal{H}^{(q,c)}$. So, following Eq. (4.26) and Theorem 2.15.1(2) we shall represent the state $\widehat{S}_0^{(q,c)}$ of the compound system at time $t = 0$ by the following mixed vector in $\mathcal{H}^{(q,c)}$

$$\boldsymbol{F}_0^{(q,c)} = \int^{\oplus} \left(\varphi \otimes f(\gamma) \right) \boldsymbol{u}(\gamma) \, dg(\gamma). \tag{4.186}$$

For simplicity let us make the assumption that the interaction begins at $t = 0$ and acts strongly and *impulsively* for a small duration T so that we can ignore the non-interaction part of the Hamiltonian generator for this duration.[59] To find the vector $\boldsymbol{F}_t^{(q,c)}$ to represent the state at time t we adopt a Schrödinger equation in the form of Eq. (4.62) in Model 4.4.3. In other words we adopt the following Schrödinger equation:

$$i\hbar \frac{\partial}{\partial t} \boldsymbol{F}_t^{(q,c)} = \rho \left(\widehat{H}_I^{(q)} \otimes \widehat{H}_I^{(c)} \right) \boldsymbol{F}_t^{(q,c)}. \tag{4.187}$$

The solution for $t \in [0, T]$ is

$$\boldsymbol{F}_t^{(q,c)} = \int^{\oplus} \Big\{ c_1 \left(\varphi_1 \otimes f(\gamma^1 - a_1 \rho t, \gamma^2) \right) \\ + c_2 \left(\varphi_2 \otimes f(\gamma^1 - a_2 \rho t, \gamma^2) \right) \Big\} \boldsymbol{u}(\gamma) d\gamma. \tag{4.188}$$

When $f(\gamma^1, \gamma^2)$ is highly localized, say around $\gamma^1 = 0$, the supports of the functions $f(\gamma^1 - a_1 \rho T, \gamma^2)$ and $f(\gamma^1 - a_2 \rho T, \gamma^2)$ can become non-overlapping for sufficiently large ρT. Then $\boldsymbol{F}_T^{(q,c)}$ becomes a mixture of two disjoint parts

$$\int^{\oplus} \left(\varphi_1 \otimes f(\gamma^1 - a_1 \rho T, \gamma^2) \right) \boldsymbol{u}(\gamma) \, dg(\gamma) \tag{4.189}$$

and

$$\int^{\oplus} \left(\varphi_2 \otimes f(\gamma^1 - a_2 \rho T, \gamma^2) \right) \boldsymbol{u}(\gamma) \, dg(\gamma) \tag{4.190}$$

with weights c_1^2 and c_2^2 respectively. The fact that $\boldsymbol{F}_t^{(q,c)}$ is a classical mixture of two disjoint parts enables us to interpret the above interaction as a

---

[59] Aharonov and Safko (1975). This amounts to taking the coupling constant as time dependent:

$$\rho(t) = \begin{cases} \rho, & 0 < t < T \\ 0, & \text{otherwise.} \end{cases}$$

The constant ρ is assumed large enough to make the product ρT significantly different from zero for experimental observation.

measurement interaction for the quantum observable $\widehat{A}^{(q)}$ of the quantum system. For the sake of argument, imagine the classical system as a classical particle with γ^1 as its position and γ^2 as its momentum.[60]

1. At the start of measurement the classical particle is positioned approximately at the origin on account of the highly localized nature of $f(\gamma^1, \gamma^2)$. If, after the interaction, the particle has moved to position $a_1\rho T$, that is, the particle has moved to a position round the peak of $f(\gamma^1 - a_1\rho T, \gamma^2)$, then we can say we have obtained the eigenvalue a_1 of $\widehat{A}^{(q)}$. A movement to $a_2\rho T$ corresponds to a measurement resulting in the eigenvalue a_2.

2. The situation is clearer if a_1 is negative and a_2 positive. Assuming a positive coupling constant, a movement of the particle to the left of the origin signals a measured value of a_1 and an opposite movement means a measured value of a_2.[61]

4.7.3 Coupling of orthodox and mixed quantum systems

The coupling of an orthodox quantum system to a mixed quantum system is even more natural and so are measurement models following from such a coupling.[62] The idea is to employ mixed quantum systems to serve as measuring devices. Let us revisit the orthodox quantum system with a two-dimensional Hilbert space $\mathcal{H}^{(q)}$ considered in the preceding section. Again we want to measure the observable $\widehat{A}^{(q)}$ which has two orthonormal eigenvectors φ_1 and φ_2 corresponding to eigenvalues a_1 and a_2 respectively. The system is assumed to be in state $\widehat{S}^{(q)} = |\varphi\rangle\langle\varphi|$ given by Eq. (4.184). Let

$$\widehat{P}_1^{(q)} = |\varphi_1\rangle\langle\varphi_1| \quad \text{and} \quad \widehat{P}_2^{(q)} = |\varphi_2\rangle\langle\varphi_2| \qquad (4.191)$$

be the projectors onto the two subspaces of $\mathcal{H}^{(q)}$ spanned by φ_1 and φ_2 respectively. We shall construct a measuring device in terms of Model 4.4.3, i.e., in terms of a mixed quantum system described by a Hilbert space $\mathcal{H}^{(m)}$ based on the three-dimensional Hilbert space $\mathcal{H}^\oplus$ in Eq. (4.50). The compound of the quantum and the measuring device is assumed to be associated with the Hilbert space

$$\mathcal{H}^{(q,m)} = \mathcal{H}^{(q)} \otimes \mathcal{H}^{(m)} = \mathcal{H}^{(q)} \otimes (\mathcal{H}_- \oplus \mathcal{H}_0 \oplus \mathcal{H}_+). \qquad (4.192)$$

---

[60] Particles here are meant in the symbolic sense. We could have an LC circuit as a second order Hamiltonian system with the flux Φ in Eq. (4.159) as the "position" of the system.

[61] Wan, Bradshaw, Trueman and Harrison (1998) has an explicit example.

[62] Wan (1980), Beltrammetti and Cassinelli (1981), van Fraassen (1991), Bub (1988).

4.7. COUPLING OF SYSTEMS OF DIFFERENT TYPES

This product space inherits a natural direct sum decomposition

$$\mathcal{H}^{(q,m)} = \left(\mathcal{H}^{(q)} \otimes \mathcal{H}_-\right) \oplus \left(\mathcal{H}^{(q)} \otimes \mathcal{H}_0\right) \oplus \left(\mathcal{H}^{(q)} \otimes \mathcal{H}_+\right). \tag{4.193}$$

We shall assume the Hilbert space of the compound system to have this particular preferred direct sum decomposition, i.e., the compound system inherits the superselection rule of the measuring device.

Now introduce a Hamiltonian generator $\widehat{H}_g^{(q,m)}$ on $\mathcal{H}^{(q,m)}$ to couple the orthodox quantum system with the measuring device:

$$\widehat{H}_g^{(q,m)} = \widehat{H}_g^{(q)} \otimes \widehat{\mathbb{I}}^{(m)} + \widehat{\mathbb{I}}^{(q)} \otimes \widehat{H}_g^{(m)} + \rho \left(\widehat{P}_1^{(q)} \otimes \widehat{L}_- + \widehat{P}_2^{(q)} \otimes \widehat{L}_+ \right), \tag{4.194}$$

where $\widehat{L}_-$ and $\widehat{L}_+$ are defined by Eqs. (4.53) and (4.54). This Hamiltonian generator is selfadjoint but not decomposable since $\widehat{L}_-$ and $\widehat{L}_+$ are not. Let

1. $\widehat{U}_t^{(q,m)} = \exp\left(-i\widehat{H}_g^{(q,m)} t\right)$ be the evolution operators generated by $\widehat{H}_g^{(q,m)}$.

2. $\boldsymbol{f}_0^{\oplus}$ be an arbitrary unit vector in $\mathcal{H}^{(m)}$ and $\boldsymbol{F}_0^{(q,m)} = \varphi \otimes \boldsymbol{f}_0^{\oplus}$.

3. $\boldsymbol{F}_t^{(q,m)} = \widehat{U}_t^{(q,m)} \boldsymbol{F}_0^{(q,m)}$.

Then $\boldsymbol{F}_t^{(q,m)}$ satisfies the following Schrödinger equation:

$$i\hbar \frac{\partial}{\partial t} \boldsymbol{F}_t^{(q,m)} = \widehat{H}_g^{(q,m)} \boldsymbol{F}_t^{(q,m)}. \tag{4.195}$$

As in the preceding section let us assume an impulsive interaction which acts strongly for a small duration T starting from $t = 0$. Ignoring the non-interaction part of the Hamiltonian generator[63] the Schrödinger equation reduces to

$$i\hbar \frac{\partial}{\partial t} \boldsymbol{F}_t^{(q,m)} = \rho \left(\widehat{P}_1^{(q)} \otimes \widehat{L}_- + \widehat{P}_2^{(q)} \otimes \widehat{L}_+ \right) \boldsymbol{F}_t^{(q,m)}. \tag{4.196}$$

An initial state $\boldsymbol{F}_0^{(q,m)}$ with $\boldsymbol{f}_0^{\oplus} = i\varphi \otimes \boldsymbol{u}_0^{\oplus}$ as in Eq. (4.57) will evolve, for $t \in [0,T]$, to

$$\begin{aligned}\boldsymbol{F}_t^{(q,m)} &= c_1 \varphi_1 \otimes \left(\sin\rho t/\hbar \, \boldsymbol{u}_-^{\oplus} + i \cos\rho t/\hbar \, \boldsymbol{u}_0^{\oplus} \right) \\ &+ c_2 \varphi_2 \otimes \left(\sin\rho t/\hbar \, \boldsymbol{u}_+^{\oplus} + i \cos\rho t/\hbar \, \boldsymbol{u}_0^{\oplus} \right).\end{aligned} \tag{4.197}$$

Suppose we adjust the interaction parameters so that $\rho T/\hbar = \pi/2$. Then at the end of the interaction at $t = T$ we have

$$\boldsymbol{F}_T^{(q,m)} = c_1 \varphi_1 \otimes \boldsymbol{u}_-^{\oplus} + c_2 \varphi_2 \otimes \boldsymbol{u}_+^{\oplus}. \tag{4.198}$$

---

[63] An alternative is to consider time evolution in the interaction picture.

The interpretation of the whole process is obvious. An initial pure state represented by the tensor product $\boldsymbol{F}_0^{(q,m)}$ of φ and $\boldsymbol{f}_0^{\oplus}$ will evolve into a state represented by the mixed vector $\boldsymbol{F}_T^{(q,m)}$. This is a classical mixture of $\varphi_1 \otimes \boldsymbol{u}_-^{\oplus}$ and $\varphi_2 \otimes \boldsymbol{u}_+^{\oplus}$. Let us see how the mixed quantum system can serve as a measuring device. At time $t = 0$ the mixed system is in pure state $\boldsymbol{u}_0^{\oplus}$. Immediately after the completion of the interaction the coupled system evolve into the classical mixture $\boldsymbol{F}_T^{(q,m)}$ with φ_1 coupled to $\boldsymbol{u}_-^{\oplus}$ and φ_2 to $\boldsymbol{u}_+^{\oplus}$. We can visualize $\boldsymbol{u}_-^{\oplus}$ and $\boldsymbol{u}_+^{\oplus}$ as corresponding to the two states of a *pointer observable* of the mixed quantum system corresponding to the pointer swinging to the left and right respectively from the neutral position represented by $\boldsymbol{u}_0^{\oplus}$. A swing to the left would imply a measured value a_1 for $\widehat{A}^{(q)}$ and an opposite swing would mean a measured value of a_2. The probabilities for these measured results are given by $|c_1|^2$ and $|c_2|^2$ respectively as expected.

4.7.4 Coupling of classical and mixed quantum systems

We can go on to set up models for coupling of a classical system with a mixed quantum system. Such a coupling enables us to construct models of measurement of classical observables using a mixed quantum system as a measuring device. At first sight this may not seem to serve any purpose at all. Surprising as it may be such a measuring model does serve an important purpose. The point is that a measuring device based on a mixed quantum system can be more effective and more accurate than any classical measuring device. The best examples are magnetometers for measuring a magnetic field. There are magnetometers based on classical physics of course. However, there are magnetometers constructed from superconducting quantum interference devices (SQUIDs for short).[64] A SQUID, which is commercially available, is a most accurate device for measuring classical magnetic fluxes and classical magnetic fields. Clearly a SQUID is not a classical system. As we shall discuss in detail in Chapter 7 a SQUID may be regarded as a mixed quantum system. So, the coupling of classical systems with mixed quantum systems is not just of academic interest. They are important for practical applications.

4.8 Concluding Remarks

There is a strong belief in some circles in what may be called *the universality of (orthodox) quantum mechanics.*[65] The idea is that quantum mechanics is universally valid so that all theories of relevant physical systems are derivable

---

[64] Feynman (1965) Chap 21, Rose-Innes and Rhoderick (1980), Gallop (1990), Wan and Harrison (1997).

[65] Ludwig (1985) Chapter XVIII, Busch, Lahti and Mittelstaedt (1996) §III 1.1 p. 28.

4.8. CONCLUDING REMARKS

from it. In particular the behaviour of macroscopic systems, including classical systems, should be derivable from orthodox quantum mechanics.[66] To the believers of the universality of quantum mechanics our Postulate UQS should be explainable in terms of Postulate OQS. In other words, we ought to be able to derive superselections rules in terms of Postulate OQS.

Of course it would be highly desirable if superselection rules can be derived from a deeper underlying theory. In §7.11 we shall explain the emergence of a superselection rule in superconductivity in terms of orthodox quantum theory. However, we shall not adopt a dogmatic view on the existence of a single universally valid orthodox quantum mechanics.[67] As already pointed out in §4.3.3 not all systems exhibiting superselection rules are macroscopic. There are molecules, both organic and inorganic exhibiting superselection rules. We regard quantum theory as a *living* theory, capable of evolution to take account of and to incorporate new discoveries and new developments, both experimentally[68] and theoretically. For example we shall devote Chapter 5 to extend quantum theory so as to be able to utilize certain symmetric operators to represent a new types of observables. However it remains possible that, just like the division of particles into Bosons and Fermions which gives rise to superselection rule in the first place, mixed quantum systems with their superselection rules may be fundamental in their own right.

Even for those who believe in the universality of orthodox quantum mechanics there is no reason not to employ our unified formalism as a practical theory to describe physical systems when appropriate, whether or not the superselection rules involved are derivable.[69]

---

[66] Here the derivation is meant to be rigorous and general. Classical behaviour in some special cases can indeed be derived from orthodox quantum mechanics. But there is as yet no rigorous and general derivation of classical behaviour from orthodox quantum mechanics.

[67] There are others not taking such a dogmatic view also, e.g., Ludwig (1985) Chapter XVIII.

[68] There have been an explosion of experimental studies of novel systems such as Bose-Einstein condensate and low-dimensional nano-structures and novel nano-electronic devices (Grabert and Devoret (1992), Schwab and Roukes (2005)), leading to many discoveries of new physics.

[69] Wightman and Glance (1989).

References

1. Aharonov, Y. and Safko, L. (1975). *Ann. Phys.* **91** 279.
2. Alanson, T. (1992). *Phys. Lett. A* **163** 41.
3. Anderson, A. (1995). *Phys. Rev. Lett.* **74** 621.
4. Araki, H. (1980). *Prog. Theor. Phys.* **66** 719.
5. Araki, H. (1986). 'A Continuous Superselection Rule as a Model of Classical Measuring Apparatus in Quantum Mechanics' in A. I. Miller, ed., *Fundamental Aspects of Quantum Theory*, a NATO ASI Series B Vol. 226, Plenum, New York.
6. Beltrametti, E. G. and Cassinelli, G. (1981). *The Logic of Quantum Mechanics*, Addison-Wesley, Reading Mass.
7. Bogolubov, N., Logunov, A. and Todorov, I. (1975). *Introduction to Axiomatic Quantum Field Theory*, Benjamin, Reading Mass.
8. Bradshaw, J. E. (1995). *Superselection Rules, Quasi-Particles and Macroscopic Quantum Systems*, St Andrews University PhD Thesis.
9. Bratteli, O. and Robinson, D. W. (1979). *Operator Algebras and Quantum Statistical Mechanics Vol. I*, Springer-Verlag, New York.
10. Bub, J. (1988). *Found. Phys.* **18** 701.
11. Busch, P., Lahti, P. J. and Mittelstaedt, P. (1996). *The Quantum Theory of Measurement*, 2nd edition, Springer-Verlag, Berlin.
12. Cisneros, C., Martines-y-Romera, R. P., Núñez-Yépez, H. N. and Salas-Brito, A. L. (1998). *Euro. J. Phys.* **19** 237.
13. Cram, J. M. and Cram D. J. (1978). *The Essence of Organic Chemistry*, Addison-Wesley, Reading, Mass.
14. Feynman, R. P., Leighton, R. B. and Sands, M. (1965). *Feynman Lectures on Physics Vol. 3*, Addison-Wesley,Reading, Mass.
15. Gallop, J. C. (1991). *SQUIDs, The Josephson Effects and Superconducting Electronics*, Adam Hilger, Bristol.
16. Ghirardi, G. C., Rimini, A. and Weber, T. (1986). *Phys. Rev. D* **34** 470.
17. Gisin, N. and Percival, I. (1992). *J. Phys. A: Math. Gen.* **25** 5677.
18. Giulini, D., Joos, E., Kiefer, C., Kupsch, J., Stamatescu, I. and Zeh, H. D. (1996). *Decoherence and the Appearance of a Classical World in Quantum Theory*, Springer-Verlag, Berlin.
19. Giulini, D., Kiefer, C. and Zeh, H. D. (1995). *Phys. Lett. A* **199** 291.
20. Grabert, H. and Devoret, M. H. (1992). *Single Charge Tunneling: Coulomb Blackage Phenomena in Nanostructures*, Plenum, New York.
21. Grigolini, P. (1993). *Quantum Mechanical Irreversibility and Measurement*, World Scientific, Singapore.
22. Hagg, R. (1992). *Local Quantum Physics*, Springer-Verlag, Berlin.
23. Haag, R. and Kastler, D. (1964). *J. Math. Phys.* **5** 848.
24. Hegstrom, R. A., and Kondepudi, D. K. (1990). *Scientific American* **263** 98.
25. Hepp, K. (1972). *Helv. Phys. Acta.* **45** 237.

26. Hughes, R. I. G. (1989). *The Structure and Interpretation of Quantum Mechanics*, Harvard University Press, Cambridge, Mass.
27. Jauch, J. M. (1964). *Helv. Phys. Acta* **37** 293.
28. Jauch, J. M. (1968). *Foundations of Quantum Mechanics*, Addison-Wesley, Reading, Mass.
29. Kastler, D. ed. (1990). *The Algebraic Theory of Superselection Sectors*, World Scientific, Singapore.
30. Koopman, B. O. (1931). *Proc. Nat. Sci.* **17** 315.
31. Kowalski, K. (1994). *Methods of Hilbert Spaces in the Theory of Nonlinear Dynamical Systems*, World Scientific, Singapore.
32. Landsman, N. P. (1991). *Int. J. Mod. Phys. A* **6** 5349.
33. Ludwig, G. (1985). *Foundations of Quantum Mechanics Vol. 2*, Springer-Verlag, New York.
34. Machida, S. and Namiki, M. (1980). *Prog Theor. Phys.* **63** 1457, **63** 1833.
35. Müler-Herold, U. (1985). *J. Chem. Education* **62** 379.
36. Namiki, M. (1988). *Found. Phys.* **18** 29.
37. Namiki, M. and Pascazio, S. (1993). *Phys. Rep.* **232** 301.
38. Núñez-Yépez, H. N., Vargas, C. A. and Salas-Brito, A. L. (1988). *J. Phys. A: Math. Gen.* **21** L651.
39. Núñez-Yépez, H. N., Vargas, C. A. and Salas-Brito, A. L. (1998). *Phys. Rev. A* **39** 4306.
40. Omnès, R. (1994). *The Interpretation of Quantum Mechanics*, Princeton University Press, Princeton.
41. Piron, C. (1976). *Foundations of Quantum Physics*, Benjamin, Reading, Mass.
42. Primas, H. and Müller-Herold, U. (1978). *Advances in Chemical Physics* **38** Chapters 1-6.
43. Reed, M. and Simon, B. (1975). *Methods of Modern Mathematical Physics Vol. 2: Fourier Analysis, Selfadjointness*, Academic Press, New York.
44. Roberts, J. E. and Roepstorff, G. (1969). *Commun. Math. Phys.* **11** 321.
45. Rose-Innes, A. C. and Rhoderick, E. H. (1980). *Introduction to Superconductivity*, Pergamon Press, Oxford.
46. Schwab, K. C. and Roukes, M. L. (2005). *Physics Today* **58** 36.
47. Sewell, G. L. (2002). *Quantum Mechanics and its Emergent Macrophysics*, Princeton University Press, Princeton, N. J.
48. van Fraassen, B. C. (1991). *Quantum Mechanics, An Empiricist View*, Clarendon Press, Oxford.
49. Wan, K. K. (1980). *Can. J. Phys.* **58** 976.
50. Wan, K. K., Bradshaw, J., Trueman, C. and Harrison, F. E. (1998). *Found. Phys.* **28** 1739.
51. Wan, K. K. and Fountain, R. H. (1996). *Found. Phys.* **26** 1165.
52. Wan, K. K., and Harrison, F. E. (1997). *J. Phys. A: Math. Gen.* **30** 4731.
53. Wan, K. K. and McLean, R. G. D. (1984a). *J. Phys. A: Math. Gen.* **17** 825. Corrigenda **17** 2363 (1984).

54. Wan, K. K. and McLean, R. G. D. (1984b). *J. Phys. A: Math. Gen.* **17** 837. Corrigenda **17** 2363 (1984).
55. Wick, G. C., Wightman, A. S. and Wigner E. P. (1952). *Phys. Rev.* **88** 101.
56. Wightman, A. S. and Glance, N. (1989). *Nucl. Phys. B (Proc Suppl)* **6** 202.
57. Williams, H., J. (1982). *Introduction to Organic Chemistry*, Wiley, New York.
58. Zurek, W H. (1981). *Phys. Rev. D* **24** 1516.
59. Zurek, W H. (1982). *Phys. Rev. D* **26** 1862.
60. Zurek, W H. (1991). *Physics Today* April 81-90.

Chapter 5

Generalized Quantum Mechanics

5.1 Introduction

Generally in a physical theory observables and states are independent quantities. We could have an abstract mathematical description of observables which may appear to have no obvious physical meaning by themselves; the same is true for the description of states. These descriptions become physically meaningful through the probability measures they generate for the measured values of observables in various states. A mere existence of probability measures is not enough though. We further require that for each given observable there should be *a sufficiently large set of states* such that the probability measures they generate should produce **finite** expectation values and **finite** variances. An infinite expectation value is not useful and a finite expectation value is not very meaningful if the associated variance is infinite.[1]

Quantum mechanics is based on the mathematical framework of a Hilbert space $\mathcal{H}$. Here observables are not directly related to states. It is therefore necessary to spell out how to generate a probability measure. In orthodox quantum mechanics this is done in Postulate OQS in §3.2.1. The mathematical basis for the postulate is provided in the discussions in §2.11.1 which tell us that a selfadjoint operator $\widehat{A}$ together with a unit vector ϕ in the domain $\mathcal{D}(\widehat{A})$ of $\widehat{A}$ gives us three pieces of physical information:

1. A unique probability measure and its associated probability function for

---

[1] The precise meaning of *a sufficently large set of states* should be made clear in any specific theory.

the values of observable $\widehat{A}$ in state ϕ:

$$\mathcal{M}_\phi^{\hat{A}}(\Lambda) = \langle \phi \mid \widehat{M}^{\hat{A}}(\Lambda)\phi \rangle, \quad \text{and} \quad \mathcal{F}_\phi^{\hat{A}}(\tau) = \langle \phi \mid \widehat{F}^{\hat{A}}(\tau)\phi \rangle. \tag{5.1}$$

2. A finite expectation value:

$$\mathcal{E}(\hat{A},\phi) = \mathcal{E}(\mathcal{F}_\phi^{\hat{A}}) = \langle \phi \mid \widehat{A}\phi \rangle < \infty. \tag{5.2}$$

3. A finite variance

$$\mathcal{V}(\hat{A},\phi) = \mathcal{V}(\mathcal{F}_\phi^{\hat{A}}) = \left\| \widehat{A}\phi \right\|^2 - \langle \phi \mid \widehat{A}\phi \rangle^2 < \infty. \tag{5.3}$$

The crucial quantity here is the PV measure $\widehat{M}^{\hat{A}}$ or its associated PV function $\widehat{F}^{\hat{A}}$ from which everything can be calculated. It is then quite appropriate and meaningful to say that an observable corresponds to a PV measure or to a PV function rather than to a selfadjoint operator. Moreover there is a dense set of states, i.e., $\mathcal{D}(\widehat{A})$, on which the observable has finite expectation values and variances. In the context of quantum mechanics a *sufficiently large set of states* means a dense set of states in the Hilbert space.

So far we have assumed measuring processes, such as those presented in §3.6, which can achieve arbitrary accuracies. It is of interest to see what would happen if we employ a measuring process having a finite resolution ϵ, i.e., the value τ produced by such a measuring process is a nominal value signifying only that the measured value lies in the range $(\tau-\epsilon, \tau+\epsilon)$.[2] Consider a position measurement of a system in one-dimension. Suppose we want to determine the position probability function $\mathcal{F}_\phi^{\hat{x}}(\tau)$ in state ϕ. This is the probability of a measured position value lying in the range $(-\infty, \tau]$. We have, according to Eq. (2.486),

$$\mathcal{F}_\phi^{\hat{x}}(\tau) = \langle \phi \mid \widehat{F}^{\hat{x}}(\tau)\, \phi \rangle = \int_{-\infty}^{\tau} |\phi(x)|^2\, dx. \tag{5.4}$$

We can carry out a yes-no experiment of the spectral projector $\widehat{F}^{\hat{x}}(\tau)$. With a measuring process of finite resolution, there is a chance that one is measuring $\widehat{F}^{\hat{x}}(\tau')$ instead, where τ' may be bigger or smaller than τ. So, a yes result produced by such a measuring process could in fact be a yes result for a position value in the range $(-\infty, \tau']$. Suppose there is a degree of randomness in this so that we can describe the situation probabilistically. In other words we assume that such a measuring process has a probability density function $f(\epsilon)$, referred to as a *confidence function*, with properties

---

[2] As pointed out in §3.6.2 (footnote) the traditional Stern-Gerlach experiment for spin measurement is one such example.

5.1. INTRODUCTION

1. $f(\epsilon)$ is an even function of ϵ, i.e., $f(\epsilon) = f(-\epsilon)$,

2. $f(\epsilon)$ peaks at $\epsilon = 0$,

3. $f(\epsilon)$ possesses a finite expectation value $\mathcal{E}(f)$ and a finite variance $\mathcal{V}(f)$,

associated with it so that an attempt to obtain the value of $\mathcal{F}_\phi^{\hat{x}}(\tau)$ would lead to various possible values $\mathcal{F}_\phi^{\hat{x}}(\tau + \epsilon)$ with a probability distribution given by $f(\epsilon)$. The average of $\mathcal{F}_\phi^{\hat{x}}(\tau+\epsilon)$ over various possible values of ϵ, to be denoted by $\mathcal{F}_\phi^{\hat{x}f}(\tau)$, is given by

$$\mathcal{F}_\phi^{\hat{x}f}(\tau) = \int_{-\infty}^{\infty} f(\epsilon)\, \mathcal{F}_\phi^{\hat{x}}(\tau + \epsilon)\, d\epsilon. \tag{5.5}$$

A measuring process capable of arbitrary accuracy would correspond to

$$f(\epsilon) = \delta(\epsilon), \quad \text{namely} \quad \mathcal{F}_\phi^{\hat{x}\delta}(\tau) = \mathcal{F}_\phi^{\hat{x}}(\tau). \tag{5.6}$$

Since $\mathcal{F}_\phi^{\hat{x}f}(\tau)$ is generally different from $\mathcal{F}_\phi^{\hat{x}}(\tau)$ we can say that a measuring process of finite resolution does not actually measure $\widehat{F}^{\hat{x}}(\tau)$. If we attempt to employ a similar measuring process of finite resolution to measure the position observable[3] we would in effect be measuring a different observable which is not represented by the selfadjoint operator $\hat{x}$. Intuitively we would expect such a measuring process of finite resolution to measure a certain *approximate observable* to $\hat{x}$. Let us denote this measuring process-dependent and as yet undefined approximate observable by $\hat{x}_f$. We say that a measuring process of finite resolution measures observable $\hat{x}_f$ in the sense that the measurement yields the probability function $\mathcal{F}_\phi^{\hat{x}f}(\tau)$.[4]

Applying the above analysis to a general observable $\widehat{A}$ we can see that an attempt to obtain the probability function $\mathcal{F}_\phi^{\hat{A}}(\tau)$ by measuring $\widehat{F}^{\hat{A}}(\tau)$ using a measuring process of finite resolution would yield a probability function

$$\mathcal{F}_\phi^{\hat{A}f}(\tau) = \int_{-\infty}^{\infty} f(\epsilon)\, \mathcal{F}_\phi^{\hat{A}}(\tau + \epsilon)\, d\epsilon. \tag{5.7}$$

Consequently a measuring process of finite resolution may be formally regarded as measuring an approximate observable $\widehat{A}_f$ to $\widehat{A}$.

---

[3] As shown in Eq. (3.314) a measurement of $\hat{x}$ is obtained by measuring the spectral projectors

$$\widehat{M}^{\hat{x}}((\tau_{j-1}, \tau_j]) = \widehat{F}^{\hat{x}}(\tau_j) - \widehat{F}^{\hat{x}}(\tau_{j-1}).$$

[4] Ali and Emch (1974), Davies (1976).

Following Eq. (2.486) for $\mathcal{F}_\phi^{\hat{A}}(\tau)$ let us rewrite Eq. (5.7) in the form of an expectation value, i.e.,

$$\mathcal{F}_\phi^{\hat{A}_f}(\tau) = \langle \phi \mid \widehat{F}_+^{\hat{A}_f}(\tau)\phi \rangle, \tag{5.8}$$

where

$$\widehat{F}_+^{\hat{A}_f}(\tau) = \int_{-\infty}^{\infty} f(\epsilon) \, \widehat{F}^{\hat{A}}(\tau + \epsilon) \, d\epsilon. \tag{5.9}$$

In other words we have a new operator-valued function $\widehat{F}_+^{\hat{A}_f}$ from which the probability function can be obtained in the same way as probability functions are obtained from a PV function. An examination reveals that generally $\widehat{F}_+^{\hat{A}_f}$ is not a PV function. Instead it is a generalized spectral function (POV function) as defined in Definition 2.9(1).[5] These results strongly suggest the following physical interpretation:

1. A measuring process of finite resolution measures an *approximate observable* characterized by a POV function $\widehat{F}_+^{\hat{A}_f}$. We may write down a formal expression for the observable as a symmetric operator having $\widehat{F}_+^{\hat{A}_f}$ as its generalized spectral function in the sense of Definition 2.10.1(1) and C2.10.1(1) C2, i.e.,

$$\widehat{A}_f = \int_{-\infty}^{\infty} \tau \, d\widehat{F}_+^{\hat{A}_f}(\tau). \tag{5.10}$$

2. The probability function for the values of $\widehat{A}_f$ in state ϕ is given by

$$\mathcal{F}_\phi^{\hat{A}_f}(\tau) = \langle \phi \mid \widehat{F}_+^{\hat{A}_f}(\tau)\phi \rangle. \tag{5.11}$$

3. The observable $\widehat{A}$ and its associated approximate observables $\widehat{A}_f$ can be shown to possess identical expectation values in the same state. The differences in the probability functions $\mathcal{F}_\phi^{\hat{A}}(\tau)$ and $\mathcal{F}_\phi^{\hat{A}_f}(\tau)$ manifest themselves in the approximate observables having a bigger variance introduced by the finite resolution of the measuring process. In other words we have

$$\mathcal{E}(\widehat{A}, \phi) = \mathcal{E}(\mathcal{F}_\phi^{\hat{A}}) \;=\; \mathcal{E}(\mathcal{F}_\phi^{\hat{A}_f}) = \mathcal{E}(\widehat{A}_f, \phi), \tag{5.12}$$

$$\mathcal{V}(\widehat{A}, \phi) = \mathcal{V}(\mathcal{F}_\phi^{\hat{A}}) \;\leq\; \mathcal{V}(\mathcal{F}_\phi^{\hat{A}_f}) = \mathcal{V}(\widehat{A}_f, \phi). \tag{5.13}$$

---

[5] Ali and Emch (1974), Davies (1976).

5.1. INTRODUCTION

The conventional Stern-Gerlach experiment for spin measurement is often cited as a paradigm of this situation. The probability distribution of the measured values for such an experiment should then be described by a POV measure rather than a PV measure.[6] Since a POV measure is associated with a symmetric operator we should then associate a symmetric operator with an approximate observable.

Our present discussion shows that symmetric operators can play a role in quantum theory. The need for an extension from PV measures to POV measures in quantum theory is recognized by many authors.[7] Many more reasons may be put forward in favour of such an extension. Some often cited ones are:

1. Many measuring processes are of a finite resolution.[8]

2. The need for an extension beyond selfadjoint operators emerges very clearly from the difficulties in quantizing incomplete momenta in §3.3.5 and §3.3.7. One would like to ask why an incomplete momentum such as the radial momentum should not have a counterpart in quantum mechanics.

3. A desire to formulate a quantum theory in phase space.[9]

4. The need for a quantum theory for massless particles. There are great difficulties in formulating a selfadjoint position operator for photons.[10]

5. The desire for a theory of joint measurement of incompatible observables.[11]

However, instead of listing specific reasons, it would be more satisfying if one can put forward an analysis of the concept of observables[12] which would automatically include symmetric operators for their description. We shall do this in the next section.

---

[6] Busch and Schreock (1989). For further motivation for an extension to POV measures the reader is referred to Busch, Grabowski and Lahti (1997).

[7] Ali and Emch (1974), Davies (1976), Kraus (1977), (1983), Holevo (1982), Ludwig (1985), Prugovečki (1984), Busch, Lahti and Mittelstaedt (1996).

[8] This alone is not sufficient to warrant a fundamental amendment to Postulate OQS in §3.2.1 to include symmetric operators, given that measuring processes capable of achieving an arbitrary accuracy exist.

[9] Prugovečki (1984), Schroeck (1996).

[10] Kraus (1977), (1983) and (1900), Comi (1980), Jordan (1978).

[11] Davies (1976) p. 16, Busch and Lahti (1984), Busch (1985), Uffink (1994).

[12] Grabowski (1989), Wan, Fountain and Tao (1995), Fountain (1995).

5.2 Maximal Symmetric Operators and Observables

5.2.1 Observables: Concept and description

Conceptually an observable is a property of a physical system which can manifest itself quantitatively in the form of numerical values when the system interacts with certain other systems, the other systems constituting the measuring device. An observable together with a state should give us the three pieces of physical information listed in Eqs. (5.1), (5.2) and (5.3) for a sufficiently large set of states. Orthodox quantum theory employs selfadjoint operators for the representation of observables for the obvious reason that a selfadjoint operator with its unique PV function unambiguously gives us the three desirable pieces of information for a dense set of states. A symmetric operator generally has no unique POV function associated with it, as pointed out in C2.10.1(1) C1, and hence no unique probability function can be generated in a given state. To avoid this non-uniqueness one may attempt to employ POV functions directly to represent observables, bypassing symmetric operators altogether. However, there is a serious problem with this due to the fact pointed out in C2.11.2(1) C3 that a general POV function may not lead to a probability function with a finite variance in any state. This contradicts the requirement set out in Eq. (5.3). We desire that every observable should have finite expectation values and finite variances in a sufficiently large set of states. An observable which has infinite variances in every state would not be physically meaningful. A natural approach is to find a restricted set of POV functions to represent observables.

Let us start with a POV function $\widehat{F}_+(\tau)$ in a Hilbert space $\mathcal{H}$. From Theorem 2.11.2(1) we know that each unit vector $\phi \in \mathcal{H}$ gives rise to a probability function $\mathcal{F}_\phi(\tau) = \langle \phi \mid \widehat{F}_+(\tau)\phi \rangle$. It follows that a family of probability functions $\mathcal{F}_\phi$, one for each unit vector ϕ in $\mathcal{H}$, is generated by a POV function. We shall denote such a set of probability functions generated by a POV function by[13]

$$\mathbf{F}(\mathcal{H}) = \{ \mathcal{F}_\phi : \|\phi\| = 1, \phi \in \mathcal{H} \}, \tag{5.14}$$

and call it

a family of probability functions on the Hilbert space $\mathcal{H}$.

---

[13] Two unit vectors differing by a constant phase factor generate the same probability function. Note that we shall confine ourselves to probability functions generated by POV functions. In orthodox quantum theory the relationship between probability functions and projectors is given by Gleason's Theorem (Isham (1995) §9.2.4).

5.2. MAXIMAL SYMMETRIC OPERATORS AND OBSERVABLES

Definition 5.2.1(1)   **Maximal families of probability functions**

- Let $\mathbf{F}(\mathcal{H})$ be a family of probability functions $\mathcal{F}_\phi$ on $\mathcal{H}$. If there exists a dense linear subset $\mathcal{D}$ in $\mathcal{H}$ such that for every unit vector ϕ in $\mathcal{D}$ we have

$$\mathcal{E}(\mathcal{F}_\phi) = \int_{-\infty}^{\infty} \tau \, d\mathcal{F}_\phi(\tau) < \infty, \tag{5.15}$$

$$\mathcal{V}(\mathcal{F}_\phi) = \int_{-\infty}^{\infty} \left(\tau - \mathcal{E}(\mathcal{F}_\phi)\right)^2 d\mathcal{F}_\phi(\tau) < \infty, \tag{5.16}$$

then $\mathbf{F}(\mathcal{H})$ is said to have finite expectation values and variances on $\mathcal{D}$. A family of probability functions having finite expectation values and variances on a dense linear subset $\mathcal{D}$ is denoted by $\mathbf{F}(\mathcal{H}, \mathcal{D})$.

- A family $\mathbf{F}(\mathcal{H}, \mathcal{D})$ of probability functions $\mathcal{F}_\phi$ on a Hilbert space $\mathcal{H}$ is called a *maximal family of probability functions* on the Hilbert space $\mathcal{H}$ if given any other family $\mathbf{F}'(\mathcal{H}, \mathcal{D})$ of probability functions $\mathcal{F}'_\phi$ on $\mathcal{H}$ with the same expectation values on the same linear dense set $\mathcal{D}$, i.e.,

$$\mathcal{E}(\mathcal{F}'_\phi) = \mathcal{E}(\mathcal{F}_\phi) \quad \forall \phi \in \mathcal{D}, \tag{5.17}$$

we have either

$$\mathcal{F}'_\phi = \mathcal{F}_\phi \quad \forall \phi \in \mathcal{D} \tag{5.18}$$

or

$$\mathcal{V}(\mathcal{F}'_\phi) \geq \mathcal{V}(\mathcal{F}_\phi) \; \forall \phi \in \mathcal{D} \quad \text{and} \quad \mathcal{V}(\mathcal{F}'_\phi) > \mathcal{V}(\mathcal{F}_\phi) \; \text{for some} \; \phi \in \mathcal{D}. \tag{5.19}$$

In view of Eqs. (5.12) and (5.13) we would naturally consider:

1. A maximal family of probability functions $\mathbf{F}(\mathcal{H}, \mathcal{D})$ as arising from an observable which possesses a finite expectation value and a corresponding finite variance in every state $\phi \in \mathcal{D}$.

2. A non-maximal family $\mathbf{F}'(\mathcal{H}, \mathcal{D})$ having the same expectation values on $\mathcal{D}$, i.e., satisfying Eq. (5.17), as generated by an approximate observable. The increase in variance can be attributed to the use of a measuring process of finite resolution as evident in Eq. (5.13).

We shall formalize this interpretation in a postulate later. Before that we want to find out what kind of POV functions will generate a maximal family of probability functions. The answer is provided in the Theorem 5.2.1(1).

Theorem 5.2.1(1) Maximal families and maximal symmetric operators[14]

- Every maximal family of probability functions is generated by the POV function $\widehat{F}_+^{\widehat{S}}(\tau)$ of a maximal symmetric operator $\widehat{S}$ by

$$\mathcal{F}_\phi^{\widehat{S}}(\tau) = \langle \phi \mid \widehat{F}_+^{\widehat{S}}(\tau) \phi \rangle. \tag{5.20}$$

- The POV function $\widehat{F}_+^{\widehat{S}}(\tau)$ of every maximal symmetric operator $\widehat{S}$ generates a maximal family of probability functions by Eq. (5.20).

Corollary 5.2.1(1) Maximal families and maximal symmetric operators

- Maximal families of probability functions on a Hilbert space $\mathcal{H}$ correspond one-to-one to maximal symmetric operators in $\mathcal{H}$.

Now it makes sense to denote a maximal family of probability function by $\mathbf{F}^{\widehat{S}}(\mathcal{H}, \mathcal{D}(\widehat{S}))$, where $\mathcal{D}(\widehat{S})$ is the domain of $\widehat{S}$. This notation signifies the unique relationship between a maximal family of probability functions and a maximal symmetric operator $\widehat{S}$. The relationship is based on the following properties:

1. Members of the family $\mathcal{F}_\phi^{\widehat{S}}$ are given by Eq. (5.20).

2. The family gives rise to a finite expectation values $\mathcal{E}(\mathcal{F}_\phi^{\widehat{S}}) = \langle \phi \mid \widehat{S}\phi \rangle$ for every $\phi \in \mathcal{D}(\widehat{S})$.

3. The family gives rise to a finite variance $\mathcal{V}(\mathcal{F}_\phi^{\widehat{S}}) = ||\widehat{S}\phi||^2 - \mathcal{E}(\mathcal{F}_\phi^{\widehat{A}})^2$ for every $\phi \in \mathcal{D}(\widehat{S})$.

Each family $\mathbf{F}^{\widehat{S}}(\mathcal{H}, \mathcal{D}(\widehat{S}))$ should correspond to an observable since the three pieces of information in Eqs. (5.1), (5.2) and (5.3) associated with an observable are embodied in $\mathbf{F}^{\widehat{S}}(\mathcal{H}, \mathcal{D}(\widehat{S}))$. For each given $\mathbf{F}^{\widehat{S}}(\mathcal{H}, \mathcal{D}(\widehat{S}))$ there may well be other non-maximal families

$$\mathbf{F}'(\mathcal{H}, \mathcal{D}(\widehat{S})), \quad \mathbf{F}''(\mathcal{H}, \mathcal{D}(\widehat{S})), \quad \mathbf{F}'''(\mathcal{H}, \mathcal{D}(\widehat{S})), \ldots \tag{5.21}$$

---

[14] In order not to interrupt the physical arguments we have placed the proof of this theorem in §5.7 near the end of this chapter.

5.2. MAXIMAL SYMMETRIC OPERATORS AND OBSERVABLES   403

of probability functions giving the same expectation values on the same dense linear subset $\mathcal{D}(\widehat{S})$. These probability functions will lead to bigger variances generally. We can interpret the difference in the variances as arising from the use of measuring processes with a finite resolution, e.g., $\mathbf{F}'(\mathcal{H}, \mathcal{D}(\widehat{S}))$ may be generated by a POV function in the form of Eq. (5.9). The family $\mathbf{F}^{\widehat{S}}(\mathcal{H}, \mathcal{D}(\widehat{S}))$ with the minimum variances corresponds to measurements made by a measuring process capable of arbitrary accuracy. We shall pursue this in more details when we discuss approximate observables in §5.3.1.

We conclude that an observable is conceptually characterizable by a maximal family of probability functions on a Hilbert space with the different probability functions in the family corresponding to different states of the system. This enables us to arrive at the following mathematical description of observables.

Mathematical representation of observables

- An observable is describable by a maximal symmetric operator $\widehat{S}$, or equivalently by its POV measure $\widehat{M}_+^{\widehat{S}}$ or its POV function $\widehat{F}_+^{\widehat{S}}$, in the sense that the corresponding maximal family $\mathbf{F}^{\widehat{S}}(\mathcal{H}, \mathcal{D}(\widehat{S}))$ of probability functions $\mathcal{F}_\phi^{\widehat{S}}$ are generated by $\widehat{F}_+^{\widehat{S}}$ in accordance with Eq. (5.20). The resulting expectation value and variance for any unit vector $\phi \in \mathcal{D}(\widehat{S})$ are obtainable directly in terms of $\widehat{S}$ by

$$\mathcal{E}(\mathcal{F}_\phi^{\widehat{S}}) = \langle \phi \mid \widehat{S}\phi \rangle \quad \text{and} \quad \mathcal{V}(\mathcal{F}_\phi^{\widehat{S}}) = ||\widehat{S}\phi||^2 - \mathcal{E}(\mathcal{F}_\phi^{\widehat{S}})^2. \tag{5.22}$$

Compared with orthodox quantum mechanics in Chapter 3 this concept of observables, taking their representation into the set of maximal symmetric operators, is a generalization. Within this set we have selfadjoint operators and strictly maximal symmetric operators. Both these operators play the same role in providing uniqueness and the same expressions for various physical quantities, e.g, expectation values and variances. For simplicity as well as clarity we shall start with a generalization of orthodox quantum mechanics without superselection rules, i.e., we want to incorporate maximal symmetric operators formally into a generalized postulate to replace Postulate OQS in §3.2.1.

Postulate GQS on generalized quantum statics

- This is as Postulate OQS in §3.2.1 except with selfadjoint operators replaced by maximal symmetric operators.

We can also replace maximal symmetric operators by their POV functions or their POV measures in the above postulate. The notion of observables here is more restrictive than the more liberal statement identifying all POV measures with observables adopted by many authors. Our main motivation here is to formulate a generalized quantum mechanics to accommodate more observables, e.g., incomplete momentum discussed in §3.3.5, so that the resulting theory can be applied to more general physical systems, e.g., quantum circuit systems. The inclusion of maximal symmetric operators as observables in Postulate GQS is sufficient for our purposes. We can now accommodate quantized incomplete momenta such as the radial momentum operator presented in E3.3.5(1) E2, the momentum of a particle confined in the half line $I\!\!E^+$ given in E3.3.5(1) E3 and the momentum for a Y-shaped quantum circuit to be discussed in §7.7 in Chapter 7. Another obvious application is the emergence of a *time operator* within the theory, a topic to be discussed later in §5.5.

We can expect some fundamental differences between observables represented by selfadjoint operators and observables described by strictly maximal symmetric operators. Before proceeding further on this let us introduce some definitions first.

Definition 5.2.1(2)   Intrinsic and extrinsic observables

- **Intrinsic observables**   Observables represented by maximal symmetric operators as stated in Postulate GQS are called *intrinsic observables*. There are two kinds of intrinsic observables:

    1. *Sharp intrinsic observables*   These observables correspond one-to-one to selfadjoint operators.
    2. *Unsharp intrinsic observables*   These observables correspond one-to-one to strictly maximal symmetric operators.[15]

- **Extrinsic observables**   These are approximate observables in §5.1 arising from measuring processes of finite resolution. These are also called *unsharp extrinsic observables*. They are describable in terms of non-maximal symmetric operators, e.g., $\widehat{A}_f$ in Eq. (5.10) and their POV functions. In §5.3.1 we shall extend this group of extrinsic observables to include approximations to observables represented by strictly maximal symmetric operators.

Bearing in mind that a maximal symmetric operator is either selfadjoint or

---

[15] Our definitions on sharpness and unsharpness here are different from the more liberal use of the terms by other authors.

5.2. MAXIMAL SYMMETRIC OPERATORS AND OBSERVABLES

strictly maximal symmetric some important physical characteristics of various types of observables are listed as follows:

1. The spectral measure of a selfadjoint operator $\widehat{A}$ consists of spectral projectors $\widehat{M}^{\widehat{A}}(\Lambda)$. It follows that a state ϕ exists such that

$$\mathcal{M}_\phi^{\widehat{A}}(\Lambda) = \langle \phi \mid \widehat{M}^{\widehat{A}}(\Lambda) \phi \rangle = 1, \quad (5.23)$$

unless $\widehat{M}^{\widehat{A}}(\Lambda) = 0$. In other words there exist states in which the observable $\widehat{S}$ possesses a value either in Λ or in its complement $\Lambda^c = \mathbb{R} - \Lambda$ with certainty.[16] It is this certainty which is the basis for the concept of *sharpness* in Definition 5.2.1(2).

2. The generalized spectral measure of a strictly maximal symmetric operator $\widehat{S}$ consists of positive operators $\widehat{M}_+^{\widehat{S}}(\Lambda)$ which are not projectors for some Borel sets Λ and for these Borel sets

$$\| \widehat{M}^{\widehat{S}}(\Lambda) \| < 1 \quad (5.24)$$

so that states satisfying Eq. (5.23) do not exist, i.e., we have

$$\mathcal{M}_\phi^{\widehat{S}}(\Lambda) = \langle \phi \mid \widehat{M}_+^{\widehat{S}}(\Lambda) \phi \rangle < 1. \quad (5.25)$$

The notion of *unsharpness* in Definition 5.2.1(2) refers to this lack of certainty in predicting a value in these Borel sets. For an unsharp observable we simply cannot prepare a state in which the observable will be found with certainty to have a value in one of these Borel sets.

3. A non-maximal symmetric operator $\widehat{S}_0$ does not determine a unique spectral function; therefore it does not by itself represent an observable in our present theory as given by Postulate GQS. However $\widehat{S}_0$ does generate observables in the form of its maximal symmetric extensions $\widehat{S}$. Moreover, $\widehat{S}_0$ can be regarded as the restriction to domain $\mathcal{D}(\widehat{S}_0)$ of its maximal extensions $\widehat{S}$ in that for states in $\mathcal{D}(\widehat{S}_0)$ we can use the symmetric operator directly to evaluate expectation values and variances, namely for all $\phi \in \mathcal{D}(\widehat{S}_0)$ we have:

$$\mathcal{E}(\mathcal{F}_\phi^S) = \langle \phi \mid \widehat{S}_0 \phi \rangle, \quad \mathcal{V}(\mathcal{F}_\phi^S) = \| \widehat{S}_0 \phi \|^2 - \langle \phi \mid \widehat{A}_0 \phi \rangle^2. \quad (5.26)$$

Generally different maximal extensions show themselves in different probability distributions since they possess distinct POV functions $\widehat{F}_+^{\widehat{S}}$.

---

[16] When $\widehat{M}^{\widehat{A}}(\Lambda) = 0$ we can find ϕ such that $\langle \phi \mid \widehat{M}^{\widehat{A}}(\Lambda^c) \phi \rangle = 1$.

4. The concept of unsharpness can be extended to cover extrinsic observables. This means that we have two kind of unsharpness:

   (a) **Intrinsic unsharpness**   This arises from strictly maximal symmetric nature of unsharp intrinsic observables. The radial momentum operator is a typical example.

   (b) **Extrinsic unsharpness**   This relates to extrinsic observables. Here unsharpness is due to experimental outcomes registered by measuring processes of finite resolution, i.e., we do not have perfect confidence in the measured values, the traditional Stern-Gerlach measurement of spin being an example. We call both extrinsic observables and their measuring processes unsharp.

5.2.2 Measurement of intrinsically unsharp observables

In §3.6 we discussed how, in principle, a sharp observable represented by a selfadjoint operator can be measured in terms of local position measurements through a process of spectral separation. We can apply the same procedure to measure an unsharp observable represented by a strictly maximal symmetric operator $\widehat{S}$. The idea is to reduce everything to the measurement of observables represented by selfadjoint operators.

The expectation value of an unsharp observable $\widehat{S}$ in state ϕ can be discretized, as done in §3.2.3, as

$$\mathcal{E}(\widehat{S}, \phi) = \int_{\infty}^{\infty} \tau \, d\langle \phi \mid \widehat{F}_+^{\hat{S}} \phi \rangle \approx \int_{-N}^{N} \tau \, d\langle \phi \mid \widehat{F}_+^{\hat{S}} \phi \rangle \qquad (5.27)$$

$$\approx \sum_j \int_{\Lambda_j} \tau \, d\langle \phi \mid \widehat{F}_+^{\hat{S}} \phi \rangle \qquad (5.28)$$

$$= \sum_j \tau_j \, \langle \phi \mid \widehat{M}_+^{\hat{S}}(\Lambda_j) \, \phi \rangle, \qquad (5.29)$$

where $\{\Lambda_j\}$ is a partition of the interval $[-N, N]$. The quantity $\langle \phi \mid \widehat{M}_+^{\hat{S}}(\Lambda_j) \, \phi \rangle$ is the expectation value of selfadjoint operator $\widehat{M}_+^{\hat{S}}(\Lambda_j)$ to which our previous measurement theory applies.[17]

---

[17] There are other ways of measuring unsharp observables. For example we could couple the system with an ancillary system to reduce the problem to measurement of projectors (Fuches (2001)).

5.3 Approximate and Related Observables

5.3.1 Approximate observables

Let $\widehat{S}$ be a strictly maximal symmetric operator representing an observable. For a unit vector $\phi \in \mathcal{D}(\widehat{S})$ we have the probability function $\mathcal{F}_\phi^{\widehat{S}}(\tau) = \langle \phi \mid \widehat{F}_+^{\widehat{S}}(\tau) \phi \rangle$ for the measured values of the observable. The analysis in §5.1 can be applied to deal with the measurement of $\widehat{S}$ using measuring processes with a finite resolution. By assuming the existence of a confidence function f with properties as specified in §5.1 for the measuring process we obtain a new probability function

$$\mathcal{F}_\phi^{\widehat{S}_f}(\tau) = \int_{-\infty}^{\infty} f(\epsilon)\, \mathcal{F}_\phi^{\widehat{S}}(\tau + \epsilon)\, d\epsilon. \tag{5.30}$$

Following Eq. (5.9) we can introduce a new POV function $\widehat{F}_+^{\widehat{S}_f}$ by

$$\widehat{F}_+^{\widehat{S}_f}(\tau) = \int_{-\infty}^{\infty} f(\epsilon)\, \widehat{F}_+^{\widehat{S}}(\tau + \epsilon)\, d\epsilon \tag{5.31}$$

so that the probability function can be obtained directly from $\widehat{F}_+^{\widehat{S}_f}(\tau)$:

$$\mathcal{F}_\phi^{\widehat{S}_f}(\tau) = \langle \phi \mid \widehat{F}_+^{\widehat{S}_f}(\tau) \phi \rangle. \tag{5.32}$$

The expectation value $\mathcal{E}(\mathcal{F}_\phi^{\widehat{S}_f})$ and the variance $\mathcal{V}(\mathcal{F}_\phi^{\widehat{S}_f})$ can also be calculated from the probability function $\mathcal{F}_\phi^{\widehat{S}_f}$ in the standard manner. The inaccuracy of the measuring process leads to an apparent change of the probability function which results in an increase in the variance. However, the properties of f means that the average value of the observable is unaffected. The resulting approximate observable $\widehat{S}_f$ may be written down symbolically as

$$\widehat{S}_f = \int_{-\infty}^{\infty} \widehat{F}_+^{\widehat{S}_f}(\tau)\, d\tau. \tag{5.33}$$

Here we can again see the motivation in Definition 5.2.1(1) of a maximal family $\mathbf{F}^{\widehat{S}}(\mathcal{H}, \mathcal{D}(\widehat{S}))$ of probability functions. With different confidence functions a host of other families $\mathbf{F}^{\widehat{S}_f}(\mathcal{H}, \mathcal{D}(\widehat{S}))$ of probability functions $\mathcal{F}_\phi^{\widehat{S}_f}$ can be generated by unit vectors in the same dense linear subset $\mathcal{D}(\widehat{S})$ on which they give the same expectation values as that of the original observable $\widehat{S}$. The original observable $\widehat{S}$ leads to the smallest variance in every state in

$\mathcal{D}(\widehat{A})$. But $\mathbf{F}^{S_f}(\mathcal{H}, \mathcal{D}(\widehat{S}))$ is not a maximal family and it therefore does not correspond to a maximal symmetric operator. In keeping with Definition 5.2.1(2) we call $\widehat{S}_f$ an extrinsic observables.

The significance or otherwise of approximate observables depends on the nature of the measuring processes involved. Even in the realm of classical physics a measuring process would have inherent inaccuracy. The situation is even more obvious in classical statistical physics where even the physical systems, e.g., thermodynamical systems, are realizable only approximately. However, the fundamental issue is not that of the existence of inaccuracy, but that of whether the inaccuracy can be arbitrarily reduced. Hence we shall regard approximate observables as a useful but less fundamental generalization of orthodox quantum mechanics. This is why extrinsic observables are not formally included in Postulate GQS.

5.3.2 Related family of observables

The fact that a measuring processes has a finite resolution also means that it may well be impossible to distinguish a related set of observables. Let us illustrate this with a simple example.

Recall the case of a particle confined to the interval $\Lambda = (a, b)$ of $I\!\!E$ by an infinite square potential well. As pointed out in §3.3.7 the particle's momentum is incomplete so that when quantized as $\widehat{p}_0(\Lambda)$ acting on $C_0^\infty(\Lambda)$ it is not essentially selfadjoint. Examples 2.5.1(1) E1 tells us that there is a one-parameter family of selfadjoint extensions $\widehat{p}_\lambda(\Lambda)$. These extensions represent a one-parameter family of distinct observables, i.e., $\widehat{p}_\lambda(\Lambda)$ and $\widehat{p}_{\lambda'}(\Lambda)$ with $\lambda \neq \lambda'$ represent two different observables with two distinct sets of eigenvalues given by Eq. (2.168). We call this a *closely related family of observables* because the spectrum of $\widehat{p}_\lambda(\Lambda)$ can be arbitrarily close to that of $\widehat{p}_{\lambda'}(\Lambda)$ as λ' gets close to λ.

Suppose one sets out to try to measure a particular member of this family $\widehat{p}_\lambda(\Lambda)$ using a measuring process with a finite resolution one would not be able to distinguish this chosen observable $\widehat{p}_\lambda(\Lambda)$ from a neighbouring one, e.g., $\widehat{p}_{\lambda'}(\Lambda)$ with λ' sufficiently close to λ. Two confidence functions have to be introduced to establish a probability function for the nominal values recorded by the measuring device. First, we have a confidence function f for the usual inaccuracy incurred, assuming $\widehat{p}_\lambda(\Lambda)$ is being measured. Secondly, we must have a new confidence function g to account for the uncertainty as to which observable, e.g., $\widehat{p}_\lambda(\Lambda)$ or $\widehat{p}_{\lambda'}(\Lambda)$, is being measured.[18]

---

[18] Wan, Fountain and Tao (1995).

5.4 Implications on Quantization

Postulate GQS broadens the horizon of quantum mechanics in several ways. First let us examine the quantization problems discussed in §3.3. There is no difficulty in achieving an initial quantization map in Eq. (3.74) from a classical observable A to a symmetric operator $\widehat{A}_0$ defined on the domain $C_0^\infty(I\!\!E^n)$ in $L^2(I\!\!E^n)$. A fundamental problem arises when this initially quantized operator is not essentially selfadjoint. Then $\widehat{A}_0$ would generally admit a family of maximal symmetric operators. Orthodox quantum mechanics will be confronted with the following progressively worsening situations:

1. Selfadjointess without uniqueness and uniqueness without selfadjointness as seen in E3.3.5(1) E1.

2. Non-selfadjointness and non-uniqueness. The initially quantized operator may admit a family of strictly maximal symmetric operators.

An examination of these two cases leads to the following conclusions:

1. It is possible to accommodate selfadjointess without uniqueness with Postulate OQS. The values of observables of a classical particle are determined locally in the neighbourhood of the particle and are not affected by any distant influence. In contrast a selfadjoint operator is generally a global quantity which can include distant boundary conditions in its very definition, $\widehat{p}_\lambda(\Lambda)$ in E2.5.1(1) E1 being an example. It is therefore conceivable that a classical observable may be quantized into different quantum observables corresponding to different distant boundary conditions. We have already discussed this problem in §3.3.7. More examples will be pressented in Chapters 6 and 7. This non-uniqueness also emerges in an attempt to establish a time observable in quantum theory.

2. The problem of non-selfadjointness is more serious and cannot be solved within Postulate OQS. This problem disappears if Postulate OQS is replaced by Postulate GQS. Explicit physical examples in Chapter 7, especially in §7.7.1, which involve strictly maximal symmetric operators serve to substantiate such a replacement.

In the next two sections we shall look at some more examples of the kind of quantities Postulate GQS can bring into the theory.

5.5 Time Operators and Uncertainty Relation

As an application of Postulate GQS let us investigate the notion of a *time operator*. There have been many attempts to introduce selfadjoint *time op-*

410   CHAPTER 5. GENERALIZED QUANTUM MECHANICS

erators within orthodox quantum mechanics without success.[19] The need for a time operator arises in the context of the energy–time uncertainty relation which is usually presented as

$$\Delta t \Delta E \geq \frac{\hbar}{2}. \tag{5.34}$$

This energy–time uncertainty relation is very useful physically, for example in the study of decay and the broadening of energy levels. A striking feature is that this relation is capable of different interpretations to fit in with different physical situations. This is because Eq. (5.34) is simply not well-defined. The crux of the matter lies in the fact that an energy–time uncertainty relation cannot be defined in the same way as the position–momentum uncertainty relation within orthodox quantum mechanics.[20] In orthodox quantum theory the uncertainty relation for two observables $\widehat{A}$ and $\widehat{B}$ states that the uncertainties in $\widehat{A}$ and in $\widehat{B}$ in any given state ϕ are related by[21]

$$\Delta(\widehat{A},\phi)\Delta(\widehat{B},\phi) \geq \frac{1}{2} |\langle \phi \,|\, [\,\widehat{A}, \widehat{B}\,]\, \phi \rangle |. \tag{5.35}$$

The position–momentum uncertainty relation is a special case of this. However, the energy–time uncertainty relation does not follows from Eq. (5.35) because of the absence of a selfadjoint time operator $\widehat{t}$ which is canonically conjugate to a positive Hamiltonian $\widehat{H}$, i.e., the absence of a selfadjoint time operator $\widehat{t}$ satisfying[22]

$$[\,\widehat{t},\,\widehat{H}\,]\,\phi = i\hbar\,\phi \tag{5.36}$$

on a certain dense set of vectors ϕ. In other words we do not have a time observable in orthodox quantum mechanics. As a result the quantity Δt in Eq. (5.34) is ill-defined. Indeed different authors have different definitions for Δt and ΔE.[23]

Our interest here is to lay the foundation for the introduction of a time observable, and hence the energy–time uncertainty relation in the context of Postulate GQS. Unlike the position and momentum operators of a particle which are fixed largely by the geometry of the physical space, the Hamiltonian

---

[19] Many other approaches to the notion of time have been proposed, from discrete time with an indivisible interval (Misra (1995)) to statistical treatment based on the relationship between time reversible dynamics and irreversible law of increase of entropy (Misra (1978), (1979), Misra, Prigogine and Courbage (1979).

[20] Aharonov and Bohm (1961), Capri (1985), Busch (1990), Peres (1993).

[21] The uncertainty $\Delta(\widehat{A},\phi)$ of an observable $\widehat{A}$ has been introduced in Definition 2.11.1(1) with a working expression given in C2.11.1(1) C3.

[22] Messiah Vol. 1 (1967) Chap IV Section 10, Allcock (1969), Recami (1977), Rayski and Rayski (1977), Price and Chissick (1977), Bauer (1983), Auletta (2001) §10.2.

[23] Messiah Vol. 1 (1967) Chap VIII §13, Rae (1981) §8.5, Capri (1985) §2.10, Merzbacher (1998) pp. 20-22.

5.5. TIME OPERATORS AND UNCERTAINTY RELATION

changes with interactions so that a time operator conjugate to the Hamiltonian should also have to change with the interactions. We have to abandon the idea that one and the same time operator is applicable to all physical systems.[24] Instead we may have different time operators for different physical systems, just like we have different Hamiltonians for different systems.[25] This can then explain the fact that the energy–time uncertainty relation has different interpretations for different systems. Each physical system has to be treated on an individual basis when it comes to the time operator. Physically this is understandable since time is about change and we conceive a notion of time, i.e., a time duration, by observing the change of certain physical systems, e.g., the movement of the hands of a clock. There is no way to measure a time interval if nothing changes in the physical world. This shows up vividly in the formulation by Mandelstam and Tamm (1945) of an uncertainty in time which we shall now describe.

Consider a system for which the Hamiltonian $\widehat{H}$ coincides with the Hamiltonian generator $\widehat{H}_g$. Let $\widehat{A}$ be a selfadjoint observable and $\Delta(\widehat{A}, \phi_t)$ be the uncertainty of $\widehat{A}$ in state ϕ_t at time t. Applying Eq. (5.35) to $\widehat{A}$ and the Hamiltonian $\widehat{H}$ we get

$$\Delta(\widehat{A}, \phi_t) \Delta(\widehat{H}, \phi_t) \geq \frac{1}{2} \left| \langle \phi_t \mid [\widehat{A}, \widehat{H}] \phi_t \rangle \right|. \tag{5.37}$$

Now introduce a new quantity $\Delta(\tau_{\widehat{A}}, \phi_t)$ in terms of the uncertainty $\Delta(\widehat{A}, \phi_t)$ and the expectation value $\mathcal{E}(\widehat{A}, \phi_t)$ of $\widehat{A}$ in state ϕ by

$$\Delta(\tau_{\widehat{A}}, \phi_t) = \frac{\Delta(\widehat{A}, \phi_t)}{\left| d\mathcal{E}(\widehat{A}, \phi_t)/dt \right|}. \tag{5.38}$$

This new quantity has the dimension of time. We may interpret:[26]

1. $d\mathcal{E}(\widehat{A}, \phi_t)/dt$ as the rate of change of $\mathcal{E}(\widehat{A}, \phi_t)$, i.e., a kind of "velocity" indicating how fast the expectation value $\mathcal{E}(\widehat{A}, \phi_t)$ changes with time.

2. $\Delta(\tau_{\widehat{A}}, \phi_t)$ as the time duration for the expectation value to change by an amount equal to the uncertainty $\Delta(\widehat{A}, \phi_t)$.

Making use of Eqs. (3.226) and (5.37) we can see that this new quantity is related to the uncertainty in the Hamiltonian by

$$\Delta(\tau_{\widehat{A}}, \phi_t) \Delta(\widehat{H}, \phi_t) \geq \frac{1}{2} \hbar. \tag{5.39}$$

---

[24] This is paradoxical since in classical non-relativistic physics, time is assumed to be absolute and independent of any motion or interaction. One could argue that different time operators may not contradict Newton's notion of absolute time though.
[25] Busch, Grabowski and Lahti (1994), Busch, Grabowski and Lahti (1997) Chapter 3 Section 4.
[26] Messiah Vol. 1 (1967) Chap VIII §13.

If we take $\Delta(\tau_{\widehat{A}}, \phi_t)$ as an uncertainty in time, then Eq. (5.38) above does represent a mathematically well-defined form of energy–time uncertainty relation which also has a definite physical meaning.

Returning to the theme of time operators let us consider the simple case of a free particle of mass m moving along $I\!\!E$. The Hamiltonian is equal to the kinetic energy, i.e., $\widehat{H}^{(0)} = \widehat{p}^2/2m$ acting in the Hilbert space $L^2(I\!\!E)$. Classically such a free particle would travel with a uniform velocity v with momentum $p = mv$. The quantity

$$t = \frac{x}{v} = m\frac{x}{p} \qquad (5.40)$$

is the *time of flight* from the origin to any position x. It has been suggested that the time operator for such a free particle should be the *quantized time of flight operator*.[27] Using the symmetrization rule of quantization given by Eq. (3.82) we formally get

$$\widehat{t} = \frac{m}{2}\left(\frac{1}{\widehat{p}}\,\widehat{x} + \widehat{x}\,\frac{1}{\widehat{p}}\right), \qquad \frac{1}{\widehat{p}} \text{ being the inverse of } \widehat{p}. \qquad (5.41)$$

This operator is well-defined since the momentum operator in $L^2(I\!\!E)$ is invertible. Moreover one can verify that $\widehat{t}$ is canonically conjugate to $\widehat{H}^{(0)}$ in the sense of Eq. (5.36). It follows that $\widehat{t}$ is a reasonable candidate for a time operator. Unfortunately $\widehat{t}$ is not selfadjoint;[28] instead it is strictly maximal symmetric.[29] This is why we do not have a time observable in orthodox quantum mechanics. However, this presents no problem in the context of Postulate GQS which accepts strictly maximal symmetric operators as intrinsic observables.[30] One can then proceed to derive an energy–time uncertainty relation which can then be interpreted in the usual way. More examples, e.g., a time operator for the simple harmonic oscillator, are available in the literature.[31]

Through its expectation value $\mathcal{E}(\widehat{t}, \phi_t)$ with respect to a non-stationary state ϕ_t, a time operator $\widehat{t}$ is related to the classical time variable t as one would expect. Using Eqs. (3.226) and (5.36) we have

$$\frac{d}{dt}\mathcal{E}(\widehat{t}, \phi_t) = \frac{1}{i\hbar}\langle \phi_t \mid [\,\widehat{t},\widehat{H}\,]\,\phi_t\rangle = 1 \quad \Rightarrow \quad \mathcal{E}(\widehat{t}, \phi_t) = t + t_0. \qquad (5.42)$$

The constant t_0 may be identified with the same quantity in Eq. (3.225).

---

[27] See, for example, Aharonov and Bohm (1961).

[28] Busch, Grabowski and Lahti (1994).

[29] To avoid interrupting the flow of arguments here we have postponed the proof of this result to §5.8.

[30] An alternative approach based POV measures are adopted by many authors, e.g., Giannitrapani (1997), Atmanspacher and Amann (1998), Busch, Grabowski and Lahti (1997).

[31] Busch, Grabowski and Lahti (1997) Chapter 3 Section 4.

5.6 Local Values in Coordinate and in Phase Spaces

5.6.1 Expectation values in terms of local values

Classical mechanics is a local theory. Physical quantities are effective in the neighbourhood of the particle in the spatial space E or in the phase space Γ.[32] The values of observables are determined locally in the neighbourhood of the particle. The question arises as to whether we can localize the values of quantum mechanical observables in some well-defined sense. There are local quantities such as local observables represented by local operators introduced in §2.12, such as local position operators χ_Λ in $L^2(E)$ which play an important role in our measurement theory in §3.6. We have also presented schemes for localizing arbitrary observables. However, it is the expectation values which are directly relevant to experimental physics. Any local nature of quantum mechanics should show up in the expectation values. To be specific consider a particle in one-dimension associated with the Hilbert space $L^2(E)$. An expectation value $\langle \phi \mid \widehat{A}\phi \rangle$ is a global quantity in the form of an integral over the entire space E. Let us pose the following question:

> *Can we regard an expectation value $\langle \phi \mid \widehat{A}\phi \rangle$ as a sum of local values distributed over either the coordinate space E or the phase space Γ?*

There are two well-known approaches to answer this question:

1. **Phase space approach** The idea originated from Wigner[33] in an attempt to express an expectation values of $\widehat{A}$ in state ϕ as the average of its classical counterpart $A(x,p)$ under a certain ϕ-dependent distribution function $w_\phi(x,p)$ on the classical phase space Γ. In other words we would express $\langle \phi \mid \widehat{A}\phi \rangle$ as the following integral over Γ:

$$\langle \phi \mid \widehat{A}\phi \rangle = \int A(x,p)\, w_\phi(x,p)\, dxdp. \qquad (5.43)$$

Two problems immediately present themselves. Firstly we may have difficulty writing down the classical counterpart $A(x,p)$, given an arbitrary selfadjoint operator $\widehat{A}$ and *vice versa*. Even if we manage to find a suitable $A(x,p)$ there is no guarantee that a probability density function $w_\phi(x,p)$ exists to make the above equation work for an arbitrary state.

---
[32] For simplicity we shall consider the particle in one-dimensional motion along E.
[33] Wigner (1932).

Indeed the function $w_\phi(x,p)$ proposed by Wigner, known as the *Wigner function*, is not even positive definite.[34]

2. **Coordinate space approach** Specific schemes exist to assign local values to certain observables in the coordinate space. For example Bohm (1952) proposed a scheme to assign a state-dependent momentum value $\pi_\phi(x)$ to a particle of mass m at each point x in the coordinate space so that the momentum expectation value $\langle \phi \mid \widehat{p}\phi \rangle$ is a sum of these local values under the probability density function $|\phi(x)|^2$ over the coordinate space $I\!\!E$. Writing the complex wave function $\phi \in L^2(I\!\!E)$ in polar form, i.e.,

$$\phi(x) = R(x)e^{iS(x)}, \tag{5.44}$$

where R and S are real-valued functions of x, Bohm makes the following state-dependent value assignments:

(a) A momentum value

$$\pi_\phi(x) = \frac{dS(x)}{dx} \tag{5.45}$$

is assigned to the particle at x. Then it is easily verified that the quantum mechanical momentum expectation value is recovered by summing this value under the probability density function $|\phi(x)|^2$, i.e.,

$$\mathcal{E}(\widehat{p}, \phi) = \langle \phi \mid \widehat{p}\phi \rangle = \int |\phi(x)|^2 \, \pi_\phi(x) \, dx. \tag{5.46}$$

(b) A kinetic energy value $K_\phi(x)$ is assigned to the particle at x given by

$$K_\phi(x) = \frac{1}{2m} \pi_\phi^2(x) + U_\phi(x), \tag{5.47}$$

where $U_\phi(x)$ is a new quantity known as the *quantum potential* determined by the wave function using the expression

$$U_\phi(x) = -\frac{\hbar^2}{2m} \left(\frac{1}{R(x)} \frac{d^2 R(x)}{dx^2} \right). \tag{5.48}$$

The quantum mechanical kinetic energy expectation value is recovered by summing this value under the probability density function $|\phi(x)|^2$, i.e.,

$$\mathcal{E}(\widehat{K}, \phi) = \langle \phi \mid \widehat{K}\phi \rangle = \int |\phi(x)|^2 \, K_\phi(x) \, dx. \tag{5.49}$$

---

[34] Freyberger *et al.* (1997) for a short review.

5.6. LOCAL VALUES IN COORDINATE AND IN PHASE SPACES

(c) A total energy value $E_\phi(x)$ is assigned to the particle at x given by

$$E_\phi(x) = \frac{1}{2m}\pi_\phi^2(x) + U_\phi(x) + V(x), \qquad (5.50)$$

where $V(x)$ is the usual potential energy appearing in the Hamiltonian $\widehat{H}$ of the system. The quantum mechanical expectation value is recovered by summing this value under the probability density function $|\phi(x)|^2$, i.e.,

$$\mathcal{E}(\widehat{H},\phi) = \langle \phi \mid \widehat{H}\,\phi \rangle = \int |\phi(x)|^2\, E_\phi(x)\, dx. \qquad (5.51)$$

A quantum potential interpretation of quantum mechanics then follows from this assignment of local momentum values.[35] A close examination of the way Bohm establishes his local momentum values reveals a lack of generality. The scheme does not apply to all states, e.g., not for wave functions which are not differentiable. Apart from the momentum, the kinetic energy and the Hamiltonian there is no prescription to produce local values for an arbitrary quantum observable in a given state.[36] Moreover, those individual local values $\pi_\phi(x), K_\phi(x)$ and $E_\phi(x)$ have no direct quantum mechanical meaning.

Despite the fundamental difficulties inherent in the schemes of Wigner and Bohm their ideas have aroused a great deal of interest and applications, especially that of Wigner.[37] Indeed local values established by various schemes have found many applications going beyond traditional physics, e.g., in physical chemistry.[38] We shall not review various schemes here. Instead, we shall demonstrate that a notion of local values in coordinate space and in phase space, which have a direct quantum mechanical meaning, can be established, provided we are prepared to extend orthodox quantum mechanics, i.e., generalizing Postulate OQS in §3.2.1 to Postulate GQS in §5.2.1.

5.6.2 Local values and semi-local observables

We aim to propose a general procedure for making local value assignment which is generally applicable to all quantum observables in such way that local values themselves can have a quantum mechanical meaning, making the whole

---

[35] Bohm (1952), Holland (1993). For more discussions of Bohm, the person, and his work see Hiley and Peat (1987), Whitaker (1996).

[36] Wan and Sumner (1988), Sumner (1988).

[37] Wigner function proves to be an extremely useful construct in many applications. Hillery, O'Connell, Scully and Wigner (1986), Kim and Noz (1991), Leonhardt (1997).

[38] Dahl and Avery (1984), Bader and Essen (1984), Deb (1984).

scheme lying within the extended quantum theory embodied by Postulate GQS.

We start by an examination of the expectation value. In $L^2(I\!\!E)$ the expectation value of an observable $\widehat{A}$ in state $\phi \in \mathcal{D}(\widehat{A})$ is expressible as an integral over $I\!\!E$:

$$\langle \phi \mid \widehat{A}\,\phi \rangle = \int_{-\infty}^{\infty} \phi(x)^* \left(\widehat{A}\phi\right)(x)\,dx. \tag{5.52}$$

The integrand is not necessarily real so that its values cannot be identified with real local values of the observable. The complex part of the integrand cannot make any global contribution to the expectation value which is real, i.e., when integrated it would give a zero value. So we can take the real part of the integrand to define a real-valued function on $I\!\!E$ by

$$f_\phi^{\widehat{A}}(x) = \frac{1}{2}\left\{\phi^*(x)\left(\widehat{A}\phi\right)(x) + \left(\widehat{A}\phi\right)^*(x)\phi(x)\right\}, \tag{5.53}$$

so that

$$\langle \phi \mid \widehat{A}\,\phi \rangle = \int_{-\infty}^{\infty} f_\phi^{\widehat{A}}(x)\,dx. \tag{5.54}$$

We interpret $f_\phi^{\widehat{A}}(x)$ as the *local value density* of $\widehat{A}$ in state ϕ at a point x. Compared with Bohm's schemes we can see that:

1. When applied to the momentum and the kinetic energy our results agree with Bohm's. To check this we shall rewrite the wave function $\phi \in L^2(I\!\!E)$ in polar form as in Eq. (5.44). Then we get:

$$f_\phi^{\widehat{p}}(x) = \frac{1}{2}\left\{\phi^*(x)\left(\widehat{p}\phi\right)(x) + \left(\widehat{p}\phi\right)^*(x)\phi(x)\right\} \tag{5.55}$$

$$= |\phi(x)|^2 \frac{dS(x)}{dx}. \tag{5.56}$$

These results agree with Eqs. (5.45) and (5.46). For the kinetic energy operator $\widehat{K}$ we can again apply Eq. (5.53) to get:

$$f_\phi^{\widehat{K}}(x) = |\phi(x)|^2 \left\{\frac{1}{2m}\left(\frac{dS(x)}{dx}\right)^2 + U_\phi(x)\right\}, \tag{5.57}$$

where U_ϕ is the quantum potential given in Eq. (5.48). Note that although the kinetic energy operator $\widehat{K}$ is basically the square of the momentum operator $\widehat{p}$ but a local value of $\widehat{K}$ is not proportional to the square of the corresponding local value of the momentum. An additional term $U_\phi(x)$ appears.[39] This is just Bohm's quantum potential.

---
[39] Wan and Sumner (1988).

5.6. LOCAL VALUES IN COORDINATE AND IN PHASE SPACES

2. Our method is generally applicable, not restricted to momentum and energy.

3. Our present scheme assigns local value densities in the sense that a global expectation value is an integral of the local value density. No probability function is required. The question of negative probability simply does not arise. The expression for $f_\phi^{\hat{p}}(x)$ shows that a probability density function $|\phi(x)|^2$ is automatically incorporated into local values and local value density in our present scheme.

We can go one step further to construct local values. Integrating the local value density over an interval Λ we get

$$\int_\Lambda f_\phi^{\hat{A}}(x)\, dx = \frac{1}{2} \int_\Lambda \left\{ \phi^*(x)(\widehat{A}\phi)(x) + (\widehat{A}\phi)^*(x)\phi(x) \right\} dx$$

$$= \frac{1}{2} \left(\langle \chi_\Lambda \phi \mid \widehat{A}\phi \rangle + \langle \widehat{A}\phi \mid \chi_\Lambda \phi \rangle \right) \tag{5.58}$$

$$= \mathrm{Re}\, \langle \chi_\Lambda \phi \mid \widehat{A}\phi \rangle, \tag{5.59}$$

which can be interpreted as the *local value* in the interval Λ of $\widehat{A}$ in state ϕ. This local value has a meaning within our quantum theory, i.e., it is the correlation term between the vector $\widehat{A}\phi$ and $\chi_\Lambda \phi$.[40] We can go even further. Let $\widehat{B}$ be a bounded observable in $L^2(E)$. Then $\widehat{B}$ gives rise to a selfadjoint semi-local operator of the form[41]

$$\widehat{B}_\Lambda^{(sl)} = \frac{1}{2} (\chi_\Lambda \widehat{B} + \widehat{B} \chi_\Lambda). \tag{5.60}$$

Let $\mathcal{E}(\widehat{B}_\Lambda^{(sl)}, \phi)$ be the expectation value of $\widehat{B}_\Lambda^{(sl)}$ in state ϕ, i.e.,

$$\mathcal{E}(\widehat{B}_\Lambda^{(sl)}, \phi) = \langle \phi \mid \widehat{B}_\Lambda^{(sl)} \phi \rangle. \tag{5.61}$$

Compared Eq. (5.58) we see that

$$\mathcal{E}(\widehat{B}_\Lambda^{(sl)}, \phi) = \int_\Lambda f_\phi^{\hat{B}}(x)\, dx. \tag{5.62}$$

It follows that we can equate the local value in the interval Λ of $\widehat{B}$ in state ϕ to the expectation value of the semi-local observable $\widehat{B}_\Lambda^{(sl)}$ in state ϕ.

Finally, let Λ_j be a set of intervals forming a partition of E, i.e.,

$$\cup_j \Lambda_j = E, \quad \Lambda_j \cap \Lambda_k = \emptyset \text{ if } j \neq k, \tag{5.63}$$

---

[40] See Eq. (3.43).
[41] See Definition 2.12.1(1). The superscripts (sl) stand for "semi local" and are used to distinguish semi-local operators from local operators.

and let $\mathcal{E}(\widehat{B}^{(sl)}_{\Lambda_j}, \phi)$ be the expectation value of $\widehat{B}^{(sl)}_{\Lambda_j}$ in state ϕ. Then we have

$$\mathcal{E}(\widehat{B}, \phi) = \sum_j \mathcal{E}(\widehat{B}^{(sl)}_{\Lambda_j}, \phi). \qquad (5.64)$$

It should be pointed out that generally these local values are generated by semi-local operators, not by local operators. This reflects the non-local nature of quantum mechanics.[42]

The situation becomes techincally more complicated for unbounded operators $\widehat{A}$, since $\widehat{A}^{(sl)}_{\Lambda}$ is generally not selfadjoint. We then have to choose a maximal symmetric extension $\bar{A}^{(sl)}_{\Lambda}$ to serve as an observable. To carry through the above analysis we have to be careful in considering the domains of all those semi-local observables $\bar{A}^{(sl)}_{\Lambda_j}$.

5.6.3 Local values in generalized phase space

We can extend local momentum values to the classical phase space Γ. Let $\widehat{F}^{\hat{p}}(p)$ be the spectral function of the momentum operator. Applying Eq. (5.53) to the momentum operator $\hat{p}$ and making use of Eqs. (2.323), (2.349) and (2.350) we can express $f^{\hat{p}}_\phi(x)$ as

$$\frac{1}{2}\left\{\phi^*(x)\left(\int_{-\infty}^\infty p\, d\widehat{F}^{\hat{p}}(p)\phi\right)(x) + \left(\int_{-\infty}^\infty p\, d\widehat{F}^{\hat{p}}(p)\phi\right)^*(x)\phi(x)\right\}, \qquad (5.65)$$

or more directly in terms of ϕ and its Fourier transform $\widetilde{\phi}(p)$ as

$$\frac{1}{2}\frac{1}{\sqrt{2\pi\hbar}}\left\{\phi^*(x)\int_{-\infty}^\infty p\, e^{ipx}\,\widetilde{\phi}(p)\,dp + \phi(x)\int_{-\infty}^\infty p\, e^{-ipx}\,\widetilde{\phi}^*(p)\,dp\right\}. \qquad (5.66)$$

This result enables us to express the momentum expectation as

$$\langle \phi \mid \hat{p}\, \phi \rangle = \int_{-\infty}^\infty f^{\hat{p}}_\phi(x)\, dx \qquad (5.67)$$

$$= \int_{-\infty}^\infty \int_{-\infty}^\infty W^{\hat{p}}_\phi(x,p)\, dx\, dp, \qquad (5.68)$$

where $W^{\hat{p}}_\phi(x,p)$ is a function on the phase space $\Gamma = \{x, p\}$:

$$W^{\hat{p}}_\phi(x,p) = \frac{p}{2\sqrt{2\pi\hbar}}\left(\phi^*(x)\, e^{ipx}\,\widetilde{\phi}(p) + \phi(x)\, e^{-ipx}\,\widetilde{\phi}^*(p)\right). \qquad (5.69)$$

---

[42] See C2.12.1(1) C2 for the relationship between local and semi-local observables.

5.6. LOCAL VALUES IN COORDINATE AND IN PHASE SPACES

The idea now is to interpret $W_\phi^{\hat{p}}(x,p)$ as the *local value density of momentum in state ϕ in the phase space*.[43] By integrating over any volume in the phase space we obtain the corresponding *local value* in phase space.

Clearly the above analysis can be extended to other observables. Associated with each observable $\hat{A}$ we can construct a *generalized phase space* as the Cartesian product of the spatial space E and the spectrum $sp(\hat{A})$ of the operator $\hat{A}$:

$$\Gamma_{\hat{A}} = E \times sp(\hat{A}) = \{(x,\tau)\}. \tag{5.70}$$

Then $\Gamma_{\hat{p}}$ for the momentum coincides with the usual classical phase space Γ. We can establish an expression for the local value density of $\hat{A}$ in the spatial space E and an expression for the local value density of $\hat{A}$ in its generalized phase space $\Gamma_{\hat{A}}$. The desired expressions are[44]

$$f_\phi^{\hat{A}}(x) = \frac{1}{2}\left\{\phi^*(x)\left(\hat{A}\phi\right)(x) + \left(\hat{A}\phi\right)^*(x)\phi(x)\right\}, \tag{5.71}$$

$$W_\phi^{\hat{A}}(x,\tau) = \frac{\tau}{2}\left\{\phi^*(x)\left(\frac{\partial \widehat{F}^{\hat{A}}(\tau)\phi(x)}{\partial \tau}\right) + \phi(x)\left(\frac{\partial \widehat{F}^{\hat{A}}(\tau)\phi(x)}{\partial \tau}\right)^*\right\}, \tag{5.72}$$

$$\langle\phi \mid \hat{A}\phi\rangle = \int_{-\infty}^{\infty} f_\phi^{\hat{A}}(x)\,dx \tag{5.73}$$

$$= \int_{-\infty}^{\infty}\int_{-\infty}^{\infty} W_\phi^{\hat{A}}(x,\tau)\,dxd\tau. \tag{5.74}$$

In the expression for $W_\phi^{\hat{A}}(x,\tau)$ we have assumed differentiability of $\widehat{F}^{\hat{A}}(\tau)\phi$ with respect to τ.

These local values add up directly to give the global expectation value without requiring any probability function. It follows that negative probabilities never come into the consideration. Even for a positive operator local values can be negative at places without causing any conceptual problem.[45]

---

[43] This expression for the local value density of the momentum in phase space is consistent with the value distribution proposed by Margenau and Hill (1961).

[44] Alternative expressions for $f_\phi^{\hat{A}}(x)$ are:

$$f_\phi^{\hat{A}}(x) = \frac{1}{2}\left\{\phi^*(x)\left(\int_{-\infty}^{\infty} \tau\, d\widehat{F}^{\hat{A}}(\tau)\phi\right)(x) + \text{complex conjugate}\right\}$$

$$= \frac{1}{2}\left\{\phi^*(x)\int_{-\infty}^{\infty}\tau\left(\frac{\partial \widehat{F}^{\hat{A}}(\tau)\phi(x)}{\partial \tau}\right)d\tau + \text{complex conjugate}\right\}.$$

[45] For more details see Wan and Sumner (1988), Wan and Sumner (1991).

5.7 Appendix on Maximal Probability Families

We shall summarize a proof of Theorem 5.2.1(1) linking maximal families of probability functions to maximal symmetric operators.[46]

Lemma 5.7(1)   **For the proof of Theorem 5.2.1(1)**

- Let $\widehat{F}'(\tau)$ be a POV function in a Hilbert space $\mathcal{H}$ generating a family of probability functions $\mathcal{F}'_\phi(\tau) = \langle \phi \mid \widehat{F}'(\tau)\phi \rangle$ on $\mathcal{H}$ having finite expectation values and variances on a dense linear subset $\mathcal{D}$. Then there exists a symmetric operator $\widehat{S}'$ in $\mathcal{H}$ with domain $\mathcal{D}$ such that

$$\langle \phi \mid \widehat{S}'\phi \rangle = \int \tau \, d\langle \phi \mid \widehat{F}'(\tau)\phi \rangle \quad \forall \phi \in \mathcal{D}, \tag{5.75}$$

$$\|\widehat{S}'\phi\|^2 \leq \int \tau^2 \, d\langle \phi \mid \widehat{F}'(\tau)\phi \rangle \quad \forall \phi \in \mathcal{D}. \tag{5.76}$$

Compared with Definition 2.10.1(1) and on account of the Eq. (5.76) which is an inequality we see that Lemma 5.7(1) does not claim that $\widehat{F}'(\tau)$ is necessarily a generalized spectral function of a symmetric operator. According to C2.10.1(1) C1 there may not be a symmetric operator which would admit $\widehat{F}'(\tau)$ as one of its generalized spectral functions. Equation (5.76) suggests that $\widehat{S}'$ together with one of its POV functions may lead to a smaller variance in some states. This enables us to relate the POV function of a maximal symmetric operator to a maximal family of probability functions on $\mathcal{H}$. This will be seen in the proof of Theorem 5.2.1(1) later.

Proof of Lemma 5.7(1)

Theorem 2.9(1) in §2.9 tells us that there exists a PV function $\widehat{F}'_{ex}(\tau)$ in a Hilbert space $\mathcal{H}_{ex}$ which contains $\mathcal{H}$ as a subspace such that for all $\phi \in \mathcal{H} \subset \mathcal{H}_{ex}$ we have

$$\widehat{F}'(\tau)\phi = \widehat{P}_{ex}\widehat{F}'_{ex}(\tau)\phi, \tag{5.77}$$

where $\widehat{P}_{ex}$ is the projector on $\mathcal{H}_{ex}$ projecting onto $\mathcal{H}$. We have, for all $\phi \in \mathcal{D}$,

$$\begin{aligned}
\langle \phi \mid \widehat{F}'(\tau)\phi \rangle &= \langle \phi \mid \widehat{P}_{ex}\widehat{F}'_{ex}(\tau)\phi \rangle = \langle \widehat{P}_{ex}\phi \mid \widehat{F}'_{ex}(\tau)\phi \rangle_{ex} \\
&= \langle \phi \mid \widehat{F}'_{ex}(\tau)\phi \rangle_{ex}, \tag{5.78} \\
\int \tau \, d\langle \phi \mid \widehat{F}'(\tau)\phi \rangle &= \int \tau \, d\langle \phi \mid \widehat{F}'_{ex}(\tau)\phi \rangle_{ex}, \tag{5.79}
\end{aligned}$$

---

[46] Wan, Fountain and Tao (1995), Fountain (1995).

5.7. APPENDIX ON MAXIMAL PROBABILITY FAMILIES

where $\langle \cdot \mid \cdot \rangle_{ex}$ signifies the scalar product in $\mathcal{H}_{ex}$. As a PV function $\widehat{F}'_{ex}(\tau)$ defines a selfadjoint operator $\widehat{A}'_{ex}$ in $\mathcal{H}_{ex}$. The domain of $\widehat{A}'_{ex}$ contains $\mathcal{D}$ since

$$\int \tau^2 \, d\langle \phi \mid \widehat{F}'_{ex}(\tau)\phi \rangle_{ex} = \int \tau^2 \, d\langle \phi \mid \widehat{F}'(\tau)\phi \rangle \quad \forall \phi \in \mathcal{D}, \tag{5.80}$$

and on account of Eqs. (2.274) and (5.16) we have, for every $\phi \in \mathcal{D}$,

$$\int \tau^2 \, d\langle \phi \mid \widehat{F}'(\tau)\phi \rangle < \infty, \tag{5.81}$$

and hence

$$\int \tau^2 \, d\langle \phi \mid \widehat{F}'_{ex}(\tau)\phi \rangle_{ex} < \infty. \tag{5.82}$$

It follows from Spectral Theorem 2.8.1(1) in §2.8.1 and Eq. (5.79) that, for every $\phi \in \mathcal{D}$,

$$\begin{aligned}
\int \tau \, d\langle \phi \mid \widehat{F}'(\tau)\phi \rangle &= \langle \phi \mid \widehat{A}'_{ex} \phi \rangle_{ex} = \langle \widehat{P}_{ex} \phi \mid \widehat{A}'_{ex} \widehat{P}_{ex} \phi \rangle_{ex} \\
&= \langle \phi \mid \widehat{P}_{ex} \widehat{A}'_{ex} \widehat{P}_{ex} \phi \rangle_{ex} \\
&= \langle \phi \mid \widehat{S}' \phi \rangle,
\end{aligned} \tag{5.83}$$

where

$$\widehat{S}' = \widehat{P}_{ex} \widehat{A}'_{ex} \widehat{P}_{ex} \tag{5.84}$$

can be taken as an operator in $\mathcal{H}$ acting on the domain $\mathcal{D}$. This operator is symmetric in $\mathcal{H}$ and satisfies Eq. (5.75).

Next we have, on $\mathcal{D}$,

$$\begin{aligned}
\int \tau^2 d\langle \phi \mid \widehat{F}'(\tau)\phi \rangle &= \int \tau^2 d\langle \phi \mid \widehat{F}'_{ex}(\tau)\phi \rangle_{ex} \\
&= \left\| \widehat{A}'_{ex} \phi \right\|_{ex}^2 = \left\| \widehat{A}'_{ex} \widehat{P}_{ex} \phi \right\|_{ex}^2.
\end{aligned} \tag{5.85}$$

Using inequality (2.32) in §2.2.1 we get

$$\left\| \widehat{S}' \phi \right\|^2 = \left\| \widehat{P}_{ex} \widehat{A}'_{ex} \widehat{P}_{ex} \phi \right\|_{ex}^2 \leq \left\| \widehat{A}'_{ex} \widehat{P}_{ex} \phi \right\|_{ex}^2. \tag{5.86}$$

This implies the desired result:

$$\left\| \widehat{S}' \phi \right\|^2 \leq \int \tau^2 \, d\langle \phi \mid \widehat{F}'(\tau)\phi \rangle. \tag{5.87}$$

Proof of Theorem 5.2.1(1)

First a family of probability functions $\mathcal{F}_\phi^{\widehat{S}}$ on $\mathcal{H}$ generated by the generalized spectral function $\widehat{F}_+^{\widehat{S}}(\tau)$ of a maximal symmetric operator $\widehat{S}$ in $\mathcal{H}$ with domain $\mathcal{D}$ is a maximal family. To show this let $\widehat{F}'(\tau)$ be another POV function which generates a family $\mathbf{F}'(\mathcal{H}, \mathcal{D})$ of probability functions $\mathcal{F}_\phi'(\tau)$ such that

$$\mathcal{E}(\mathcal{F}_\phi') = \mathcal{E}(\mathcal{F}_\phi^{\widehat{S}}) \quad \forall \phi \in \mathcal{D}. \tag{5.88}$$

Then by Lemma 5.7(1), there exists a symmetric operator $\widehat{S}'$ with domain $\mathcal{D}$ such that

$$\langle \phi \mid \widehat{S}'\phi \rangle = \int \tau \, d\langle \phi \mid \widehat{F}'(\tau)\phi \rangle. \tag{5.89}$$

Since

$$\langle \phi \mid \widehat{S}\phi \rangle = \mathcal{E}(\mathcal{F}_\phi^{\widehat{S}}) = \mathcal{E}(\mathcal{F}_\phi') = \langle \phi \mid \widehat{S}'\phi \rangle \quad \forall \phi \in \mathcal{D} \tag{5.90}$$

we have, according to Eq. (2.42),

$$\widehat{S}'\phi = \widehat{S}\phi \quad \forall \phi \in \mathcal{D}. \tag{5.91}$$

By Lemma 5.7(1) we have, on $\mathcal{D}$,

$$\int \tau^2 \, d\langle \phi \mid \widehat{F}'(\tau)\phi \rangle \geq \|\widehat{S}'\phi\|^2 = \|\widehat{S}\phi\|^2 \tag{5.92}$$

$$\Rightarrow \quad \mathcal{V}(\mathcal{F}_\phi') \geq \mathcal{V}(\mathcal{F}_\phi^{\widehat{S}}). \tag{5.93}$$

This means that the variance arising from this new probability function $\mathcal{F}_\phi'$ is generally bigger in some states. A maximal symmetric operator possesses a unique POV function so that the equality sign in the above expressions holds for all $\phi \in \mathcal{D}$ only if $\widehat{F}'(\tau) = \widehat{F}_+^{\widehat{S}}(\tau)$. This is because the equality sign in Eq. (5.92) implies that $\widehat{F}'(\tau)$ is a spectral function of $\widehat{S}'$, and hence of $\widehat{S} = \widehat{S}'$. It follows that the spectral function of a maximal symmetric operator gives rise to the smallest variances. In other words it generates a maximal family of probability functions.

Next let $\mathbf{F}(\mathcal{H}, \mathcal{D})$ be a maximal family of probability functions on $\mathcal{H}$, and let $\widehat{F}(\tau)$ be the POV function which generates $\mathbf{F}(\mathcal{H}, \mathcal{D})$. According to Lemma 5.7(1) there is a symmetric operator $\widehat{S}'$ satisfying Eqs. (5.75) and (5.76). As a symmetric operator $\widehat{S}'$ possesses at least one POV function $\widehat{F}^{\widehat{S}'}(\tau)$ which in turn generates a new family $\mathbf{F}^{\widehat{S}'}(\mathcal{D})$ of probability functions $\mathcal{F}_\phi^{\widehat{S}'}$ with $\mathcal{E}(\mathcal{F}_\phi^{\widehat{S}'}) = \mathcal{E}(\mathcal{F}_\phi)$ on $\mathcal{D}$. We have, again by Lemma 5.7(1),

$$\int \tau^2 \, d\langle \phi \mid \widehat{F}(\tau)\phi \rangle \geq \|\widehat{S}'\phi\|^2 = \int \tau^2 \, d\langle \phi \mid \widehat{F}^{\widehat{S}'}(\tau)\phi \rangle \tag{5.94}$$

$$\Rightarrow \quad \mathcal{V}(\mathcal{F}_\phi) \geq \mathcal{V}(\mathcal{F}_\phi^{\widehat{S}'}). \tag{5.95}$$

This is a contradiction since $\mathbf{F}(\mathcal{H}, \mathcal{D})$ is maximal, unless $\widehat{F}(\tau) = \widehat{F}^{\hat{S}'}(\tau)$. It follows that $\widehat{F}^{\hat{S}'}(\tau)$ has to be the spectral function of $\widehat{S}'$. The same reasoning also implies that $\widehat{S}'$ cannot admit two distinct spectral functions. It follows from Theorem 2.10.2(1) that $\widehat{S}'$ is maximal symmetric.

5.8 Appendix on Time Operators

In some treatment of the energy–time uncertainty relation the time variable is taken to be simply the numerical variable t and the Hamiltonian operator taken as $i\hbar\,\partial/\partial t$ so as to carry out a formal analogy with the position–momentum uncertainty relation. This treatment does not make mathematical sense since t and $i\hbar\partial/\partial t$ are not operators in the Hilbert space $\mathcal{H}$ in the coordinate representation, e.g., $\mathcal{H} = L^2(I\!\!E)$. However, as pointed out in §2.8.4, it can make mathematical sense to treat the Hamiltonian, which is a selfadjoint operator in $\mathcal{H}$ as a multiplication operator in a spectral representation space of the operator. This opens up the possibility of treating time as a differential operator in that space.

For simplicity consider a free particle in one-dimensional motion in $I\!\!E$. The Hamiltonian is equal to the kinetic energy $\widehat{K}$. From Eq. (2.416) and (2.426) we know that the kinetic energy operator is transformed into the multiplication operator $\widetilde{\widetilde{K}}_+ = \epsilon$ in the kinetic energy representation space $\widetilde{\widetilde{L}}_+^2(\widetilde{\widetilde{I\!\!E}})$. We can define a new operator $\widetilde{\widetilde{t}}_+$ in $\widetilde{\widetilde{L}}_+^2(\widetilde{\widetilde{I\!\!E}})$ by

$$\widetilde{\widetilde{t}}_+ \widetilde{\widetilde{\phi}}_+(\epsilon) = i\hbar \frac{d}{d\epsilon} \widetilde{\widetilde{\phi}}_+(\epsilon) \tag{5.96}$$

on the domain

$$\left\{ \widetilde{\widetilde{\phi}}_+(\epsilon) \in S_{2,1}(\widetilde{\widetilde{I\!\!E}}) : \widetilde{\widetilde{\phi}}_+(0) = 0 \right\}, \tag{5.97}$$

where $S_{2,1}$ is a Sobolev space introduced in C2.5.1(2) C4. Noting that the energy variable ϵ is restricted to the range $(0, \infty)$ and following the analysis of $\widehat{p}(I\!\!E^+)$ in Eq. (2.209) we can deduce that $\widetilde{\widetilde{t}}_+$ is strictly maximal symmetric with deficiency indices (1,0). This operator is also canonically conjugate to $\widetilde{\widetilde{K}}_+$ in the sense that the following commutation relation

$$[\widetilde{\widetilde{t}}_+, \widetilde{\widetilde{K}}_+]\widetilde{\widetilde{\phi}}_+ = i\hbar \widetilde{\widetilde{\phi}}_+ \tag{5.98}$$

holds on a dense set of elements $\widetilde{\widetilde{\phi}}_+$ in $\widetilde{\widetilde{L}}_+^2(\widetilde{\widetilde{I\!\!E}})$.

To gain a physical appreciation of this new operator we need to carry out a transformation back to the subspace $L_+^2(I\!\!E)$ in the coordinate representation.

First we can transform back to the momentum representation space $\widetilde{L}_+^2(\widetilde{I\!\!E})$. Using Eqs. (2.404), (2.407) and (2.410) we have

$$\begin{aligned}
\frac{d}{d\epsilon}\widetilde{\phi}_+(\epsilon) &= \frac{dp}{d\epsilon}\frac{d}{dp}\left\{\left(\frac{m}{2\epsilon}\right)^{1/4}\widetilde{\phi}_+(p)\right\} \quad (5.99)\\
&= \left(\frac{m}{p}\right)\frac{d}{dp}\left\{\left(\frac{m}{p}\right)^{1/2}\widetilde{\phi}_+(p)\right\}\\
&= \left(\frac{m}{p}\right)\left\{\left(\frac{m}{p}\right)^{1/2}\frac{d}{dp} - \left(\frac{m}{p}\right)^{1/2}\frac{1}{2p}\right\}\widetilde{\phi}_+(p)\\
&= \left(\frac{m}{p}\right)^{1/2} m\left\{\frac{1}{p}\frac{d}{dp} - \frac{1}{2p^2}\right\}\widetilde{\phi}_+(p)\\
&= \left(\frac{m}{p}\right)^{1/2}\frac{m}{2}\left\{\frac{1}{p}\frac{d}{dp} + \frac{d}{dp}\frac{1}{p}\right\}\widetilde{\phi}_+(p).\\
i\hbar\frac{d}{d\epsilon}\widetilde{\phi}_+(\epsilon) &= \left(\frac{m}{p}\right)^{1/2}\frac{m}{2}\left\{\frac{1}{p}\left(i\hbar\frac{d}{dp}\right) + \left(i\hbar\frac{d}{dp}\right)\frac{1}{p}\right\}\widetilde{\phi}_+(p). \quad (5.100)
\end{aligned}$$

Using Eqs. (2.410), (2.411) and (5.96) we get

$$\widetilde{t}_+\widetilde{\phi}_+(\epsilon) \mapsto \frac{m}{2}\left\{\frac{1}{p}\left(i\hbar\frac{d}{dp}\right) + \left(i\hbar\frac{d}{dp}\right)\frac{1}{p}\right\}\widetilde{\phi}_+(p). \quad (5.101)$$

The corresponding time operator $\widetilde{t}_+$ in the momentum representation space $\widetilde{L}_+^2(\widetilde{I\!\!E})$ is

$$\begin{aligned}
\widetilde{t}_+ &= \frac{m}{2}\left\{\frac{1}{p}\left(i\hbar\frac{d}{dp}\right) + \left(i\hbar\frac{d}{dp}\right)\frac{1}{p}\right\} \quad (5.102)\\
&= \frac{m}{2}\left\{\frac{1}{\widetilde{p}}\widetilde{x} + \widetilde{x}\frac{1}{\widetilde{p}}\right\} \quad (5.103)
\end{aligned}$$

in the notation of Eqs. (2.388) and (2.389). Using Eqs. (2.387), (2.388) and (2.389) we finally obtain the time operator $\widehat{t}_+$

$$\widehat{t}_+ = \frac{m}{2}\left\{\frac{1}{\widehat{p}}\widehat{x} + \widehat{x}\frac{1}{\widehat{p}}\right\} = \frac{m}{2}\left\{\widehat{x}\frac{1}{\widehat{p}} + \frac{1}{\widehat{p}}\widehat{x}\right\} \quad (5.104)$$

in the coordinate representation space $L_+^2(I\!\!E)$.[47] This operator satisfies the

---

[47]There is a confusion about the sign of the time operator and of the time-energy commutation relation in that some authors (Busch, Grabowski and Lahti (1997)) have adopted $\widehat{t}_+$ in Eq. (5.104) as the time operator with a time–energy commutation relation $[\widetilde{t}_+, \widetilde{H}_+] = -i\hbar$. Our results are consistent with that of Rayski and Rayski (1977).

following time–energy commutation relation[48]

$$[\widehat{t}_+, \widehat{K}_+] = i\hbar. \tag{5.105}$$

The operator $\widehat{t}_+$, which may be interpreted as the *time-of-flight operator* in $L_+^2(I\!E)$, is strictly maximal symmetric since it is obtained by unitary transformations from $\widetilde{t}_+$. We can repeat the process in the subspace $L_-^2(I\!E)$ of negative momentum to obtain a corresponding time operator $\widehat{t}_-$.

Finally we form the direct sum operator $\widehat{t}_- \oplus \widehat{t}_+$ in $L^2(I\!E) = L_-^2(I\!E) \oplus L_+^2(I\!E)$. However, this direct sum operator would have a more obscure interpretation. For an arbitrary wave function which is a superposition of positive and negative momentum parts, i.e., when the particle has a probability of moving right and a probability of moving left it is not clear what it means by "time-of-flight". Mathematically this is reflected in the fact that $\widetilde{t}_- \oplus \widetilde{t}_+$ have many selfadjoint extensions in $\widetilde{L}_-^2(I\!E) \oplus \widetilde{L}_+^2(I\!E)$, as shown by Theorem 2.14.1(1) in §2.14.1. One may go on to examine the physics of these selfadjoint extensions.

Our conclusion is that for a free quantum particle there exists a meaningful time-of-flight operator canonically conjugate to the free Hamiltonian $\widehat{H}^{(0)} = \widehat{K}$ in each of the subspaces $L_-^2(I\!E)$ and $L_+^2(I\!E)$. It follows that an uncertainty relation can be established in the usual manner for $\widehat{t}$ and $\widehat{H}^{(0)}$. To admit $\widehat{t}_\pm$, which are strictly maximal symmetric, as observables we have to adopt Postulate GQS in §5.2.1 rather than Postulate OQS in §3.2.1. It is clear from our analysis here that time operator does depend on the Hamiltonian of the system, i.e., different physical systems with different Hamiltonians may well have different conjugate time operators.[49]

5.9 Concluding Remarks

We have incorporated strictly maximal symmetric operators into orthodox quantum mechanics by extending Postulate OQS to Postulate GQS. We can also carry out the same extension in the context of Postulate UQS put forward in §4.2.1. In other words we would use decomposable strictly maximal symmetric operators in the direct integral Hilbert space $\mathcal{H}^\oplus$ to represent observables. The necessity and usefulness of such an extension to Postulate UQS are demonstrated in the physical examples represented in §7.7 in Chapter 7. As pointed out in §2.14.1 and seen in the discussion on time operators in the

---

[48] $\widehat{K}_+$ is the kinetic operator acting on $L_+^2(I\!E)$ introduced in Eq. (2.403). This is also the Hamiltonian for a free particle.

[49] For more discussions on the concept of time see Zeh (1992).

preceding section, the direct sum or the direct integral of strictly maximal symmetric operators are not necessarily strictly maximal symmetric. So, care has to be taken when we build up an observable in $\mathcal{H}^{\oplus}$ in terms of strictly maximal symmetric operators within each supersector.

References

1. Aharonov, Y. and Bohm, D. (1961). *Phys. Rev.* **122** 1649.
2. Ali, S. T. and Emch, G. G. (1974). *J. Math. Phys.* **15** 176.
3. Allcock, G. R. (1969). *Ann. Phys.* **53** 311.
4. Atmanspacher, H. and Amann, A. (1998). *Int. J. Theor. Phys.* **37** 629.
5. Bader, R. and Essen, H. (1984). 'The Mechanics of and an Equation for the Electron Charge Density' in J. Dahl and A. Avery, eds., *Local Density Approximations in Quantum Chemistry and Solid State Physics*, Plenum, New York.
6. Bauer, M. (1983). *Ann. Phys.* **150** 1.
7. Bohm, D. (1952). *Phys. Rev.* **85** 166.
8. Busch, P. (1990). *Found. Phys.* **20** 1.
9. Busch, P. (1985). *Int. J. Theor. Phys.* **24** 63.
10. Busch, P., Lahti, P. and Mittelstaedt, P. (1996). *The Quantum Theory of Measurement*, 2nd edition, Springer-Verlag, Berlin.
11. Busch, P., Grabowski, M. and Lahti, P. J. (1994). *Phys Lett. A* **191** 357.
12. Busch, P., Grabowski, M. and Lahti, P. J. (1997). *Operational Quantum Physics*, Springer-Verlag, Berlin.
13. Busch, P and Lahti, P. J. (1984). *Phys. Rev. D* **29** 1634.
14. Busch, P. and Schroeck, F. E., Jr. (1989). *Found. Phys.* **19** 807.
15. Capri, A. Z. (1985). *Nonrelativistic Quantum Mechanics*, Benjamin/Cummings, Menlo Park, California.
16. Comi, M. (1980). *IL Nuovo Cimento A* **56** 299.
17. Dahl, J. P. and Avery, J. eds. (1984). *Local Density Approximations in Quantum Chemistry and Solid State Physics*, Plenum, New York.
18. Davies, E. B. (1976). *Quantum Theory of Open Systems*, Academic Press, London.
19. Deb, B. M. (1984). 'Some Aspects of the Role of Single-Particle Density in Chemistry' in J. Dahl and A. Avery, eds., *Local Density Approximations in Quantum Chemistry and Solid Sate Physics*, Plenum, New York.
20. Fountain, R. H. (1995). *Observables, Maximal Symmetric Operators, POV Measures and their Applications in Quantum Mechanics*, St Andrews University PhD Thesis.

REFERENCES

21. Freyberger, M., Bardroff, P., Leichtle, C., Schrade, G. and Schleich, W. (1997). *Physics World* **10** 42.
22. Fuches, C. A. (2001). 'Quantum Foundations in the Light of Quantum Information', arXiv: quant-ph/0106166.
23. Giannitrapani, R. (1997). *Int. J. Theor. Phys.* **36** 1575.
24. Grabowsi, M. (1989). *Found. Phys.* **19** 923.
25. Hiley, B. and Peat, F. D. eds. (1987). *Quantum Implications: Essays in Honour of David Bohm*, Routledge & Kegan Paul, London.
26. Hillery, M., O'Connell, R., Scully, M. and Wigner, E. (1984). *Physics Reports* **106** 121.
27. Holevo, A. S. (1982). *Probabilistic and Statistical Aspects of Quantum Theory*, North-Holland, Amsterdam.
28. Holland, P. R. (1993). *The Quantum Theory of Motion*, Cambridge University Press, Cambridge.
29. Isham, C. J. (1995). *Lectures on Quantum Theory*, Imperial College Press, London.
30. Jordan, T. F. (1978). *J. Math. Phys.* **19** 1382.
31. Kim, Y. S. and Noz, M. E. (1991). *Phase Space Picture of Quantum Mechanics: Group Theoretical Approach*, World Scientific, Singapore.
32. Kraus, K. (1977). 'Position Observables of the Photon' in W. C. Price and S. S. Chissick, eds., *The Uncertainty Principle and the Foundations of Quantum Mechanics*, Wiley, New York.
33. Kraus, K. (1984). *States, Effects and Operations*, Springer-Verlag, Berlin.
34. Leonhardt, U. (1997). *Measuring the Quantum State of Light*, Cambridge University Press, Cambridge.
35. Ludwig, G. (1985). *Foundations of Quantum Mechanics Vol. 2*, Springer-Verlag, New York.
36. Mandelstam L. and Tamm, I. (1945). *J. Phys.* (USSR) **9** 249-254.
37. Margenau, H. and Hill, R. N. (1961). *Prog. Theor. Phys.* **26** 722.
38. Merzbacher, E. (1998). *Quantum Mechanics*, 3rd edition, Wiley, New York.
39. Messiah, A. (1967). *Quantum Mechanics Vol. 1*, North-Holland, Amsterdam.
40. Misra, B. (1978). *Proc. Natl. Acad. Sci. USA* **75** 1627.
41. Misra, B. (1979). *J. Stat. Phys.* **48** 1925.
42. Misra, B. (1995). *Found. Phys.* **25** 1087.
43. Misra, B., Prigogine, I. and Courbage, M. (1979). *Physica A* **98** 1.
44. Peres, A. (1995). *Quantum Theory: Concepts and Methods*, Kluwer, Dordrecht.
45. Price, W. C. and Chissick, S. S. eds. (1977). *The Uncertainty Principle and the Foundations of Quantum Mechanics*, Wiley, New York.
46. Prugovečki, E. (1984). *Stochastic Quantum Mechanics and Quantum Spacetime*, Reidel, Dordrecht.
47. Rae, A. I. M. (1981). *Quantum Mechanics*, MaGraw-Hill, London.
48. Rayski, J. and Rayski, J. M., Jr. (1977). 'On the Meaning of the Time-Energy Uncertainty Relation' in W. C. Price and S. S. Chissick, eds., *The Uncertainty Principle and the Foundations of Quantum Mechanics*, Wiley, New York.

49. Recami, E. (1977). 'A Time Operator and the Time-Energy Uncertainty Relation' in W. C. Price and S. S. Chissick, eds., *The Uncertainty Principle and the Foundations of Quantum Mechanics*, Wiley, New York.
50. Schroeck, F. E., Jr. (1996). *Quantum Mechanics on Phase Space*, Kluwer, Dordrecht.
51. Sumner, P. (1988). *Spatial and Generalized Phase Space Distributions of Observable Values in Quantum Mechanics*, St Andrews University PhD Thesis.
52. Uffink, J. (1994). *Int. J. Theor. Phys.* **33** 199.
53. Wan, K. K., Fountain, R. H. and Tao, Z. Y. (1995). *J. Phys. A: Math. Gen.* **28** 2379.
54. Wan, K. K. and Sumner, P. (1988). *Phys. Lett. A* **128** 458.
55. Wan, K. K. and Sumner, P. (1991). *Nuovo Cimento B* **106** 593.
56. Wigner, E. (1932). *Phys. Rev.* **40** 749.
57. Whitaker, A. (1996). *Einstein, Bohr and the Quantum Dilemma*, Cambridge University Press, Cambridge.
58. Zeh, H. D. (1992). *The Physical Basis of the Direction of Time*, Springer-Verlag, Berlin.

Part III

Point Interactions, Macroscopic Quantum Systems and Superselection Rules

Chapter 6

Point Interactions

6.1 Introduction

In many physical applications one often considers potentials that are strongly localized. Point interaction models are an extreme idealization of this and describe physical systems where the potentials are constant except at a countable number of isolated points. By modifying these basic idealizations by various perturbations and approximations we can describe more complicated and realistic interactions. The simplest and the most familiar point interaction is that of a one-dimensonal model of a potential step at the origin along the x-axis, i.e., the potential is zero left of the origin and rises abruptly to a constant value $V_0 > 0$ to the right of the origin. Another example is that of a δ-function potential well; the associated Hamiltonian is commonly written in the form

$$-\frac{\hbar^2}{2m}\frac{d^2}{dx^2} - V_0\,\delta(x), \quad V_0 > 0. \tag{6.1}$$

This can only be a formal expression since a δ-function does not constitute an operator in $L^2(I\!E)$. Given any $\phi(x) \in L^2(I\!E)$ the function $\delta(x)\,\phi(x)$ is not square-integrable, and is hence not in $L^2(I\!E)$.

It is possible to describe such δ-function potentials or other point interactions rigorously in terms of well-defined maximal symmetric operators. As will be seen in §6.5.2 we can find a selfadjoint operator in $L^2(I\!E)$, denoted by $\widehat{H}^{[2-]}$ in §6.5.2, which has the effect of what we would expect of the above formal expression of a Hamiltonian with a δ-function potential well. The way to achieve this is rather simple in principle. All we need is to proceed as follows:

1. To start with we would avoid all those points of discontinuity of the potential, e.g., the point $x = 0$ for the above δ-potential, by setting up

a symmetric operator which does not act on those discontinuity points.

2. Next, work out all the selfadjoint and strictly maximal symmetric extensions to the above symmetric operator.

3. Finally select one of those extension operators to produce the effect of the point interaction in question, e.g., the effect of a δ-function potential well. This is how $\widehat{H}_{-}^{[2-]}$ is obtained and selected in §6.5.2

The whole problem of point interactions then becomes intrinsically related to the subject of selfadjoint extensions of symmetric operators.[1] Since the selfadjoint extensions of a large class of differential operators are specifiable in terms of conditions at the points of discontinuity we can classify single point interactions by boundary conditions at the points of discontinuity of the potential.

In this chapter we aim to review a scheme based on the above idea, to be referred to as *quantization by parts*, for a systematic treatment of point interaction in one dimension.[2] This method together with the results obtained in this chapter will be extended to study some specific macroscopic quantum systems, i.e., superconducting circuit systems, in the next chapter. Point interactions also play a role in the study of other macroscopic quantum systems. The Gross-Pitaevskii equation for the description of Bose-Einstein condensate of dilute gases is derived using a δ-function potential.[3] Such point interactions are also employed in studying possible interference devices based on ring-shaped Bose-Einstein condensates.[4]

It is worth stressing that we can turn the whole problem of point interactions on its head in the following manner. Take the example of the selfadjoint operator $\widehat{H}_{-}^{[2-]}$ mentioned earlier. Orthodox quantum mechanics asserts that, being selfadjoint $\widehat{H}_{-}^{[2-]}$ can serve as an observable. The question then becomes what intuitive physical interpretation we should give $\widehat{H}_{-}^{[2-]}$ if we are to employ it as a Hamiltonian. The answer is that $\widehat{H}_{-}^{[2]}$ behaves like a Hamiltonian with a δ-function potential well. The moral is that quantum mechanics can deal with potential discontinuities in a mathematically rigorous manner to enable these potentials to be incorporated into a selfadjoint Hamiltonian operator. In classical mechanics Newtonian equations are unable to deal with discontinuous potential functions, e.g., a square potential barrier, because the force becomes

---

[1] Albeverio, Gesztesy, Hoegh-Krohn and Holden (1988), Albeverio, Gesztesy and Holden (1992).

[2] Exner and Seba (1986), (1989a), (1989b), Exner, Seba and Stovicek (1989), Blank, Exner and Havlicek (1994) §14.6, Wan and Fountain (1996), (1998), Trueman (1999), Trueman and Wan (2000). See Giamarchi (2004) for quantum physics in one-dimensional systems.

[3] Gross (1961), (1963), Pitaevskii (1961). For a review see Dalfovo and Giorgini (1999) p. 474.

[4] Anderson, Dholakia and Wright (2003).

infinite or generally undefinable at a potential discontinuity. Such potentials are then perceived as unphysical. However, these seemingly unphysical quantities in the classical sense acquire a new legitimacy in quantum theory. In the context of Postulate OQS a square potential barrier can be treated rigorously along with other continuous potentials. In fact Hamiltonians incorporating point interactions with consequent discontinuity in their eigenfunctions turn out to lead to new physics such as Josephson effect in superconductivity.[5]

We should point out that point interactions in the form of point sources and point scatterers are also employed in classical physics, e.g., point charges can be described in terms of a δ-function charge density[6] and point scatterers can also be described by a δ-function.[7] However, point interactions can lead to a much richer range of interactions and properties in quantum theory. To rule out point interactions, as some people may do on the ground of its being intuitively unphysical, would amount to ruling out a large class of selfadjoint operators as well as directly contradicting Postulate OQS of orthodox quantum mechanics.[8]

6.2 Extensions of Symmetric Operators

When applying von Neumann's formula to $\widehat{p}_0$ in $L^2(\Lambda)$ in E2.5.2(1) E1 we find that all the selfadjoint extensions $\widehat{p}_\lambda$ can be specified by boundary conditions. It turns out that this is generally true for differential operators. To highlight this let us recall operator $\widehat{K}_0(\Lambda)$ in E2.5.1(1) E2. This operator is symmetric and it possesses many selfadjoint extensions. The selfadjoint extension $\widehat{K}^\infty(\Lambda)$ is one example. This particular extension is specified by its domain containing functions ϕ in the Sobolev space $S_{2,2}(\Lambda)$ satisfying boundary conditions $\phi(0) = \phi(a) = 0$. It turns out that all the selfadjoint extensions of $\widehat{K}_0(\Lambda)$ can be specified by certain conditions on the values of ϕ and its derivative $\phi' = d\phi/dx$ at the boundaries $x = 0$ and $x = a$. We shall spell out everything explicitly here; the results presented will also be directly relevant to later applications.

Let $\underline{\alpha} = \{\alpha'_2, \alpha_2, \alpha'_1, \alpha_1\}$ and $\underline{\beta} = \{\beta'_2, \beta_2, \beta'_1, \beta_1\}$ be two sets of complex numbers subject to the following conditions:

(C1) The first set is not a multiple of the second set, i.e., there does not exist a number w such that

$$\alpha'_2 = w\,\beta'_2, \ \alpha_2 = w\,\beta_2, \ \alpha'_1 = w\,\beta'_1 \ \text{and} \ \alpha_1 = w\,\beta_1. \tag{6.2}$$

---

[5] This will be discussed in Chapter 7.
[6] Jackson (1999).
[7] de Vries (1998) for a review of point scatterers for classical waves.
[8] The quantum concept of spin also seems to be intuitively unphysical in classical physics.

(C2) These complex numbers are related by
$$\alpha_2'^* \alpha_2 - \alpha_2^* \alpha_2' = \alpha_1'^* \alpha_1 - \alpha_1^* \alpha_1', \quad \beta_2'^* \beta_2 - \beta_2^* \beta_2' = \beta_1'^* \beta_1 - \beta_1^* \beta_1', \tag{6.3}$$
$$\alpha_2'^* \beta_2 - \alpha_2^* \beta_2' = \alpha_1'^* \beta_1 - \alpha_1^* \beta_1', \quad \beta_2'^* \alpha_2 - \beta_2^* \alpha_2' = \beta_1'^* \alpha_1 - \beta_1^* \alpha_1'. \tag{6.4}$$

Let $\mathcal{D}_{\underline{\alpha},\underline{\beta}}$ be a subset of the Sobolev space $S_{2,2}(\Lambda)$ consisting of all the functions ϕ satisfying the following boundary conditions

$$\alpha_2' \phi_2' - \alpha_2 \phi_2 = \alpha_1' \phi_1' - \alpha_1 \phi_1, \tag{6.5}$$
$$\beta_2' \phi_2' - \beta_2 \phi_2 = \beta_1' \phi_1' - \beta_1 \phi_1, \tag{6.6}$$

where[9]

$$\phi_2 = \phi(0), \quad \phi_1 = \phi(a), \tag{6.7}$$
$$\phi_2' = \left[\frac{d\phi}{dx}\right]_{x=0}, \quad \phi_1' = \left[\frac{d\phi}{dx}\right]_{x=a}. \tag{6.8}$$

Then all selfadjoint extensions of $\widehat{K}_0(\Lambda)$ are given by the following theorem.[10]

Theorem 6.2(1)   **On selfadjoint extensions of $\widehat{K}_0(\Lambda)$**

- The operator $\widehat{K}_{\underline{\alpha},\underline{\beta}}(\Lambda)$ in $L^2(\Lambda)$ defined on the domain $\mathcal{D}_{\underline{\alpha},\underline{\beta}}$ by

$$\widehat{K}_{\underline{\alpha},\underline{\beta}}(\Lambda) \phi = -\frac{\hbar^2}{2m} \frac{d^2 \phi}{dx^2} \quad \forall \phi \in \mathcal{D}_{\underline{\alpha},\underline{\beta}} \tag{6.9}$$

is selfadjoint and conversely every selfadjoint extension of $\widehat{K}_0(\Lambda)$ is of this form.

The Hamiltonian of an infinite square potential well $\widehat{K}^\infty(\Lambda)$ is the selfadjoint extension corresponding to the following choice of $\underline{\alpha}$ and $\underline{\beta}$:

$$\alpha_2' = 0, \quad \alpha_2 \neq 0, \quad \alpha_1' = 0, \quad \alpha_1 = 0, \tag{6.10}$$
$$\beta_2' = 0, \quad \beta_2 = 0, \quad \beta_1' = 0, \quad \beta_1 \neq 0. \tag{6.11}$$

Such a choice renders functions in the domain vanishing at $x = 0$ and $x = a$ as required physically by the infinite square potential well.

Another example is the family of extensions $\widehat{K}_\lambda = \widehat{p}_\lambda^2/2m$ constructed directly from the selfadjoint operators $\widehat{p}_\lambda$ in E2.5.1(1) E1. These extensions correspond to the choice

$$\alpha_2' = 0, \quad \alpha_2 = 1, \quad \alpha_1' = 0, \quad \alpha_1 = e^{i\lambda}, \tag{6.12}$$
$$\beta_2' = 1, \quad \beta_2 = 0, \quad \beta_1' = e^{i\lambda}, \quad \beta_1 = 0. \tag{6.13}$$

---

[9] The reason for labelling $\phi(0)$ as ϕ_2 rather than ϕ_1 will become apparent in §6.8.2.
[10] Hudson and Pym (1980) Theorem 10.5.3 on p. 269.

6.3 Extensions of Direct Sum Operators

6.3.1 Direct sums and their selfadjoint extensions

We shall have many occasions to employ direct sums of differential operators. It turns out that for a large number of cases, boundary conditions can again be used to characterize the selfadjoint extensions of direct sum operators. We have already seen an example in Theorem 2.14.1(1) where the direct sum of two first-order differential operators $\widehat{p}^-$ and $\widehat{p}^+$ is considered. In this section we are interested in the direct sum of second order differential operators, i.e., $\widehat{K}_0(E^-)$ and $\widehat{K}_0(E^+)$ introduced in E2.5.1(1) E7.[11] For later applications we shall re-label various quantities as follows:

$$\mathcal{H}_1 = L^2(E^-), \quad \widehat{p}_1 = \widehat{p}^-, \quad \widehat{K}_{10} = \widehat{K}_0(E^-), \quad (6.14)$$
$$\mathcal{H}_2 = L^2(E^+), \quad \widehat{p}_2 = \widehat{p}^+, \quad \widehat{K}_{20} = \widehat{K}_0(E^+). \quad (6.15)$$

As mentioned in E2.5.1(1) E7 operator $\widehat{K}_{20}$ in $L^2(E^+)$ is symmetric with deficiency indices $(1,1)$. It follows that $\widehat{K}_{20}$ admits a one-parameter family of selfadjoint extensions. The situation is the same for $\widehat{K}_{10}$ in $L^2(E^-)$. Now consider the direct sum operator

$$\widehat{K}_0^{(1,2)} = \widehat{K}_{10} \oplus \widehat{K}_{20} \quad \text{defined on} \quad \mathcal{D}(\widehat{K}_0^{(1,2)}) = \mathcal{D}(\widehat{K}_{10}) \oplus \mathcal{D}(\widehat{K}_{20}). \quad (6.16)$$

The deficiency indices of $\widehat{K}_0^{(1,2)}$ are $(2,2)$ according to Eq. (2.630). It follows that $\widehat{K}_0^{(1,2)}$ possesses a four-parameter family of selfadjoint extensions. We can work out these extensions using von Neumann's formula in Theorem 2.5.2(1). Let $\bar{K}_{10}$ and $\bar{K}_{20}$ be the closures of $\widehat{K}_{10}$ and $\widehat{K}_{20}$ respectively. Then $\bar{K}_{10} \oplus \bar{K}_{20}$ is closed, according to a property of direct sum of operators stated in §2.14.1. Hence the closure of $\widehat{K}_0^{(1,2)}$ is

$$\bar{K}_0^{(1,2)} = \bar{K}_{10} \oplus \bar{K}_{20}. \quad (6.17)$$

Let $\mathcal{N}^{(+i)}$ and $\mathcal{N}^{(-i)}$ be the deficiency subspaces of $\bar{K}_0^{(1,2)}$, i.e.,

$$\mathcal{N}^{(+i)} = \left\{ \phi^{(+i)} \in \mathcal{H}^{(1,2)} : (\bar{K}_0^{(1,2)\dagger} - i)\phi^{(+i)} = 0 \right\}, \quad (6.18)$$
$$\mathcal{N}^{(-i)} = \left\{ \phi^{(-i)} \in \mathcal{H}^{(1,2)} : (\bar{K}_0^{(1,2)\dagger} + i)\phi^{(-i)} = 0 \right\}. \quad (6.19)$$

Then $\mathcal{N}^{(+i)}$ is two-dimensional with normalized basis functions:[12]

$$g_1 = ce^{\rho x_1} \oplus 0, \quad g_2 = 0 \oplus ce^{-\rho x_2}, \quad \rho = \exp\left(-\frac{1}{4}i\pi\right), \quad (6.20)$$

---

[11] Fountain (1995), Wan and Fountain (1996).
[12] Reed and Simon Vol. 2 (1975) p. 144. For simplicity we have chosen $\hbar^2/2m = 1$ here so that $\widehat{K}_{20}$ and $\widehat{K}_{10}$ have the expression $-d^2/dx^2$. But the results obtained in this section are valid whether we have made this simplification or not.

where c is a normalization constant. Similarly we have normalized basis functions in $\mathcal{N}^{(-i)}$:

$$f_1 = ce^{\rho^* x_1} \oplus 0, \quad f_2 = 0 \oplus ce^{-\rho^* x_2}, \quad \rho^* = \exp\left(\frac{1}{4}i\pi\right). \quad (6.21)$$

Let $\widehat{K}^{(1,2)}$ denote a selfadjoint extension of $\bar{K}_0^{(1,2)}$.[13] Then, by von Neumann's formula, the domain of $\widehat{K}^{(1,2)}$ must be of the form

$$\mathcal{D}(\widehat{K}^{(1,2)}) = \mathcal{D}(\bar{K}_0^{(1,2)}) + \left\{\varphi + \widehat{\mathcal{U}}\varphi : \varphi \in \mathcal{N}^{(+i)}\right\}, \quad (6.22)$$

where $\widehat{\mathcal{U}}$ is a unitary mapping of $\mathcal{N}^{(+i)}$ onto $\mathcal{N}^{(-i)}$. When acting on

$$\Phi = \phi + \varphi + \widehat{\mathcal{U}}\varphi \in \mathcal{D}(\widehat{K}^{(1,2)}) \quad (6.23)$$

we have

$$\widehat{K}^{(1,2)}\Phi = \bar{K}_0^{(1,2)}\phi + i\varphi - i\widehat{\mathcal{U}}\varphi. \quad (6.24)$$

Using basis functions $\{g_1, g_2\}$ of $\mathcal{N}^{(+i)}$ and $\{f_1, f_2\}$ of $\mathcal{N}^{(-i)}$ we can write down the following expressions:

$$\varphi = \sum_k c_k g_k \in \mathcal{N}^{(+i)}, \quad k=1,2 \quad (6.25)$$

$$u_{jk} = \langle f_j | \widehat{\mathcal{U}} g_k \rangle, \quad (6.26)$$

$$\widehat{\mathcal{U}} g_k = \sum_j u_{jk} f_j, \quad \text{and} \quad (6.27)$$

$$\Phi = \phi + \sum_k c_k \left(g_k + \sum_j u_{jk} f_j\right). \quad (6.28)$$

The 2×2 matrix (u_{jk}) is unitary and hence parameterizable by four real parameters. This is consistent with our earlier statement on the existence of a four-parameter family of selfadjoint extensions.

We already know a family of selfadjoint extensions:

$$\widehat{K}_\lambda^{(1,2)} = \frac{1}{2m}\left(\widehat{P}_\lambda^{(1,2)}\right)^2, \quad (6.29)$$

where $\widehat{P}_\lambda^{(1,2)}$ is given in Theorem 2.14.1(1). For later physical applications we need some other extensions. Let us narrow down the four-parameter family to a two-parameter family by a symmetry requirement. Consider the Euclidean space $\mathbb{E}$ with the origin removed. What is left is the union $\mathbb{E}^- \cup \mathbb{E}^+$. The

---

[13] We have three related yet different operators: $\widehat{K}_0^{(1,2)}$, $\bar{K}_0^{(1,2)}$ and $\widehat{K}^{(1,2)}$.

6.3. EXTENSIONS OF DIRECT SUM OPERATORS

direct sum $\mathcal{H}^{(1,2)}$ is a Hilbert space consisting of square-integrable functions on $I\!\!E^- \cup I\!\!E^+$. Let $\varphi(x_1)$ be a function in $L^2(I\!\!E^-)$ and $\psi(x_2)$ be a function in $L^2(I\!\!E^+)$. Then $\varphi(x_1) \oplus \psi(x_2)$ is a function in $\mathcal{H}^{(1,2)}$. Moreover, we can see that $\varphi(-x_2)$ becomes a function in $L^2(I\!\!E^+)$ and $\psi(-x_1)$ becomes a function in $L^2(I\!\!E^-)$. We can see that $\varphi(-x_2)$ is a mirror image of $\varphi(x_1)$ and $\psi(-x_1)$ is a mirror image of $\psi(x_2)$, the mirror being situated at the coordinate origin in both cases. It follows that $\psi(-x_1) \oplus \varphi(-x_2)$ is also a member of $\mathcal{H}^{(1,2)}$. To effect such a mirror reflection we can introduce a *parity operator* $\widehat{\wp}$ on $\mathcal{H}^{(1,2)}$ by

$$\widehat{\wp}\{\varphi(x_1) \oplus \psi(x_2)\} = \psi(-x_1) \oplus \varphi(-x_2). \tag{6.30}$$

For examples we have

$$\widehat{\wp} g_1 = g_2, \quad \widehat{\wp} g_2 = g_1, \tag{6.31}$$

$$\widehat{\wp} f_1 = f_2, \quad \widehat{\wp} f_2 = f_1, \tag{6.32}$$

$$\widehat{\wp}\{e^{\rho x_1} \oplus e^{-\rho^* x_2}\} = e^{\rho^* x_1} \oplus e^{-\rho x_2}, \tag{6.33}$$

$$\widehat{\wp} \eta_{\lambda,p}^{(1,2)} = e^{i\lambda} e^{-ipx_1} \oplus e^{-ipx_2}. \tag{6.34}$$

where, as in Theorem 2.14.1(1),[14]

$$\eta_{\lambda,p}^{(1,2)} = e^{ipx_1} \oplus e^{i\lambda} e^{ipx_2}. \tag{6.35}$$

What we want here is to find all the selfadjoint extensions $\widehat{K}^{(1,2)}$ to $\bar{K}_0^{(1,2)}$ which have a mirror reflection symmetry. Since $\widehat{K}_{10}$ and $\widehat{K}_{20}$ are mirror images of each other their closures would also have a mirror reflection symmetry. In other words $\mathcal{D}(\bar{K}_0^{(1,2)})$ is invariant under $\widehat{\wp}$ and the operator expression for $\bar{K}_0^{(1,2)}$, being a second order differential operator without the first order term, is also unchanged under $\widehat{\wp}$, i.e., we have

$$\widehat{\wp} \mathcal{D}(\bar{K}_0^{(1,2)}) = \mathcal{D}(\bar{K}_0^{(1,2)}), \tag{6.36}$$

$$\widehat{\wp} \bar{K}_0^{(1,2)} \phi = \bar{K}_0^{(1,2)} \widehat{\wp} \phi, \quad \phi \in \mathcal{D}(\bar{K}_0^{(1,2)}). \tag{6.37}$$

In view of Eq. (6.31) the space $\mathcal{N}^{(+i)}$ is invariant under $\widehat{\wp}$. For $\widehat{K}^{(1,2)}$ to possess a reflection symmetry in the above sense we require $\widehat{\wp}$ to commute with $\widehat{\mathcal{U}}$, i.e., we want

$$\widehat{\wp} \widehat{\mathcal{U}} \varphi = \widehat{\mathcal{U}} \widehat{\wp} \varphi, \quad \varphi \in \mathcal{N}^{(+i)}. \tag{6.38}$$

In terms of matrix elements we require

$$\langle f_j | \widehat{\mathcal{U}} \widehat{\wp} g_k \rangle = \langle f_j | \widehat{\wp} \widehat{\mathcal{U}} g_k \rangle. \tag{6.39}$$

---

[14] $\eta_{\lambda p}^{(1,2)}$, introduced in Theorem 2.14.1(1), is a generalized eigenfunction of $\widehat{P}_\lambda^{(1,2)}$.

For example we must have[15]

$$\langle f_1 | \widehat{\mathcal{U}} \widehat{\wp} g_1 \rangle = \langle f_1 | \widehat{\wp} \widehat{\mathcal{U}} g_1 \rangle \Rightarrow \langle f_1 | \widehat{\mathcal{U}} g_2 \rangle = \langle f_2 | \widehat{\mathcal{U}} g_1 \rangle, \quad (6.40)$$
$$\langle f_1 | \widehat{\mathcal{U}} \widehat{\wp} g_2 \rangle = \langle f_1 | \widehat{\wp} \widehat{\mathcal{U}} g_2 \rangle \Rightarrow \langle f_1 | \widehat{\mathcal{U}} g_1 \rangle = \langle f_2 | \widehat{\mathcal{U}} g_2 \rangle. \quad (6.41)$$

It follows that
$$u_{12} = u_{21} = v, \quad u_{11} = u_{22} = u, \quad (6.42)$$

or
$$u_{jk} = u\,\delta_{jk} + v\,(1 - \delta_{jk}). \quad (6.43)$$

Being a unitary matrix we also have
$$|u|^2 + |v|^2 = 1 \quad \text{and} \quad uv^* + vu^* = 0. \quad (6.44)$$

These constraints reduce the four-parameter family to a two-parameter family specified by two real parameters. The family of extensions $\widehat{K}_\lambda^{(1,2)}$ with $\lambda \neq 0, \pi$ does not have the required mirror reflection symmetry, e.g., the domain of $\widehat{K}_\lambda^{(1,2)}$ is not invariant under $\widehat{\wp}$ as shown in Eq. (6.34). These extensions are therefore not included here.

We can now rewrite Eq. (6.28) for $\Phi \in \mathcal{D}(\widehat{K}^{(1,2)}) \subset \mathcal{H}^{(1,2)}$ in the form

$$\Phi = \Phi_1 + \Phi_2, \quad (6.45)$$

where
$$\Phi_1 = \phi_1 + c_1 g_1 + (c_1 u + c_2 v)\,f_1, \quad (6.46)$$
$$\Phi_2 = \phi_2 + c_2 g_2 + (c_1 v + c_2 u)\,f_2, \quad (6.47)$$

and
$$\phi_1 = \phi^- \oplus 0, \quad \phi^- \in \mathcal{D}(\bar{K}_0^-), \quad (6.48)$$
$$\phi_2 = 0 \oplus \phi^+, \quad \phi^+ \in \mathcal{D}(\bar{K}_0^+). \quad (6.49)$$

Being in the domain $\mathcal{D}(\bar{K}_0^+)$ the function ϕ^+ and its derivative vanish at $x = 0$ according to Eq. (2.219) in E2.5.1 E7. Similarly ϕ^- and its derivative also vanish at $x = 0$. Together with Eqs. (6.20) and (6.21) these results enable us to relate various values at the boundary:

$$\begin{aligned} \phi_{10} &= \phi'_{10} = \phi_{20} = \phi'_{20} = 0, \\ g_{10} &= f_{10} = g_{20} = f_{20} = c, \end{aligned} \quad (6.50)$$

where a dash signifies a derivative with respect to x and the subscript 0 denotes values evaluated at $x = 0$ as limits from the left and from the right appropriately. Substituting into Eqs. (6.46) and (6.47) we get

$$\Phi_{10} = c\,c_1 + c\,(c_1 u + c_2 v), \quad \Phi'_{10} = c\,c_1 \rho + c\,\rho^*(c_1 u + c_2 v), \quad (6.51)$$
$$\Phi_{20} = c\,c_2 + c\,(c_1 v + c_2 u), \quad \Phi'_{20} = -c\,c_2 \rho - c\,\rho^*(c_1 v + c_2 u). \quad (6.52)$$

---

[15] The parity operator is both unitary and selfadjoint.

6.3. EXTENSIONS OF DIRECT SUM OPERATORS

6.3.2 Selfadjoint extensions in terms of boundary conditions

From Eqs. (6.51) and (6.52) we can see that the effect of the unitary matrix can be incorporated into *boundary conditions* on the values of $\Phi \in \mathcal{D}(\widehat{K}^{(1,2)})$ at the boundary at $x = 0$. These conditions can then be used to characterize the corresponding selfadjoint extensions of $\bar{K}_0^{(1,2)}$. There are two convenient forms of expressing these boundary conditions to exhibit the correlations between the two sides of the origin.

Boundary Conditions relating Φ_{10}, Φ_{20} to Φ'_{10}, Φ'_{20}

We can rewrite Eqs. (6.51) and (6.52) in terms of two parameters α and β in the form:

$$\Phi_{10} = \alpha \Phi'_{10} - \beta \Phi'_{20}, \tag{6.53}$$

$$\Phi_{20} = \beta \Phi'_{10} - \alpha \Phi'_{20}. \tag{6.54}$$

The reason for rewriting the relations in terms of α and β is to highlight the linkage between the left and right hand sides of the origin. Clearly β acts like a kind of coupling constant linking both sides of the origin. To show that we can indeed do this let us substitute Eqs. (6.51) and (6.52) into these equations and equate the coefficients of arbitrary constants c_1, c_2. We obtain two equations relating α, β to u, v:

$$1 + u = \alpha(\rho + \rho^* u) + \beta \rho^* v, \tag{6.55}$$

$$v = \alpha \rho^* v + \beta(\rho + \rho^* u). \tag{6.56}$$

We can express u and v in terms of α and β and vice versa:

$$u = \frac{(\rho + \rho^*)\alpha - \alpha^2 + \beta^2 - 1}{(1 - \alpha \rho^*)^2 - (\beta \rho^*)^2}, \tag{6.57}$$

$$v = \frac{(\rho - \rho^*)\beta}{(1 - \alpha \rho^*)^2 - (\beta \rho^*)^2}, \tag{6.58}$$

$$\alpha = \frac{(1 + u)(\rho + \rho^* u) - \rho^* v^2}{(\rho + \rho^* u)^2 - (\rho^* v)^2}, \tag{6.59}$$

$$\beta = \frac{(\rho - \rho^*)v}{(\rho + \rho^* u)^2 - (\rho^* v)^2}. \tag{6.60}$$

Any pair of real parameters α, β define a corresponding pair of values of u, v since the common denominator in Eqs. (6.58) and (6.57) does not vanish for any real values of α and β. Also a pair of real parameters α, β do exist for any given u and v, provided the denominator common in Eqs. (6.59) and

(6.60) does not vanish. We can demonstrate this as follows. Multiplying Eq. (6.55) by v^* and Eq. (6.56) by u^* and adding them together we get, using the properties of u, v given by Eq. (6.44),

$$v^* = \alpha \rho v^* + \beta \rho u^* + \beta \rho^*. \tag{6.61}$$

Compare this with the complex conjugate of Eq. (6.56) we get

$$(\alpha^* - \alpha) v^* + (\beta^* - \beta) u^* = -i (\beta^* - \beta), \tag{6.62}$$

and

$$(\alpha - \alpha^*) v + (\beta - \beta^*) u = i (\beta - \beta^*). \tag{6.63}$$

These equations are consistent if

$$\alpha - \alpha^* = \beta - \beta^* = 0, \tag{6.64}$$

i.e., if α and β are real. Multiplying Eqs. (6.62) and (6.63) and using the properties of u and v given in Eq. (6.44) we can show that

$$|\alpha - \alpha^*|^2 = |\beta - \beta^*|^2. \tag{6.65}$$

From this we can deduce from Eq. (6.63) that $u = i \pm v$, a result which contradicts the premise that the common denominator in Eqs. (6.59) and (6.60) does not vanish.

When the denominator common in Eqs. (6.59) and (6.60) vanishes different boundary conditions are required. There are three cases to consider:

Case (1) $\rho + \rho^* u = \rho^* v$. This is equivalent to $u = i + v, v \neq 0$. Equations (6.51) and (6.52) give rise to the following boundary conditions:

$$\Phi_{20} + \Phi_{10} = \gamma (\Phi'_{20} - \Phi'_{10}), \tag{6.66}$$
$$\Phi'_{20} + \Phi'_{10} = 0, \tag{6.67}$$

with

$$\gamma = -\frac{1 + (1 - i) v}{\sqrt{2}\, v}, \quad \text{provided } v \neq 0. \tag{6.68}$$

An example is the extension obtained by setting $\gamma = 0$. This corresponds to

$$v = -\frac{1 + i}{2}, \quad u = -\frac{1 - i}{2}. \tag{6.69}$$

This choice of u, v satisfies Eq. (6.44). The above boundary conditions reduce to

$$\Phi_{20} + \Phi_{10} = 0, \quad \Phi'_{20} + \Phi'_{10} = 0. \tag{6.70}$$

6.3. EXTENSIONS OF DIRECT SUM OPERATORS

The resulting extension is $\widehat{K}_\lambda^{(1,2)}$ in Eq. (6.29) with $\lambda = \pi$.

Case (2) $\rho + \rho^* u = -\rho^* v$. This is equivalent to $u = i - v, v \neq 0$.
Equations (6.51) and (6.52) give rise to the following boundary conditions:

$$\Phi_{20} - \Phi_{10} = -\zeta \Phi'_{10}, \tag{6.71}$$
$$\Phi'_{20} - \Phi'_{10} = 0, \tag{6.72}$$

with

$$\zeta = -2 \frac{1 - (1-i)v}{\sqrt{2}\, v}, \quad \text{provided } v \neq 0. \tag{6.73}$$

An obvious example is the extension obtained by setting $\zeta = 0$. This corresponds to

$$v = \frac{1+i}{2}, \quad u = -\frac{1-i}{2}. \tag{6.74}$$

The above boundary conditions reduce to the continuity of the wave function and its derivative across the origin. The resulting extension is simply the usual kinetic energy operator $\widehat{K}(I\!E)$ for a particle moving in $I\!E$ introduced in E2.4.1 E4. This is the same as $\widehat{K}_\lambda^{(1,2)}$ with $\lambda = 0$. Note that the choice of this pair of u, v satisfies Eq. (6.44).

Case (3) $v = 0, \rho + \rho^* u = 0$. This is equivalent to $v = 0, u = i$ allowed by Eqs. (6.44), (6.46) and (6.47). These values are not compatible with, and hence not realizable by, boundary conditions (6.53) and (6.54) since they contradict Eqs. (6.55) and (6.56). These values are realizable as the *Neumann condition* on the half lines,[16] i.e.,

$$\Phi'_{20} = \Phi'_{10} = 0. \tag{6.75}$$

This amounts to taking the selfadjoint extension:[17]

$$\frac{1}{2m} \left(\widehat{p}_1 \widehat{p}_1^\dagger \oplus \widehat{p}_2 \widehat{p}_2^\dagger \right). \tag{6.76}$$

From E2.5.1 E6 we know that $\widehat{p}_2^\dagger$ acts on $S_{2,1}(I\!E^+)$ and $\widehat{p}_2$ acts on functions ϕ in $S_{2,1}(I\!E^+)$ satisfying boundary condition $\phi(0) = 0$. It follows that the product operator $\widehat{p}_2 \widehat{p}_2^\dagger$ must act on functions satisfying boundary condition $\phi'(0) = 0$ because of the general definition of the domain of a product operator as given in Eq. (2.30).

---

[16] See §3.3.7. While the Neumann condition requires the vanishing of the derivative the Dirichlet condition requires the vanishing of the function at the boundary. See also Blank, Exner and Havlicek (1994) p. 137.

[17] Reed and Simon Vol. 2 (1975) p. 181.

Boundary Conditions relating Φ'_{10}, Φ'_{20} to Φ_{10}, Φ_{20}

Conditions (6.53) and (6.54) may be transformed into the form of

$$\Phi'_{10} = a\Phi_{10} + b\Phi_{20}, \tag{6.77}$$
$$\Phi'_{20} = -a\Phi_{20} - b\Phi_{10}, \tag{6.78}$$

where

$$a = \frac{\alpha}{\alpha^2 - \beta^2}, \quad b = \frac{-\beta}{\alpha^2 - \beta^2}, \tag{6.79}$$
$$\alpha = \frac{a}{a^2 - b^2}, \quad \beta = -\frac{b}{a^2 - b^2}. \tag{6.80}$$

provided $\alpha^2 \neq \beta^2$. Here the coupling between both sides of the origin are again transparent with b playing the role of a coupling constant linking both sides of the origin. For each pair of a, b we have a selfadjoint extension to $\widehat{K}^{(1,2)}$ which we shall denote by $\widehat{K}^{(1,2)}_{a,b}$.

When $\alpha^2 = \beta^2$ we have the following cases:

Case (1) We could have $\alpha = \beta = 0$ which is equivalent to $u = -1, v = 0$. This special case corresponds to the Dirichlet condition on the half lines, i.e.,

$$\Phi_{20} = \Phi_{10} = 0, \tag{6.81}$$

which in turn amounts to taking the following selfadjoint extension[18]

$$\frac{1}{2m}\left(\widehat{p}_1^\dagger \widehat{p}_1 \oplus \widehat{p}_2^\dagger \widehat{p}_2\right). \tag{6.82}$$

Case (2) We could have $\alpha = \beta \neq 0$ which implies that Φ is continuous across the junction with

$$\Phi_{20} = \Phi_{10} = \alpha\left(\Phi'_{10} - \Phi'_{20}\right). \tag{6.83}$$

Case (3) We could have $\alpha = -\beta \neq 0$ which implies that Φ is discontinuous across the junction with

$$\Phi_{20} = -\Phi_{10} = -\alpha\left(\Phi'_{10} + \Phi'_{20}\right). \tag{6.84}$$

---
[18]Reed and Simon Vol. 2 (1975) p. 181.

6.4 Quantization by Parts and Point Interactions

In the absence of any potential a particle of mass m moving in one-dimension along $I\!\!E$ has associated with it the Hilbert space $L^2(I\!\!E)$ on $I\!\!E$; the Hamiltonian of the system is equal to its kinetic energy. We are interested in potentials that are strongly localized. Our object is to model these highly localized potentials by point interactions. The traditional method involves writing down the Schrödinger equation with a Hamiltonian containing the various point interaction potentials put in by hand, e.g., Eq. (6.1) in §6.1. Here we shall present a scheme of *quantization by parts* which will bring out the various types of point interactions automatically.

Mathematically the geometry of a one-dimensional system with a point interaction may be idealized as the one-dimensional Euclidean space $I\!\!E$ broken into two halves, $I\!\!E^-$ and $I\!\!E^+$. The potential abruptly changes at the point $x = 0$. This is referred to as a *branch point* or a *junction* which separates the real line into two branches $\boldsymbol{B}_1 = I\!\!E^-$ and $\boldsymbol{B}_2 = I\!\!E^+$. As before, functions in $\mathcal{H}_1 = L^2(I\!\!E^-)$ are denoted by the subscript 1, e.g., ψ_1 and functions in $\mathcal{H}_2 = L^2(I\!\!E^+)$ are denoted by the subscript 2, e.g., ψ_2. Classically the particle is free when it is on $\boldsymbol{B}_1$ or on $\boldsymbol{B}_2$ and when restricted to these branches the Hamiltonian would agree with the kinetic energy, apart from some additive constant. The idea of quantization by parts is to quantize in each branch separately, and then combine the results to establish the quantum observables for the particle moving across these branches. For example, we shall set up the momentum operators for the particle on branches $\boldsymbol{B}_1$ and $\boldsymbol{B}_2$ separately first, and then combine the results to obtain the momentum operator for the particle existing in the geometric space formed by the two branches together with the branch point. In analogy with classical electrical circuits with branches and branch points we shall call the geometric space formed by the branches and branch points a *circuit*. A circuit may have many branches and branch points. Our present geometry may be called a two-branch circuit.

A striking feature of such a quantization process involves the role played by the kinetic energy. Take the familiar example of a particle confined by an infinite square potential well discussed in E2.5.1 E2. We start with the second order differential operator $\widehat{K}_0(\Lambda)$ acting on domain $C_0^\infty(\Lambda)$ with an aim of obtaining a kinetic energy operator. However, when we look for selfadjointness we find that $\widehat{K}_0(\Lambda)$ has many different selfadjoint extensions in $L^2(\Lambda)$. For example, we have the selfadjoint extension $\widehat{K}^\infty(\Lambda)$ defined by boundary conditions in Eq. (2.173). These conditions embody the infinite and confining nature of the square well potential. A kinetic energy operator should not incorporate any potential. It follows that $\widehat{K}^\infty(\Lambda)$ is no longer a kinetic energy operator. Instead it is the Hamiltonian resulting from an infinite square potential well. Such a state of affair, already discussed in some detail in §3.3.7 in Chapter 3, will also show up in many of the examples to be presented in later sections. Typically when we combine separately quantized symmetric kinetic energy operators on the branches together we will find that the resulting selfadjoint operators are no longer merely kinetic energy operators. Instead they acquire the role of a Hamiltonian with point interactions. This is a case of the "sum" is greater than its parts. Moreover, these selfadjoint extensions enable us to define Hamiltonians with point interactions which are not expressible as a sum of a kinetic energy operator and a potential energy operator, as in Eq. (6.1). Generally we shall call a quantity defined on a branch a *partial quantity*, e.g., we can have a *partial momentum operator* in branch B_1 and we call those separately quantized symmetric kinetic energy operators *partial Hamiltonians*. Different ways of combining these partial operators on the branches result in different physical systems corresponding to different point interactions. Let us now set out the quantization processes in detail.

Quantization by parts: A three-stage quantization scheme

Stage 1: Partial quantization This involves quantities on each branch separately:

1. *Partial Hilbert spaces* On branch B_1 the Hilbert space is taken to be $\mathcal{H}(B_1) = L^2(I\!E^-)$. A similar treatment applies on branch B_2 where the Hilbert space is $\mathcal{H}(B_2) = L^2(I\!E^+)$.

2. *Partial observables* Physical quantities on B_1 are quantized as symmetric operators in $\mathcal{H}(B_1)$, and a similar treatment applies on branch B_2. For a start we can take the two strictly maximal symmetric operators $\widehat{p}_1 = \widehat{p}^-$ and $\widehat{p}_2 = \widehat{p}^+$ in E2.4.1 E5 as the *partial momentum operators* in branches B_1, and B_2 respectively. For the Hamiltonian we shall take, as the *partial Hamiltonian* on B_1, the differential operator $\widehat{H}_{10}$ acting on the domain

$$\mathcal{D}(\widehat{H}_{10}) = C_0^\infty(B_1) = C_0^\infty(I\!E^-). \tag{6.85}$$

In other words we have

$$\widehat{H}_{10}\phi = -\frac{\hbar^2}{2m}\frac{d^2\phi}{dx^2} \quad \forall \phi \in C_0^\infty(B_1). \tag{6.86}$$

Similarly on B_2 the partial Hamiltonian $\widehat{H}_{20}$ is defined on $C_0^\infty(I\!E^+)$. Note that we have not defined partial Hamiltonians in terms of selfadjoint operators. The fact that $\widehat{H}_{10}$ and $\widehat{H}_{20}$ are symmetric and not selfadjoint is of crucial importance in what follows.

6.4. QUANTIZATION BY PARTS AND POINT INTERACTIONS 445

Stage 2: Composite quantization   This involves combining the partial quantities on the branches into a single quantity for the circuit as a whole. This consists of taking the direct sum of the partially quantized quantities:

1. *The composite Hilbert space* $\mathcal{H}^{(1,2)}$ The Hilbert space for the system as a whole, referred to as the *composite Hilbert space*, is taken to be the direct sum of the partial Hilbert spaces for the branches, i.e.,

$$\mathcal{H}^{(1,2)} = \mathcal{H}(\boldsymbol{B}_1) \oplus \mathcal{H}(\boldsymbol{B}_2). \tag{6.87}$$

2. *Composite observables*   Observables for the system as a whole are referred to as *composite observables*. To obtain composite observables we start by forming the direct sums of their corresponding partial quantities in the branches. These direct sums are generally not maximal symmetric, let alone selfadjoint. We must proceed to find their selfadjoint extensions and then use these extensions to represent the observables of the system. If selfadjoint extensions do not exist, we would accept strictly maximal symmetric extensions in the spirit of Postulate GQS presented in §5.2.1.[19] For the momentum and the Hamiltonian we have the following direct sums:

    (a) *Composite momentum*   We shall first form the direct sum of the partial momenta, i.e.,

    $$\widehat{P}_0^{(1,2)} = \widehat{p}_1 \oplus \widehat{p}_2. \tag{6.88}$$

    This operator is symmetric with a one-parameter family of selfadjoint extensions $\widehat{P}_\lambda^{(1,2)}$, according to Theorem 2.14.1(1). We shall assume that one of these selfadjoint extensions can serve as the momentum of the system.

    (b) *Composite Hamiltonian*   We form the direct sum of the partial Hamiltonians, i.e.,

    $$\widehat{H}_0^{(1,2)} = \widehat{H}_{10} \oplus \widehat{H}_{20}. \tag{6.89}$$

    This direct sum $\widehat{H}_0^{(1,2)}$ is again symmetric with many selfadjoint extensions. We shall assume that one of these selfadjoint extensions can serve as the Hamiltonian of the system.

A few remarks are warranted here before proceeding to the final stage of quantization. Clearly composite quantization generally gives rise to non-uniqueness, e.g., we have a host of selfadjoint Hamiltonian operators to choose from. However, this non-uniqueness is a blessing in disguise. It turns out that

---

[19] In §7.7 we shall encounter an example of such a situation in a three-branch circuit.

different selfadjoint extensions to $\widehat{H}_0^{(1,2)}$ may be identified with different point interactions. So, by working out all the selfadjoint extensions we would obtain a description of all point interactions. We can then select an appropriate one to describe a physical system with a particular point interaction.

One may attempt to circumvent this non-uniqueness by taking a selfadjoint extension of $\widehat{H}_{10}$ within $\mathcal{H}(\boldsymbol{B}_1)$ and one for $\widehat{H}_{20}$ within $\mathcal{H}(\boldsymbol{B}_2)$ to form a direct sum. This sum would be automatically selfadjoint and we would have obtained a single selfadjoint direct sum operator. But this does not avoid the non-uniqueness of composite quantization since there is non-uniqueness in the choice of selfadjoint extensions of $\widehat{H}_{10}$ and $\widehat{H}_{20}$ in the first place. More importantly, such a procedure would have eliminated many selfadjoint extensions of $\widehat{H}_0^{(1,2)}$, especially those involving correlations between the wave functions between the branches. This will be evident from the boundary conditions worked out in details for various point interactions in the next section.

Finally we would point out that in the case of one-branch circuits such as a circular coil this composite stage would consist of taking selfadjoint extensions of partially quantized quantities within the branch without having to construct any direct sums.

Stage 3: Correlative quantization We *may* have to correlate or match several cognate observables on physical grounds in order to achieve a consistent final theory applicable to a given physical system. To appreciate this let us look at a two-branch circuit where a large number of possible Hamiltonian and momentum operators are obtained in composite quantization. We may well have to select a matching pair of Hamiltonian and momentum to describe a given physical system. For example this matching process will be required to establish a theory of Josephson junction in superconductivity, a subject to be discussed in the next chapter. For some other systems this stage may not be needed.

6.5 Classification of Point Interactions in $I\!\!E$

Consider the generalized eigenfunction of the momentum operator $\widehat{p}$ in $L^2(I\!\!E)$, i.e., the familiar plane wave

$$\eta_p(x) = \frac{1}{\sqrt{[\ell]}} e^{ipx} = \frac{1}{\sqrt{[\ell]}} e^{ikx}, \quad p = \hbar k \qquad (6.90)$$

where $[\ell]$ denotes a unit length, i.e., it has the value 1 and the dimension of length.[20] Such a plane wave is formally a one-particle wave function since

---

[20] Generally a wave function $\phi(x) \in L^2(I\!\!E)$ is taken to have the dimension of [length$^{1/2}$] so that the normalization integral $\int |\phi(x)|^2\, dx$ is dimensionless. The expression for $\eta_p(x)$ first

6.5. CLASSIFICATION OF POINT INTERACTIONS IN $I\!E$

it is a function of a single position variable, and yet it is not normalizable to 1 and hence not a member of $L^2(I\!E)$. A formal calculation of its associated probability current density, using the standard formula in quantum mechanics

$$J_\phi = \frac{1}{2m}\left\{(\widehat{p}\phi)^*\phi + \phi^*(\widehat{p}\phi)\right\} \tag{6.91}$$

for a normalized wave function $\phi \in L^2(I\!E)$ gives a value for the above plane wave to be

$$J_{\eta_p} = p/m[\ell]. \tag{6.92}$$

This result suggests that, when investigating scattering problems, we can employ the plane wave to represent a beam of particles, each of momentum p, with beam density one particle per length.[21] This suggestion arises from the intuition that a beam of particles moving with velocity $v = p/m$ with beam intensity one particle per unit length will have a probability current density, i.e., particle flux, passing any fixed point equal to $p/m[\ell]$ per second.[22] For a free particle in $I\!E$ the Hamiltonian operator $\widehat{H}^{(0)}(I\!E)$, also denoted by $\widehat{H}^{(0)}$ for short, and the kinetic energy operator $\widehat{K}(I\!E) = \widehat{p}^2/2m$ coincide. They admit generalized eigenfunctions of the form

$$\eta(x) = c_1 e^{ikx} + c_2 e^{-ikx}, \quad x \in I\!E, \tag{6.93}$$

where k is a real constant. Such an eigenfunction represents two beams of particles travelling in opposite directions with beam intensity ratio $|c_1/c_2|^2$. If the beam travelling to the right is of intensity one particle per unit length we can rewrite the above function as

$$\eta(x) = \frac{1}{\sqrt{[\ell]}}\left(e^{ikx} + R e^{-ikx}\right), \quad x \in I\!E, \tag{6.94}$$

for some constant R.

We are interested in interactions localized at the origin which leads to a Hamiltonian with a mirror reflection symmetry about the origin, i.e., a Hamiltonian commuting with the parity operator $\widehat{\wp}$ defined by Eq. (6.30). We want to know how many physically distinct types of such point interactions there are in one dimension. According to our quantization by parts scheme this

---

introduced in Eq. (2.210) is dimensionless. For a discussion of dimensions and dimensional analysis see Giancoli (1988) pp. 6-9.

[21] Dicke and Wittke (1960) §3.5. For motion in $I\!E^3$ the plane wave would be $[\ell]^{-3/2}\exp i\boldsymbol{p}.\boldsymbol{x}$ which has the dimension of $[\text{length}^{3/2}]$. The beam intensity would then be counted as being one per unit volume.

[22] In $I\!E$ the quantity $p/m[\ell]$ has the dimension of $[\text{time}^{-1}]$. The particle flux also has the dimension of $[\text{time}^{-1}]$. In $I\!E^3$ the particle flux has the dimensions of $[\text{length}^2 \times \text{time}^{-1}]$, the same as that of $p/m[\ell]^2$ calculated from Eq. (6.91) using the plane wave in $I\!E^3$.

reduces to the classification of selfadjoint extensions of $\widehat{H}_0^{(1,2)}$. Each class of selfadjoint extensions then act as a class of Hamiltonians representing a particular type of point interaction with distinctive physical properties. According to the discussions in §6.3.1 and §6.3.2 these extensions are characterizable in terms of boundary conditions either of the form of Eqs. (6.53) and (6.54) or of the form of Eqs. (6.77) and (6.78). Under a point interaction at the origin a particle of mass m would move freely from the far left along $I\!E$ until it reaches the origin $x = 0$ where there is an abrupt change of potential. On reaching the origin, a particle would generally have a probability of being reflected as well as being transmitted through the point potential to move over to the right of the origin.

The presence of a point interaction at the origin, such as a step function potential, implies that the Hamiltonian for the system $\widehat{H}$ is not equal to the kinetic energy operator $\widehat{K}(I\!E)$. However, apart from a possible additive constant, $\widehat{H}$ acts as a second order differential operator away from the origin where the potential is either zero or a constant. For a scattering situation with particles incident from the left, the standard physical argument leads us to consider generalized eigenfunctions of the Hamiltonian of the form:[23]

$$\eta(x) = \begin{cases} \frac{1}{\sqrt{[\ell]}} \left(e^{ik_1 x} + R e^{-ik_1 x} \right), & x < 0 \\ \frac{1}{\sqrt{[\ell]}} \left(T e^{ik_2 x} \right), & x > 0 \end{cases} \quad (6.95)$$

where k_1 is real and positive, but R, T and k_2 could be complex. For example, we could have $k_2 = i\kappa$, $\kappa \in I\!R$, then we have in effect an exponentially decaying wave function right of the origin. These constants are related to the point interaction potential. Some point interactions also admit square integrable eigenfunctions of the forms

$$\phi(x) = \begin{cases} \sqrt{\kappa} \, e^{\kappa x}, & x < 0 \\ \sqrt{\kappa} \, e^{-\kappa x}, & x > 0 \end{cases}, \quad (6.96)$$

or

$$\varphi(x) = \begin{cases} \sqrt{\kappa} \, e^{\kappa x}, & x < 0 \\ -\sqrt{\kappa} \, e^{-\kappa x}, & x > 0 \end{cases}, \quad (6.97)$$

where κ is real and positive. At $x = 0$ the function $\phi(x)$ is continuous with a discontinuous derivative while $\varphi(x)$ is discontinuous with a continuous derivative.

For later applications we need to introduce *transfer matrices* to relate the wave function and its spatial derivative at two neighbouring points.[24] For

---
[23] We have assumed an incident beam of intensity of one particle per unit length. For $x > 0$ there is no further potential to reflect the wave. Hence there is only one term.

[24] Pearson (1988) pp. 490-497, Merzbacher (1998) p. 169.

6.5. CLASSIFICATION OF POINT INTERACTIONS IN $I\!E$

the generalized eigenfunction $\eta_p(x)$ in Eq. (6.90) we can introduce a transfer matrix $\mathcal{M}_k^{(0)}(x_2, x_1)$ to relate

$$\eta_p(x_1) \quad \text{and} \quad \eta_p'(x_1) \quad \text{to} \quad \eta_p(x_2) \quad \text{and} \quad \eta_p'(x_2) \qquad (6.98)$$

by[25]

$$\mathcal{M}_k^{(0)}(x_2, x_1) \begin{pmatrix} \eta_p(x_1) \\ \eta_p'(x_1) \end{pmatrix} = \begin{pmatrix} \eta_p(x_2) \\ \eta_p'(x_2) \end{pmatrix}, \qquad (6.99)$$

where

$$\mathcal{M}_k^{(0)}(x_2, x_1) = \begin{pmatrix} \cos k(x_2 - x_1) & k^{-1} \sin k(x_2 - x_1) \\ -k \sin k(x_2 - x_1) & \cos k(x_2 - x_1) \end{pmatrix}. \qquad (6.100)$$

Note that this matrix depends only on the separation $(x_2 - x_1)$, not on the starting point x_1. When the wave function and its derivative are continuous as in the present case, the transfer matrix tends to the 2×2 identity matrix $\mathcal{I}$ as $x_1 \to x_2$, i.e.,

$$\mathcal{M}_k^{(0)}(x_2, x_1) \to \mathcal{I} \quad \text{as} \quad x_1 \to x_2. \qquad (6.101)$$

The idea of transfer matrices can be applied to a general wave function. When the wave function or its derivative are discontinuous at x_1 or x_2 the above transfer matrix is meant to relate

$$\psi(x_{1+}) \quad \text{and} \quad \psi'(x_{1+}) \quad \text{to} \quad \psi(x_{2-}) \quad \text{and} \quad \psi'(x_{2-}), \qquad (6.102)$$

where $\psi(x_{1+})$ and $\psi'(x_{1+})$ signify limiting values as $x \to x_1$ from the right and $\psi(x_{2-})$ and $\psi'(x_{2-})$ denote limiting values as $x \to x_2$ from the left. For later applications we shall introduce four different kinds of transfer matrices, i.e., $\mathcal{M}(x_2, x_1)$, $\mathcal{M}\{x_2, x_1\}$, $\mathcal{M}\{x_2, x_1)$ and $\mathcal{M}\{x_2, x_1]$, which relate $\psi(x_1)$ to $\psi(x_2)$ in four different ways:

1. To $\psi(x_{2-})$, $\psi'(x_{2-})$ from $\psi(x_{1+})$, $\psi'(x_{1+})$ by $\mathcal{M}(x_2, x_1)$.

2. To $\psi(x_{2+})$, $\psi'(x_{2+})$ from $\psi(x_{1-})$, $\psi'(x_{1-})$ by $\mathcal{M}\{x_2, x_1\}$.

3. To $\psi(x_{2+})$, $\psi'(x_{2+})$ from $\psi(x_{1+})$, $\psi'(x_{1+})$ by $\mathcal{M}\{x_2, x_1)$.

---

[25] A prime denotes a spatial derivative, i.e., $\eta_p'(x) = d\eta_p(x)/dx$. Here x_1 and x_2 refer to two arbitrary points on $I\!E$ with $x_1 < x_2$. Equation (6.104) shows that when we go from x_1 to x_2 and then to x_3, the transfer matrix from x_1 to x_2 acts first before the transfer matrix from x_2 to x_3. We have therefore chosen the notation $\mathcal{M}_k^{(0)}(x_2, x_1)$, instead of $\mathcal{M}_k^{(0)}(x_1, x_2)$, to match the order in which transfer matrices act. $\mathcal{M}_k^{(0)}(x_2, x_1)$ also applies to $\eta(x)$ in Eqs. (6.94) and (6.94).

4. To $\psi(x_{2+})$, $\psi'(x_{2+})$ from $\psi(x_1)$, $\psi'(x_1)$ by $\mathcal{M}\{x_2, x_1\}$, when $\psi(x)$ and $\psi'(x)$ are continuous at $x = x_1$.

The wave function and its derivative may not be continuous at the origin. The transfer matrix at $x = 0$ is defined to be

$$\lim_{\epsilon \to 0} \mathcal{M}\{-\epsilon, \epsilon\}, \qquad \epsilon > 0 \tag{6.103}$$

which is generally not an identity matrix. We should mention a useful property, referred to as the *multiplicative property*, of transfer matrices, i.e., the transfer matrices linking successive points can be multiplied together to obtain the transfer matrix linking the beginning and the end points. For example, if $x_3 > x_2 > x_1$ then

$$\mathcal{M}\{x_3, x_1\} = \mathcal{M}\{x_3, x_2] \times \mathcal{M}(x_2, x_1\}. \tag{6.104}$$

When there is a point interaction at $x = x_2$ care has to be taken in choosing the limiting points at $x = x_2$.

We are now ready to classify point interactions into 12 different types.

6.5.1 Type 1 (BC1): The step potential

The simplest point interaction arises from a step potential, e.g.,

$$U(x) = \begin{cases} 0, & \text{if } x < 0 \\ U_0, & \text{if } x > 0 \end{cases}, \tag{6.105}$$

where U_0 is a numerical constant. This potential can be added on to the usual free Hamiltonian $\widehat{H}^{(0)}$ to obtain a Hamiltonian

$$\widehat{H}^{[1]}(E) = \widehat{H}^{(0)} + V(x). \tag{6.106}$$

The boundary conditions imposed on the functions in the domain of $\widehat{H}^{[1]}(E)$, to be referred to as (BC1), are

$$\Psi_{20} = \Psi_{10}, \quad \Psi'_{20} = \Psi'_{10}. \tag{6.107}$$

These conditions mean that the generalized eigenfunctions and their derivatives must be continuous across the origin.

In what follows we shall investigate point interactions not expressible in the form of a sum of $\widehat{H}^{(0)}$ and a potential function.[26] There are eleven types of such interactions resulting in eleven types of selfadjoint Hamiltonians $\widehat{H}^{[2]}(E)$, $\widehat{H}^{[3]}(E), \ldots, \widehat{H}^{[12]}(E)$. We often abbreviate $\widehat{H}^{[2]}(E)$ by $\widehat{H}^{[2]}$ and so on.

---
[26] Many do not even admit symbolic expressions like Eq. (6.1).

6.5.2 Type 2 (BC2): δ-interaction as high-pass filters

There are two selfadjoint extensions of $\widehat{H}_0^{(1,2)}$ in composite quantization which can relate to a δ-function potential, one for an attractive δ-function potential the other for a repulsive one.

Case 1   Attractive δ-interaction as high-pass filters and (BC2_)

This interaction is traditionally described by a Hamiltonian with an attractive δ-function potential written in the form of Eq. (6.1). The corresponding eigenvalue equation is written as:

$$\left(-\frac{\hbar^2}{2m}\frac{d^2}{dx^2} - V_0\,\delta(x)\right)\xi^{[2-]} = E^{[2-]}\xi^{[2-]}, \quad V_0 > 0. \tag{6.108}$$

The constant V_0, to be referred to as the strength of the δ-interaction, shall denote a positive constant throughout this chapter. This equation can be rewritten as

$$\frac{\hbar^2}{2m}\left(-\frac{d^2}{dx^2} - k_0\,\delta(x)\right)\xi^{[2-]} = E^{[2-]}_{k_0}\xi^{[2-]}, \tag{6.109}$$

where

$$k_0 = \frac{2m}{\hbar^2}V_0 \quad \text{or} \quad V_0 = \frac{\hbar^2}{2m}k_0. \tag{6.110}$$

The constant k_0, which is proportional to the product mV_0, plays a significant role in a number of different point interactions. We shall call k_0 the *characteristic wave number of the point interaction*.[27] The above equation admits a solution of the form of Eq. (6.96), i.e.,[28]

$$\phi^{[2-]}_{k_0}(x) = \begin{cases} \sqrt{(k_0/2)}\,\exp\{(k_0/2)\,x\}, & \text{if } x \leq 0 \\ \sqrt{(k_0/2)}\,\exp\{-(k_0/2)\,x\}, & \text{if } x > 0 \end{cases}. \tag{6.111}$$

The corresponding eigenvalue can be obtained by evaluating the eigenvalue equation away from the origin where the potential term vanishes. We have

$$E^{[2-]}_{-,k_0} = -\frac{1}{8m}\left(\hbar k_0\right)^2 = -\frac{m}{2\hbar^2}V_0^2. \tag{6.112}$$

---

[27] The parameter k in Eq. (6.90) is known as the wave number of the plane wave. While the constant U_0 in Eq. (6.105) has the dimension of energy the constant V_0 in Eq. (6.108) has the dimension [energy × length] with the delta function taken as having the dimension of [length^{-1}] since the integral $\int \delta(x)\,dx$ is taken as dimensionless. It follows that $V_0/\hbar$ has the dimension of velocity. The constant k_0, which is taken to be positive throughout this chapter, has the dimension of [length^{-1}].

[28] Merzbacher (1998) p. 107. Greiner (1989) p. 93.

To verify these assertions we simply differentiate $\phi_{k_0}^{[2-]}(x)$ to get

$$\frac{d\phi_{k_0}^{[2-]}(x)}{dx} = \begin{cases} \frac{1}{2} k_0 \sqrt{(k_0/2)} \; \exp\{(k_0/2)\,x\}, & \text{if } x \leq 0 \\ -\frac{1}{2} k_0 \sqrt{(k_0/2)} \; \exp\{-(k_0/2)\,x\}, & \text{if } x > 0 \end{cases}. \quad (6.113)$$

Following Eq. (2.249) we get

$$\frac{d^2 \phi_{k_0}^{[2-]}(x)}{dx^2} = -k_0 \sqrt{(k_0/2)}\, \delta(x) + \frac{1}{4} k_0^2 \, \phi_{k_0}^{[2-]}(x). \quad (6.114)$$

Making use of the formal substitution

$$\delta(x)\phi_{k_0}^{[2-]}(x) = \delta(x)\phi_{k_0}^{[2-]}(0) = \delta(x)\sqrt{(k_0/2)} \quad (6.115)$$

we get

$$\frac{\hbar^2}{2m}\left(-\frac{d^2}{dx^2} - k_0\,\delta(x)\right)\phi_{k_0}^{[2-]}(x)$$

$$= \frac{\hbar^2}{2m}\left(-\frac{d^2 \phi_{k_0}^{[2-]}(x)}{dx^2} - k_0\,\delta(x)\phi_{k_0}^{[2-]}(x)\right)$$

$$= \frac{\hbar^2}{2m}\left(k_0\sqrt{(k_0/2)}\,\delta(x) - \frac{1}{4} k_0^2 \,\phi_{k_0}^{[2-]}(x) - k_0\,\delta(x)\sqrt{(k_0/2)}\right)$$

$$= -\frac{(\hbar k_0)^2}{8m}\,\phi_{k_0}^{[2-]}(x),$$

leading to the same eigenvalue $E_{-,k_0}^{[2-]}$. Eigenfunction $\phi_{k_0}^{[2-]}$ is normalized and represents a bound state. Note that while $\phi_{k_0}^{[2-]}$ is continuous across the origin its derivative is not. Indeed $\phi_{k_0}^{[2-]}$ satisfies the following boundary conditions:

$$\Phi_{20} - \Phi_{10} = 0, \quad \Phi'_{20} - \Phi'_{10} = -k_0 \, \Phi_{10}. \quad (6.116)$$

We shall call these boundary conditions (BC2$_-$). Here Φ_{10} and Φ_{20} are the values of the wave function obtained as x tends to 0 from the left and from the right respectively, and Φ'_{10} and Φ'_{10} are the values of its derivative obtained similarly. The corresponding transfer matrix at $x = 0$ is

$$\mathcal{M}_{k_0}^{[2-]} = \begin{pmatrix} 1 & 0 \\ -\frac{2m}{\hbar^2}V_0 & 1 \end{pmatrix} = \begin{pmatrix} 1 & 0 \\ -k_0 & 1 \end{pmatrix}. \quad (6.117)$$

Moreover, by integrating over a vanishing interval containing the origin, we can show that the operator expression given formally by Eq. (6.1) acts meaningfully only on wave functions satisfying (BC2$_-$).

6.5. CLASSIFICATION OF POINT INTERACTIONS IN $I\!E$

As it stands Eq. (6.1) does not define an operator in $L^2(I\!E)$ since it contains a delta function[29] at the origin and care has to be taken in an ad hoc manner to make sense of the eigenvalue Eq. (6.108). If we now scan the set of selfadjoint extensions of $\widehat{H}_0^{(1,2)}$ and their defining boundary conditions given in §6.3.2 we will reach the following conclusion:

A point interaction with an attractive δ-function potential is described by a Hamiltonian $\widehat{H}_{k_0}^{[2-]}$ defined by a selfadjoint extension of $\widehat{H}_0^{(1,2)}$ specified by boundary conditions (BC2$_-$) obtained from Eqs. (6.53) and (6.54) with $\alpha = \beta = k_0^{-1}$. Alternatively we can simply define such a point interaction directly in terms of $\widehat{H}_{k_0}^{[2-]}$ without reference to Eq. (6.1).[30]

Apart from the above bound state $\widehat{H}_{k_0}^{[2-]}$ also admits generalized eigenfunctions of the form of Eq. (6.95) with $k_1 = k_2 = k$ and the following expressions for R and T in terms of m, V_0 and k:

$$R_k^{[2-]} = -\frac{mV_0}{mV_0 + i\hbar^2 k} = -\frac{k_0/2}{k_0/2 + ik}, \qquad (6.118)$$

$$T_k^{[2-]} = \frac{i\hbar^2 k}{mV_0 + i\hbar^2 k} = \frac{ik}{k_0/2 + ik}. \qquad (6.119)$$

It is interesting to note that

$$k = \frac{1}{2}k_0 \;\Rightarrow\; \left|R_k^{[2-]}\right|^2 = \left|T_k^{[2-]}\right|^2 = \frac{1}{2}. \qquad (6.120)$$

An attractive δ-function potential can serve as a *high-pass filter* with negligible phase shift for the transmitted wave since[31]

$$k \gg k_0 \;\Rightarrow\; R_k^{[2-]} \approx 0, \quad T_k^{[2-]} \approx 1. \qquad (6.121)$$

$$k \ll k_0 \;\Rightarrow\; R_k^{[2-]} \approx -1, \quad T_k^{[2-]} \approx 0. \qquad (6.122)$$

---

[29] δ-functions do not represent wave functions. They are not square-integrable.

[30] The Hilbert space used in §6.3.2 is $L^2(I\!E^-) \oplus L^2(I\!E^+)$. This is unitarily equivalent to $L^2(I\!E)$ since we can convert a function on $I\!E_0^+ \cup I\!E_0^-$ to a function on $I\!E$ by assigning a value to the function at $x = 0$. The choice of value for the function at $x = 0$ is arbitrary. As pointed out in C2.6.1(1) C3 a function in the Hilbert space $L^2(I\!E)$ is only defined up to a set of Lebesgue measure zero so that the choice of its value at any point does not affect the function as an element of $L^2(I\!E)$.

[31] The corresponding reflection and transmission coefficients are $|R_k|^2$ and $|T_k|^2$ respectively.

Case 2   Repulsive δ-interaction as high-pass filters and (BC2$_+$)

A *repulsive δ-interaction* is traditionally given by the following formal Hamiltonian

$$-\frac{\hbar^2}{2m}\frac{d^2}{dx^2} + V_0\,\delta(x) = \frac{\hbar^2}{2m}\left(-\frac{d^2}{dx^2} + k_0\,\delta(x)\right). \qquad (6.123)$$

This operator expression acts meaningfully only on wave functions satisfying the following boundary conditions (BC2$_+$):

$$\Phi_{20} - \Phi_{10} = 0, \quad \Phi'_{20} - \Phi'_{10} = k_0\,\Phi_{10}, \qquad (6.124)$$

where k_0 is related to V_0 by Eq. (6.110). The corresponding transfer matrix at $x = 0$ is

$$\mathcal{M}^{[2+]}_{k_0} = \begin{pmatrix} 1 & 0 \\ \frac{2m}{\hbar^2}V_0 & 1 \end{pmatrix} = \begin{pmatrix} 1 & 0 \\ k_0 & 1 \end{pmatrix}. \qquad (6.125)$$

We can conclude that:

> A point interaction with a repulsive δ-function potential is described by a Hamiltonian $\widehat{H}^{[2+]}_{k_0}$ defined by a selfadjoint extension of $\widehat{H}^{(1,2)}_0$ specified by boundary conditions (BC2$_+$) obtained from Eqs. (6.53) and (6.54) with $\alpha = \beta = -k_0^{-1}$. Indeed we can define such a point interaction directly in terms of $\widehat{H}^{[2+]}_{k_0}$ without reference to Eq. (6.123).

This Hamiltonian also admits generalized eigenfunctions of the form given by Eq. (6.95) with $k_1 = k_2 = k$ and

$$R^{[2+]}_k = -\frac{k_0/2}{k_0/2 - ik}, \quad T^{[2+]}_k = -\frac{ik}{k_0/2 - ik}. \qquad (6.126)$$

This interaction behaves as a high-pass filter with negligible phase shift for the transmitted wave since

$$k \gg k_0 \;\Rightarrow\; R^{[2+]}_k \approx 0, \quad T^{[2+]}_k \approx 1, \qquad (6.127)$$

$$k \ll k_0 \;\Rightarrow\; R^{[2+]}_k \approx -1, \quad T^{[2+]}_k \approx 0, \qquad (6.128)$$

again with

$$k = \frac{1}{2}k_0 \;\Rightarrow\; \left|R^{[2+]}_k\right|^2 = \left|T^{[2+]}_k\right|^2 = \frac{1}{2}. \qquad (6.129)$$

We call this interaction repulsive because $\widehat{H}^{[2+]}_{k_0}$ admits no bound states within the Hilbert space. This terminology applies to all the potentials labelled repulsive in this chapter.

6.5. CLASSIFICATION OF POINT INTERACTIONS IN $I\!\!E$

Finally the δ-potential could be situated at $x = \epsilon \neq 0$, i.e., we could have a δ-function $\delta(x - \epsilon)$ in Eq. (6.123), but the transfer matrix at $x = \epsilon$ linking the wave function and its derivative from ϵ_- to ϵ_+ will be the same as $\mathcal{M}_{k_0}^{[2+]}$. The same remark applies to $\mathcal{M}_{k_0}^{[2-]}$ for the attractive δ-potential.

6.5.3 Type 3 (BC3): δ'-interaction as low-pass filters

Most point interactions cannot be described by an explicit potential function in a simple manner, not even in a formal sense. The so-called δ'-interaction is an example. These interactions can be properly described by Hamiltonians obtained from appropriate selfadjoint extensions of $\widehat{H}_0^{(1,2)}$.

Case 1   Attractive δ'-interaction as low-pass filters and (BC3$_-$)

The so-called δ'-interaction is often symbolically represented by the following formal Hamiltonian with a δ'-function potential:[32]

$$\frac{\hbar^2}{2m}\left(-\frac{d^2}{dx^2} - k_0\,\delta'_{[\ell]}(x)\right), \quad \delta'_{[\ell]}(x) = [\ell]\,\delta'(x) = [\ell]\,\frac{d\delta(x)}{dx}. \tag{6.130}$$

This is very appealing at first sight since a δ'-potential can be visualized as a kind of dipole potential formed by a pair of attractive and repulsive δ-potentials centered around the origin, i.e.,

$$\delta'(x) = \frac{d\delta(x)}{dx} = \lim_{\epsilon \to 0} \frac{\delta(x+\epsilon) - \delta(x-\epsilon)}{2\epsilon}. \tag{6.131}$$

Unfortunately Eqs. (6.130) and (6.131) do not lead to a definition of a selfadjoint Hamiltonian.[33] We can again scan the set of selfadjoint extensions of $\widehat{H}_0^{(1,2)}$ to try to obtain a Hamiltonian which may in some restricted sense be pictured as corresponding to a δ'-function potential. This search results in a Hamiltonian $\widehat{H}_{k_0}^{[3-]}$ specified by the following boundary conditions (BC3$_-$):[34]

$$\Phi_{20} - \Phi_{10} = -\frac{1}{k_0}\,\Phi'_{10}, \quad \Phi'_{20} - \Phi'_{10} = 0, \tag{6.132}$$

with transfer matrix at $x = 0$ given by

$$\mathcal{M}_{k_0}^{[3-]} = \begin{pmatrix} 1 & -k_0^{-1} \\ 0 & 1 \end{pmatrix}. \tag{6.133}$$

---

[32] Here k_0 is a constant having the dimension $[\text{length}^{-1}]$ as before, and $[\ell]$ is a unit length inserted purely for dimensional reasons.
[33] Seba (1986), (1987), Coutinho, Nogami and Perez (1997).
[34] Blank, Exner and Havlicek (1994) p. 478, Albeverio, Gesztesy, Hoegh-Krohn and Holden (1988).

Conditions (BC3_) arise from Eqs. (6.71) and (6.72) with $\zeta = k_0^{-1}$. The eigenvalue equation

$$\widehat{H}_{k_0}^{[3-]} \xi^{[3-]}(x) = E_{k_0}^{[3-]} \xi^{[3-]}(x) \tag{6.134}$$

admits a solution of the form of Eq. (6.97), i.e.,

$$\varphi_{k_0}^{[3-]}(x) = \begin{cases} \sqrt{2k_0}\ \exp\{(2k_0)\,x\}, & x < 0 \\ -\sqrt{2k_0}\ \exp\{-(2k_0)\,x\}, & x > 0 \end{cases}. \tag{6.135}$$

Clearly $\varphi_{k_0}^{[3-]}(x)$ satisfies boundary conditions (BC3_) and represents a bound state corresponding to the eigenvalue

$$E_{-,k_0}^{[3-]} = -\frac{1}{2m}(2\hbar k_0)^2 = -\frac{8m}{\hbar^2} V_0^2. \tag{6.136}$$

Hamiltonian $\widehat{H}_{k_0}^{[3-]}$ also admits generalized eigenfunctions of the form of Eq. (6.95) with $k_1 = k_2 = k$ and

$$R_k^{[3-]} = \frac{ik}{2k_0 + ik}, \qquad T_k^{[3-]} = \frac{2k_0}{2k_0 + ik}. \tag{6.137}$$

An attractive δ'-interaction can serve as a *low-pass filter* with negligible phase shift for the transmitted wave since

$$k \gg k_0 \quad \Rightarrow \quad R_k^{[3-]} \approx 1, \quad T_k^{[3-]} \approx 0, \tag{6.138}$$

$$k \ll k_0 \quad \Rightarrow \quad R_k^{[3-]} \approx 0, \quad T_k^{[3-]} \approx 1, \tag{6.139}$$

$$k = 2k_0 \quad \Rightarrow \quad \left|R_k^{[3-]}\right|^2 = \left|T_k^{[3-]}\right|^2 = \frac{1}{2}. \tag{6.140}$$

The appearance of the factor $2k_0$, rather than $k_0/2$, may be attributed to the link between a δ'-interaction with four δ-interactions, as shown in Fig. 6.5.3(1) later.

In contrast to a δ-interaction $\widehat{H}^{[3-]}$ operates on a domain consisting of wave functions which are discontinuous. This discontinuity causes a problem for the traditional Eq. (6.130) using $\delta'(x)$ as a potential function. Take for example the eigenfunction $\varphi_{k_0}^{[3-]}(x)$. To carry out a formal calculation according to Eq. (2.249) we get

$$\varphi_{k_0}^{[3-]\prime}(x) = -2\sqrt{2k_0}\,\delta(x) + \begin{cases} (2k_0)^{\frac{3}{2}}\ \exp\{(2k_0)\,x\}, & \text{if } x < 0 \\ (2k_0)^{\frac{3}{2}}\ \exp\{-(2k_0)\,x\}, & \text{if } x > 0 \end{cases}$$

$$\varphi_{k_0}^{[3-]\prime\prime}(x) = -2\sqrt{2k_0}\,\delta'(x) + (2k_0)^2\,\varphi_{k_0}^{[3-]}(x). \tag{6.141}$$

6.5. CLASSIFICATION OF POINT INTERACTIONS IN $\mathbb{E}$

Hence, we have

$$\frac{\hbar^2}{2m}\left(-\frac{d^2}{dx^2} - k_0\,[\,\ell\,]\delta'(x)\right)\varphi_{k_0}^{[3-]}(x) \tag{6.142}$$

$$= \frac{\hbar^2}{2m}\left(-(2k_0)^2\,\varphi_{k_0}^{[3-]}(x) + 2\sqrt{2k_0}\,\delta'(x) - k_0\,[\,\ell\,]\,\delta'(x)\,\varphi_{k_0}^{[3-]}(x)\right). \tag{6.143}$$

The last term on the right which contains a δ'-function together with a discontinuous $\varphi_{k_0}^{[3-]}(x)$ is undefined and is not formally cancelled out by the middle term which also contains a δ'-function.[35] So Eq. (6.130) fails to lead to the equivalence of eigenvalue Eq. (6.134). Also one cannot obtain boundary conditions (BC3$_-$) by a formal integration of Eq. (6.134) as we can in the case of a δ-potential. A number of investigations reveal a link between boundary conditions (BC3$_-$) and a δ'-potential in a complicated manner including a renormalization process.[36] It is therefore misleading to interpret $\widehat{H}_{k_0}^{[3-]}$ and boundary conditions (BC3$_-$) as literally equivalent to a δ'-potential. Failure to appreciate this will lead to a lot of confusion.[37] We shall return to give an intuitive understanding to a δ'-interaction in term of a limiting family of δ-interactions after a brief discussion of a repulsive δ'-interaction.

Case 2   **Repulsive δ'-interaction as low-pass filters and (BC3$_+$)**

We can introduce a repulsive δ'-interaction described by a selfadjoint extension $\widehat{H}_{k_0}^{[3+]}$ of $\widehat{H}_0^{(1,2)}$ specified by boundary conditions (BC3$_+$):

$$\Phi_{20} - \Phi_{10} = \frac{1}{k_0}\Phi'_{10}, \quad \Phi'_{20} - \Phi'_{10} = 0 \tag{6.144}$$

with transfer matrix at $x = 0$ given by

$$\mathcal{M}_{k_0}^{[3+]} = \begin{pmatrix} 1 & k_0^{-1} \\ 0 & 1 \end{pmatrix}. \tag{6.145}$$

Extension $\widehat{H}_{k_0}^{[3+]}$ admits generalized eigenfunctions of the form of Eq. (6.95) with $k_1 = k_2 = k$ and

$$R_k^{[3+]} = -\frac{ik}{2k_0 - ik}, \quad T_k^{[3+]} = \frac{2k_0}{2k_0 - ik}. \tag{6.146}$$

---

[35]If we try to use the prescription in Papoulis (1962) p. 274
$$f(x)\delta'(x) = f(0)\delta'(x) - f'(0)\delta(x)$$
we can see the difficulty due to the discontinuity of $\varphi_{k_0}^{[3-]\prime}(x)$ at $x = 0$.

[36]Seba (1986), (1987), Albeverio, Gesztesy and Holden (1992).

[37]Zhao (1992), Albeverio, Gesztesy and Holden (1992), Coutinho, Nogami and Perez (1997).

This point interaction can serve as a low pass filter with negligible phase shift for the transmitted wave since

$$k \gg k_0 \quad \Rightarrow \quad R_k^{[3+]} \approx 1, \quad T_k^{[3+]} \approx 0, \tag{6.147}$$

$$k \ll k_0 \quad \Rightarrow \quad R_k^{[3+]} \approx 0, \quad T_k^{[3+]} \approx 1, \tag{6.148}$$

$$k = 2k_0 \quad \Rightarrow \quad \left|R_k^{[3+]}\right|^2 = \left|T_k^{[3+]}\right|^2 = \frac{1}{2}. \tag{6.149}$$

Limiting families of δ-function potentials

We have pointed out that an attempt to understand a δ'-interaction in the form of a dipole potential proves to be inappropriate. However, we can have an understanding of $\widehat{H}_{k_0}^{[3-]}$ and $\widehat{H}_{k_0}^{[3+]}$ in terms of *a limiting family of δ-function potentials* whose meaning will become clear in what follows. Such an understanding will be seen to apply to a large class of point interactions.

Asymmetric quadrapole point potential for repulsive δ'-interaction

Let ϵ be a small distance much less than a unit length, n be a positive integer, and let

$$v_{n\epsilon}^+(x) = g_\epsilon^+ \, \delta(x - n\epsilon), \quad g_\epsilon^+ = \frac{\hbar^2}{2m\epsilon}\left(\frac{1}{\sqrt{k_0\epsilon}} - 1\right). \tag{6.150}$$

When ϵ is sufficiently small g_ϵ^+ is positive and $v_{n\epsilon}^+(x)$ represents a repulsive δ-function potential at $x = n\epsilon$. For an attractive potential at $x = n\epsilon$ we introduce

$$v_{n\epsilon}^-(x) = g_\epsilon^- \, \delta(x - n\epsilon), \quad g_\epsilon^- = \frac{\hbar^2}{2m\epsilon}\left(\sqrt{k_0\epsilon} - 1\right), \tag{6.151}$$

since g_ϵ^- is negative for small ϵ. Note that g_ϵ^- and g_ϵ^+ are independent of n.

We could have $v_\epsilon^-(x)$, $v_{2\epsilon}^+(x)$, $v_{3\epsilon}^+(x)$ and $v_{4\epsilon}^-(x)$ representing four δ-function potentials, two attractive and two repulsive, located at $x = \epsilon, 2\epsilon, 3\epsilon$ and 4ϵ from the origin. Combining these together we obtain a new potential

$$V_\epsilon^{[3+]}(x) = v_\epsilon^-(x) + v_{2\epsilon}^+(x) + v_{3\epsilon}^+(x) + v_{4\epsilon}^-(x). \tag{6.152}$$

The corresponding formal Hamiltonian is

$$\begin{aligned}\widehat{H}_\epsilon^{[3+]} &= -\frac{\hbar^2}{2m}\frac{d^2}{dx^2} + V_{+,\epsilon}^{[3]}(x) \tag{6.153}\\ &= \frac{\hbar^2}{2m}\Big(-\frac{d^2}{dx^2} + k_\epsilon^-\delta(x-\epsilon) + k_\epsilon^+\delta(x-2\epsilon) \\ &\qquad + k_\epsilon^+\delta(x-3\epsilon) + k_\epsilon^-\delta(x-4\epsilon)\Big), \tag{6.154}\end{aligned}$$

6.5. CLASSIFICATION OF POINT INTERACTIONS IN $I\!E$

where
$$k_\epsilon^- = \frac{2m}{\hbar^2} g_\epsilon^- \quad \text{and} \quad k_\epsilon^+ = \frac{2m}{\hbar^2} g_\epsilon^+. \tag{6.155}$$

This potential is symbolically shown in Fig. 6.5.3(1).

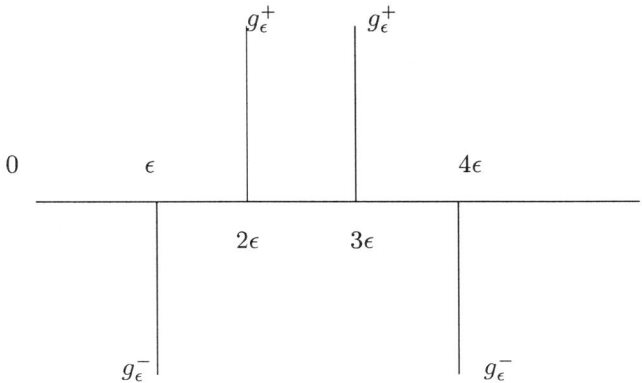

Fig. 6.5.3(1) $V_\epsilon^{[3+]}$ for a repulsive δ'-potential.

The corresponding formal Hamiltonian is

$$\widehat{H}_\epsilon^{[3+]} = -\frac{\hbar^2}{2m} \frac{d^2}{dx^2} + V_{+,\epsilon}^{[3]}(x) \tag{6.156}$$

$$= \frac{\hbar^2}{2m} \left(-\frac{d^2}{dx^2} + k_\epsilon^- \delta(x-\epsilon) + k_\epsilon^+ \delta(x-2\epsilon) \right.$$
$$\left. + k_\epsilon^+ \delta(x-3\epsilon) + k_\epsilon^- \delta(x-4\epsilon) \right), \tag{6.157}$$

where
$$k_\epsilon^- = \frac{2m}{\hbar^2} g_\epsilon^- \quad \text{and} \quad k_\epsilon^+ = \frac{2m}{\hbar^2} g_\epsilon^+. \tag{6.158}$$

We shall now show that this formal Hamiltonian offers an understanding of a repulsive δ'-interaction. The argument[38] is based on the fact that the transfer matrix $\mathcal{M}_{k_0,\epsilon}^{[3+]}\{4\epsilon, 0\}$ linking the wave function Φ in the domain of $\widehat{H}_\epsilon^{[3+]}$ and its derivative Φ' between $x = 0_-$ and $x = 4\epsilon_+$ tends to $\mathcal{M}_{k_0}^{[3+]}$ as $\epsilon \to 0$. Consequently $\widehat{H}_\epsilon^{[3+]}$ would act on wave functions satisfying boundary conditions (BC3$_+$) in the limit as $\epsilon \to 0$. To verify the argument we first

---
[38] Following Pearson (1988) pp. 490-497.

write down the transfer matrix $\mathcal{M}^{[3+]}_{k_0,\epsilon}\{4\epsilon,0\}$ in terms of the transfer matrices between the following pairs of positions:

$$4\epsilon_+, 4\epsilon_-; \quad 4\epsilon_-, 3\epsilon_+; \quad 3\epsilon_+, 3\epsilon_-; \quad 3\epsilon_-, 2\epsilon_+; \quad (6.159)$$

$$2\epsilon_+, 2\epsilon_-; \quad 2\epsilon_-, \epsilon_+; \quad \epsilon_+, \epsilon_-; \quad \epsilon_-, 0_-. \quad (6.160)$$

There are only two distinct types of matrices involved here. The transfer matrices between $4\epsilon_-, 3\epsilon_+$, between $3\epsilon_-, 2\epsilon_+$, between $2\epsilon_-, \epsilon_+$, and between $\epsilon_-, 0_-$ are for free evolution and for $\eta(x)$ in Eq. (6.93). These matrices are of the form of Eq. (6.100). We shall denote these matrices by:

$$\mathcal{M}^{(0)}_k(4\epsilon,3\epsilon), \quad \mathcal{M}^{(0)}_k(3\epsilon,2\epsilon), \quad \mathcal{M}^{(0)}_k(2\epsilon,\epsilon), \quad \mathcal{M}^{(0)}_k(\epsilon,0\}. \quad (6.161)$$

These transfer matrices can be joined up with the transfer matrices for the δ-interactions at $x = \epsilon, 2\epsilon, 3\epsilon$ and 4ϵ. The attractive interactions at ϵ and 4ϵ have the same transfer matrix $\mathcal{M}^{[2]}_{k_\epsilon^-}$ given, according to Eq. (6.117), by[39]

$$\mathcal{M}^{[2]}_{k_\epsilon^-} = \begin{pmatrix} 1 & 0 \\ k_\epsilon^- & 1 \end{pmatrix}, \quad (6.162)$$

while the repulsive interactions at 2ϵ and 3ϵ have the same transfer matrix $\mathcal{M}^{[2]}_{k_\epsilon^+}$ given, according to Eq. (6.125), by

$$\mathcal{M}^{[2]}_{k_\epsilon^+} = \begin{pmatrix} 1 & 0 \\ k_\epsilon^+ & 1 \end{pmatrix}. \quad (6.163)$$

We obtain the desired matrix $\mathcal{M}^{[3+]}_{k_0,\epsilon}\{4\epsilon,0\}$ by multiplying the above matrices together as we do in Eq. (6.104), i.e., we have

$$\begin{aligned}\mathcal{M}^{[3+]}_{k_0,\epsilon}\{4\epsilon,0\} &= \mathcal{M}^{[2]}_{k_\epsilon^-} \times \mathcal{M}^{(0)}_k(4\epsilon,3\epsilon) \times \mathcal{M}^{[2]}_{k_\epsilon^+} \times \mathcal{M}^{(0)}_k(3\epsilon,2\epsilon) \\ &\times \mathcal{M}^{[2]}_{k_\epsilon^+} \times \mathcal{M}^{(0)}_k(2\epsilon,\epsilon) \times \mathcal{M}^{[2]}_{k_\epsilon^-} \times \mathcal{M}^{(0)}_k(\epsilon,0\}. \end{aligned} \quad (6.164)$$

As $\epsilon \to 0$ the matrices in (6.161) reduce to the form

$$\begin{pmatrix} 1 + O(\epsilon^2) & \epsilon + O(\epsilon^3) \\ -k^2\epsilon + O(\epsilon^3) & 1 + O(\epsilon^2) \end{pmatrix}. \quad (6.165)$$

Here $O(\epsilon^2)$ denotes terms of second order in ϵ and above and $O(\epsilon^3)$ denotes those of third order and above. A lengthy calculation yields the following expression, up to order of ϵ, for $\mathcal{M}^{[3+]}_{k_0,\epsilon}\{4\epsilon,0\}$:

$$\mathcal{M}^{[3+]}_{k_0,\epsilon}\{4\epsilon,0\} = \begin{pmatrix} 1+O(\epsilon) & k_0^{-1}+O(\epsilon^{1/2}) \\ O(\epsilon) & 1+O(\epsilon) \end{pmatrix}. \quad (6.166)$$

---

[39]Here k_ϵ^- is negative and k_ϵ^+ is positive.

6.5. CLASSIFICATION OF POINT INTERACTIONS IN $I\!E$

As $\epsilon \to 0$ the potentials collapse to the origin and we get

$$\lim_{\epsilon \to 0} \mathcal{M}^{[3+]}_{k_0,\epsilon}\{4\epsilon, 0\} = \mathcal{M}^{[3+]}_{k_0}. \tag{6.167}$$

We conclude that $\widehat{H}^{[3+]}_{k_0}$ with boundary conditions (BC3$_+$) may be visualized as corresponding to a Hamiltonian with a limiting family of δ-function potentials. This family does not correspond to a dipole potential shown in Eq. (6.131) since it consists of attractive and repulsive potentials of different strengths. We shall call this family of two asymmetric dipoles of δ-interactions an *asymmetric quadrapole point interaction* and this set of δ-function potentials as an *asymmetric quadrapole point potential*. One can now appreciate the appearance of the factor $2k_0$ in various places, in contrast to the appearance of $\frac{1}{2}k_0$ for a single δ-interaction.

It turns out that such an asymmetric quadrapole point interaction forms a basic unit which can be combined with additional δ-interactions and additonal asymmetric quadrapole point interactions to produce the effect of many other point interactions. We shall demonstrate this in the case of an attractive δ'-interaction presently. More examples will be presented later on.

Note that the choice of a limiting family of δ-interactions is not unique, e.g., some of the spacings could be changed without affecting the limit and in some later examples even the strength of the interactions involved can be changed. It is also possible to have the above family of δ-interactions starting from any point, say from $x = b > 0$. Then we will have

$$\begin{aligned}\mathcal{M}^{[3+]}_{k_0,\epsilon}\{b+4\epsilon, b\} &= \mathcal{M}^{[2]}_{k_\epsilon^-} \times \mathcal{M}^{(0)}_k(b+4\epsilon, b+3\epsilon) \times \mathcal{M}^{[2]}_{k_\epsilon^+} \\ &\quad \times \mathcal{M}^{(0)}_k(b+3\epsilon, b+2\epsilon) \times \mathcal{M}^{[2]}_{k_\epsilon^+} \times \mathcal{M}^{(0)}_k(b+2\epsilon, b+\epsilon) \\ &\quad \times \mathcal{M}^{[2]}_{k_\epsilon^-} \times \mathcal{M}^{(0)}_k(b+\epsilon, b\}. \end{aligned} \tag{6.168}$$

Asymmetric quadrapole point potentials for attractive δ'-interaction

The attractive δ'-interaction is not obtained in terms of a family of potentials of the form $-V^{[3+]}_\epsilon(x)$ as one might expect. A rather complicated combination of asymmetric quadrapole point potentials and δ-potentials is required. Let us consider the following family of point interactions:

1. an attractive δ-interaction of strength V_0 with $k_0 = 2mV_0/\hbar^2$ at the origin, followed by

2. an asymmetric quadrapole point potential starting at the origin and ending at $x = 4\epsilon$, followed by

3. an attractive δ-interaction of strength V_0 at $x = 5\epsilon$, followed by

4. an asymmetric quadrapole point potential starting at $x = 5\epsilon$ and ending at $x = 9\epsilon$, followed by

5. an attractive δ-interaction of strength V_0 at $x = 10\epsilon$, followed by

6. an attractive δ-interaction of strength $2V_0$ at $x = 11\epsilon$, followed by

7. an asymmetric quadrapole point potential starting at $x = 11\epsilon$ and ending at $x = 15\epsilon$, followed by

8. an attractive δ-interaction of strength $2V_0$ at $x = 16\epsilon$, followed by

9. an asymmetric quadrapole point potential starting at $x = 16a$ and ending at $x = 20\epsilon$.

The corresponding potential is

$$\begin{aligned} V_\epsilon^{[3]-}(x) &= -V_0\,\delta(x) + V_\epsilon^{[3+]}(x) - V_0\,\delta(x-5\epsilon) \\ &+ V_\epsilon^{[3+]}(x-5\epsilon) - V_0\,\delta(x-10\epsilon) \\ &- 2V_0\,\delta(x-11\epsilon) + V_\epsilon^{[3+]}(x-11\epsilon) \\ &- 2V_0\delta(x-16\epsilon) + V_\epsilon^{[3+]}(x-16\epsilon). \end{aligned} \qquad (6.169)$$

To justify our claim we can examine the transfer matrix linking the wave function and its derivative for the resulting Hamiltonian. It is[40]

$$\begin{aligned} &\mathcal{M}_{k_0,\epsilon}^{[3-]}\{20\epsilon,0\} \\ &= \mathcal{M}_{k_0,\epsilon}^{[3+]}\{20\epsilon,16\epsilon\} \times \mathcal{M}_{2k_0}^{[2-]} \times \mathcal{M}_k^{(0)}(16\epsilon,15\epsilon) \times \mathcal{M}_{k_0,\epsilon}^{[3+]}\{15\epsilon,11\epsilon\} \\ &\times \mathcal{M}_{2k_0}^{[2-]} \times \mathcal{M}_k^{(0)}(11\epsilon,10\epsilon) \times \mathcal{M}_{k_0}^{[2-]} \times \mathcal{M}_k^{(0)}(10\epsilon,9\epsilon) \times \mathcal{M}_{k_0,\epsilon}^{[3+]}\{9\epsilon,5\epsilon\} \\ &\times \mathcal{M}_{k_0}^{[2-]} \times \mathcal{M}_k^{(0)}(5\epsilon,4\epsilon) \times \mathcal{M}_{k_0,\epsilon}^{[3+]}\{4\epsilon,0\} \times \mathcal{M}_{k_0}^{[2-]}. \end{aligned} \qquad (6.170)$$

To the first order of ϵ we get

$$\mathcal{M}_{k_0,\epsilon}^{[3]-}\{20\epsilon,0\} = \begin{pmatrix} 1 + O(\epsilon^{1/2}) & -k_0^{-1} + O(\epsilon^{1/2}) \\ O(\epsilon^{1/2}) & 1 + O(\epsilon^{1/2}) \end{pmatrix}. \qquad (6.171)$$

As $\epsilon \to 0$ the potentials collapse to a point and

$$\lim_{\epsilon \to 0} \mathcal{M}_{k_0,\epsilon}^{[3-]}\{20\epsilon,0\} = \begin{pmatrix} 1 & -k_0^{-1} \\ 0 & 1 \end{pmatrix}, \qquad (6.172)$$

which is the transfer matrix for an attractive δ'-interaction.

A diagramatic illustration of $V_\epsilon^{[3]-}(x)$ is sketched in Fig. 6.5.3(2).

---

[40]In Eq. (6.164) if the last matrix $\mathcal{M}_k^{(0)}(\epsilon,0\}$ is replaced by $\mathcal{M}_k^{(0)}(\epsilon,0)$ to exclude the origin, the resulting product matrix will be denoted by $\mathcal{M}_{k_0,\epsilon}^{[3+]}\{4\epsilon,0)$.

6.5. CLASSIFICATION OF POINT INTERACTIONS IN $\mathbb{E}$

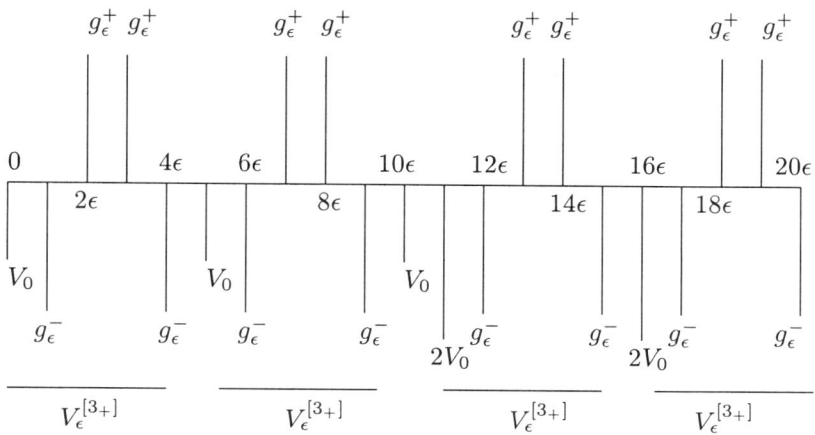

Fig. 6.5.3(2) $V_\epsilon^{[3-]}$ for an attractive δ'-potential.

6.5.4 Type 4 (BC4): Perfect reflector

There is a selfadjoint extension $\widehat{H}^{[4]}$ corresponding to a rigid and impenetrable partition at the origin separating the real line into two uncorrelated branches. This extension is specified by boundary conditions (BC4)

$$\Phi_{20} = 0, \quad \Phi_{10} = 0. \tag{6.173}$$

This is the Dirichlet condition arising from Eqs. (6.53) and (6.54) with $\alpha = \beta = 0$ and $\widehat{H}^{[4]}$ is given explicitly earlier by Eq. (6.82). There is no transfer matrix for (BC4) since the two branches are unrelated. The generalized eigenfunctions of $\widehat{H}^{[4]}$ on branch $\boldsymbol{B}_1$ are of the form of a standing wave[41]

$$\eta_k^{[4]}(x) = \frac{1}{\sqrt{[\ell]}} \left(e^{ikx} - e^{-ikx}\right), \quad x < 0 \tag{6.174}$$

where k is real. Hamiltonian $\widehat{H}^{[4]}$ with the point interaction at the origin acts as a *perfect reflector* for incident plane waves of all wave lengths. An incoming plane wave from the left undergoes a π-phase change on reflection. A similar situation holds on branch $\boldsymbol{B}_2$.

---

[41] We employ the notation φ to denote normalizable state, e.g., bound states, and η to denote an unnormalizable state, e.g., generalized eigenfunctions.

6.5.5 Type 5 (BC5): Elastic reflectors

The following boundary conditions (BC5)

$$\Phi_{20} = -\alpha\,\Phi'_{20}, \quad \Phi_{10} = \alpha\,\Phi'_{10}, \quad \alpha \in I\!R \text{ and } \alpha \neq 0, \tag{6.175}$$

are known as *elastic boundary conditions*[42] which arise from Eqs. (6.53) and (6.54) with $\beta = 0$. There is no transfer matrix since the wave functions on both sides of the origin are not directly related; they are not totally unrelated either since they share a common parameter α. Conditions (BC5) specify a set of selfadjoint extensions $\widehat{H}_\alpha^{[5]}$ which admit generalized eigenfunctions of the form

$$\eta_k^{[5]}(x) = \begin{cases} \frac{1}{\sqrt{[\ell]}}\left(e^{ikx} + R_k^{[5]}\,e^{-ikx}\right), & x < 0, \\ \frac{1}{\sqrt{[\ell]}}\left(R_k^{[5]}\,e^{ikx} + e^{-ikx}\right), & x > 0, \end{cases} \tag{6.176}$$

where k is real and

$$R_k^{[5]} = (i\alpha k - 1)/(i\alpha k + 1), \quad |R_k^{[5]}| = 1. \tag{6.177}$$

As a result an incident wave from the left or the right is totally reflected. We can rewrite $R_k^{[5]}$ as

$$R_k^{[5]} = e^{i\gamma}, \quad \gamma = \tan^{-1}\left[\frac{-2\alpha k}{1-(\alpha k)^2}\right]. \tag{6.178}$$

This means that there is a flexibility in the phase shift on reflection,[43] i.e., the phase shift is adjustable and not fixed to be π as in the perfect reflector, hence the name *elastic reflector*.

6.5.6 Type 6 (BC6): Open end

The following conditions (BC6):

$$\Phi'_{10} = 0, \quad \Phi'_{20} = 0, \tag{6.179}$$

---

[42] Bratteli and Robinson (1981) p. 65.

[43] A wave function Φ is said to have a phase shift λ across the point $x = 0$ if Φ satisfies the following boundary condition

$$\Phi_{20} = e^{i\lambda}\,\Phi_{10}, \quad \lambda > 0.$$

A wave function suffers a *phase retardation* or *phase delay* if

$$\Phi_{20} = e^{-i\lambda}\,\Phi_{10}, \quad \lambda > 0.$$

The term phase shift is also used as a general term for any phase change on transmission and on reflection at a boundary.

specify a selfadjoint extension $\widehat{H}^{[6]}$. This is the Neumann condition given by Eq. (6.75) and $\widehat{H}^{[6]}$ is explicitly given by Eq. (6.76). There is no transfer matrix. In contrast to the perfect reflector we have the derivative of the wave function vanishing as it tends to the origin from both sides. This corresponds to the boundary conditions on a vibrating semi-infinite classical string with an open end.[44] Generalized eigenfunctions of $\widehat{H}^{[6]}$ on the left of the origin are

$$\eta_k^{[6]}(x) = \frac{1}{\sqrt{[\ell]}} \left(e^{ikx} + e^{-ikx} \right), \quad x < 0, \tag{6.180}$$

where k is real. We have a standing wave with a vanishing probability current; there is no transfer of probability across the origin. The wave undergoes no phase change on reflection. As a result the incident and reflected waves reinforce each other at the open end.

6.5.7 Type 7 (BC7): Ideal π-phase shifters

In optics there are *phase shifters* or *phase retarders*, e.g., wave and phase plates, which shift the phase of a light wave on transmission.[45] Ideal phase shifters are transparent without causing any reflection.[46] For quantum waves we can have phase shifters provided by point interactions. There are several types of point interactions resulting in different phase shifts, e.g., a π-phase shift, a $\frac{1}{2}\pi$-phase shift or an arbitrary phase shift. We shall start with a π-phase shifter. The point interaction having the effect of a perfectly transparent π-phase shifter at the origin corresponds to the Hamiltonian $\widehat{H}^{[7]}$ defined by boundary conditions (BC7):

$$\Phi_{20} + \Phi_{10} = 0, \quad \Phi'_{20} + \Phi'_{10} = 0. \tag{6.181}$$

These arise from Eqs. (6.66) and (6.67) with $\gamma = 0$. The corresponding transfer matrix is

$$\mathcal{M}^{[7]} = \begin{pmatrix} -1 & 0 \\ 0 & -1 \end{pmatrix}. \tag{6.182}$$

For phase shifters in optics the amount of phase shift generally depends on the frequency of the incoming light waves. In contrast our present $\widehat{H}^{[7]}$ causes a π-phase shift of an incoming wave and its derivative of all frequencies of incident waves without loss, hence the name *ideal π-phase shifters*. This becomes

---

[44] Crawford (1968) p. 68, Coulson (1965) p. 91.
[45] In optics we have objects which are transparent but have the effect of causing a phase shift. These are called *phase objects* (Longhurst (1963) p. 294).
[46] Longhurst (1963) p. 463, Born and Wolf (1999) p. 820.

obvious when we write down its generalized eigenfunctions

$$\eta_k^{[7]}(x) = \begin{cases} \frac{1}{\sqrt{[\ell]}} e^{ikx}, & x < 0 \\ \frac{1}{\sqrt{[\ell]}} e^{i\pi} e^{ikx}, & x > 0 \end{cases}. \tag{6.183}$$

Compared with Eq. (6.95) we can see that $\eta^{[7]}$ corresponds to having

$$R_k^{[7]} = 0, \quad T_k^{[7]} = -1 \quad \text{for all } k. \tag{6.184}$$

We can visualize this point interaction in terms of the following family of δ-interactions:

1. An attractive δ-interaction of strength $2V_0$ at the origin, followed by

2. an asymmetric quadrupole point interaction from the origin and to $x = 4\epsilon$, followed by

3. an attractive δ-interaction of strength $2V_0$ at $x = 5\epsilon$, followed by

4. an asymmetric quadrupole point interaction starting at $x = 5\epsilon$ and ending at $x = 9\epsilon$.

The corresponding potential is

$$V_\epsilon^{[7]}(x) = -2V_0\, \delta(x) + V_\epsilon^{[3+]}(x) - 2V_0\, \delta(x - 5\epsilon) + V_\epsilon^{[3+]}(x - 5\epsilon). \tag{6.185}$$

The transfer matrix linking the wave function and its derivative between $x = 0$ and 9ϵ is

$$\begin{aligned}\mathcal{M}_\epsilon^{[7]}\{9\epsilon, 0\} &= \mathcal{M}_{k_0,\epsilon}^{[3+]}\{9\epsilon, 5\epsilon\} \times \mathcal{M}_{2k_0}^{[2-]} \times \mathcal{M}_k^{(0)}(5\epsilon, 4\epsilon) \\ &\quad \times \mathcal{M}_{k_0,\epsilon}^{[3+]}\{4\epsilon, 0\} \times \mathcal{M}_{2k_0}^{[2-]}\end{aligned} \tag{6.186}$$

which reduces to

$$\mathcal{M}_\epsilon^{[7]}\{9\epsilon, 0\} = \begin{pmatrix} -1 + O(\epsilon^{1/2}) & O(\epsilon^{1/2}) \\ O(\epsilon^{1/2}) & -1 + O(\epsilon^{1/2}) \end{pmatrix}. \tag{6.187}$$

As $\epsilon \to 0$ the potentials collapse to the origin and we have

$$\lim_{\epsilon \to 0} \mathcal{M}_\epsilon^{[7]}\{0, 9\epsilon\} = \mathcal{M}^{[7]}. \tag{6.188}$$

A diagramatic illustration of $V_\epsilon^{[7]}(x)$ is sketched in Fig. 6.5.7.

6.5. CLASSIFICATION OF POINT INTERACTIONS IN $\mathbb{E}$

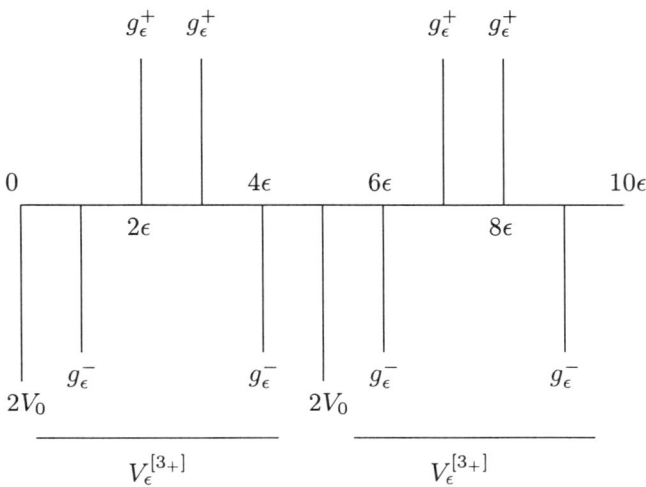

Fig. 6.5.7 $V_\epsilon^{[7]}$ for the ideal π-phase shifter.

6.5.8 Type 8 (BC8): High-pass π-phase shifters

Case 1   Attractive high-pass π-phase shifters and (BC8$_-$)

It is possible to have non-transparent phase shifters. For example we could have phase shifters, to be referred to as *high-pass π-phase shifter*, which are transparent to high frequencies but opaque to low frequencies. Consider boundary conditions (BC8$_-$):

$$\Phi_{20} + \Phi_{10} = 0, \quad \Phi'_{20} + \Phi'_{10} = k_0 \Phi_{10}, \tag{6.189}$$

arising from Eqs. (6.53) and (6.54) with $\alpha = -\beta \neq 0$. The resulting selfadjoint extension $\widehat{H}_{k_0}^{[8_-]}$ admits eigenfunction of the form of Eq. (6.97), i.e.,

$$\varphi_{k_0}^{[8_-]}(x) = \begin{cases} \sqrt{(k_0/2)} \ \exp\{(k_0/2)\, x\}, & x < 0 \\ -\sqrt{(k_0/2)} \ \exp\{-(k_0/2)\, x\}, & x > 0 \end{cases} \tag{6.190}$$

corresponding to an eigenvalue

$$E_{-,k_0}^{[8_-]} = -\frac{1}{8m}\left(\hbar k_0\right)^2. \tag{6.191}$$

The transfer matrix for (BC8$_-$) is

$$\mathcal{M}_{k_0}^{[8-]} = \begin{pmatrix} -1 & 0 \\ k_0 & -1 \end{pmatrix} = \mathcal{M}^{[7]} \times \mathcal{M}_{k_0}^{[2-]}. \tag{6.192}$$

This shows a link with a Type 2 attractive δ-potential which acts as a high-pass filter. The present Hamiltonian acts as a high-pass π-phase shifter in the sense that $\widehat{H}_{k_0}^{[8-]}$ admits generalized eigenfunctions of the form of Eq. (6.95) with $k_1 = k_2 = k$ and coefficients $R_{k_0}^{[8-]}$ and $T_{k_0}^{[8-]}$ related to those for an attractive δ-potential in Eqs. (6.118) and (6.119) by

$$R_k^{[8-]} = R_k^{[2-]}, \quad T_k^{[8-]} = -T_k^{[2-]}. \tag{6.193}$$

Both types of interactions can act as a high-pass filter. The difference between the present interaction and a δ-interaction lies in the following result

$$k \gg k_0 \quad \Rightarrow \quad T_k^{[2-]} \approx 1, \quad T_k^{[8-]} \approx -1. \tag{6.194}$$

Consequently the present interaction causes an additional π-phase shift between the transmitted and incident waves.

In view of the relationship between the present interaction with a Type 2 attractive δ-potential and the Type 7 ideal π-phase shifter shown in Eq. (6.192) we can appreciate that the limiting family of δ-potentials for $\widehat{H}_{k_0}^{[8-]}$ contains an attractive δ-potential of strength V_0 at the origin followed by $V_\epsilon^{[7]}(x - \epsilon)$ starting at $x = \epsilon$, i.e.,

$$V_\epsilon^{[8-]}(x) = -V_0\,\delta(x) + V_\epsilon^{[7]}(x - \epsilon). \tag{6.195}$$

The transfer matrix is then

$$\mathcal{M}_{k_0,\epsilon}^{[8-]}\{10\epsilon, 0\} = \mathcal{M}_\epsilon^{[7]}\{10\epsilon, \epsilon\} \times \mathcal{M}_k^{(0)}(\epsilon, 0) \times \mathcal{M}_{k_0}^{[2-]} \tag{6.196}$$

$$\Rightarrow \quad \lim_{\epsilon \to 0} \mathcal{M}_{k_0,\epsilon}^{[8-]}\{10\epsilon, 0\} = \mathcal{M}_{k_0}^{[8-]}. \tag{6.197}$$

Case 2 **Repulsive high-pass π-phase shifters and (BC8$_+$)**

Another extension is $\widehat{H}_{k_0}^{[8]}$ specified by boundary conditions (BC8$_+$):

$$\Phi_{20} + \Phi_{10} = 0, \quad \Phi'_{20} + \Phi'_{10} = -k_0\,\Phi_{10}, \tag{6.198}$$

with transfer matrix

$$\mathcal{M}_{k_0}^{[8+]} = \begin{pmatrix} -1 & 0 \\ -k_0 & -1 \end{pmatrix} = \mathcal{M}^{[7]} \times \mathcal{M}_{k_0}^{[2+]}. \tag{6.199}$$

6.5. CLASSIFICATION OF POINT INTERACTIONS IN $\mathbb{E}$

This Hamiltonian also acts as a high-pass π-phase shifter since $\widehat{H}_{k_0}^{[8+]}$ admits generalized eigenfunctions of the form of Eq. (6.95) with $k_1 = k_2 = k$ and

$$R_k^{[8+]} = R_k^{[2+]}, \quad T_k^{[8+]} = -T_k^{[2+]}. \tag{6.200}$$

This extension differs from $\widehat{H}_{k_0}^{[2+]}$ in that we now have an additional π-phase shift between the transmitted and incident particles.

The corresponding limiting family of δ-interactions contains a repulsive δ-potential of strength V_0 at the origin followed by $V_\epsilon^{[7]}(x - \epsilon)$, i.e.,

$$V_\epsilon^{[8+]}(x) = V_0\,\delta(x) + V_\epsilon^{[7]}(x - \epsilon). \tag{6.201}$$

The transfer matrices involved are related by

$$\mathcal{M}_{k_0,\epsilon}^{[8+]}\{0, 10\epsilon\} = \mathcal{M}_\epsilon^{[7]}\{10\epsilon, \epsilon\} \times \mathcal{M}_k^{(0)}(\epsilon, 0) \times \mathcal{M}_{k_0}^{[2+]} \tag{6.202}$$

$$\Rightarrow \quad \lim_{\epsilon \to 0} \mathcal{M}_{k_0,\epsilon}^{[8+]}\{10\epsilon, 0\} = \mathcal{M}_{k_0}^{[8+]}. \tag{6.203}$$

6.5.9 Type 9 (BC9): Low-pass π-phase shifters

Case 1   Attractive low-pass π-phase shifters and (BC9$_-$)

It is possible to have phase shifters, referred to as low-pass π-phase shifters, which are opaque to high frequencies but transparent to low frequencies. For such an attractive low-pass π-phase shifter the Hamiltonian $\widehat{H}_{k_0}^{[9-]}$ is determined by boundary conditions (BC9$_-$):

$$\Phi_{20} + \Phi_{10} = \frac{1}{k_0} \Phi_{10}', \quad \Phi_{20}' + \Phi_{10}' = 0. \tag{6.204}$$

These arise from Eqs. (6.66) and (6.67) with $\gamma \neq 0$. The eigenvalue equation of $\widehat{H}_{k_0}^{[9-]}$

$$\widehat{H}_{k_0}^{[9-]} \xi^{[9-]}(x) = E^{[9-]}\xi^{[9-]}(x) \tag{6.205}$$

admits a solution of the form of Eq. (6.96), i.e.,

$$\phi_{k_0}^{[9-]}(x) = \begin{cases} \sqrt{(2k_0)} \, \exp\{(2k_0)\,x\}, & x < 0 \\ \sqrt{(2k_0)} \, \exp\{-(2k_0)\,x\}, & x > 0 \end{cases}. \tag{6.206}$$

This represents a bound state corresponding to the eigenvalue

$$E_{-,k_0}^{[9-]} = -\frac{2}{m}(\hbar k_0)^2. \tag{6.207}$$

The corresponding transfer matrix is

$$\mathcal{M}_{k_0}^{[9-]} = \begin{pmatrix} -1 & k_0^{-1} \\ 0 & -1 \end{pmatrix} = \mathcal{M}_{k_0}^{[3-]} \times \mathcal{M}^{[7]}. \tag{6.208}$$

This shows a link with a Type 3 attractive δ'-interaction which acts as a low-pass filter. This point interaction acts as a low-pass filter with a π-phase shift since $\widehat{H}_{k_0}^{[9-]}$ admits generalized eigenfunctions of the form of Eq. (6.95) with $k_1 = k_2 = k$ and

$$R_k^{[9-]} = R_k^{[3-]}, \quad T_k^{[9-]} = -T_k^{[3-]}. \tag{6.209}$$

Compared with Eqs. (6.138), (6.139) and (6.140) we can see that $\widehat{H}_{k_0}^{[9-]}$ causes an additional π-phase shift between the transmitted and incident particles.

The corresponding limiting family of δ-interactions consists of $V_\epsilon^{[7]}(x)$ and $V_\epsilon^{[3-]}(x - 10\epsilon)$, i.e.,

$$V_\epsilon^{[9-]}(x) = V_\epsilon^{[7]}(x) + V_\epsilon^{[3-]}(x - 10\epsilon). \tag{6.210}$$

The transfer matrix is

$$\mathcal{M}_{k_0,\epsilon}^{[9-]}\{30\epsilon, 0\} = \mathcal{M}_{k_0,\epsilon}^{[3-]}\{30\epsilon, 10\epsilon\} \times \mathcal{M}_k^{(0)}(10\epsilon, 9\epsilon) \times \mathcal{M}_\epsilon^{[7]}\{9\epsilon, 0\} \tag{6.211}$$

$$\Rightarrow \quad \lim_{\epsilon \to 0} \mathcal{M}_{k_0,\epsilon}^{[9-]}\{30\epsilon, 0\} = \mathcal{M}_{-,k_0}^{[9]}. \tag{6.212}$$

Case 2   **Repulsive low-pass π-phase shifters and (BC9$_+$)**

The repulsive case is given by the Hamiltonian $\widehat{H}_{k_0}^{[9+]}$ fixed by boundary conditions (BC9$_+$):

$$\Phi_{20} + \Phi_{10} = -\frac{1}{k_0} \Phi_1'(0), \quad \Phi_{20}' + \Phi_{10}' = 0. \tag{6.213}$$

with transfer matrix

$$\mathcal{M}_{k_0}^{[9+]} = \begin{pmatrix} -1 & -k_0^{-1} \\ 0 & -1 \end{pmatrix} = \mathcal{M}_{k_0}^{[3+]} \times \mathcal{M}^{[7]}. \tag{6.214}$$

This potential acts as a low-pass filter with an additional π-phase since $\widehat{H}_{k_0}^{[9+]}$ admits generalized eigenfunctions of the form of Eq. (6.95) with $k_1 = k_2 = k$ and

$$R_{k_0}^{[9+]} = R_{k_0}^{[3+]}, \quad T_{k_0}^{[9+]} = -T_{k_0}^{[3+]}. \tag{6.215}$$

The corresponding limiting family of δ-interactions consists of $V_\epsilon^{[7]}(x)$ and $V_\epsilon^{[3+]}(x - 9\epsilon)$, i.e.,

$$V_\epsilon^{[9+]}(x) = V_\epsilon^{[7]}(x) + V_\epsilon^{[3+]}(x - 9\epsilon). \tag{6.216}$$

6.5. CLASSIFICATION OF POINT INTERACTIONS IN $\mathbb{E}$ 471

The transfer matrices involved are related by

$$\mathcal{M}_{k_0,\epsilon}^{[9+]}\{13\epsilon, 0\} = \mathcal{M}_{k_0,\epsilon}^{[3+]}(13\epsilon, 9\epsilon\} \times \mathcal{M}_{\epsilon}^{[7]}\{9\epsilon, 0\} \quad (6.217)$$

$$\Rightarrow \lim_{\epsilon \to 0} \mathcal{M}_{+,k_0,\epsilon}^{[9]}\{13\epsilon, 0\} = \mathcal{M}_{k_0}^{[9+]}. \quad (6.218)$$

6.5.10 Type 10 (BC10): Ideal mid-pass $\frac{1}{2}\pi$-phase shifters

Case 1 Attractive mid-pass $-\frac{1}{2}\pi$-phase shifter and (BC10$_-$)

In addition to π-shifters which are either high-pass or low-pass we can also have $\frac{1}{2}\pi$-phase shifters which are mid-pass. The attractive case is described by Hamiltonian $\widehat{H}_{k_0}^{[10-]}$ specified by boundary conditions (BC10$_-$) arising from Eqs. (6.53) and (6.54) with $\alpha = 0$:

$$\Phi_{20} = -\frac{1}{k_0}\Phi'_{10}, \quad \Phi_{10} = \frac{1}{k_0}\Phi'_{20}. \quad (6.219)$$

The corresponding transfer matrix is

$$\mathcal{M}_{k_0}^{[10-]} = \begin{pmatrix} 0 & -k_0^{-1} \\ k_0 & 0 \end{pmatrix}. \quad (6.220)$$

The eigenvalue equation of $\widehat{H}^{[10-]}$ yields an eigenfunction of the form of Eq. (6.96), i.e.,

$$\phi_{k_0}^{[10-]}(x) = \begin{cases} \sqrt{k_0}\, e^{k_0 x}, & x < 0 \\ \sqrt{k_0}\, e^{-k_0 x}, & x > 0 \end{cases}. \quad (6.221)$$

This represents a bound state corresponding to an eigenvalue

$$E_{-,k_0}^{[10-]} = -\frac{1}{2m}(\hbar k_0)^2. \quad (6.222)$$

This Hamiltonian also admits generalized eigenfunctions of the form of Eq. (6.95) with $k_1 = k_2 = k$ and

$$R_k^{[10-]} = \frac{k^2 - k_0^2}{k^2 + k_0^2}, \quad T_k^{[10-]} = -\frac{2ik_0 k}{k^2 + k_0^2}. \quad (6.223)$$

Since

$$k \gg k_0 \Rightarrow R_k^{[10-]} \approx 1, \quad T_k^{[10-]} \approx 0, \quad (6.224)$$

$$k \ll k_0 \Rightarrow R_k^{[10-]} \approx -1, \quad T_k^{[10-]} \approx 0, \quad (6.225)$$

472   CHAPTER 6. POINT INTERACTIONS

this interaction effectively forbids any transmission for extremely low and high values of k. When $k = k_0$ we have

$$k = k_0 \quad \Rightarrow \quad R_k^{[10-]} = 0, \ T_k^{[10-]} = -i. \tag{6.226}$$

This implies a complete transmission with a $-\frac{1}{2}\pi$-phase shift in the transmitted wave relative to the incident wave, hence we have an ideal attractive mid-pass $-\frac{1}{2}\pi$-phase shifter. Here the minus sign means a retardation of the phase by $\frac{1}{2}\pi$ across $x = 0$ as seen in the corresponding generalized eigenfunction

$$\eta_{k_0}^{[10-]}(x) = \begin{cases} \frac{1}{\sqrt{[\ell]}} e^{ik_0 x}, & x < 0 \\ \frac{1}{\sqrt{[\ell]}} e^{-i\frac{\pi}{2}} e^{ik_0 x}, & x > 0 \end{cases}. \tag{6.227}$$

An appropriate limiting family of δ-interactions for this Hamiltonian consists of an attractive δ'-interaction and two repulsive δ-interactions:

$$V_\epsilon^{[10-]}(x) = V_0\, \delta(x) + V_\epsilon^{[3-]}(x - \epsilon) + V_0\, \delta(x - 22\epsilon). \tag{6.228}$$

The corresponding transfer matrix is

$$\mathcal{M}_{k_0,\epsilon}^{[10-]}\{22\epsilon, 0\} = \mathcal{M}_{k_0}^{[2+]} \times \mathcal{M}_k^{(0)}(22\epsilon, 21\epsilon)$$
$$\times \ \mathcal{M}_{k_0,\epsilon}^{[3-]}(21\epsilon, \epsilon) \times \mathcal{M}_k^{(0)}(\epsilon, 0) \times \mathcal{M}_{k_0}^{[2+]}$$
$$\Rightarrow \quad \lim_{\epsilon \to 0} \mathcal{M}_{k_0,\epsilon}^{[10-]}\{22\epsilon, 0\} = \mathcal{M}_{k_0}^{[10-]}. \tag{6.229}$$

Case 2   **Repulsive mid-pass $\frac{1}{2}\pi$-phase shifter and (BC10$_+$)**

The repulsive case is described by a Hamiltonian $\widehat{H}_{k_0}^{[10+]}$ specified by boundary conditions (BC10$_+$):

$$\Phi_{20} = \frac{1}{k_0} \Phi'_{10}, \quad \Phi_{10} = -\frac{1}{k_0} \Phi'_{20} \tag{6.230}$$

with a transfer matrix

$$\mathcal{M}_{k_0}^{[10+]} = \begin{pmatrix} 0 & k_0^{-1} \\ -k_0 & 0 \end{pmatrix}. \tag{6.231}$$

This Hamiltonian admits generalized eigenfunctions of the form of Eq. (6.95) with

$$R_{k_0}^{[10+]} = \frac{k^2 - k_0^2}{k^2 + k_0^2}, \quad T_{k_0}^{[10+]} = \frac{2ik_0 k}{k^2 + k_0^2}. \tag{6.232}$$

6.5. CLASSIFICATION OF POINT INTERACTIONS IN $I\!\!E$

When $k = k_0$ we have a complete transmission with $\frac{1}{2}\pi$-phase shift in the transmitted wave relative to the incident wave as shown in the following generalized eigenfunction of $\widehat{H}_{k_0}^{[10+]}$:

$$\eta_{k_0}^{[10+]}(x) = \begin{cases} \frac{1}{\sqrt{[\ell]}} e^{ik_0 x}, & x < 0 \\ \frac{1}{\sqrt{[\ell]}} e^{i\frac{\pi}{2}} e^{ik_0 x}, & x > 0 \end{cases}. \qquad (6.233)$$

This extension also effectively forbids any transmission for extremely low and high values of k, hence an ideal mid-pass $\frac{1}{2}\pi$-phase shifter.

An appropriate family of δ-interactions for this Hamiltonian consists of two attractive δ-interactions and a repulsive δ'-interaction:

$$V_\epsilon^{[10+]}(x) = -V_0\,\delta(x) + V_\epsilon^{[3+]}(x) - V_0\,\delta(x - 5\epsilon). \qquad (6.234)$$

The corresponding transfer matrix is

$$\mathcal{M}_{k_0,\epsilon}^{[10+]}\{5\epsilon, 0\} = \mathcal{M}_{k_0}^{[2-]} \times \mathcal{M}_k^{(0)}(5\epsilon, 4\epsilon) \times \mathcal{M}_{k_0,\epsilon}^{[3+]} \times \mathcal{M}_{k_0}^{[2-]}$$
$$\Rightarrow \lim_{\epsilon \to 0} \mathcal{M}_{k_0,\epsilon}^{[10+]}\{5\epsilon, 0\} = \mathcal{M}_{k_0}^{[10+]}. \qquad (6.235)$$

6.5.11 Type 11 (BC11): Partial mid-pass filter

Point interactions can produce more varied effects than those mentioned in the preceding sections. Let us go back to the general boundary conditions spelled out in Eqs. (6.53) and (6.54). Introduce a new type of selfadjoint extensions $\widehat{H}^{[11]}$ specified by boundary conditions (BC11):

$$\Phi_{10} = \alpha \Phi'_{10} - \beta \Phi'_{20}, \qquad \Phi_{20} = \beta \Phi'_{10} - \alpha \Phi'_{20}, \qquad (6.236)$$

$$\alpha \neq 0 \neq \beta, \qquad \alpha^2 > \beta^2. \qquad (6.237)$$

The conditions placed on α and β here distinguish this particular type of extension from previous ones. Since $\alpha^2 \neq \beta^2$ we can recast Eq. (6.236) into the form of Eqs. (6.77) and (6.78), i.e., we can rewrite the above boundary conditions as

$$\Phi'_{10} = a\,\Phi_{10} + b\,\Phi_{20}, \qquad \Phi'_{20} = -a\,\Phi_{20} - b\,\Phi_{10}, \qquad (6.238)$$

$$a \neq 0 \neq b, \qquad a^2 > b^2. \qquad (6.239)$$

Here a and b are related to α and β by Eqs. (6.79) and (6.80). The corresponding transfer matrix at $x = 0$ is

$$\mathcal{M}_{a;b}^{[11]} = \frac{1}{b}\begin{pmatrix} -a & 1 \\ a^2 - b^2 & -a \end{pmatrix}. \qquad (6.240)$$

474   CHAPTER 6. POINT INTERACTIONS

Extension $\widehat{H}^{[11]}$ admits generalized eigenfunctions of the form of Eq. (6.95) with $k_1 = k_2 = k$ and

$$R_k^{[11]} = \frac{b^2 - a^2 - k^2}{(a+ik)^2 - b^2}, \quad T_k^{[11]} = -\frac{2ibk}{(a+ik)^2 - b^2}. \tag{6.241}$$

Since

$$a^2 - b^2 > 0 \quad \Rightarrow \quad (a+b)(a-b) > 0, \tag{6.242}$$

We can divide Type 11 into two cases, depending on the sign of $a+b$ and $a-b$. Case 1 is characterized by

$$(a+b) > 0 \quad \text{and} \quad (a-b) > 0. \tag{6.243}$$

In this case Eq. (6.243) implies $a > 0$. Case 2 is characterized by

$$(a+b) < 0 \quad \text{and} \quad (a-b) < 0. \tag{6.244}$$

In this case Eq. (6.243) implies $a < 0$.

Conditions given by Eqs. (6.238), (6.239) and (6.243) are referred to as (BC11$_+$) while conditions given by Eqs. (6.238), (6.239) and (6.244) are referred to as (BC11$_-$). Here the subscripts $\pm$ refer to the sign of a. Let us study these two cases separately.

Case 1 The corresponding selfadjoint extension $\widehat{H}_{a,b}^{[11+]}$ admits two bound states corresponding to the normalized eigenfunctions of the forms of Eqs. (6.96) and (6.97):

1. Eigenfunction $\phi_{\kappa_1}^{[11+]}$:

$$\phi_{\kappa_1}^{[11+]}(x) = \begin{cases} \sqrt{\kappa_1}\, e^{\kappa_1 x}, & x < 0 \\ \sqrt{\kappa_1}\, e^{-\kappa_1 x}, & x > 0 \end{cases}, \tag{6.245}$$

where

$$\kappa_1 = a + b, \tag{6.246}$$

which is positive. The corresponding energy eigenvalue is

$$E_{-,\kappa_1}^{[11+]} = -\frac{1}{2m}(\hbar \kappa_1)^2 = -\frac{\hbar^2}{2m}(a+b)^2. \tag{6.247}$$

2. Eigenfunction $\varphi_{\kappa_2}^{[11+]}$:

$$\varphi_{\kappa_2}^{[11+]}(x) = \begin{cases} \sqrt{\kappa_2}\, e^{\kappa_2 x}, & x < 0 \\ -\sqrt{\kappa_2}\, e^{-\kappa_2 x}, & x > 0 \end{cases}, \tag{6.248}$$

6.5. CLASSIFICATION OF POINT INTERACTIONS IN $I\!\!E$

where
$$\kappa_2 = a - b, \tag{6.249}$$
which is again positive. The corresponding energy eigenvalue is
$$E^{[11+]}_{-,\kappa_2} = -\frac{1}{2m}(\hbar\kappa_2)^2 = -\frac{\hbar^2}{2m}(a-b)^2. \tag{6.250}$$

A corresponding limiting family of δ-interactions for $b > 0$ is then
$$V^{[11+]}_{a,b>0,\epsilon}(x) = -V_1\,\delta(x) + V^{[3+]}_\epsilon(x) - V_1\,\delta(x - 5\epsilon), \tag{6.251}$$
where
$$V_0 = \frac{\hbar^2}{2m}b, \quad k_0 = \frac{2m}{\hbar^2}V_0 = b, \quad V_1 = \frac{\hbar^2}{2m}(a+b). \tag{6.252}$$

The corresponding transfer matrix is:
$$\begin{aligned}
& \mathcal{M}^{[11+]}_{a,b>0,\epsilon}\{5\epsilon, 0\} \\
& = \mathcal{M}^{[2-]}_{k_{10}} \times \mathcal{M}^{(0)}_k(5\epsilon, 4\epsilon) \times \mathcal{M}^{[3+]}_{k_0,\epsilon}\{4\epsilon, 0\} \times \mathcal{M}^{[2-]}_{k_{10}},
\end{aligned} \tag{6.253}$$
where
$$k_{10} = \frac{2m}{\hbar^2}V_1 = a + b. \tag{6.254}$$

We have
$$\mathcal{M}^{[11+]}_{a,b>0,\epsilon}\{0, 5\epsilon\} \to \mathcal{M}^{[11+]}_{a,b>0} \quad \text{as } \epsilon \to 0. \tag{6.255}$$

Equation (6.241) for transmission and reflection applies here and we can identify the following features:

1. $k \to 0 \implies R^{[11+]}_k \to -1, \quad T^{[11+]}_k \to 0.$

2. $k \to \infty \implies R^{[11+]}_k \to 1, \quad T^{[11+]}_k \to 0.$

3. It follows that maximum transmission occurs at some intermediate value of k. We cannot achieve a complete transmision with a zero reflection since we cannot have k^2 equal to $b^2 - a^2$ which is negative, and hence the name *partial mid-pass filter*. Maximum transmission occurs when
$$k = \sqrt{a^2 - b^2} \implies T^{[11+]}_k = -b/a. \tag{6.256}$$

We can have a maximum transmission with no phase shift when b/a is negative, a feature which distinguishes the present interaction from Type 10 interaction.[47]

Case 2 The Hamiltonian corresponding to (BC11_−) is denoted by $\widehat{H}^{[11-]}_{a,b}$. The results for this case can also be worked out.

---

[47]Note that b could be positive or negative. The sign of b determines whether there is a phase shift of π or not.

6.5.12 Type 12 (BC12): Ideal tunable phase shifters

Finally consider boundary conditions (BC12):

$$\Phi_{10} = \alpha \Phi'_{10} - \beta \Phi'_{20}, \qquad \Phi_{20} = \beta \Phi'_{10} - \alpha \Phi'_{20}, \qquad (6.257)$$

$$\alpha \neq 0 \neq \beta, \qquad \alpha^2 < \beta^2. \qquad (6.258)$$

We can recast Eq. (6.257) into the form of Eq. (6.77) and (6.78):

$$\Phi'_{10} = a\,\Phi_{10} + b\,\Phi_{20}, \qquad \Phi'_{20} = -a\,\Phi_{20} - b\,\Phi_{10}, \qquad (6.259)$$

$$a \neq 0 \neq b, \qquad a^2 < b^2. \qquad (6.260)$$

Here a and b are related to α and β by Eqs. (6.79) and (6.80). The corresponding transfer matrix $\mathcal{M}^{[12]}_{a,b}$ is the same as $\mathcal{M}^{[11]}_{a,b}$ in Eq. (6.240).[48]

The seemingly innocent condition $a^2 < b^2$ produces quite a different physical effect, i.e., the corresponding Hamiltonian can act as an ideal phase shifter with of a complete transmission. This is why we separate this from Type 11. Since

$$b^2 - a^2 > 0 \quad \Rightarrow \quad (b+a)(b-a) > 0 \qquad (6.261)$$

we can again separate Type 12 into two cases, depending on the sign of $(b+a)$ and $(b-a)$. Case 1 is characterized by

$$(b+a) > 0 \quad \text{and} \quad (b-a) > 0. \qquad (6.262)$$

We note that Eq. (6.262) implies $b > 0$. Case 2 is characterized by

$$(b+a) < 0 \quad \text{and} \quad (b-a) < 0. \qquad (6.263)$$

In this case Eq. (6.262) implies $b < 0$.

Conditions given in Eqs. (6.259), (6.260) and (6.262) are referred to as (BC12$_+$) and conditions given in Eqs. (6.259), (6.260) and (6.263) are referred to as (BC12$_-$).

Case 1 The corresponding selfadjoint extension $\widehat{H}^{[12+]}_{a,b}$ admits a bound state corresponding to a normalized eigenfunction of the form of Eq. (6.96):

$$\phi^{[12+]}_\kappa(x) = \begin{cases} \sqrt{\kappa}\, e^{\kappa x}, & x < 0 \\ \sqrt{\kappa}\, e^{-\kappa x}, & x > 0 \end{cases}, \qquad (6.264)$$

where

$$\kappa = b + a, \qquad (6.265)$$

---

[48]The limiting family of δ-interactions in Eq. (6.251) and its corresponding family of transfer matrices in Eq. (6.251) also apply here.

6.5. CLASSIFICATION OF POINT INTERACTIONS IN $\mathbb{E}$

which is positive. The corresponding eigenvalue is

$$E_{-,\kappa}^{[12+]} = -\frac{1}{2m}(\hbar\kappa)^2 = -\frac{\hbar^2}{2m}(b+a)^2. \tag{6.266}$$

A distinctive feature emerges arising from the condition $b^2 - a^2 > 0$ when we examine generalized eigenfunctions of the form of Eq. (6.95) with $k_1 = k_2 = k$. The coefficients $R^{[12]}$ and $T^{[12]}$ agree with $R^{[11]}$ and $T^{[11]}$ given by Eq. (6.241). But in contrast to Type 11 we can now achieve a complete transmission with a zero reflection coefficient. This is achieved when

$$k = \sqrt{b^2 - a^2}. \tag{6.267}$$

We then have

$$R_k^{[12+]} = 0, \quad T_k^{[12+]} = e^{i\lambda}, \tag{6.268}$$

where

$$\lambda = \tan^{-1}\left(-\frac{\sqrt{b^2 - a^2}}{a}\right). \tag{6.269}$$

The generalized eigenfunction corresponding to such a complete transmission is

$$\eta_k^{[12+]}(x) = \begin{cases} \frac{1}{\sqrt{[\ell]}} e^{ikx}, & x < 0 \\ \frac{1}{\sqrt{[\ell]}} e^{i\lambda} e^{ikx}, & x > 0 \end{cases}. \tag{6.270}$$

We can examine the relationship of various quantities further by substituting this eigenfunction into the boundary conditions given by Eq. (6.259). On taking the real and imaginary parts of the resulting equations we obtain the following constraints relating a, b, λ and k:

$$a = -b\cos\lambda, \tag{6.271}$$

$$k = b\sin\lambda. \tag{6.272}$$

So, given a and b we have a perfect transmission when the incoming wave has a wave number $k = \sqrt{b^2 - a^2}$. The transmitted wave will have its phase shifted by an amount λ determined by Eq. (6.271). By changing a and b we can change or tune the wave number for a complete transmission and the phase of the transmitted wave.

Note that some phase shifts are forbidden on account of Eq. (6.271):

1. When $\lambda = \frac{1}{2}\pi$ parameter a would have to vanish, contradicting (BC12).

2. When $\lambda = \pi$ we have $k = 0$ and $a = b$ which violates Eq. (6.260).

3. When $\lambda = 0$ we have $k = 0$ and $a^2 = b^2$ which violates Eq. (6.260).

Case 2 This resulting Hamiltonian is denoted by $\widehat{H}_{a,b}^{[12-]}$. Again we can also work out the properties of this Hamiltonian.

In the next chapter the results of this and previous sections will be applied directly to the studies of point interactions in superconducting circuits, e.g., Eqs. (6.271) and (6.272) are related to the Josephson equation in a superconducting circuit.

6.6 Remarks on Quantization by Parts

We have seen a dozen distinct types of single point interactions in $I\!E$ which are symmetrical about the origin. Even though each of the interactions mentioned are different physically, they are all obtained by dividing the physical space $I\!E$ into two branches and by using the method of quantization by parts. The method of quantization by parts for dealing with point interactions can be extended to other geometric configurations such as circles and other multi-branch configuration. Some of these will be discussed in the next chapter.

Often the unspoken aim of traditional quantization schemes is to produce a unique quantized theory. The present work demonstrates the fallacy of such an aim. A quantum mechanical description of a physical system is generally more complex than the corresponding classical description. It is then natural that seemingly the same classical system in a given geometry can be quantized into quite distinct quantum systems. We can summarize our results as follows:

1. We start with partial quantization in the branches to establish partially quantized quantities in the branches. The composition of partial Hamiltonians gives rise to a dozen types of selfadjoint extensions to serve as the Hamiltonians for the corresponding point interactions, each having a distinct physical characteristic.

2. Each type of point interaction is determinable by a corresponding set of boundary conditions.

3. The non-uniqueness of composite quantization is a general feature of our scheme. This demonstrates the richness and complexity of point interactions as well as the non-uniqueness of quantization in general.

4. The basic building block for point interactions is the δ-interaction which can be visualized as effected by a δ-function potential. Many other point interactions can then be visualized in terms of a limiting family of δ-interactions. In particular we can appreciate a repulsive δ'-interaction

6.6. REMARKS ON QUANTIZATION BY PARTS

as a limiting family of four δ-interactions, i.e., an asymmetric quadrapole interaction. Most of the rest of point interactions can then be pictured in terms of a finite combination of δ-interactions and asymmetric quadrapole interactions.

5. Point interactions are idealizations of highly localized potentials. For example an attractive δ-interaction is an idealization of a very high and narrow square potential well while a repulsive δ-interaction corresponds to high and narrow square potential barrier.[49] More physical examples will be presented in the next chapter.

6. There are other types of point interactions which are not symmetric at the junction. An example is $\widehat{K}_\lambda^{(1,2)}$ given by Eq. (6.29).

Symmetric operators and the studies of their extensions are not often discussed in physics texts. As seen in relation to point interactions they are well-worth exploring for their physical applications.

It should be pointed out that a quantization is not just a mathematical process. We have to incorporate the physics involved, and it is the mathematical non-uniqueness which enables us to incorporate physics into the correlative quantization process which proves to be of fundamental importance in deriving some novel properties of macroscopic quantum system, a subject matter of the next chapter. To gain a flavour of things to come, let us take a look at the quantization of momentum. Recall that on branches B_1 and B_2 we have partial momenta $\widehat{p}_1$ and $\widehat{p}_2$ respectively. In composite quantization we combine these partial momenta into the direct sum $\widehat{p}_1 \oplus \widehat{p}_2$. This direct sum is symmetric with a one-parameter family of selfadjoint extensions $\widehat{P}_\lambda^{(1,2)}$. Our scheme then assumes that one of these selfadjoint extensions will serve as the momentum for the system as a whole. A momentum for the system obtained this way may not be compatible with an arbitrary point interaction Hamiltonian. It all depends on the compatibility of boundary condition (2.639) which fixes $\widehat{P}_\lambda^{(1,2)}$ and those specifying the Hamiltonian, e.g., Eqs. (6.259) and (6.260) for $\widehat{H}_{a,b}^{[12+]}$. We may want to require $\widehat{P}_\lambda^{(1,2)}$ to be compatible with $\widehat{H}_{a,b}^{[12+]}$ in some well-defined sense. An example is the constraints in Eqs. (6.271) and (6.272). As the studies in the next chapter will show, such a compatibility requirement may have a severe consequence on the nature of the resulting theory.

---

[49] Merzbacher (1998) p. 98, p. 107.

6.7 Charged Particles in Circular Motion

6.7.1 Charged particles constrained to move in a circle

Consider a classical system comprising a particle of mass m_q and electric charge q constrained to move in a circle $\mathcal{C}$ of radius r in the x-y plane centred at the origin in the presence of a constant and uniform external magnetic field of magnitude B along the positive z-axis. In cylindrical coordinates the corresponding vector potential is $\vec{A} = (0, \frac{1}{2}Br, 0)$, which has a magnitude $A(r) = \frac{1}{2}Br$. We shall treat the system as one-dimensional, i.e., as motion of a particle in a circle $\mathcal{C}$ of radius r. In keeping with the notation in E1.5.1(1) E2 we shall employ the polar coordinate variable θ. The arc length from the point $\theta = 0$ is then $s = r\theta$ which may be called the *linear position variable* to contrast the *angular position variable* θ. In E1.6.4(1) E5 we introduce the momentum $P_\theta(\mathcal{C})$ conjugate to the angle variable θ. This is a complete momentum with its Hamiltonian vector field given by Eq. (1.241). The linear momentum $P_s(\mathcal{C})$ conjugate to the linear position variable s is related to $P_\theta(\mathcal{C})$ by $P_s(\mathcal{C}) = P_\theta(\mathcal{C})/r$. In the presence of an external magnetic field described above the classical Hamiltonian becomes[50]

$$H(\mathcal{C}) = \frac{1}{2m_q}\left(P_s(\mathcal{C}) - qA_{ex}\right)^2, \quad A_{ex} = \frac{1}{2}Br. \quad (6.273)$$

This Hamiltonian, serving also as the Hamiltonian generator, describes a classical particle going round the constraining circle of radius r with a constant speed which manifests itself in the appearance of an electric current flowing round the ring. For brevity we shall rewrite $P_s(\mathcal{C})$ simply as $P(\mathcal{C})$ from now on.

To quantize we first associate the quantized system with the Hilbert space $L^2(\mathcal{C})$ of square-integrable functions on the circle with respect to the measure $r\,d\theta$. The state of the quantized system is fully described by an element ϕ of $L^2(\mathcal{C})$. The linear momentum $P(\mathcal{C})$ is quantized as a selfadjoint operator defined by the operator expression

$$\widehat{P}(\mathcal{C}) = -i\hbar \frac{1}{r}\frac{d}{d\theta} \quad (6.274)$$

with a domain consisting of absolutely continuous functions $\phi(\theta)$ on $\mathcal{C}$ satisfying the following periodic boundary condition[51]

$$\phi(\theta) = \phi(\theta + 2\pi). \quad (6.275)$$

---
[50]Goldstein (1950) §7.3. Dicke and Wittke (1960) p. 85.
[51]Martin (1981) pp. 46-47.

6.7. CHARGED PARTICLES IN CIRCULAR MOTION

The quantized Hamiltonian, serving also as the Hamiltonian generator, is

$$\widehat{H}(\mathcal{C}) = \frac{1}{2m_q} \left(\widehat{P}(\mathcal{C}) - qA_{ex}\right)^2, \quad A_{ex} = \frac{1}{2}Br. \tag{6.276}$$

Operators $\widehat{P}(\mathcal{C})$ and $\widehat{H}(\mathcal{C})$ possess the following eigenvalues

$$p_n = \frac{1}{r} n\hbar = \frac{1}{2\pi r} nh \quad \text{and} \quad E_n = \frac{1}{2m_q}(p_n - qA_{ex})^2 \tag{6.277}$$

with corresponding normalized eigenfunctions

$$\varphi_n(\theta) = c_n \, e^{\frac{i}{\hbar} p_n r\theta} = c_n \, e^{in\theta}, \tag{6.278}$$

where

$$n = 0, \pm 1, \pm 2, \ldots, \quad \text{and} \quad c_n = \frac{1}{\sqrt{2\pi r}}. \tag{6.279}$$

We can introduce a probability current $\jmath_\phi$ associated with a normalized wave function ϕ for a charged particle in an external magnetic field, in accordance with orthodox quantum mechanics, by[52]

$$\jmath_\phi = \frac{1}{2m_q} \left\{ \left[\left(\widehat{P}(\mathcal{C}) - qA_{ex}\right)\phi\right]^* \phi + \phi^* \left[\left(\widehat{P}(\mathcal{C}) - qA_{ex}\right)\phi\right] \right\}. \tag{6.280}$$

When ϕ is not an eigenfunction of $\widehat{P}(\mathcal{C})$ the probability current $\jmath_\phi$ is generally θ-dependent, i.e., we can write it as $\jmath_\phi(\theta)$. This probability current is not an observable represented by a well-defined selfadjoint operator in $L^2(\mathcal{C})$. It is possible to contrive an operator expression, say,[53]

$$\widehat{\jmath}(\theta_0) = \frac{1}{2m_q} \left(\left(\widehat{P}(\mathcal{C}) - qA_{ex}\right)\delta(\theta - \theta_0) + \delta(\theta - \theta_0)\left(\widehat{P}(\mathcal{C}) - qA_{ex}\right)\right) \tag{6.281}$$

and call it the *probability current density operator* at the point $\theta = \theta_0$ with the understanding that the probability current in state ϕ at the point $\theta = \theta_0$ is equal to the formal expectation value

$$\begin{aligned}
\langle \phi \mid \widehat{\jmath}(\theta_0)\phi \rangle &= \frac{1}{2m_q} \int \phi^*(\theta) \left(\left(\widehat{P}(\mathcal{C}) - qA_{ex}\right)\delta(\theta - \theta_0) \right. \\
&\quad \left. + \delta(\theta - \theta_0)\left(\widehat{P}(\mathcal{C}) - qA_{ex}\right)\right) \phi(\theta) \, rd\theta \\
&= \frac{1}{2m_q} \left\{\left(\left(\widehat{P}(\mathcal{C}) - qA_{ex}\right)\phi\right)^* \phi + \phi^* \left(\left(\widehat{P}(\mathcal{C}) - qA_{ex}\right)\phi\right)\right\}_{\theta=\theta_0} \\
&= [\jmath_\phi(\theta)]_{\theta=\theta_0}. \tag{6.282}
\end{aligned}$$

---

[52] Feynmann, Leighton and Sands (1965) §21-2.
[53] Merzbacher (1998) p. 49, p. 553, Feynman (1972) pp. 294-295.

However, $\hat{\jmath}_{\theta_0}$ is not a properly defined operator in $L^2(\mathcal{C})$.

For the momentum eigenfunction φ_n we get a θ-independent value of the probability current, i.e.,

$$J_{\varphi_n} = |c_n|^2 \left(\frac{1}{m_q}\left(p_n - qA_{ex}\right)\right). \tag{6.283}$$

This value of the probability current depends on the normalization constant c_n in the following manner:

1. The current is due to a single particle going round the circle if we choose $c_n = 1/\sqrt{2\pi r}$, i.e., if we choose the normalized eigenfunction $\varphi_n(\theta)$.

2. The current is due to a beam of particles circling round with beam intensity of one particle per unit length if we choose[54] $c_n = 1/\sqrt{[\ell]}$, i.e., if we choose the following eigenfunction

$$\eta_n(\theta) = \frac{1}{\sqrt{[\ell]}} e^{ip_n r\theta} = \frac{1}{\sqrt{[\ell]}} e^{in\theta}. \tag{6.284}$$

For consistency with the plane wave in Eq. (6.90) and the study of superconducting systems in the next chapter we shall take $c_n = 1/\sqrt{[\ell]}$ and use momentum eigenfunctions of the form of $\eta_n(\theta)$. We would have a corresponding probability current

$$J_{\eta_n} = \frac{1}{m_q[\ell]}\left(p_n - qA_{ex}\right). \tag{6.285}$$

We interpret η_n as representing a beam of particles going round the circle, each of momentum p_n, with beam density one particle per unit length.

A probability current gives rise to an electrical current which can now be taken to be[55]

$$j_{\varphi_n} = q\,J_{\varphi_n} = \frac{q|c_n|^2}{m_q}\left(p_n - qA_{ex}\right) \tag{6.286}$$

or

$$j_{\eta_n} = q\,J_{\eta_n} = \frac{q}{m_q[\ell]}\left(p_n - qA_{ex}\right). \tag{6.287}$$

This electric current in turn generates a magnetic field. Following classical circuit theory we shall assume that the resulting magnetic flux enclosed by the ring to be proportional to the current. In other words when the system is in

---

[54] As in Eq. (6.90) the symbol $[\ell]$ denotes a unit length and is inserted here for dimensional reason.

[55] Feynman, Leighton and Sands (1965) p. 21-6.

6.7. CHARGED PARTICLES IN CIRCULAR MOTION

an eigenstate η_n, the total magnetic flux Φ_{η_n} enclosed within the ring should be given by

$$\Phi_{\eta_n} = \Phi_{ex} + L j_{\eta_n}, \qquad (6.288)$$

where L is the self-inductance[56] and

$$\Phi_{ex} = \pi r^2 B = 2\pi r A_{ex} \qquad (6.289)$$

is the external flux applied to the ring. The self-inductance of a classical circuit depends on the geometry of the circuit. For the present quantized situation we assume the self-inductance to be dependent on the circuit geometry as well as the density of the particle beam.[57] To obtain the result needed later for a corresponding superconducting circuit we shall assume the following expression

$$L = L_{\eta_n} = 2\pi r [\ell] \frac{m_q}{q^2} \qquad (6.290)$$

for a beam of particles with unit beam density as represented by state η_n. From Eq. (6.288) we obtain the value of Φ_{η_n} for the total enclosed flux to be[58]

$$\Phi_{\eta_n} = n\frac{h}{q} = n\Phi_0, \qquad (6.291)$$

where

$$\Phi_0 = \frac{h}{q}. \qquad (6.292)$$

The externally applied magnetic field does not affect the values of Φ_{η_n} and Φ_0.

If we were to choose the normalization constant to be $c_n = 1/\sqrt{2\pi r}$ we would use a different expression for the self-inductance, i.e.,[59]

$$L_{\varphi_n} = \frac{2\pi r}{c_n^2} \frac{m_q}{q^2} = (2\pi r)^2 \frac{m_q}{q^2}. \qquad (6.293)$$

A substitution of j_{φ_n} and its associated self-inductance L_{φ_n} into Eq. (6.288) gives rise to the same set of values $n\Phi_0$ for the total enclosed flux.

The conclusion is that the values of the total enclosed flux is independent of the applied field and the normalization of the eigenfunctions. It follows that physically we can imagine the quantized flux as due to a single charged particle going round the circle, and we can also picture the situation as due

---

[56] Bleaney and Bleaney (1957) p. 147.
[57] Wan and Harrison (1993), Wan and Saglam (2005).
[58] Equation (6.291) for the quantized flux is consistent with the result obtained by a semiclassical approach seen in Kittel (1996) pp. 255-257 using Bohr-Sommerfeld quantization rule (Landau and Lifshits (2002) p. 171).
[59] The unit length $[\ell]$ in Eq. (6.290) ensures that L_{η_n} has the same dimemsion as L_{φ_n}.

to a beam of particles going round. Because of our chosen normalization for η_n the current j_{η_n} and the flux Φ_{η_n} are due to a beam of $2\pi r/[\ell]$ particles going round the circle.[60] Operators $\widehat{P}(\mathcal{C})$ in Eq. (6.274) and $\widehat{H}(\mathcal{C})$ in Eq. (6.276) are the momentum and the Hamiltonian of a single particle. To get the corresponding quantities for the beam we have to multiply the factor $2\pi r$, e.g., the Hamiltonian for the beam $\widehat{H}_{bm}(\mathcal{C})$ is taken to be

$$\widehat{H}_{bm}(\mathcal{C}) = \frac{2\pi r}{[\ell]} \widehat{H}(\mathcal{C}). \tag{6.294}$$

The eigenvalues of $\widehat{H}_{bm}(\mathcal{C})$ are expressible in terms of Φ_{η_n} and Φ_{ex} as

$$E_{bm,n} = \frac{2\pi r}{[\ell]} E_n = \frac{1}{2L} \left(\Phi_{\eta_n} - \Phi_{ex} \right)^2. \tag{6.295}$$

Note that a stable and θ-independent current, and hence a stable enclosed flux for the quantized system, can be introduced only when the state is described by an eigenfunction of both the momentum and the Hamiltonian.[61] To pursue this point further we should emphasize the following:

1. Suppose the state is an eigenfunction of the momentum but not of the Hamiltonian. This becomes a possibility when the Hamiltonian is not given by Eq. (6.276) or Eq. (6.294) and instead it contains an additional interaction term which does not commute with the momentum operator. Then the initial probability current will become unstable, i.e., it will change in time, as the state evolves under the Hamiltonian.

2. Suppose the state is not an eigenfunction of the momentum. For the present system consider the following example:

$$\eta_{nn'}(\theta) = \frac{1}{\sqrt{2}} \left(\eta_n(\theta) + \eta_{n'}(\theta) \right) \tag{6.296}$$

is not an eigenfunction of $\widehat{P}(\mathcal{C})$. Serving as an initial state it will evolve under the Hamiltonian $\widehat{H}(\mathcal{C})$ as

$$\eta_{nn',t}(\theta) = \frac{1}{\sqrt{2}} \left(\eta_n(\theta) e^{-i\omega_n t} + \eta_{n'}(\theta) e^{-i\omega_{n'} t} \right), \tag{6.297}$$

where
$$\omega_n = E_n/\hbar, \quad \omega_{n'} = E_{n'}/\hbar. \tag{6.298}$$

---
[60] Dividing by $[\ell]$ renders $2\pi r/[\ell]$ dimensionless.
[61] We assume that $\widehat{H}(\mathcal{C})$ acts as the Hamiltonian generator as well.

6.7. CHARGED PARTICLES IN CIRCULAR MOTION

Substituting $\eta_{nn',t}$ into Eq. (6.280) leads to a probability current

$$j_{\eta_{nn'}}(\theta,t) = \frac{1}{2m[\ell]} \left(\frac{\hbar}{r}(n+n') - qA_{ex} \right)$$
$$\times \left(1 + \cos\left[(n-n')\theta - (\omega_n - \omega_{n'})t \right] \right) \qquad (6.299)$$

which is different at different points on the circle. This θ-dependent and time dependent probability current makes it difficult to establish a magnetic flux trapped by the ring. Equation (6.288) would not be applicable, e.g., if a θ-dependent and unstable electric current $j_{\eta_{nn'}}(\theta,t) = q\,j_{\eta_{nn'}}(\theta,t)$ is subsituted for j_{η_n} in Eq. (6.288) we would obtain a value for a total enclosed flux $\Phi_{\eta_{nn'}}$ which is θ-dependent, and makes no sense.

3. Suppose the state is an eigenfunction of the Hamiltonian but not of the momentum. Any eigenfunction of the Hamiltonian would correspond to a stationary state but it may not correspond to any θ-independent current flowing in the circuit if it is not an eigenfunction of the momentum operator. This is the case when $n' = -n$ in Eq. (6.296). The resulting current is θ-dependent, even though it is time-independent.[62]

4. The cause of the problem lies in the fact that electric current[63] is not an observable represented by a selfadjoint operator in the Hilbert space $L^2(\mathcal{C})$. It follows that we cannot claim that $j_{\eta_{nn'}}(\theta,t)$ is the measured values of an electric current observable in state $\eta_{nn',t}$ in the sense of Postulate OQS in §3.2.1.

This section is a precursor to the study of superconducting systems in the next chapter. For instance we shall investigate electrical current flow in a superconducting ring and the quantized nature of the magnetic flux enclosed in the ring in §7.3. We shall see that a superconducting ring is not an orthodox quantum system; it is a mixed quantum system with a superselection rule as allowed by Postulate UQS in §4.2. We shall endeavour to establish electric current and enclosed flux as physical observables for a superconducting ring and other superconducting systems. Before proceeding further we shall investigate point interactions in a circle since these point interactions are directly relevant to our study of superconductivity in the next chapter.

---

[62] In the presence of a magnetic field, the canonical momentum is no longer proportional to the velocity even in classical mechanics (Goldstein (1950) p. 49). This explains why there is still a non-zero probability current when $n' = -n$. For more discussions see Feynman, Leighton and Sands (1965) pp. 21-3 to 21-8.

[63] As in classical circuit theory, we are considering a stable current which is a global quantity, having the same value throughout the circle or any branch of a multi-branch circuit.

6.7.2 Charged particles in 3-dimensions

The analysis in §6.7.1 can be extended to motion in 3-dimensions.[64] Consider the spherically symmetric case of the electron in a hydrogen. Let $\phi_{n\ell m}(r, \theta, \varphi)$ be a normalized eigenfunction of the hydrogen Hamiltonian, where n, ℓ and m are respectively the principal, the angular momentum and the magnetic quantum numbers. The electron may be visualized to execute a circular motion of radius a about the z-axis when m is not zero in the sense that there is a non-zero probability current density circulating the z-axis of radius a defined by

$$\jmath(r, \theta, \varphi) = -\frac{i\hbar}{2m_e}\left(\psi_{n\ell m}^* \frac{\partial \psi_{n\ell m}}{a\,\partial \varphi} - \frac{\partial \psi_{n\ell m}^*}{a\,\partial \varphi}\psi_{n\ell m}\right) \quad (6.300)$$

$$= \frac{L_z}{m_e a}\left|\psi_{n\ell m}(r, \theta, \varphi)\right|^2, \quad (6.301)$$

where m_e is the mass of electron and $L_z = m\hbar$ is the z-component angular momentum eigenvalue. The probability current across an elementary surface area $rdrd\theta$ perpendicular to the current flow is[65]

$$\jmath(r, \theta, \varphi)\,rdrd\theta = \frac{L_z}{m_e a}\left|\psi_{n\ell m}(r, \theta, \varphi)\right|^2 rdrd\theta. \quad (6.302)$$

This probability current gives rise to an electric current[66]

$$j(r, \theta, \varphi)rdrd\theta = -\frac{eL_z}{m_e a}\left|\psi_{n\ell m}(r, \theta, \varphi)\right|^2 rdrd\theta, \quad (6.303)$$

circling the z-axis in a circle of radius $a = r\sin\theta$ flowing perpendicularly across surface area $rdrd\theta$.

Classically an electrical current of magnitude I going round a circular loop of radius a gives rise to a magnetic moment

$$M = \pi a^2 I, \quad (6.304)$$

and a classical particle of charge q and mass m_q circling the z-axis with speed v along a circle of radius a on the x-y plane generates an electric current

$$I = \frac{qv}{2\pi a} = \frac{qL_{cz}}{2\pi m_q a^2}, \quad L_{cz} = m_q v a. \quad (6.305)$$

---

[64] Wan and Saglam (2005).
[65] See Spiegel (1974) Fig. 7-27 (b) on p. 153.
[66] Here the negative sign shows that the electric current flows in an opposite direction to that of the probability current due to the negative nature of the electron charge.

6.7. CHARGED PARTICLES IN CIRCULAR MOTION

Note that L_{cz} is the angular momentum along the z-axis.[67] This current in turn gives rise to a magnetic moment M_z along the z-axis of magnitude[68]

$$M_z = \frac{1}{2} qav = \frac{q}{2m_q} L_{cz}. \tag{6.306}$$

To establish the quantum magnetic moment of an orbiting electron we may carry out a formal quantization of Eq. (6.306) to obtain a *magnetic moment operator* along the z-direction

$$\widehat{M}_z = -\frac{e}{2m_e} \widehat{L}_z. \tag{6.307}$$

For a more satisfying and explicit analysis we can proceed as follows:

1. An orbiting electron gives rise to an electric current element circulating the z-axis with a radius $a = r \sin\theta$, i.e., $j(r,\theta,\varphi)\,rdrd\theta$.

2. Each current element $j(r,\theta,\varphi)\,rdrd\theta$ gives rise to a magnetic moment element $d\mu_z^{(o)}$ given, in accordance with Eq. (6.304), by

$$d\mu_z^{(o)} = \pi a^2 j(r,\theta,\varphi)\,rdrd\theta \tag{6.308}$$

$$= -\pi a^2 \frac{eL_z}{m_e a} \left|\psi_{n\ell m}(r,\theta,\varphi)\right|^2 rdrd\theta \tag{6.309}$$

$$= -\frac{\pi e L_z}{m_e} \left|\psi_{n\ell m}(r,\theta,\varphi)\right|^2 r^2 \sin\theta\,drd\theta. \tag{6.310}$$

3. The total magnetic moment is then

$$\mu_z^{(o)} = \int_0^\infty \int_0^\pi -\frac{\pi e L_z}{m_e} |\psi_{n\ell m}(r,\theta,\varphi)|^2 r^2 \sin\theta\,drd\theta \tag{6.311}$$

$$= -\frac{e}{2m_e} L_z, \tag{6.312}$$

where we have used the normalized nature of the energy eigenfunction.[69]

---

[67] The symbol L_{cz} is for a classical angular momentum and is different from the angular momentum eigenvalue L_z in Eq. (6.301).

[68] Jackson (1999) Eq. (5.59) on p. 187.

[69] $|\psi_{n\ell m}(r,\theta,\varphi)|^2$ being independent of the angle variable φ, the normalization of the eigenfunction implies

$$\int_0^\infty \int_0^\pi 2\pi |\psi_{n\ell m}(r,\theta,\varphi)|^2 r^2 \sin\theta\,drd\theta = 1.$$

4. The operator which would admit $\mu_z^{(o)}$ as eigenvalues and $\psi_{n\ell m}$ as eigenfunctions is[70]

$$\widehat{\mu}_z^{(o)} = -\frac{e}{2m_e}\widehat{L}_z. \qquad (6.313)$$

This agrees with $\widehat{M}_z$ in Eq. (6.307).

We can also examine the magnetic flux enclosed by a circulating current. The magnetic flux enclosed by an electron in an orbital motion in an external magntic field is discussed in Kittel's well-known book.[71] We can extend the study to the orbital motion of an electron in an atom. In Bohr's intuitive model of the hydrogen atom an electron would go round the nucleus in a circular orbit. The radius and the speed of each allowed orbit can be calculated within Bohr's model. Such an orbital motion of the electron will cause a circulating electric current. In addition to generating a magnetic moment such a current should also enclose a magnetic flux. Using the expression of self-inductance in Eq. (6.293) we can evaluate the magnetic flux which turns out to have values equal to multiples of h/e. We can pursue this further to consider the quantum mechanical model of the hydrogen atom in three-dimensions. We shall again use the expression for the self-inductance in Eq. (6.293). To avoid confusion we shall re-label the self-inductance as $L_e(a)$ for a current round a circular path of radius a.[72] Then we have:

1. A magnetic flux element $d\Phi_z^{(o)}(r,\theta,\varphi)$ generated and enclosed by each current element $j_e(r,\theta,\varphi)\,rdrd\theta$:

$$\begin{aligned}d\Phi_z^{(o)}(r,\theta,\varphi) &= L_e(a)\,j_e(r,\theta,\varphi)\,rdrd\theta \qquad (6.314)\\ &= m_e\left(\frac{2\pi a}{e}\right)^2\left(-\frac{eL_z}{m_e a}\right)\left|\psi_{n[\ell]m}(r,\theta,\varphi)\right|^2 rdrd\theta\\ &= \left(-\frac{2\pi}{e}L_z\right)2\pi\left|\psi_{n[\ell]m}(r,\theta,\varphi)\right|^2 r^2\sin\theta\,drd\theta. \quad (6.315)\end{aligned}$$

2. The total magnetic flux generated and enclosed by an orbiting electron

---

[70] The superscript signifies the fact that the magnetic moment arises from orbital motion.

[71] Kittel (1996) pp. 255-257. Further studies of magnetic flux and electron motion including a proposal that an electron spin magnetic moment gives rise to an intrinsic magnetic flux associated with electron spin are presented by Saglam and Boyacioglu (2002a), Saglam and Boyacioglu (2002b), Wan and Saglam (2005).

[72] The expression of self-inductance in Eq. (6.293) is dependent only on the mass, the charge and the radius a of the circulating current, i.e., we have

$$L_e(a) = (2\pi a)^2\frac{m_e}{e^2}.$$

in an energy eigenstate $\psi_{n\ell m}(r,\theta,\varphi)$:

$$\Phi_z^{(o)} = \int_0^\infty \int_0^\pi d\Phi_z^{(o)}(r,\theta,\varphi) \tag{6.316}$$

$$= -\frac{2\pi}{e} L_z \int_0^\infty \int_0^\pi 2\pi \left|\psi_{n[\ell]m}(r,\theta,\varphi)\right|^2 r^2 \sin\theta\, drd\theta$$

$$= -\frac{2\pi}{e} L_z. \tag{6.317}$$

3. A corresponding total magnetic flux operator:

$$\widehat{\Phi}_z^{(o)} = -\frac{2\pi}{e} \widehat{L}_z. \tag{6.318}$$

With $L_z = m\hbar$ we get

$$\Phi_z^{(o)} = -m\,\Phi_e^{(o)}, \qquad \Phi_e^{(o)} = h/e. \tag{6.319}$$

We call $\Phi_z^{(o)}$ an *electron orbital magnetic flux* in the z-direction. Our result shows that an electron orbital magnetic flux is quantized into a multiple of $\Phi_e^{(o)} = h/e$. We shall call $\Phi_e^{(o)}$ the *electron orbital magnetic flux quantum*. This is to contrast the situation in a superconductor where the current is due to pairs of electrons known as Cooper pairs. As will be discussed in §7.3 the corresponding *Cooper pair magnetic flux quantum* in a superconducting ring has the value $h/2e$.

6.8 Point Interactions in a Circle

Consider the motion of a particle of mass m in a circle $\mathcal{C}$. Clearly we can have point interactions in a circle, e.g., a δ-function interaction at some point in the circle $\mathcal{C}$. To be specific let us consider point interactions localized at $\theta = 0$ which is symmetric to clockwise and anti-clockwise motion. To tackle a point interaction at $\theta = 0$ we shall first introduce a circle with a cut at $\theta = 0$, i.e., with the point $\theta = 0$ removed. We shall adhere to the notation in E1.5.1(1) E2 by denoting a circle with a cut by $\mathcal{C}_c$ and the coordinate in $\mathcal{C}_c$ by $\vartheta \in (0, 2\pi)$.

The Hilbert space for our present one-branch geometry is $L^2(\mathcal{C}_c)$ of square-integrable functions on $\mathcal{C}_c$ with respect to the measure $r\,d\vartheta$. Following the method of quantization by parts we shall start with partially quantized operators, e.g., differential operators acting on $C_0^\infty(\mathcal{C}_c)$. There is no need to take any direct sum in the composite quantization stage for a one-branch geometry. All we need is to construct selfadjoint extensions of the partially quantized operators.

Since $\mathcal{C}_c$ is topologically equivalent to the interval $(0, 2\pi r)$ in $I\!\!E$ we can identify $L^2(\mathcal{C}_c)$ with $L^2(0, 2\pi r)$.[73] This allows us to utilize E2.5.1(1) E1 and Theorem 6.2(1) to write down all the selfadjoint extensions for the momentum and the Hamiltonian operators.

6.8.1 Momentum operators

Let us start with the differential operator

$$\widehat{P}_0(\mathcal{C}_c) = -i\hbar \frac{1}{r} \frac{d}{d\vartheta} \qquad (6.320)$$

defined on the domain

$$\mathcal{D}(\widehat{P}_0(\mathcal{C}_c)) = C_0^\infty(\mathcal{C}_c) \qquad (6.321)$$

as the partially quantized momentum. As dicussed in E2.5.1(1) E1 this is a symmetric operator, and there is a one-parameter family of selfadjoint extensions

$$\widehat{P}_\lambda(\mathcal{C}_c) = -i\hbar \frac{1}{r} \frac{d}{d\vartheta} \qquad (6.322)$$

parameterized by $\lambda \in (-\pi, \pi]$ and defined on domain[74]

$$\mathcal{D}(\widehat{P}_\lambda(\mathcal{C}_c)) = \left\{ \Psi : \Psi \in S_{2,1}(\mathcal{C}_c), \text{ and } \Psi(0) = e^{i\lambda}\Psi(2\pi) \right\}. \qquad (6.323)$$

In composite quantization the momentum is quantized as $\widehat{P}_\lambda(\mathcal{C}_c)$. This operator possesses a discrete spectrum

$$p_{\lambda,n}(\mathcal{C}_c) = \frac{\hbar}{r} \left(n - \frac{\lambda}{2\pi} \right). \qquad (6.324)$$

In line with Eq. (6.284) we can write down the corresponding eigenfunctions:

$$\eta_{\lambda,n}(\vartheta) = \frac{1}{\sqrt{[\ell]}} e^{i p_{\lambda,n} r \vartheta} = \frac{1}{\sqrt{[\ell]}} e^{i(n-\frac{\lambda}{2\pi})\vartheta}, \quad n = 0, \pm 1, \pm 2, \ldots. \qquad (6.325)$$

The value of λ cannot be determined at this stage; it has to be chosen on physical ground based on a more detailed knowledge of the physical system in question. Explicit examples of this will be discussed later, e.g., in §7.4.3.

---

[73] The circumference of $\mathcal{C}_c$ is $2\pi r$. If we introduce a coordinate s in the interval $(0, 2\pi r)$ then an element f of $L^2(0, 2\pi r)$ is a function of s, i.e., $f = f(s)$, $s \in (0, 2\pi r)$. We can identify $f(s) \in L^2(0, 2\pi r)$ with $f(r\vartheta) \in L^2(\mathcal{C}_c)$. This identification is unitary, bearing in mind that $s = r\vartheta$ and the measure or volume element for integration in $L^2(\mathcal{C}_c)$ is $r\,d\vartheta$.

[74] $S_{2,1}(\mathcal{C}_c)$ is the Sobolev space $S_{2,1}(\Lambda)$, $\Lambda = \{r\vartheta \in (0, 2\pi r)\}$. Since r is a constant we can regard functions in $L^2(\mathcal{C}_c)$, $C_0^\infty(\mathcal{C}_c)$ and $S_{2,1}(\mathcal{C}_c)$ as functions of $\vartheta \in (0, 2\pi)$ without having to write down r explicitly every time.

6.8. POINT INTERACTIONS IN A CIRCLE

Although the mathematics so far is the same as those in E2.5.1(1) E1 and in E2.5.2(1) E2 relating to an interval the physics is rather different. For example we can associate an eigenfunction $\eta_{\lambda,n}$ in $L^2(\mathcal{C}_c)$ with a probability current, in accordance with Eq. (6.285),

$$j_{\lambda,n}(\mathcal{C}_c) = \frac{1}{m[\ell]} p_{\lambda,n} = \frac{\hbar}{mr[\ell]} \left(n - \frac{\lambda}{2\pi} \right) \tag{6.326}$$

flowing round and round $\mathcal{C}_c$ in the same way as we do for $\mathcal{C}$. The cut at $\theta = 0$ represents a point interaction which a current may tunnel through. In the case of an interval $\Lambda = (0, a)$ in E2.5.1(1) E1 it is difficult to visualize a probability current going out at point $x = a$ and then returning to enter the interval from $x = 0$.

6.8.2 Hamiltonians with reflection symmetry

The partially quantized Hamiltonian is

$$\widehat{H}_0(\mathcal{C}_c) = -\frac{\hbar^2}{2mr^2}\frac{d^2}{d\vartheta^2} = \frac{1}{2m}\widehat{P}_0^2(\mathcal{C}_c) \tag{6.327}$$

acting on $C_0^\infty(\mathcal{C}_c)$. Again, mathematically the situation is like that of an interval, and we can make use of Theorem 6.2(1) to establish a host of selfadjoint extensions of $\widehat{H}_0(\mathcal{C}_c)$, i.e., we can identify each $\widehat{K}_{\underline{\alpha},\underline{\beta}}$ in Theorem 6.2(1) with a selfadjoint extension of $\widehat{H}_0(\mathcal{C}_c)$. Since these extensions would play the role of Hamiltonians we shall re-label them as $\widehat{H}_{\underline{\alpha},\underline{\beta}}$. We are interested only in motions which are symmetrical in the clockwise and anti-clockwise directions. In other words there should be no preferred direction of current flow either clockwise or anti-clockwise.[75] So, we require the Hamiltonian to have a reflection symmetry about the cut, i.e., Hamiltonians which are unchanged by the transformation $\vartheta \to 2\pi - \vartheta$, and it should therefore commute with the parity operator $\widehat{\wp}(\mathcal{C}_c)$ defined on $L^2(\mathcal{C}_c)$ by

$$\left(\widehat{\wp}(\mathcal{C}_c)\Psi\right)(\vartheta) = \Psi(2\pi - \vartheta). \tag{6.328}$$

The domain of the Hamiltonian must also be invariant with respect to this parity operator. This requirement rules out a number of selfadjoint extensions to $\widehat{H}_0(\mathcal{C}_c)$. Obvious examples are selfadjoint extensions of the form

$$\widehat{H}_\lambda(\mathcal{C}_c) = \frac{1}{2m}\widehat{P}_\lambda^2(\mathcal{C}_c), \quad \lambda \neq 0, \pi. \tag{6.329}$$

---

[75] We confine ourselves here to systems whose Hamiltonian can be identified with their Hamiltonian generator for time evolution.

Clearly $\eta_{\lambda,n}$ in Eq. (6.325) is in the domain $\mathcal{D}(\widehat{H}_\lambda(\mathcal{C}_c))$ of $\widehat{H}_\lambda(\mathcal{C}_c)$. But

$$\left(\widehat{\wp}(\mathcal{C}_c)\eta_{\lambda,n}\right)(\vartheta) = \eta_{\lambda,n}(2\pi - \vartheta) \qquad (6.330)$$

fails to satisfy the boundary condition in Eq. (6.323).[76] Hence these extensions are not included in our consideration in what follows. The two cases with $\lambda = 0$ and $\lambda = 2\pi$ not included in Eq. (6.329) are worth mentioning here:[77]

1. When $\lambda = 0$ we have the periodic boundary condition (6.275) which results in the free Hamiltonian for a particle going round the circle $\mathcal{C}$ without a cut.

2. When $\lambda = \pi$ we have boundary condition

$$\Psi(0) = e^{i\pi}\,\Psi(2\pi) = -\Psi(2\pi). \qquad (6.331)$$

Compared with Eq. (6.183) we anticipate that this would correspond to a point interaction at the cut acting as a π-phase shifter.

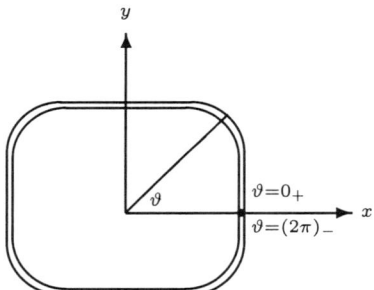

Fig. 6.8.2   Circle with the point $\vartheta = 0$ removed.

We can carry out a classification of all these selfadjoint extensions with a reflection symmetry into a number of types in terms of boundary conditions on the wave function Ψ at $\vartheta = 0_+, (2\pi)_-$ similar to those presented in §6.5. We shall start by spelling out the sign convention and relevant notation on the circle with a cut shown in Fig. 6.8.2:

---

[76]Expression

$$\left(\widehat{\wp}(\mathcal{C}_c)\eta_{\lambda,n}\right)(2\pi) = \eta_{\lambda,n}(0) = e^{i\lambda}\eta_{\lambda,n}(2\pi) = e^{i\lambda}\left(\widehat{\wp}(\mathcal{C}_c)\eta_{\lambda,n}\right)(0)$$

is not the same as the boundary condition in Eq. (6.323). When $\lambda = \pi/2$, for example, $\jmath_{\lambda,n}(\mathcal{C}_c)$ gives rise to a set of clockwise currents not the same as the set of anti-clockwise currents.

[77]These two cases are compatible with our symmetry requirement, and are included in our list of point interactions, to be classified later. See §6.9.1 and §6.9.7.

6.8. POINT INTERACTIONS IN A CIRCLE

1. The anti-clockwise direction is taken as positive, e.g., a positive current flows from $\vartheta = (2\pi)_-$ across the cut to $\vartheta = 0_+$, and a current flowing from $\vartheta = 0_+$ across the cut to $\vartheta = (2\pi)_-$ is regarded as negative.

2. Boundary values of the wave function are labelled as follows:[78]

$$\Psi_{10} = \Psi(2\pi_-), \qquad \Psi_{20} = \Psi(0_+), \qquad (6.332)$$

$$\Psi'_{10} = \left[\frac{1}{r}\frac{d\Psi(\vartheta)}{d\vartheta}\right]_{\vartheta=2\pi_-}, \qquad \Psi'_{20} = \left[\frac{1}{r}\frac{d\Psi(\vartheta)}{d\vartheta}\right]_{\vartheta=0_+}. \qquad (6.333)$$

The effect of the parity operator on these boundary values are as follows:[79]

$$\left[\widehat{\wp}(\mathcal{C}_c)\Psi\right]_{20} = \Psi_{10}, \qquad (6.334)$$

$$\left[\widehat{\wp}(\mathcal{C}_c)\Psi\right]_{10} = \Psi_{20}, \qquad (6.335)$$

$$\left[\left(\widehat{\wp}(\mathcal{C}_c)\Psi\right)'\right]_{20} = -\Psi'_{10}, \qquad (6.336)$$

$$\left[\left(\widehat{\wp}(\mathcal{C}_c)\Psi\right)'\right]_{10} = -\Psi'_{20}. \qquad (6.337)$$

Now let us return to consider the classification of all those $\widehat{H}_{\underline{\alpha},\underline{\beta}}$ which have a reflection symmetry about the cut. This hinges on a detailed study of the domains of the operators in question. As mentioned earlier we require the domain $\mathcal{D}(\widehat{H}_{\underline{\alpha},\underline{\beta}})$ of any $\widehat{H}_{\underline{\alpha},\underline{\beta}}$ to be invariant with respect to the parity operator $\widehat{\wp}(\mathcal{C}_c)$, i.e., we require

$$\Psi \in \mathcal{D}(\widehat{H}_{\underline{\alpha},\underline{\beta}}) \quad \Rightarrow \quad \widehat{\wp}(\mathcal{C}_c)\Psi \in \mathcal{D}(\widehat{H}_{\underline{\alpha},\underline{\beta}}). \qquad (6.338)$$

In other words $\widehat{\wp}\Psi$ must satisfy Eqs. (6.5) and (6.6). This leads to the following additional conditions on Ψ:

$$-\alpha'_1\Psi'_{20} - \alpha_1\Psi_{20} = -\alpha'_2\Psi'_{10} - \alpha_2\Psi_{10}, \qquad (6.339)$$

$$-\beta'_1\Psi'_{20} - \beta_1\Psi_{20} = -\beta'_2\Psi'_{10} - \beta_2\Psi_{10}. \qquad (6.340)$$

---

[78] This is the same as the notation used to write Theorem 6.2(1). A dash represents a spatial derivative, i.e., a derivative with respect to $r\vartheta$, not with respect to ϑ.

[79] Note that

$$\frac{d}{d\vartheta}\left(\widehat{\wp}(\mathcal{C}_c)\Psi\right)(\vartheta) = \frac{d}{d\vartheta}\Psi(2\pi - \vartheta) = -\frac{d\Psi(2\pi - \vartheta)}{d(2\pi - \vartheta)}.$$

$$\left[\left(\widehat{\wp}(\mathcal{C}_c)\Psi\right)'\right]_{20} = \left[d\left(\widehat{\wp}(\mathcal{C}_c)\Psi(\vartheta)\right)/r\,d\vartheta\right]_{\vartheta=0_+} = -\left[\frac{d\Psi(\vartheta)}{r\,d\vartheta}\right]_{\vartheta=2\pi_-}.$$

$$\left[\left(\widehat{\wp}(\mathcal{C}_c)\Psi\right)'\right]_{10} = \left[d\left(\widehat{\wp}(\mathcal{C}_c)\Psi(\vartheta)\right)/r\,d\vartheta\right]_{\vartheta=2\pi_-} = -\left[\frac{d\Psi(\vartheta)}{r\,d\vartheta}\right]_{\vartheta=0_+}.$$

Compared with Eqs. (6.5) and (6.6) we see that these equations correspond to two new sets of parameters

$$\tilde{\underline{\alpha}} = \{-\alpha'_1, \alpha_1, -\alpha'_2, \alpha_2\}, \quad \tilde{\underline{\beta}} = \{-\beta'_1, \beta_1, -\beta'_2, \beta_2\}. \tag{6.341}$$

To comply with condition (C1) for Theorem 6.2(1) we must have[80]

$$\text{either} \quad \tilde{\underline{\alpha}} = \underline{\alpha}, \ \tilde{\underline{\beta}} = \underline{\beta} \quad \text{or} \quad \tilde{\underline{\alpha}} = \underline{\beta}, \ \tilde{\underline{\beta}} = \underline{\alpha}. \tag{6.342}$$

The first alternative is immediately contradicted by the periodic boundary condition in Eq. (6.275). The correct choice is then

$$\tilde{\underline{\beta}} = \{-\beta'_1, \beta_1, -\beta'_2, \beta_2\}. \tag{6.343}$$

In other words we have

$$\alpha'_2 = -\beta'_1, \quad \alpha_2 = \beta_1, \quad \alpha'_1 = -\beta'_2, \quad \alpha_1 = \beta_2. \tag{6.344}$$

Eqs. (6.5) and (6.6) become

$$\alpha'_2 \Psi'_{20} - \alpha_2 \Psi_{20} = \alpha'_1 \Psi'_{10} - \alpha_1 \Psi_{10}, \tag{6.345}$$
$$\alpha'_1 \Psi'_{20} + \alpha_1 \Psi_{20} = \alpha'_2 \Psi'_{10} + \alpha_2 \Psi_{10}. \tag{6.346}$$

These can be re-arranged into the form

$$\Psi'_{10} = a\Psi_{10} + b\Psi_{20}, \tag{6.347}$$
$$\Psi'_{20} = -a\Psi_{20} - b\Psi_{10}, \tag{6.348}$$

where the constants are related by

$$a = \frac{\alpha_2 \alpha'_2 + \alpha_1 \alpha'_1}{\alpha'^2_1 - \alpha'^2_2}, \quad b = -\frac{\alpha_2 \alpha'_1 + \alpha_1 \alpha'_2}{\alpha'^2_1 - \alpha'^2_2}, \tag{6.349}$$

provided $\alpha'^2_1 - \alpha'^2_2 \neq 0$. All these are very similar to the conditions in Eqs. (6.77) and (6.78) for point interactions in $I\!E$. We shall regard these boundary conditions as arising from some point interactions at the cut and proceed to classify these point interactions in terms of boundary conditions. Since the operator expression for any selfadjoint extension $\widehat{H}_{\underline{\alpha},\underline{\beta}}$ away from the the cut at $\vartheta = 0$ is of the form of Eq. (6.327) its eigenfunctions away from the cut will be of one of the following forms:

1. Trigonometric functions:

$$\eta(\vartheta) = c_k \left(e^{ikr\vartheta} + R\, e^{-ikr\vartheta} \right), \tag{6.350}$$

   where k is real and R is a constant, possibly k-dependent. When $R = \pm 1$ the eigenfunction becomes cosine and sine functions respectively.

---

[80] We have set the proportionality constants to 1 for simplicity.

6.9. CLASSIFICATION OF POINT INTERACTIONS IN $\mathcal{C}$

2. Hyperbolic functions:

$$\psi(\vartheta) = c_\kappa \left(e^{\kappa r \vartheta} + R e^{-\kappa r \vartheta}\right), \qquad (6.351)$$

where κ is real.[81] When $R = \pm 1$ it becomes hyperbolic cosine and sine functions respectively.

6.9 Classification of Point Interactions in $\mathcal{C}$

6.9.1 Type 1 (BCC1): Free motion

The simplest selfadjoint extension[82] of $\widehat{H}_0(\mathcal{C}_c)$ is $\widehat{H}^{(1)}(\mathcal{C}_c)$ which is determined by its domain consisting of functions satisfying the following periodic boundary conditions (BCC1):[83]

$$\Psi_{20} = \Psi_{10}, \quad \Psi'_{20} = \Psi'_{10}. \qquad (6.352)$$

These conditions are obtained by setting

$$\alpha'_2 = \alpha'_1, \quad \alpha_2 = \alpha_1 \neq 0 \qquad (6.353)$$

in Eqs. (6.345) and (6.346). The resulting Hamiltonian is the same as $\widehat{H}(\mathcal{C})$ in Eq. (6.276) with the magnetic vector potential $A = 0$, or $\widehat{H}_\lambda(\mathcal{C}_c)$ in Eq. (6.329) with $\lambda = 0$. Physically this extension corresponds to free motion in the circle without any point interaction. When there is no risk of confusion we shall simply denote $\widehat{H}^{(1)}(\mathcal{C}_c)$ by $\widehat{H}^{(1)}$.

6.9.2 Type 2 (BCC2): δ-interaction

This selfadjoint extension is given by boundary conditions (BCC2_):

$$\Psi_{20} - \Psi_{10} = 0, \quad \Psi'_{20} - \Psi'_{10} = -k_0 \Psi_{10}, \qquad (6.354)$$

and the extension operator is denoted by $\widehat{H}^{(2-)}(\mathcal{C}_c)$. As before k_0 will denote a positive constant throughout this chapter. These conditions are obtained by setting

$$\alpha'_2 = \alpha'_1 \neq 0, \quad \alpha_2 \neq \alpha_1 \qquad (6.355)$$

---

[81] As before k and κ have the dimension of [length^{-1}]. The radius r is inserted to produce a dimensionless exponential.

[82] We use $\widehat{H}^{(1)}$, ... to distinguish from the corresponding selfadjoint extensions $\widehat{H}^{[1]}$, ... for point interactions on $I\!\!E$. (BCC1) can apply to a torus (Gustafson and Sigal (2003) p. 68).

[83] The notation BCC represents a set of boundary conditions on the circle $\mathcal{C}_c$.

in Eqs. (6.345) and (6.346).

As in §6.5.2 we can symbolically write a Hamiltonian with an attractive δ-function potential in terms of coordinate variable ϑ_π defined by Eq. (1.237) in E1.5.1(1) E2. Recall that ϑ_π varies continuously in the range $(-\pi, \pi)$ and that $\vartheta_\pi = 0$ corresponds to $\theta = 0$.[84] In terms of ϑ_π we have the following formal operator expression for $\widehat{H}^{(2-)}$:

$$-\frac{\hbar^2}{2m}\frac{1}{r^2}\frac{d^2}{d\vartheta_\pi^2} - V_0\,\delta(r\vartheta_\pi) = \frac{\hbar^2}{2m}\left(-\frac{1}{r^2}\frac{d^2}{d\vartheta_\pi^2} - k_0\,\delta(r\vartheta_\pi)\right), \qquad (6.356)$$

showing an attractive δ-interaction; k_0 is related to V_0 by Eq. (6.110).

This Hamiltonian is attractive in that $\widehat{H}^{(2-)}$ admits a negative eigenvalue corresponding to an eigenfunction of the form given by Eq. (6.351):

$$\psi^{(2-)}_{\kappa,\text{ch}}(\vartheta) = c_\kappa \left(e^{\kappa r\vartheta} + e^{2\pi\kappa r}\,e^{-\kappa r\vartheta}\right) \qquad (6.357)$$
$$= 2c_\kappa\,e^{\pi\kappa r}\cosh\kappa r(\vartheta - \pi), \qquad (6.358)$$

with a negative eigenvalue $E^{(2-)}_\kappa(\mathcal{C}_c)$ given by an expression of the form[85]

$$E^{(2-)}_{-,\kappa}(\mathcal{C}_c) = -\frac{1}{2m}(\hbar\kappa)^2. \qquad (6.359)$$

The values of κ are fixed, according to (BCC2$_-$), by

$$2\kappa\tanh\kappa r\pi = k_0. \qquad (6.360)$$

We can see graphically, i.e., plotting $(\tanh \kappa r\pi)$ against (κ^{-1}), that there are two solutions to κ of the same absolute value with one positive and the other negative. However, these two values lead to the same eigenvalue $E^{(2-)}_{-,\kappa}$ and the same eigenfunction $\psi^{(2-)}_{\kappa,\text{ch}}$ when the normalization constant c_κ is included.

This Hamiltonian also admits eigenfunctions $\eta^{(2-)}_k(\vartheta)$ of the form of Eq. (6.350). Since $\Psi_{10} = \Psi_{20}$ the eigenfunction reduces to

$$\eta^{(2-)}_{k,\text{cos}}(\vartheta) = c_k\left(e^{ikr\vartheta} + e^{i2\pi kr}\,e^{-ikr\vartheta}\right) \qquad (6.361)$$
$$= 2c_k\,e^{i\pi kr}\cos kr(\vartheta - \pi). \qquad (6.362)$$

The remaining condition in (BCC2$_-$) imposes a link between k and k_0:

$$2k\tan kr\pi = -k_0. \qquad (6.363)$$

---

[84]E1.5.1(1) E2 introduces three related angle variables, namely θ, ϑ and ϑ_π.

[85]There is a minus sign in the subscript to indicate an attractive interaction with a negative energy eigenvalue. The need for a minus sign arises because $\widehat{H}^{(2-)}$ also admits eigenfunctions with positive eigenvalues.

6.9. CLASSIFICATION OF POINT INTERACTIONS IN $\mathcal{C}$

This gives rise to a discrete set of values for k.[86] The corresponding eigenvalues $E^{(2_-)}_{+,k}$ are positive given by an expression of the form

$$E^{(2_-)}_{+,k}(\mathcal{C}_c) = \frac{1}{2m}(\hbar k)^2. \tag{6.364}$$

Replacing k_0 and V_0 by $-k_0$ and $-V_0$ we obtain a repulsive δ-interaction which would admit no negative eigenvalues. This statement applies to many of the interactions presented below and we shall not keep repeating such a statement in what follows.

6.9.3 Type 3 (BCC3): δ'-interaction

Selfadjoint extension $\widehat{H}^{(3_-)}(\mathcal{C}_c)$ is specified by conditions (BCC3$_-$):

$$\left(\Psi_{20} - \Psi_{10}\right) = -\frac{1}{k_0}\Psi'_{10}, \quad \Psi'_{20} - \Psi'_{10} = 0. \tag{6.365}$$

These conditions are obtained from Eqs. (6.345) and (6.346) by setting

$$\alpha'_2 \neq \alpha'_1, \quad \alpha_2 = \alpha_1 \neq 0. \tag{6.366}$$

In analogy with (BC3$_-$) in §6.5.3 we consider (BCC3$_-$) as defining an attractive δ'-interaction when $k_0 > 0$ since $\widehat{H}^{(3_-)}$ admits an eigenfunction of the form

$$\psi^{(3_-)}_{\kappa,\text{sh}}(\vartheta) = c_\kappa \sinh \kappa r(\vartheta - \pi), \tag{6.367}$$

with a negative eigenvalue $E^{(3_-)}_{-,\kappa}(\mathcal{C}_c)$ given by Eq. (6.359), where κ is linked to k_0 by

$$\kappa \coth \kappa r\pi = 2k_0. \tag{6.368}$$

This Hamiltonian also admits eigenfunctions of the form

$$\eta^{(3_-)}_{k,\sin}(\vartheta) = c_k \sin kr(\vartheta - \pi), \tag{6.369}$$

where k is related to k_0 by

$$k \cot kr\pi = 2k_0. \tag{6.370}$$

The corresponding eigenvalues $E^{[3]}_{+,k}(\mathcal{C}_c)$ are given by Eq. (6.364).

---

[86] This is similar to results obtained from the study of a finite square potential well. See Merzbacher (1998) p. 105 and Greiner (1989) p. 92.

6.9.4 Type 4 (BCC4): Perfect reflector

Selfadjoint extension $\widehat{H}^{(4)}(\mathcal{C}_c)$ is specified by the Dirichlet condition (BCC4):

$$\Phi_{20} = 0, \quad \Phi_{10} = 0, \tag{6.371}$$

obtained from Eqs. (6.345) and (6.346) by setting

$$\text{either} \quad \alpha_2' = \alpha_1' = \alpha_2 = 0, \; \alpha_1 \neq 0 \tag{6.372}$$
$$\text{or} \quad \alpha_2' = \alpha_1' = \alpha_1 = 0, \; \alpha_2 \neq 0 \tag{6.373}$$

in Eqs. (6.5) and (6.6), i.e., Eqs. (6.10) and (6.11). Eigenfunctions of $\widehat{H}^{(4)}$ are of the form:

$$\eta_{n,\sin}^{(4)}(\vartheta) = c_n \sin \frac{1}{2} n\vartheta, \quad n = \pm 1, \pm 2, \ldots, \tag{6.374}$$

corresponding to eigenvalues

$$E_n^{(4)}(\mathcal{C}_c) = \frac{\hbar^2 n^2}{8mr^2}. \tag{6.375}$$

This can be compared with that of an infinite square potential well of width $2\pi r$ in Eq. (2.175).

6.9.5 Type 5 (BCC5): Elastic reflector

Selfadjoint extension $\widehat{H}_\varepsilon^{(5)}(\mathcal{C}_c)$ is specified by conditions (BCC5):

$$\Phi_{20} = -\varepsilon\, \Phi'_{20}, \quad \Phi_{10} = \varepsilon\, \Phi'_{10}, \tag{6.376}$$

obtained from Eqs. (6.345) and (6.346) by setting

$$\text{either} \quad \alpha_2 = \alpha_2' = 0, \quad \alpha_1 \neq 0 \neq \alpha_1' \tag{6.377}$$
$$\text{or} \quad \alpha_1 = \alpha_1' = 0, \quad \alpha_2 \neq 0 \neq \alpha_2'. \tag{6.378}$$

We call this an *elastic reflector*. The Hamiltonian is attractive and admits negative eigenvalues with eigenfunctions of the forms:[87]

1. $\psi_{\kappa,\text{sh}}^{(5)}(\vartheta) = c_\kappa \sinh \kappa r(\vartheta - \pi)$, if ϵ is such that a κ exists satisfying

$$\tanh \kappa r \pi = \epsilon \kappa. \tag{6.379}$$

---
[87] The constants κ and k in the following equations are meant to be real, not complex.

6.9. CLASSIFICATION OF POINT INTERACTIONS IN $\mathcal{C}$

2. $\psi_{\kappa,\text{ch}}^{(5)}(\vartheta) = c_\kappa \cosh \kappa r(\vartheta - \pi)$, if ϵ is such that a κ exists satisfying

$$\coth \kappa r \pi = \epsilon \kappa. \tag{6.380}$$

The corresponding negative eigenvalues $E_{-,\kappa}^{(5)}$ are given by Eq. (6.359).
This Hamiltonian also admits eigenfunctions of the forms:

1. $\eta_{k,\sin}^{(5)}(\vartheta) = c_k \sin kr(\vartheta - \pi)$, if ϵ is such that a k exists satisfying

$$\tan kr\pi = \epsilon k. \tag{6.381}$$

2. $\eta_{k,\cos}^{(5)}(\vartheta) = c_k \cos kr(\vartheta - \pi)$, if ϵ is such that a k exists satisfying

$$\cot kr\pi = -\epsilon k. \tag{6.382}$$

The corresponding eigenvalues $E_{+,k}^{(5)}$ are given by Eq. (6.364).

6.9.6 Type 6 (BCC6): Open end

Selfadjoint extension $\widehat{H}^{(6)}(\mathcal{C}_c)$ corresponds to the Neumann condition (BCC6):

$$\Phi'_{10} = 0, \quad \Phi'_{20} = 0, \tag{6.383}$$

obtained from Eqs. (6.345) and (6.346) by setting either

$$\alpha'_2 = \alpha_1 = \alpha_2 = 0, \quad \alpha'_1 \neq 0, \tag{6.384}$$

or

$$\alpha'_1 = \alpha_1 = \alpha_2 = 0, \quad \alpha'_2 \neq 0. \tag{6.385}$$

Its eigenfunctions are of the form:

$$\eta_{n,\cos}^{(6)}(\vartheta) = c_n \cos \frac{1}{2} n\vartheta, \quad n = 0, \pm 1, \pm 2, \ldots, \tag{6.386}$$

corresponding to eigenvalues

$$E_n^{(6)} = \frac{\hbar^2 n^2}{8mr^2}. \tag{6.387}$$

6.9.7 Type 7 (BCC7): Ideal dynamic π-phase shifter

Selfadjoint extension $\widehat{H}^{(7)}(\mathcal{C}_c)$ corresponds to conditions (BCC7):

$$\Phi_{20} + \Phi_{10} = 0, \quad \Phi'_{20} + \Phi'_{10} = 0 \tag{6.388}$$

obtained from Eqs. (6.345) and (6.346) by setting

$$\alpha'_2 = -\alpha'_1, \quad \alpha_2 = \alpha_1. \tag{6.389}$$

This Hamiltonian admits plane waves as eigenfunctions

$$\eta_n^{(7)}(\vartheta) = \frac{1}{\sqrt{[\ell]}} e^{i(n-\frac{1}{2})\vartheta}, \quad n = 0, \pm 1, \pm 2, \ldots \tag{6.390}$$

with eigenvalues

$$E_n^{(7)} = \frac{\hbar^2}{2mr^2}\left(n - \frac{1}{2}\right)^2. \tag{6.391}$$

These eigenfunctions incur a π-phase shift across the cut. This Hamiltonian is the same as $\widehat{H}_\lambda(\mathcal{C}_c)$ in Eq. (6.329) with $\lambda = \pi$ and $\eta_n^{(7)}(\vartheta)$ is the same as $\eta_{\lambda,n}$ of Eq. (6.325) with $\lambda = \pi$.

This point interaction is fundamentally different to many other point interactions in $\mathcal{C}_c$. The point interactions discussed previously admit real eigenfunctions which correspond to a zero probability current in the absence of external magnetic field. The present eigenfunctions are complex and they can give rise to a non-zero probability current throughout the circle.[88] We may call the previous point interactions as static ones[89] and the present interaction as a dynamic one on account of its eigenfunctions describing a current flow round the circle. We shall return to these important results in our study of superconducting systems in §7.4.

6.9.8 Type 8 (BCC8): Static π-phase shifter

Selfadjoint extension $\widehat{H}^{(8-)}(\mathcal{C}_c)$ corresponds to conditions (BCC8$_-$):

$$\Phi_{20} + \Phi_{10} = 0, \quad \Phi'_{20} + \Phi'_{10} = k_0\,\Phi_{10} \tag{6.392}$$

obtained from Eqs. (6.345) and (6.346) by setting

$$\alpha'_2 = -\alpha'_1, \quad \alpha_2 \neq \alpha_1. \tag{6.393}$$

---

[88]The expression for the probability current will depend on the choice of the momentum operator. The momentum to match $\widehat{H}^{(7)}$ would be $\widehat{P}_\lambda(\mathcal{C}_c)$ in Eq. (6.322) with $\lambda = \pi$. We shall return to this in §7.4.

[89]Type 1 Hamiltonian does admit complex eigenfunctions with a probability current flow. But there is no point interaction there.

6.9. CLASSIFICATION OF POINT INTERACTIONS IN $\mathcal{C}$

As before k_0 is assumed positive. This Hamiltonian is attractive since $\widehat{H}^{(8-)}$ admits negative eigenvalue given by Eq. (6.359) corresponding to eigenfunction

$$\psi_{\kappa,\text{sh}}^{(8-)}(\vartheta) = c_\kappa \sinh \kappa r(\vartheta - \pi), \quad (6.394)$$

where κ is related to k_0 by

$$2\kappa \coth \kappa r\pi = k_0. \quad (6.395)$$

All these eigenfunctions incur a π-phase shift across the cut and being real they correspond to a static situation without giving rise to a non-zero probability current.

This Hamiltonian also admits eigenfunctions of the form

$$\eta_{k,\sin}^{(8-)}(\vartheta) = c_k \sin kr(\vartheta - \pi), \quad (6.396)$$

where k is related to k_0 by

$$2k \cot kr\pi = k_0. \quad (6.397)$$

The corresponding eigenvalues $E_{+,k}^{(8-)}$ are given by Eq. (6.364).

6.9.9 Type 9 (BCC9): Gradient π-phase shifter

Selfadjoint extension $\widehat{H}^{(9-)}(\mathcal{C}_c)$ corresponds to conditions (BCC9$_-$):

$$\Phi_{20} + \Phi_{10} = \frac{1}{k_0} \Phi_{10}', \quad \Phi_{20}' + \Phi_{10}' = 0, \quad (6.398)$$

obtained from Eqs. (6.345) and (6.346) by setting

$$\alpha_2' \neq \alpha_1', \quad \alpha_2 = -\alpha_1 \neq 0. \quad (6.399)$$

The second condition in (BCC9$_-$) imposes a π-phase shift in the gradient of the wave function. This Hamiltonian is attractive since it admits an eigenfunction

$$\psi_{\kappa,\text{ch}}^{(9-)}(\vartheta) = c_\kappa \cosh \kappa r(\vartheta - \pi), \quad (6.400)$$

where κ are related to k_0 by

$$\kappa \tanh \kappa r\pi = 2k_0. \quad (6.401)$$

The corresponding eigenvalue given by Eq. (6.359) is negative.

This Hamiltonian also admits eigenfunctions of the form

$$\eta_{k,\cos}^{(9-)}(\vartheta) = c_k \cos kr(\vartheta - \pi), \quad (6.402)$$

where $k \in \mathbb{R}$ are related to V_0 by

$$k \tan kr\pi = -2k_0. \quad (6.403)$$

The corresponding eigenvalues $E_{+,k}^{(9-)}$ are given by Eq. (6.364).

6.9.10 Type 10 (BCC10): Ideal $\frac{1}{2}\pi$-phase shifter

Selfadjoint extension $\widehat{H}^{(10_-)}(\mathcal{C}_c)$ corresponds to conditions (BCC10$_-$):

$$\Phi_{20} = -\frac{1}{k_0}\Phi'_{10}, \quad \Phi_{10} = \frac{1}{k_0}\Phi'_{20}, \qquad (6.404)$$

obtained from Eqs. (6.345) and (6.346) by setting

$$\text{either} \quad \alpha_1 = \alpha'_2 = 0 \quad \text{or} \quad \alpha_2 = \alpha'_1 = 0. \qquad (6.405)$$

This Hamiltonian, which correlates the wave function directly to its gradient across the cut, is attractive since it admits the following eigenfunctions

$$\psi_{\kappa,\text{sh}}^{(10_-)}(\vartheta) = c_\kappa \sinh \kappa r(\vartheta - \pi), \quad \text{where} \quad \kappa \coth \kappa r\pi = k_0, \qquad (6.406)$$

$$\psi_{\kappa,\text{ch}}^{(10_-)}(\vartheta) = c_\kappa \cosh \kappa r(\vartheta - \pi), \quad \text{where} \quad \kappa \tanh \kappa r\pi = -k_0. \qquad (6.407)$$

The corresponding negative eigenvalues $E_{-,\kappa}^{(10_-)}$ are given by Eq. (6.359).

This Hamiltonian also admits eigenfunctions of the forms:

$$\eta_{k,\sin}^{(10_-)}(\vartheta) = c_k \sin kr(\vartheta - \pi), \quad \text{where} \quad k \cot kr\pi = k_0, \qquad (6.408)$$

$$\eta_{k,\cos}^{(10_-)}(\vartheta) = c_k \cos kr(\vartheta - \pi), \quad \text{where} \quad k \tan kr\pi = k_0. \qquad (6.409)$$

The corresponding eigenvalues $E_{+,k}^{(10_-)}$ are positive and given by Eq. (6.364).

This point interaction is called an ideal $\frac{1}{2}\pi$-phase shifter since for certain appropriate values of k_0 the Hamiltonian admits an eigenfunction of the form $\eta_{\lambda,n}$ in Eq. (6.325) with $\lambda = \frac{1}{2}\pi$, i.e., we have eigenfunctions of the form

$$\eta_{\pi/2,n}^{(10_-)}(\vartheta) = \frac{1}{\sqrt{[\ell]}} e^{i(n-\frac{1}{4})\vartheta}, \qquad (6.410)$$

which can give rise to a non-zero probability current flow across the cut. Substituting this function into (BCC10$_-$) we obtain the following link between k_0 and an integer n:

$$rk_0 = -(n - 1/4). \qquad (6.411)$$

Since k_0 is positive, the integer n must assume a value of $0, -1, -2, \ldots$. These eigenfunctions suffer a phase shift of $\frac{1}{2}\pi$ across the cut in an anti-clockwise direction. For the case of $n = 0$ we have

$$\eta_{\pi/2,0}(\vartheta) = \frac{1}{\sqrt{[\ell]}} e^{-i\frac{1}{4}\vartheta}, \quad rk_0 = \frac{1}{4}. \qquad (6.412)$$

6.9.11 Type 11 (BCC11): Static junction correlator

When $\alpha_1'^2 - \alpha_2'^2 \neq 0$ we have Eqs. (6.347) and (6.348) which can be taken as the basis of new boundary conditions (BCC11):

$$\Psi_{10}' = a\Psi_{10} + b\Psi_{20}, \qquad (6.413)$$
$$\Psi_{20}' = -a\Psi_{20} - b\Psi_{10}, \qquad (6.414)$$

where

$$a \neq 0 \neq b, \qquad a^2 > b^2. \qquad (6.415)$$

Here the constants α and β are related to a and b by Eqs. (6.79) and (6.80). These conditions determine a new type of selfadjoint extensions. There are two cases depending on the sign of $a+b$ and $a-b$.

Case 1 $\quad (a+b) > 0 \text{ and } (a-b) > 0$

This is equivalent to $a > 0$. A selfadjoint extension of this type, denoted by $\widehat{H}_{a,b}^{(11+)}(\mathcal{C}_c)$, admits eigenfunctions of the form

$$\psi_{\kappa,\text{ch}}^{(11+)} = c_\kappa \cosh \kappa r(\vartheta - \pi), \quad \text{where} \quad \kappa \tanh \kappa r\pi = a+b, \qquad (6.416)$$

corresponding to negative eigenvalues given by Eq. (6.359).

This Hamiltonian also possesses the following eigenfunctions

$$\eta_{k,\cos}^{(11+)} = c_k \cos kr(\vartheta - \pi), \quad \text{where} \quad k \tan kr\pi = a+b \qquad (6.417)$$

corresponding to positive eigenvalues given by Eq. (6.364).

Case 2 $\quad (a+b) < 0 \text{ and } (a-b) < 0$

This is equivalent to $a < 0$. A selfadjoint extension of this type, denoted by $\widehat{H}_{a,b}^{(11-)}(\mathcal{C}_c)$, admits eigenfunctions[90]

$$\psi_{\kappa,\text{sh}}^{(11+)} = c_\kappa \sinh \kappa r(\vartheta - \pi), \quad \text{where} \quad \kappa \coth \kappa r\pi = -(a-b) \qquad (6.418)$$

corresponding to negative eigevalues given by Eq. (6.359).

This Hamiltonian also possesses the following eigenfunctions

$$\eta_{k,\sin}^{(11+)} = c_k \sin kr(\vartheta - \pi), \quad \text{where} \quad k \cot kr\pi = -(a-b) \qquad (6.419)$$

corresponding to positive eigenvalues given by Eq. (6.364).

This Hamiltonian does not admit plane wave $\eta_{\lambda,n}(\vartheta)$ as eigenfunctions.[91]

---

[90] Eigenfunctions in Case 1 are continuous across the junction while the eigenfunctions in Case 2 are not.

[91] The reason is that $\eta_{\lambda,n}(\vartheta)$ violates (BCC11). On a substitution of $\eta_{\lambda,n}(\vartheta)$ into (BCC11) we get Eq. (6.423) which implies $a^2 < b^2$.

6.9.12 Type 12 (BCC12): Ideal tunable phase shifters

This type of extension is specified by boundary conditions (BCC12):

$$\Psi'_{10} = a\Psi_{10} + b\Psi_{20}, \tag{6.420}$$
$$\Psi'_{20} = -a\Psi_{20} - b\Psi_{10}, \tag{6.421}$$

where

$$a \neq 0 \neq b, \qquad a^2 < b^2. \tag{6.422}$$

There are two cases depending on the sign of $a+b$ and $a-b$.

Let us concentrate on the case with $(b+a) > 0$ and $(b-a) > 0$ This is equivalent to $b > 0$. A selfadjoint extension of this type, denoted by $\widehat{H}^{(12+)}_{a,b}(\mathcal{C}_c)$, possesses positive eigenvalues corresponding to eigenfunctions of the form of Eqs. (6.417) and (6.419) respectively. We can also have negative eigenvalues corresponding to eigenfunctions of the form of Eqs. (6.416) and (6.418).

In view of the similarity between boundary conditions (BCC12) and (BC12) in §6.5.12 we expect this Hamiltonian to admit complex eigenfunctions of the form of $\eta_{\lambda,n}(\vartheta)$ in Eq. (6.325), albeit subject to some additional constraints on λ and n due to boundary conditions (BCC12). This is indeed the case and we can obtain these constraints by substituting $\eta_{\lambda,n}(\vartheta)$ into Eqs. (6.420) and (6.421), and then taking the real and imaginary parts of the resulting equations. The constraints are:

$$a = -b\cos\lambda, \quad \lambda \neq 0, \frac{1}{2}\pi, \tag{6.423}$$

$$\frac{1}{r}\left(n - \frac{\lambda}{2\pi}\right) = b\sin\lambda. \tag{6.424}$$

For any given a and b such an eigenfunction entails a phase shift λ determined by Eq. (6.423) and a value of $p_{\lambda,n}$ given by

$$p_{\lambda,n} = \frac{\hbar}{r}\left(n - \frac{\lambda}{2\pi}\right), \tag{6.425}$$

or

$$p_{\lambda,n} = \hbar b \sin\lambda. \tag{6.426}$$

The corresponding energy eigenvalue is

$$E_n^{(12+)} = \frac{1}{2m} p_{\lambda,n}^2, \tag{6.427}$$

or

$$E_n^{(12+)} = \frac{1}{2m} p_{\lambda,n}^2 = \frac{\hbar^2}{2mr^2}\left(n - \frac{\lambda}{2\pi}\right)^2. \tag{6.428}$$

6.10. CURRENT AND STATIONARY STATES IN A CIRCLE

This eigenfunction describes a dynamic situation with a non-zero current flowing round the circle across the cut. The phase shift across the cut can be tuned by adjusting the values of a and b. Note that Eq. (6.423) implies $a^2 < b^2$. It follows that a Type 11 junction correlator cannot admit an eigenfunction of this form.

A closer examination of Eqs. (6.423) and (6.424) reveals a possible inconsistency. In view of the relevance of this in later applications we shall investigate this a little more. Given any a, b the first equation fixes two values of the phase shift, i.e., $\pm \lambda$. However, it is not certain that there exist two matching integers, i.e., $\pm n$, to satisfy the second equation. So, the existence of a complex eigenfunction of the form $\eta_{\lambda,n}(\vartheta)$ requires a careful choice of a and b. A matching pair of a and b can be established as follows:

1. Choose parameter b and an integer n at the start. Then obtain possible values of λ from Eq. (6.424), consistent with the chosen b and n. As can be seen graphically there could be more than one possible value,[92] for an appropriately chosen n, i.e., if $|n|$ is too large no such value of λ can be found.

2. Then, substitute each possible value $\lambda_1, \lambda_2, \ldots$ of the phase shift obtained above into Eq. (6.423) to work out the corresponding value of parameter a, e.g., a_1 from λ_1 and a_2 from λ_2.

3. For each momentum eigenvalue, e.g., $p_{\lambda_1,n}$ there corresponds a set of parameter values, i.e., a pre-determined b with a matching $a = a_1$. It follows that a momentum eigenvalue corresponds to a Hamiltonian $\widehat{H}_{a,b}^{(12+)}(\mathcal{C}_c)$. This Hamiltonian would admit $\eta_{\lambda,n}$ as an eigenfunction without inconsistency.

Similar to the Hamiltonian in $I\!E$ in §6.5.12 specified by boundary conditions (BC12) our present Hamiltonian can also admit a plane wave eigenfunction if the parameters a, b are suitably chosen. The constant b is seen to act like a coupling constant of the junction linking the wave function on opposite sides of the junction.

6.10 Current and Stationary States in a Circle

The analysis in the preceding section demonstrates that the method of quantization by parts applies equally well to a circular geometry. Although point

---

[92]Possible values of λ can be obtained from the intersections of the following curves

$$y = rb \sin \lambda, \quad y = n - \lambda/2\pi$$

in the y-λ plane.

interactions in $I\!\!E$ and in $\mathcal{C}$ appear similar in terms of the boundary conditions, the physical interpretation of the resulting Hamiltonians can be quite different.

In the absence of an external magnetic field most of the interactions lead to stationary eigenstates in $\mathcal{C}$ which are static without leading to a probability current flow on account of the Hamiltonian eigenfunctions being real. For applications requiring stationary states with a current flow, i.e., with a non-zero probability current, we will have to select those Hamiltonians which admit plane waves as eigenfunctions. Apart from the free Hamiltonian $\widehat{H}(\mathcal{C})$ there are just three cases in which stationary states can lead to a current flow:

1. *The ideal dynamic π-phase shifter.* The Hamiltonian $\widehat{H}^{(7)}(\mathcal{C}_c)$ admits complex eigenfunctions $\eta_n^{(7)}(\vartheta)$. Comparing with Eq. (6.325) we can see that these are also the eigenfunctions of the momentum operator $\widehat{P}_{\lambda=\pi}(\mathcal{C}_c)$.

2. *The ideal $\frac{1}{2}$-phase shifter.* The Hamiltonian $\widehat{H}_-^{[10]}(\mathcal{C}_c)$ admits complex eigenfunctions $\eta_{\pi/2,n}^{(10)}(\vartheta)$ given by Eq. (6.410) which is also the eigenfunctions of the momentum operator $\widehat{P}_{\lambda=\pi/2}(\mathcal{C}_c)$.

3. *The ideal tunable phase shifters.* The Hamiltonian $\widehat{H}_{a;}^{(12)}(\mathcal{C}_c)$ admits $\eta_{\lambda,n}(\vartheta)$ in Eq. (6.325) as eigenfunctions.

These results have important applications as will be seen in the next chapter.

References

1. Albeverio, S., Gesztesy, F., Hoegh-Krohn, R. and Holden, H. (1988). *Solvable Models in Quantum Mechanics*, Springer-Verlag, New York Inc.
2. Albeverio, S., Gesztesy, F. and Holden, H. (1992). *J. Phys. A: Math. Gen.* **26** 3903.
3. Anderson, B. P., Dholakia, K. and Wright, E. M. (2003). *Phys. Rev. A* **67** 033601.
4. Blank, J., Exner, P. and Havlicek, M. (1994). *Hilbert Space Operators in Quantum Physics*, American Institute of Physics Press, New York.
5. Bleaney, B. I. and Bleaney, B. (1957). *Electricity and Magnetism*, Clarendon Press, Oxford.

REFERENCES

6. Born, M. and Wolf, E. (1999). *Principles of Optics*, Cambridge University Press, Cambridge.
7. Bratteli, O. and Robinson, D. W. (1981). *Operator Algebras and Quantum Statistical Mechanics II*, Springer-Verlag, New York.
8. Coulson, C. A . (1965). *Waves*, Oliver and Boyd, Edinburgh.
9. Coutinho, F., Nogami, Y. and Perez, J. (1997). *J. Phys. A: Math. Gen.* **30** 3937.
10. Crawford, F. S., Jr. (1968). *Waves - Berkeley Physics Course Vol. 3*, MaGraw-Hill, New York.
11. Dalfovo, F. and Giorgini, S. (1999). *Rev. Mod. Phys.* **71** 463.
12. de Vries, P. (1998). *Rev. Mod. Phys.* **70** 447.
13. Dicke, R. H. and Wittke, J. P. (1963). *Introduction to Quantum Mechanics*, Addison-Wesley, Reading, Mass. World Student Series Edition.
14. Exner, P. and Seba, P. (1987). *J. Math. Phys.* **28** 386.
15. Exner, P. and Seba, P. (1989a). *Rep. Math. Phys.* **28** 7.
16. Exner, P. and Seba, P. (1989b). 'Quantum Junctions and the Self-Adjoint Extensions Theory' in P. Exner and P. Seba, eds., *Applications of Self-Adjoint Extensions in Quantum Physics*, Springer-Verlag, Berlin.
17. Exner, P., Seba, P. and Stovicek, P. (1989). 'Quantum Waveguides' in P. Exner and P. Seba, eds., *Applications of Self-Adjoint Extensions in Quantum Physics*, Springer-Verlag, Berlin.
18. Feynman, R. P. (1972). *Statistical Mechanics*, Benjamin, Reading, Mass.
19. Feynman, R. P., Leighton, R. B. and Sands, M. (1965). *The Feynman Lectures on Physics Vol. 3*, Addison-Wesley, Reading, Mass.
20. Fountain, R. H. (1995). *Observables, Maximal Symmetric Operators, POV Measures and their Applications in Quantum Mechanics*, St Andrews University PhD Thesis.
21. Gesztesy, F. and Holden, H. (1987). *J. Phys. A: Math. Gen.* **20** 5157.
22. Giamarchi, T. (2004) *Quantum Physics in One Dimension*, Clarendon Press, Oxford.
23. Giancoli, D. C. (1988). *Physics for Scientists and Engineers*, Prentice-Hall, Englewood Cliffs, New Jersey. 2nd edition.
24. Goldstein, H. (1950). *Classical Mechanics*, Addison-Wesley, Reading, Mass.
25. Greiner, W. (1989). *Quantum Mechanics Vol. 1 - An Introduction*, Springer-Verlag, Berlin.
26. Gross, E. P. (1961). *Nuovo Cimento* **20** 454.
27. Gross, E. P. (1963). *J. Math. Phys.* **4** 195.
28. Gustafson, S. J. and Sigal, I. M. (2003). *Mathematical Concepts of Quantum Mechanics*, Springer-Verlag, Berlin.
29. Hudson, V. and Pym, S. J. (1980). *Applications of Functional Analysis and Operator Theory*, Academic Press, London.
30. Jackson, J. D. (1999). *Classical Electrodynamics*, Wiley, New York.
31. Kittel, C. (1996) *Introduction to Solid State Physics*, Wiley, New York. 7th edition.
32. Landau, L. D. and E. M. Lifshitz, (2002). *Quantum Mechanics*, 3rd edition, Butterworth-Heinemann, Oxford.

33. Longhurst, R. S. (1963). *Geometric and Physical Optics*, Longmans, London.
34. Martin, J. L. (1981). *Basic Quantum Mechanics*, Clarendon Press, Oxford.
35. Merzbacher, E. (1998). *Quantum Mechanics*, 3rd edition, Wiley, New York.
36. Papoulis, A. (1962). *The Fourier Integral and Its Applications*, McGraw-Hill, New York.
37. Pearson, D. B. (1988). *Quantum Scattering and Spectral Theory*, Academic Press, London.
38. Pitaevskii, L. P. (1961). *Sov. Phys. JETP* **13** 451.
39. Reed, M. and Simon, B. (1975). *Methods of Modern Mathematical physics Vol. 2: Fourier Analysis, Selfadjointness*, Academic Press, New York.
40. Saglam, M. and Boyacioglu, B. (2002a). *Int. J. Mod. Phys. B* **16** 607.
41. Saglam, M. and Boyacioglu, B. (2002b). *Phys. Stat. Sol. Phys. B* **230** 133.
42. Seba, P. (1986). *Rep. Math. Phys.* **24** 111.
43. Seba, P. (1987). *Ann. Phys.* **44** 323.
44. Spiegel, M. R. (1974). *Advanced Calculus*, a Schaum's Outline Series, McGraw-Hill, New-York.
45. Trueman, C. (1999). *Superselection Rules, Quantization by Parts and Point Interactions*, St Andrews University PhD Thesis.
46. Trueman, C. and Wan, K. K. (2000). *J. Math. Phys.* **41** 1.
47. Wan, K. K. and Fountain, R. H. (1996). *Found. Phys.* **26** 1165.
48. Wan, K. K. and Fountain, R.H. (1998). *Int. J. Theor. Phys.* **37** 2153.
49. Wan, K. K. and Saglam, M. (2005). *Intrinsic Magnetic Flux of the Electron's Orbital and Spin Motion*, preprint.
50. Zhao, B-H. (1992). *J. Phys. A: Math. Gen.* **25** L617.

Chapter 7

Macroscopic Quantum Systems

7.1 Single-Particle Representation

There has been a lot of interest in macroscopic quantum systems such as superconductors, superfluid helium and Bose-Einstein condensates. Such systems are macroscopic both in their spatial dimensions and in the number of particles involved, and yet they do not behave classically. Superconducting systems are often constructed in the form of an electrical circuit through which a resistanceless current, known as a *supercurrent*, flows. Such circuits are referred to as *superconducting quantum circuits* or simply *quantum circuits*, since the behaviour of a supercurrent is quite different from a traditional electric current which obeys classical circuit theory such as Ohm's law. The simplest circuit consists of a single *branch* which is idealized as either a line of finite or infinite length or a circular coil. Generally a circuit consists of several branches linked together at some points called *branch points*. The linking together of two or more branches at a branch point need not be a physical joining of the branches. One can form a linkage by placing the ends of two branches sufficiently close together without actually joining them, e.g., one could leave a gap of about a few nanometers. Such a linkage is called a *junction*. A superconducting current can tunnel through the insulating gap without needing an applied voltage across the gap. These systems are physically very important. Our aim here is to establish a description of these systems within the framework of quantum mechanics and to investigate any fundamental differences between traditional microscopic quantum systems and these macroscopic quantum systems. We shall argue that macroscopic quantum systems can behave differently from orthodox quantum systems. They could exhibit some very distinctive features, e.g., they could behave like mixed quantum systems with a superselection rule in the context of Postulate UQS in §4.2.1. To make our discussion more definite we shall concentrate on superconductivity in low temperatures, e.g., in the limit as the temperature tends to zero. We shall investigate flux quantization as well as the Josephson equation for the tunnelling of a supercurrent through a Josephson junction in a superconducting circuit. We shall also examine a number of different circuit configurations to clarify some conceptual problems in quantum mechanics.

There is a generally accepted microscopic theory of low temperature superconductivity, namely the BCS theory which is based on the interactions between the conduction electrons and the crystal lattice.[1] According to the BCS theory, superconductivity becomes possible because at low temperatures the balance of all interactions enables electrons in a superconductor to form pairs, known as Cooper pairs, which would behave like Bosons and "condense" into the ground state, i.e., all the Cooper pairs may exist in the ground state of the Hamiltonian of the system. Such a collection of Cooper pairs is referred to as a *condensate*. It is the motion of this condensate as a whole which produces the various phenomena of superconductivity. We shall give a brief review of BCS theory later in §7.11. A theory which can describe every aspect of such a many-particle system would be very complicated. However, a surprisingly simple approach has proved successful in modelling a number of superconducting phenomena. The idea is to consider the condensate as a single system behaving like a single quasi-particle and hence describable by a one-particle wave function.

In classical mechanics it is often possible to describe certain behaviour of a many-body system in terms of a single particle, e.g., the description of the translational motion of a rigid body by a single particle situated in the center of mass of the body.[2] A similar situation holds in quantum theory. When one is considering the translational motion one can often ignore the motion of its constituent particles and describe the system in terms of a single particle with an appropriate effective mass. Such a single-particle representation of a compound system helps us to appreciate certain properties of the system revealed by experiments, without having to study explicitly the complex interactions of the constituent components of the compound system. Take the example of an experiment on the wave property of fullerene atoms C_{60}.[3] A diffraction

---

[1] Bardeen, Cooper and Schreiffer (1957).

[2] To continue with an analogy with rigid bodies in classical mechanics a condensate can be visualized as a rigid body in the momentum space rather than in the coordinate space (Tilley and Tilley (1991) p. 40).

[3] Arndt *et al.* (1999). See Tonomura (1989) for experiments with electrons and see Bonse and Rauch (1979) for experiments with neutrons.

7.1. SINGLE-PARTICLE REPRESENTATION

pattern is observed when a beam of fullerene atoms are sent through a diffraction grating consisting slits of width 50 nm and 100 nm apart. A fullerene atom of dimension 1 nm is massive compared with its constituent electrons, protons and neutrons. For translational motion involved in the interference experiment it turns out that one can treat a C_{60} atom as a single particle having the mass M of the whole atom moving with a speed V in the sense that the atom as a whole can be given a de Broglie wavelength $\lambda = h/MV$. The observed interference pattern is consistent with a beam of single particles of de Broglie wavelength $\lambda = h/MV$.[4]

There is no *a priori* reason why a single-particle quantum mechanics should work so well. To appreciate this one may consider a single-particle approach as a *phenomenalogical theory*. An analogy is the relationship between thermodynamics and statistical mechanics. Like the former a single-particle quantum mechanical approach can give a number of experimentally verifiable relations and results, e.g., flux quantization and Josephson effects in superconductivity, but it is not the complete story. It does not provide further detailed information of the system, e.g., how Cooper pairs and condensate are formed in the case of superconductivity. For a complex system in certain equilibrium states which can be characterized by a relatively *small number of parameters* a phenomenological theory can prove to be perfectly adequate as well as being very useful. An ideal gas in thermal equilibrium is specified by its volume, temperature and pressure. This is possible because at equilibrium the temperature and pressure are the same through out the gas. When the gas is not in equilibrium such a simple description fails; there is no longer a single temperature or pressure for the system as a whole.

So, we have a peculiar situation where quantum mechanics, often regarded as the most fundamental of all physical theories, can also play the role of a phenomenological theory. As a phenomenological theory it does not delve too deeply into the exact details of the physical systems.[5] Instead it concentrates on a small number of most important physical properties of the system. This is one of the reasons for the successes and general applicability of quantum theory. In each specific case one can of course go deeper into the reason why a single-particle approach works.[6]

---

[4]For a fullerene atom the de Broglie wavelength $\lambda = h/MV$ is approximately equal to 2.5 pm for a speed $V \approx 220$ ms^{-1}.

[5]This is in contrast with certain hidden variable theories which try to work out every details, e.g., the trajectory of motion for each particle.

[6]The reason this approach works for superconducting systems will be commented on in §7.11.

7.2 Macroscopic Wave Function Hypothesis

Let us adopt a single-particle representation of the condensate based on the following hypothesis championed by Feynman.[7]

Macroscopic wave function hypothesis:

- The condensate as a whole can be described by a single-particle wave function corresponding to a *quasi-particle* of mass m_c and charge $-q_c$ twice those of an electron, i.e.,[8]

$$q_c = 2e, \quad m_c = 2m_e, \qquad (7.1)$$

where e is the elementary charge (taken as positive) and m_e is the electron mass. This single-particle wave function shall be referred to as a *macroscopic wave function*.

Many important properties of a condensate can be obtained from studying the macroscopic wave function associated with the system under an appropriate one-particle Hamiltonian. In adopting this approach one has to bear in mind that the model is naturally an idealized one, useful for describing only some specific properties.

It should be emphasized that while a macroscopic wave function is formally a function of a single position variable x the wave function is not meant to represent just a single quasi-particle. This is similar to the study of scattering problems in orthodox quantum mechanics where one is interested in the scattering of a beam of particles which are all in the same state. One can represent such a beam of particles without invoking a multi-particle wave function. For motion in one dimension in coordinate space $I\!E$ a plane wave in Eq. (6.90) introduced in §6.5 is formally a one-particle wave function since it is a function of a single position variable x. But such a plane wave also represents a beam of particles of the same momentum p. A similar treatment shall be adopted here to describe a condensate. In other words we shall employ such a plane wave to represent a beam of quasi-particles each of momentum p with beam density one particle per unit volume. Such a beam may be regarded as consisting of particles moving with velocity $v_c = p/m_c$ and it gives rise to a probability current j_p and an electrical current j_p in the coordinate space related to the

---

[7]London (1961), Feynman, Leighton and Sands (1965) §21-9, Feynman (1972) p. 304, Barone and Paterno (1982) p. 2, pp. 9-10, p. 18, p. 23, Tilley and Tilley (1990) pp. 38-41. We shall return in §7.11 to discuss how this hypothesis can be related to the BCS theory, following a brief review of the BCS theory there.

[8]The subscript shows the resemblance of these quasi-particles to the Cooper pairs. Note that q_c is positive so that a Cooper pair possesses a negative charge of $-q_c = -2e$.

7.3. UNIFORMLY THICK SUPERCONDUCTING RINGS

quasi-particle momentum p by[9]

$$J_p = \frac{p}{m_c[\ell]}, \quad j_p = -q_c \frac{p}{m_c[\ell]}. \tag{7.2}$$

A physically relevant example is the motion of a particle in a circle discussed in §6.7, §6.8 and §6.9 where we have a beam of particles all of the same momentum going round the circle contributing to a probability current and an electric current going round the circle.

7.3 Uniformly Thick Superconducting Rings

7.3.1 Physical properties

Let us start with a simple one-branch quantum circuit, i.e., a uniformly thick superconducting ring, abbreviated as TSCR, as shown in Fig. 7.3.1.[10]

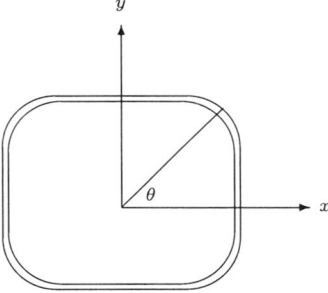

Fig. 7.3.1   Uniformly thick superconducting ring.

The properties of a TSCR in the presence of a uniform and constant external magnetic field are well-known. First we have the Meissner effect which prevents an external magnetic field from penetrating into the bulk of a superconductor.[11] A typical procedure to circumvent the Meissner effect is to place the ring in an external magnetic field at room temperature. The magnetic field will enter the ring, i.e., the ring will enclose a magnetic flux. One then lowers the temperature until the ring becomes superconducting. Generally a

---
[9]As before, the symbol $[\ell]$ denotes a unit length, i.e., it has the value 1 and the dimension of length.
[10]The ring is meant to be circular.
[11]We shall confine ourselves to Type 1 superconductors to which the Meissner effect applies.

dc supercurrent will be induced to flow in the ring without any applied voltage.[12] The ring will enclose a magnetic flux. As already shown in E1.6.2(1) E1 in §1.6.2 the magnetic flux enclosed in a resistanceless ring will not change, i.e., changing the external magnetic field will not alter the enclosed flux.[13] In other words the total magnetic flux Φ enclosed in a TSCR will not change as long as superconductivity is maintained, even if the external magnetic field is changed or removed altogether. This is made possible by a corresponding change of the supercurrent, i.e., when the external magnetic field changes the supercurrent will automatically alter so as to keep the total enclosed flux Φ unchanged. One says that no flux can enter or leave the ring on account of the Meissner effect which prevents a magnetic field from entering the bulk of a superconductor. One can change Φ in a TSCR only by destroying superconductivity, and hence the superconducting system, e.g., by raising the temperature for example.

The total magnetic flux Φ enclosed by the ring can be directly measured, without destroying superconductivity in the process. It is found to be quantized in units of $\Phi_c = h/2e$, i.e., the measured flux values are multiples of Φ_c, i.e.,[14]

$$0, \pm\Phi_c, \pm 2\Phi_c, \pm 3\Phi_c, \ldots, \qquad \Phi_c = h/q_c. \tag{7.3}$$

The value Φ_c is referred to as the *Cooper pair magnetic flux quantum*, or the *flux quantum* for short.

7.3.2 Superconducting rings: Preliminaries

Before we set out a formal model theory let us spell out the following preliminaries:

1. The system and the macroscopic wave function hypothesis We are interested in a TSCR in the presence of a uniform and time-independent external magnetic field. We shall consider the general situation that the ring contains a non-zero amount of magnetic flux and that a supercurrent circulates the ring. We shall assume the macroscopic wave function hypothesis stated in §7.2, i.e., we shall associate a TSCR with a Hilbert space and consider the state of the condensate representable by a vector and observables by appropriate selfadjoint operators in the Hilbert space.

---

[12] The term dc is meant to signify a direct current as opposed to an alternative current. The latter will be referred to as an ac current. Note also that a dc current is meant to be time-independent and stable as in an equilibrium classical circuit, i.e., transient situations are excluded.
[13] Rose-Inness and Rhoderick (1978) §1.3.
[14] Deaver and Fairbank (1961), Doll and Nabauer (1961).

7.3. UNIFORMLY THICK SUPERCONDUCTING RINGS

2. A preliminary model It is natural to adopt the theory of charged particles in a circle presented in §6.7 to model our quasi-particles. In other words, we simply replace m and q there by m_c and $-q_c$ to arrive at the following description of our quasi-particles:

1. Idealize the ring as a circle $\mathcal{C}$ of radius r and take the Hilbert space to be $L^2(\mathcal{C})$.

2. Take the momentum operator to be $\widehat{P}(\mathcal{C})$ in Eq. (6.274). Following Eq. (6.276) the Hamiltonian is taken to be:[15]

$$\widehat{H}(\mathcal{C}) = \frac{1}{2m_c}\left(\widehat{P}(\mathcal{C}) + q_c A_{ex}\right)^2. \tag{7.4}$$

3. The plane wave $\eta_n(\theta)$ in Eq. (6.284) can be used to represent a beam of quasi-particles of density one particle per unit length of the circle, each having momentum p_n and energy E_n given in Eq. (6.277), i.e.,

$$p_n = \frac{1}{r}n\hbar = \frac{1}{2\pi r}nh \quad \text{and} \quad E_n = \frac{1}{2m_c}\left(p_n + q_c A_{ex}\right)^2. \tag{7.5}$$

In accordance with Eq. (6.287) such a beam of quasi-particles generates a dc current round the ring:

$$j_n = -\frac{q_c}{m_c[\ell]}\left(p_n + q_c A_{ex}\right). \tag{7.6}$$

4. The total enclosed magnetic flux is

$$\Phi_n = \Phi_{ex} + L_c j_n. \tag{7.7}$$

Here the external flux applied to the ring Φ_{ex} is the same as in Eq. (6.288), i.e., $\Phi_{ex} = 2\pi r A_{ex}$. The self-inductance L_c is assumed to be given by Eq. (6.290) with m and q replaced by m_c and q_c, i.e.,[16]

$$L_c = (2\pi r[\ell])\frac{m_c}{q_c^2}. \tag{7.8}$$

As a result we obtain:

$$\Phi_n = -n\Phi_c, \tag{7.9}$$

where

$$\Phi_c = h/q_c \tag{7.10}$$

is the *Cooper pair magnetic flux quantum*. These agree with experimentally observed values.

---
[15] We have replaced m and q in Eq. (6.276) with m_c and $-q_c$.
[16] The subscript in L signifies the self-inductance of a current due to the circular motion of quasi-particles of mass m_c and charge q_c.

It should be pointed out that we could have chosen to employ the wave function $\varphi_n(\theta)$ in Eq. (6.278) to represent a beam of quasi-particles. This has the apparent advantage of being able to take into account Cooper pairs density in a given superconductor, i.e., by identifying c_n^2 with the density of Cooper pairs. Then the expressions for the current and the self-inductance will both depend on c_n^2 as shown in Eqs. (6.283) and (6.293). However, these changes do not affect the final outcome for the total enclosed flux.

Note that both the flux and the supercurrent are *global quantities*, i.e., they are independent of the angular position variable θ on the ring. The flux is directly measurable. We can also regard the supercurrent as measurable in the sense that it can be deduced from the difference between the total enclosed flux and the external flux, if we know the density of Cooper pairs.[17] They should therefore be represented by properly defined selfadjoint operators in the Hilbert space $L^2(\mathcal{C})$, not one of those improperly defined operators like $\hat{\jmath}_{\theta_0}$ in Eq. (6.281). We shall set out a number of preliminary steps necessary to achieve this aim in what follows.

3. Current and flux as observables Following Eqs. (7.6) and (7.7) we introduce the supercurrent operator by

$$\hat{J}(\mathcal{C}) = -\frac{q_c}{m_c\,[\ell]}\left(\hat{P}(\mathcal{C}) + q_c A_{ex}\right) \qquad (7.11)$$

and the total enclosed flux operator by

$$\hat{\Phi}(\mathcal{C}) = \Phi_{ex} + L_c\,\hat{J}(\mathcal{C}). \qquad (7.12)$$

The vector potential A_{ex} in Φ_{ex} and in $\hat{J}(\mathcal{C})$ cancel out to reduce the flux operator to

$$\hat{\Phi}(\mathcal{C}) = -\frac{2\pi r}{q_c}\,\hat{P}(\mathcal{C}) = -\frac{2\pi}{q_c}\,\hat{L}_z(\mathcal{C}), \qquad \hat{L}_z(\mathcal{C}) = r\hat{P}(\mathcal{C}). \qquad (7.13)$$

Here $\hat{L}_z(\mathcal{C})$ is identifiable with the angular momentum operator for the angular motion round the circle. These current and flux operators are selfadjoint in $L^2(\mathcal{C})$ with $\eta_n(\theta)$ in Eq. (6.284) as eigenfunctions and j_n and Φ_n given by Eqs. (7.6) and (7.9) as their respective eigenvalues. These would be the measured values of the supercurrent and total enclosed flux, according to orthodox quantum mechanics. The flux operator is independent of the external magnetic field and the radius of the circle. Following Eqs. (6.294) and (6.295) we can introduce the Hamiltonian for a beam of $2\pi r/[\ell]$ particles going round

---

[17] See also Feynman, Leighton and Sands (1965) pp. 21-18.

7.3. UNIFORMLY THICK SUPERCONDUCTING RINGS

the ring. This Hamiltonian can be rewritten as:

$$\widehat{H}_{bm}(\mathcal{C}) = \frac{2\pi r}{[\ell]} \widehat{H}(\mathcal{C}) = \frac{1}{2L_c} \left(\widehat{\Phi}(\mathcal{C}) - \Phi_{ex} \right)^2. \tag{7.14}$$

4. Current and flux stability We shall assume that the Hamiltonian $\widehat{H}(\mathcal{C})$ also serves as the Hamiltonian generator. Since $\widehat{H}(\mathcal{C})$ commutes with $\widehat{J}(\mathcal{C})$ and $\widehat{\Phi}(\mathcal{C})$ we conclude that $\eta_n(\theta)$ represents a stationary state. Such a stationary state would correspond to a stable dc current j_n and flux Φ_n. We shall denote $\widehat{H}(\mathcal{C})$, $\widehat{P}(\mathcal{C})$, $\widehat{J}(\mathcal{C})$ and $\widehat{\Phi}(\mathcal{C})$ by $\widehat{H}$, $\widehat{P}$, $\widehat{J}$ and $\widehat{\Phi}$ in what follows for brevity.

Apart from the introduction of the supercurrent and enclosed flux operators we have more or less repeated §6.7 here. In other words we have so far treated the quasi-particles as an orthodox quantum system just like quantized charged particles in a circle in §6.7. In what follows we shall present physical arguments to show that superconducting systems are not orthodox quantum systems.

5. Boundedness of observables and the Hilbert space Superconductivity is destroyed simply by heating the ring up above the critical temperature, by increasing the external magnetic field to above a certain critical value B_c, or by increasing the supercurrent above a critical value j_c. Physically we can see that increasing the temperature causes an increase in the net momenta of the Cooper pairs, until it becomes energetically favourable for them to split up, thereby breaking up the condensate and destroying superconductivity. Similarly both the critical field B_c and the critical supercurrent j_c are associated with a maximum critical value of the momenta of the Cooper pairs, and hence of the quasi-particle representing the condensate.[18] It follows that we must exclude states of excessively high momentum in our model. This exclusion is achieved by taking the Hilbert space for a TSCR to be a subspace of $L^2(\mathcal{C})$ in the following manner:

1. The appropriate Hilbert space for a TSCR is a subspace of $L^2(\mathcal{C})$ spanned by momentum eigenfunctions $\{\eta_n(\theta)\}$ for $n = 0, \pm 1, \pm 2, \ldots, \pm N$ up to some finite positive integer N, in order to restrict the momentum to an appropriate range of values to maintain superconductivity. The integer N is related j_c and B_c.[19] We shall denote this subspace by $L_N^2(\mathcal{C})$.

---

[18] Changing the external magnetic field has the effect of changing the supercurrent in order to maintain the quantized flux.

[19] The critical field B_c and the critical supercurrent j_c are related by the Silsbee's Rule (Fujita and Godoy (1996) p. 289), i.e., j_c is equal to the superconducting current generated by B_c. For example, the maximum value of the enclosed flux cannot exceed $\pi r^2 B_c$, i.e., N is the largest integer satisfying $N < (r^2 e/\hbar) B_c$.

2. Observables of the system must correspond to selfadjoint operators on $L_N^2(\mathcal{C})$,[20] rather than in $L^2(\mathcal{C})$. The natural thing to do is to try to reduce our previous operators in $L^2(\mathcal{C})$ to $L_N^2(\mathcal{C})$ in the sense of Definition 2.3(2). Not every selfadjoint operator in $L^2(\mathcal{C})$ admits $L_N^2(\mathcal{C})$ as an invariant subspace onto which it has a selfadjoint reduction. Fortunately $L_N^2(\mathcal{C})$ completely reduces $\widehat{P}$, $\widehat{H}$, $\widehat{\Phi}$ and $\widehat{J}$ so that their reductions to $L_N^2(\mathcal{C})$, denoted respectively by $\widehat{P}_N$, $\widehat{H}_N$, $\widehat{\Phi}_N$ and $\widehat{J}_N$, are selfadjoint on $L_N^2(\mathcal{C})$. On the other hand both the angular and linear position operators $\widehat{\theta} = \theta$ and $\widehat{s} = r\theta$ have no reduction on $L_N^2(\mathcal{C})$, let alone selfadjoint ones. It follows that they are not observables.

Here we have one striking example of physical systems whose basic observables are described by bounded operators. This is also a new feature of the present superconducting system which is different from those quantized charged particles on a circle presented in §6.7.

6. The breakdown of the superposition principle What we have so far is a quantum system which, when in states represented by plane waves η_n, possesses a finite range of quantized values of momentum, energy, current and total enclosed flux. Suppose the superposition principle operates here, i.e., a linear combination

$$\varphi = c_n \eta_n + c_k \eta_k \qquad (7.15)$$

of states η_n and η_k is allowed to represent a pure state of the system. Such a state would correspond to the total enclosed flux not having a definite value, and a measurement will yield either $n\Phi_c$ or $k\Phi_c$. This is the same as the familiar situation that a superposition of z-component spin-up and spin-down states defines a new state without a definite z-component spin value prior to a spin measurement, and that a measurement will yield either $-\hbar/2$ or $\hbar/2$. However, we have already pointed out that once established the enclosed flux cannot be altered without destroying the superconducting system, i.e., no flux can enter or leave the ring as long as the ring remains superconducting. It follows that any measured flux value must be the value enclosed by the TSCR before the measurement, i.e., the condensate is always in a state with a definite flux value $n\Phi_c$, the same before and after measurement. Moreover there is no experimental evidence of any stable dc superconducting state for a TSCR corresponding to a superposition of different flux eigenstates. For example, when the ring is cooled down to reach superconductivity in the presence of an external magnetic field, the induced dc supercurrent will automatically ensure

---

[20]Generally we accept strictly maximal symmetric operators in accordance with Postulate GQS presented in §4.2.1. Indeed we shall consider a circuit configuration which necessitates the use of strictly maximal symmetric operators in §7.7.

7.3. UNIFORMLY THICK SUPERCONDUCTING RINGS

that the condensate is in a flux eigenstate so that the enclosed flux takes on a value $n\Phi_c$. We conclude that the existence of a pure state φ obtained by a coherent superposition of η_n and η_k would contradict the physical properties of a TSCR. This leads us to two physical assumptions for the description of the quasi-particles set out in the next item.

7. **Physical Assumptions on dc superconductivity**

   - **(PAS1)** A dc superconducting state corresponding to an established dc supercurrent is describable by an eigenfunction η_n of the supercurrent operator $\widehat{J}_N$ with the current equal to the corresponding eigenvalue j_n.

   - **(PAS2)** A dc superconducting state corresponding to an established dc supercurrent must also correspond to an eigenfunction of the Hamiltonian $\widehat{H}_N$ of the system.[21]

Let us clarify these assumptions with the following comments:

1. Statement (PAS1) serves to define dc superconducting states, i.e., they correspond to eigenvectors η_n of $\widehat{J}_N$. By definition a superconducting system exists only in a superconducting state. It follows that not every vector in the Hilbert space $L_N^2(\mathcal{C})$ can represent a state of a superconducting system with a dc supercurrent. In particular there is no superposition of different superconducting states, since such a superposition is not an eigenvector of $\widehat{J}_N$. Assumption (PAS1) directly contradicts Postulate OQS of orthodox quantum statics in §3.2.1. In other words (PAS1) amounts to the assertion that:

   our present superconducting system is not an orthodox quantum system.

2. Statement (PAS2) ensures the stability of the dc supercurrent in a superconducting state. In the present case this condition is satisfied. As we shall see later, for a general quantum circuit an eigenfunction of the current operator is not automatically an eigenfunction of the Hamiltonian. A process of correlative quantization, the third stage of the quantization by parts scheme, is required to render (PAS2) compatible with (PAS1).

3. Statements (PAS1) and (PAS2) are meant to apply to some specific superconducting systems like a TSCR under a uniform and static magnetic field. They are not meant to apply to all superconducting systems, e.g., those having ac supercurrents.

---
[21] The Hamiltonian generator and the Hamiltonian are assumed to be the same.

8. Environment on dc superconductivity A superconducting circuit is often placed in an external magnetic field. We consider this external field as the *environment* the superconducting circuit is coupled to. A superconducting system is meant to be the circuit plus its environment. When a superconducting circuit is placed in a different environment, i.e., in a different magnetic field, we have in effect a different physical system. This is not a novel concept. Particles under different external potentials constitute different physical systems, even in classical mechanics.

7.3.3 Superconducting rings as equilibrium mixed quantum systems

Our preliminary analysis shows that a TSCR in a dc superconducting state formulated in the Hilbert space $L_N^2(\mathcal{C})$ violates the superposition principle and Postulate OQS of §3.2.1. Therefore it is not an orthodox quantum system. Instead it is a mixed quantum system satisfying Postulate UQS of §4.2.1.

Let $\mathcal{H}_n(\mathcal{C})$ be a subspace in $L_N^2(\mathcal{C})$ spanned by the eigenfunction η_n, and let us decompose $L_N^2(\mathcal{C})$ as a direct sum of $\mathcal{H}_n(\mathcal{C})$ and denote the resulting direct sum by $\mathcal{H}_N^\oplus(\mathcal{C})$, i.e.,

$$\mathcal{H}_N^\oplus(\mathcal{C}) = \oplus_n \mathcal{H}_n(\mathcal{C}), \quad n \in [-N, N]. \tag{7.16}$$

Decomposable operators on $\mathcal{H}_N^\oplus(\mathcal{C})$ are of the form

$$\widehat{A}^\oplus(\mathcal{C}) = \oplus_n \widehat{A}_n(\mathcal{C}), \tag{7.17}$$

where $\widehat{A}_n(\mathcal{C})$ is an operator on $\mathcal{H}_n(\mathcal{C})$. For examples the operators $\widehat{P}_N$, $\widehat{J}_N$, $\widehat{H}_N$ and in particular $\widehat{\Phi}_N$ defined in the preceding section are all decomposable. To emphasize this we shall rewrite $\widehat{P}_N$, $\widehat{J}_N$, $\widehat{H}_N$ and $\widehat{\Phi}_N$ as $\widehat{P}_N^\oplus(\mathcal{C})$, $\widehat{J}_N^\oplus(\mathcal{C})$, $\widehat{H}_N^\oplus(\mathcal{C})$ and $\widehat{\Phi}_N^\oplus(\mathcal{C})$.

In line with the notation of C2.13.2(3) C2 and §4.4.4 we shall employ a boldface symbol $\boldsymbol{\eta}_n$ when we want to refer to η_n as an element of Hilbert space $\mathcal{H}_n(\mathcal{C})$. We shall denote its extension to $\mathcal{H}_N^\oplus(\mathcal{C})$ in accordance with Eqs. (2.591) and (2.592) by $\boldsymbol{\eta}_n^\oplus$. Operators $\widehat{P}_N^\oplus(\mathcal{C})$, $\widehat{J}_N^\oplus(\mathcal{C})$, $\widehat{\Phi}_N^\oplus(\mathcal{C})$ and $\widehat{H}_N^\oplus(\mathcal{C})$ have the following direct sum decomposition:

$$\widehat{P}_N^\oplus(\mathcal{C}) = \oplus_n p_n |\boldsymbol{\eta}_n\rangle\langle\boldsymbol{\eta}_n|, \tag{7.18}$$

$$\widehat{J}_N^\oplus(\mathcal{C}) = \oplus_n j_n |\boldsymbol{\eta}_n\rangle\langle\boldsymbol{\eta}_n|, \tag{7.19}$$

$$\widehat{H}_N^\oplus(\mathcal{C}) = \oplus_n E_n |\boldsymbol{\eta}_n\rangle\langle\boldsymbol{\eta}_n|, \tag{7.20}$$

$$\widehat{\Phi}_N^\oplus(\mathcal{C}) = \oplus_n \Phi_n |\boldsymbol{\eta}_n\rangle\langle\boldsymbol{\eta}_n|. \tag{7.21}$$

7.3. UNIFORMLY THICK SUPERCONDUCTING RINGS

They also have the following spectral decompositions:

$$\widehat{P}_N^\oplus(\mathcal{C}) = \sum_n p_n |\boldsymbol{\eta}_n^\oplus\rangle\langle\boldsymbol{\eta}_n^\oplus|, \qquad (7.22)$$

$$\widehat{J}_N^\oplus(\mathcal{C}) = \sum_n j_n |\boldsymbol{\eta}_n^\oplus\rangle\langle\boldsymbol{\eta}_n^\oplus|, \qquad (7.23)$$

$$\widehat{H}_N^\oplus(\mathcal{C}) = \sum_n E_n |\boldsymbol{\eta}_n^\oplus\rangle\langle\boldsymbol{\eta}_n^\oplus|, \qquad (7.24)$$

$$\widehat{\Phi}_N^\oplus(\mathcal{C}) = \sum_n \Phi_n |\boldsymbol{\eta}_n^\oplus\rangle\langle\boldsymbol{\eta}_n^\oplus|. \qquad (7.25)$$

We are now in a position to summarize our model into the following postulate:

Postulate for TSCR

- A TSCR in a uniform and static external magnetic field has associated with it a Hilbert space which has a preferred direct sum decomposition, i.e., $\mathcal{H}_N^\oplus(\mathcal{C})$, and all its observables are represented by decomposable selfadjoint operators in $\mathcal{H}_N^\oplus(\mathcal{C})$, e.g., $\widehat{P}_N^\oplus(\mathcal{C})$, $\widehat{J}_N^\oplus(\mathcal{C})$, $\widehat{H}_N^\oplus(\mathcal{C})$ and $\widehat{\Phi}_N^\oplus(\mathcal{C})$, in accordance of Postulate UQS in §4.2.1.

- The dynamics is given in accordance of Postulate UQD(P) in §4.4.2, i.e., by a preserving unitary evolution group

$$\widehat{U}_t = \exp(-i\widehat{H}_g t) \qquad (7.26)$$

where the Hamiltonian generator $\widehat{H}_g$ is taken to coincide with the Hamiltonian $\widehat{H}_N^\oplus(\mathcal{C})$. In other words, the time evolution of a superconducting state is explicitly given by

$$\boldsymbol{\eta}_{n,t}^\oplus(\theta) = \boldsymbol{\eta}_n^\oplus(\theta)\, e^{-iE_n t} = \frac{1}{\sqrt{[\ell]}}\, e^{in\theta}\, e^{-iE_n t}. \qquad (7.27)$$

This evolution corresponds to the following Schrödinger equation:

$$i\hbar \frac{\partial}{\partial t} \boldsymbol{\eta}_{n,t}^\oplus = \widehat{H}_N^\oplus(\mathcal{C}) \boldsymbol{\eta}_{n,t}^\oplus. \qquad (7.28)$$

This postulate tells us that a TSCR is a mixed quantum system possessing the following properties:

CHAPTER 7. MACROSCOPIC QUANTUM SYSTEMS

1. The system possesses a superselection rule with $\mathcal{H}_n(\mathcal{C})$ serving as supersectors.[22] These supersectors are one-dimensional so that all selfadjoint decomposable operators, e.g., $\widehat{P}_N^\oplus(\mathcal{C})$, $\widehat{H}_N^\oplus(\mathcal{C})$, $\widehat{J}_N^\oplus(\mathcal{C})$, $\widehat{\Phi}_N^\oplus(\mathcal{C})$, mutually commute. Hence a TSCR serves as a non-trivial example of a system which is not classical despite the fact that all its observables mutually commute.

2. Supersectors $\mathcal{H}_n(\mathcal{C})$ correspond one-to-one to pure states, e.g., a pure vector $\eta_n^\oplus$ describes a pure state. A linear combination of pure vectors belonging to different supersectors, e.g.,

$$\eta^\oplus = \frac{1}{\sqrt{2}}\left(\eta_n^\oplus + \eta_k^\oplus\right) \tag{7.29}$$

corresponds to a mixed state in the sense of Definition 4.3.1(1) Case (1) in §4.3.1 since

$$\langle \eta_n^\oplus \mid \widehat{A}^\oplus \, \eta_k^\oplus \rangle = 0, \quad n \neq k \tag{7.30}$$

for all observables $\widehat{A}^\oplus(\mathcal{C})$. This explains why there are no pure states which are superpositions of different flux states.

3. Any one of the observables $\widehat{P}_N^\oplus(\mathcal{C})$, $\widehat{J}_N^\oplus(\mathcal{C})$, $\widehat{H}_N^\oplus(\mathcal{C})$ and $\widehat{\Phi}_N^\oplus(\mathcal{C})$, can serve as a superselection operator. All these operators are bounded.

4. Neither of the position operators $\widehat{\theta} = \theta$ and $\widehat{s} = a\theta$ represents an observable of the quasi-particle since they are neither decomposable nor selfadjoint on $\mathcal{H}_N^\oplus(\mathcal{C})$. Physically this corresponds to the understanding that the condensate is *delocalized* and that any attempt to localize it in order to measure its position would destroy the superconducting system.

Eq. (7.27) tells us that a superconducting state $\eta_{nt}^\oplus$ is stationary and it evolves within one and the same supersector. We can conclude that a TSCR in a static environment with a dc supercurrent is an equilibrium mixed quantum system in the sense of Definition 4.6.3(1) and that it is fundamentally different from an orthodox quantum particle going round a circle described in §6.7.

7.4 Superconducting Rings with a Junction

7.4.1 Josephson junction and dc Josephson effect

A physically important system is that of a thick superconducting ring interrupted by a thin insulating layer, typically of the order of nanometers, known as a *Josephson junction* or JJ for short. This is depicted in Fig. 7.4.1.

---
[22]See also Sewell (2002) pp. 233-234.

7.4. SUPERCONDUCTING RINGS WITH A JUNCTION

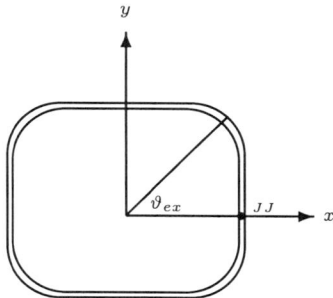

Fig. 7.4.1   Thick ring interrupted by a JJ.

A junction could be achieved physically by a cut, with or without an insulating oxide layer.[23] A dc supercurrent, albeit weaker than before, can still go round the ring without needing any voltage applied across the junction. This phenomenan is known as a *dc Josephson effect*. This effect was predicted by Josephson, using BCS theory with a tunnelling Hamiltonian.[24] Intuitively we can imagine the junction as a potential barrier and that Cooper pairs can tunnel through the barrier without loss of momentum to produce a supercurrent.[25] A supercurrent could be set up initially with the help of an external magnetic field in a similar fashion as for a TSCR. A new feature here is that the Meissner effect loses its rigid hold on the magnetic flux enclosed by the ring as magnetic flux can leak in and out of the ring through the junction. One therefore would not expect the total enclosed magnetic flux to be quantized as exact multiples of the Cooper pair flux quantum Φ_c.

Let us start our investigation of such a dc Josephson effect with the external magnetic field removed. The idea is to treat the junction as the source of a point interaction. In §6.9 we have classified point interactions for a particle of mass m and charge q moving in a circle. It is now a matter of applying the results to our present superconducting system. This involves

1. replacing m and charge q in §6.8 and §6.9 by m_c and charge $-q_c$,

2. introducing a supercurrent operator, and

3. finding a Hamiltonian which is compatible with the current operator in the sense of (PAS1) and (PAS2) in item 7 of §7.3.2.

The details are set out in §7.4.2 to §7.4.7.

---

[23] It is possible to alter the uniformity of the ring by a constriction at a chosen point. Such a constriction is known as a *weak link*.
[24] Josephson (1962).
[25] Mandl (1992) p. 35, §2.2.

7.4.2 Supercurrent and magnetic flux operators

The Hilbert space for the system is initially taken to be $L^2(\mathcal{C}_c)$. The first new feature to emerge is the non-uniqueness of momentum operator in $L^2(\mathcal{C}_c)$, i.e., we have a one-parameter family of possible momentum operators $\widehat{P}_\lambda(\mathcal{C}_c)$ defined by Eqs. (6.322) and (6.323). As in Eqs. (7.11) and (7.12), we shall take the corresponding supercurrent and total enclosed magnetic flux operators to be

$$\widehat{J}_\lambda(\mathcal{C}_c) = -\frac{q_c}{m_c[\ell]} \widehat{P}_\lambda(\mathcal{C}_c), \quad \widehat{\Phi}_\lambda(\mathcal{C}_c) = L_c \widehat{J}_\lambda(\mathcal{C}_c). \tag{7.31}$$

Setting $\widehat{L}_{\lambda,z}(\mathcal{C}_c) = r\widehat{P}_\lambda(\mathcal{C}_c)$ and assuming the self-inductance L_c to be given by Eq. (7.8) we get

$$\widehat{\Phi}_\lambda(\mathcal{C}_c) = -\frac{2\pi r}{q_c} \widehat{P}_\lambda(\mathcal{C}_c) = -\frac{2\pi}{q_c} \widehat{L}_{\lambda,z}(\mathcal{C}_c), \tag{7.32}$$

as in Eqs. (7.11) and (7.13). These operators admit $\eta_{\lambda,n}(\vartheta)$ in Eq. (6.325) as eigenfunctions with

$$j_{\lambda,n}(\mathcal{C}_c) = -\frac{q_c}{m_c[\ell]} p_{\lambda,n}(\mathcal{C}_c) = -\frac{q_c \hbar}{m_c r[\ell]} \left(n - \frac{\lambda}{2\pi}\right) \tag{7.33}$$

and

$$\Phi_{\lambda,n}(\mathcal{C}_c) = -\left(n - \frac{\lambda}{2\pi}\right) \Phi_c \tag{7.34}$$

as their respective eigenvalues.

In the present case we have assumed that the external magnetic field has been removed and that $\lambda \neq 0$ for a TSRC with a JJ. The magnetic flux $\Phi_{\lambda,n}(\mathcal{C}_c)$ is generated by the circulating supercurrent. Clearly $\lambda = 0$ corresponds to the periodic boundary condition for a continuous ring without a junction. Intuitively we can see that for an interrupted ring there is no compelling reason to impose the periodic boundary condition to maintain the continuity of the eigenfunction at the junction, i.e., there is no reason why the wave function should not have a discontinuity across the junction. Indeed, according to Josephson, the wave function for the Josephson effect should suffer a discontinuous phase shift λ across the junction and that this phase shift is directly related to the supercurrent j by the following equation

$$j = j_c \sin \lambda, \tag{7.35}$$

known as the *Josephson equation*. The maximum current which can pass the junction is known as the *critical current* of the junction, to be denoted by j_c. Note that a critical current is meant to be a maximum current either

7.4. SUPERCONDUCTING RINGS WITH A JUNCTION

anti-clockwise or clockwise. Our task is to find a Hamiltonian which would incorporate the interaction caused by the junction, and then use (PAS1) and (PAS2) to relate the current with the phase shift so as to derive the Josephson equation.

7.4.3 The Hamiltonian: Preliminary results

We start with the assumption that the effect of a junction is describable by a point interaction. It follows that we must choose a Hamiltonian from the list of point interaction Hamiltonians given in §6.9. According to (PAS1) a dc superconducting state is described by an eigenfunction $\eta_{\lambda,n}$ of the supercurrent operator, and according to (PAS2) a dc supercurrent must also correspond to an eigenfunction of the Hamiltonian. This means that we must choose a point interaction Hamiltonian which admits $\eta_{\lambda,n}$ as an eigenfunction. Compared with a more general type of functions given by Eq. (6.95) we can see that a superconducting state implies a total transmission with no reflection at the junction. Any reflection would correspond to an electrical resistance.

Now, for $\eta_{\lambda,n}$ to be an eigenfunction of the Hamiltonian we must require $\eta_{\lambda,n}$ to satisfy the boundary conditions of the chosen Hamiltonian. This rules out the δ-interactions, the δ'-interactions, the reflector, the elastic reflectors, the open end, the static and gradient π-phase shifters, since these interactions only admit eigenfunctions which will lead to a zero probability current and hence a zero supercurrent. We are left with three possibilities:

1. A ideal dynamic π-phase shifter in §6.9.7: $\widehat{H}^{(7)}(\mathcal{C}_c)$ admits plane waves as eigenfunctions as shown in Eq. (6.390).

2. An ideal $\frac{1}{2}\pi$-phase shifter in §6.9.10: $\widehat{H}^{(10-)}(\mathcal{C}_c)$ also admits plane waves as eigenfunctions as shown in Eq. (6.410).

3. An ideal tunable phase-shifter in §6.9.12: $\widehat{H}^{(12+)}_{a,b}(\mathcal{C}_c)$ also admits plane waves as eigenfunctions.

Before going into the two special situations of a π-phase shifter and a $\frac{1}{2}\pi$-phase shifter we shall examine Hamiltonian $\widehat{H}^{(12+)}_{a,b}(\mathcal{C}_c)$ first.

Hamiltonian $\widehat{H}^{(12+)}_{a,b}(\mathcal{C}_c)$ is determined by two real parameters a and b. From the discussion in §6.9.12 we know that $\widehat{H}^{(12+)}_{a,b}(\mathcal{C}_c)$ can admit $\eta_{\lambda,n}$ as an eigenfunction provided a and b are chosen in such a way that Eqs. (6.423) and (6.424) are satisfied for some λ and n. For ease of reference let us write things down explicitly here. First, the state corresponding to a supercurrent $j_{\lambda,n}$ is described by

$$\eta_{\lambda,n} = \frac{1}{\sqrt{[\ell]}} e^{i(n-\frac{\lambda}{2\pi})\vartheta}. \qquad (7.36)$$

The constants involved are related by

$$a = -b\cos\lambda, \tag{7.37}$$

$$\frac{1}{r}\left(n - \frac{\lambda}{2\pi}\right) = b\sin\lambda, \tag{7.38}$$

$$p_{\lambda,n}(\mathcal{C}_c) = \hbar b\sin\lambda. \tag{7.39}$$

The following interpretation presents itself:

1. Eq. (7.39) gives rise to the Josephson equation, i.e., by setting

$$j(\mathcal{C}_c) = j_{\lambda,n}(\mathcal{C}_c) = -\frac{q_c}{m_c[\ell]}p_{\lambda,n}(\mathcal{C}_c), \quad j_c(\mathcal{C}_c) = -\frac{q_c}{m_c[\ell]}\hbar b \tag{7.40}$$

we obtain the Josephson equation

$$j(\mathcal{C}_c) = j_c(\mathcal{C}_c)\sin\lambda. \tag{7.41}$$

While the current can be positive or negative, depending on λ, the value of $j_c(\mathcal{C}_c)$ here is negative since $b > 0$.[26] If we wish to render the relation between j and λ unique we have to restrict the range of phase shift to

$$\lambda \in (-\pi/2, 0) \cup (0, \pi/2). \tag{7.42}$$

2. The magnetic flux enclosed by the ring assumes the value $\Phi_{\lambda,n}$ given by Eq. (7.34). As the phase shift changes the magnetic flux $\Phi_{\lambda,n}$ and the current would change accordingly. This is interpreted as due to the magnetic flux leaking in or out of the ring through the junction.

3. Note that $\lambda = \pm\pi/2$ is not achievable by Hamiltonian $\widehat{H}_{a,b}^{(12+)}(\mathcal{C}_c)$ since Eq. (7.37) renders $a = 0$, violating boundary conditions (BCC12). Also $\lambda = 0, \pi$ renders $a^2 = b^2$ violating (BCC12). Nevertheless we can have a superconducting state with a phase shift of $\pi/2$ or π across the junction. We shall return to discuss these cases in §7.4.5 and §7.4.6 in terms of an ideal $\pi/2$-phase shifter and an ideal dynamic π-phase shifter.

4. In §6.9.12 we mention a possible problem concerning Eqs. (7.37), (7.38) and (7.39), i.e., parameters a and b have to be chosen very carefully to avoid inconsistency. Here we shall assume that for all TSCRs with a JJ capable of carrying a supercurrent their parameters a and b do indeed satisfy Eqs. (7.37), (7.38) and (7.39). In conjunction with the discussion in §6.9.12 we would offer the following understanding of the different roles played by a and b:

---

[26] The sign of $j_c(\mathcal{C}_c)$ is of no fundamental importance.

7.4. SUPERCONDUCTING RINGS WITH A JUNCTION

(a) A given TSCR with a JJ determines the radius r and the parameter b which acts as the coupling constant characterizing the junction.

(b) Parameter a is left to be flexible and determined by the physical processes which help to set up the supercurrent. In other words, in setting up a supercurrent one is setting up the values of n and λ; the value of a is determined in the process and is related to λ and b in accordance with Eq. (7.37). A change of the phase shift from λ to a different value λ' would cause a corresponding change in the parameter value, i.e., from a to a'. This will become clearer later in §7.4.7 when we bring in an external magnetic field. An external magnetic field serving as the environment of the ring will be seen to play a part in fixing the value of a.

(c) As discussed in §6.9.12 a given momentum eigenvalue determines the Hamiltonian $\widehat{H}_{a,b}^{(12+)}(\mathcal{C}_c)$. Similarly a supercurrent also determines $\widehat{H}_{a,b}^{(12+)}(\mathcal{C}_c)$.

(d) It follows from the discussion above that there is only one momentum operator $\widehat{P}_\lambda(\mathcal{C}_c)$, and hence one supercurrent operator $\widehat{J}_\lambda(\mathcal{C}_c)$ and one enclosed flux operator $\widehat{\Phi}_\lambda(\mathcal{C}_c)$, compatible with $\widehat{H}_{a,b}^{(12+)}(\mathcal{C}_c)$ in the sense that they share one common eigenfunction $\eta_{\lambda,n}$. In accordance with (PAS1) and (PAS2) this eigenfunction can describe a superconducting state. In such a superconducting state the current satisfies the Josephson equation.

This is not the end of the story. There are two unusual problems:

1. A superconducting state $\eta_{\lambda,n}$ automatically singles out a λ, and the operators $\widehat{P}_\lambda$, $\widehat{J}_\lambda$, $\widehat{\Phi}_\lambda$ and $\widehat{H}_{a,b}^{(12+)}$. A different state $\eta_{\lambda',n}$ together with its current $j_{\lambda',n}$ would mean a different phase shift and a different set of operators, i.e., $\widehat{P}_{\lambda'}$, $\widehat{J}_{\lambda'}$, $\widehat{\Phi}_{\lambda'}$ and $\widehat{H}_{a',b}^{(12+)}$. This is not a standard situation, since a change of state usually does not necessitate a change of operators.

2. The supercurrent operator $\widehat{J}_\lambda$ possesses other eigenfunctions $\eta_{\lambda,n'}$ and eigenvalues $j_{\lambda,n'}$. But these eigenfunctions are not capable of representing a superconducting state since its associated eigenvalue $j_{\lambda,n'}$ cannot satisfy the Josephson equation with given b and λ, i.e., Eqs. (7.40) and (7.41) tell us that
$$j_{\lambda,n'} \neq j_c \sin \lambda = j_{\lambda,n}. \tag{7.43}$$
A new current $j_{\lambda,n'}$ should correspond to a new phase shift λ' in accordance with a corresponding Josephson equation, i.e.,
$$j_{\lambda',n'} = j_c \sin \lambda'. \tag{7.44}$$

528   CHAPTER 7. MACROSCOPIC QUANTUM SYSTEMS

This new phase shift will in turn entail a new current operator $\widehat{J}_{\lambda'}$ which possesses further new eigenvalues. This process can go on to generate a chain of new current eigenvalues and new current operators.

The question now is whether we should assign the current operator $\widehat{J}_\lambda$ to the system with an established current $j = j_{\lambda,n}$. If we do so, what is the status of the other eigenvalues $j_{\lambda,n'}$, given that these values are not related to the phase shift λ by the Josephson equation?

One may be tempted to get out of this difficulty by claiming that not all eigenvalues of the current operator $\widehat{J}_\lambda$ can represent a supercurrent value. This would avoid generating new phase shifts and hence new current operators. However, such a claim, as it stands, is arbitrary and it would contradict the basic tenets of orthodox quantum mechanics. To overcome this difficulty we have to reject $\widehat{J}_\lambda$ in favour of a different operator to serve as the current operator and consider a TSCR with a JJ as a mixed quantum system. We shall do this in the next section.

7.4.4 Superconducting ring with a Josephson junction as an equilibrium mixed quantum system

A resolution of the difficulty presented in the preceding section lies in the recognition that the condensate in a TSCR with a JJ is a mixed quantum system in the same sense the condensate in a TSCR without a JJ is. The argument runs as follows:

1. Let $\mathcal{H}_{\lambda,n}$ be the one-dimensional subspace of $L^2(\mathcal{C}_c)$ spanned by the eigenfunction $\eta_{\lambda,n}$ of $\widehat{J}_\lambda$. Then $\mathcal{H}_{\lambda,n}$ is an invariant subspace of $\widehat{J}_\lambda$ and we have a selfadjoint reduction of $\widehat{J}_\lambda$ on $\mathcal{H}_{\lambda,n}$, to be denoted by $\widehat{J}_{\lambda,n}$. Similarly let the reductions of $\widehat{P}_\lambda$, $\widehat{\Phi}_\lambda$ to $\mathcal{H}_{\lambda,n}$ be denoted by $\widehat{P}_{\lambda,n}$, $\widehat{\Phi}_{\lambda,n}$ respectively. On each $\mathcal{H}_{\lambda,n}$ we have a Hamiltonian $\widehat{H}_{\lambda,n}^{(12+)}$ obtained by the reduction of $\widehat{H}_{a,b}^{(12+)}$ for a given b and an appropriate a. On account of Eq. (6.428) we have, on $\mathcal{H}_{\lambda,n}$, the following relation:

$$\widehat{H}_{\lambda,n}^{(12+)} = \frac{1}{2m_c} \widehat{P}_{\lambda,n}^2. \qquad (7.45)$$

2. For sufficiently large values of b there will be a finite set of matching pairs of n and λ satisfying Eq. (7.38). Physically this means that a TSCR with an appropriate radius r and coupling constant b at the junction can admit a finite set of superconducting states $\eta_{\lambda,n}$ for some matching pairs of n and λ satisfying Eq. (7.38), i.e., the ring allows a set of supercurrent $j_{\lambda,n}$ to flow round it.

7.4. SUPERCONDUCTING RINGS WITH A JUNCTION

3. Each allowed current $j_{\lambda,n}$ corresponds to a subspace $\mathcal{H}_{\lambda,n}$. It follows that the state space for the ring is the direct sum over these subspaces:

$$\mathcal{H}^{\oplus}(\mathcal{C}_c) = \oplus_{\lambda,n} \mathcal{H}_{\lambda,n}, \tag{7.46}$$

where the sum is over some matching pairs of n and λ satisfying Eq. (7.38). This is then the Hilbert space associated with the ring in the sense of Postulate UQS in §4.2. In other words $\eta_{\lambda,n}$ corresponds to a pure superconducting state. As before, to highlight the fact that $\eta_{\lambda,n}$ is an element of $\mathcal{H}^{\oplus}(\mathcal{C}_c)$ we shall rewrite $\eta_{\lambda,n}$ in boldface as $\boldsymbol{\eta}^{\oplus}_{\lambda,n}$.

4. Observables are decomposable operators on $\mathcal{H}^{\oplus}$. Clearly these correspond to the following direct sums

$$\widehat{P}^{\oplus}(\mathcal{C}_c) = \oplus_{\lambda,n} \widehat{P}_{\lambda,n}, \tag{7.47}$$

$$\widehat{J}^{\oplus}(\mathcal{C}_c) = \oplus_{\lambda,n} \widehat{J}_{\lambda,n} = -\frac{q_c}{m_c[\ell]} \widehat{P}^{\oplus}(\mathcal{C}_c), \tag{7.48}$$

$$\widehat{\Phi}^{\oplus}(\mathcal{C}_c) = \oplus_{\lambda,n} \widehat{\Phi}_{\lambda,n} = -\frac{2\pi r}{q_c} \widehat{P}^{\oplus}(\mathcal{C}_c), \tag{7.49}$$

$$\widehat{H}^{\oplus}(\mathcal{C}_c) = \oplus_{\lambda,n} \widehat{H}^{(12+)}_{\lambda,n} = \frac{1}{2m_c} \widehat{P}^{\oplus}(\mathcal{C}_c)^2. \tag{7.50}$$

The sum is over matching pairs of n and λ satisfying Eq. (7.38).

We can now see that an attempt to represent a Josephson junction in terms of a point interaction entails a superselection rule in the form of a preferred direct sum decomposition of the Hilbert space in the sense of Postulate UQS. We would therefore regard the condensate in the TSCR with a JJ with a stipulated range of currents as a mixed quantum system. Properties similar to those stated for a TSCR in the preceding section, such as no superposition of states of different currents, the delocalized nature of the condensate and the dynamics, apply here. This is also an equilibrium mixed quantum system with its Hamiltonian generator identified with the Hamiltonian. We can write the time-dependence of a superconducting state explicitly as:

$$\boldsymbol{\eta}^{\oplus}_{\lambda,n,t} = \boldsymbol{\eta}^{\oplus}_{\lambda,n} e^{-iE_{\lambda,n}t}, \tag{7.51}$$

where $E_{\lambda,n}$ is the eigenvalue of $\widehat{H}^{\oplus}(\mathcal{C}_c)$ corresponding to the eigenfunction $\boldsymbol{\eta}^{\oplus}_{\lambda,n}$. Because of the junction magnetic flux can leak in and out of the ring so that the total enclosed flux is not confined to a multiple of the flux quantum Φ_c.

7.4.5 Superconducting ring with a π-junction

An interesting situation arises when the phase difference at the junction λ is equal to π. When we set $\lambda = \pi$ the Josephson equation leads to a zero current. It is wrong to conclude that a superconducting ring with a JJ cannot admit a superconducting state with a π-phase shift across the junction. It is just that the point interaction embodied by $\widehat{H}_{a;b}^{(12+)}(\mathcal{C}_c)$ does not apply since boundary conditions (BCC12) are violated, i.e., a substitution of $\eta_{\pi,n}$ into Eqs. (6.420) and (6.420) leads to contradictions. So, the Josephson equation simply does not apply.

In order to have a π-phase shift a different point interaction must come into play. The ideal dynamic π-phase shifter $\widehat{H}^{(7)}(\mathcal{C}_c)$ with eigenfunction $\eta_n^{(7)}(\vartheta)$ given by Eq. (6.390) would work here, i.e., we can employ $\widehat{H}^{(7)}(\mathcal{C}_c)$ as the Hamiltonian and $\widehat{J}_{\lambda=\pi}$ as the supercurrent operator. Then $\widehat{H}^{(7)}(\mathcal{C}_c)$ and $\widehat{J}_{\lambda=\pi}(\mathcal{C}_c)$ share eigenfunctions $\eta_n^{(7)}$. So, we could indeed have a junction which would allow a supercurrent to flow through with the corresponding superconducting state $\eta_n^{(7)}$ suffering a π phase shift at the junction. We can also consider the system as a mixed quantum system with each superconducting state $\eta_n^{(7)}$ spanning a supersector denoted by $\mathcal{H}_{\pi,n}$. This result is consistent with π-junction discovered in superconductivity and superfluid.[27] Note that a π-junction is not a simple continuation of a Josephson junction with λ tends to π. A π-junction is physically quite different. A π-junction gives rise to an equilibrium mixed quantum system described here by the following:

1. A Hilbert space having the following preferred direct sum decomposition:

$$\mathcal{H}_\pi^\oplus(\mathcal{C}_c) = \oplus_n \mathcal{H}_{\pi,n}. \qquad (7.52)$$

2. Observables are decomposable operators on $\mathcal{H}_\pi^\oplus(\mathcal{C}_c)$. We have

$$\widehat{H}_\pi^\oplus(\mathcal{C}_c) = \oplus_n \widehat{H}_{\pi,n}, \qquad (7.53)$$
$$\widehat{J}_\pi^\oplus(\mathcal{C}_c) = \oplus_n \widehat{J}_{\pi,n}, \qquad (7.54)$$
$$\widehat{\Phi}_\pi^\oplus(\mathcal{C}_c) = \oplus_n \widehat{\Phi}_{\pi,n}, \qquad (7.55)$$

where $\widehat{H}_{\pi,n}$, $\widehat{J}_{\pi,n}$ and $\widehat{\Phi}_{\pi,n}$ are respectively the reductions on the supersector $\mathcal{H}_{\pi,n}$ of operators $\widehat{H}^{(7)}(\mathcal{C}_c)$, $\widehat{J}_{\lambda=\pi}(\mathcal{C}_c)$ and $\widehat{\Phi}_{\lambda=\pi}(\mathcal{C}_c)$.

7.4.6 Superconducting ring with a $\frac{1}{2}\pi$-junction

As mentioned earlier the Josephson equation derived using $\widehat{H}_{a;b}^{(12+)}(\mathcal{C}_c)$ is not applicable for a phase shift of $\frac{1}{2}\pi$. From the discussion in §7.9.10 we see that

---
[27] Backhaus (1998).

7.4. SUPERCONDUCTING RINGS WITH A JUNCTION

for suitable values of k_0 an deal $\frac{1}{2}\pi$-phase shifter $\widehat{H}^{(10-)}(\mathcal{C}_c)$ can admit a superconducting state in the form of $\eta_{\pi/2,n}^{(10)}(\vartheta)$ and $\eta_{\pi/2,0}^{(10)}(\vartheta)$ given by Eqs. (6.410) and (6.412) with a phase shift of $\frac{1}{2}\pi$ and a supercurrent

$$j = -\frac{q_c}{m_c[\ell]}\hbar k_0. \quad (7.56)$$

We can go on to construct a direct sum space similar to the constructs in Eq. (7.52) and treat the system as a mixed quantum system.

7.4.7 Superconducting ring with a Josephson junction in an external magnetic field

In the presence of a uniform and time-independent external magnetic field specified by a vector potential having the value A_{ex} constant along the circumference of the ring the system becomes more complicated. We shall consider this case briefly. In line with Eq. (7.11) the partially quantized current operator is

$$\widetilde{J}_0(\mathcal{C}_c, A_{ex}) = -\frac{q_c}{m_c[\ell]}\left(\widetilde{P}_0(\mathcal{C}_c) + q_c A_{ex}\right). \quad (7.57)$$

In line with Eq. (7.12) the corrresponding total enclosed flux operator is

$$\widetilde{\Phi}_0(\mathcal{C}_c, A_{ex}) = \Phi_{ex} + L_c \widetilde{J}_0(\mathcal{C}_c, A_{ex}), \quad (7.58)$$

where $\widetilde{P}_0(\mathcal{C}_c)$ is defined by Eqs. (6.320) and (6.321), L_c is the self-inductance as given in Eq. (7.8) and Φ_{ex} is the externally applied flux ($\Phi_{ex} = 2\pi r A_{ex}$). These operators are defined on $C_0^\infty(\mathcal{C}_c)$. The quantized quantities correspond to the selfadjoint extensions of the partially quantized quantities., i.e.,

$$\widehat{J}_\lambda(\mathcal{C}_c, A_{ex}) = -\frac{q_c}{m_c[\ell]}\left(\widehat{P}_\lambda(\mathcal{C}_c) + q_c A_{ex}\right). \quad (7.59)$$

The total enclosed flux operator is

$$\widehat{\Phi}_\lambda(\mathcal{C}_c, A_{ex}) = \Phi_{ex} + L_c \widehat{J}_\lambda(\mathcal{C}_c, A_{ex}), \quad (7.60)$$

$$= -\frac{2\pi r}{q_c}\widehat{P}_\lambda(\mathcal{C}_c). \quad (7.61)$$

The supercurrent and the total enclosed flux operators possess the following eigenvalues in state $\eta_{\lambda,n}$ given by Eq. (7.36):

$$j_{\lambda,n}(\mathcal{C}_c, A_{ex}) = -\frac{q_c}{m_c[\ell]}\left\{\frac{\hbar}{r}\left(n - \frac{\lambda}{2\pi}\right) + q_c A_{ex}\right\}, \quad (7.62)$$

$$\Phi_{\lambda,n}(\mathcal{C}_c, A_{ex}) = -\left(n - \frac{\lambda}{2\pi}\right)\Phi_c. \quad (7.63)$$

The current and the flux eigenvalues are related by

$$\Phi_{\lambda,n}(\mathcal{C}_c, A_{ex}) = \Phi_{ex} + L_c\, j_{\lambda,n}(\mathcal{C}_c, A_{ex}). \tag{7.64}$$

Substituting the values for L_c and $j_{\lambda,n}(\mathcal{C}_c, A_{ex})$ in Eq. (7.62) into Eq. (7.64) we will obtain a value of $\Phi_{\lambda,n}(\mathcal{C}_c, A_{ex})$ agreeing with Eq. (7.63).

The Josephson equation relating the current to the phase holds, i.e.,[28]

$$j_{\lambda,n}(\mathcal{C}_c, A_{ex}) = j_c(\mathcal{C}_c, A_{ex})\sin\lambda, \tag{7.65}$$

where

$$j_c(\mathcal{C}_c, A_{ex}) = j_c(\mathcal{C}_c) = -\frac{q_c}{m_c[\ell]}\hbar b, \tag{7.66}$$

we get

$$\Phi_{\lambda,n}(\mathcal{C}_c, A_{ex}) - \Phi_{ex} = -rb\,\Phi_c\sin\lambda, \tag{7.67}$$

or[29]

$$\left(n - \frac{\lambda}{2\pi}\right)\Phi_c + \Phi_{ex} = rb\,\Phi_c\sin\lambda. \tag{7.68}$$

There are different cases we can consider depending on the values of the parameters. A straightforward situation occurs if the parameter b satisfies $2\pi rb < 1$. Then the total enclosed flux $\Phi_{\lambda,n}(\mathcal{C}_c, A_{ex})$ and hence the current $j_{\lambda,n}(\mathcal{C}_c, A_{ex})$ are a single-valued function of the external magnetic flux Φ_{ex}.[30] Take the simple case of $n = 0$. Eq. (7.68) reduces to

$$2\pi\Phi_{ex}/\Phi_c = \lambda + 2\pi rb\sin\lambda, \quad 2\pi rb < 1. \tag{7.69}$$

The external flux Φ_{ex} is seen to determine λ which in turn fixes the supercurrent and the total enclosed flux.[31] The phase in turn determines the operators, and hence their reductions in the supersector spanned by $\eta_{\lambda,0}^{\oplus}$. We arrive at a situation that the environment, i.e., Φ_{ex}, determines the state and the operators of the system. As we vary the environment we will effectively obtain

---

[28] Tilley and Tilley (1990) p. 270.

[29] Eq. (7.68) is consistent with the following boundary condition:

$$\Psi'_{10} + iq_c A_{ex}\Psi_{10} = a\Psi_{10} + b\Psi_{20}.$$

A substitution of $\eta_{\lambda,n}$ into this boundary condition would yield Eq. (7.68).

[30] See Tiley and Tiley (1990) pp. 270-271, especially Fig. 7.9, and Gallop (1991) p. 53.

[31] Assume that λ_1 and λ_2 satisfy Eq. (7.69) for a given Φ_{ex} and that $\lambda_2 > \lambda_1$. By a subtraction we get $0 = (\lambda_2 - \lambda_1) + 2\pi rb(\sin\lambda_2 - \sin\lambda_1)$, which is a contradiction when $2\pi rb < 1$, unless $\lambda_2 = \lambda_1$ since

$$\left|\sin\lambda_2 - \sin\lambda_1\right| = \left|\int_{\lambda_1}^{\lambda_2}\cos\lambda\,d\lambda\right| \leq \int_{\lambda_1}^{\lambda_2} d\lambda = \lambda_2 - \lambda_1.$$

a set of different systems.[32] We can incorporate such a set of systems into a single mixed quantum system with a continuous superselection rule. Each supersector corresponds to a specific value of Φ_{ex} and the Hilbert space is a direct integral of these supersectors with respect to the continuous parameter Φ_{ex}. More explicit details are available in Wan and Harrison (1993).

We can go on to establish the theory for a π-junction and a $\frac{1}{2}\pi$-junction in an external magnetic field.

7.5 Feynman's Derivation of Josephson's Equation

Consider a long superconductor divided into two halves by a junction as shown in Fig. 7.5.

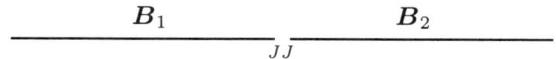

Fig. 7.5   dc Josephson effect.

We shall consider the case where the junction is symmetrical about its center. When the metal is in a superconducting state a direct current can tunnel through the junction as a supercurrent, without needing any applied voltage across the insulator junction and without generating a voltage across the junction either. Mathematically the geometry of the system may be idealized as the real line $I\!\!E$ broken into two half lines, with $\boldsymbol{B}_1 = I\!\!E^-$ representing branch 1 and $\boldsymbol{B}_2 = I\!\!E^+$ representing branch 2. The condensate is to be represented by a wave function on this broken line $I\!\!E_0 = I\!\!E^- \cup I\!\!E^+ = \boldsymbol{B}_1 \cup \boldsymbol{B}_2$.

Feynman gives a derivation of the Josephson equation in this situation using a macroscopic wave function.[33] In Feynman's derivation the macroscopic wave function on the left hand side of the junction, i.e., on $\boldsymbol{B}_1$, is denoted by ψ_1 and on the right hand side, i.e., on $\boldsymbol{B}_2$, it is denoted by ψ_2. Feynman then assumes that there should be a coupling between the left and the right hand sides and that this coupling manifests itself in two equations relating ψ_1 and ψ_2, namely

$$i\hbar \frac{\partial \psi_1}{\partial t} = A_1 \psi_1 + B\psi_2, \tag{7.70}$$

$$i\hbar \frac{\partial \psi_2}{\partial t} = A_2 \psi_2 + B\psi_1, \tag{7.71}$$

---

[32] We pointed out earlier that a different environment meant a different system.
[33] Feynman, Leighton and Sands (1965) §21-9.

where A_1, A_2, B are some numerical constants.

Feynman further assumes a potential difference V across the junction. By relating A_1, A_2 to V he arrives at the following equations:

$$i\hbar \frac{\partial \psi_1}{\partial t} = \frac{q_c V}{2}\psi_1 + B\psi_2, \tag{7.72}$$

$$i\hbar \frac{\partial \psi_2}{\partial t} = -\frac{q_c V}{2}\psi_2 + B\psi_1. \tag{7.73}$$

With a few more assumptions on the expression for the current and eventually setting the potential V to zero Feynman obtains Josephson's equation describing the supercurrent for a dc Josephson effect to be

$$j = j_0 \sin \lambda, \tag{7.74}$$

where λ is the phase difference between ψ_1 and ψ_2 at the junction.

Feynman's equations are equivalent to two separate equations in the bulk of the two disjoint lengths of superconductor and two boundary conditions at the junction, namely we have in the bulk of the superconductors

$$i\hbar \frac{\partial \psi_1(x,t)}{\partial t} = \frac{q_c V}{2}\psi_1(x,t) \quad \text{on} \quad \boldsymbol{B}_1, \tag{7.75}$$

$$i\hbar \frac{\partial \psi_2(x,t)}{\partial t} = -\frac{q_c V}{2}\psi_2(x,t) \quad \text{on} \quad \boldsymbol{B}_2, \tag{7.76}$$

and at the junction, i.e., at $x = 0$, we have

$$i\hbar \left[\frac{\partial \psi_1(x,t)}{\partial t}\right]_{x=0} = \frac{q_c V}{2}\psi_1(0,t) + B\psi_2(0,t), \tag{7.77}$$

$$i\hbar \left[\frac{\partial \psi_2(x,t)}{\partial t}\right]_{x=0} = -\frac{q_c V}{2}\psi_2(0,t) + B\psi_1(0,t). \tag{7.78}$$

The coupling between the left and the right hand sides is effected by boundary conditions (7.77) and (7.78) with B playing the role of a coupling constant. It is these boundary conditions which eventually lead to the desired Josephson equation. There have been attempts to recast Feynman's equations into the form of a standard Schrödinger equation.[34] We shall not pursue Feynman's derivation in what follows. Instead we shall consider the junction as a point interaction to derive the Josephson equation.

---
[34] Barone and Paterno (1982) p. 2, pp. 10-11, p. 18, p. 23.

7.6 Superconducting Wire with a Junction

7.6.1 Point interactions

What we have is a two-branch quantum circuit with the junction as the branch point. We shall now treat the junction as a point interaction and set up a description of the flow of a dc supercurrent in the circuit. We shall derive the Josephson equation based on our method of quantization by parts. The geometry here is the same as that for point interactions discussed in §6.4 and §6.5 so that we can make use of many of the results obtained there. Since this is a two branch circuit we need to carry out all the three stages of our quantization by parts scheme set out in §6.4.

7.6.2 Momentum and supercurrent operators

Partial quantization We shall follow §6.3.1 to set up various quantities in the branches. On $\boldsymbol{B}_2$ the associated Hilbert space is $\mathcal{H}_2 = L^2(I\!\!E^+)$. The momentum is the strictly maximal symmetric operator $\widehat{p}_2 = \widehat{p}^+$ introduced in E2.5.1(1) E6. Similarly we have the Hilbert space $\mathcal{H}_1 = L^2(I\!\!E^-)$ and momentum operator $\widehat{p}_1 = \widehat{p}^-$ in $\boldsymbol{B}_1$.

Composite quantization We shall take the direct sum $\mathcal{H}^{(1,2)} = \mathcal{H}_1 \oplus \mathcal{H}_2$ as the Hilbert space for the two branch circuit as a whole. We then form the direct sum operator $\widehat{p}_1 \oplus \widehat{p}_2$ in $\mathcal{H}^{(1,2)}$. This direct sum operator is not selfadjoint, but it possesses a one-parameter family of selfadjoint extensions $\widehat{P}_\lambda^{(1,2)}$. Generalized eigenfunctions of $\widehat{P}_\lambda^{(1,2)}$ corresponding to eigenvalue p are given by Eq. (6.35), i.e., $\eta_{\lambda,p}^{(1,2)}$. So, for the momentum operator of the system as a whole we shall have to choose a selfadjoint extension $\widehat{P}_\lambda^{(1,2)}$. We shall decide which selfadjoint extension to take later in correlation with the choice of a Hamiltonian.

The corresponding compositely quantized supercurrent operator for the system is

$$\widehat{J}_\lambda^{(1,2)} = -\frac{q_c}{m_c[\ell]} \widehat{P}_\lambda^{(1,2)} \tag{7.79}$$

which admits $\eta_{\lambda,p}^{(1,2)}$ as generalized eigenfunctions with eigenvalue

$$j_{\lambda,p} = -\frac{q_c}{m_c[\ell]} p. \tag{7.80}$$

7.6.3 Hamiltonian operator 1: π-junction

Partial quantization Following §6.4 we shall start with partially quantized Hamiltonians on the branches

$$\widehat{H}_{0s} = -\frac{\hbar^2}{2m}\frac{d^2}{dx^2} \quad \text{acting on } C_0^\infty(\boldsymbol{B}_s), \ s = 1, 2. \tag{7.81}$$

Composite quantization It is shown in §6.5 that the direct sum operator $\widehat{H}_{01} \oplus \widehat{H}_{02}$ in $\mathcal{H}^{(1,2)}$ possesses a dozen different types of possible selfadjoint extensions. These extensions acting as Hamiltonians would represent a dozen different point interactions. Not all these Hamiltonians are suitable for the description of superconductivity. We have to make a choice.

Correlative quantization To describe a dc supercurrent we must have a matching pair of momentum and Hamiltonian to comply with assumptions (PAS1) and (PAS2) on dc superconductivity spelled out in item 7 in §7.3.2. The situation is similar to the discussion in §7.4.3. We desire the selected Hamiltonian to admit $\eta_{\lambda,p}^{(1,2)}$ as eigenfunctions for some range of values of λ and p. This requires the boundary conditions for the chosen Hamiltonian be satisfiable by $\eta_{\lambda,p}^{(1,2)}$. This rules out the δ-interactions, the δ'-interactions, the reflector, the elastic reflectors and the open end. We shall ignore the free Hamiltonian, despite its compatibility with $\eta_{\lambda,p}^{(1,2)}$ for $\lambda = 0$, since this corresponds to free motion with no junction at all.

We then come to an ideal π-phase shifter represented by Hamiltonian $\widehat{H}^{[7]}$ described in §6.5.7. By comparing $\eta_p^{[7]}$ in Eq. (6.183) with $\eta_{\lambda,p}^{(1,2)}$ we can see that $\eta_{\lambda,p}^{(1,2)}$ for $\lambda = \pi$, i.e.,

$$\eta_{\pi,p}^{(1,2)} = \frac{1}{\sqrt{[\ell]}}\left(e^{ipx_1} \oplus e^{i\pi}e^{ipx_2}\right), \quad x_1 \in \boldsymbol{E}^-, \ x_2 \in \boldsymbol{E}^+, \tag{7.82}$$

does satisfy boundary conditions (BC7) of $\widehat{H}^{[7]}$. Following the discussion in §7.4 we conclude that an ideal π-phase shifter can act as the Hamiltonian of a long superconducting wire with a junction, to be called a π-junction. This particular point interaction allows a supercurrent to flow across the junction with the corresponding superconducting state $\eta_{\pi,p}^{(1,2)}$ suffering a π-phase shift across the junction.

There are other point interactions admitting supercurrent across the junction. We shall explore these in the next two sections.

7.6.4 Hamiltonian operator 2: $\frac{1}{2}\pi$-junction

If we continue the list of Hamiltonians in §6.5 we will come to high-pass and low pass π-phase shifters. These can be ruled out immediately as they do not

7.6. SUPERCONDUCTING WIRE WITH A JUNCTION

admit $\eta_{\lambda,p}^{(1,2)}$ as eigenfunctions. Next on the list is an ideal mid-pass $\frac{1}{2}\pi$-phase shifter. For definiteness let us consider the Hamitonian $\widehat{H}^{[10+]}$ determined by boundary conditions (BC10$_+$).[35] We can see that

$$\eta_{\frac{\pi}{2},p}^{(1,2)} = \frac{1}{\sqrt{[\ell]}} \left(e^{ipx_1} \oplus e^{i\frac{\pi}{2}} e^{ipx_2} \right), \quad x_1 \in I\!\!E^-, \quad x_2 \in I\!\!E^+ \qquad (7.83)$$

would satisfy boundary conditions (BC10$_+$) since this function is of the same form as $\eta_{k_0}^{[10+]}$ in Eq. (6.233). Substituting $\eta_{\frac{\pi}{2},p}^{(1,2)}$ into boundary conditions (BC10$_+$) produces a matching value of the momentum, $p = p_0$, given by

$$p_0 = \hbar k_0 \qquad (7.84)$$

with an accompanying current

$$j = j_{\frac{\pi}{2},p_0} = -\frac{q_c}{m_c[\ell]} \hbar k_0. \qquad (7.85)$$

This equation may be taken as the expression linking the critical current j_c to the momentum p_0 which can be identified with the maximum momentum a Cooper pair can possess.

There are only two more types of point interactions. A Type 11 interaction, which represents a partial mid-pass filter, does not admit $\eta_{\lambda,p}^{(1,2)}$ as an eigenstate, but a Type 12 ideal tunable phase shifter does. As in the case of a superconducting ring discussed in §7.4.3 we shall show in what follows that a Type 12 interaction can also describe a Josephson junction here.

7.6.5 Hamiltonian operator 3: Josephson junction

Recall the discussion in §6.5.12 on ideal tunable phase shifters corresponding to a Hamitonian $\widehat{H}_{a,b}^{[12+]}$. By comparing $\eta_k^{[12+]}$ in Eq. (6.270) with $\eta_{\lambda,p}^{(1,2)}$ we see that $\widehat{H}_{a,b}^{[12+]}$ and $\widehat{P}_\lambda^{(1,2)}$ can be compatible in the sense of (PAS1) and (PAS2) of §7.3.2 provided that Eqs. (6.271) and (6.272) are satisfied and that the momentum eigenvalue p is related to k by $p = \hbar k$.[36] In other words $\eta_{\lambda,p}^{(1,2)}$ is also an eigenfunction of $\widehat{H}_{a,b}^{[12+]}$ if

$$p = \hbar k = \hbar b \sin \lambda, \qquad (7.86)$$

where λ is related to a and b by[37]

$$a = -b \cos \lambda. \qquad (7.87)$$

---

[35] We can also use an ideal mid-pass $-\frac{1}{2}\pi$-phase shifter with Hamiltonian $\widehat{H}^{[10-]}$.
[36] Recall Eq. (6.267) which states that $k = \sqrt{b^2 - a^2}$. This is consistent with Eqs. (6.271) and (6.272), and with Eqs. (7.86) and (7.87).
[37] As in §7.4.3 $\lambda = 0, \pi$ are excluded. When $\lambda = \pi$ we have a contradiction with the condition $a^2 \neq b^2$ of (BC12). When $\lambda = 0$ we have free motion in $I\!\!E$ without a junction.

The following interpretation now presents itself:

1. A state like $\eta_{\lambda,p}^{(1,2)}$ represents a total transmission across the junction without reflection, hence the superconducting nature of the state.

2. By setting
$$j = -\frac{q_c}{m_c[\ell]} p, \quad j_c = -\frac{q_c}{m_c[\ell]} \hbar b \tag{7.88}$$
we obtain from Eq. (7.86) the Josephson equation
$$j = j_c \sin \lambda. \tag{7.89}$$
If we wish to render the relation between j and λ unique we can restrict the range of phase shift to
$$\lambda \in (-\pi/2, 0) \cup (0, \pi/2). \tag{7.90}$$

3. Parameter b serves as a coupling constant between the two branches, effectively characterizing the junction. Physically b can be determined from the maximum current j_c. This current can be obtained, to an arbitrary degree of accuracy, from the maximum supercurrent achievable in a suitable experiment.[38] As discussed in §7.3.3 the value of a is fixed by b and λ. Since λ is fixed by the current j through the Josephson equation we can regard λ and a as functions of j.

4. A superconducting state $\eta_{\lambda,p}^{(1,2)}$ corresponds to an energy eigenvalue
$$E_p = E_{a,b}^{[12+]} = \frac{1}{2m} p^2 = \frac{\hbar^2 b^2}{2m} \sin^2 \lambda. \tag{7.91}$$

5. As in the case of TSCR the π-junction and $\frac{1}{2}\pi$-junction are not special cases of the present Josephson junction.

Using the analysis in §7.3.3 we can conclude that a long superconductor interrupted by a Josephson junction is not an orthodox quantum system describable in the Hilbert space $\mathcal{H}^{(1,2)}$ with a Hamiltonian $\widehat{H}_{a,b}^{[12+]}$. Firstly, a different value of the current implies different values of λ and a, and hence a different Hamiltonian. More seriously, each $\widehat{H}_{a,b}^{[12+]}$ possesses other generalized eigenfunctions of the form of Eq. (6.95). These functions cannot describe

---

[38] The exact value of the maximum current j_c is not experimentally realizable with a Type 12 point interaction since this requires $\lambda = \pi/2$. This renders $a = 0$ contradicting (BC12). However, an experiment with a Type 12 point interaction can approach the value j_c to give a value of b with high accuracy.

7.6. SUPERCONDUCTING WIRE WITH A JUNCTION

a superconducting state since they are not eigenfunctions of the supercurrent operator and physically these functions represent a typical scattering situation with non-zero reflection. In line with the discussions in §7.3.4 we can resolve all these problems by treating a long superconducting wire with a Josephson junction as a mixed quantum system. This time we have to deal with a Hilbert space having a preferred continuous direct integral decomposition because the current and the phase shift can assume a continuous set of values.

Recall E2.4(1) E1 and C2.4(1) C7 where the notion of a momentum space $\widetilde{I\!E}$ is introduced which leads to a continuous direct integral decomposition of $\widetilde{L}^2(\widetilde{I\!E})$ in the momentum space in E2.13.2(1) E3. One can consider such an integral decomposition in terms of a decomposition into a continuous set of singular subspace introduced in E2.13.2(2) E2. In the same fashion we can consider $\eta^{(1,2)}_{\lambda,p}$ formally spanning a singular subspace $\mathcal{H}^{(1,2)}_{\lambda,p}$ in $\mathcal{H}^{(1,2)}$. When in a superconducting state corresponding to a steady dc current j the system is described by $\eta^{(1,2)}_{\lambda,p}$. The supercurrent operator and the Hamiltonian should respectively be the selfadjoint reductions of $\widehat{J}^{(1,2)}_\lambda$ and $\widehat{H}^{[12+]}_{a,b}$ on the singular subspace $\mathcal{H}^{(1,2)}_{\lambda,p}$. To accommodate other current values we can form the direct integral space from subspaces $\mathcal{H}^{(1,2)}_{\lambda,p}$ and construct a mixed quantum system. This we shall do in the following section.

7.6.6 Superconducting wire with a Josephson junction as a mixed equilibrium quantum system

Since the current j determines both λ and p we can re-label

$$\eta^{(1,2)}_{\lambda,p} \quad \text{as} \quad \boldsymbol{\eta}(j) \quad \text{and} \quad \mathcal{H}_{\lambda,p} \quad \text{as} \quad \mathcal{H}(j), \tag{7.92}$$

dropping the superscripts for brevity in the process.[39] We can also re-label p as $p(j)$ to indicate its dependence on j. As mentioned in C2.14.2(2) C2 the momentum operator $\widehat{P}^{(1,2)}_\lambda$ has a formal reduction to $\mathcal{H}(j)$ as a multiplication operator $\widehat{P}(j)$. Similar reductions of the supercurrent operator $\widehat{J}^{(1,2)}_\lambda$ and the Hamiltonian $\widehat{H}^{[12+]}_{a,b}$ on $\mathcal{H}(j)$ are labelled as $\widehat{J}(j)$ and $\widehat{H}(j)$. The effects of these reduced operators on $\mathcal{H}(j)$ are:

$$\widehat{P}(j)\,\mathcal{H}(j) = p(j)\,\mathcal{H}(j), \tag{7.93}$$
$$\widehat{J}(j)\,\mathcal{H}(j) = j\,\mathcal{H}(j), \tag{7.94}$$
$$\widehat{H}(j)\,\mathcal{H}(j) = \frac{1}{2m}p^2(j)\,\mathcal{H}(j). \tag{7.95}$$

---

[39] In line with previous notation, e.g., C2.13.2(3) C2, §4.4.4 and §7.2.3 we employ a bold face symbol to highlight the fact that $\boldsymbol{\eta}(j)$ represents a vector in the Hilbert space $\mathcal{H}(j)$.

We can now form the direct integral space for the dc Josephson effect:

$$\mathcal{H}^\oplus = \int^\oplus \mathcal{H}(j)\,dj, \quad j \in (-|j_c|, 0) \cup (0, |j_c|). \tag{7.96}$$

Elements of $\mathcal{H}^\oplus$ are of the form

$$\phi^\oplus = \int^\oplus \phi(j)\,\boldsymbol{\eta}(j)\,dj, \quad \psi^\oplus = \int^\oplus \psi(j)\,\boldsymbol{\eta}(j)\,dj, \tag{7.97}$$

with scalar product

$$\langle \phi^\oplus \mid \psi^\oplus \rangle_\oplus = \int \phi^*(j)\psi(j)\,dj. \tag{7.98}$$

We can also form the direct integral operators of $\widehat{P}(j)$, $\widehat{J}(j)$ and $\widehat{H}(j)$ on $\mathcal{H}^\oplus$ by:

$$\widehat{P}^\oplus = \int^\oplus \widehat{P}(j)\,dj \quad \text{and} \quad \widehat{P}^\oplus \phi^\oplus = \int^\oplus p(j)\phi(j)\,\boldsymbol{\eta}(j)\,dj, \tag{7.99}$$

$$\widehat{J}^\oplus = \int^\oplus \widehat{J}(j)\,dj \quad \text{and} \quad \widehat{J}^\oplus \phi^\oplus = \int^\oplus j\,\phi(j)\,\boldsymbol{\eta}(j)\,dj, \tag{7.100}$$

$$\widehat{H}^\oplus = \int^\oplus \widehat{H}(j)\,dj \quad \text{and} \quad \widehat{H}^\oplus \phi^\oplus = \int^\oplus \frac{1}{2m} p^2(j)\,\phi(j)\,\boldsymbol{\eta}(j)\,dj. \tag{7.101}$$

We shall summarize our results for the description of a long superconductor interrupted by a Josephson junction as follows:[40]

1. A long superconductor exhibiting a dc Josephson effect is a mixed quantum system possessing a continuous superselection rule. Its associated Hilbert space is given by the direct integral $\mathcal{H}^\oplus$ in Eq. (7.96).

2. Physical observables are represented by decomposable selfadjoint operators in $\mathcal{H}^\oplus$. In particular, the momentum, the supercurrent and the Hamiltonian operators are represented by multiplication operators $\widehat{P}^\oplus$, $\widehat{J}^\oplus$ and $\widehat{H}^\oplus$ respectively. As before there is no position observable.

3. This is an equilibrium system with the Hamiltonian generator $\widehat{H}_g$ identified with the Hamiltonian $\widehat{H}^\oplus$. Time evolution lies within the same supersector $\mathcal{H}(j)$ with the time-dependence of a superconducting state given by:

$$\boldsymbol{\eta}_t(j) = \boldsymbol{\eta}(j)\,e^{-iE_p t} = \frac{1}{\sqrt{[\ell]}}\left(e^{ipx_1} \oplus e^{i\lambda}e^{ipx_2}\right)e^{-iE_p t}. \tag{7.102}$$

Here p and λ are functions of j.

---
[40] Wan and Harrison (1993), Wan and Fountain (1995), Trueman and Wan (2000).

7.6. SUPERCONDUCTING WIRE WITH A JUNCTION

In view of the link between the current j and the phase shift λ we can also parameterize the supersectors in terms of the phase shift λ. We can re-label

$$\eta^{(1,2)}_{\lambda,p} \quad \text{as} \quad \boldsymbol{\eta}_\lambda \quad \text{and} \quad \mathcal{H}_{\lambda,p} \quad \text{as} \quad \mathcal{H}_\lambda, \tag{7.103}$$

and rewrite $\mathcal{H}^\oplus$ as

$$\mathcal{H}^\oplus = \int^\oplus \mathcal{H}_\lambda \, d\lambda. \tag{7.104}$$

Following C2.13.2(3) C4 we can express such a pure singular vector as:

$$\boldsymbol{\eta}^\oplus_\lambda = \int^\oplus \delta(\lambda' - \lambda) \, \boldsymbol{\eta}_{\lambda'} \, d\lambda'. \tag{7.105}$$

Observables can be similarly re-labelled. For examples we can rewrite the momentum operator, the supercurrent operator and the Hamiltonian as

$$\widehat{P}^\oplus = \int^\oplus \widehat{P}_\lambda \, d\lambda, \quad \widehat{J}^\oplus = \int^\oplus \widehat{J}_\lambda \, d\lambda, \quad \widehat{H}^\oplus = \int^\oplus \widehat{H}_\lambda \, d\lambda. \tag{7.106}$$

We can express their action on a singular state $\boldsymbol{\eta}^\oplus_\lambda$ as

$$\widehat{P}^\oplus \boldsymbol{\eta}^\oplus_\lambda = p \, \boldsymbol{\eta}^\oplus_\lambda, \quad \widehat{J}^\oplus \boldsymbol{\eta}^\oplus_\lambda = j \, \boldsymbol{\eta}^\oplus_\lambda, \quad \widehat{H}^\oplus \boldsymbol{\eta}^\oplus_\lambda = E_p \, \boldsymbol{\eta}^\oplus_\lambda. \tag{7.107}$$

More intuitively and in terms of the reduced operators in the supersector $\mathcal{H}_\lambda$ we have

$$\widehat{P}_\lambda \boldsymbol{\eta}_\lambda = p(\lambda) \, \boldsymbol{\eta}_\lambda, \quad \widehat{J}_\lambda \boldsymbol{\eta}_\lambda = j(\lambda) \, \boldsymbol{\eta}_\lambda, \quad \widehat{H}_\lambda \boldsymbol{\eta}_\lambda = E_p \, \boldsymbol{\eta}_\lambda, \tag{7.108}$$

where E_p is related to λ by Eq. (7.91). State $\boldsymbol{\eta}^\oplus_\lambda$ evolves to $\boldsymbol{\eta}^\oplus_{\lambda,t}$ in accordance with the Schrödinger equation with $\widehat{H}^\oplus$ serving as the Hamiltonian generator. In terms of quantities in the supersector $\mathcal{H}_\lambda$ we have

$$i\hbar \frac{d}{dt} \boldsymbol{\eta}_{\lambda,t} = \widehat{H}_\lambda \boldsymbol{\eta}_{\lambda,t} \quad \text{with} \quad \boldsymbol{\eta}_{\lambda,t} = e^{-\tfrac{i}{\hbar}E_p t} \boldsymbol{\eta}_\lambda. \tag{7.109}$$

In terms of quantities in $\mathcal{H}^\oplus$ we have the following formal equation:

$$i\hbar \frac{d}{dt} \boldsymbol{\eta}^\oplus_{\lambda,t} = \widehat{H}^\oplus \boldsymbol{\eta}^\oplus_{\lambda,t}, \tag{7.110}$$

where

$$\boldsymbol{\eta}^\oplus_{\lambda,t} = \int^\oplus \delta(\lambda' - \lambda) \, e^{-\tfrac{i}{\hbar}E_{p'} t} \, \boldsymbol{\eta}_{\lambda'} \, d\lambda'. \tag{7.111}$$

Note that we can follow C2.14.4(2) C3 to employ a singular statistical operator $\widehat{S}^\oplus_s$ to represent a singular state such as $\boldsymbol{\eta}^\oplus_\lambda$.

7.7 Y-Shape Circuits

7.7.1 Momentum and supercurrent operators: Special cases

New features can emerge as the circuits become more complicated. Here we shall investigate a three branch circuit in the form of a Y-shape configuration shown in Fig. 7.7.1.

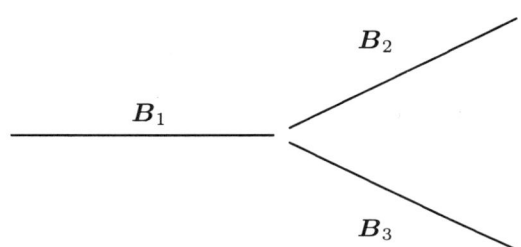

Fig. 7.7.1   Y-shape circuit with a Josephson junction.

A Josephson junction is located at the branch point preventing a direct contact of the three branches. A dc supercurrent is fed into branch B_1 from the left. To carry out a quantization by parts we shall first write down the Hilbert spaces for the branches, i.e.,

$$\mathcal{H}_1 = L^2(I\!\!E^-), \quad \mathcal{H}_2 = L^2(I\!\!E^+), \quad \mathcal{H}_3 = L^2(I\!\!E^+) \tag{7.112}$$

for branches B_1, B_2 and B_3 respectively. The corresponding partially quantized momenta are given respectively by

$$\widehat{p}_1 = \widehat{p}_- \text{ in } \mathcal{H}_1, \quad \widehat{p}_2 = \widehat{p}_+ \text{ in } \mathcal{H}_2 \text{ and } \quad \widehat{p}_3 = \widehat{p}_+ \text{ in } \mathcal{H}_3. \tag{7.113}$$

According to E2.5.1(1) E6 these three operators are all strictly maximal symmetric with deficiency indices (0,1), (1,0) and (1,0) respectively. Their direct sum is

$$\widehat{p}_1 \oplus \widehat{p}_2 \oplus \widehat{p}_3 \quad \text{in} \quad \mathcal{H}^{(1,2,3)} = \mathcal{H}_1 \oplus \mathcal{H}_2 \oplus \mathcal{H}_3, \tag{7.114}$$

which has deficiency indices (2,1), according to Eq. (2.630). It follows that this direct sum operator has no selfadjoint extensions. Theorem 2.5.1(1) and C2.5.1(2) C1 tell us that there are many strictly maximal symmetric extensions. The momentum observable for the system would then be a strictly maximal symmetric operator, rather than a selfadjoint one, chosen from the

7.7. Y-SHAPE CIRCUITS

set of strictly maximal symmetric extensions. As before different choices of extensions correspond to different physical systems. By first considering branches B_1 and B_2 which are separated by a Josephson junction we can construct an obvious family of extensions as follows:

1. We know that $\hat{p}_1 \oplus \hat{p}_2$ in $\mathcal{H}_1 \oplus \mathcal{H}_2$ possesses a one-parameter family of selfadjoint extensions $\hat{P}_\lambda^{(1,2)}$ in $\mathcal{H}_1 \oplus \mathcal{H}_2$.

2. This can then be combined with $\hat{p}_3$ to form a one-parameter family of extensions of the form

$$\hat{P}_\lambda^{(1,2;3)} = \hat{P}_\lambda^{(1,2)} \oplus \hat{p}_3 \quad \text{in} \quad \mathcal{H}_1 \oplus \mathcal{H}_2 \oplus \mathcal{H}_3. \tag{7.115}$$

These operators have deficiency indices (1,0), according to Eq. (2.630). Hence they are strictly maximal symmetric.

3. The corresponding supercurrent operator will then be

$$\hat{J}_\lambda^{(1,2;3)} = -\frac{q_c}{m_c[\ell]} \hat{P}_\lambda^{(1,2;3)}. \tag{7.116}$$

This choice of extensions has the striking physical consequence of allowing the supercurrent fed from the left into B_1 to tunnel across the junction into B_2 and forbidding any current flowing into B_3. The reason for this lies in the form of the generalized eigenfunctions of $\hat{J}_\lambda^{(1,2;3)}$ and our previous assumption (PAS1) in §7.3.2 that a superconducting state for a steady dc supercurrent should correspond to a generalized eigenfunction of the supercurrent operator. Clearly $\hat{P}_\lambda^{(1,2)}$ in $\mathcal{H}_1 \oplus \mathcal{H}_2$ admits generalized eigenfunctions of the form $\eta_{\lambda,p}^{(1,2)}$ given by Eq. (6.35). We also know from E2.5.1(1) E6 that the maximal symmetric operator $\hat{p}_3$ in $\mathcal{H}_3$ possesses no generalized eigenfunctions.[41] It follows that the generalized eigenfunctions of the supercurrent operator $\hat{J}_\lambda^{(1,2;3)}$ are of the form

$$\eta_{\lambda,p}^{(1,2;3)} = \eta_{\lambda,p}^{(1,2)} \oplus 0_3 = \frac{1}{\sqrt{[\ell]}} \left(e^{ipx_1} \oplus e^{i\lambda} e^{ipx_2} \oplus 0_3 \right), \tag{7.117}$$

where 0_3 is the zero element in $\mathcal{H}_3$. For a state represented by $\eta_{\lambda,p}^{(1,2;3)}$ there is no current flowing in B_3; the supercurrent flows from B_1 to B_2 without any splitting up into B_3 at the junction. One way of looking at this would be to visualize our present Y-shape circuit behaving as follows:

1. The tunnelling of the condensate takes place only between two branches, i.e., from B_1 to B_2.

2. The condensate as a single entity in B_1 does not split up to tunnel into two disjoint branchs B_2 and B_3 simultaneously.

---

[41] Messiah (1967) p. 346.

544   CHAPTER 7. MACROSCOPIC QUANTUM SYSTEMS

Clearly we can repeat the process to establish a momentum operator $\widehat{P}_\lambda^{(1,3;2)}$ and a corresponding supercurrent operator $\widehat{J}_\lambda^{(1,3;2)}$ with generalized eigenfunction

$$\eta_{\lambda,p}^{(1,3;2)} = \frac{1}{\sqrt{[\ell]}} \left(e^{ipx_1} \oplus 0_2 \oplus e^{i\lambda} e^{ipx_3} \right) \tag{7.118}$$

which allows the current to flow from $\boldsymbol{B}_1$ to $\boldsymbol{B}_3$ with no current flow in $\boldsymbol{B}_2$. We shall return to discuss the implications of all this in the next few sections.

7.7.2 Hamiltonian operators: Special cases

It is obvious which Hamiltonian to choose to match the choice of the supercurrent operator $\widehat{J}_\lambda^{(1,2;3)}$. Physically we can visualize a Josephson junction connecting branches $\boldsymbol{B}_1$ and $\boldsymbol{B}_2$ given by a point interaction arising from the Hamiltonian $\widehat{H}_{a,b}^{[12+]}$ in §7.6.5 and §7.6.6. Branch $\boldsymbol{B}_3$ can be left independent with a selfadjoint Hamiltonian $\widehat{H}_3 = \widehat{p}_3^\dagger \widehat{p}_3 / 2m_c$.[42] In other words we have

$$\widehat{H}_{a,b}^{(1,2;3)} = \widehat{H}_{a,b}^{[12+]} \oplus \frac{1}{2m_c} \widehat{p}_3^\dagger \widehat{p}_3. \tag{7.119}$$

Then the analysis of §7.6.5 and §7.6.6 can be applied here to produce a supercurrent satisfying the Josephson equation flowing from $\boldsymbol{B}_1$ to $\boldsymbol{B}_2$. There is no current flowing into $\boldsymbol{B}_3$.

A similar situation holds when $\widehat{J}_\lambda^{(1,3;2)}$ is chosen as the supercurrent operator. The matching Hamiltonian will be[43]

$$\widehat{H}_{a,b}^{(1,3;2)} = \widehat{H}_{a,b}^{[12+]} \oplus \frac{1}{2m_c} \widehat{p}_2^\dagger \widehat{p}_2. \tag{7.120}$$

We have a supercurrent following from $\boldsymbol{B}_1$ to $\boldsymbol{B}_3$. There is no current flowing into $\boldsymbol{B}_2$.

7.7.3 Physics of strictly maximal symmetric operators

This Y-shape circuit illustrates two important points:

1. There is a need for the inclusion of strictly maximal symmetric operators as legitimate observables in quantum mechanics, otherwise we will not be able to describe important observables in many quantum systems, e.g., there would be no momentum or supercurrent operators for the above Y-shape circuit system. This is why we need to generalize Postulate OQS in §3.2.1 to Postulate GQS in §4.2.1.

---
[42] From §2.10.3 we know that the square of a strictly maximal symmetric operator, e.g., $\widehat{p}_3^2$, is not selfadjoint. This is why we choose the Friedrichs extension $\widehat{p}_3^\dagger \widehat{p}_3$ which is selfadjoint.
[43] Following the procedure of §7.6.6 we have to take its reduction to $\mathcal{H}(j)$.

2. There is a fundamental difference in the physical properties between selfadjoint observables and strictly maximal symmetric ones. This arises from the existence of a complete orthonormal set of eigenfunctions, albeit generalized ones in many cases, for selfadjoint operators and the possible non-existence of eigenfunctions of strictly maximal symmetric operators in general. For example, $\widehat{p}_3$ in $\mathcal{H}_3$ admits no generalized eigenfunctions satisfying the required boundary condition and it is this fact that leads to the particular form of the eigenfunctions for the composite supercurrent operator $\widehat{J}_\lambda^{(1,2;3)}$ which implies the absence of current flow in $\boldsymbol{B}_3$. Compared with the investigation in §2.9 and §2.10 we can see that this is related to the nature of generalized spectral functions of strictly maximal symmetric operators.

There are other possibilities arising from different choices of the momentum and the supercurrent operators. We could have a π-junction at the branch point for example. Some examples of these more general cases are given in the following two sections

7.7.4 Momentum and supercurrent operators: General cases

We can work out the maximal symmetric extensions of

$$\widehat{p}_1 \oplus \widehat{p}_2 \oplus \widehat{p}_3 \quad \text{in} \quad \mathcal{H}_1 \oplus \mathcal{H}_2 \oplus \mathcal{H}_3. \tag{7.121}$$

First, we note that all these extensions are restrictions of the adjoint of $\widehat{p}_1 \oplus \widehat{p}_2 \oplus \widehat{p}_3$.[44] It can be shown that these strictly maximal symmetric extensions are characterizable by restrictions of the adjoint operator to domains satisfying certain boundary conditions on the wave functions at the junction at $x^s = 0$, $s = 1, 2, 3$. These boundary conditions are determined by three real parameters α, λ_2, λ_3 in the following manner:[45]

$$\Phi_{20} = \sqrt{1-\alpha^2}\, e^{i\lambda_2}\, \Phi_{10}, \tag{7.122}$$
$$\Phi_{30} = \alpha\, e^{i\lambda_3}\, \Phi_{10}, \tag{7.123}$$

where

$$\alpha \in [0,1], \quad \lambda_2, \lambda_3 \in (-\pi, \pi], \tag{7.124}$$

and

$$\Phi_s(x_s) \in \mathcal{H}_s, \quad \Phi_{s0} = \Phi_s(0), \quad \text{and} \quad s = 1, 2, 3. \tag{7.125}$$

---

[44]The domain of the adjoint operator is $\mathcal{D}(\widehat{p}_1^\dagger) \oplus \mathcal{D}(\widehat{p}_2^\dagger) \oplus \mathcal{D}(\widehat{p}_3^\dagger)$.
[45]Wan and Fountain (1998).

In other words we have a 3-parameter family of strictly maximal symmetric extensions. Generalized eigenfunctions of these extension operators are the form

$$\frac{1}{\sqrt{[\ell]}} \left(e^{ipx_1} \oplus \sqrt{1-\alpha^2}\, e^{i\lambda_2}\, e^{ipx_2} \oplus \alpha\, e^{i\lambda_3}\, e^{ipx_3} \right) \tag{7.126}$$

which clearly satisfy the above boundary conditions. There are two obvious cases to consider:

1. **Symmetric case** If the circuit is symmetrical in B_2 and B_3 we may seek extensions which are invariant with respect to the interchange of B_2 and B_3. Clearly such extensions correspond to the choice

$$\alpha = 2^{-1/2}, \quad \lambda_2 = \lambda_3 = \lambda. \tag{7.127}$$

   The boundary conditions become

$$\Phi_{20} = \frac{1}{\sqrt{2}} e^{i\lambda} \Phi_{10}, \tag{7.128}$$

$$\Phi_{30} = \frac{1}{\sqrt{2}} e^{i\lambda} \Phi_{10}. \tag{7.129}$$

   The resulting strictly maximal symmetric extensions, to be denoted by $\widehat{P}_\lambda^{(1:2,3)}$, possess generalized eigenfunctions of the form

$$\eta_{\lambda,p}^{(1:2,3)} = \frac{1}{\sqrt{[\ell]}} \left(e^{ipx_1} \oplus \frac{1}{\sqrt{2}} e^{i\lambda} e^{ipx_2} \oplus \frac{1}{\sqrt{2}} e^{i\lambda} e^{ipx_3} \right). \tag{7.130}$$

2. **Asymmetrical case** There are two obviously asymmetrical arrangements:

   (a) When $\alpha = 0$ the boundary conditions reduce to a one-parameter family determined by $\lambda = \lambda_2$ with λ_3 playing no role, i.e.,

$$\Phi_{20} = e^{i\lambda} \Phi_{10}, \tag{7.131}$$

$$\Phi_{30} = 0. \tag{7.132}$$

   The resulting eigenfunction coincides with $\eta_{\lambda,p}^{(1,2;3)}$ and the resulting extension to the supercurrent operator coincides with our previous $\widehat{J}_\lambda^{(1,2;3)}$ in Eq. (7.116).

   (b) When $\alpha = 1$ the boundary conditions reduce

$$\Phi_{20} = 0, \tag{7.133}$$

$$\Phi_{30} = e^{i\lambda} \Phi_{10}. \tag{7.134}$$

   The resulting extension to the supercurrent operator coincides with our previous $\widehat{J}_\lambda^{(1,3;2)}$ with eigenfunctions $\eta_{\lambda,p}^{(1,3;2)}$ in Eq. (7.118).

7.7.5 Hamiltonian operators: General cases

The partially quantized Hamiltonians in the branches are

$$\widehat{K}_{0s} = -\frac{\hbar^2}{2m}\frac{d^2}{d(x_s)^2} \quad \text{defined on} \quad C_0^\infty(\boldsymbol{B}_s), \quad s = 1, 2, 3. \tag{7.135}$$

For composite quantization we construct the direct sum

$$\widehat{H}_{01} \oplus \widehat{H}_{02} \oplus \widehat{H}_{03} \quad \text{defined on} \quad \mathcal{D}(\widehat{H}_{01}) \oplus \mathcal{D}(\widehat{H}_{02}) \oplus \mathcal{D}(\widehat{H}_{03}). \tag{7.136}$$

The deficiency indices of this direct sum operator are $(3,3)$. So, according to Theorem 2.5.1(1), this operator possesses a 9-parameter family of selfadjoint extensions.[46]

It would be tedious, as well as pointless, to list all the extensions here. It is sufficient to consider only those extensions which can be correlated with the momentum and current operators obtained earlier. Let us confine ourselves to the symmetric and asymmetric cases discussed in §7.7.4:

1. **Symmetrical case** We would choose those extensions which are invariant with respect to the interchange of $\boldsymbol{B}_2$ and $\boldsymbol{B}_3$. This reduces the 9-parameter family to a 2-parameter family of operators. There is a 2-parameter family specified by boundary conditions on the wave functions at the junction at $x_s = 0$ in terms of two real parameters a, b on the domains of the extension operators, i.e.,[47]

$$\Phi'_{10} = a\,\Phi_{10} + b\,(\Phi_{20} + \Phi_{30}), \tag{7.137}$$
$$\Phi'_{20} + \Phi'_{30} = -a\,(\Phi_{20} + \Phi_{30}) - 2b\,\Phi_{10}, \tag{7.138}$$

   where a dash represents a differentiation with respect to an appropriate position variable. We shall denote the resulting extensions by $\widehat{H}^{(1:2,3)}_{a,b}$.

2. **Asymmetrical case** Comparing with the discussion in §7.7.2 we can see that the corresponding Hamiltonian operators are $\widehat{H}^{(1,2;3)}_{a,b}$ and $\widehat{H}^{(1,3;2)}_{a,b}$ given by Eqs. (7.119) and (7.120).

7.7.6 Correlation

We shall continue with the examination of the two cases of interest:

---

[46] Blank, Exner and Havlicek (1994) p. 480.
[47] Wan and Fountain (1998).

1. **Symmetrical case**   Substituting $\eta^{(1;2,3)}_{\lambda,p}$ into boundary conditions given by Eqs. (7.137) and (7.138) we obtain[48]

$$ip = a + \sqrt{2}\,b\,e^{i\lambda}, \tag{7.139}$$

$$ip\,e^{i\lambda} = -a\,e^{i\lambda} - \sqrt{2}\,b, \quad \text{provided } \lambda \neq 0, \pi. \tag{7.140}$$

Equating real and imaginary parts of the above equations yields

$$0 = a + \sqrt{2}\,b\cos\lambda, \tag{7.141}$$

$$p = \hbar\sqrt{2}\,b\sin\lambda. \tag{7.142}$$

These result in a Josephson equation relating the phase shift across the junction and the supercurrent flowing from B_1 equally into B_2 and B_3 according to

$$j = j_c \sin\lambda, \tag{7.143}$$

where

$$j = -\frac{q_c}{m_c[\ell]}\,p, \quad j_c = -\frac{q_c}{m_c[\ell]}\,\hbar\sqrt{2}\,b. \tag{7.144}$$

The critical current j_c is characteristic of the junction and independent of λ. The incoming current from B_1 splits up equally and flows into B_2 and B_3.[49] Note that there is an enhancement of the critical current by the factor $\sqrt{2}$ when compared with the case of a long superconducting wire interrupted by a JJ.

2. **Asymmetrical case**   We either have the Hamiltonian $\widehat{H}^{(1,2;3)}_{a,b}$ matching the supercurrent operator $\widehat{J}^{(1,2;3)}_{\lambda}$ or the Hamiltonian $\widehat{H}^{(1,3;2)}_{a,b}$ matching the supercurrent operator $\widehat{J}^{(1,3;2)}_{\lambda}$. The system is acting like a two-branch circuit since there is no current in the third branch.

7.7.7 Superselection rules

To complete the theory a superselection rule has to be introduced. In the two asymmetric cases which are essentially the same as two-branch circuits we have already established the necessary superselection rule.

---

[48] When $\lambda = 0, \pi$ the function $\eta^{(1;2,3)}_{\lambda,p}$ does not satisfy boundary conditions (BC3) and should be excluded.

[49] We are talking about current flow in an algebraic sense. The direction of the current is opposite to the direction of the motion of the Cooper pairs. So, a negative current means a (positive) flow of Cooper pairs from B_1 into B_2 and B_3.

7.7. Y-SHAPE CIRCUITS

The symmetrical case follows similarly. Let us confine ourselves to a situation in which the phase shift λ in Josephson equation Eq. (7.143), has a restricted range, i.e.,

$$\lambda \in (0, \pi/2). \tag{7.145}$$

Then the supercurrent is uniquely related to λ. The analysis set out for the two-branch circuit in §7.6.5 which establishes a superselection rule applies here. The result is the same as before in that a continuous superselection rule exists parameterized by the current j.[50]

To sum up we have three descriptions of the circuit, one symmetrical and two asymmetrical. It may seem natural to assume that the former is the correct description of the circuit. However, analysis presented in the next section reveals that this may not necessarily be the case.

7.7.8 Condensate in a pure or in a mixed state

An experiment to examine how a supercurrent fed into $\boldsymbol{B}_1$ will flow down the circuit could have the following three probable outcomes:

1. Outcome 1   The current from $\boldsymbol{B}_1$ flows entirely into $\boldsymbol{B}_2$ with no current in $\boldsymbol{B}_3$. This means that the condensate is in a pure state described by $\eta_{\lambda,p}^{(1,2;3)}$ in Eq. (7.117).

2. Outcome 2   The current from $\boldsymbol{B}_1$ flows entirely into $\boldsymbol{B}_3$ with no current in $\boldsymbol{B}_2$. The condensate is in a pure state described by $\eta_{\lambda,p}^{(1,3;2)}$ in Eq. (7.118).

3. Outcome 3   The current from $\boldsymbol{B}_1$ splits up and flows equally into $\boldsymbol{B}_2$ and $\boldsymbol{B}_3$ with an enhancement of the critical current by a factor of $\sqrt{2}$.

The first two cases above are unambiguous. Supposing that an experiment confirms case 3 above, namely that the current from $\boldsymbol{B}_1$ splits up and flows equally into $\boldsymbol{B}_2$ and $\boldsymbol{B}_3$. The questions are whether

1. the state for outcome 3 is described by $\eta_{\lambda,p}^{(1:2,3)}$ in Eq. (7.130), and

2. whether such a state is pure or mixed.

If the state for outcome 3, whatever it is, represents a pure state then the Cooper pairs forming the condensate will all be in the same pure state. We are then confronted with the conceptual problem of how each Cooper pair gets

---
[50]Wan and Fountain (1998). We have assumed a positive value for b so that the current is negative for $\lambda \in (0, \pi/2)$ corresponding to a flow of Cooper pairs from $\boldsymbol{B}_1$ into $\boldsymbol{B}_2$ and $\boldsymbol{B}_3$.

divided at the junction to flow downstream into branches B_2 and B_3. This situation is more acute than the usual query as to how an electron physically passes through the double-slit in a double-slit diffraction experiment. Quite unlike the situation in an interferometer configuration, to be discussed in §7.9 later, the Cooper pairs going down B_2 and B_3 will physically separate further and further apart and they do not meet up together. If each Cooper pair were to "split up" into "two components" and these "components" never met up again we would be confronted with a de Broglie type paradox,[51] namely we are faced with the problem of not knowing what happens to each Cooper pair after the splitting. In particular we do not know where a Cooper pair is, e.g., if it is in B_2 or B_3. We can avoid the above difficulty by arguing that the state for outcome 3 should be a mixture,[52] i.e., a mixture of $\eta_{\lambda,p}^{(1,2;3)}$ and $\eta_{\lambda,p}^{(1,3;2)}$. In other words the condensate consists of a mixture of two parts, one part corresponds to $\eta_{\lambda,p}^{(1,2;3)}$ and the other part corresponds to $\eta_{\lambda,p}^{(1,3;2)}$.[53] This means that the incoming current in B_1 consists of two components with one component represented by $\eta_{\lambda,p}^{(1,2;3)}$ flowing into B_2 and the other component described by $\eta_{\lambda,p}^{(1,3;2)}$ flowing into B_3. Cooper pairs in B_1 are correspondingly divided into two groups described separately by $\eta_{\lambda,p}^{(1,2;3)}$ and $\eta_{\lambda,p}^{(1,3;2)}$; each group forms a current component. Cooper pairs are not having to "split up" at the junction.[54] The current in B_1 flowing into B_2 and B_3 is not accompanied by the splitting up of each Cooper pair.[55] The situation would be quite different if B_2 and B_3 were to meet up to form an interference circuit. We shall continue our conceptual discussion in §7.12.2 by comparing a Y-shape circuit with a similar configuration for a photon or an electron experiment with a beam splitter.

As far as performing experiments is concerned it is probably easier to employ a continuous Y-Shape circuit presented in the next section.

---

[51] Selleri and Tarozzi (1981), Wan (1988). The de Broglie paradox will be discussed in §8.2.
[52] Wan and Fountain (1996), Wollman (1993).
[53] Wan and Fountain (1996).
[54] Because of the *intrinsic global nature* of the condensate in dc effects the kind of delayed choice experiments described in Wheeler (1983) pp. 182-213, Hellmuth, Zajonc and Walther (1985) pp. 417-422, Greenstein and Zajonc (1997), Auletta (2001) §26 are not appropriate here.
[55] In a classical circuit, when a classical current flows into a junction and subsequently into different branches, each charge carrier, i.e., an electron, will flow into one of the branches, rather than "splitting itself up" into different branches.

7.8 Continuous Y-Shape Circuit

A simple but non-trivial configuration is that of a Y-shape circuit where the three branches join up at the branch point to eliminate the Josephson junction. For such a continuous circuit one may be tempted to adopt the following continuity conditions at the branch point:

$$\Phi_{20} = \Phi_{10}, \quad \Phi_{30} = \Phi_{10}. \tag{7.146}$$

However these conditions are inappropriate since they contradict general boundary conditions in Eqs. (7.122) and (7.123) for maximal symmetric extensions of the composite momentum operator. For simplicity let us confine our attention to the symmetrical and asymmetric cases:

1. A symmetrical continuous Y-shape circuit is a special case of Y-shape circuits corresponding to boundary conditions given by Eq. (7.127) with continuity of the phase at the branch point, i.e.,

$$\lambda_2 = \lambda_3 = 0. \tag{7.147}$$

We have

$$\Phi_{10} = \frac{1}{\sqrt{2}} (\Phi_{20} + \Phi_{30}). \tag{7.148}$$

The resulting composite momentum denoted by $\widehat{P}_{\lambda=0}^{(1:2,3)}$ admits eigenfunctions

$$\eta_{\lambda=0,p}^{(1:2,3)} = \frac{1}{\sqrt{[\ell]}} \left(e^{ipx_1} \oplus \frac{1}{\sqrt{2}} e^{ipx_2} \oplus \frac{1}{\sqrt{2}} e^{ipx_3} \right). \tag{7.149}$$

To determine the Hamiltonian we impose a further boundary condition on the symmetric nature of the derivative:[56]

$$\Phi'_{10} = \frac{1}{\sqrt{2}} (\Phi'_{20} + \Phi'_{30}) \tag{7.150}$$

which, together with Eq. (7.148), determines a Hamiltonian $\widehat{H}_{\lambda=0}^{(1:2,3)}$ admitting $\eta_{\lambda=0,p}^{(1:2,3)}$ as eigenfunctions.

We have an interesting situation where the wave function still suffers a discontinuity at the branch point despite the continuity of the circuit at the branch point. The incoming current in B_1 splits equally and flows into B_2 and B_3.

---

[56] The present condition on the derivatives is not a special case of Eqs. (7.137) and (7.138). Otherwise we would arrive at the Josephson equation with zero current on account of $\lambda = 0$. A detailed derivation of Eqs. (7.137) and (7.138) and Eq. (7.150) is available in Wan and Fountain (1998).

2. Asymmetrical case with boundary conditions $\alpha = 0, \lambda_2 = 0$ amounts to a continuity between branches $\boldsymbol{B}_1$ and $\boldsymbol{B}_2$ while ignoring $\boldsymbol{B}_3$, i.e.

$$\Phi_{10} = \Phi_{20}, \quad \Phi_{30} = 0. \tag{7.151}$$

The resulting composite momentum is given by

$$\widehat{P}^{(1,2;3)}_{\lambda_2=0} = \widehat{p} \oplus \widehat{p}_3, \tag{7.152}$$

where $\widehat{p}$ is the usual momentum in $L^2(\mathbb{R}) = \mathcal{H}_1 \oplus \mathcal{H}_2$. $\widehat{P}^{(1,2;3)}_{\lambda_2=0}$ admits eigenfunctions of the form

$$\eta^{(1,2;3)}_{\lambda_2=0,p} = \frac{1}{\sqrt{[\ell]}} \left(e^{ipx_1} \oplus e^{ipx_2} \oplus 0_3 \right). \tag{7.153}$$

The supercurrent operator is

$$\widehat{J}^{(1,2;3)}_{\lambda_2=0} = -\frac{q_c}{m_c[\ell]} \widehat{P}^{(1,2;3)}_{\lambda_2=0}. \tag{7.154}$$

Let $\widehat{H} = \widehat{p}^2/2m_c$ be the usual Hamiltonian operator in $L^2(\mathbb{R}) = \mathcal{H}_1 \oplus \mathcal{H}_2$. Then we can take the composite Hamiltonian as

$$\widehat{H}^{(1,2;3)}_{\lambda_2=0} = \widehat{H} \oplus \widehat{H}_3, \quad \text{where} \quad \widehat{H}_3 = \frac{1}{2m_c} \widehat{p}_3^\dagger \widehat{p}_3. \tag{7.155}$$

Physically this describes an incoming current in $\boldsymbol{B}_1$ going straight through to $\boldsymbol{B}_2$ with no current flowing into $\boldsymbol{B}_3$.

3. Asymmetric case with the boundary conditions $\alpha = 0, \lambda_3 = 0$ is the same as the case above except with $\boldsymbol{B}_2$ and $\boldsymbol{B}_3$ interchanged.

The discussion in the preceding section on whether the condensate is in a pure or a mixed state applies here.

7.9 Superconducting Quantum Interference Devices

Our method of quantization by parts can be applied to multi-branch circuits. Examples are four-branch circuits consisting of two leads connecting to a thick superconducting ring, five-branch circuits consisting of two leads connecting to a thick superconducting ring interrupted by a Josephson junction (the so-called rf-SQUID configuration) and six-branch circuits consisting of two leads

7.9. SUPERCONDUCTING QUANTUM INTERFERENCE DEVICES 553

connecting to a thick superconducting ring interrupted by two Josephson junctions, the so-called dc-SQUID configuration, a circuit depicted in Fig. 7.9.[57]

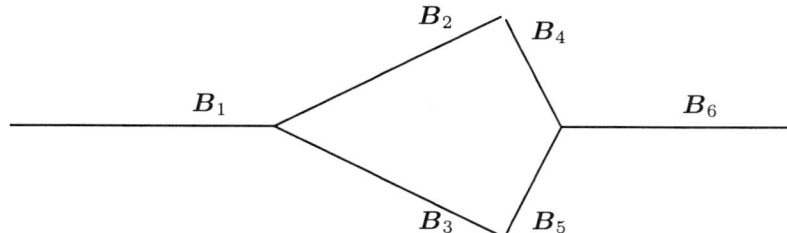

Fig. 7.9   Superconducting quantum interference device.

The circuit configuration is similar to the geometry of a typical interferometer experiment for light or an electron beam shown in Fig. 7.12.2(4) in §7.12.2. The matching of a superconducting state at the *incoming branch point* and the *outgoing branch point* is similar to the interference of the wave function from the two paths of an interferometer. So, the circuit is called a *superconducting quantum interference device*, or a SQUID. Operating with a dc current it is known as a dc-SQUID. In many applications a magnetic field is applied to influence the current flow in the circuit. The circuit, i.e., branches B_2, B_3, B_4 and B_5, will then enclose an external flux Φ_{ex}.

A dc-SQUID is a six-branch circuit with two Josephson junctions. One can reduce this to a four-branch circuit without the junctions. One can also have asymmetric branches. A dc supercurrent j can be fed into B_1 from the far left. This current will split at the incoming branch point to flow into B_2 and B_3, and then converge again at the outgoing branch point on the right to flow into B_6. Moreover, we can place such a circuit in an external magnetic field.

Clearly the method of quantization by parts can be applied to such a circuit. In view of the number of branches involved a detailed analysis is rather lengthy, and is available in Harrison and Wan (1997). We shall not present such a lengthy analysis here. Instead we shall summarize certain features of such circuits:

1. The initially quantized observables on the branches must match up at the Josephson junctions and the two branch points in the composite and correlative quantization stages in order to produce global quantities for the system as a whole from the partially quantized quantities. An eigenfunction for the global supercurrent operator representing a superconducting state will have components in various branches. The components on

---
[57]Gallop (1991).

the branches must match up at the incoming and the outgoing branch points. Such matching can lead to some striking characteristic behaviour of the system. When the circuit is placed in an external magnetic field the maximal supercurrent j_{max} which can be passed through the circuit from B_1 is related to the externally applied magnetic flux enclosed by the circuit Φ_{ex} by:

$$j_{max} = 2j_c \cos \frac{\pi \Phi_{ex}}{\Phi_c}, \qquad (7.156)$$

where j_c is the critical current of one of the Josephson junctions. This result can be confirmed by our quantization by parts scheme. We see that j_{max} is an oscillatory function of Φ_{ex} with two neighbouring peaks separated by a single flux quantum Φ_c. As pointed out before a supercurrent is an observable. In particular the current j_{max} is measurable. So, by measuring j_{max} we can detect any variation of the external magnetic flux up to a fraction of a Cooper pair flux quantum Φ_c. This provides an accurate measurement of the external flux and the magnetic field, not easily achievable by a classical magnetometer.[58] Magnetometers constructed from a superconducting circuit such as the one we have just studied are known as SQUID magnetometers. They are now widely used.

2. In previous models we have emphasized the existence of a superselection rule, and the one-dimensional nature of the supersectors. Like previous circuits a dc SQUID is a mixed quantum system. However, we must point out that all this does not preclude interference, e.g., there is effective interference at the out going branch point in a SQUID.[59]

3. Mixed quantum systems possessing both quantum and classical properties are not just of interest in the theoretical study of the foundations of quantum mechanics; they can prove to be extremely useful in practical applications as seen in a SQUID serving as a magnetometer.

7.10 Non-Equilibrium Mixed Quantum System

Equilibrium mixed quantum systems could be turned into non-equilibrium systems under different physical conditions. As an illustration consider a long superconductor interrupted by a Josephson junction as shown in Fig. 7.10. A dc supercurrent can tunnel through the JJ without needing or developing a voltage across the junction. When a small and constant voltage V is applied

---

[58] Feynman, Leighton and Sands (1965) §21-18.
[59] Harrison and Wan (1997).

7.10. NON-EQUILIBRIUM MIXED QUANTUM SYSTEM

across the junction a new phenomenan, known as an *ac Josephson effect*, occurs, i.e., an ac (alternating) supercurrent will flow across the junction.[60]

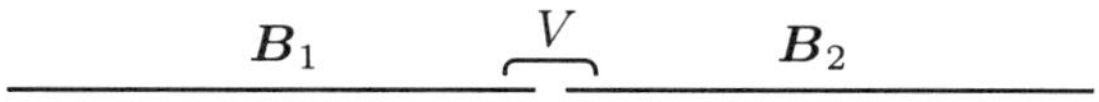

$$\underline{\qquad B_1 \qquad \overset{V}{\frown} \qquad B_2 \qquad}$$

Fig. 7.10   AC Josephson effect.

The time dependence of the current and the phase shift across the junction are related to the voltage by the following Josephson equation:[61]

$$j(t) = j_c \sin \lambda(t), \quad \lambda(t) = \lambda(0) - \frac{q_c V}{\hbar} t, \qquad (7.157)$$

or

$$j(t) = j_c \sin\left(\lambda(0) - \frac{q_c V}{\hbar} t\right). \qquad (7.158)$$

The dc Josephson equation is often regarded as a special case when $V = 0$. Another example is a TSCR with a Josephson junction. When placed in an external magnetic field which varies with time, an ac Josephson effect can occur and we again have a non-equilibrium situation.[62]

The physical argument for the ac Josephson equation is well known.[63] The applied voltage generates a potential difference at the branch point which would bring about a change in the phase shift. If we assume that the rate of change of λ is proportional to the applied voltage, or more specifically if we

---

[60] Fujita and Godoy (1996) p. 252. There are other ways to cause a voltage to appear across the junction. A common method is to drive a dc current greater than the critical current $|j_c|$ through a Josephson junction from an external current source. A dc voltage would appear across the junction (Rose-Innes and Rhoderick (1980) p. 167). One can also irradiate the junction with microwaves to cause a voltage to appear across the junction (Williams (1970) p. 177). An ac Josephson effect is physically complicated and generally it may involve a supercurrent component as well as a normal current component, together with microwave absorption or radiation at the junction. We shall consider a very simplified situation in this section. A more complex situation with the junction having a capacitive effect will be discussed in §8.4.

[61] de Bruyn Ouboter (1978) pp. 42-47. Depending on the sign convention used for various quantities some authors, e.g., Feynman, Leighton and Sands (1965), would adopt an equation of the form

$$j(t) = j_0 \sin \lambda(t), \quad \lambda(t) = \lambda(0) + \frac{q_c V}{\hbar} t.$$

[62] This ac Josephson effect further highlights the distinction between classical and macroscopic quantum circuits, i.e., an applied dc voltage across a resistor in a classical circuit does not produce an ac current in the resistor.

[63] Feynman, Leighton and Sands (1965). de Bruyn Ouboter (1978) pp. 42-47. Also see a relevant footnote in §8.4.1 for a reasoning based on Faraday's law.

assume that the phase factor satisfies a Schrödinger-type equation of the form

$$i\hbar \frac{d}{dt} e^{i\lambda} = q_c V e^{i\lambda}, \qquad (7.159)$$

we get

$$\frac{d\lambda}{dt} = -\frac{q_c}{\hbar} V. \qquad (7.160)$$

This agrees with the time-dependence of the phase shift in Eq. (7.157).

Before the voltage is applied we have an equilibrium system in a singular pure state $\eta_{\lambda,t}^\oplus$ given by Eq. (7.111). This state evolves within the supersector $\mathcal{H}_\lambda$. The Hamiltonian generator is identified with the Hamiltonian leading to Eq. (7.110) as the Schrödinger equation. The application of a voltage across the junction amounts to the creation of a new Hamiltonian generator. As a result the system will evolve differently. At each moment in time we still have the Josephson equation. This means that the supercurrent, the momentum and the phase shift will maintain their previous relations but they all become time dependent. The system is forced out of equilibrium to move from one supersector to another, i.e., the wave function would appear to oscillate from one supersector to another in time on account of the time-dependence of λ. Note that Eq. (7.157) applies even when the phase shift $\lambda(t)$ assumes (1) the values 0 and π for which the current vanishes, and (2) the values $\pm\frac{1}{2}\pi$ for which the current reaches its maximum. Moreover, we can formally extend the range of λ outside $[-\pi, \pi]$ in a periodic manner to accommodate $t \in (-\infty, \infty)$.[64]

Let us denote the resulting extended direct integral space by $\mathcal{H}^{(ac)\oplus}$. For the ac Josephson effect a pure singular state at any instance of time would still be representable by a pure singular vector of the form $\eta_\lambda^\oplus$ given in Eq. (7.105). The time dependence of the phase in the Josephson equation represents a transition of the state vector at time $t=0$

$$\eta_0^{(ac)\oplus} = \int^\oplus \delta\bigl(\lambda - \lambda(0)\bigr)\, \eta_\lambda \, d\lambda, \qquad (7.161)$$

to

$$\eta_t^{(ac)\oplus} = \int^\oplus \delta\Bigl(\lambda - \Bigl[\lambda(0) - \frac{q_c V}{\hbar} t\Bigr]\Bigr)\, \eta_\lambda \, d\lambda, \qquad (7.162)$$

where the superscript ac signifies an ac Josephson effect.

---

[64] We can extend the range of λ to include $0, \pm\frac{1}{2}\pi$ and $\pm\pi$. For these values we shall identify η_λ in Eq. (7.103) as follows: (1) $\eta_{\lambda=0} = \eta_{0,0}^{(1,2)}$ in Eq. (6.35), (2) $\eta_{\lambda=\pi} = \eta_{\pi,p}^{(1,2)}$ in Eq. (7.82), (3) $\eta_{\lambda=\pi/2} = \eta_{\frac{\pi}{2},p}^{(1,2)}$ in Eq. (7.83) for a $\frac{1}{2}\pi$-junction and so on. We can formally extend the range of λ from $[-\pi, \pi]$ to $(-\infty, \infty)$.

7.10. NON-EQUILIBRIUM MIXED QUANTUM SYSTEM

We can express the ac Josephson equation (7.158) in terms of the time dependence of the state vector within the general formalism presented in §4.4. Since the ac Josephson equation resembles an equation for a classical ac current it follows that the motion across supersectors can be formally described in a similar way as a classical current is described in §4.5.1. In other words we can use Eq. (4.120) to write the general form of a time-dependent singular state vector of this system. First we note that the time-dependence of λ generates a translation group T acting on λ given by[65]

$$T_t \lambda = \lambda - \frac{q_c V}{\hbar} t \quad \text{or} \quad T_t^{-1} \lambda = \lambda + \frac{q_c V}{\hbar} t \qquad (7.163)$$

with

$$\frac{\partial (T_t^{-1}\lambda)}{\partial \lambda} = 1. \qquad (7.164)$$

It follows from Eq. (4.120) that a pure singular state $\eta_0^{(ac)\oplus}$ with a phase shift $\lambda(0)$ at $t=0$ will evolve as

$$\eta_t^{(ac)\oplus} = \int^\oplus \delta\!\left(T_t^{-1}\lambda - \lambda(0)\right) \eta_\lambda \, d\lambda. \qquad (7.165)$$

This agrees with Eq. (7.162).

We can then go on to establish an apparent Schrödinger-type equation. Differentiating with respect to time we get

$$i\hbar \frac{\partial}{\partial t} \eta_t^{(ac)\oplus} = \int^\oplus \left\{ i\hbar \frac{\partial}{\partial t} \delta\!\left(\lambda - \left[\lambda(0) - \frac{q_c V}{\hbar} t\right]\right) \right\} \eta_\lambda \, d\lambda \qquad (7.166)$$

$$= \int^\oplus \left\{ iq_c V \, \delta'\!\left(\lambda - \left[\lambda(0) - \frac{q_c V}{\hbar} t\right]\right) \right\} \eta_\lambda \, d\lambda, \qquad (7.167)$$

where δ' is the derivative of the δ-function with respect to its argument. Now we can construct the Hamiltonian generator $\widehat{H}_g^{(ac)}$ in terms of the vector field for the group T which is, in accordance with Eq. (1.136) and Eqs. (3.125) to (3.127),

$$\widehat{\mathcal{X}} = -\frac{q_c V}{\hbar} \frac{\partial}{\partial \lambda} \quad \text{since} \quad \frac{d(T_t \lambda)}{dt} = -\frac{q_c V}{\hbar}. \qquad (7.168)$$

It follows from Eq. (4.121) that the Hamiltonian generator $\widehat{H}_g^{(ac)}$ is given by

$$\widehat{H}_g^{(ac)} \eta_t^{(ac)\oplus} = \int^\oplus \left\{ iq_c V \frac{\partial}{\partial \lambda} \delta\!\left(\lambda - \left[\lambda(0) - \frac{q_c V}{\hbar} t\right]\right) \right\} \eta_\lambda \, d\lambda$$

$$= \int^\oplus \left\{ iq_c V \, \delta'\!\left(\lambda - \left[\lambda(0) - \frac{q_c V}{\hbar} t\right]\right) \right\} \eta_\lambda \, d\lambda. \qquad (7.169)$$

---

[65] We have extended the range of λ to $(-\infty, \infty)$.

As a result we have an equation of motion:

$$\widehat{H}_g^{(ac)} \, \boldsymbol{\eta}_t^{(ac)\oplus} = i\hbar \frac{\partial}{\partial t} \boldsymbol{\eta}_t^{(ac)\oplus}. \tag{7.170}$$

The present example serves to illustrate how an equilibrium system can become non-equilibrium due to external factors. The Hamiltonian generator $\widehat{H}_g^{(ac)}$ is not the same as the Hamiltonian of the system.[66]

The theory presented in this section can be formulated from a different point of view. The idea is to regard the junction as a Hamiltonian system with the phase shift λ as the 'position variable',[67] Alternatively we may regard the non-equilibrium time dependence of λ as the motion of a Hamiltonian system with λ as the canonical position variable. We then introduce a phase shift operator $\widehat{\lambda}^\oplus$ as a multiplication operator acting in the direct integral Hilbert space $\mathcal{H}^{(ac)\oplus}$, i.e.,

$$\widehat{\lambda}^\oplus \boldsymbol{f}^\oplus = \int^\oplus \{\lambda f(\lambda)\} \boldsymbol{\eta}_\lambda \, d\lambda. \tag{7.171}$$

This operator is decomposable. The next step is to introduce a 'momentum operator' $\widehat{\varpi}$ canonically conjugate to $\widehat{\lambda}^\oplus$. We can define $\widehat{\varpi}$ by

$$\widehat{\varpi} \boldsymbol{f}^\oplus = \int^\oplus \left\{ -i\hbar \frac{\partial}{\partial \lambda} f(\lambda) \right\} \boldsymbol{\eta}_\lambda \, d\lambda. \tag{7.172}$$

Then $\widehat{\varpi}$ is canonically conjugate to $\widehat{\lambda}^\oplus$ in the sense of the following commutation relation:

$$\left[\widehat{\lambda}^\oplus, \widehat{\varpi}\right] \boldsymbol{f}^\oplus = i\hbar \, \boldsymbol{f}^\oplus. \tag{7.173}$$

The Hamiltonian generator in Eq. (7.169) is now seen to be proportional to $\widehat{\varpi}$, i.e., formally we have $\widehat{H}_g^{(ac)} = -(q_c V/\hbar)\widehat{\varpi}$. It is not surprising that the Hamiltonian generator is linear in the canonical momentum, since the variable in question, i.e., λ, varies linearly in time.[68] This particular approach is applicable to more complicated junction systems, e.g., capacitive junctions. A detailed discussion is given in §8.4.1 in Chapter 8.

7.11 BCS Theory and Superselection Rules

The unified formalism in Chapter 4 makes use of superselection rules to provide a formal characterization of classical properties. In §4.3.3 we also discuss whether these superselection rules can always be derived from some more

---

[66] For a discussion of energy see Rose-Inness and Rhoderick (1978) pp. 167-168, Fujita and Godoy (1996) p. 175, pp. 253-254.

[67] Leggett (1987), Berkley et at. (2003), Johnson et at. (2003). The value of λ can be deduced from the supercurrent.

[68] Percival and Richard (1982) p. 60 for an example in classical mechanics.

7.11. BCS THEORY AND SUPERSELECTION RULES

fundamental theory. Much work has of course been done on trying to derive superselection rules from orthodox quantum mechanics, particularly in a measurement context. In superconducting circuits superselection rules can be derived within a macroscopic wave function approach. But we want to see if these superselection rules as well as the macroscopic wave function approach can be justified by the standard microscopic theory of superconductivity of Bardeen, Cooper and Schrieffer.[69] The BCS theory of superconductivity is founded on the idea that the electrons in a superconductor exist in pairs, due to a net attraction between them caused by interaction with phonons. There are a number of different treatments of the situation. We shall follow that of Bogoliubov, for superconductivity at zero temperature.[70] In this idealized and formal treatment in terms of a standard formalism of quantum field theory, the electrons are initially confined in a cubic box of volume V which will be allowed to tend to infinity. Within the box the electrons interact with the Hamiltonian

$$\widehat{H}^{(0)}_{BCS} = \sum_f \eta_f \widehat{a}^\dagger_f \widehat{a}_f - \frac{1}{2V} \sum_{f,f'} \xi_f \xi_{f'} \widehat{a}^\dagger_f \widehat{a}^\dagger_{-f} \widehat{a}_{f'} \widehat{a}_{-f'} \qquad (7.174)$$

where

1. $\widehat{a}^\dagger_f, \widehat{a}_f$ are respectively the free electron creation and annihilation operators. The subscript f stands for momentum p and spin s, i.e., $f = (p, s)$ with $-f = (-p, -s)$,

2. η_f and ξ_f are real-valued functions of f satisfying certain convergence criteria in the limit of infinite volume V.[71]

Clearly $\widehat{a}^\dagger_f \widehat{a}^\dagger_{-f}$ will create a pair of electrons of opposite momentum and spin from the vacuum while $\widehat{a}_f \widehat{a}_{-f}$ will annihilate such a pair. It follows that this Hamiltonian describes an interacting system in which electrons are created and annihilated in pairs of opposite momentum and spin, known as *Cooper pairs*, and it is therefore referred to as the BCS *pairing Hamiltonian*. Our first interest lies in the system of electrons in the ground state, known as a *condensate*.[72] For an observer in the laboratory frame the condensate is, by definition, a "single-state system" described by the ground state of the BCS pairing Hamiltonian $\widehat{H}^{(0)}_{BCS}$. In such a BCS theory excited states of the Hamiltonian are found to separate from the ground state by an energy gap.[73]

---

[69] Bardeen, Cooper and Schrieffer (1957). This theory, known as the BCS theory, can be found in various monographs on superconductivity.
[70] Bogoliubov (1970), Bogoliubov and Bogoliubov (1992) pp. 359-366.
[71] Bogoliubov (1970) p. 78.
[72] Tilley and Tilley (1990) p. 38.
[73] As discussed earlier, the condensate corresponds to the ground state.

An approximate ground state, known as the *BCS ground state*, is the form

$$|\Psi_g\rangle = \prod_f \left(u_f + v_f \hat{a}_f^\dagger \hat{a}_{-f}^\dagger\right) |0\rangle, \qquad (7.175)$$

where $|0\rangle$ is the vacuum state of the many-particle system, u_f and v_f are numerical functions of f. This approximate solution to the ground state can be shown to be asymptotically exact in the limit of infinite volume. This state also corresponds to a zero total momentum,[74] an unsurprising result since electrons are created in pairs of opposite momentum. This means that there is no current flowing in the superconductor when in ground state, i.e., the condensate is at rest.

To obtain a superconducting state with a non-zero supercurrent a net momentum must be imparted to the condensate. Feynman pointed out an intuitive way of visualizing a current carrying state, namely that a condensate in a state $|\Psi_{P=0}\rangle$ of zero net momentum as seen by an observer at rest in the laboratory frame will be seen to be carrying a current by another observer in a moving inertial frame, or vice versa.[75] We can formalize this idea (for non-relativistic velocities) by using a Galilean transformation to introduce a net momentum to the condensate. Consider the case of a long and straight superconducting wire. The Cooper pairs in a supercurrent are assumed to move in a straight line, i.e., in the Euclidean space $\mathbb{E}$. Suppose the condensate is in a state of zero net momentum as seen by an observer in a frame moving to the right. In the stationary laboratory frame each Cooper pair will be seen to have a net momentum $\Delta p = m_c v$ in the same direction, giving rise to a current flowing to the left. In other words the state in the stationary frame is given by a state $|\Psi_{P\neq 0}\rangle$ having a non-zero net momentum $P \neq 0$, i.e.,[76]

$$|\Psi_{P\neq 0}\rangle = \prod_f \left(u_f + v_f \hat{a}^\dagger_{(f+\frac{\Delta p}{2})} \hat{a}^\dagger_{(-f+\frac{\Delta p}{2})}\right) |0\rangle_s, \qquad (7.176)$$

where

$$\pm f + \frac{\Delta p}{2} = (\pm p + \frac{\Delta p}{2}, s). \qquad (7.177)$$

There would also be a corresponding transformed Hamiltonian $\hat{H}_{BCS}^{(\Delta p)}$.[77] So, each current-carrying superconducting state in the stationary frame is isomorphic to an inertial frame moving with speed v relative to the laboratory frame.

---

[74] Bogoliubov (1970) p. 82.
[75] Feynman (1972) p. 291.
[76] Each electron would acquire a momentum of $\Delta p/2$.
[77] See Wan, Bradshaw, Trueman and Harrison (1998) §7 and Appendices D and E for explicit expressions.

7.11. BCS THEORY AND SUPERSELECTION RULES

One can then argue that the incoherence of superconducting states corresponding to different current follows from the incoherence of the corresponding inertial frames. Thus the superselection rule between different superconducting states derived in our macroscopic wave function model in §7.6.6 is justifiable by the standard microscopic theory.[78]

In the simple situation discussed above a condensate carrying a given steady dc current exists as a single-state system. This also sheds light on why a simple one-particle macroscopic wave function approach is effective: a single state of the condensate can be mapped one-to-one to a macroscopic wave function.[79]

We should point out that this argument based on a transformation to and from a moving inertial frame does not produce any superselection rules when applied to an orthodox quantum system. As an example, consider a free particle in a stationary frame and in a moving frame. For clarity let us denote quantities in the moving frame by a prime, e.g., in the stationary frame we have spatial coordinate x, wave functions $\phi(x,t)$ which form a Hilbert space $\mathcal{H}$ and in the moving frame we have coordinate x', and wave functions $\varphi'_v(x',t)$ which form a Hilbert space $\mathcal{H}'$. There is a mapping between states in the two frames by a Galilean transformation.[80] Let $\varphi_v(x,t)$ be the function in the stationary frame obtained from a Galilean transformation from $\varphi'_v(x',t)$. Then $\varphi_v(x,t)$ is an element of $\mathcal{H}$ and we can employ the superposition principle in orthodox quantum mechanics to form a coherent superposition of $\varphi_v(x,t)$ and any given state $\phi(x,t)$ in $\mathcal{H}$. No superselection rules appear. This is in marked contrast to the condensate which is a single-state system. The transformed state is generally not a state of the original stationary system with the BCS Hamiltonian $\widehat{H}_{BCS}^{(0)}$,[81] and is therefore incoherent from the original state. In the case of the condensate in a long straight superconductor, the transformed state $|\Psi_{P\neq 0}\rangle$ cannot be coherently added to $|\Psi_{P=0}\rangle$ to form a new pure state.

---

[78] A note of caution is warranted here. We have seen that superconductors can be used to construct complex circuits. There are circuit configurations for which the simple arguments based on a continuous and straight superconducting wire does not hold. In other words there are circuits in which there may well be coherent superpositions of different superconducting states. We shall investigate such a possibility in §8.4 in relation to Schrödinger's cat states.
[79] There is also a corresponding transformation of operators such as the Hamiltonian generator (Kuper (1968) p. 141).
[80] Merzbacher (1998) Eqs. (4.111) and (4.106) on pp. 76-77.
[81] The transformed state would correspond to the Hamiltonian $\widehat{H}_{BCS}^{(\Delta p)}$.

7.12 Conceptual Analyses

7.12.1 Non-uniqueness of quantization

Our analysis shows that when quantized a classical observable does not generally correspond to a unique selfadjoint operator. First, one has to replace the rigid selfadjointness requirement by a maximal symmetry requirement, and secondly one has to allow for non-uniqueness. To proceed further one has to regard the physical conditions or environment the system is subjected to as an integral part of the quantization problem and use these conditions to correlate cognate observables. Only then can a consistent and quantized theory emerge. For example different physical conditions can mean a different choice of boundary conditions leading to a different quantized system. The non-uniqueness arising should be regarded as desirable since it is precisely this non-uniqueness which enables us to accommodate different physical conditions or environments. Our attempt to deal with quantum circuits by the method of quantization by parts serves to illustrate all these points.

7.12.2 Y-shape circuits, equilibrium mixed quantum systems and non-locality

A number of interesting conceptual problems on the non-local nature of a condensate in a Y-shape circuit and in a SQUID configuration can be compared with similar features of light under similar geometric constraints. To be specific let us consider the following experiments:

Photon beam splitter experiment A beam of monochromatic light L_1 travelling along the x-axis from the far left is intercepted by a beam splitter S_1 situated at the origin. The splitter divides L_1 into two beams, L_2 and L_3, of equal intensity with L_2 reflected vertically upwards along the y-axis and L_3 transmitted horizontally to the right along the x-axis. Two photon detectors D_2 and D_3 are placed at some distance away from the beam splitter with D_2 in the path of L_2 and D_3 in the path of L_3 as shown in Fig. 7.12.2(1).

When the intensity of the incoming beam L_1 is so low that the beam consists of one photon crossing the beam splitter at a time we find that only one of the two detectors will fire at a time since there is just one photon to be detected at a time. We can tell by which path a particular photon had travelled. This is direct evidence that an individual photon does not split up into two halves at the beam splitter in this particular experiment. On average there will be an equal number of counts registered by the two detectors so that when the intensity of L_1 increases sufficiently we see an incoming beam L_1 splitting up simultaneously into two equal beams L_2 and L_3. No interference

7.12. CONCEPTUAL ANALYSES

between the two beams L_2 and L_3 takes place. In other words the incoming photon beam consists of a mixture of two beams with one reflected up to form L_2 and the other one passing though to form L_3.

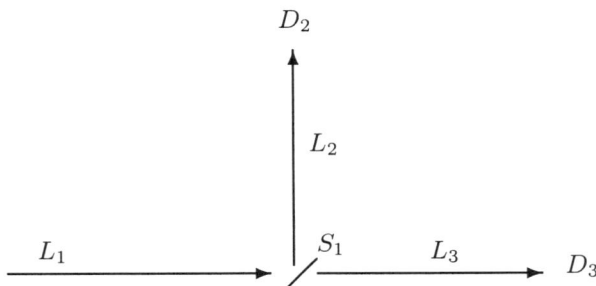

Fig. 7.12.2(1)   Photon beam splitter experiment with a beam splitter.

Y-shape circuit experiment   One can see an analogy here with a superconducting Y-shape circuit shown in Fig. 7.7.1. For the Y-shape circuit with a supercurrent j fed into branch $\boldsymbol{B}_1$ from the left we expect, based on the discussions §7.7.8, one of the following three outcomes to occur:

1. **Outcome 1**   The supercurrent j flows from $\boldsymbol{B}_1$ to $\boldsymbol{B}_2$ with no current flow in $\boldsymbol{B}_3$. In other words there is a total absence of current in $\boldsymbol{B}_3$.

2. **Outcome 2**   The supercurrent j flows from $\boldsymbol{B}_1$ to $\boldsymbol{B}_3$ with no current flow in $\boldsymbol{B}_2$. The conceptual situation is identical to Outcome 1 above.

Outcomes 1 and 2 resemble a photon beam splitter experiment with a single photon. This photon will go into the path of one detector with no photon detection in the other detector. We can also compare this situation with a double-slit experiment of electrons. Let us consider two experimental arrangements.

   (a) Experimental arrangement 1:   An electron beam is directed towards a double-slit, as shown in Fig. 7.12.2(2). We can observe an interference pattern on the screen which demonstrates the splitting up of the wave function through the slits and the recombining of the wave function to interfere at the screen.

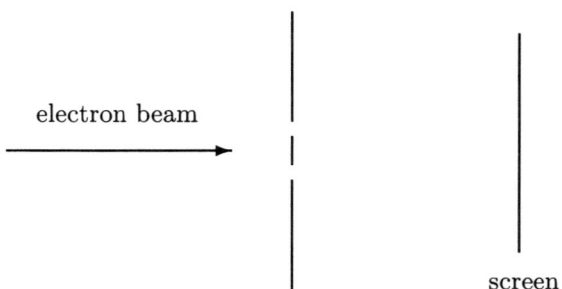

Fig. 7.12.2(2)   Standard double-slit experiment.

(b) Experimental arrangement 2:   Place a partition behind the double-slit, as shown in Fig. 7.12.2(3):

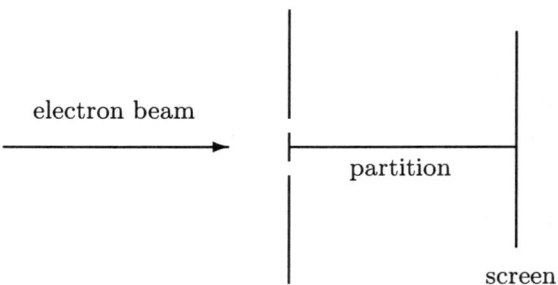

Fig. 7.12.2(3)   Double-slit experiment with a partition.

The partition is designed to prevent any recombination of the wave function behind the double-slit. One would then expect each electron to go through only one of the slits and no double-slit interference pattern is expected to emerge. We would argue that the original electron beam would divide into a mixture of two parts, with one part going through the upper slit and the remaining part going through the lower slit. There is no interference between the two parts.

7.12. CONCEPTUAL ANALYSES

3. **Outcome 3**   The supercurrent j fed in from the left to branch $\boldsymbol{B}_1$ flows equally into both $\boldsymbol{B}_2$ and $\boldsymbol{B}_3$, a result which resembles a photon beam splitter experiment with a strong incident beam L_1. Cooper pairs will tunnel into either $\boldsymbol{B}_2$ or $\boldsymbol{B}_3$ in equal numbers to produce equal currents in the two branches. To reconcile with our theory we can assume the condensate to be in a mixed state, i.e., a mixture of

   - state $\eta_{\lambda,p}^{(1,2;3)}$ corresponding to the Cooper pairs going into $\boldsymbol{B}_2$, and
   - state $\eta_{\lambda,p}^{(1,3;2)}$ corresponding to the Cooper pairs going into $\boldsymbol{B}_3$.

   As with the photon beam splitter experiment and the doublt-slit experiment for electrons depicted in Fig. 7.12.2(2), these two parts of the condensate never meet up again so that the question of their coherence and interference simply does not arise. We are confronted with no conceptual difficulty since no Cooper pair has to split up at the junction to travel simultaneously down $\boldsymbol{B}_2$ or $\boldsymbol{B}_3$.

Photon interferometer experiment[82]   Let us now modify a photon beam splitter experiment by adding two perfectly reflecting mirrors M_2 and M_3 in the paths of L_2 and L_3 respectively and a further beam splitter at the place where beams L_2 and L_3 meet up again. This way we have constructed an interferometer configuration so that L_2 and L_3 are brought back together as shown in Fig. 7.12.2(4).

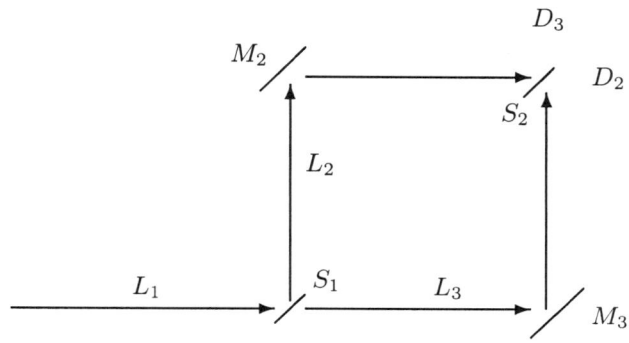

Fig. 7.12.2(4)   Interferometer experiment.

In the usual language used in optics we would say that beams L_2 and L_3 are coherent and would interfere with each other when they meet up at the second

---
[82] Greenstein and Zajonc (1997) Fig. 2-8 on p. 39.

beam splitter S_2. Let us consider the motion of individual photons and their detection by the detectors. The fact that beams L_2 and L_3 would interfere with each other is a strong evidence that each photon comes to S_2 by both paths, as it were. When a photon is detected by a detector, say D_2, after the interference, we cannot tell by which path a photon had travelled.[83] This is in sharp contrast to the beam splitter experiment shown in Fig. 7.12.2(1) where there is no interference so that each photon arriving at the detector goes through only one path. In §7.9 we also discussed a similar experiment with superconductors, i.e., SQUIDs.

When looking at a beam splitter in an interferometer experiment we often encounter the question as to how an incoming photon in L_1 knows what to do at the beam splitter S_1, e.g., "to split" or not "to split", since the photon cannot know what experimental configuration comes after the beam splitter. This question has been investigated in the form of the so-called *delayed-choice experiment*.[84] We will gain an insight into this question by an analysis of this question in the context of superconducting circuits in the following section.

7.12.3 Equilibrium states, globalization and non-locality

As with a photon interferometer experiment one may ask the usual naive non-locality question on the motion of Cooper pairs in a dc SQUID in Fig. 7.9:

> *How would each Cooper pair know exactly how to "split up" at the incoming branch point in a way as to produce the correct currents flowing into branches B_2 and B_3,[85] given that these currents depend on the circuit configuration downstream?*

Based on the theory presented so far we would offer the following analysis:

1. A superconducting system with a dc current is an equilibrium system; a superconducting state represents the condensate as a whole in the form of a beam of Cooper pairs. Physical observables, including the Hamiltonian, are established by global consideration, i.e., by considering the entire circuit configuration. These quantities determine the global behaviour of the system, i.e., behaviour throughout the entire circuit such as current distribution in different branches.

---

[83]Herzog *et al.* (1995), Rarity (1996). The knowledge of which path a quantum particle travels and the emergence of an interference effect are known to be mutually exclusive. For information on neutron interferometry see Bonse and Rauch (1979).

[84]Hellmuth, Zajonc and Walther (1985) pp. 417-422, Greenstein and Zajonc (1997).

[85]Generally a SQUID circuit is not symmetrical, i.e., branches B_3 and B_5 do not have to be symmetrical to B_2 and B_4. In other words the current in B_3 may not be the same as the current in B_2.

7.12. CONCEPTUAL ANALYSES

2. When we naively picture an individual Cooper pair arriving at the junction we can imagine the following two scenarios:

   (a) An individual Cooper pair has the "ability to sense" in a non-local manner the rest of the circuit configuration to enable it to behave accordingly at the junction.

   (b) There is no question of Cooper pairs having the "ability to sense" the entire circuit configuration in a way described above. The condensate is in equilibrium, and an equilibrium superconducting state is established by considering the global circuit configuration, and is then able to affect the behaviour of the Cooper pairs throughout the circuit.

3. An equilibrium state is global in nature. Properties of a system in equilibrium states are generally global in nature as well. To reach an equilibrium state the system must be in a stable global environment, and then allowed to evolve with its global environment. A gas in contact with a wildly fluctuating heat source will not reach any equilibrium state. If a gas contained in a given volume is placed in thermal contact with a stable heat bath of a given temperature the gas will in time reach an equilibrium state of the same temperature. At an equilibrium state a knowledge of the temperature at one location enables one to claim the same knowledge of temperature at another distant location. This is a kind of *globalization* of a thermodynamical system which presents no conceptual problem as this involves no instantaneous transfer of information from one point to another. There is no causality or non-locality problem either.

4. Another example is that of classical electric circuits. At first sight one is confronted with an apparent non-locality problem: when a current comes to a branch point how does it know how to divide up to flow into different branches, given that such a division depends on the circuit configuration down stream? The answer is simple. When the dc voltage source in a circuit is switched on, a transient current is generated which rises from zero to a steady dc current. The growth of the transient is a dynamical process determined by the global circuit configuration as the transient current develops down the circuit. The final dc current represents an equilibrium state.[86] No non-locality problem arises. A similar argument may be applied to a superconductung circuit. Consider the case of a SQUID. Being in an interferometer circuit configuration a supercurrent divides and flows into two branches and emerges at the

---
[86] Bleaney and Bleaney (1957) pp. 152-157.

other end to interfere. This is again an equilibrium state. There is no causality problem involved.

7.12.4 Quantum/Classical divide 2

In the discussions in this chapter we know that a compound system can behave quantum mechanically and that a single-particle representation of a compound quantum system is capable of describing many physically observable results, from superconductivity to the diffraction of a beam of fullerene atoms. The question then arises as to whether there is a limitation to all this. One might be tempted to claim that, as technology improves, one can succeed in diffracting bigger and more massive molecules and even viruses and biological cells in the sense of being able to produce and detect an interference effect. Against this is the generally accepted criterion for the disappearance of the interference effect, i.e., if we can in principle determine which path an individual particle in the beam takes in going through the diffraction grating, we will not be able to produce an interference pattern in the experiment. A particle can reveal its path by emission of radiation or by interaction with a probing radiation. However, the wavelength involved must be smaller than the separation between slits in the grating in order to provide unambiguous information on which slit the particle went through. It follows that if we send a beam of marbles through a grating made of iron bars several centimeters apart we will not observe any interference effect because we can determine, through visible light, the trajectory of each marble as it goes through the iron bars. So, it appears that size really matters and we cannot expect to observe an interference effect for arbitrarily large and complex systems. As the object gets bigger and becomes more massive there will come a time when we can determine its trajectory in principle, and hence effectively remove any interference effect. The object then loses its wave property and behaves classically. However, size is not everything. A superconducting condensate consists of a macroscopic number of constituent particles. Moreover, the transition between quantum and classical regimes is not necessarily a gradual and asymptotic one. For example, the transition from normal to superconductivity occurs rather abruptly as the temperature goes down and passes a certain critical value.

The conclusion must be that the quantum/classical divide is very complicated and it cannot be universally characterized by size alone nor by a certain universal asymptotic limit. Every case has to be dealt with individually. The emergence of superselection rules should be regarded as a general feature of macroscopic quantum systems, although we are not claiming that all macroscopic quantum systems must possess a superselection rule.

7.13 Orthodox Quantum Systems

The method of quantization by part can be applied to orthodox quantum systems under geometric constraints.[87] The difference here is that (PAS1) and (PAS2) in §7.3.2 for superconducting systems no longer apply. As a result superselection rules inherent in some superconducting systems no longer apply.

Here we shall illustrate how the method of quantization by parts can provide a description of an orthodox quantum particle, say an electron, passing through a double-slit configuration shown in Fig. 7.13.

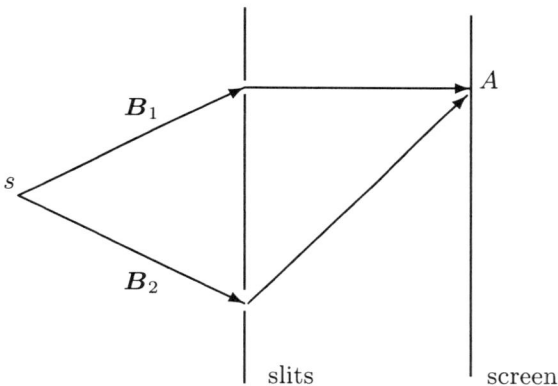

Fig. 7.13   Double-slit diffraction experiment.

We start with a classical particle which leaves the point source s and arrives at a point A on the screen by taking either one or the other of the two continuous paths shown as $\boldsymbol{B}_1$ and $\boldsymbol{B}_2$ in Fig. 7.13. Let us treat the two paths $\boldsymbol{B}_1$ and $\boldsymbol{B}_2$ as continuous lines.[88] Next let us introduce a linear position variable x_1 along $\boldsymbol{B}_1$. This position variable x_1 is designed to vary from 0 to a_1 with $x_1 = 0$ at the point source s and $x_1 = a_1$ ending at point A on the screen. A second position variable x_2 is introduced for path $\boldsymbol{B}_2$ in a similar way, i.e., we have $x_2 \in [0, a_2]$ with $x_2 = 0$ coinciding with the point source s and $x_2 = a_2$ ending up at point A.

When we come to quantization we can view the geometry in Fig. 7.13 for the motion from s to A as an effective two-branch circuit. The scheme of quantization by parts can then be carried out. For branch $\boldsymbol{B}_1$ the appropriate

---
[87] Exner and Seba (1989), Fountain (1995).
[88] Path $\boldsymbol{B}_1$ forms a continuous line from s to A. Path $\boldsymbol{B}_2$ is similarly continuous. The fact that the paths are not straight does not matter here.

Hilbert space is $\mathcal{H}_1 = L^2(\boldsymbol{B}_1, dx_1)$ and on branch $\boldsymbol{B}_2$ the Hilbert space is $\mathcal{H}_2 = L^2(\boldsymbol{B}_2, dx_2)$.

Partial quantization

1. The momentum on $\boldsymbol{B}_1$ is quantized as the symmetric operator introduced in E2.5.1(1) E1, i.e., $\widehat{p}_{01}(A)$ acting in $\mathcal{H}_1$.[89] This operator admits a one-parameter family of selfadjoint extensions $\widehat{p}_{\lambda_1}(A)$ with eigenfunctions

$$\eta^{(1)}_{\lambda_1,p_1}(x_1) = \frac{1}{\sqrt{a_1}} e^{ip_1 x_1}, \quad p_1 = \hbar \left(\frac{2n_1\pi - \lambda_1}{a_1}\right). \tag{7.178}$$

Similarly we have a partially quantized momentum $\widehat{p}_{\lambda_2}(A)$ in $\mathcal{H}_2$ with eigenfunctions

$$\eta^{(2)}_{\lambda_2,p_2}(x_2) = \frac{1}{\sqrt{a_2}} e^{ip_2 x_2}, \quad p_2 = \hbar \left(\frac{2n_2\pi - \lambda_2}{a_2}\right). \tag{7.179}$$

2. We can take the operator introduced in E2.5.1(1) E2 to serve as a partially quantized Hamiltonian, i.e.,[90]

$$\widehat{H}_{01}(A) = \frac{1}{2m} \widehat{p}^2_{01}(A) \text{ on } \boldsymbol{B}_1, \tag{7.180}$$

$$\widehat{H}_{02}(A) = \frac{1}{2m} \widehat{p}^2_{02}(A) \text{ on } \boldsymbol{B}_2. \tag{7.181}$$

As shown in Theorem 6.2(1) these operators possess a large number of selfadjoint extensions. Obvious examples are

$$\widehat{H}_{\lambda_1}(A) = \frac{1}{2m} \widehat{p}^2_{\lambda_1}(A), \quad \widehat{H}_{\lambda_2}(A) = \frac{1}{2m} \widehat{p}^2_{\lambda_2}(A). \tag{7.182}$$

Composite quantization Some simple choices are as follows:

1. For the momentum we take

$$\widehat{p}^{(1,2)}_{\underline{\lambda}}(A) = \widehat{p}_{\lambda_1}(A) \oplus \widehat{p}_{\lambda_2}(A), \tag{7.183}$$

where $\underline{\lambda}$ denotes the pair (λ_1, λ_2).

---

[89] Following E2.5.1(1) E1 we should have used the notation $\widehat{p}_{01}(\Lambda_1)$, $\Lambda_1 = [0, a_1]$. But it is more transparent to used the end point A on the screen, instead of the interval Λ_1, to indicate our aim to derive an interference pattern on the screen.

[90] We assume that the double-slit geometry has the effect of dividing the motion between points s and A into two branches without introducing any additional potential along the branches.

7.13. ORTHODOX QUANTUM SYSTEMS

2. For the Hamiltonian we take

$$\widehat{H}_{\underline{\lambda}}^{(1,2)}(A) = \widehat{H}_{\lambda_1}(A) \oplus \widehat{H}_{\lambda_2}(A) \tag{7.184}$$

where $\underline{\lambda} = (\lambda_1, \lambda_2)$ are the same as those for the momenta above. We shall also identify the Hamiltonian generator for the dynamics of the system with $\widehat{H}_{\underline{\lambda}}^{(1,2)}(A)$.

3. The composite momentum and the composite Hamiltonian operators

$$\widehat{p}_{\underline{\lambda}}^{(1,2)}(A) \quad \text{and} \quad \widehat{H}_{\underline{\lambda}}^{(1,2)}(A) \tag{7.185}$$

share the following normalized eigenfunctions

$$\eta_{\underline{\lambda},p}^{(1,2)}(x_1, x_2) = \frac{1}{\sqrt{2}} \eta_{\lambda_1,p}^{(1)}(x_1) \oplus \frac{1}{\sqrt{2}} \eta_{\lambda_2,p}^{(2)}(x_2). \tag{7.186}$$

Here we have set $p_1 = p_2 = p$, i.e.,[91]

$$\frac{2\pi n_1 - \lambda_1}{a_1} = \frac{2\pi n_2 - \lambda_2}{a_2}. \tag{7.187}$$

For a double-slit experiment the particle source emits a particle in a momentum eigenstate with a given momentum value. In our present setup this corresponds to $\eta_{\underline{\lambda},p}^{(1,2)}$ in Eq. (7.186). The interference effect at the point A on the screen is due to the superposition of the branch functions $\eta_{\lambda_1,p}^{(1)}(x_1)$ and $\eta_{\lambda_2,p}^{(2)}(x_2)$ overlapping at the point A, i.e., at $x_1 = a_1$ for $\eta_{\lambda_1,p}^{(1)}$ and $x_2 = a_2$ for $\eta_{\lambda_2,p}^{(2)}$, to give a value of the composite wave function $\eta_{\underline{\lambda},p}^{(1,2)}$ at A to be

$$\eta_{\underline{\lambda},p}^{(1,2)}(A) = \frac{1}{\sqrt{2}} \left(\eta_{\lambda_1,p}^{(1)}(a_1) + \eta_{\lambda_2,p}^{(2)}(a_2) \right). \tag{7.188}$$

Its absolute value square is

$$\left| \eta_{\underline{\lambda},p}^{(1,2)}(A) \right|^2$$

$$= \frac{1}{2} \left| \eta_{\lambda_1,p}^{(1)}(a_1) \right|^2 + \frac{1}{2} \left| \eta_{\lambda_2,p}^{(2)}(a_2) \right|^2 + \text{Re}\left\{ \left(\eta_{\lambda_1,p}^{(1)}(a_1) \right)^* \eta_{\lambda_2,p}^{(2)}(a_2) \right\}$$

$$= \frac{1}{2a_1} + \frac{1}{2a_2} + \frac{1}{\sqrt{a_1 a_2}} \cos k(a_2 - a_1), \quad k = p/\hbar. \tag{7.189}$$

---

[91] We can have
$$n_1 = n_2 = n = 0, \quad \frac{\lambda_1}{\lambda_2} = \frac{a_1}{a_2}.$$
In many practical experimental arrangements we would have $a_2 - a_1 \ll a_2$.

Taking the approximation $a_2 \approx a_1 = a$ we finally arrive at the familiar result:

$$\left| \eta_{\underline{\lambda},p}^{(1,2)}(A) \right|^2 \approx I \left(1 + \cos k(a_2 - a_1) \right), \tag{7.190}$$

Here $I = 1/a$ may be interpreted as the intensity of the incoming wave and $k(a_2 - a_1)$ is just the phase difference between the two branches corresponding to the path difference.[92] We can interpret $\left| \eta_{\underline{\lambda},p}^{(1,2)}(A) \right|^2$ as a position probability density at A on the screen.

What is said above applies to all points on the screen. As A moves up and down the screen $\left| \eta_{\underline{\lambda},p}^{(1,2)}(A) \right|^2$ varies from a minimun value of zero to a maximum value of $2I$. Since $\eta_{\underline{\lambda},p}^{(1,2)}$ is also an eigenfunction of the Hamiltonian $\widehat{H}_{\underline{\lambda}}^{(1,2)}(A)$, which also serves as the Hamiltonian generator, we have a stationary situation. A steady interference pattern will emerge on the screen to show a position probability distribution corresponding to the probability density $\left| \eta_{\underline{\lambda},p}^{(1,2)}(A) \right|^2$ as a sufficient number of particles are emitted by the source s.

The following remarks help to clarify the status of our discussions here:

1. Our present treatment of interference at A obtained by considering two possible electron paths is different from Feynman's sum-over-paths approach. In the sum-over-paths picture, interference at A is interpreted as a weighted sum over all the alternative paths from s to A.[93]

2. Despite the similarity between the geometry of Fig. 7.9 and Fig. 7.13 our present double-slit diffraction situation is quite different from a dc SQUID. For the double-slit diffraction experiment the point A can move about the screen and we have constructive interference at places and destructive interference at some other places.

An exciting achievement in electronic circuitry is the development of nanostructures, i.e., structures of the dimensions of nanometers.[94] It is possible to form electrical circuits in nanostructures. We call these nano-circuits. Current flows in nano-circuits can be formed by the motion of a few electrons. The study of electronics in *nano-circuits* may be referred to as *nano-electronics* or *single-electronics*. A nano-circuit is formed by conducting wires of diameter of a few nanometers. An example is that of a single-wall carbon nano-tube.[95] To visualize this we can consider a single layer or a single sheet of carbon atoms arranged in a two-dimensional hexagonal lattice structure, i.e., a graphite sheet,

---

[92] Note that with a different normalization determined by an actual experiment the intensity I will be different from $1/a$.

[93] See Baym (1969) pp. 69-74 for a highly readable exposition of Feynman's work.

[94] See Schwab and Roukes (2005) for a brief review of nanoelectromechanical structures and quantum mechanics.

[95] Mintmire (1992) for a picture of a nanotube made of fullerene molecules.

and then roll this sheet into a small tube of diameter of about one nanometer. The motion of an electron along the wall of the tube can be separated into a circular component round the tube and a translational component along the length of the tube. The motion round the circumference of the tube has already been used to observe the Aharonov-Bohm effect. This is done by threading magnetic field through the hollow of the tube and passing current through leads connected to diametrically opposite points across the tube.[96] When the circular motion can be ignored, we have an effective one-dimensional conductor, serving as a *quantum wire*.[97] Electrical circuits can then be constructed. We could have a single electron going throw the circuit at a time. Quantum properties arise from the fact that the wave function can remain coherent over a distance of over 100 nm, i.e., over a considerable part of the circuit, so that interference experiments can be performed. The situation is different from that involving a supercurrent. Here we can have one electron going through the circuit at a time, while a supercurrent consists of massive number of Cooper pairs.

7.14 Prospects and Other Approaches

Our present method of quantization by parts can be extended to other circuit configurations. We shall mention two briefly here. Firstly we can include a capacitor in what is known as a *capacitive junction* in a superconducting circuit. This will be dicussed in §8.4.1. Secondly we can have a *point contact* circuit formed by a long superconducting wire linked to a flat (two-dimensional) superconductor through a Josephson junction.[98] This configuration can be idealized mathematically as the union of two branches, i.e., $I\!\!E^- \cup I\!\!E_0^2$ with the branch point at the chosen coordinate origin of $I\!\!E^2$.[99] Such a circuit for the motion of an electron has been discussed by a number of people;[100] the situation there is quite different since the momentum operator and its eigenfunctions do not play such a central role for electrons. We can deal with such a configuration for superconductivity by quantization by parts, using Hilbert spaces $L^2(I\!\!E^-), L^2(I\!\!E_0^2)$ and their direct sum $L^2(I\!\!E^-) \oplus L^2(I\!\!E_0^2)$. Now, consider the momentum operator. For the branch $I\!\!E^-$ along which the supercurrent is fed in, the partially quantized momentum is just $\widehat{p}^-$ introduced in E2.5.1(1) E6. For the plane $I\!\!E_0^2$ the radial momentum operator $\widehat{p}_r(I\!\!E_0^2)$ in $L^2(I\!\!E_0^2)$ has the

---

[96] Cobden (1999).
[97] Razavy (1997).
[98] Barone and Paternò (1982).
[99] $I\!\!E_0^2$ is the space $I\!\!E^2$ with the origin removed.
[100] Exner and Seba (1987).

expression[101]

$$\widehat{p}_r(I\!\!E_0^2) = -i\hbar \left(\frac{\partial}{\partial r} + \frac{1}{2r} \right) \quad (7.191)$$

in polar coordinates (r, θ) in the plane. This operator is strictly maximal symmetric with deficiency indices (0,1). The composite quantization would amount to taking a selfadjoint extension of the direct sum $\widehat{p}^- \oplus \widehat{p}_r$ in $L^2(I\!\!E^-) \oplus L^2(I\!\!E_0^2)$. Selfadjoint extensions exist since the deficiency indices of $\widehat{p}^- \oplus \widehat{p}_r$ are (1,1). We can go on to establish corresponding supercurrent operators.

Superfluid helium is known to be describable by a macroscopic wave function.[102] Moreover, superfluid helium flow through a weak link also exhibits Josephson effects, although the geometric constraints involved are quite different.[103] Our present theory should also be applicable in such situations.

Finally we should mention another phenomenological approach to superconducting circuits. From E1.6.3(1) E3 in Chapter 1 we know that a classical resistanceless LC circuit is a Hamiltonian system with the charge Q on the capacitor and the magnetic flux Φ through the inductor as canonical variables. One can quantize such a circuit system in the same way as a classical harmonic oscillator is quantized:

1. Take the flux Φ as a "coordinate variable" assuming values from $-\infty$ to ∞.

2. Define a Hilbert space $\mathcal{H}_{LC}$ of complex-valued functions of Φ, square-integrable with respect to the measure $d\Phi$.

3. Define the flux and charge operators in $\mathcal{H}_{LC}$ by the following operator expressions

$$\widehat{\Phi} = \Phi, \quad \widehat{Q} = -i\hbar \frac{d}{d\Phi}. \quad (7.192)$$

Operating on the usual domains these give rise to two selfadjoint operators in $\mathcal{H}_{LC}$.

4. Define the Hamiltonian operator as

$$\widehat{H}_{LC} = \frac{\widehat{Q}^2}{2C} + \frac{\widehat{\Phi}^2}{2L}. \quad (7.193)$$

It is clear that a quantized LC circuit behaves like a quantized harmonic oscillator, with $\widehat{Q}^2/2C$ as the 'kinetic energy' and $\widehat{\Phi}^2/2L$ as the potential energy.

---

[101] This is different from the radial momentum operator in $I\!\!E^3$ given by Eqs. (2.456) and (2.457)
[102] Tilley and Tilley (1990).
[103] For a review of superfluid helium weak links see Davis and Packard (2002).

It is possible to modify this Hamiltonian, e.g., by inserting additional interaction terms, to model the behaviour of certain superconducting circuits.[104] Many important recent experiments on superconducting circuits are based on theories using this approach. The reason is that many junctions are constructed to make them behave like a capacitor. As with a capacitor, when a voltage is applied across such a junction charges will accumulate on the two sides of the junction. The system will then behave very differently. A description of the charge has to be introduced in the theory. We shall return in §8.4 to relate this approach to the theory presented in this chapter.

References

1. Arndt, M., Nairz, O.,Vos-Andreae, J., Keller, C., van der Zouw, G. and Zeilinger, A. (1999). *Nature* **401** 680.
2. Backhaus, S., Pereversev, S., Simmonds, R. W., Loshak, A., Davies, J. C. and Packard., R. E. (1998). *Nature* **392** 687.
3. Bardeen, J., Cooper, L. N. and Schreiffer, J. R. (1957). *Phys. Rev.* **108** 1175.
4. Barone, A. and Paternò, G. (1982). *Physics and Applications of the Josephson Effect*, Wiley, New York.
5. Baym, G. (1969). *Lectures on Quantum Mechanics*, Benjamin, New York.
6. Beltrametti, E. G. and Cassinelli, G. (1981). *The Logic of Quantum Mechanics*, Addison-Wesley, Reading, Mass.
7. Blank, J., Exner, P. and Havlíček, M. (1994). *Hilbert Space Operators in Quantum Physics*, American Institute of Physics Press, New York.
8. Bleaney, B. I. and Bleaney, B. (1957). *Electricity and Magnetism*, Clarendon Press, Oxford.
9. Bogoliubov, N. N. (1970). *Lectures on Quantum Statistics Vol. 2*, MacDonald Technical and Scientific, London. Part 2.
10. Bogoliubov, N. N. and Bogoliubov, N. N., Jr. (1992). *An Introduction to Quantum Statistical Mechanics*, Gordon and Breach, Lausanne, Switzerland.
11. Bonse, U. and Rauch, H. (1979). *Neutron Interferometry*, Clarendon Press, Oxford.
12. Clark, T. D. (1987). 'Macroscopic Quantum Objects' in B. J. Hiley and E. D. Peat, eds., *Quantum Implications, Essays in honour of David Bohm*, Routledge & Kegan, London.

---

[104] Leggett (1980), (1987), Srivastava and Widom (1987), Clark (1987) pp. 121-150, Clark, Prance, Prance and Spiller (1990), Spiller, Clark, Prance and Widom (1992), Wan and Fountain (1996) Appendix D.

13. Clark, T. D., Prance, H., Prance, R. J. and Spiller, T. P. eds. (1991). *Macroscopic Quantum Phenomena*, World Scientific, Singapore.
14. Cobden, D. H. (1999). *Nature* **397** 648.
15. Davis, J. C. and Packard, R. E. (2002). *Rev. Mod. Phys.* **74** 741.
16. Deaver, B., Jr. and Fairbank, W. (1961). *Phys. Rev. Lett.* **7** 43.
17. de Bruyn Ouboter, R. (1977). 'Macroscopic Quantum Phenomena in Superconductors' in B. B. Schwartz and S. Foner, eds., *Superconductor Applications: SQUIDs and Machines*, NATO ASI Series B Vol. 21, Plenum, New York.
18. de Gennes, P. G. (1963). *Phys. Lett.* **5**, 22.
19. de Gennes, P. G. (1964). *Rev. Mod. Phys.* **36** 225.
20. Doll, R. and Nabauer, M. (1961). *Phys. Rev. Lett.* **7** 51.
21. Exner, P. and Seba, P. (1987). *J. Math. Phys.* **28** 386.
22. Exner, P. and Seba, P. (1989). *Rep. Math. Phys.* **28** 7.
23. Exner, P. and Seba, P. (1989). 'Quantum Junctions and the Self-Adjoint Extensions Theory', also P. Exner, P. Seba and P. Stovicek, 'Quantum Waveguides', both in P. Exner and P. Seba, eds., *Applications of Self-Adjoint Extensions in Quantum Physics*, Springer-Verlag, Berlin.
24. Feynman, R. P. (1972). *Statistical Mechanics*, Benjamin, Reading, Mass.
25. Feynman, R. P., Leighton, R. B. and Sands, M. (1965). *The Feynman Lectures on Physics Vol. 3*, Addison-Wesley, New York.
26. Fujita, S. and Godoy, S. (1996). *Quantum Statistical Theory of Superconductivity*, Plenum, New York.
27. Gallop, J. C. (1991). *SQUIDS, the Josephson Effects and Superconducting Electronics*, Adam Hilger, Bristol.
28. Greenstein, G. and Zajonc, A. G. (1997). *The Quantum Challenge*, Jones and Bartlett Publishers, Sudbury, Mass.
29. Harrison, F. E. and Wan, K. K. (1997). *J. Phys. A: Math. Gen.* **30** 4731.
30. Harrison, F. E. (1993). *Superselection rules, Quantum Measurement Problems and Macroscopic Quantum Systems*, St Andrews University PhD Thesis.
31. Hegstrom, R. A. and Sols, F. (1995). *Found. Phys.* **25** 681.
32. Hellmuth, T., Zajonc, A. G. and Walther, H. (1985). 'Realization of a "Delayed-Choice" Mach-Zehnder Interferometer' in P. Lahti and P. Mittelstaedt, eds., *Symposium in the Foundations of Modern Physics*, World Scientific, Singapore.
33. Herzog, T. J., Kwiat, P. G., Weinfurter, H. and Zeilinger, A. (1995). *Phys. Rev. Lett.* **75** 3034.
34. Josephson, B. D. (1962). *Phys. Lett.* **1** 251.
35. Kuper, C. G. (1968). *An Introduction to the Theory of Superconductivity*, Clarendon Press, Oxford.
36. Leggett, A. J. (1980). *Prog. Theor. Phys.* Supplement **69** 80.
37. Leggett, A. J. (1987). 'Quantum Mechanics at the Macroscopic Level' in J. Souletie and J. Vannimenus, eds., *Chance and Matter*, Elsevier, Amsterdam.
38. London, F. (1961). *Superfluids Vol. 1 and Vol. 2*, Dover, New York.

REFERENCES

39. Mandl, F. (1992). *Quantum Mechanics*, Wiley, Chichester.
40. Merzbacher, E. (1998). *Quantum Mechanics*, 3rd edition, Wiley, New York.
41. Mintmire, J. W., Dunlap, B. I. and White, C. T. (1992). *Phys. Rev. Lett.* **68** 631.
42. Rarity, J. (1996). *Physics World* **9** February 19.
43. Razavy, M. (1997). *Phys. Rev. A* **55** 4102.
44. Rose-Innes, A. C. and Rhoderick, E. H. (1980). *Introduction to Superconductivity*, Pergamon Press, London.
45. Schwab, K. C. and Roukes, M. L. (2005). *Physics Today* **58** 36.
46. Selleri, F. and Tarozzi, G. (1981). *La Rivita del Nuovo Cimento* **4** 1.
47. Sewell, G. L. (2002). *Quantum Mechanics and its Emergent Macrophysics*, Princeton University Press, Princeton, N. J.
48. Spiller, T. P., Clark, T. D., Prance, R. J. and Widom, A. (1992). 'Quantum Phenomena in Circuits at Low Temperature' in E. D. Brewer, ed., *Progress in Low Temperature Physics Vol. XIII*, North-Holland, Amsterdam.
49. Srivastava Y. and Widom, A. (1987). *Physics Reports* **148** 1-65.
50. Tilley, D. R. and Tilley, J. (1990). *Superfluidity and Superconductivity*, Adam Hilger, Bristol.
51. Tonomura, A., Endo, J., Matsuda, T. and Kawasaki, T. (1989). *Am. J. Phys.* **57** 117.
52. Trueman, C. and Wan, K. K. (2000). *J. Math. Phys.* **41** 1.
53. Wan, K. K. (1988). *Found. Phys.* **18** 887.
54. Wan, K. K., Bradshaw, J., Trueman, C. and Harrison, F. E. (1998). *Found. Phys.* **28** 1739.
55. Wan, K. K. and Harrison, F. E. (1993). *Phys. Lett. A* **174** 1.
56. Wan, K.K., Fountain, R. H. and Tao, Z. Y. (1995). *J. Phys. A: Math. Gen.* **28** 2379.
57. Wan, K.K. and Fountain, R. H. (1996). *Found. Phys.* **26** 1165.
58. Wan, K.K. and Fountain, R. H. (1998). *Int. J. Theor. Phys.* **37** 2153.
59. Wheeler, J. A. (1983). 'Law without Law' in J. A. Wheeler and W. H. Zurek, eds. *Quantum Theory of Measurement*, Princeton University Press, Princeton, N. J.
60. Widom, A. (1979). *J. Low. Temp. Phys.* **37** 449.
61. Widom, A. (1991). 'Implications of Superconducting Circuits for Relativistic Quantum Electrodynamics' in T. D. Clark, H. Prance, R. J. Prance and T. P. Spiller, eds., *Macroscopic Quantum Phenomena*, World Scientific, Singapore.
62. Williams, J. E. C., (1970). *Superconductivity and its Applications*, Pion Limited, London.
63. Wollman, D. A. et al., (1993). *Phys. Rev. Lett.* **71** 2134.

Part IV
Asymptotic Disjointness, Asymptotic Separability, Quantum Mechanics on Path Space and Superselection Rules

Chapter 8

Separability and Decoherence

8.1 Introduction

A theory formulated in precise mathematical terms is often an idealization as far as physics is concerned. In classical physics the very notion of a solid object, e.g., a billiard ball, of a precise mass m is an idealization since a billiard ball travelling in a vacuum is constantly losing atoms on its surface, and hence its mass. A thermodynamic system as an infinite system is an idealization of a physically finite system. We can often construct a relatively simple mathematical model theory as an idealization to approximate an actual physical system. But there is a fundamental difference between classical and quantum mechanical idealizations.

Consider two model systems in one spatial dimension $I\!E$ in classical mechanics characterized by their Hamiltonians H_{sho} and H_{spr} given below:

$$H_{sho} = \frac{1}{2m}p^2 + \frac{1}{2}kx^2, \qquad H_{spr} = \frac{1}{2m}p^2 + V(x), \qquad (8.1)$$

where $V(x)$ is a smooth function on $I\!E$ which is equal to $\frac{1}{2}kx^2$ for $|x| \leq x_p$ and flattens out to a certain finite value as $x \to \pm\infty$. Here H_{sho} describes an ideal harmonic oscillator. Mechanically a harmonic oscillator is realizable in the form of a particle of mass m attached to an elastic spring. For any physical spring the harmonic potential will fail beyond a certain point, i.e., there is a distance, known as the *proportionality limit*, beyond which the potential begins to deviate from the harmonic form. The Hamiltonian H_{spr} is a more realistic description of a physical oscillator, with x_p identifiable with the proportionality limit of the spring. As the spring is further stretched there comes a point, known as the *elastic limit* x_e, beyond which the restoring force of the spring fails completely and the particle will no longer return to the origin.[1] Classical mechanics is a *local theory* in the sense that the motion of the particle is influenced by the potential within an immediate neighbourhood of the particle. For example, the potential outside the proportionality limit x_p has no effect on the motion of the particle within x_p. With a suitable choice of initial conditions it is possible to confine the motion entirely within the proportionality limit for all times. With such initial conditions the motion under H_{spr} is identical to the motion under H_{sho}. It follows that although the actual Hamiltonian of an oscillator should be of the form of H_{spr} we can employ H_{sho} as the Hamiltonian for motion within the proportionality limit without incurring any approximation. It is in this sense that a simple harmonic oscillator, in the form of a particle attached to a spring, oscillating with a certain finite amplitude less than the proportionality limit of the spring can be described by H_{sho}.

The situation is fundamentally different in quantum theory. The Hamiltonians H_{sho} and H_{spr} become operators $\widehat{H}_{sho}$ and $\widehat{H}_{spr}$ acting in $L^2(I\!\!E)$. These are *global operators* whose properties are affected by the potential over the entire space $I\!\!E$. Any tailing off the potential $V(x)$ even at large distances away will still have an effect on the properties of $\widehat{H}_{spr}$. Generally $\widehat{H}_{sho}$ and $\widehat{H}_{spr}$ would possess different energy eigenvalues and eigenfunctions which would show that they correspond to two distinct quantum systems. A harmonic potential is often used to describe the vibrational potential of a diatomic molecule. For such a molecule the harmonic potential will eventually break down as the distance between the two atoms increases. Indeed as the distance keeps increasing the molecule will eventually break up resulting in the two atoms separating permanently and resuming their respective identities, analogous with a classical oscillator moving beyond the elastic limit of the spring. Fortunately, $\widehat{H}_{sho}$ produces results which are often a good enough approximation to that of $\widehat{H}_{spr}$ for low lying eigenstates to enable us to employ $\widehat{H}_{sho}$ to describe the vibration of diatomic molecules.[2] One can see that even very simple and apparently mathematically exact models like harmonic oscillators in quantum theory are inherently approximate in nature. Moreover, one cannot always treat physical conditions at a long distance away as

---

[1] A harmonic potential is often an approximation used for small oscillatory motion about an equilibrium point of a more general potential on account of the Taylor expansion of potential about the equilibrium point (Merzbacher (1998) p. 79).

[2] Although the eigenvalues and eigenfunctions of $\widehat{H}_{sho}$ and $\widehat{H}_{spr}$ do not coincide, some low lying eigenvalues and their corresponding eigenfunctions for $\widehat{H}_{sho}$ and $\widehat{H}_{spr}$ may be very close to each other. The motion of these eigenstates can approximate each other, but are not identical as in the classical case.

8.1. INTRODUCTION

a small perturbation. Take the examples of point interactions discussed in §6. The behaviour of the particle over the entire space, including the existence or otherwise of energy eigenstates, is affected by the physical conditions at a single point. The behaviour of an entire supercurrent flowing in a long wire is influenced by the presence of a single junction, i.e., the Josephson junction, whose effect cannot be eliminated by making the wire longer.

The global nature of quantum systems can also be seen when we consider two spatially separated states. Let $\phi_1(x)$ and $\phi_2(x)$ be two wave functions localized in separated and disjoint regions in space. These two wave functions may well obey the superposition principle stated in Definition 3.3.2(1). In other words $\phi_1(x)$ and $\phi_2(x)$ may be coherent, i.e., there may be the possibility of observing the correlation or interference between these two wave functions. Mathematically this means the existence of physical observables, represented by selfadjoint operators $\widehat{A}$, such that $\langle \phi_1 \mid \widehat{A} \phi_2 \rangle \neq 0$, despite the spatial separation of the two wave functions. An obvious question to ask is whether the correlations between two states would decrease as their spatial separation increases. Theorem 3.5.2(1) suggests the answer is yes. We shall follow this with the introduction of a notion of *equivalence FAPP* and *disjointness FAPP* to highlight a process in which the coherence of quantum states is progressively obliterated. Explicit examples will be presented in the next section.[3] Again approximation plays an important part here.

As discussed in §3.2.5, we can only measure a finite set $\mathcal{O}_{qm}^n$ of independent bounded observables with a certain upper bound of inaccuracy ε_{inc} in the sense of Eq. (3.58) in any experiment for the system in a limited or restricted set $\mathcal{S}_{qm}^{res}$ of states. As before let us denote such an experiment by $\boldsymbol{E}_{xp}(\mathcal{O}_{qm}^n, \mathcal{S}_{qm}^{res}, \varepsilon_{inc})$. There are two special cases of interest:

1. We want to perform an experiment to measure a finite set $\mathcal{O}_{qm}^n$ of observables, to an upper bound of inaccuracy ε_{inc}, in order to try to distinguish the two given states. We shall denote such an experiment by $\boldsymbol{E}_{xp}(\mathcal{O}_{qm}^n, \varepsilon_{inc})$.

2. We want to perform an experiment with respect to a restricted set $\mathcal{S}_{qm}^{res}$ of states, to an upper bound of inaccuracy ε_{inc}, in order to try to distinguish to two observables. We shall denote such an experiment by $\boldsymbol{E}_{xp}(\mathcal{S}_{qm}^{res}, \varepsilon_{inc})$.

For convenience we shall summarize the analysis based on a notion of

equivalence for all practical purposes   or   *equivalence FAPP*

described first in §3.2.5 into a couple of definitions.

---
[3] We are talking about states and observables in $L^2(I\!\!E^n)$. For spin, see Wan and Jackson (1985) and also §8.6.2 later.

Definition 8.1(1)   On equivalence FAPP of observables

- Two observables $\widehat{A}_1$ and $\widehat{A}_2$ are said to be equivalent FAPP with respect to a set of states, $\mathcal{S}_{qm}^{res}$, to an inaccuracy ε_{inc}, if a small ε_{inc} defined by

$$\varepsilon_{inc} = \text{supremum} \left\{ \left| \mathcal{E}(\widehat{A}_1, \phi) - \mathcal{E}(\widehat{A}_2, \phi) \right| \; \forall \phi \in \mathcal{S}_{qm}^{res} \right\} \tag{8.2}$$

exists.[4] We write[5]

$$\widehat{A}_1 \equiv \widehat{A}_2 \;\; \text{FAPP}(\mathcal{S}_{qm}^{res}, \varepsilon_{inc}). \tag{8.3}$$

The two Hamiltonians $\widehat{H}_{she}$ and $\widehat{H}_{spr}$ in Eq. (8.1) are equivalent FAPP for an appropriate set $\mathcal{S}_{qm}^{res}$ of low lying eigenstates and their linear combinations. Other examples are approximate and related family of observables introduced in §5.3.1 and §5.3.2.

Definition 8.1(2)   On equivalence and disjointness FAPP of states

- Two pure states ϕ_1 and ϕ_2 are said to be equivalent FAPP with respect to a set of observables, $\mathcal{O}_{qm}^n$, to an upper bound of inaccuracy ε_{inc}, if a small ε_{inc} defined by[6]

$$\varepsilon_{inc} = \text{supremum} \left\{ \left| \mathcal{E}(\widehat{A}, \phi_1) - \mathcal{E}(\widehat{A}, \phi_2) \right| \; \forall \widehat{A} \in \mathcal{O}_{qm}^n \right\} \tag{8.4}$$

exists. We write

$$\phi_1 \equiv \phi_2 \;\; \text{FAPP}(\mathcal{O}_{qm}^n, \varepsilon_{inc}). \tag{8.5}$$

- Two pure states ϕ_1 and ϕ_2 are said to be *disjoint* FAPP with respect to a set of observables, $\mathcal{O}_{qm}^n$, to an upper bound of inaccuracy ε_{inc}, if a small ε_{inc} defined by

$$\varepsilon_{inc} = \text{supremum} \left\{ \left| \langle \phi_1 \mid \widehat{A} \, \phi_2 \rangle \right| \; \forall \widehat{A} \in \mathcal{O}_{qm}^n \right\} \tag{8.6}$$

exists.

---

[4] In practice ε_{inc} would be assumed to be small in some definite sense, e.g., small compared with certain known quantities. The smallness of ε_{inc} is often fixed by experimental inaccuracy.

[5] Often the relevant quantities involved, i.e., the set of states $\mathcal{S}_{qm}^{res}$, are obvious so that we would simply write $\widehat{A}_1 \equiv \widehat{A}_2$ FAPP.

[6] $\mathcal{O}_{qm}^n$ is a finite set. If $\widehat{A}$ is in $\mathcal{O}_{qm}^n$ it does not mean that $a\widehat{A}$ is in $\mathcal{O}_{qm}^n$ for any arbitrarily large number a, or we would have an arbitrarily large ε_{inc} due to an arbitrarily large a.

8.1. INTRODUCTION

- Let c_1, c_2 be two complex numbers such that $|c_1|^2 + |c_2|^2 = 1$, and let

$$\phi = c_1\,\phi_1 + c_2\,\phi_2, \tag{8.7}$$

and

$$\widehat{D} = |c_1|^2\,\widehat{P}_{\phi_1} + |c_2|^2\,\widehat{P}_{\phi_2}, \tag{8.8}$$

where

$$\widehat{P}_{\phi_1} = |\phi_1\rangle\langle\phi_1|, \;\; \widehat{P}_{\phi_2} = |\phi_2\rangle\langle\phi_2|. \tag{8.9}$$

Then[7] the two states represented respectively by ϕ and $\widehat{D}$ are said to be equivalent FAPP with respect to a set of observables, $\mathcal{O}_{qm}^n$, to an upper bound of inaccuracy ε_{inc}, if a small ε_{inc} defined by

$$\varepsilon_{inc} = \text{supremum}\left\{\, |\mathcal{E}(\widehat{A}, \phi) - \mathcal{E}(\widehat{A}, \widehat{D})|\, \big|\, \forall \widehat{A} \in \mathcal{O}_{qm}^n \right\} \tag{8.10}$$

exists. We write[8]

$$\phi \equiv \widehat{D} \;\; \text{FAPP}(\mathcal{O}_{qm}^n, \varepsilon_{inc}). \tag{8.11}$$

From Eq. (3.41) we can see that the disjointness of two states, e.g., ϕ_1 and ϕ_2 in Eq. (8.6), means a lack of correlation between the two states. Two pure states being disjoint is clearly related to their linear combinations being equivalent to mixtures.[9] It should be emphasized that generally whether a given linear combination of pure states has the significance of a coherent superposition or not depends on the set of operators which are admissible as observables. We have the following three cases:

1. For an orthodox quantum system Postulate OQS in §3.2.1 operates. Two pure states ϕ_1 and ϕ_2 cannot be disjoint with respect to **every** finite set $\mathcal{O}_{qm}^n$ of observables.

2. For a mixed quantum system Postulate UQS in §4.2.1 operates. The system possesses superselection rules. Pure states belonging to different supersectors are disjoint.

3. In practice the situation is less clear cut. It is possible within orthodox quantum mechanics to have a linear combination of pure states identifiable FAPP with a mixture, leading to a breakdown of the superposition

---
[7] Isham (1995) p. 151.
[8] We would simply write $\phi \equiv \widehat{D}$ FAPP when the observables $\mathcal{O}_{qm}^n$ involved are obvious.
[9] The inaccuracies involved in various definitions need not be the same.

principle. This is related to the disjointness of the constituent states since we can substitute Eq. (8.6) into Eq. (8.10) to get[10]

$$\left| \mathcal{E}(\widehat{A}, \phi) - \mathcal{E}(\widehat{A}, \widehat{D}) \right| \leq 2 \left| c_1^* c_2 \right| \left| \langle \phi_1 \mid \widehat{A} \phi_2 \rangle \right| \leq 2 \left| c_1^* c_2 \right| \varepsilon_{inc} \quad (8.12)$$

for every $\widehat{A} \in \mathcal{O}_{qm}^n$. Our present discussions can be compared with the concept of selection rules in orthodox quantum mechanics.[11]

In what follows we shall examine a number of examples within orthodox quantum mechanics to illustrate case 3 above.

8.2 Scattering Systems and de Broglie Paradox

8.2.1 Scattering systems

From §3.4.3 we know that a simple scattering system possesses bound states and scattering states. Let us denote a given bound state at time $t = 0$ by ϕ_b and similarly a scattering state at $t = 0$ by ϕ_s. At time $t > 0$ these states will evolve under a propagator $\widehat{\mathcal{U}} = \{\widehat{U}_t : t \in (-\infty, \infty)\}$ to states $\widehat{U}_t \phi_b$ and $\widehat{U}_t \phi_s$ respectively. While the bound state wave function $\widehat{U}_t \phi_b$ will remain around a bounded region at all times, the scattering state wave function $\widehat{U}_t \phi_s$ will move to spatial infinity. As a result the overlap between $\widehat{U}_t \phi_b$ and $\widehat{U}_t \phi_s$ over any finite spatial region tends to zero as $t \to \infty$. Let $\mathcal{B}(\mathcal{H})$ denote the set of all bounded operators on the Hilbert space $\mathcal{H}$ associated with the system. Theorem 3.5.2(1) tells that

$$\lim_{t \to \infty} \langle \widehat{U}_t \phi_b \mid \widehat{A} \widehat{U}_t \phi_s \rangle = 0 \quad \forall \widehat{A} \in \mathcal{B}(\mathcal{H}). \quad (8.13)$$

So, for any finite set $\mathcal{O}_{qm}^n$ of bounded observables $\widehat{A}_j, j = 1, \ldots, n$, the following correlation term between $\widehat{U}_t \phi_b$ and $\widehat{U}_t \phi_s$

$$\langle \widehat{U}_t \phi_b \mid \widehat{A}_j \widehat{U}_t \phi_s \rangle \quad (8.14)$$

become arbitrarily small as $t \to \infty$. We conclude that for sufficiently large times $\widehat{U}_t \phi_b$ and $\widehat{U}_t \phi_s$ are disjoint FAPP. It follows that at large times a linear combination

$$\phi_t = c_1 \widehat{U}_t \phi_b + c_2 \widehat{U}_t \phi_s \quad (8.15)$$

is equivalent FAPP to the mixture

$$\widehat{D}_t = |c_1|^2 \, |\widehat{U}_t \phi_b\rangle\langle\widehat{U}_t \phi_b| + |c_2|^2 \, |\widehat{U}_t \phi_s\rangle\langle\widehat{U}_t \phi_s|. \quad (8.16)$$

---

[10] A reference to Eqs. (3.39), (3.40), (3.41) and (3.42) would be help to understand the argument here.
[11] Phillips (2003).

If we trace the time from $-\infty$ to ∞ we will see a change of the relation between $\widehat{U}_t \phi_s$ and $\widehat{U}_t \phi_b$ from being disjoint at large negative times, through to a coherent regime for small times, and finally back to being disjoint at large positive times.

8.2.2 de Broglie paradox

The following thought experiment in quantum mechanics is known as the de Broglie paradox.[12] Imagine an experimental apparatus in a laboratory in Paris consisting of a box B with rigid walls which can be divided into two halves B_1 and B_2 by a sliding wall in the middle of B. Suppose that there is an electron inside box B and that the state of the electron is represented by a linear combination of two wave functions ϕ_1 and ϕ_2 with ϕ_1 localized in B_1 and ϕ_2 localized B_2. The box is then divided into B_1 and B_2 by inserting the sliding wall, possibly trapping ϕ_1 in B_1 and ϕ_2 in B_2. Now, move B_2 to Tokyo while keeping B_1 in Paris. Suppose one then opens box B_1 in Paris and finds the electron inside. The question is whether the electron was inside B_1 all along before one opened the box and found the electron inside. If this were the case, then the wave function should be identically zero in B_2 all along so as to give a zero probability of the electron in B_2, i.e., the process of dividing box B would also collapse the entire electron wave function into B_1.[13] Suppose this were not the case, i.e., when box B was divided the electron wave function did not collapse entirely into B_1. In other words the electron wave function had a component in B_1 and a component in B_2 all along before B_1 was opened. Given that the two boxes B_1 and B_2 are so far apart the question is how the electron wave function which included a component in B_2 could possibly collapse into B_1 by the act of opening box B_1. In order to carry out a quantitative analysis of such a thought experiment we have to set up suitable mathematical models.

A model in terms of translation operators

Consider two wave functions $\phi_1(x)$ and $\phi_2(x)$ in $L^2(I\!E)$ localized near the coordinate origin. Let us move $\phi_2(x)$ to the right by a distance a using the translation operator $\widehat{T}_a$ introduced in C2.4(2) C2, i.e.,

$$\left(\widehat{T}_a \phi_2\right)(x) = \phi_2(x-a), \qquad \widehat{T}_a = e^{-ia\hat{p}}. \qquad (8.17)$$

Then the overlap between $\phi_1(x)$ and $\widehat{T}_a \phi_2(x)$ becomes vanishingly small as

---

[12] de Broglie (1959). Selleri and Tarozzi (1981).
[13] The discussion in §8.3.9 tells us that a rigid sliding wall acting as an infinite potential barrier of finite and non-zero width can render the state into a classical mixture of ϕ_1 and ϕ_2. A finite potential barrier which is infinitely wide also has the same effect.

a tends to infinity. We would expect

$$\lim_{a \to \infty} \langle \phi_1 \mid \widehat{T}_a \phi_2 \rangle = 0. \tag{8.18}$$

This indeed is the case, i.e., let $\xi(x)$ and $\eta(x)$ be any two members of $L^2(I\!\!E)$, then

$$\lim_{a \to \infty} \langle \xi \mid \widehat{T}_a \eta \rangle = 0. \tag{8.19}$$

To prove this we first recall the Fourier transform operator $\widehat{U}_F$ in $L^2(I\!\!E)$ introduced in E2.4(1) E1. In view of the unitary nature of the Fourier transform operator and Eq. (2.132) we have

$$\begin{aligned}\langle \xi \mid \widehat{T}_a \eta \rangle &= \langle \widehat{U}_F \xi \mid \widehat{U}_F \widehat{T}_a \eta \rangle = \left\langle \widehat{U}_F \xi \mid \left(\widehat{U}_F \widehat{T}_a \widehat{U}_F^\dagger \right) \widehat{U}_F \eta \right\rangle \\ &= \int_{-\infty}^{\infty} \widetilde{\xi}^*(p) \, e^{-ipa} \, \widetilde{\eta}(p) \, dp \\ &\to 0 \quad \text{as } a \to \infty.\end{aligned}$$

The last step is due to the Riemann-Lebesgue lemma.[14] This result also implies that for any bounded selfadjoint operator $\widehat{A}$ we have

$$\lim_{a \to \infty} \langle \xi \mid \widehat{A} \widehat{T}_a \eta \rangle = \lim_{a \to \infty} \langle \varphi \mid \widehat{T}_a \eta \rangle = 0, \tag{8.20}$$

where $\varphi = \widehat{A}\xi$ is independent of a.

Now, suppose we have a particle whose state is initially described by a linear combination of $\phi_1(x)$ and $\phi_2(x)$, and suppose after the translation the linear combination becomes

$$\Phi_a = c_1 \phi_1 + c_2 \widehat{T}_a \phi_2. \tag{8.21}$$

We then have a de Broglie paradox situation since ϕ_1 and $\widehat{T}_a \phi_2$ are far apart. But the paradox can be resolved. Equation (8.20) tells us that:

1. ϕ_1 and $\widehat{T}_a \phi_2$ becomes disjoint as $a \to \infty$.

2. A linear combination of ϕ_1 and $\widehat{T}_a \phi_2$ is equivalent FAPP to a classical mixture of ϕ_1 and $\widehat{T}_a \phi_2$.

Consequently when we come to ask where the particle is, the answer is either somewhere round the origin or somewhere round the support of the function $\widehat{T}_a \phi_2$ some distance away. There is no question of Φ_a, regarded as a coherent superposition, collapsing instantaneously into ϕ_1 or $\widehat{T}_a \phi_2$ on a position measurement.

---

[14] Papoulis (1962) p. 278.

A model in terms of scattering systems

The above mathematical description is rather artificial and difficult to realize in practice. However, we can model the essence of de Broglie's thought experiment realistically in terms of a scattering system. Suppose the state of the system is initially a coherent superposition of a bound state and a scattering state, i.e., we have

$$\phi = \frac{1}{\sqrt{2}}\left(\phi_b + \phi_s\right). \tag{8.22}$$

Let

$$\widehat{U}_t\phi = \frac{1}{\sqrt{2}}\left(\widehat{U}_t\phi_b + \widehat{U}_t\phi_s\right) \tag{8.23}$$

and

$$\widehat{D}_t = \frac{1}{2}\left(|\widehat{U}_t\phi_b\rangle\langle\widehat{U}_t\phi_b| + |\widehat{U}_t\phi_s\rangle\langle\widehat{U}_t\phi_s|\right). \tag{8.24}$$

According to the discussions in §8.2.1, we have at large times the following equivalence FAPP:

$$\widehat{U}_t\phi \equiv \widehat{D}_t \quad \text{FAPP}. \tag{8.25}$$

To relate to de Broglie's thought experiment we can imagine the initial state ϕ to be enclosed in a large box B in a laboratory in Paris. As time evolves we divide the box into a part in Paris where the bound state $\widehat{U}_t\phi_b$ remains and another part which follows the scattering state to a far away region in Tokyo. One then asks the question where the particle is at large times, given that $\widehat{U}_t\phi_b$ remains in Paris and $\widehat{U}_t\phi_s$ lies in a region far away in Tokyo. At sufficiently large times the state $\widehat{U}_t\phi$ is equivalent FAPP to a classical mixture $\widehat{D}_t$ of $\widehat{U}_t\phi_b$ and $\widehat{U}_t\phi_s$. This result provides an answer, i.e., the particle is either in Paris where $\widehat{U}_t\phi_b$ remains or in Tokyo where $\widehat{U}_t\phi_s$ lies. In the limit as $t \to \infty$, no experiment can detect any correlation between $\widehat{U}_t\phi_b$ and $\widehat{U}_t\phi_s$. This is similar to the ionization process in the quantum measurement model discussed in §3.6.

8.3 Schrödinger's Cat States

8.3.1 Classical-like states

In 1935 Schrödinger published a paper which introduced the Schrödinger's cat paradox.[15] This paradox is to do with coherent superpositions of *classically distinguishable states*, or *classical-like states* or *macroscopically distinguishable states* of a quantum system, often referred to as Schrödinger's cat states. Here

---

[15] Schrödinger (1935). Jauch (1968) p. 185. There are quite a number of qualitative books on this subject, e.g., Gribbin (1984), Rae (1986).

classical or macroscopic distinguishability of states and classical-likeness of states are recognizable on an intuitive and individual basis, rather than on precisely defined criteria.[16] The object of this section is to provide a study of the concept of various "cats", their states, and different notions of disjointness. The harmonic oscillator is often used in the discussion of Schrödinger's cat states. So, we shall provide some relevant material on the harmonic oscillator here.

The motion of a classical harmonic oscillator of mass m and angular frequency ω is given by the time dependence of its position x_c and momentum p_c according to:

$$x_c(t) = \alpha \cos(\omega t + \theta), \qquad p_c(t) = -m\omega\alpha \sin(\omega t + \theta), \qquad (8.26)$$

where θ is the initial phase and α the amplitude of oscillation. Here the subscript c signifies classical quantities. For the corresponding quantized oscillator the Hamiltonian is

$$\widehat{H}_{sho} = \frac{1}{2m}\,\widehat{p}^2 + \frac{1}{2}\,m\omega^2\,\widehat{x}^2, \quad \widehat{p} = -i\hbar d/dx, \quad \widehat{x} = x. \qquad (8.27)$$

The following normalized wave packet is a well-known solution of the time-dependent Schrödinger equation for the quantized oscillator:[17]

$$\phi_{zt}^{(c)}(x) = c(t) \left(\frac{m\omega}{\pi\hbar}\right)^{\frac{1}{4}} \exp\left[\tfrac{i}{\hbar}p_c(t)x - \frac{m\omega}{2\hbar}\left(x - x_c(t)\right)^2\right], \qquad (8.28)$$

where $c(t)$ satisfies

$$i\hbar\,\frac{\dot{c}(t)}{c(t)} = \frac{1}{2m}\,p_c^2(t) - \frac{1}{2}\,m\omega^2 x_c^2(t) + \frac{1}{2}\,\hbar\omega, \qquad (8.29)$$

or more explicitly

$$c(t) = \exp\left[\tfrac{i}{\hbar}\left(\frac{1}{4}m\alpha^2\omega \sin 2(\omega t + \theta) - \frac{1}{2}\hbar\omega t\right)\right]. \qquad (8.30)$$

A subscript z is attached to the function because $\phi_{zt}^{(c)}$ is an eigenfunction of the annihilation operator

$$\widehat{a} = \frac{1}{\sqrt{2m\hbar\omega}}\left(m\omega\widehat{x} + i\widehat{p}\right) \qquad (8.31)$$

---

[16] Davisovish et al. (1996) for various experimental schemes on quantum optical systems.
[17] Yurke and Stoler (1986), Howard and Roy (1987), Monroe et at. (1996), Gerry and Knight (1991), de Lange and Raab (1991) p. 50.

8.3. SCHRÖDINGER'S CAT STATES

associated with the quantized oscillator corresponding to the time-dependent[18] complex eigenvalue z given by

$$z = \left(\frac{m\omega}{2\hbar}\right)^{\frac{1}{2}} \left(x_c(t) + \frac{i}{m\omega} p_c(t)\right). \quad (8.32)$$

Such a wave packet is known as a *coherent state*, hence the superscript (c).[19] A coherent state may be considered *classical-like* in the following sense:

1. The corresponding position probability density function is

$$|\phi_{zt}^{(c)}(x)|^2 = \left(\frac{m\omega}{\pi\hbar}\right)^{1/2} \exp\left[-\frac{m\omega}{\hbar}(x - x_c(t))^2\right]. \quad (8.33)$$

   This represents a wave packet oscillating about the origin without distortion like a classical oscillator.

2. The position and momentum expectation values oscillate like classical values:

$$\langle \phi_{zt}^{(c)} | \hat{x} \phi_{zt}^{(c)} \rangle = x_c(t), \quad \langle \phi_{zt}^{(c)} | \hat{p} \phi_{zt}^{(c)} \rangle = p_c(t). \quad (8.34)$$

   The amplitude of oscillation can be made arbitrarily large by increasing α.

3. For large oscillation amplitudes the energy expectation value tends to the classical value:[20]

$$\langle \phi_{zt}^{(c)} | \hat{H} \phi_{zt}^{(c)} \rangle \approx \frac{1}{2} k\alpha^2, \quad k = m\omega^2. \quad (8.35)$$

   Here $\frac{1}{2} k\alpha^2$ is the total energy of a classical oscillator vibrating with amplitude α.

These coherent states are not mutually orthogonal since[21]

$$|\langle \phi_{zt}^{(c)} | \phi_{z't}^{(c)} \rangle| = e^{-\frac{1}{2}|z-z'|^2}. \quad (8.36)$$

When $z' = -z$ we have, writing $\phi_{z't}^{(c)}$ as $\phi_{-zt}^{(c)}$,

$$\langle \phi_{zt}^{(c)} | \phi_{-zt}^{(c)} \rangle = e^{-2|z|^2} = e^{-(m\omega/\hbar)\alpha^2} \to 0 \quad \text{as } |z|, \alpha \to \infty. \quad (8.37)$$

---

[18] The magnitude of z is time independent, i.e.,

$$|z|^2 = \left(\frac{m\omega}{2\hbar}\right)\alpha^2.$$

[19] Klauder and Skagarstam (1985), de Lange and Raab (1991), Merzbacher (1998).
[20] Schiff (1955) §13.
[21] de Lange and Raab (1991).

Let $\widehat{\wp}$ be the parity operator defined in the usual way by $\widehat{\wp}\phi(x) = \phi(-x)$. Then we have,

$$\widehat{\wp}\phi^{(c)}_{zt} = \phi^{(c)}_{-zt}, \qquad \widehat{\wp}\phi^{(c)}_{-zt} = \phi^{(c)}_{zt}. \tag{8.38}$$

Clearly $\phi^{(c)}_{zt}$ and $\phi^{(c)}_{-zt}$ are mirror images of each other so that

$$\langle\phi^{(c)}_{zt} \mid \widehat{\wp}\phi^{(c)}_{-zt}\rangle = \langle\phi^{(c)}_{zt} \mid \phi^{(c)}_{zt}\rangle = 1. \tag{8.39}$$

Consequently

$$\langle\phi^{(c)}_{zt} \mid \widehat{A}\,\phi^{(c)}_{-zt}\rangle \neq 0, \quad \text{even in the limit as } |z| \to \infty \tag{8.40}$$

for some observables represented by selfadjoint operators $\widehat{A}$, parity $\widehat{\wp}$ being an obvious example. It follows that $\phi^{(c)}_{zt}$ and $\phi^{(c)}_{-zt}$ can form a coherent superposition with non-vanishing correlation even for large $|z|$, i.e., for large α.

It should be pointed out that being classical-like and being disjoint are two different concepts, and one does not follow from the other. Coherent states are classical-like but are not mutually disjoint. Moreover, classical-like states are identified on an individual basis in a qualitative manner, and are not precisely defined.

8.3.2 Classical cats and their states

Let us begin the discussion with classical cats.[22] We are interested in the well-being of such cats. Taking an extreme view we will classify the well-being of a classical cat into two categories: *live* or *dead*. In other words a cat is assumed to exist in one of only two pure states, being live or being dead. By definition these two states are not coherent, i.e., there is no coherent superposition of a live and a dead state (a state of limbo). Within our general formalism of physical theory embodied in Postulate UQS of §4.2.1 we can describe such a classical cat in terms of a two-dimensional Hilbert space spanned by two orthonormal vectors ψ_ℓ and ψ_d, with ψ_ℓ representing the pure state in which the cat is live and ψ_d representing the pure state in which the cat is dead. The absence of any state of limbo is achieved by a superselection rule which forbids any coherent superposition of ψ_ℓ and ψ_d. In other words the Hilbert space has a preferred direct sum decomposition as

$$\mathcal{H}^\oplus_{cc} = \mathcal{H}_\ell \oplus \mathcal{H}_d, \tag{8.41}$$

where $\mathcal{H}_\ell$ is spanned by ψ_ℓ and $\mathcal{H}_d$ is spanned by ψ_d. All observables (on the well being) of a classical cat are represented by selfadjoint decomposable

---
[22]Wan, Green and Trueman (2000).

8.3. SCHRÖDINGER'S CAT STATES

operators acting on $\mathcal{H}_{cc}^{\oplus}$, i.e.,

$$\widehat{A}_{cc}^{\oplus} = a_\ell |\psi_\ell\rangle\langle\psi_\ell| \oplus a_d |\psi_d\rangle\langle\psi_d|, \tag{8.42}$$

for some real numbers a_ℓ, a_d. Operators of this form are unable to correlate the live and dead states, i.e.,

$$\langle\psi_\ell \mid \widehat{A}_{cc}^{\oplus} \psi_d\rangle = 0. \tag{8.43}$$

So, any linear combination

$$\psi_{cc} = c_\ell \psi_\ell \oplus c_d \psi_d, \quad |c_\ell|^2 + |c_d|^2 = 1, \tag{8.44}$$

is equivalent to a classical mixture described by the statistical operator[23]

$$\widehat{S}_{cc}^{\oplus} = |c_\ell|^2 |\psi_\ell\rangle\langle\psi_\ell| \oplus |c_d|^2 |\psi_d\rangle\langle\psi_d|, \tag{8.45}$$

with ψ_{cc} and $\widehat{S}_{cc}^{\oplus}$ giving identical expectation values to all observables of the form $\widehat{A}_{cc}^{\oplus}$. Since any linear combination ψ_{cc} will not have the significance of a coherent superposition we will not be confronted with the conceptual paradox of a state of limbo for classical cats.

8.3.3 Quantum cats and their states

Here we are interested in the well being of quantum cats. Again there are two states on the well being of a quantum cat, i.e., the pure state ϕ_ℓ for the cat being live and the pure state ϕ_d for the cat being dead. These two orthonormal pure states span a two-dimensional Hilbert space $\mathcal{H}_{qc}$. In contrast to classical cats a quantum cat is by definition an orthodox quantum system satisfying Postulate OQS of §3.2.1. No superselection rule is assumed to exist for quantum cats so that all selfadjoint operators on $\mathcal{H}_{qc}$ represent observables. There will be observables $\widehat{A}_{qc}$ on $\mathcal{H}_{qc}$ capable of correlating ϕ_ℓ and ϕ_d, i.e.,

$$\langle\phi_\ell \mid \widehat{A}_{qc} \phi_d\rangle \neq 0. \tag{8.46}$$

Consequently any linear combination $\phi_{qc} = c_\ell \phi_\ell + c_d \phi_d$ represents a new pure state describing a state of limbo with the cat being partially live and partially dead, a situation quite distinct from a classical mixture of ϕ_ℓ and ϕ_d. A quantum cat is hence no different from any two-level orthodox quantum system.

---

[23] We can rewrite ψ_{cc} and $\widehat{D}_{cc}^{\oplus}$ as

$$\psi_{cc} = c_\ell \psi_\ell^{\oplus} + c_d \psi_d^{\oplus}, \quad \text{where} \quad \psi_\ell^{\oplus} = \psi_\ell \oplus 0, \ \psi_d^{\oplus} = 0 \oplus \psi_d,$$
$$\widehat{S}_{cc}^{\oplus} = |c_\ell|^2 |\psi_\ell^{\oplus}\rangle\langle\psi_\ell^{\oplus}| + |c_d|^2 |\psi_d^{\oplus}\rangle\langle\psi_d^{\oplus}|.$$

8.3.4 Disjointness and Schrödinger's cat states

A quantized oscillator is an orthodox quantum system with no superselection rules and no disjoint states. Let us now relax the definition of disjointness to arrive at a weaker but useful concept based on the notion of FAPP.

Definition 8.3.4(1)   Schrödinger's cat states in the strong sense

- Let φ_{1z0}, φ_{2z0} be a family of pairs of states of a system at time $t = 0$, parameterized by a parameter z. In time these states will evolve to states φ_{1zt} and φ_{2zt} respectively. Let

$$\Psi_{zt} = c_1 \varphi_{1zt} + c_2 \varphi_{2zt}, \quad |c_1|^2 + |c_2|^2 = 1 \qquad (8.47)$$

and

$$\widehat{D}_{zt} = |c_1|^2 |\varphi_{1zt}\rangle\langle\varphi_{1zt}| + |c_2|^2 |\varphi_{2zt}\rangle\langle\varphi_{2zt}|. \qquad (8.48)$$

Suppose there is a range Δt of times and a range Δz of values of z and a certain finite set of observables $\mathcal{O}_{qm}^n$ such that for all times $t \in \Delta t$ and for all $z \in \Delta z$ we have, for any given small ε_{inc}.

$$\Psi_{zt} \equiv \widehat{D}_{zt} \quad \text{FAPP}(\mathcal{O}_{qm}^n, \varepsilon_{inc}). \qquad (8.49)$$

Then φ_{1zt}, φ_{2zt} are called *disjoint* $\text{FAPP}(\mathcal{O}_{qm}^n, \varepsilon_{inc})$ for $t \in \Delta t$ and $z \in \Delta z$. These states φ_{1zt}, φ_{2zt} are called

disjoint FAPP in the strong sense

if φ_{1zt}, φ_{2zt} are $\text{FAPP}(\mathcal{O}_{qm}^n, \varepsilon_{inc})$ for **every** finite set $\mathcal{O}_{qm}^n$ of observables and any preset upper bound of inaccuracy ε_{inc}.

- A linear combination Ψ_{zt} of φ_{1zt}, φ_{2zt} is called a *Schrödinger's cat state in the strong sense* if φ_{1zt}, φ_{2zt} are disjoint FAPP in the strong sense. We may call the system in a Schrödinger's cat state as a *Schrödinger's cat*.

Note that Δt and Δz could be finite or unbounded, and both Δt and Δz generally depend on both $\mathcal{O}_{qm}^n$ and ε_{inc}. We can also replace $t \in \Delta t$ by $t \to \infty$, and replace $z \in \Delta z$ by $z \to \infty$; the disjointness is then achieved in the limit as t or z tends to infinity. We shall consider several examples below, showing that disjointness may be achieved by either z or t alone. Our present concepts of disjointness are less rigid than the notion of disjointness introduced in Definition 4.3.3(1) in relation to superselection rules.

8.3. SCHRÖDINGER'S CAT STATES

8.3.5 Scattering systems and Schrödinger's cat states

Schrödinger's cat states can arise from simple scattering systems. First we need to introduce a one-parameter family of pairs of states. Given a pair of bound state $\widehat{U}_t \phi_b$ and scattering state $\widehat{U}_t \phi_s$ of a scattering system we can formally define one-parameter family of pairs of states

$$\phi_{bzt} = \widehat{U}_t \phi_b \quad \text{and} \quad \phi_{szt} = \widehat{U}_t \phi_s. \tag{8.50}$$

Definition 8.3.4(1) then formally applies.[24] We can conclude that $\widehat{U}_t \phi_b$ and $\widehat{U}_t \phi_s$ are disjoint FAPP in the strong sense for all z in the limit as $t \to \infty$.

The state ϕ_t in Eq. (8.15) is a Schrödinger's cat state in the strong sense in the limit as $t \to \infty$.

8.3.6 Quantized oscillator and Schrödinger's cat states

Let $\varphi_n(x)$ be a normalized eigenfunction of the quantized harmonic oscillator Hamiltonian $\widehat{H}_{sho}$ corresponding to eigenvalue $(n+\frac{1}{2})\hbar\omega$. The corresponding time dependent eigenfunction is

$$\varphi_{nt}(x) = \varphi_n(x) e^{-i(n+\frac{1}{2})\omega t}, \quad n = 0, 1, 2, \ldots. \tag{8.51}$$

Let us first establish a thereom.

Theorem 8.3.6   Coherent states and asymptotic disjointness

- Let $\phi_{zt}^{(c)}$ be a coherent state given by Eq. (8.28) and let $\widehat{A}$ be any bounded operator. We have

$$\lim_{|z| \to \infty} \langle \phi_{zt}^{(c)} \mid \widehat{A} \varphi_{nt} \rangle = 0. \tag{8.52}$$

- Let ψ_t be a finite linear combination of $\varphi_{nt}(x)$, i.e.,

$$\psi_t = \sum_{n=0}^{N} c_n \varphi_{nt}, \tag{8.53}$$

then we have

$$\lim_{|z| \to \infty} \langle \phi_{zt}^{(c)} \mid \widehat{A} \psi_t \rangle = 0. \tag{8.54}$$

---

[24] In this case the pairs are independent of z.

Proof First we shall show that for any given vector $\xi \in L^2(E)$ and t we have
$$\lim_{|z|\to\infty} \langle \phi_{zt}^{(c)} \mid \widehat{A}\xi \rangle = 0. \tag{8.55}$$
To prove this we first observe that
$$\langle \phi_{zt}^{(c)} \mid \xi \rangle$$
$$= \int_{-\infty}^{\infty} c^*(t) \exp\left[-ip_c(t)x - \frac{m\omega}{2\hbar}(x - x_c(t))^2\right] \xi(x)\, dx$$
$$= c^*(t) \int_{-\infty}^{\infty} \exp\left[-ip_c(t)x\right] \exp\left[-\frac{m\omega}{2\hbar}(x - x_c(t))^2\right] \xi(x)\, dx. \tag{8.56}$$
Now, let
$$\eta(x) = \exp\left[-\frac{m\omega}{2\hbar} x^2\right], \tag{8.57}$$
and let $\widehat{T}_a$ be a translation operator with $a = x_c(t)$. Then we have
$$\exp\left[-\frac{m\omega}{2\hbar}(x - x_c(t))^2\right] = \widehat{T}_a \eta(x). \tag{8.58}$$
For any given time t such that $x_c(t) \neq 0$ we have
$$|\langle \phi_{zt}^{(c)} \mid \xi \rangle|$$
$$= \left| c^*(t) \int_{-\infty}^{\infty} \exp\left[-ip_c(t)x\right] \exp\left[-\frac{m\omega}{2\hbar}(x - x_c(t))^2\right] \xi(x)\, dx \right|$$
$$\leq \left(\frac{m\omega}{\pi\hbar}\right)^{\frac{1}{4}} \left| \int_{-\infty}^{\infty} \exp\left[-\frac{m\omega}{2\hbar}(x - x_c(t))^2\right] \xi(x)\, dx \right| \tag{8.59}$$
$$= \left(\frac{m\omega}{\pi\hbar}\right)^{\frac{1}{4}} |\langle \widehat{T}_a \eta \mid \xi \rangle| \tag{8.60}$$
$$\to 0 \quad \text{as } a \to \infty \quad \text{by Eq. (8.19)}. \tag{8.61}$$
Moreover, since
$$a = x_c(t) \to \infty \quad \Leftrightarrow \quad \alpha \to \infty \quad \Leftrightarrow \quad |z| \to \infty, \tag{8.62}$$
we deduce that
$$\lim_{|z|\to\infty} |\langle \phi_{zt}^{(c)} \mid \xi \rangle| = 0. \tag{8.63}$$
This result remains valid if we replace ξ by $\widehat{A}\xi$, i.e., we have
$$\lim_{|z|\to\infty} \langle \phi_{zt}^{(c)} \mid \widehat{A}\xi \rangle = 0, \tag{8.64}$$

because $\widehat{A}\xi$ is just another member of $L^2(I\!\!E)$.

When $x_c(t) = 0$, i.e., $\cos(\omega t + \theta) = 0$, the above argument in terms of $\widehat{T}_a$ fails. However, Eq. (8.55) remains valid. From Eq. (8.26) we know that $x_c(t) = 0$ implies $p_c(t) = \mp m\omega a$. It follows that

$$\langle \phi_{zt}^{(c)} \mid \xi \rangle$$
$$= \int_{-\infty}^{\infty} c^*(t) \exp\left[-ip_c(t)x - \frac{m\omega}{2\hbar}(x - x_c(t))^2\right] \xi(x)\,dx$$
$$= c^*(t) \int_{-\infty}^{\infty} \exp\left[\pm im\omega a x\right] \exp\left[-\frac{m\omega}{2\hbar}x^2\right] \xi(x)\,dx \quad (8.65)$$
$$\to 0 \quad \text{as } \alpha \to \infty, \quad (8.66)$$

on account of the Riemann-Lebesgue lemma.[25] Equation (8.52) follows from the fact that for any given t we have

$$|\langle \phi_{zt}^{(c)} \mid \widehat{A}\,\varphi_{nt}\rangle| = |\langle \phi_{zt}^{(c)} \mid \widehat{A}\,\varphi_n\rangle|. \quad (8.67)$$

The final result in the form of Eq. (8.54) then follows immediately.

As an example consider the following linear combination

$$\psi_t = \frac{1}{\sqrt{2}}\left(\varphi_{0t} + \varphi_{1t}\right) = \frac{1}{\sqrt{2}}\left(\varphi_0\,e^{-\frac{i}{2}\omega t} + \varphi_1\,e^{-\frac{3i}{2}\omega t}\right), \quad (8.68)$$

which represents a wave packet with an oscillating position expectation value, albeit not keeping its shape as it oscillates, i.e.,

$$\langle \psi_t \mid \widehat{x}\,\psi_t\rangle = \left(\frac{\hbar}{2m\omega}\right)^{\frac{1}{2}} \cos\omega t. \quad (8.69)$$

Theorem 8.3.6 tells us that $\phi_{zt}^{(c)}$ and ψ_t are disjoint FAPP for large $|z|$, and hence $c_1\phi_{zt}^{(c)} + c_2\psi_t$ is a Schrödinger's cat state in the strong sense.

8.3.7 Weak Schrödinger's cat states

One way to formulate a concept of classical distinguishability of states is to go back to classical mechanical systems of a finite degree of freedom where two distinct states of a classical particle do not form a coherent superposition. It is then tempting to identify classical or macroscopic distinguishability with disjointness. The problem here is that distinct states of some classical systems can form coherent superposition. For example, two classical harmonic waves

---

[25] Papoulis (1962) p. 278.

of different frequencies and amplitudes can interfere with each other to form a coherent superposition. Hence one cannot employ two classical waves to represent the live and dead states of a classical cat. The moral of all this is that care has to be taken when one uses the term *classical*. It is misleading to identify classical distinguishability with disjointness, and this is why we have avoided such an identification. But one has no objection to using the term classical distinguishability to mean disjointness on an *ad hoc* basis when it makes good sense.

Many authors consider a linear combination of two coherent states as a Schrödinger's cat state. An example is[26]

$$\Phi_{zt} = \frac{1}{\sqrt{2}} \left(\phi_{zt}^{(c)} + \phi_{-zt}^{(c)} \right). \tag{8.70}$$

This would amount to a weaker definition of Schrödinger's cat states since Φ_{zt} does not satisfy Definition 8.3.4(1) on Schrödinger's cat states. The coherent states involved are not disjoint FAPP with respect to every finite set of observables, because of Eq. (8.40). However, $\phi_{zt}^{(c)}$ and $\phi_{-zt}^{(c)}$ are indeed disjoint FAPP with respect to some chosen set of observables. It follows that treating a linear combination of $\phi_{zt}^{(c)}$ and $\phi_{-zt}^{(c)}$ as a Schrödinger's cat state amounts to the following definition:

Definition 8.3.7(1)   **Schrödinger's cat state in weak senses**

- A linear combination Ψ_{zt} of two states φ_{1zt}, φ_{2zt} described in Definition 8.3.4(1) is called a Schrödinger's cat state in the weak sense 1 if φ_{1zt}, φ_{2zt} are disjoint FAPP$(\mathcal{O}_{qm}^n, \varepsilon_{inc})$ for $t \in \Delta t$ and $z \in \Delta z$ and a suitably chosen $\mathcal{O}_{qm}^n$ and ε_{inc}.

- A linear combination of two states which are classically or macroscopically distinguishable in some specific sense is called a Schrödinger's cat state in the weak sense 2.

These are much weaker definitions. In weak sense 1 the disjointness is required only with respect to a specific set of observables. Of course this set of observables need to be non-trival in some physical sense which should be made clear in each case. For example, Φ_{zt} in Eq. (8.70) is a Schrödinger's cat state in the weak sense 1 with respect to a set of observables containing the Hamiltonian.[27] We can see explicitly how disjointness of its constituent states

---

[26] Yurke and Stoler (1986), Sivakumar (1998).

[27] We have not used $\phi_{z^*t}^{(c)}$, where z^* is the complex conjugate of z, employed by some authors (Brune *et al.* (1996)).

8.3. SCHRÖDINGER'S CAT STATES

is achieved asymptotically.[28] Take for example the Hamiltonian $\widehat{H}$ which is expressible in terms the annihilation operator $\widehat{a}$ and its adjoint $\widehat{a}^\dagger$ as

$$\widehat{H} = \hbar\omega \left(\widehat{a}^\dagger \widehat{a} + \frac{1}{2} \right). \tag{8.71}$$

We have, using Eq. (8.36),

$$\langle \phi_{zt}^{(c)} | \widehat{H} \phi_{-zt}^{(c)} \rangle = \langle \phi_{zt}^{(c)} | \hbar\omega \widehat{a}^\dagger \widehat{a} \phi_{-zt}^{(c)} \rangle + \frac{1}{2} \hbar\omega \langle \phi_{zt}^{(c)} | \phi_{-zt}^{(c)} \rangle \tag{8.72}$$

$$= \hbar\omega \langle \widehat{a} \phi_{zt}^{(c)} | \widehat{a} \phi_{-zt}^{(c)} \rangle + \frac{1}{2} \hbar\omega \langle \phi_{zt}^{(c)} | \phi_{-zt}^{(c)} \rangle \tag{8.73}$$

$$= \hbar\omega \left(z^*(-z) + \frac{1}{2} \right) e^{-2|z|^2} \tag{8.74}$$

$$\to \quad 0 \quad \text{as} \quad |z| \to \infty. \tag{8.75}$$

Although Schrödinger's cat states in the weak sense 2 appear to be even weaker and vague, this definition is physically important. In many specific cases it is possible to give a well-defined understanding to the meaning of classical distinguishability, e.g., the two states are in the form of spatially non-overlapping wave packets. Many experimental investigations of Schrödinger's cat states are based on this weaker definition. We shall see an explicit example in §8.3.9 in an ammonia molecule.

8.3.8 Periodic Schrödinger's cat states

We aim to look for a cat that lies between the two extreme cases of being either entirely classical or being entirely quantum, i.e., we seek a *periodic Schrödinger's cat* possessing states which are

1. quantum sometimes, capable of exhibiting coherent superpositions,

2. classical sometimes in the sense of being disjoint, exhibiting no coherent superpositions, and

3. periodically quantum and classical.

To demonstrate the existence of such a cat let us consider the motion of a quantum particle having the following Hamiltonian

$$\widehat{H} = \widehat{H}_{sho} + \frac{1}{2} m\Omega g(t) \widehat{x}, \tag{8.76}$$

---

[28]Myatt *et al.* (2002).

with

$$g(t) = \alpha (\Omega + 2\omega) \cos(\Omega + \omega)t + \alpha (\Omega - 2\omega) \cos(\Omega - \omega)t, \qquad (8.77)$$

where α, Ω are two positive constants. We can view this system as a charged oscillator under a time-dependent external electric field. Let

$$X_c(t) = \alpha \cos \Omega t \cos \omega t, \qquad (8.78)$$
$$P_c(t) = m\dot{X}_c = -m\alpha\left(\Omega \sin \Omega t \cos \omega t + \omega \cos \Omega t \sin \omega t\right), \qquad (8.79)$$

and let

$$Z = \left(\frac{m\omega}{2\hbar}\right)^{\frac{1}{2}} \left(X_c(t) + \frac{i}{m\omega} P_c(t)\right). \qquad (8.80)$$

Then a solution of the time-dependent Schrödinger equation with this Hamiltonian is

$$\Phi_{Zt}(x) = C(t) \left(\frac{m\omega}{\pi\hbar}\right)^{\frac{1}{4}} \exp\left[i P_c(t)x - \frac{m\omega}{2\hbar}(x - X_c(t))^2\right], \qquad (8.81)$$

where $C(t)$ is a solution of

$$i\hbar \frac{\dot{C}}{C} = \frac{1}{2m} P_c^2 - \frac{1}{2} m\omega^2 X_c^2 + \frac{1}{2} \hbar\omega. \qquad (8.82)$$

We can have a factorizable solution of the form

$$C(t) = c(t) \, c_1(t) \, c_2(t) \, c_3(t) \, c_4(t) \qquad (8.83)$$

where $c(t)$ is the corresponding factor in Eq. (8.30) for the unperturbed oscillator, and

$$c_1(t) = \exp\left[-i\frac{m\alpha^2}{8}\left(\Omega^2 t + \omega \sin 2\omega t\right)\right], \qquad (8.84)$$

$$c_2(t) = \exp\left[i\frac{m\alpha^2}{16}\left(\Omega \sin 2\Omega t - \frac{\Omega^2}{\omega} \sin 2\omega t\right)\right], \qquad (8.85)$$

$$c_3(t) = \exp\left[i\frac{m\alpha^2}{32}\left(\Omega^2 - 2\Omega\omega + 2\omega^2\right) \frac{\sin 2(\Omega - \omega)t}{\Omega - \omega}\right], \qquad (8.86)$$

$$c_4(t) = \exp\left[i\frac{m\alpha^2}{32}\left(\Omega^2 + 2\Omega\omega + 2\omega^2\right) \frac{\sin 2(\Omega + \omega)t}{\Omega + \omega}\right]. \qquad (8.87)$$

Clearly $C(t)$ in Eq. (8.83) reduces to $c(t)$ when $\Omega = 0$. The corresponding spatial probability density function is

$$|\Phi_{Zt}(x)|^2 = \left(\frac{m\omega}{\pi\hbar}\right)^{1/2} \exp\left[-\frac{m\omega}{\hbar}(x - X_c(t))^2\right]. \qquad (8.88)$$

8.3. SCHRÖDINGER'S CAT STATES

We see that Eq. (8.88) represents a wave packet which oscillates without change of shape with its center oscillating according to $X_c(t)$.

When $\Omega << \omega$ we may regard the oscillation in $X_c(t)$ as consisting of a harmonic oscillation $\cos \omega t$ with angular frequency ω and a slowly varying amplitude $|\alpha \cos \Omega t|$. The oscillation amplitude varies slowly and periodically between α and 0 according to $|\alpha \cos \Omega t|$. It follows that given a very large α the wave packet $\Phi_{Zt}(x)$ will oscillate with small amplitudes sometimes and with large amplitudes at other times. Therefore $\Phi_{Zt}(x)$ and $\Phi_{-Zt}(x)$ will be periodically disjoint FAPP in the weak sense 1 as the oscillation amplitude $|\alpha \cos \Omega t|$ varies slowly and periodically.

It would be interesting to see an experimental realization of those periodic cat states and examine the effect and any possible applications of a pair of states which behave in such a periodic manner. Results on harmonic oscillators can be converted to single mode optical systems.[29]

8.3.9 Double-well potentials and chiral molecules

Double-well potentials and their applications to model certain molecules are well-known.[30] The ammonia molecule NH_3 has three hydrogen atoms sited in the corners of an equilateral triangle lying in the y-z plane, as shown.[31]

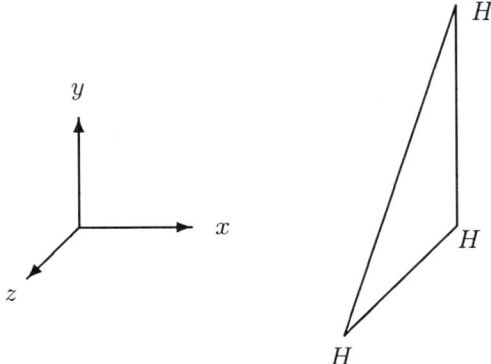

Fig. 8.3.9(1)   Ammonia molecule.

One would imagine an ammonia molecule to have a pyramidal structure with the nitrogen atom positioned in the apex on the x-axis either on the

---
[29]Yurke and Stoler (1986), Gerry and Knight (1997).
[30]Lévy-Leblond (1990) pp. 355-368, Merzbacher (1988) pp. 149-159, Feynman, Leighton and Sands (1965) §8-6 and §9-1.
[31]See also Fig. 8.3.9(2) and Fig. 8.3.9(3).

left or on the right of the triangle formed by the hydrogen atoms on the y-z plane.[32] However, the situation turns out to be more complicated as the molecule can exist in many different states. The motion of the nitrogen atom along the x-axis is governed by the potential generated by the hydrogen atoms. We can model the motion of the nitrogen atom by that of a particle in one dimension along the x-axis under the influence of a double-well potential, a square double-well potential for example. As we shall see shortly, an ammonia molecule can shape like a "fuzzy diamond" in its stationary ground state. It can also be in a non-stationary state describing an oscillation of the nitrogen atom between left and right of the plane formed by the hydrogen atoms.[33]

A square double-well potential in one dimension of depth $V_0 > 0$, width of each well a and separation $2b$, located in

$$\Lambda_\ell = \bigl(-(a+b), -b\bigr) \quad \text{and} \quad \Lambda_r = \bigl(b, a+b\bigr), \tag{8.89}$$

and symmetrically on either sides of the origin is given by the potential:

$$V(x) = \begin{cases} V_0, & x \leq -(a+b), \\ 0, & x \in \Lambda_\ell, \\ V_0, & x \in [-b, b], \\ 0, & x \in \Lambda_r, \\ V_0, & x \geq (a+b). \end{cases} \tag{8.90}$$

Case 1 Infinite square double-well potential   When $V_0 = \infty$ we have two infinite square wells. The eigenfunctions and eigenvalues of the Hamiltonian with such an infinite square double-well potential are directly obtainable from that of a single infinite square potential well presented in E2.5.1(1) E2. Each eigenvalue is degenerate with degeneracy 2.[34] In particular, we have, corresponding to the ground state energy eigenvalue E_1, two degenerate ground state eigenfunctions $\phi_{1\ell}$ and ϕ_{1r} localized respectively in Λ_ℓ and Λ_r. The infinite nature of the potential wells over an extended region[35] $[-b, b]$ cuts off any correlation between $\phi_{1\ell}$ and ϕ_{1r}. To build this feature into a formal theory we shall take the Hilbert space associated with the system to be

$$\mathcal{H}^\oplus = L^2(\Lambda_\ell) \oplus L^2(\Lambda_r) \tag{8.91}$$

so that $\phi_{1\ell} \in L^2(\Lambda_\ell)$ and $\phi_{1r} \in L^2(\Lambda_r)$. A superselection rule operates with $L^2(\Lambda_\ell)$ and $L^2(\Lambda_r)$ as supersectors. Observables are then of the form

$$\widehat{A}^\oplus = \widehat{A}_\ell \oplus \widehat{A}_r, \tag{8.92}$$

---

[32] Bransden and Joachain (1992) p. 649.
[33] See Fig. 8.3.9(2) for ammonia molecules and Fig. 8.3.9(3) for chiral molecules.
[34] Lévy-Leblond (1990) p. 357.
[35] A potential, infinite at isolated points like a δ-potential, does not necessarily cut off all correlations.

8.3. SCHRÖDINGER'S CAT STATES

where $\widehat{A}_\ell$ acts in $L^2(\Lambda_\ell)$ and $\widehat{A}_r$ acts in $L^2(\Lambda_r)$. It follows that there is no correlation between $\phi_{1\ell}$ and ϕ_{1r}.

We can form a symmetric ground state eigenfunction and an anti-symmetric ground state eigenfunction by taking the following combinations:

$$\phi_1^{\oplus s} = \frac{1}{\sqrt{2}}\left(\phi_{1\ell}^\oplus + \phi_{1r}^\oplus\right), \quad \phi_1^{\oplus a} = \frac{1}{\sqrt{2}}\left(\phi_{1\ell}^\oplus - \phi_{1r}^\oplus\right), \tag{8.93}$$

where

$$\phi_{1\ell}^\oplus = \phi_{1\ell} \oplus 0_r, \quad \phi_{1r}^\oplus = 0_\ell \oplus \phi_{1r}. \tag{8.94}$$

These are stationary states. Moreover, the superselection rule renders these states equivalent to the following classical mixture:[36]

$$\widehat{S}^\oplus = \frac{1}{2}\left(|\phi_{1\ell}^\oplus\rangle\langle\phi_{1\ell}^\oplus| + |\phi_{1r}^\oplus\rangle\langle\phi_{1r}^\oplus|\right). \tag{8.95}$$

Case 2 Finite square double-well potential If the potential V_0 is finite the degeneracy is lifted. In place of the ground state energy E_1 for the infinite double-well we shall have two non-degenerate energy levels: E_1^s and E_1^a with $E_1^s < E_1^a$. These energy levels correspond respectively to a symmetric eigenfunction ψ_1^s and an anti-symmetric eigenfunction ψ_1^a.[37] These are real-valued functions[38] and they represent stationary states with time evolution given by

$$\psi_{1t}^s = \psi_1^s\, e^{-iE_1^s t}, \quad \psi_{1t}^a = \psi_1^a\, e^{-iE_1^a t}. \tag{8.96}$$

Intuitively it is clear that Case 2 reduces to Case 1 as V_0 becomes arbitrarily large. Less obvious is the transition from Case 2 to an effective Case 1 as the well separation $2b$ tends to infinity in the sense that we can recover the energy eigenvalues and their degeneracy of an infinite square double-well potential.[39] For large but finite b the energy gap $\epsilon > 0$ between E_1^s and E_1^a becomes small and we also have[40]

$$E_1^s = E_1 - \frac{1}{2}\epsilon, \quad E_1^a = E_1 + \frac{1}{2}\epsilon. \tag{8.97}$$

In contrast to Eq. (8.93) the following superpositions

$$\Psi_1^\ell = \frac{1}{\sqrt{2}}\left(\psi_1^s + \psi_1^a\right), \quad \Psi_1^r = \frac{1}{\sqrt{2}}\left(\psi_1^s - \psi_1^a\right) \tag{8.98}$$

---

[36] See the second footnote in §8.2.2 on de Broglier paradox.
[37] See Lévy-Leblond (1990) Fig. 6.3.3 on p. 362 for a diagramatic illustration.
[38] Merzbacher (1988) p. 46.
[39] Lévy-Leblond (1990) p. 360. This is another example of spatial separation giving rise to a superselection rule.
[40] Lévy-Leblond (1990) p. 361. Here E_1 is the ground state energy eigenvalue of an infinite square well.

represent two non-stationary states. As the notation suggests Ψ_1^ℓ lies mainly round the left well at time $t = 0$, and Ψ_1^r lies mainly round the right well.[41] In time Ψ_1^ℓ will evolve to Ψ_{1t}^ℓ given by

$$\Psi_{1t}^\ell = \frac{1}{\sqrt{2}} \left(\psi_{1t}^s + \psi_{1t}^a \right) \quad (8.99)$$

$$= \frac{1}{\sqrt{2}} e^{-iE_1 t} \left(\psi_1^s e^{i\omega t} + \psi_1^a e^{-i\omega t} \right), \quad \omega = \epsilon/2\hbar \quad (8.100)$$

$$= e^{-iE_1 t} \left(\Psi_1^\ell \cos\omega t + i\, \Psi_1^r \sin\omega t \right). \quad (8.101)$$

The position probability density function

$$\left| \Psi_{1t}^\ell(x) \right|^2 = \left| \Psi_1^\ell(x) \, \cos\omega t \right|^2 + \left| \Psi_1^r(x) \, \sin\omega t \right|^2. \quad (8.102)$$

oscillates from left to right and back, i.e., from $|\Psi_1^\ell|^2$ on the left to $|\Psi_1^r|^2$ on the right and back, with a period

$$T = 2\pi\hbar/\epsilon, \quad (8.103)$$

a result known as *quantum beats*.[42] Note that this period agrees with the photon frequency emitted or absorbed for transitions between states Ψ^ℓ and Ψ^r. Moreover the energy expectation value in state Ψ_t^ℓ coincides with E_1. Another useful expression is the correlation term between $\Psi_{t_2}^\ell$ and $\Psi_{t_1}^\ell$ defined by Eq. (3.43), i.e.,

$$\langle \Psi_{1t_2}^\ell \mid \Psi_{1t_1}^\ell \rangle = e^{iE_1(t_2-t_1)} \cos\omega(t_2 - t_1), \quad (8.104)$$

giving a transition probability

$$\left| \langle \Psi_{1t_2}^\ell \mid \Psi_{1t_1}^\ell \rangle \right|^2 = \left| \cos\omega(t_2 - t_1) \right|^2. \quad (8.105)$$

Quantum beat can be understood as a tunnelling of the wave function through the potential barrier, i.e., tunnelling through the region $(-b, b)$ separating the two potential wells in Λ_ℓ and Λ_r. When the separation $2b$ becomes arbitrarily large the tunnelling will die away, as in the case of an infinite double-well.

Let us now consider how our present square double-well potential can model ammonia and other molecules of a similar structure:

Ammonia molecules[43]   The idea is to model the potential governing the motion of the nitrogen atom by a finite square double-well potential with an

---

[41] See Lévy-Leblond (1990) Fig. 6.3.4 on p. 363 for a diagramatic illustration.
[42] Lévy-Leblond (1990) p. 232, Merzbacher (1988) p. 154.
[43] Wightman and Glance (1989), Feynman, Leighton and Sands (1965) Chapters 8 and 9, Bransden and Joachain (1992) pp. 649-654. More details are available in Herzberg (1945).

8.3. SCHRÖDINGER'S CAT STATES

appropriate separation $2d$. The motion of the nitrogen atom is then describable by one of the wave functions mentioned in Case 2 above. The ground state ψ_1^s is a symmetric function of x, i.e., symmetric with respect to the plane of the hydrogen atoms. If the molecule is in its stationary ground state ψ_{1t}^s then it would have a "fuzzy diamond", rather than a pyramidal, shape with the nitrogen atom having an equal probability of being on the right and the left of the y-z plane where the hydrogen atoms are situated. An ammonia molecule can be excited into other states. An example is Ψ_{1t}^ℓ which is a superposition of stationary states ψ_{1t}^s and ψ_{1t}^a. This superposition produces a non-stationary state in which the nitrogen atom would oscillate from the left to the right of the y-z plane and back again as shown explicitly in Eqs. (8.101) and (8.102). For an ammonia molecule the energy gap ϵ is of the order of 10^{-4} electron volt, giving a frequency $\nu \approx 24000$ MHz and an oscillation period $T \approx 4 \times 10^{-11}$ s. This corresponds to microwaves when we consider emission and absorption of radiation.[44]

The wave function Ψ_{1t}^ℓ is also seen to be a linear superposition of Ψ_1^ℓ and Ψ_1^r which can be regarded as classically distinguishable on account of their minimal overlap. It follows that Ψ_t^ℓ is a Schrödinger's cat state in the weak sense 2 given in Definition 8.3.7(1). The situation is schematically shown in Fig. 8.3.9(2) where the two broken lines on either side of the triangular base lying in the y-z plane indicates a superposition of Ψ^ℓ to Ψ^r.

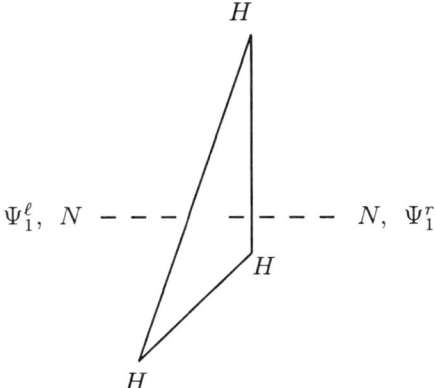

Fig. 8.3.9(2)   Superposition of Ψ^ℓ and Ψ^r.

Chiral molecules   Equation (8.103) tells us that the oscillation period T can become arbitrarily large if the energy gap ϵ is arbitrarily small. Consequently the oscillation can become negligible or undetectable. Typically T

---
[44]Indeed this oscillation is utilized to construct *masers* (microwave amplification by stimulated emission of radiation). Feynman, Leighton and Sands (1965) §9.1.

can range from years to over the lifetime of the universe![45] This is the case for a group of larger and heavier molecules known as chiral molecules.[46] Typically these molecules are formed by four groups G_1, G_2, G_3, G_4 of (possibly different) atoms of which the first three groups, G_1, G_2, G_3 are sited at the corners of a triangular base lying in, say, the y-z plane with the fourth group G_4, which may consist of a single atom, localizing itself on the left **or** the right of the triangular base. Such an arrangement results in the molecule assuming a definite pyramidal (tetrahedron) shape, as shown in Fig. 8.3.9(3) in contrast to the fuzzy diamond or oscillating shapes of an ammonia molecule shown in Fig. 8.3.9(2).

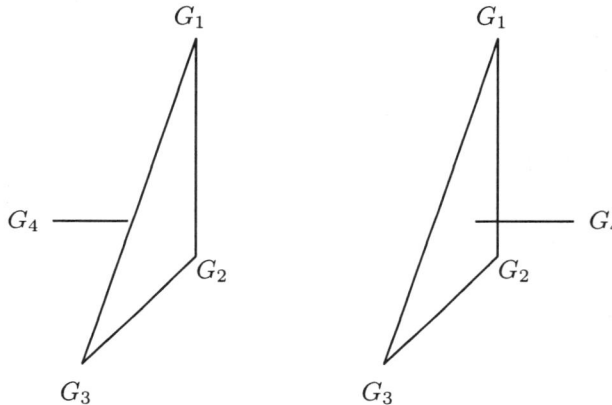

Fig. 8.3.9(3)   Mirror images and disjoint chiral states.

In terms of our square double-well model this means setting

$$\epsilon = 0 \text{ FAPP} \quad \text{and} \quad \omega = 0 \text{ FAPP} \tag{8.106}$$

so that Ψ_{1t}^ℓ becomes stationary, i.e., Eq. (8.101) becomes

$$\Psi_{1t}^\ell = \Psi_1^\ell \, e^{-iE_1 t} \qquad \text{FAPP,} \tag{8.107}$$

leading to a state with G_4 remaining indefinitely on the left of the triangular base.[47] Similarly if G_4 is initially on the right of the triangular base it will remain so indefinitely. These two positions of G_4 and hence the two shapes of the molecules are distinguishable if the four groups of atoms are different. As already mentioned in §4.3.3, for a group of molecules, collectively known

---

[45]Lévy-Leblond (1990) p. 366, Merzbacher (1988) p. 159.
[46]Cram and Cram (1978) pp. 167-178, Williams (1982) p. 240.
[47]This corresponds to an infinite double-well potential mentioned in Case 1 earlier.

8.3. SCHRÖDINGER'S CAT STATES

as *optical isomers*, the two different structural shapes lead to different optical properties. In terms of the language introduced in §4.6.1 these molecules form a mixed quantum system. A superselection rule operates rendering these two states Ψ_1^ℓ and Ψ_1^r disjoint. The situation resembles that of a classical cat.

The two positions the atom on the left or the right of the triangular base are mirror images of each other. Objects which are not identical to their mirror images are said to possess *chirality* or *handedness*. Many biological systems clearly exhibit chirality, such as snails with right-handed spiralling shells, plants climbing up spiralling in a right handed manner. Most strikingly many molecules such as DNA and protein molecules crucial to life are chiral. Various reasons such as symmetry breaking and the weak nuclear force have been advanced to account for the phenomenon of chirality.[48]

8.3.10 Dynamic and asymptotic decoherence

In §4.3.3 we mention a number of origins of supereselection rules. These can also serve as sources of disjointness and decoherence of states. It is widely known that decoherence can be achieved by coupling to an appropriate external environment.[49] The discussions in the preceding sections highlight the situation that a coherent superposition is not such an absolute concept, and its significance in representing a pure state could be eroded gradually either in time or by other means.[50] The correlations between states effected by a given set of observables can become so small as to be undetectable; such a situation can be achieved as a result of a normal dynamical process without further external intervention. We call a process which reduces correlations to arbitrarily small amounts in the sense of Eq. (8.6) and the corresponding erosion of coherence and transition from a coherent superposition to a mixture in the sense of Eq. (8.10) or Definition 8.3.4(1) as a *decoherent process*. A number of examples in the preceding sections serve to demonstrate how dynamical processes can achieve decoherence asymptotically. In these examples we consider some simple single-particle systems, and the decoherent process is a unitary dynamical process determined by a given Hamiltonian generator. We call such processes *dynamical decoherence processes* or *asymptotic decoherence processes*.

Spatial separation could be a cause for decoherence, as in an ionization process and in the case of a square double-well potential with sufficiently large separation of the two wells. As in the case involving coherent states a de-

---

[48]Hegstrom and Kondepudi (1990), Janoschek (1991), Pfeifer (1980), Amann (1988).
[49]Myatt *et al.* (2000).
[50]See Anatopoulos (2002) for a short review of some general issues involved in the phenomenon of decoherence, and a series articles in the monograph *Decoherence and the Appearance of a Classical World in Quantum Theory* by Giulini *et al.* (1996).

coherence process does not have to be achieved by spatial separation. The example of a periodic Schrödinger's cat shows that decoherence is not always a one-way process. Generally decoherence is achieved asymptotically rather than abruptly. In our simple and idealized examples this manifests itself in a residual correlation which goes to zero only in an asymptotic limit. This appears to suggest that there will never be complete decoherence. We should not take this literally since those undetectable residual correlations can easily be totally destroyed by environmental and other factors in the real physical world.

Decoherence is not always to do with a transition between classical and quantum regimes. An ionization process is not a transition between classical and quantum regimes; both the outgoing electron and the remaining ion are quantum systems in their own right.

8.4 Superconducting Schrödinger's Cat States

8.4.1 Breakdown of superselection rules and capacitive junction

As discussed in §7.10 a superconductor is capable of carrying an alternating current. In the presence of a Josephson junction such an alternative current is accompanied by the appearance of a voltage across the junction. Also a changing current mean a changing phase shift λ across the junction. As stated in Eq. (7.160) the rate of change of the phase shift λ across the junction is proportional to the voltage across the junction.[51]

So far we have dealt with idealized superconducting circuits. Practical superconducting circuits operating under different conditions can behave quite differently. Physically it is possible to construct a junction which acts like a

---

[51] There are various derivations this result, e.g., de Bruyn Ouboter (1978) pp. 42-47, Feynman, Leighton and Sands (1965), Barone and Paterno (1982) §1.4. We may also treat Eq. (7.160) as a postulate in a formal theory. We can gain an intuition of this result for a superconducting ring with a Josephson junction in an external magnetic field. One way to change the current in such a ring is to vary the external magnetic flux applied to the ring. As a result the total enclosed flux would also change in time. A voltage V will be induced across the junction according to Faraday's law (Leggett (1987) p. 421):

$$V = -\frac{d\Phi}{dt}.$$

Combining this with Eq. (7.63) we get

$$V = -\frac{\Phi_c}{2\pi}\frac{d\lambda}{dt} = -\frac{\hbar}{q_c}\frac{d\lambda}{dt},$$

agreeing with Eq. (7.160). Here we have assumed that the change in Φ is within the same n in the context of Eq. (7.63).

8.4. SUPERCONDUCTING SCHRÖDINGER'S CAT STATES

capacitor in the sense that a voltage V across the junction will cause charges to accumulate on the opposite sides of the junction, and if Q is the total charge accumulated on a chosen side of the junction then Q is proportional to V, i.e., we have the usual relationship $V = CQ$ of a capacitor with the proportionality constant C referred to as the capacitance of the junction.[52]

So far we have not introduced any mathematical representation of the charge on the junction within our present theoretical framework. The object of this section is to remedy this situation. We shall confine our attention to two specific cases.

Case 1 **Superconducting wire interrupted by a capacitive junction** The configuration is schematically shown in the figure below:

$$\underline{\qquad\qquad B_1 \qquad\qquad \overset{V}{\overbrace{\quad\;\;}}\qquad B_2 \qquad\qquad}$$
$$C$$

Fig. 8.4.1(1)   Superconducting wire with a capacitive junction.

In §7.6.6 we are able to describe the system with a dc supercurrent as a mixed quantum system with a direct integral Hilbert space $\mathcal{H}^\oplus$ given by Eq. (7.104). An element of $\mathcal{H}^\oplus$ is of the form

$$\boldsymbol{f}^\oplus = \int^\oplus f(\lambda)\,\boldsymbol{\eta}_\lambda\,d\lambda, \tag{8.108}$$

where $\boldsymbol{\eta}_\lambda$ is specified by Eq. (7.103). As mentioned in §7.10 an addition of a voltage across the junction or feeding a current bigger than the critical current through the junction will change the situation. The system becomes nonequilibrium and an ac Josephson effect emerges. More fundamental changes of the system occur when the junction becomes an effective capacitor. Let us consider a standard situation when an external current source drive a constant current j_{ex} through the junction to cause an ac Josephson effect.[53] Following an approach in §7.10 we shall treat this effective capacitive junction as a Hamiltonian system with λ as the 'position variable'. The present system is quite different so that the Hamiltonian used in §7.10 is no longer applicable here. Let us start afresh and use a standard expression for the Hamiltonian. In other words we take the Hamiltonian H^w_{cap} for the present system to consist of an effective kinetic energy term K^w_{cap} and a potential energy term V^w_{cap}.[54]

---

[52] Leggett (1987) p. 439 for illustrations.
[53] Rose-Innes (1980) pp. 162-168. Leggett (1987) p. 418.
[54] The superscript and the subscripts together signify the present system of a long wire with a capacitive junction, to constrast to the ring system to be considered later.

A generally accepted potential energy term is given by[55]

$$V_{cap}^w = -\frac{|j_c|\hbar}{q_c}\cos\lambda - \frac{j_{ex}\hbar}{q_c}\lambda. \qquad (8.109)$$

Let us express the Hamiltonian as[56]

$$H_{cap}^w = K_{cap}^w + V_{cap}^w \qquad (8.110)$$

$$= \frac{1}{2\varrho}\varpi^2 - \frac{|j_c|\hbar}{q_c}\cos\lambda - \frac{j_{ex}\hbar}{q_c}\lambda, \qquad (8.111)$$

where ϖ is the canonical momentum conjugate to λ and ϱ plays the role of 'mass'. If we recast the term K_{cap}^w in the standard form for an LC circuit, i.e., $Q^2/2C$ as in Eq. (7.193), by setting[57]

$$\frac{1}{2\varrho}\varpi^2 = \frac{1}{2C}Q^2, \qquad (8.112)$$

we can rewrite the Hamiltonian as

$$H_{cap}^w = \frac{1}{2C}Q^2 - \frac{|j_c|\hbar}{q_c}\cos\lambda - \frac{j_{ex}\hbar}{q_c}\lambda. \qquad (8.113)$$

---

[55]Leggett (1987) Eq. (2.23) p. 420. Classically a potential energy arises from the work done by the voltage V on the current j across the junction. The work done by the voltage over a time interval dt is $Vj\,dt$ (Rose-Innes (1980) pp. 167-168, Barone and Paterno (1982) §1.6). We have assumed an electrical potential difference V in a direction such that Vj is positive initially for small values of λ. It follows that a positive work is done by the voltage which can then converted into the potential energy of the junction, as there is no resistive dissipation of energy here. Using Eq. (7.157) and (7.160) for an ac Josephson effect we get

$$\int_0^t Vj\,dt = \int_0^\lambda \left(|j_c|\sin\lambda\right)\frac{\hbar}{q_c}d\lambda = \frac{|j_c|\hbar}{q_c}\left(1-\cos\lambda\right).$$

The first term which is independent of λ can be dropped without affecting the dynamics of the subsequent Hamiltonian system. The second term in V_{cap}^w arises from the external current, i.e., it is due to

$$\int_0^t Vj_{ex}\,dt = j_{ex}\int_0^t V\,dt = -\frac{j_{ex}\hbar}{q_c}\lambda.$$

These analyses are meant to give an intuitive understanding to the terms involved and should not be taken literally. As Fujita and Godoy ((1996) p. 175, p. 254) pointed out, a condensate is not accelerated by an electric force, and therefore a junction voltage should cause no change of the energy of the condensate. Fujita and Godoy ((1996) p. 254) attribute the potential energy term to the microwave associated with an ac Josephson effect. Recall that microwave can give rise to an ac Josephson effect (see footnote to §7.7).

[56]Berkley et al. (2003), Johnson et al. (2003). Also see Eqs. (7.172) and (7.173) for ϖ.

[57]Berkley et al. (2003). Here C is the junction capacitance and Q is the charge on the junction capacitor.

8.4. SUPERCONDUCTING SCHRÖDINGER'S CAT STATES

By setting[58]

$$\varrho = \left(\frac{\hbar}{q_c}\right)^2 C, \tag{8.114}$$

equation (8.112) then leads to[59]

$$\varpi = \pm\frac{\hbar}{q_c} Q. \tag{8.115}$$

If we agree that charge and the total energy are physical observables we should be able to represent these quantities as operators in the Hilbert space associated with the system. The Hilbert space associated with the system, denoted by $\mathcal{H}_{cap}^{w\oplus}$, should be obtained from $\mathcal{H}^{\oplus}$ for the original non-capacitive junction. As discussed in §7.6.6 we can express the Hilbert space $\mathcal{H}^{\oplus}$ as

1. a direct integral over a range of supercurrent $j \in (-|j_c|, 0) \cup (0, |j_c|)$ as in Eq. (7.96), or

2. a direct integral over a range of phase shift $\lambda \in (-\frac{\pi}{2}, 0) \cup (0, \frac{\pi}{2})$ as in Eq. (7.104).

The restricted ranges are taken to maintain a one-to-one relationship between j and λ. If we want to work with a direct integral of spaces $\mathcal{H}_\lambda$ and not worry about having to transform back to a direct integral of $\mathcal{H}(j)$ as in §7.6.6 we can extend the range of λ to $(-\pi, \pi)$ to arrive at a direct integral space[60] which can serve as the Hilbert space for our present capacitive junction, i.e., we have

$$\mathcal{H}_{cap}^{w\oplus} = \int_{-\pi}^{\oplus \pi} \mathcal{H}_\lambda \, d\lambda. \tag{8.116}$$

This Hilbert space consists of elements $\boldsymbol{f}_{cap}^{w\oplus}$ of the form of Eq. (8.108) where the integral is over $\lambda \in (-\pi, \pi)$. The phase shift λ then corresponds to the multiplication operator

$$\widehat{\lambda}_{cap}^{w\oplus} = \int_{-\pi}^{\oplus \pi} \lambda \, \widehat{\boldsymbol{I}}(\lambda) \, d\lambda. \tag{8.117}$$

One would then expect the canonical momentum ϖ to correspond to a differential operator $\widehat{\varpi}_{cap}^w$ given, as in Eq. (7.172), by

$$\widehat{\varpi}_{cap}^w \boldsymbol{f}_{cap}^{w\oplus} = \int_{-\pi}^{\oplus \pi} -i\hbar \frac{\partial f(\lambda)}{\partial \lambda} \, \boldsymbol{\eta}_\lambda \, d\lambda \tag{8.118}$$

---

[58] Berkley et at. (2003).
[59] Johnson et at. (2003). The charge operator will then have the desired eigenvalues.
[60] In such an extension we shall take: (1) $\boldsymbol{\eta}_0$ to correspond to $\eta_{\lambda=0,p=0}^{(1,2)}$ for a zero current, (2) $\boldsymbol{\eta}_{\pm\frac{\pi}{2}}$ to correspond to $\eta_{\lambda=\frac{\pi}{2},p_0}$ of a $\frac{\pi}{2}$-junction in Eq. (7.83) for a maximum current, and (3) $\boldsymbol{\eta}_{\pm\pi}$ to $\eta_{\lambda=\pm\pi,p=0}^{(1,2)}$ for a zero current.

so that the following commutation relation[61]

$$\left[\widehat{\lambda}_{cap}^{w\oplus}, \widehat{\varpi}_{cap}^{w}\right] \boldsymbol{f}_{cap}^{w\oplus} = i\hbar \, \boldsymbol{f}_{cap}^{w\oplus} \tag{8.119}$$

is satisfied. To render $\widehat{\varpi}_{cap}^{w}$ selfadjoint we must impose some boundary conditions on the functions in the domain of the operator. A natural choice would simply be the usual periodic boundary condition $f(-\pi) = f(\pi)$.[62] Choosing the positive sign in Eq. (8.115) the charge operator should then be given by[63]

$$\widehat{Q}_{cap}^{w} \boldsymbol{f}_{cap}^{w\oplus} = \frac{q_c}{\hbar} \widehat{\varpi}_{cap}^{w} \boldsymbol{f}_{cap}^{w\oplus} \tag{8.120}$$

$$= \int_{-\pi}^{\oplus \pi} -iq_c \frac{\partial f(\lambda)}{\partial \lambda} \boldsymbol{\eta}_\lambda \, d\lambda. \tag{8.121}$$

This makes sense since $\widehat{Q}_{cap}^{w}$ possesses a desired set of eigenvalues.[64] We can write down the eigenfunctions and eigenvalues explicitly. Let $\boldsymbol{f}_{cap,n}^{w\oplus}$ be defined by

$$f_n(\lambda) = \frac{1}{\sqrt{2\pi}} e^{in\lambda}, \quad n = 0, \pm 1, \pm 2, \ldots. \tag{8.122}$$

Then we have:

$$\widehat{Q}_{cap}^{w} \int_{-\pi}^{\oplus \pi} f_n(\lambda) \boldsymbol{\eta}_\lambda \, d\lambda = \int_{-\pi}^{\oplus \pi} -iq_c \frac{\partial f_n(\lambda)}{\partial \lambda} \boldsymbol{\eta}_\lambda \, d\lambda \tag{8.123}$$

$$= nq_c \int_{-\pi}^{\oplus \pi} f_n(\lambda) \boldsymbol{\eta}_\lambda \, d\lambda \tag{8.124}$$

$$\Rightarrow \quad \widehat{Q}_{cap}^{w} \boldsymbol{f}_{cap,n}^{w\oplus} = nq_c \boldsymbol{f}_{w,n}^{\oplus}. \tag{8.125}$$

---

[61] Because of the periodic boundary conditions imposed on the domain of $\widehat{Q}_{cap}^{w}$ we have to be careful about the domain of the commutator $[\widehat{\lambda}_{cap}^{w\oplus}, \widehat{\varpi}_{cap}^{w}]$ to avoid getting into trouble. See Fano (1971) pp. 407-408.

[62] The domain of $\widehat{\varpi}_{cap}^{w}$ consists of $\boldsymbol{f}_{cap}^{w\oplus}$ defined by $f(\lambda)$ which are absolutely continuous in λ and satisfy the periodic boundary condition $f(-\pi) = f(\pi)$.

[63] It is possible to choose the negative sign. From the Hamilton's equations given rise by the Hamiltonian H_{cap}^{w} we get $\varpi/\varrho = d\lambda/dt$. Comparing with Eq. (7.160) we get $\varpi/\varrho = -(q_c/\hbar)V = -(q_c/\hbar)Q/C$. Equating $\varpi^2/2\varrho = Q^2/2C$ leads back to Eq. (8.114). Despite this intuitive argument it is more convenient to choose the positive sign to arrive at the operator $\widehat{Q}_{cap}^{w}$ in Eq. (8.121). It will become apparent later that this would fit in better with the notion that charge and flux are canonical variables. It is the square of the charge which appears in the Hamiltonian. So, the choice of the sign of the charge operator does not make any difference.

[64] This also justifies the choice of the periodic boundary condition used to define $\widehat{\varpi}_{cap}^{w}$. It is possible to proceed with a direct integral over a range of $\lambda \in (-\frac{\pi}{2}, \frac{\pi}{2})$. Then we have to change the charge operator with a corresponding change of $f_n(\lambda)$ in order to satisfy the periodic boundary condition and produce the same eigenvalues.

8.4. SUPERCONDUCTING SCHRÖDINGER'S CAT STATES

The charge on the capacitor is therefore a multiple of q_c.[65]

We shall refer to the phase shift and the charge as *canonically conjugate observables at the junction* or as *canonical junction observables*. Since $\widehat{Q}^w_{cap}$ does not commute with $\widehat{\lambda}^{w\oplus}_{cap}$ which is a diagonalizable operator in $\mathcal{H}^{w\oplus}_{cap}$ we can infer from Theorem 2.14.2(2) that $\widehat{Q}^w_{cap}$ is not decomposable in $\mathcal{H}^{w\oplus}_{cap}$. Hence, the acceptance of $\widehat{Q}^w_{cap}$ as an observable means that the superselection rule set up for the original non-capacitive junction described in §7.6.5 and §7.6.6 is broken, i.e., the admission of charge as an observable implies the

breakdown of the superselection rule

for our present system.

The Hamiltonian operator representing the capacitive effect of the junction in $\mathcal{H}^{w\oplus}_{cap}$ can now be written as

$$\widehat{H}^w_{cap} = \frac{1}{2C}\left(\widehat{Q}^w_{cap}\right)^2 - \frac{|j_c|\hbar}{q_c}\cos\widehat{\lambda}^{w\oplus}_{cap} - \frac{j_{ex}\hbar}{q_c}\widehat{\lambda}^{w\oplus}_{cap}, \qquad (8.126)$$

which is again non-decomposable. More explicitly we have

$$\widehat{H}^w_{cap}\int_{-\pi}^{\oplus\pi} f(\lambda)\,\eta_\lambda\,d\lambda$$

$$= \int_{-\pi}^{\oplus\pi}\left\{-\frac{q_c^2}{2C}\frac{\partial^2 f(\lambda)}{\partial\lambda^2} - \frac{|j_c|\hbar}{q_c}\cos\lambda\, f(\lambda) - \frac{j_{ex}\hbar}{q_c}\lambda\right\}\eta_\lambda\,d\lambda. \qquad (8.127)$$

For practical calculations we can simply take an effective Hilbert space formed by functions $f(\lambda)$ of λ which are square-integrable with respect to the measure $d\lambda$ over the range $(-\pi,\pi)$,[66] and represent the operators as[67]

$$\widehat{\lambda}^{w\oplus}_{cap} = \lambda, \qquad (8.128)$$

$$\widehat{Q}^w_{cap} = -iq_c\frac{\partial}{\partial\lambda}, \qquad (8.129)$$

$$\widehat{H}^w_{cap} = -\frac{q_c^2}{2C}\frac{\partial^2}{\partial\lambda^2} - \frac{|j_c|\hbar}{q_c}\cos\lambda - \frac{j_{ex}\hbar}{q_c}\lambda. \qquad (8.130)$$

Case 2 **A TSCR interrupted by a capacitive junction** The circuit configuration is schematically shown in Fig. 8.4.1(2). A constant and uniform external magnetic field is applied to the ring in a direction perpendicular to the plane of the ring.[68]

---

[65] If desired, it is possible to change the range of the variable λ so that the resulting charge operator would admit multiples of e as eigenvalues.

[66] This can be compared with $L^2(\mathcal{C})$ in §6.7.1 and $L^2(\mathcal{C}_c)$ in §6.8.

[67] We may call this a *phase representation*.

[68] As Leggett ((1987) pp. 420-421) pointed out, a ring interrupted by a capacitive junction

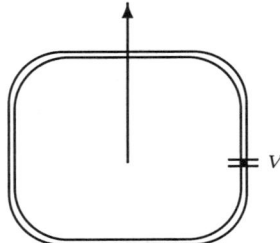

Fig. 8.4.1(2)   Think ring interrupted by a capacitive junction.

We shall make use of the notation and results of §7.4, e.g., the external magnetic flux enclosed by the ring is denoted by Φ_{ex}.

Let us first establish the Hilbert space for the system. Recall the function $\eta_{\lambda,n} \in L^2(\mathcal{C}_c)$ given by Eq. (7.36). Let $\mathcal{H}_{\lambda,n}$ denote the one-dimensional Hilbert space spanned by $\eta_{\lambda,n}$. To highlight the fact that $\eta_{\lambda,n}$ is to be regarded as a vector of a Hilbert space we shall rewrite it in bold as $\boldsymbol{\eta}_{\lambda,n}$. For each n let us introduce a direct integral Hilbert space $\mathcal{H}^{r\oplus}_{cap,n}$ with elements of the form[69]

$$\boldsymbol{f}^{r\oplus}_{cap,n} = \int_{-\pi}^{\oplus \pi} f_n(\lambda)\, \boldsymbol{\eta}_{\lambda,n}\, d\lambda. \tag{8.131}$$

Following Eqs. (8.117) and (8.121) we can introduce a phase shift operator and a charge operator acting in $\mathcal{H}^{r\oplus}_{cap,n}$, i.e.,

$$\widehat{\lambda}^{r\oplus}_{cap,n}\, \boldsymbol{f}^{r\oplus}_{cap,n} = \int_{-\pi}^{\oplus \pi} \{\lambda f_n(\lambda)\}\, \boldsymbol{\eta}_{\lambda,n}\, d\lambda, \tag{8.132}$$

$$\widehat{Q}^{r}_{cap,n}\, \boldsymbol{f}^{r\oplus}_{cap,n} = \int_{-\pi}^{\oplus \pi} \left\{ -iq_c \frac{\partial f_n(\lambda)}{\partial \lambda} \right\}\, \boldsymbol{\eta}_{\lambda,n}\, d\lambda. \tag{8.133}$$

When acting on $f(\lambda)$ as the wave function Eqs. (8.128) and (8.129) apply.

In addition we have a flux operator to represent the total magnetic flux enclosed by the ring. In view of Eq. (7.34) we can define a flux operator in terms of the phase operator on $\mathcal{H}^{r\oplus}_{cap,n}$ by

$$\widehat{\Phi}^{r\oplus}_{cap,n} = -\left(n - \frac{1}{2\pi}\widehat{\lambda}^{r\oplus}_{cap,n}\right)\Phi_c, \quad \Phi_c = \frac{h}{q_c}, \tag{8.134}$$

---

is conceptually simpler than a long wire interrupted by a similar junction. This because the former is a self-contained system, while the latter involves an external current.

[69] The superscript r and subscripts cap indicate quantities associated with a capacitive junction in the ring. In the absence of an external magnetic field the phase shift and n are related because of the Josephson equation, resulting in a discrete set of values for λ, and a subsequent direct sum Hilbert space, as shown in §7.4.4. In the presence of an external magnetic field the situation changes. As the external magnetic field varies continuously, the corresponding phase shift also varies continuously in step (see §7.4.7). We can therefore construct a direct integral space $\mathcal{H}^{r\oplus}_{cap,n}$.

8.4. SUPERCONDUCTING SCHRÖDINGER'S CAT STATES

or

$$\widehat{\lambda}^{r\oplus}_{cap,n} = 2\pi n + 2\pi \frac{\widehat{\Phi}^{r\oplus}_{cap,n}}{\Phi_c}. \tag{8.135}$$

This flux operator is diagonalizable, i.e., we have

$$\widehat{\Phi}^{r\oplus}_{cap,n} f^{r\oplus}_{cap,n} = \int_{-\pi}^{\oplus\pi} \left\{ -\left(n - \frac{\lambda}{2\pi}\right) \Phi_c f_n(\lambda) \right\} \eta_{\lambda,n} \, d\lambda \tag{8.136}$$

$$= \int_{-\pi}^{\oplus\pi} \Phi_n(\lambda) f_n(\lambda) \eta_{\lambda,n} \, d\lambda, \tag{8.137}$$

where

$$\Phi_n(\lambda) = -\left(n - \frac{\lambda}{2\pi}\right) \Phi_c \quad \text{or} \quad \lambda = 2\pi \left(n + \frac{\Phi_n}{\Phi_c}\right). \tag{8.138}$$

We can express various quantities in terms of the flux operator instead of the phase operator:

1. Commutation relation:

$$\left[\widehat{\Phi}^{r\oplus}_{cap,n}, \widehat{Q}^{r}_{cap,n} \right] = i\hbar. \tag{8.139}$$

The flux and the charge are seen as canonically conjugate observables.

2. Potential energy:

   (a) First we have

   $$-\frac{|j_c|\Phi_c}{2\pi} \cos \widehat{\lambda}^{r\oplus}_{cap,n} = -\frac{|j_c|\Phi_c}{2\pi} \cos \left(2\pi \frac{\widehat{\Phi}^{r\oplus}_{cap,n}}{\Phi_c} \right). \tag{8.140}$$

   This corresponds to V^w_{cap} for the junction in a long wire without an external current.

   (b) Next, Our present system possesses an additional energy due to the magnetic flux, i.e., we have[70]

   $$\frac{1}{2L_c} \left(\widehat{\Phi}^{r\oplus}_{cap,n} - \Phi_{ex} \right)^2. \tag{8.141}$$

Comparing with Eq. (8.126) we can conclude that the Hamiltonian incorporating the capacitive effect and the flux is[71]

$$\widehat{H}^{r}_{cap,n} = \frac{1}{2C} \left(\widehat{Q}^{r}_{cap,n} \right)^2 - \frac{|j_c|\Phi_c}{2\pi} \cos \left(2\pi \frac{\widehat{\Phi}^{r\oplus}_{cap,n}}{\Phi_c} \right)$$

$$+ \frac{1}{2L_c} \left(\widehat{\Phi}^{r\oplus}_{cap,n} - \Phi_{ex} \right)^2. \tag{8.142}$$

---

[70] For energy due to enclosed flux see Eq. (7.14).
[71] Leggett (1987) p. 422.

When acting on $\boldsymbol{f}^{r\oplus}_{cap,n}$ we have

$$\widehat{H}^r_{cap,n} \int_{-\pi}^{\oplus \pi} f_n(\lambda)\, \boldsymbol{\eta}_{\lambda,n}\, d\lambda$$

$$= \int_{-\pi}^{\oplus \pi} \left\{ -\frac{q_c^2}{2C} \frac{\partial^2 f_n(\lambda)}{\partial \lambda^2} - \left(\frac{|j_c|\Phi_c}{2\pi} \cos \lambda\right) f_n(\lambda) \right.$$

$$\left. + \frac{1}{2L_c} \left(-\left(n - \frac{\lambda}{2\pi}\right)\Phi_c - \Phi_{ex}\right)^2 f_n(\lambda) \right\} \boldsymbol{\eta}_{\lambda,n} d\lambda. \quad (8.143)$$

In practical applications we are directly interested in the enclosed flux rather than the phase shift. Hence a *flux representation* based on the treatment of a quantum LC circuit in §7.4 in Chapter 7 is often adopted. This is obtained as follows:

1. Choose an appropriate range of flux values relevant to a particular application or experiment. As an example let us consider the range from 0 to Φ_c. This means that we would collect an appropriate set of functions $\eta_{\lambda,n}$ which correspond to the total enclosed flux $\Phi = \Phi_{\lambda,n} \in (0, \Phi_c)$.[72] We shall label these functions in terms of Φ, i.e., we rewrite $\eta_{\lambda,n}$ as η_Φ and the corresponding Hilbert spaces $\mathcal{H}_{\lambda,n}$ as $\mathcal{H}(\Phi)$. Each η_Φ spans a Hilbert spaces $\mathcal{H}(\Phi)$. We can now form the direct integral of these Hilbert spaces:

$$\mathcal{H}^{r\oplus}_{fr} = \int_0^{\Phi_c} \mathcal{H}(\Phi)\, d\Phi. \quad (8.144)$$

Elements of $\mathcal{H}^{r\oplus}_{fr}$ are of the form[73]

$$\boldsymbol{f}^{r\oplus}_{fr} = \int_0^{\Phi_c} f(\Phi)\, \boldsymbol{\eta}_\Phi\, d\Phi. \quad (8.145)$$

The flux representation space $\mathcal{H}^{r\oplus}_{fr}$ is not the same as $\mathcal{H}^{r\oplus}_{cap,n}$, e.g., when $n = 0$ we have $\mathcal{H}^{r\oplus}_{cap,0}$ corresponding to the total enclosed flux lying in the range $(-\frac{1}{2}\Phi_c, \frac{1}{2}\Phi_c)$.

---

[72]For example we may choose

$$n = 0, \quad \lambda \in (0, \pi) \quad \Rightarrow \quad \Phi \in \left(0, \frac{1}{2}\Phi_c\right),$$

and

$$n = -1, \quad \lambda \in (-\pi, 0) \quad \Rightarrow \quad \Phi \in \left(\frac{1}{2}\Phi_c, \Phi_c\right).$$

The corresponding set of $\eta_{\lambda,n}$ would correspond to $\Phi \in (0, \Phi_c)$.
[73]The subscripts fr stand for flux representation.

8.4. SUPERCONDUCTING SCHRÖDINGER'S CAT STATES

2. The flux and charge operators then act on $\boldsymbol{f}_{fr}^{\oplus}$ as[74]

$$\widehat{\Phi}_{fr}^{r\oplus} \, \boldsymbol{f}_{fr}^{r\oplus} = \int_0^{\Phi_c} \{\Phi f(\Phi)\} \, \boldsymbol{\eta}_\Phi \, d\Phi, \tag{8.146}$$

$$\widehat{Q}_{fr} \, \boldsymbol{f}_{fr}^{\oplus} = \int_0^{\Phi_c} \left\{ -i\hbar \frac{\partial f(\Phi)}{\partial \Phi} \right\} \boldsymbol{\eta}_\Phi \, d\Phi. \tag{8.147}$$

3. The Hamiltonian in Eq. (8.142) becomes

$$\widehat{H}_{fr}^r = \frac{1}{2C}\left(\widehat{Q}_{fr}^r\right)^2 - \frac{|j_c|\Phi_c}{2\pi}\cos\left(2\pi\frac{\widehat{\Phi}_{fr}^\oplus}{\Phi_c}\right) + \frac{1}{2L_c}\left(\widehat{\Phi}_{fr}^\oplus - \Phi_{ex}\right)^2. \tag{8.148}$$

4. Acting on $f(\Phi)$ as a wave function we have[75]

$$\widehat{\Phi}_{fr}^{r\oplus} = \Phi, \tag{8.149}$$

$$\widehat{Q}_{fr}^r = -i\hbar\frac{\partial}{\partial \Phi}, \tag{8.150}$$

$$\widehat{H}_{fr}^r = -\frac{\hbar^2}{2C}\frac{\partial^2}{\partial \Phi^2} - \frac{|j_c|\Phi_c}{2\pi}\cos\left(2\pi\frac{\Phi}{\Phi_c}\right) + \frac{1}{2L_c}\left(\Phi - \Phi_{ex}\right)^2. \tag{8.151}$$

The Hamiltonian is seen to consist of an effective kinetic energy term

$$\widehat{K}_{fr}^r = -\frac{\hbar^2}{2C}\frac{\partial^2}{\partial \Phi^2}, \tag{8.152}$$

and an effective potential term

$$V_{fr}^{r\oplus}(\Phi) = -\frac{|j_c|\Phi_c}{2\pi}\cos\left(2\pi\frac{\Phi}{\Phi_c}\right) + \frac{1}{2L_c}\left(\Phi - \Phi_{ex}\right)^2 \tag{8.153}$$

$$= -\frac{|j_c|\Phi_c}{2\pi}\cos\left(2\pi\frac{\Phi}{\Phi_c}\right) + \frac{\Phi_c^2}{2L_c}\left(\frac{\Phi}{\Phi_c} - \frac{\Phi_{ex}}{\Phi_c}\right)^2. \tag{8.154}$$

---

[74] We need to impose the periodic boundary condition $f(0) = f(\Phi_c)$ to achieve selfadjointness. For an intuitive comparison with the charge operator $\widehat{Q}_{cap,n}^r$ in $\widehat{H}_{cap,n}^{r\oplus}$ we note that

$$-iq_c\frac{\partial}{\partial \lambda} = -iq_c\left(\frac{\Phi_c}{2\pi}\frac{\partial}{\partial \Phi}\right) = -i\hbar\frac{\partial}{\partial \Phi}.$$

[75] $\widehat{Q}_{fr}^r$ possesses the expected eigenvalues nq_c with eigenfunctions constructed from $f(\Phi)$ proportional to $\exp(i2\pi n\Phi/\Phi_c)$. These functions satisfy the periodic boundary condition.

8.4.2 Schrödinger's cat states in superconducting systems

Let us adopt the flux representation to describe a TSCR interrupted by a capacitive junction. We shall confine our attention to the space $\mathcal{H}_{fr}^{r\oplus}$ with the total enclosed flux lying in the range $(0, \Phi_c)$. Now, consider the case when the external flux Φ_{ex} is equal to the value of a half flux quantum, i.e., $\Phi_{ex} = \frac{1}{2}\Phi_c$. Introduce a new variable ρ by

$$\rho = 2\pi \left(\frac{\Phi}{\Phi_c} - \frac{1}{2} \right), \tag{8.155}$$

then

$$\Phi \in (0, \Phi_c) \quad \Rightarrow \quad \rho \in (-\pi, \pi). \tag{8.156}$$

We can rewrite the potential $V_{fr}^{r\oplus}$ in Eq. (8.154) in terms of ρ as

$$V_{fr}^{r\oplus}(\rho) = \frac{\Phi_c^2}{(2\pi)^2 L_c} \left\{ \frac{2\pi L_c |j_c|}{\Phi_c} \cos\rho + \frac{1}{2}\rho^2 \right\}. \tag{8.157}$$

This potential is an even function of ρ. It assumes a local maximum at $\rho = 0$, provided L_c and $|j_c|$ are sufficiently large, flanked by two local minima symmetrical about $\rho = 0$.[76] These local minima occur at

$$\rho_+ > 0, \text{ i.e., } \Phi_+ > \tfrac{1}{2}\Phi_c, \text{ and } \rho_- < 0, \text{ i.e., } \Phi_- < \tfrac{1}{2}\Phi_c. \tag{8.158}$$

At these minima we have

$$\rho_\pm = \frac{2\pi L_c |j_c|}{\Phi_c} \sin\rho_\pm, \tag{8.159}$$

or more explicitly

$$\left(\Phi_\pm - \frac{1}{2}\Phi_c \right) = L_c |j_c| \sin\frac{2\pi}{\Phi_c} \left(\Phi_\pm - \frac{1}{2}\Phi_c \right) \tag{8.160}$$

---

[76] Friedman et al. (2000), Leggett (1987) pp. 422-423.

$$V_{fr}^{r\oplus\prime} = \frac{dV_{fr}^{r\oplus}(\rho)}{d\rho} = \frac{\Phi_c^2}{(2\pi)^2 L_c}\left(-\frac{2\pi L_c|j_c|}{\Phi_c}\sin\rho + \rho \right),$$

$$V_{fr}^{r\oplus\prime\prime} = \frac{d^2 V_{fr}^{r\oplus}(\rho)}{d\rho^2} = \frac{\Phi_c^2}{(2\pi)^2 L_c}\left(-\frac{2\pi L_c|j_c|}{\Phi_c}\cos\rho + 1 \right).$$

$\rho = 0 \Rightarrow V_{fr}^{r\oplus\prime} = 0$ but $V_{fr}^{r\oplus\prime\prime} < 0$ only if $\frac{2\pi L_c|j_c|}{\Phi_c} > 1$.

Since $V_{fr}^{r\oplus}(\rho)$ is an even function of ρ the two neighbouring minima are symmetric about $\rho = 0$, occurring at $\rho_+ > 0$ and $\rho_+ < 0$.

8.4. SUPERCONDUCTING SCHRÖDINGER'S CAT STATES

which can be reduced to a Josephson equation in the form of Eq. (7.64):[77]

$$\Phi_\pm - \Phi_{ex} = L_c \, |j_c| \sin \lambda_\pm \qquad (8.161)$$

where

$$\Phi_{ex} = \frac{1}{2}\Phi_c \quad \text{and} \quad \lambda_\pm = \frac{2\pi}{\Phi_c}\left(\Phi_\pm - \frac{1}{2}\Phi_c\right). \qquad (8.162)$$

The supercurrent $j = j_c \sin \lambda_\pm$ flows in the clockwise direction at the minimum Φ_+ and in the anti-clockwise direction at Φ_- round the ring.[78]

To conclude we see that the potential $V_{fr}^{r\oplus}$ is of the form of a finite double-well potential. Consequently a discussion similar to that in §8.3.9 can apply here.[79] In other words we can have:

1. The two lowest energy eigenstates of the Hamiltonian $\widehat{H}_{fr}^r$ may have only a slight energy difference similar to the two eigenstates in Eq. (8.96).

2. The two asymmetrical superpositions of these energy eigenstates similar to those oscillating states in Eq. (8.98) may be formed. The existence or otherwise of these asymmetrical superpositions needs to be verified experimentally as in the case of ammonia molecules. This would confirm the breakdown of the superselection rule as stated in §8.4.1

3. Since the oscillating states in Eq. (8.98) may be regarded as classically distinguishable, the experimental confirmation of these states, as in the case of ammonia molecules, would demonstrate the existence of superposition of classically distinguishable states in superconducting systems. This is particularly interesting since we have two energy eigenstates one of which corresponds to a clockwise current flow and the other to an anticlockwise current flow. A superposition of these two energy eigenstates could be interpreted as having currents flowing through a superconducting ring in opposite directions simultaneously.[80]

4. Experiments by Friedman et al. (2000) in Stoney Brook and van der Wal et al. (2000) in the Netherlands show strong evidence for the existence of such superpositions.

---

[77] Wan and Harrison (1993).
[78] At Φ_+ the phase shift λ_+ lies in $(0, \pi)$ so that $\sin \lambda_+ > 0$, while at Φ_- the phase shift λ_- lies in $(-\pi, 0)$ so that $\sin \lambda_+ < 0$. Note that j_c is negative.
[79] Leggett (1986) and (2000).
[80] The resulting currents could be in the order of microamps which can be deemed macroscopic and classically distinguishable (Leggett (2000)).

8.5 Asymptotically Separable Quantum Theory

8.5.1 Motivation

Our discussion in §8.2 tells us that an experiment may not be able to detect any correlations between a state localized nearby and a state localized arbitrarily far away since the correlations caused by any finite number of observables $\widehat{A}$ would be arbitrarily small. So far we have focused our attention on states and their spatial separation. We can equally look into the situation from the point of view of observables. Recall the discussion in §2.12 on the mathematics of local observables. The physical motivation for studying local observables emerges in §3.6 on quantum measurement, especially in §3.6.1 and §3.6.2. The reasoning is that generally it is not feasible to measure global observables, not directly in any case. We can measure local position observables χ_Λ directly, not the global position observable $\widehat{x} = x$. Moreover, any measurement can be reduced to local position measurement in an asymptotic manner.

The question then arises as to what kind of quantum mechanics will result if we confine ourselves to local observables. This kind of question was asked quite sometimes ago by workers in quantum field theory and statistical mechanics.[81] An analysis of this question results in an algebraic approach to quantum field theory based on local observables.[82] There is now a well-established formulation of algebraic quantum field theory based on C^*-algebras of observables. Once a C^*-algebra of observables is identified a formal theory can be set up algebraically. The idea is to associate a C^*-algebra with a physical system such that selfadjoint elements of the C^*-algebra are taken to correspond to physical observables with normalized positive linear functionals on the algebra corresponding to states. Such an algebraic approach to quantum field theory and to infinite quantum systems such as thermodynamical systems is well-established.[83] This approach, when applied to quantum mechanics, results in an asymptotically separable theory.

8.5.2 Asymptotically separable quantum mechanics

We shall give an outline of the theory here; more details are available in Wan and McLean (1984a), (1984b), and Wan and Jackson (1985). Consider a quantum system with its associated Hilbert space $\mathcal{H} = L^2(I\!\!E^n)$. Let $\mathcal{B}(\mathcal{H})$ denote

---

[81] Haag and Kastler (1964).

[82] Guenin (1966), Emch (1972), Bratteli and Robinson (1979), (1981), Primas and Müller-Herold (1978), Haag (1992), Landsman (1998), Sewell (1986), (2002).

[83] For a readily readable account see Bogoliubov (1975) and Sewell (1986).

8.5. ASYMPTOTICALLY SEPARABLE QUANTUM THEORY

the set of all bounded operators on $\mathcal{H}$. We know from §2.12.3 that $\mathcal{B}(\mathcal{H})$ possesses a subalgebra $\mathcal{A}_L$ of local operators $\widehat{A}_\Lambda$. Let us investigate what would happen if we restrict all bounded physical observables to selfadjoint members of $\mathcal{A}_L$, rather than $\mathcal{B}(\mathcal{H})$. Intuitively one can see that these local observables are unable to correlate states which are spatially separated by an arbitrary distance. To make this notion more precise let us consider a free quantum particle in $I\!E^3$. Theorem 3.4.2(2) on asymptotic localization in §3.4.2 tells us that two wave functions ϕ and ψ corresponding to momentum values respectively in $m\Delta v$ and $m\Delta w$ will evolve into spatial regions $t\Delta v$ and $t\Delta w$ at large times. Suppose $m\Delta v$ and $m\Delta w$ are disjoint. Then ϕ and ψ will localize asymptotically into disjoint spatial regions $t\Delta v$ and $t\Delta w$ whose separation will become arbitrarily large. These two wave functions are asymptotically separable in accordance with Definition 3.4.2(1). We also call the corresponding evolved wave functions ϕ_t and ψ_t two wave functions *infinitely separated* in the limit as $t \to \infty$. It is obvious that[84]

$$\lim_{t\to\infty} \langle \phi_t \mid \widehat{A}_\Lambda \psi_t \rangle = 0 \quad \forall \widehat{A}_\Lambda \in \mathcal{A}_L. \tag{8.163}$$

It follows that as far as local observables are concerned decoherence will set in asymptotically for states with disjoint momentum values. At large times T we have ϕ_T and ψ_T becoming disjoint FAPP, i.e.,

$$\Phi_T = \frac{1}{\sqrt{2}}(\phi_T + \psi_T) \equiv \widehat{D}_T = \frac{1}{2}\Big(|\phi_T\rangle\langle\phi_T| + |\psi_T\rangle\langle\psi_T|\Big) \text{ FAPP.} \tag{8.164}$$

The restriction of observables to local ones is crucial here. Without this restriction one cannot deduce decoherence of infinitely separated states. In other words Eq. (8.163) is not true if $\mathcal{A}_L$ is replaced by $\mathcal{B}(\mathcal{H})$. For example, for a free particle in one-dimensional motion in $I\!E$ a wave function ϕ_t moving to the right and its mirror image $\varphi_t = \widehat{\wp}\phi_t$, where $\widehat{\wp}$ is the parity operator[85] defined by Eq. (6.30), will increasingly separate in time but we have

$$\lim_{t\to\infty} \langle \phi_t \mid \widehat{A}\varphi_t \rangle \neq 0 \quad \text{for some } \widehat{A} \in \mathcal{B}(\mathcal{H}). \tag{8.165}$$

The parity operator $\widehat{\wp}$ for instance is a bounded selfadjoint operator in $\mathcal{B}(\mathcal{H})$ but $\widehat{\wp}$ is not a member of $\mathcal{A}_L$. If we let $\widehat{A} = \widehat{\wp}$, then we have,

$$\lim_{t\to\infty} \langle \phi_t \mid \widehat{A}\varphi_t \rangle = \lim_{t\to\infty} \langle \phi_t \mid \widehat{\wp}\varphi_t \rangle = \lim_{t\to\infty} \langle \phi_t \mid \phi_t \rangle = 1, \tag{8.166}$$

due to the fact that $\widehat{\wp}^2 = \widehat{I\!I}$. So, in the algebraic formulation of quantum mechanics we can obtain different quantum theories by taking different sets

---
[84] Recall that $\widehat{A}_\Lambda$ is of the form $\chi_\Lambda \widehat{A} \chi_\Lambda$. Since ϕ_t and ψ_t are moving to regions far away the scalar product $\langle \phi_t \mid \widehat{A}_\Lambda \psi_t \rangle = \langle \phi_t \mid \chi_\Lambda \widehat{A} \chi_\Lambda \psi_t \rangle$ would tend to zero as $t \to \infty$.
[85] $\widehat{\wp}$ commutes with free Hamiltonian (Roman (1965) p. 528).

of operators as observables, i.e., the choice of the C^*-algebra is crucial in determining the nature of the resulting theory. We shall consider two examples to illustrate the situation.

Orthodox quantum mechanics   If we choose

1. the set of all bounded operators $\mathcal{B}(\mathcal{H})$ as the C^*-algebra associated with the system, and

2. the set of statistically normalized positive linear functionals introduced in Definition 4.4.4(1) as states,

then we will recover orthodox quantum statics embodied in Postulate OQS of §3.2.1.

Asymptotically separable quantum mechanics   If we choose

1. the quasi-local algebra $\bar{\mathcal{A}}_L$ of operators introduced in §2.12.3 as the C^*-algebra associated with the system,[86] and

2. the set of statistically normalized positive linear functionals introduced in Definition 4.4.4(1) as the state,

then we would produce a non-orthodox theory which is asymptotically separable in the sense of Eqs. (8.163) and (8.164). It follows that as far as quasi-local observables are concerned decoherence will set in asymptotically for states with disjoint momentum values.

One can go one step further to introduce local Hamiltonian generators, such as those of the form of Eq. (2.552) in §2.12.6, which will not lead to the usual spreading of the wave packet arising from evolution generated by global Hamiltonian generators.[87]

Finally we should mention that a number of people have pointed out that an asymptotically separable theory would violate some well-established conservation laws, e.g., the angular momentum conservation law.[88] We shall return to discuss this in §8.6

8.6   Entanglement and Decoherence

There has been a great deal of interest in many-particle systems exhibiting what has come to be known as *entanglement* ever since the problem was first

---

[86]The choice of $\bar{\mathcal{A}}_L$ does not alter the local nature of the observables involved. Every $\bar{A} \in \bar{\mathcal{A}}_L$ is the uniform limit of a sequence of local observables $\widehat{A}_{\Lambda_n}$, i.e., $\lim_{n\to\infty} ||\bar{A} - \widehat{A}_{\Lambda_n}|| = 0$. In other words $\bar{A}$ can be approximated by $\mathcal{A}_{\Lambda_n}$ with arbitrary accuracy which is independent of states.

[87]Wan (1988).

[88]Selleri and Tarozzi (1981) p. 10.

8.6. ENTANGLEMENT AND DECOHERENCE

raised in 1935 by Einstein, Podolsky and Rosen (EPR).[89] Some recent experiments show strong evidence of entanglement even for macroscopic quantum systems, e.g., experiments on the entanglement of two superconducting rings with a capacitive junction.[90] Here we shall look into the entanglement and subsequent decoherence of two massive particles in the context of asymptotically separable quantum mechanics. The de Broglie paradox is about the correlations between two asymptotically separable components of a single-particle wave functions. The EPR problem is to do with distant correlations between two or more particles. We shall not consider the theory of massless particles like photons here.[91] This book deals with non-relativistic theories and massless particles are necessarily relativistic in nature. There are fundamentally new issues to be addressed in any relativistic theory. For example, the very concept of photons requires a very careful consideration. Even the definition of a position observable of a massless particle is controversial.[92] Any questions on the size or distances travelled by photons are difficult to answer. Consider the simplest situation in Bohr's theory of the hydrogen atom. A photon is emitted by a hydrogen atom when an orbiting electron makes a transition to a lower energy state. The atom is of a linear dimension of 0.1 nanometer. What is the size of the photon emitted? For the sake of argument it seems to be intuitively reasonable to assume a photon size to be of the order of its wave length which has a linear dimension of several hundred nanometers. Then one has a peculiar situation of a tiny atom emitting something several thousand times bigger than itself. Even harder to imagine is the absorption process, i.e., a tiny atom has to swallow something several thousand times bigger than itself! Similar problems emerge when considering how a photon travels in matter. In condense matter the inter-atomic distances are of the order of nanometers. It is not easy to visualize how a photon, which is huge compared with inter-atomic distances, travels in the free space between atoms and how one can measure how fast such a huge photon would travel between two atoms separated by only a fraction of the size of the photon. The problem becomes less serious as the distances becomes much bigger, i.e., when the distances involved are much bigger than the 'size' (wave length) of the photon.

8.6.1 Distinguishable particles

Let us start with two distinguishable and massive particles. The Hilbert space for the compound system of these two particles is assumed to be the tensor

---

[89]Einstein, Podolsky and Rosen (1935). For a book popularizing entanglement see Aczel (2002).
[90]Berkley *et al.* (2003), Johnson *et al.* (2003).
[91]Unfortunately most experiments on entanglement are performed on photons.
[92]Kraus (1977), Prugovečki (1984), Schroeck (1996), Bacry (1988).

624   CHAPTER 8. SEPARABILITY AND DECOHERENCE

product space $\mathcal{H}^{(c)} = \mathcal{H}_1 \otimes \mathcal{H}_2$ of the Hilbert spaces $\mathcal{H}_1$ of particle 1 and $\mathcal{H}_2$ of particle 2. Let us suppose that the two particles interact and as a result of the interaction the two-particle wave function cannot be factorized into a single tensor product of single-particle wave functions. For example, at time t we could have a two-particle wave function of the form of a linear combination of tensor products:

$$\Psi_t^{(c)} = \frac{1}{\sqrt{2}} (\psi_t \otimes \xi_t + \phi_t \otimes \eta_t), \qquad (8.167)$$

where ψ_t, ϕ_t are orthogonal unit vectors in $\mathcal{H}_1$ and similarly ξ_t, η_t are orthogonal unit vectors in $\mathcal{H}_2$. Note that orthogonality is preserved under unitary evolution so that ψ_t, ϕ_t are orthogonal for all t and so are ξ_t, η_t. Then a measurement at a later time $t = T$ on one of the particles will immediately influence the state of the other particle in the system, a conclusion now referred to as *entanglement*. States which cannot be decomposed into a single tensor product of one-particle states, e.g., states of the form of Eq. (8.167) are called *entangled states*. The following scenario serves to highlight the situation:

1. Suppose that at the time of measurement, T, the two particles are spatially very far apart and that the measurement in question obeys von Neumann's projection postulate discussed in §3.5.1.

2. The state of the system just before the measurement is

$$\Psi_T^{(c)} = \frac{1}{\sqrt{2}} (\psi_T \otimes \xi_T + \phi_T \otimes \eta_T), \qquad (8.168)$$

   and hence the state of particle 2 immediately before the measurement is neither ξ_T nor η_T.

3. If the measurement made on particle 1 results in projecting the particle's state to ψ_T, then particle 2 would have to be in state ξ_T right after the measurement. This implies a change of the state of particle 2.

4. Since particle 2 is far away, an immediate change of its state appears to be an action-at-a-distance effect. This causes a serious conceptual problem, commonly referred to as the *EPR paradox*.[93]

So, how are we to understand the apparent phenomenon that a measurement performed on a distant particle appears to have a physical effect on a particle nearby to the extent of influencing its state? Let us look into this carefully. At time T the composite state $\Psi_T^{(c)}$ appears to be a coherent superposition of its

---
[93] See Torozzi and van der Merwe (1980) and Selleri (1990) for more discussion.

8.6. ENTANGLEMENT AND DECOHERENCE

constituents $\psi_T \otimes \xi_T$ and $\phi_T \otimes \eta_T$. We like to know what kind of observables are capable of showing up the coherent nature of its constituents. Let $\widehat{B}_1$ and $\widehat{I\!\!I}_1$ be respectively a bounded observable and the identity on $\mathcal{H}_1$, and similarly for $\widehat{B}_2$ and $\widehat{I\!\!I}_2$ on $\mathcal{H}_2$. Let us consider the following two types of observables:

1. **One-particle observables**   Observables in $\mathcal{H}^{(c)}$ of the form

$$\widehat{B}_1^{(c)} = \widehat{B}_1 \otimes \widehat{I\!\!I}_2, \quad \widehat{B}_2^{(c)} = \widehat{I\!\!I}_1 \otimes \widehat{B}_2 \qquad (8.169)$$

are called *one-particle observables* as they relate to only one of the two particles in the composite system and their expectation value can be obtained from the relevant one-particle states.[94] We have

$$\langle \Psi_T^{(c)} \mid \widehat{B}_1^{(c)} \Psi_T^{(c)} \rangle = \frac{1}{2} \left(\langle \psi_T \mid \widehat{B}_1 \psi_T \rangle + \langle \phi_T \mid \widehat{B}_1 \phi_T \rangle \right), \qquad (8.170)$$

showing clearly that the expectation value $\langle \Psi_T^{(c)} \mid \widehat{B}_1^{(c)} \Psi_T^{(c)} \rangle$ is obtainable from the measurement of $\widehat{B}_1$ in states ϕ_T and ψ_T separately on particle 1 without reference to the states of the second particle, and similarly for $\langle \Psi_T^{(c)} \mid \widehat{B}_2^{(c)} \Psi_T^{(c)} \rangle$. This can also be seen from the fact that $\widehat{B}_1^{(c)}$ and $\widehat{B}_2^{(c)}$ commute, hence simultaneously and independently measurable. Moreover these one-particle observables are incapable of producing any correlation between the constituents $\psi_T \otimes \xi_T$ and $\phi_T \otimes \eta_T$ in $\Psi_T^{(c)}$.

2. **Two-particle observables**   Let us examine observables of the form

$$\widehat{B}^{(c)} = \widehat{B}_1^{(c)} \widehat{B}_2^{(c)} = \widehat{B}_1 \otimes \widehat{B}_2. \qquad (8.171)$$

For this observable we have[95]

$$\langle \Psi_T^{(c)} \mid \widehat{B}^{(c)} \Psi_T^{(c)} \rangle = \frac{1}{2} \langle \psi_T \mid \widehat{B}_1 \psi_T \rangle \langle \xi_T \mid \widehat{B}_2 \xi_T \rangle$$

$$+ \frac{1}{2} \langle \phi_T \mid \widehat{B}_1 \phi_T \rangle \langle \eta_T \mid \widehat{B}_2 \eta_T \rangle + \mathrm{Re}\, \langle \psi_T \mid \widehat{B}_1 \phi_T \rangle \langle \xi_T \mid \widehat{B}_2 \eta_T \rangle. \qquad (8.172)$$

---

[94] Jauch (1968) §11.8, Peres (1995) Chapter 5.

[95] The factor $\langle \psi_T \mid \widehat{B}_1 \phi_T \rangle$ in the correlation terms is expressible in terms of expectation values involving particle 1 only, e.g., using Eq. (2.41) we obtain

$$\langle \psi_T \mid \widehat{B}_1 \phi_T \rangle \;=\; \tfrac{1}{2} \left\{ \langle \psi_T + \phi_T \mid \widehat{B}_1 (\psi_T + \phi_T) \rangle - i \langle \psi_T + i\phi_T \mid \widehat{B}_1 (\psi_T + i\phi_T) \rangle \right.$$
$$\left. + \; i \langle \psi_T \mid \widehat{B}_1 \psi_T \rangle + i \langle \phi_T \mid \widehat{B}_1 \phi_T \rangle - \langle \psi_T \mid \widehat{B}_1 \psi_T \rangle - \langle \phi_T \mid \widehat{B}_1 \phi_T \rangle \right\}.$$

It follows that expectation value $\langle \Psi_T^{(c)} \mid \widehat{B}^{(c)} \Psi_T^{(c)} \rangle$ can be obtained by measurements performed on the two particles separately.

We can see that it is the correlations term $\text{Re}\langle\psi_T \mid \widehat{B}_1\, \phi_T\rangle\langle\xi_T \mid \widehat{B}_2\, \eta_T\rangle$ which "entangles" the constituents $\psi_T \otimes \xi_T$ and $\phi_T \otimes \eta_T$ in $\Psi_T^{(c)}$. In other words it is the presence of this correlation term due to a two-particle observable $\widehat{B}^{(c)}$ which tells us that $\Psi_T^{(c)}$ is a coherent superposition of its constituents $\psi_T \otimes \xi_T$ and $\phi_T \otimes \eta_T$.

However, the situation changes if ψ_T and ϕ_T move apart spatially, and similarly for ξ_T and η_T, as T gets bigger.[96] Then the correlation terms vanish asymptotically. As a result we can regard Ψ_T^c in Eq. (8.168) as a mixture rather than a coherent superposition, i.e., the two particles are either in state $\phi_T \otimes \xi_T$ or state $\psi_T \otimes \eta_T$. In this particular case we have a way out of the EPR paradox in the sense that at large times the nature of the state $\Psi_T^{(c)}$ changes into a mixture FAPP before the measurement.[97]

8.6.2 Identical Fermions and Pauli exclusion principle

When dealing with two identical fermions Pauli exclusion principle comes into play which tacitly implies a correlation between the two particles for a start.

Let $\mathcal{H}$ be the one-fermion Hilbert space. Then for a system of two identical fermions the Hilbert space $\mathcal{H}_f$ is well-known to be a subspace of $\mathcal{H} \otimes \mathcal{H}$ consisting of states which are anti-symmetric with respect to the interchange of the two particles. For example a state at time t may be of the form

$$\Psi_{ft} = \frac{1}{\sqrt{2}}(\psi_t \otimes \phi_t - \phi_t \otimes \psi_t), \qquad (8.173)$$

where ψ_t and ϕ_t are orthonormal one-fermion wave functions. Physical observables of the system must then be symmetric with respect to the interchange of the two particles, e.g., of the form[98]

$$\widehat{C}_f = \widehat{A} \otimes \widehat{B} + \widehat{B} \otimes \widehat{A}, \qquad (8.174)$$

where $\widehat{A}$ and $\widehat{B}$ are observables of the individual particles. Note that Ψ_{ft} is a coherent superposition of $\psi_t \otimes \phi_t$ and $\phi_t \otimes \psi_t$. The exclusion principle entails a correlation between the particles in that identical fermions would seem to 'know' the existence of each other so as to 'arrange' the state of the system as a whole to be anti-symmetric. Pauli realized this and he qualified the principle with the following statement:[99]

---

[96] In other words ψ_t and ϕ_t are asymptotically separable and so are ξ_T and η_T. It would be helpful to compare with the measurement model in §3.6 at this point.
[97] In §8.7 a notion of chronological disordering is introduced to give a qualitative view of the matter.
[98] Jauch (1968) §15.3.
[99] Pauli (1973) §36 on p. 168.

8.6. ENTANGLEMENT AND DECOHERENCE

From a superficial consideration of the exclusion principle, it might be thought that a sort of action-at-a-distance is being postulated, as a result of which even two widely separated particles are aware of one another ("sign of a contract"). However, this is not so, because the exclusion principle is only valid as long as the wave packets of the two particles overlap.

We have to understand Pauli's statement in an asymptotic sense. Let us examine how the exclusion principle can become invalid. Consider the expectation value

$$\langle \Psi_{ft} \mid \widehat{C}_f \Phi_{ft} \rangle$$
$$= \langle \psi_t \mid \widehat{A} \psi_t \rangle \langle \phi_t \mid \widehat{B} \phi_t \rangle + \langle \phi_t \mid \widehat{A} \phi_t \rangle \langle \psi_t \mid \widehat{B} \psi_t \rangle + I, \quad (8.175)$$

where

$$I = -\langle \psi_t \mid \widehat{A} \phi_t \rangle \langle \phi_t \mid \widehat{B} \psi_t \rangle - \langle \phi_t \mid \widehat{A} \psi_t \rangle \langle \psi_t \mid \widehat{B} \phi_t \rangle. \quad (8.176)$$

As far as any one-fermion observables $\widehat{A}$ and $\widehat{B}$ are concerned the correlation term I involving both ψ_t and ϕ_t in the above expression will vanish asymptotically as ψ_t and ϕ_t moves apart independently so that

$$\langle \Psi_{ft} \mid \widehat{C}_f \Phi_{ft} \rangle$$
$$= \langle \psi_t \mid \widehat{A} \psi_t \rangle \langle \phi_t \mid \widehat{B} \phi_t \rangle + \langle \phi_t \mid \widehat{A} \phi_t \rangle \langle \psi_t \mid \widehat{B} \psi_t \rangle \quad \text{FAPP}. \quad (8.177)$$

As a result the anti-symmetrized state Φ_{ft} is effective the same as the product state $\psi_t \otimes \phi_t$, i.e., we have

$$\langle \Psi_{ft} \mid \widehat{C}_f \Psi_{ft} \rangle = \langle \psi_t \otimes \phi_t \mid \widehat{C}_f \psi_t \otimes \phi_t \rangle \quad \text{FAPP}. \quad (8.178)$$

Consequently, Pauli exclusion principle looses its constraining power on states, and can be disregarded for all practical purposes, e.g., there is no point in formally writing down an anti-symmetrized state for an electron here on earth and one on some galaxy on the other side of the universe when we are dealing with observables localized here on earth![100]

It is common to illustrate the EPR problem in terms of two spin-$\frac{1}{2}$ particles. Let us consider the states of a single spin-$\frac{1}{2}$ particle of the form $\psi \otimes \alpha^\uparrow$ and $\phi \otimes \alpha^\downarrow$, where ψ and ϕ are orthonormal and $\alpha^\uparrow$ and $\alpha^\downarrow$ are spin states representing spin-up and spin-down as in §3.5.5. For a two-particle system we could have a state at time t of the form

$$\Phi_{ft}^{(s)} = \frac{1}{\sqrt{2}} \Big((\psi_t \otimes \alpha^\uparrow) \otimes (\phi_t \otimes \alpha^\downarrow) - (\phi_t \otimes \alpha^\downarrow) \otimes (\psi_t \otimes \alpha^\uparrow) \Big). \quad (8.179)$$

---
[100] Gasiorowicz (2003) pp. 204-206.

CHAPTER 8. SEPARABILITY AND DECOHERENCE

So, for observables of the form

$$\widehat{C}_f^{(s)} = (\widehat{A} \otimes \widehat{S}_i) \otimes (\widehat{B} \otimes \widehat{S}_j) + (\widehat{B} \otimes \widehat{S}_j) \otimes (\widehat{A} \otimes \widehat{S}_i), \qquad (8.180)$$

where $\widehat{S}_j$, $j = 1, 2, 3$, are one-particle spin operators,[101] we have

$$\begin{aligned}
\langle \Phi_{ft}^{(s)} \mid \widehat{C}_f^{(s)} \Phi_{ft}^{(s)} \rangle \\
= &+ \langle \psi_t \mid \widehat{A} \psi_t \rangle \langle \alpha^\uparrow \mid \widehat{S}_i \alpha^\uparrow \rangle \langle \phi_t \mid \widehat{B} \phi_t \rangle \langle \alpha^\downarrow \mid \widehat{S}_j \alpha^\downarrow \rangle \\
&+ \langle \phi_t \mid \widehat{A} \phi_t \rangle \langle \alpha^\downarrow \mid \widehat{S}_i \alpha^\downarrow \rangle \langle \psi_t \mid \widehat{B} \psi_t \rangle \langle \alpha^\uparrow \mid \widehat{S}_j \alpha^\uparrow \rangle \\
&- \langle \psi_t \mid \widehat{A} \phi_t \rangle \langle \alpha^\uparrow \mid \widehat{S}_i \alpha^\downarrow \rangle \langle \phi_t \mid \widehat{B} \psi_t \rangle \langle \alpha^\downarrow \mid \widehat{S}_j \alpha^\uparrow \rangle \\
&- \langle \phi_t \mid \widehat{A} \psi_t \rangle \langle \alpha^\downarrow \mid \widehat{S}_i \alpha^\uparrow \rangle \langle \psi_t \mid \widehat{B} \phi_t \rangle \langle \alpha^\uparrow \mid \widehat{S}_j \alpha^\downarrow \rangle. \qquad (8.181)
\end{aligned}$$

At large times T when the two particles are far apart the correlation terms involving ψ_T and ϕ_T vanish asymptotically and we arrive at

$$\begin{aligned}
\langle \Phi_{fT}^{(s)} \mid \widehat{C}_f^{(s)} \Phi_{fT}^{(s)} \rangle \\
= &+ \langle \psi_T \mid \widehat{A} \psi_T \rangle \langle \alpha^\uparrow \mid \widehat{S}_i \alpha^\uparrow \rangle \langle \phi_T \mid \widehat{B} \phi_T \rangle \langle \alpha^\downarrow \mid \widehat{S}_j \alpha^\downarrow \rangle \\
&+ \langle \phi_T \mid \widehat{A} \phi_T \rangle \langle \alpha^\downarrow \mid \widehat{S}_i \alpha^\downarrow \rangle \langle \psi_T \mid \widehat{B} \psi_T \rangle \langle \alpha^\uparrow \mid \widehat{S}_j \alpha^\uparrow \rangle \quad \text{FAPP.} \qquad (8.182)
\end{aligned}$$

Let us illustrate this result with an explicit example. Let Δ_1 and Δ_2 be two disjoint regions in the physical space $I\!\!E^n$. Suppose at large times T the wave function ψ_T localizes asymptotically to region $T\Delta_1$ and ϕ_T to $T\Delta_2$. Now let[102]

$$\widehat{A} = \chi_{T\Delta_1}, \quad \widehat{B} = \chi_{T\Delta_2}, \quad \widehat{S}_i = \widehat{S}_z, \quad \widehat{S}_j = \widehat{S}_z. \qquad (8.183)$$

Then

$$\langle \psi_T \mid \widehat{A} \psi_T \rangle \approx 1, \quad \langle \phi_T \mid \widehat{B} \phi_T \rangle \approx 1. \qquad (8.184)$$

It immediately follows that

$$\begin{aligned}
\langle \Phi_{fT}^{(s)} \mid \widehat{C}_f^{(s)} \Phi_{fT}^{(s)} \rangle \\
\approx &\ \langle \alpha^\uparrow \mid \widehat{S}_z \alpha^\uparrow \rangle \langle \alpha^\downarrow \mid \widehat{S}_z \alpha^\downarrow \rangle \qquad (8.185) \\
= &\ -\frac{1}{4} \hbar^2. \qquad (8.186)
\end{aligned}$$

No correlation is caused by $\widehat{C}_f^{(s)}$ here. In other words a simultaneous measurement at time T of local position observable $\chi_{T\Delta_1}$ and spin $\widehat{S}_z$ for one of the particles and, $\chi_{T\Delta_2}$ and spin $\widehat{S}_z$ for the other particle reveals no correlation, and $\Phi_{fT}^{(s)}$ is equivalent to a mixture as far as observable $\widehat{C}_f^{(s)}$ is concerned.

---

[101] $\widehat{S}_j$, $j = 1, 2, 3$, are the spin components along the x, y, and z directions respectively.
[102] Here $\chi_{T\Delta_1}$ and $\chi_{T\Delta_2}$ are local position observables introduced in Eq. (3.309) in §3.6.1.

8.7 Chronological Disordering

8.7.1 The concept of chronological disordering

In an asymptotically separable theory, like those presented in §8.5, the correlation between two distant states would gradually weaken as their spatial separation increases on account of the choice of local observables and Eq. (8.163). Discussions in §8.6 show that a similar situation also occurs for some entangled states of two-particles systems. To appreciate this conclusion in physical terms we shall examine a typical experimental set-up to measure correlation between two distinguishable particles in an entangled state which subsequently move freely apart, i.e., particle 1 moving left freely and particle 2 moving right freely. To establish a correlation we need to have expectation values. To obtain expectation values experimentally we require two beams of such particles, with a beam moving left and another beam moving right. Let us start by examining how a particle beam is generated.

Consider an experimenter located at a chosen coordinate origin equiped with a tunable particle source, a source similar to those discussed in §3.5.3 except that the source can be tuned to emit particles of predetermined states. Let us suppose that he sends out a particle of mass m in a known state φ_0 at $t = 0$. Here φ_0 corresponds to positive momentum values in a preset range $m\Delta v = m[v_1, v_2] \subset (0, \infty)$, i.e., the Fourier transform of φ_0 vanishes outside the range $m\Delta v$. The particle then evolves freely moving to the right with its wave function at time t denoted by φ_t. After a short interval δ the experimenter would send out another such particle, i.e., a particle in state φ_0 is sent out at time $t = \delta$. The process is repeated n times, n being a large number. After a time interval of $n\delta$ the experimenter would have generated a beam of n particles moving to the right. An observer O_r situated on the far right would in time receive a total of n particles, one by one.

In accordance with Theorem 3.4.2(1) in §3.4.2 the first particle let off by the experimenter will asymptotically localize, at sufficiently large times T, in the region $T\Delta v$ on the far right. At the same time the second particle sent out by the experimenter at the origin will localize in the region $(T - \delta)\Delta v$. Since the interval δ is pre-determined by the particle source at the origin while T can be arbitrarily large we have $(T - \delta)\Delta v \approx T\Delta v$. This means that the two particles, while localizing in the regions $T\Delta v$ and $(T - \delta)\Delta v$ respectively would have a substantial overlap of their wave functions. For some simple model wave functions it is not difficult to compute the amount of overlap of their wave funcitons.[103] Being identical with overlapping wave functions these two particles should be described by an anti-symmetric two-particle wave function, assuming they are fermions. These two fermions would

---
[103]Wan and Timson (1985).

then lose their individual labels, i.e., if an observer O_r situated on the far right detects the arrival of a particle she cannot tell whether it is the first one let off by the experimenter at the origin or the second one. Another way of looking at this is to realize that a particle detection amounts to a local position measurement. With two overlapping and spread-out wave functions and the probabilistic nature of position measurement the detection of the arrival of a particle provides no information on whether this is the first or the second particle sent out by the experimenter at the origin.

We can conclude that when observer O_r detected the first particle she could not be sure that it was the first particle sent out by the experimenter at the origin. Consequently her first measured value b_1 of an observable $\widehat{B}$ on this particle cannot be attributed with certainty to that of the first particle sent out by the experimenter. For sufficiently large T the same arguments apply to a whole beam of particles, i.e., the first particle detected by O_r from which she obtains the value listed as b_1 could be the jth particle sent out by the experimenter at the origin. This analysis tells us that

> *the chronological order in which particles are detected by an observer far away from the particle source will not necessarily coincide with the chronological order in which the particles are sent out from the particle source.*

We call such a situation *chronological disordering*.[104] Suppose O_r measures observable $\widehat{B}$ of the particles reaching her and lists her results $b_1, b_2, \ldots, b_n$ in the order she obtains them. As a result of chronological disordering the order $b_1, b_2, \ldots, b_n$ of the data obtained by O_r become somewhat arbitrary and random, i.e., this order does not necessarily coincide with the chronological order in which the particles are sent out from the particle source.

The phenomenon of chronological disordering will not affect the measurement of expectation values, i.e., the sum

$$\langle \varphi_T \mid \widehat{B}\varphi_T \rangle \approx \frac{1}{n}\left(b_1 + b_2 + \cdots + b_n\right) \tag{8.187}$$

is the same, independent of the order of the data. However, the phenomenon of chronological disordering will have a dramatic effect when we consider multi-particle systems.

8.7.2 Two-particle correlation and conservation laws

Let us re-examine two-particle entanglement in terms of the following experimental set-up. Equipped with a particle source an experimenter in the coor-

---

[104] Wan and Timson (1985), Wan (1986).

8.7. CHRONOLOGICAL DISORDERING

dinate origin generates a pair of non-identical particles at $t = 0$ and then let the particles evolve apart freely. Let us suppose that:

1. The two particles are generated in an entangled state $\Psi_0^{(c)}$ of the form of Eq. (8.167), i.e., we have, at $t = 0$ a two-particle state of the form

$$\Psi_0^{(c)} = \frac{1}{\sqrt{2}}\left(\psi_0 \otimes \xi_0 + \phi_0 \otimes \eta_0\right). \tag{8.188}$$

2. The constituent one-particle states in $\Psi_0^{(c)}$ are such that

   (a) ψ_0 and ϕ_0 correspond to positive momentum values, and

   (b) ξ_0 and η_0 correspond to negative momentum values.

Then the two particles would evolve freely to the right and to the left respectively. At a later time their wave function will be denoted by $\Psi_t^{(c)}$ as in Eq. (8.167).

After a short interval δ the experimenter generates another pair of such particles, i.e., a pair of particles in state $\Psi_0^{(c)}$ at $t = \delta$, and the process is repeated n times. Eventually we have two beams of particles moving in opposite directions, n of them moving to the left and the same number moving to the right. A total of n particles would reach an observer O_r on the far right and the same total of n particles would also reach an observer O_l on the far left.

We are interested in discovering some evidence of entanglement when the particles are far apart. As seen in Eq. (8.172) such evidence can be obtained in terms of the expectation values of appropriate observables of the two-particle system. So, let $\widehat{A}$ be an observable of the particles moving to the left and $\widehat{B}$ be an observable of the particles moving to the right. Let us measure the observable $\widehat{A} \otimes \widehat{B}$ of the two-particle system when the two particles are far apart. An observer O_l on the far left can make a measurement of observable $\widehat{A}$ of each of the particles he received, and list his results $a_1, a_2, \ldots, a_n$ chronologically. An observer O_r on the far right would measure observable $\widehat{B}$ of the particles reaching her one by one and then list her results $b_1, b_2, \ldots, b_n$ chronologically. One would then conclude that the expectation value of $\widehat{A} \otimes \widehat{B}$ is

$$\langle \Psi_T^{(c)} \mid \widehat{A} \otimes \widehat{B}\, \Psi_T^{(c)} \rangle \approx \frac{1}{n}\left(a_1 b_1 + a_2 b_2 + \cdots + a_n b_n\right). \tag{8.189}$$

This is similar to the data analysis used in the Bell Inequalities.[105]

Now consider the beam of particles moving to the right. At sufficiently large times T chronological disordering will set in so that the chronological

---
[105] Isham (1995) §9.3.2, especially Eq. (9.31). Redhead (1987) §4.1.

order in which the data $b_1, b_2, \ldots, b_n$ are listed may not correspond to the order in which the particles are sent out by the experimenter at the origin. Chronological disordering would also happen to the beam of particles moving to the left so the data $a_1, a_2, \ldots, a_n$ listed by O_l are similarly randomized. As a result there will be no proper pairing between the two sets of data $a_1, a_2, \ldots, a_n$ and $b_1, b_2, \ldots, b_n$. In other words the pair a_1, b_1 may not come from any pair of particles sent out by the experimenter.[106] This would make the sum in Eq. (8.189) for the expectation value $\langle \Psi_T^{(c)} \mid \widehat{A} \otimes \widehat{B} \, \Psi_T^{(c)} \rangle$ rather meaningless. A repetition of the experiment may well produce two sets of data listed in different orders; these new data would produce a different value for the sum in Eq. (8.189).

We can now see clearly how an attempt to gather evidence of entanglement would fail. Indeed when the particles are sufficiently far apart observers O_l and O_r can only measure one-particle observables like those in Eq. (8.169) because their expectation value calculation does not require a proper pairing of the data from O_l and O_r. It follows that any conservation law of an observable whose measurement requires a proper pairing of the experimental data from two widely separated observers would become less and less meaningful as the spatial separation of the two observers increases. The conclusion is that the usual argument using conservation laws as an insurmountable obstacle against any asymptotically separable quantum theory is not valid.[107]

It should be pointed out that chronological disordering does not show up if there is no spreading of the wavepacket. Electromagetic wavepacket travelling in a vacuum at the velocity of light will keep its shape and does not spread out. No chronological disordering would occur for photons in different wavepackets. However, in a dispersive medium an electromagetic wavepacket will spread out[108] and chronological disordering may well occur for photons. It would be interesting to design an experiment to test the phenomenon of chronological disordering for massive particles as well as for photons in a dispersive medium.[109]

---

[106] The value a_1 may come from a particle from the 3rd pair of particles sent out by the experimenter while b_1 may come from a particle in the 5th pair let off by the experimenter.

[107] Wan and Timson (1985), Ghirardi, Rimini and Weber (1976), Selleri and Tarozzi (1981).

[108] Jackson (1999) §7.9.

[109] The effect of chronological disordering on entanglement can be compared with the effect of finite coherence length on the traditional interferometer experiments for massive particles (Kaiser, Werner and George (1983), Klein, Opat and Hamilton (1983)) and waves in a dispersive medium (Hamilton, Klein and Opat (1983)).

References

1. Aczel, A. (2002). *Entanglement: The Greatest Mystery in Physics*, Wiley, New York.
2. Amann, A. (1988). 'Chirality as a Classical Observable in Algebraic Quantum Mechanics' in A. Amann et al., ed., *Fractals, Quasicrystals, Chaos, Knots and Algebraic Quantum Mechanics*, Kluwer, Dordrecht.
3. Anatopoulos, C. (2002). *Int. J. Theor. Phys.* **41** 1573.
4. Bacry, H. (1988). 'Localizability and Space in Quantum Physics' (*Lecture Notes in Physics*), Springer-Verlag, Berlin.
5. Barone, A. and Paternò, G. (1982). *Physics and Applications of the Josephson Effect*, Wiley, New York.
6. Berkley, A., Xu, H., Ramos, R., Gubrud, M., Strouch, F., Johnson, P., Anderson, J., Dragt, A., Lobb, C. and Wellstood, F. (2003). *Science* **300** 1548.
7. Bleaney, B. I. and Bleaney, B. (1957). *Electricity and Magnetism*, Clarendon Press, Oxford.
8. Bogolubov, N., Logunov, A. and Todorov, I. (1975). *Introduction to Axiomatic Quantum Field Theory*, Benjamin, Reading, Mass. Part Six.
9. Bransden, B. H. and Joachain, C. J. (1992). *Introduction to Quantum Mechanics*, Wiley, New York.
10. Bratteli, O. and Robinson, D. W. (1979). *Operator Algebras and Quantum Statistical Mechanics I*, Springer-Verlag, New York.
11. Bratteli, O. and Robinson, D. W. (1981). *Operator Algebras and Quantum Statistical Mechanics II*, Springer-Verlag, New York.
12. Brune, M., Hagley, E., Dreyer, J., Maitre, X., Maali, A., Wunderlich, C., Raimond, J. M. and Horoche, S. (1996). *Phys. Rev. Lett.* **77** 4887.
13. Cram, J. M. and Cram D. J. (1978). *The Essence of Organic Chemistry*, Addison-Wesley, Reading, Mass.
14. Davisovish, L., Brune, M., Raimond, J. M. and Horoche, S. (1996). *Phys. Rev. A* **53** 1295.
15. de Broglie, L. V. (1959). *J. Phys. Radium* **20** 936.
16. de Bruyn Ouboter, R. (1977). 'Macroscopic Quantum Phenomena in Superconductors' in B. B. Schwartz and S. Foner, eds., *Superconductor Applications: SQUIDs and Machines*, NATO ASI Series B Vol. 21, Plenum, New York.
17. de Lange, O. L. and Raab, R. E. (1991). *Operator Methods in Quantum Mechanics*, Oxford University Press, Oxford.
18. Eckern, U. (1986). *Nature* **319** 726.
19. Einstein, A., Podolsky, B. and Rosen, N. (1935). *Phys. Rev.* **47** 777.
20. Emch, G. G. (1972). *Algebraic Methods in Statistical Mechanics and Quantum Field Theory*, Wiley, New York.
21. Fano, G. (1971). *Mathematical Methods of Quantum Mechanics*, McGraw-Hill, New York.
22. Feynman, R. P., Leighton, R. B. and Sands, M. (1965). *The Feynman Lectures on Physics Vol. 3*, Addison-Wesley, Reading, Mass.

23. Friedman, J. R., Patel, V., Chen, W., Tolpygo, S. K. and Lukens, J. E. (2000). *Nature* **406** 43.
24. Gasiorowicz, S. (2003). *Quantum Physics*, Wiley International Edition.
25. Gerry, C. C. and Knight, P. L. (1997). *Am. J. Phys.* **65** 964.
26. Ghirardi, G., Rimini, A. and Weber, T. (1976). *Nuovo Cimento B* **36** 97.
27. Giulini, D., Joos, E., Kiefer, C., Kupsch, J., Stamatescu, I. and Zeh, H. D. (1996). *Decoherence and the Appearance of a Classical World in Quantum Theory*, Springer-Verlag, Berlin.
28. Gribbin, J. (1984). *In Search of Schrödinger's Cat*, Corgo Books, Reading.
29. Guenin, M. (1966). 'Algebraic Methods in Quantum Field Theory' in W. E. Witten, A. O. Barut and M. Guenin, eds., *Boulder Lectures in Theoretical Physics Vol. IXA: Mathematical Methods of Theoretical Physics*, Gordan and Breach, New York.
30. Haag, R. (1992). *Local Quantum Physics*, Springer-Verlag, Berlin.
31. Hagg, R. and Kastler, D. (1964). *J. Math. Phys.* **5** 848.
32. Hamilton, W. A., Klein, A. G. and Opat, G. I. (1983). *Phys. Rev. A* **28** 3149.
33. Hegstrom, R. A. and Kondepudi, D. K. (1990). *Scientific American* **263** 98.
34. Herzberg, H. (1945). *Molecular Spectra and Molecular Structure II: Infrared and Raman Spectra of Polyatomic Molecules*, Van Nostrand Reinhold, New York.
35. Howard, S. and Roy, S. K. (1987). *Am. J. Phys.* **55** 1109.
36. Isham, C. J. (1995). *Lectures on Quantum Theory*, Imperial College Press, London.
37. Jackson, J. D. (1999). *Classical Electrodynamics*, Wiley, New York.
38. Janoschek, R. (1991) 'Chirality: From Weak Bosons to the α-Helix' in R. Janoschek, ed., *Theories on the Origin of Biomolecular Homochirality*, Springer-Verlag, Heiselberg.
39. Jauch, J. M. (1968). *Foundations of Quantum Mechanics*, Addison-Wesley, Reading, Mass.
40. Johnson, P. R., Strauch, F. W., Gragt, A. J., Ramos, R. C., Lobb, C. J., Anderson, J. R. and Wellstood, F. C. (2003). *Phys. Rev. B* **67** 020509-1.
41. Kaiser, H., Werner, S. A. and George, E. A. (1983). *Phys. Rev. Lett.* **50** 560.
42. Klauder, J. R. and Skagarstam, B. eds. (1985). *Coherent States - Applications in Physics and Mathematics*, World Scientific, Singapore.
43. Klein, A. G., Opat, G. I. and Hamilton, W. A. (1983). *Phys. Rev. Lett.* **50** 563.
44. Kraus, K. (1977). *Position Observables of the Photon* in W. C. Price and S. S. Chissick, eds. *The Uncertainty Principle and the Foundations of Quantum Mechanics*, Wiley, New York.
45. Landsman, N. P. (1998). *Mathematical Topics Between Classical and Quantum Mechanics*, Springer-Verlag, New York.
46. Leggett, A. J. (1986). 'Quantum Mechanics at the Macroscopic Level' in J. de Boer, E. Dal and O. Ulfbeck, eds., *The Lesson of Quantum Theory*, Elsevier Science Publishers, Amsterdam.
47. Leggett, A. J. (1987). 'Quantum Mechanics at the Macroscopic Level' in J. Souletie and J. Vannimenus, eds., *Chance and Matter*, Elsevier, Amsterdam.
48. Leggett, A. J. (2000). *New Life for Schrödinger's Cat* in *Physics World* **13** No 8, 23.

49. Leonhardt, U. (1997). *Measuring the Quantum State of Light*, Cambridge University Press, Cambridge.
50. Lévy-Leblond, J. (1990). *Quantics*, North-Holland, Amsterdam.
51. Merzbacher, E. (1998). *Quantum Mechanics*, 3rd edition, Wiley, New York.
52. Monroe, C., Meekhof, D. M., King, B. E. and Wineland, D. J. (1996). *Science* **272** 1131.
53. Myatt, C., King, B., Turchette, Q., Sackett, C., Kielpinski, D., Itano, W., Monroe, C. and Wineland, D. (2000). *Nature* **403** 269.
54. Papoulis, A. (1962). *The Fourier Integral and Its Applications*, McGraw-Hill, New York.
55. Pauli, W. (1973). *Pauli Lectures on Physics Vol. 5 - Wave Mechanics*, the MIT Press, Cambridge, Mass., edited by C P Enz and translated by H R Lewis and S Margulies.
56. Peres, A. (1995). *Quantum Theory: Concepts and Methods*, Kluwer Academic Publishers, Dordrecht.
57. Pfeifer, P. (1980). *Chiral Molecules - A Superselection Rule Induced by the Radiation field*, Dissertation ETH Zürich.
58. Phillips, A. C. (2003). *Introduction to Quantum Mechanics*, Wiley, New York.
59. Primas, H. and Müller-Herold, U. (1978). *Advances in Chemical Physics* **38** Chapters 1-6.
60. Prugovečki, E. (1984). *Stochastic Quantm Mechanics and Quantum Spacetime*, Reidel, Dorchecht.
61. Rae, A. I. M. (1986). *Quantum Physics: Illusion or Reality?*, Cambridge Univeristy Press, Cambridge.
62. Redhead, M. (1987). *Incompleteness, Nonlocality and Realism*, Clarendon Press, Oxford.
63. Roman, P. (1965). *Advanced Quantum Theory*, Addison-Wesley, Reading, Mass.
64. Rose-Innes, A. C. and Rhoderick, E. H. (1980). *Introduction to Superconductivity*, Pergamon Press, London.
65. Schiff, L. I. (1955). *Quantum Mechanics*, McGraw-Hill, New York.
66. Schrödinger, E. (1935). *Naturwissenschaften*, **23** 807; 823; 844. English translation in J. A. Wheeler and W. Zurek. eds., *Quantum Theory of Measurement*, Princeton University Press, Princeton, N. J. (1983).
67. Schroeck, F. E., Jr. (1996). *Quantm Mechanics on Phase Space*, Kluwer Academic Publishers, Dordrecht.
68. Selleri, F. (1990). *Quantum Paradoxes and Physical Reality*, A. van der Merwe, ed., Kluwer Academic Publishers, Dorchecht.
69. Selleri, F. and Tarozzi, G. (1981). *La Rivita del Nouvo Cimento*, **4** 1.
70. Sewell, G. L. (1986). *Quantum Theory of Collective Phenomena*, Clarendon Press, Oxford.
71. Sewell, G. L. (2002). *Quantum Mechanics and its Emergent Macrophysics*, Princeton University Press, Princeton. N. J.
72. Sivakumar, S. (1998). *Phys. Rev. A* **58** 717.

73. Torozzi, G. and van der Merwe, A. eds. (1980). *Open Questions in Quantum Physics*, Reidel, Dordrecht.
74. van der Wal, C. H., ter Haar, A. C. J., Wilhelm, F. K., Schouten, R. N., Harmans, C. J. P. M, Orlando, T. P., Lloyd, S. and Mooij, J. E. (2000). *Science* **290** 773.
75. Wan, K. K. (1980). *Can. J. Phys.* **58** 976.
76. Wan, K. K. (1986). 'Chronological Disordering and the Absence of Correlations between Infinitely Separated States' in V. Gorini and A. Grigerio, eds., *Fundamental Aspects of Quantum Theory*, NATO ASI Series B Vol. 144, Plenum, New York.
77. Wan, K. K. (1988). *Found. Phys.* **18** 887.
78. Wan, K.K. and Fountain, R. H. (1996). *Found. Phys.* **26** 1165.
79. Wan, K. K., Green, R. and Trueman, C. (2000). *J. Opt. B.* **2** 165.
80. Wan, K. K. and Harrison, F. E. (1993). *Phys. Lett. A* **174** 1.
81. Wan, K. K. and Jackson, T. D. (1985). *Phys. Lett. A* **111** 223.
82. Wan, K. K. and McLean, R. G. D. (1984a). *J. Phys. A: Math. Gen.* **17** 825. Corrigenda **17** 2363 (1984).
83. Wan, K. K. and McLean, R. G. D. (1984b). *J. Phys. A: Math. Gen.* **17** 837. Corrigenda **17** 2363 (1984).
84. Wan. K. K. and Timson, D. (1985). *Phys. Lett. A* **111** 165.
85. Wightman, A. S. and Glance, N. (1989). *Nucl. Phys. B (Proc Suppl)* **6** 202.
86. Williams, H. J. (1982). *Introduction to Organic Chemistry*, Wiley, New York.
87. Yurke, B. and Stoler, D. (1986). *Phys. Rev. Lett.* **57** 13.

Chapter 9

Quantum Mechanics on Path Space

9.1 Introduction

We have seen the emergence of superselection rules for superconducting systems from two different perspectives. Firstly, in the macroscopic wave function approach within the formalism of single-particle quantum mechanics, the introduction of physically motivated assumptions (PAS1) and (PAS2) in §7.3.2 on superconducting states lead to a superselection rule. Secondly we can also arrive at a superselection rule in §7.11 from the many-body BCS theory of superconductivity. As pointed out in §4.3.3 there are diverse origins and derivations of superselection rules. In this chapter we shall illustrate this diversity with an explicit study of what may be called a *path space formulation of quantum mechanics*.

In 1931 Dirac pointed out a striking ambiguity in the wave function $\phi(x)$, namely that the position probability density function $|\phi(x)|^2$ only determines the wave function up to an arbitrary phase factor. Writing the wave function in the form

$$\phi(x) = R(x)\, e^{i\, S(x)}, \tag{9.1}$$

where $R(x)$ and $S(x)$ are real-valued functions of x, it would seem that adding a real constant to the function $S(x)$ produces no physically observable consequences. It is phase difference which is physically significant, not the actually value of the phase, a situation similar to the potential function in classical mechanics, i.e., it is the potential difference which is significant and adding a constant term to the potential is not going to make any difference to the dynamics of the system. So, the wave function is generally multi-valued in nature. The question is whether this multi-valued nature of the wave function can in some circumstances lead to observable consequences, despite what has been said. There have been many studies of this question. First, we need to be able to set up multi-valued functions in a systematic manner. An approach is to introduce a path space; functions defined on the path space may be multi-valued on the physical space. We shall endeavour to keep the mathematics to a bare minimum by avoiding general and abstract discussions. Instead we shall confine ourselves to concrete examples which have direct physical applications, making full use of some of the existing general formulations.[1]

9.2 Physical Space and Path Space

Let $I\!M^n$ be an n-dimensional Riemannian manifold and m a point in $I\!M^n$. In the usual formulation of quantum mechanics with $I\!M^n$ as the physical space we would first introduce complex-valued functions $\psi(m)$ on $I\!M^n$ which are square-integrable with respect to a given volume element $d\mu(m)$ on $I\!M^n$, i.e.,

$$\int_{I\!M^n} \psi^*(m)\psi(m)\, d\mu(m) \leq \infty. \tag{9.2}$$

These functions have a natural vector space structure. With the usual definition of scalar product

$$\langle \psi \mid \phi \rangle = \int_{I\!M^n} \psi^*(m)\phi(m)\, d\mu(m), \tag{9.3}$$

these functions form a scalar product space which can be completed to form a Hilbert space $L^2(I\!M^n, d\mu)$. A function ψ in $L^2(I\!M^n, d\mu)$ is normalized if

$$\int_{I\!M^n} \psi^*(m)\psi(m)d\mu(m) = 1. \tag{9.4}$$

Normalized functions ψ in $L^2(I\!M^n, d\mu)$ give rise to probability density functions $|\psi(m)|^2$.

Any two functions ψ_1, ψ_2 in $L^2(I\!M^n, d\mu)$ are said to differ by a *phase factor* if

$$\psi_1(m) = e^{i\Theta_{1,2}(m)}\psi_2(m), \tag{9.5}$$

where $\Theta_{1,2}$ is a real-valued function on $I\!M^n$, to be referred to as a *phase function* or simply a *phase*. The multiplicative factor $\exp i\Theta_{1,2}(m)$ is called a *phase factor*. A phase is called *global* if it is a constant independent of m and

---
[1] Zaccoria et al. (1983), Balachandran (1989), Horvathy, Morandi and Sudarshan (1989), Morandi (1993).

9.2. PHYSICAL SPACE AND PATH SPACE

the corresponding phase factor is also called global. Otherwise they are called *local*. For a quantum particle moving in $I\!M^n$ the standard approach is to regard a function ψ in $L^2(I\!M^n, d\mu)$ as a possible wave function with $|\psi(m)|^2$ interpreted as the position probability density function of the particle moving in $I\!M^n$. There is no observational distinction between wave functions differing by an arbitrary global phase factor since it does not affect any expectation values. A local phase factor will generally have an observable effect in terms of certain expectation values, albeit not in the position probability density function. However, all this can be affected by the topological properties of $I\!M^n$. In his 1931 paper Dirac already realized this possibility and proposed a kind of path space formulation of quantum mechanics.

Given a manifold $I\!M^n$ choose some point $m_f \in I\!M^n$ which is to be held fixed from now on. We assume that $I\!M^n$ is connected so that any point m in $I\!M^n$ may be joined to m_f by a differentiable curve, referred to as a *path* for short, in the manifold. Let σ_m be any path from m_f to m. There is an infinite set of paths

$$\Pi_m(I\!M^n) = \left\{ \sigma_m, \sigma'_m, \sigma''_m, \ldots \right\} \tag{9.6}$$

linking the fixed point m_f to every m. We shall denote the set of all the paths from m_f to all the points $m \in I\!M^n$ by

$$\Pi(I\!M^n) = \left\{ \Pi_m(I\!M^n) : m \in I\!M^n \right\} \tag{9.7}$$

and refer to $\Pi(I\!M^n)$ as the *path space* on the manifold $I\!M^n$. Such a path space would have a very complex geometric structure for a general manifold $I\!M^n$. Things are much simpler in manifolds of one and two dimensions. Fortunately we are only interested in manifolds of one and two dimensions here, since many nano-structures and quantum circuits are of such low dimensions.

Let us start with a two-dimensional manifold $I\!M^2$. Our first task is to classify paths in $I\!M^2$. Two paths ending at the same point are said to be *homotopic* or *homotopically equivalent* if they can be continuously deformed into one another.[2] Whether two paths are homotopic depends on the topological nature of $I\!M^2$. If $I\!M^2$ is the plane $I\!E^2$ then any two paths with the same end points are clearly continuously deformable into each other, and are hence homotopic. This is illustrated in Fig. 9.2.1(1) where we have:

1. $\sigma_{m,0}$ denotes a path from m_f to m passing round points a and b in an anti-clockwise direction.

---

[2] We shall adopt an intuitive approach here to avoid mathematical rigour such as having to define continuous deformation of curves. For more details see Lipschutz (1965), Nash and Sen (1983).

2. $\sigma'_{m,0}$ denotes a path from m_f to m passing round point c, again in an anti-clockwise direction.

3. $\sigma''_{m,0}$ denotes a path from m_f directly to m.

All these paths are homotopic since they can be continuously deformed into one another.[3] These paths are also homotopic to any path from m_f to m in a clockwise direction.

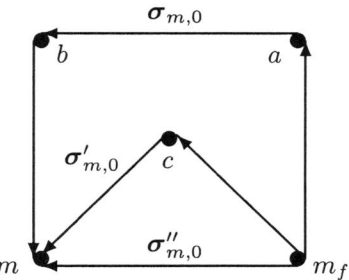

Fig. 9.2.1(1)   Paths in $I\!\!E^2$.

Now denote by $I\!\!E_h^2$ the Euclidean plane $I\!\!E^2$ with a hole, i.e., with an open and circular region removed. The hole is shown as a circle with the letter h inside in Fig. 9.2.1(2).[4]

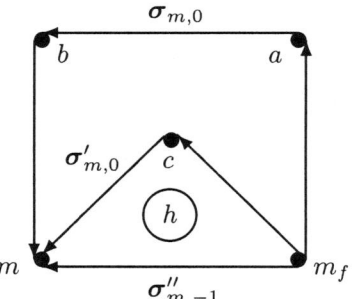

Fig. 9.2.1(2)   Paths in $I\!\!E_h^2$.

---

[3] All paths are supposed to be smooth, i.e., $\sigma_{m,0}$ goes round a and b smoothly and $\sigma'_{m,0}$ goes round c smoothly.

[4] Paths are labelled in terms of integer subscripts, e.g., 0, 1, -1 and so on. The reason for such a labelling will become apparent later on.

9.2. PHYSICAL SPACE AND PATH SPACE

Then not all the paths are homotopic. Paths $\sigma_{m,0}$ and $\sigma'_{m,0}$ are homotopic as before. However, the hole stops $\sigma_{m,0}$ from continuously deforming into $\sigma''_{m,-1}$. So, $\sigma_{m,0}$ and $\sigma''_{m,-1}$ are not homotopic. Paths $\sigma'_{m,0}$ and $\sigma''_{m,-1}$ are also not homotopic for the same reason. The hole spoils the topology of the plane so that not all paths are continuously deformable into one another. We shall devote the rest of this section to a classification of paths in the manifold $I\!E_h^2$.

To determine whether two paths are homotopic we have to examine how the paths go from the fixed point m_f to m. In particular we need to know whether the path wraps round the hole a few times before ending up at m. Figure 9.2.1(3) below serves to illustrate the situation. We shall call a closed curve, i.e., a curve starting from m_f and ending also at m_f, a *loop*. Figure 9.2.1(3) shows three types of loops:

1. $\mathcal{L}_1$ denotes a loop round the hole in an anti-clockwise direction, i.e., $\mathcal{L}_1$ starts from m_f, goes round c and d and finally returns to end at m_f.

2. $\mathcal{L}_{-1}$ denotes a loop round the hole in a clockwise direction, i.e., $\mathcal{L}_{-1}$ starts from m_f, goes round d and c and finally returns to end at m_f.

3. $\mathcal{L}_0$ denotes a loop which does not enclose the hole. All these loops, whether they go clockwise or anti-clockwise, are homotopic to each other as they can be shrunk to point m_f.[5]

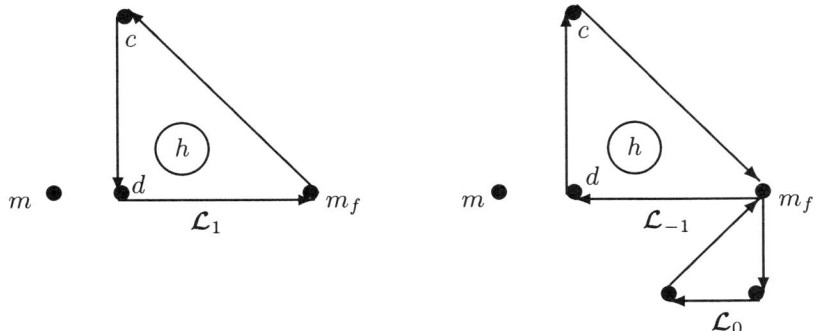

Fig. 9.2.1(3)   Loops $\mathcal{L}_1$, $\mathcal{L}_{-1}$ and $\mathcal{L}_0$ in $I\!E_h^2$.

By combining those loops in Fig. 9.2.1(3) with those simple paths in Fig. 9.2.1(2) we can construct more complex paths. For example, in Fig. 9.2.1(4), the path $\sigma_{m,1}$ is made up of sections:

---
[5]The path $\mathcal{L}_0$ is said to be contractable to a point (Lipschutz (1965) p. 186).

1. Starting from m_f the path traces out the loop $\mathcal{L}_1$ in an anti-clockwise direction.
2. The curve continues by following path $\sigma_{m,0}$ in Fig. 9.2.1(2) to end finally at m.

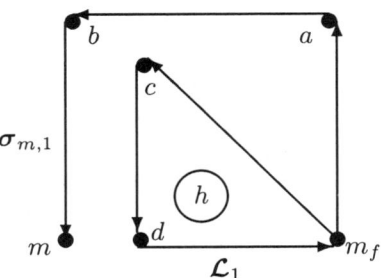

Fig. 9.2.1(4)   Loop $\mathcal{L}_1$ and path $\sigma_{m,1}$ in $I\!\!E_h^2$.

In other words $\sigma_{m,1}$ wraps round the hole in a loop in an anti-clockwise direction once before going up to a, then b to end at m, and this explains the subscript 1. In Fig. 9.2.1(2) path $\sigma_{m,0}$ does not wrap round the hole and it is hence denoted by the subscript 0. Clearly $\sigma_{m,0}$ in Fig. 9.2.1(2) is not homotopic to the path $\sigma_{m,1}$ in Fig. 9.2.1(4) since the loop $\mathcal{L}_1$ cannot be deformed away due to the presence of the hole.

It is possible to trace out the apparently same curve in a different direction as shown in Fig. 9.2.1(5).

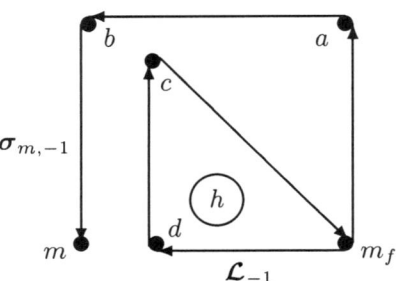

Fig. 9.2.1(5)   Loop $\mathcal{L}_{-1}$ and path $\sigma_{m,-1}$ in $I\!\!E_h^2$.

Path $\sigma_{m,-1}$ in Fig. 9.2.1(5) is made of a loop $\mathcal{L}_{-1}$, which starts from m_f and goes around the hole once in a clockwise direction to return to m_f, followed

9.2. PHYSICAL SPACE AND PATH SPACE

by a path going from m_f to m via a and b. As before this path cannot deform itself into $\sigma_{m,0}$ on account of the hole.

It turns out that all paths from m_f to m may be constructed from a section consisting of loops and a direct path from m_f to m. So, we need to introduce a classification of the loops starting and ending at m_f in $\mathbb{E}_h^2$ as a precursor to the classification of all the paths on $\mathbb{E}_h^2$. A natural classification is as follows:

Class 0 $[\mathcal{L}_0]$ = {all loops not enclosing the hole}
Class 1 $[\mathcal{L}_1]$ = {all loops circling the hole once anti-clockwise}
Class -1 $[\mathcal{L}_{-1}]$ = {all loops circling the hole once clockwise}

$\vdots$

Class ℓ $[\mathcal{L}_\ell]$ = $\begin{cases}\text{all loops round the hole } \ell \text{ times anti-clockwise if } \ell > 0 \\ \text{all loops circling the hole } |\ell| \text{ times clockwise if } \ell < 0\end{cases}$

$\vdots$

These are called *homotopic classes* of loops, and the integer ℓ of a class $[\mathcal{L}_\ell]$ is known as the *winding number* of the class. Intuitively we can see that two different loops belonging to the same class are homotopic. A member of $[\mathcal{L}_\ell]$ will be denoted by $\mathcal{L}_\ell$. We shall use these loops to classify paths.

Let $[\sigma_{m,0}]$ denote the set of all paths from m_f to m curving around the hole in an anti-clockwise direction but not enclosing the hole. Path $\sigma_{m,0}$ in Fig. 9.2.1(2) is an element of this set, and $[\sigma_{m,0}]$ consists of all paths in $\mathbf{\Pi}_m(\mathbb{E}_h^2)$ homotopic to $\sigma_{m,0}$. A general path in $\mathbf{\Pi}_m(\mathbb{E}_h^2)$ will either be in the set $[\sigma_{m,0}]$, or it will circle the hole a number of times before ending up at the point m. For example, in Fig. 9.2.1(4) path $\sigma_{m,1}$ consists of two consecutive sections: $\mathcal{L}_1 \in [\mathcal{L}_1]$ followed by $\sigma_{m,0} \in [\sigma_{m,0}]$. We shall denote this symbolically by

$$\sigma_{m,1} = \sigma_{m,0} * \mathcal{L}_1. \tag{9.8}$$

Similarly for the path $\sigma_{m,-1}$ in Fig. 9.2.1(5) we have[6]

$$\sigma_{m,-1} = \sigma_{m,0} * \mathcal{L}_{-1}. \tag{9.9}$$

It follows that any path in $\mathbf{\Pi}_m(\mathbb{E}_h^2)$ which is homotopic to $\sigma_{m,1}$ is of the form of two consecutive sections, one from $[\mathcal{L}_1]$ followed by one from $[\sigma_{m,0}]$. All the paths homotopic to $\sigma_{m,1}$ can be collected to form a class

$$[\sigma_{m,1}] = [\sigma_{m,0}] * [\mathcal{L}_1]. \tag{9.10}$$

---

[6]A path curving round clockwise from m_f to m directly without looping round the hole, e.g., $\sigma''_{m,-1}$ in Fig. 9.2.1(2), is considered homotopic to $\sigma_{m,-1}$.

Generally, we can divide all the paths in $\Pi_m(E_h^2)$ into such classes since they are homotopic to two consecutive sections, one from $[\mathcal{L}_\ell]$ followed by one from $[\sigma_{m,0}]$. Such classes are to be denoted by

$$[\sigma_{m,\ell}] = [\sigma_{m,0}] * [\mathcal{L}_\ell], \quad \ell = 0, \pm 1, \pm 2, \ldots, \tag{9.11}$$

with a member of the class denoted by $\sigma_{m,\ell}$. So, two paths $\sigma_{m,\ell}$ and $\sigma'_{m,\ell'}$ are homotopic if and only if $\ell = \ell'$.

For the case $M^2 = E^2$ without any holes all paths from m_f to the same end point m are homotopic, and we shall simply denote a path by σ_m in E^2.

9.3 Functions on Path Space

Consider a mapping f of the set of paths on E_h^2 to the complex numbers:[7]

$$f : \Pi(E_h^2) \to \mathbb{C} \quad \text{by} \quad \sigma_{m,\ell} \to f(\sigma_{m,\ell}) \in \mathbb{C}. \tag{9.12}$$

Such a mapping may be regarded as a complex-valued function on the path space $\Pi(E_h^2)$, i.e., we have a complex-valued function f on $\Pi(E_h^2)$. The value of such a function $f(\sigma_{m,\ell})$ depends on the end point m as well as the path $\sigma_{m,\ell}$. Hence $f(\sigma_{m,\ell})$ may be viewed as a multi-valued function on E_h^2. We can generate single-valued functions on E_h^2 in terms of functions on $\Pi(E_h^2)$. One way to achieve this is to choose a single path $\sigma_{m,0}$ in the class $[\sigma_{m,0}]$ of zero winding number to link m_f directly to each m. So, given a function f on $\Pi(E_h^2)$ we can define a function ψ on E_h^2 by

$$\psi(m) = f(\sigma_{m,0}). \tag{9.13}$$

Since there is only one path chosen for each end point m, this function ψ is single-valued[8] on E_h^2 with a unique value $f(\sigma_{m,0})$ at each point $m \in E_h^2$. We may call ψ a *restriction* of f to E_h^2 with respect to a set of chosen paths $\{\sigma_{m,0} : m \in E_h^2\}$.

It is interesting to investigate the reverse process. Given a (single-valued) function ψ on E_h^2 we like to know how one could generate a function f on $\Pi(E_h^2)$ such that its restriction to E_h^2 with respect to the set of chosen paths $\{\sigma_{m,0} : m \in E_h^2\}$ would coincide with ψ, i.e., $f(\sigma_{m,0}) = \psi(m)$. One way to generate such a function on $\Pi(E_h^2)$ is to annex a *path dependent phase factor* to ψ:

$$f(\sigma_{m,\ell}) = e^{i \Theta(\sigma_{m,\ell})} \psi(m), \tag{9.14}$$

---

[7] All the paths are from a single chosen fixed point m_f to another point m.
[8] To avoid confusion we should say here that unless otherwise is explicitly stated a function on E_h^2 means a single-valued function on E_h^2.

9.3. FUNCTIONS ON PATH SPACE

where the *path dependent phase* $\Theta(\sigma_{m,\ell})$ is a real-valued function on the path space $\Pi(I\!E_h^2)$ which vanishes on the set of chosen paths $\{\sigma_{m,0} : m \in I\!E_h^2\}$, i.e.,

$$\Theta(\sigma_{m,0}) = 0 \quad \forall \sigma_{m,0} \in \{\sigma_{m,0} : m \in I\!E_h^2\}. \tag{9.15}$$

There is still a great deal of arbitrariness. For instance we may have $f(\sigma_{m,0}) \neq f(\sigma'_{m,0})$, i.e., homotopic paths $\sigma_{m,0}$ and $\sigma'_{m,0}$ give rise to different values of the function. For physical applications we need to narrow down the set of functions on $\Pi(I\!E_h^2)$ considerably, e.g., we would really like to have functions having the same value on homotopic paths. This is done by imposing the following five physically motivated conditions, starting with a condition which confines the functions to those given by Eq. (9.14):[9]

C1 *Single-valuedness up to a phase as functions on $I\!E_h^2$* A function f on $\Pi(I\!E_h^2)$ is said to be *single-valued up to a phase as a function on $I\!E_h^2$* if its modulus $|f(\sigma_{m,\ell})|$ depends only on the end point m, not on the path linking m_f to m. This condition ensures that $f(\sigma_{m,\ell})$ can give rise to a (single-valued) probability density function $|f(\sigma_{m,\ell})|^2$ on $I\!E_h^2$, whenever $|f(\sigma_{m,\ell})|^2$ is integrable and normalized. This condition restricts the path dependence of $f(\sigma_{m,\ell})$ to the phase. In other words these are functions given by Eq. (9.14). In what follows we shall confine ourselves to functions f on $\Pi(I\!E_h^2)$ which are single-valued up to a phase as a function on $I\!E_h^2$.

C2 *Homotopic functions* A function f on $\Pi(I\!E_h^2)$ is called *homotopic* if its values for homotopic paths coincide, i.e., at every point $m \in I\!E_h^2$ we have[10]

$$f(\sigma'_{m,\ell}) = f(\sigma_{m,\ell}) \tag{9.16}$$

for homotopic paths $\sigma'_{m,\ell}$ and $\sigma_{m,\ell}$. In particular we have

$$f(\sigma'_{m,0}) = f(\sigma_{m,0}). \tag{9.17}$$

This enables us to define a unique restriction to $I\!E_h^2$ by[11]

$$\psi(m) = f(\sigma_{m,0}), \tag{9.18}$$

independent of the choice of any particular $\sigma_{m,0}$. Conversely we can express any homotopic function $f(\sigma_{m,\ell})$ which is single-valued up to a phase as a function on $I\!E_h^2$ in terms of ψ in the form[12]

$$f(\sigma_{m,\ell}) = e^{i\Theta(\sigma_{m,\ell})} \psi(m), \tag{9.19}$$

---

[9] Zaccoria et al. (1983), Balachandran (1989), Morandi (1993).
[10] Recall that paths $\sigma_{m,\ell}$ of the same winding number ℓ are homotopic.
[11] We have taken $\Theta(\sigma_{m,0}) = 0$ as in Eq. (9.15), since the value of this phase factor can always be absorbed into $\psi(m)$.
[12] We have assumed $\Theta(\sigma_{m,0}) = 0$ as in Eq. (9.18).

showing that the path dependence lies in the phase only. Since the function is homotopic the path dependent phase satisfies

$$e^{i\left[\Theta(\sigma'_{m,\ell})-\Theta(\sigma_{m,\ell})\right]} = 1, \tag{9.20}$$

or equivalently,

$$\Theta(\sigma'_{m,\ell}) - \Theta(\sigma_{m,\ell}) = 2n\pi \quad \text{for some integer } n. \tag{9.21}$$

Generally a homotopic function f is multi-valued as a function on $I\!\!E_h^2$. The path dependent phase of f at each point m in $I\!\!E_h^2$ changes as we change the winding number ℓ of the path. If the manifold is the plane $I\!\!E^2$ without any holes, the situation simplifies considerably, since all paths σ_m from m_f to m are homotopic in $I\!\!E^2$. It follows that the value of a homotopic function f at each point $m \in I\!\!E^2$ is independent of any particular path. In other words, f itself may be regarded as a (single-valued) function on the physical space $I\!\!E^2$. There is then no point in generalizing functions on $I\!\!E^2$ to those on the path space $\Pi(I\!\!E^2)$. Quantum mechanics formulated in terms of homotopic wave functions on the path space $\Pi(I\!\!E^2)$ will be identical to the usual theory in terms of (single-valued) wave functions on the physical space $I\!\!E^2$.

Since $I\!\!E_h^2$ does not have the topology of $I\!\!E^2$ functions defined on the path space $\Pi(I\!\!E_h^2)$ are multi-valued when regarded[13] as functions on $I\!\!E_h^2$. A theory based on functions on the path space $\Pi(I\!\!E_h^2)$ will have new features. Before we can investigate the physical implications of these new features we need to introduce further conditions to further narrow down those path space functions.

C3 *Universal path dependent phase* To establish a set of functions on $\Pi(I\!\!E_h^2)$ which has the properties of a vector space we need to introduce a universal path dependent phase. Two homotopic functions f_1 and f_2 on $\Pi(I\!\!E_h^2)$ are said to share a path dependent phase $\Theta(\sigma_{m,\ell})$ if

$$f_1(\sigma_{m,\ell}) = e^{i\Theta(\sigma_{m,\ell})}\psi_1(m), \quad \psi_1(m) = f_1(\sigma_{m,0}), \tag{9.22}$$

$$f_2(\sigma_{m,\ell}) = e^{i\Theta(\sigma_{m,\ell})}\psi_2(m), \quad \psi_2(m) = f_2(\sigma_{m,0}). \tag{9.23}$$

A path dependent phase is said to be *universal* to a set of functions on $\Pi(I\!\!E_h^2)$ if this path dependent phase is shared by all the functions in the set.

Let $\mathcal{F}(\Pi(I\!\!E_p^2))$ be a set of functions on $\Pi(I\!\!E_h^2)$ which satisfy conditions C1, C2 and C3, i.e., they are homotopic functions which are single-valued up to a phase as functions $I\!\!E_h^2$ and they share the same path dependent phase Θ. Then these functions have a vector space structure. To appreciate why

---

[13] Do not confuse with the restrictions of these functions on $I\!\!E_h^2$.

9.3. FUNCTIONS ON PATH SPACE

we require a set of functions to share a universal path dependent phase we need to consider the sum of two functions. Let $\boldsymbol{f}_1$ and $\boldsymbol{f}_2$ be two homotopic functions on $\Pi(I\!\!E_p^2)$ with their respective path dependent phases Θ_1 and Θ_2, i.e., we have

$$\boldsymbol{f}_1(\sigma_{m,\ell}) = e^{i\Theta_1(\sigma_{m,\ell})}\,\psi_1(m), \qquad (9.24)$$

$$\boldsymbol{f}_2(\sigma_{m,\ell}) = e^{i\Theta_2(\sigma_{m,\ell})}\,\psi_2(m). \qquad (9.25)$$

Now define their sum, $\boldsymbol{f} = \boldsymbol{f}_1 + \boldsymbol{f}_2$, as a function on $\Pi(I\!\!E_h^2)$, by

$$(\boldsymbol{f}_1 + \boldsymbol{f}_2)(\sigma_{m,\ell}) = \boldsymbol{f}_1(\sigma_{m,\ell}) + \boldsymbol{f}_2(\sigma_{m,\ell}) \qquad (9.26)$$

$$= e^{i\Theta_1(\sigma_{m,\ell})}\,\psi_1(m) + e^{i\Theta_2(\sigma_{m,\ell})}\,\psi_2(m). \qquad (9.27)$$

To have a vector space structure we must require this sum to satisfy C1. Let us work out the square of the modulus of $\boldsymbol{f}(\sigma_{m,\ell})$:

$$|\boldsymbol{f}(\sigma_{m,\ell})|^2 = |\psi_1(m)|^2 + |\psi_2(m)|^2$$
$$+ 2\,\mathrm{Re}\left\{\left(e^{i[\Theta_2(\sigma_{m,\ell})-\Theta_1(\sigma_{m,\ell})]}\right)\psi_1^*(m)\psi_2(m)\right\}. \qquad (9.28)$$

Clearly $|\boldsymbol{f}(\sigma_{m,\ell})|^2$ is path dependent and is unable to give rise to a single-valued probability density function on $I\!\!E_h^2$ unless

$$\Theta_2(\sigma_{m,\ell}) - \Theta_1(\sigma_{m,\ell}) = 2n\pi. \qquad (9.29)$$

In other words we must have

$$\Theta_1(\sigma_{m,\ell}) = \Theta_2(\sigma_{m,\ell}) = \Theta(\sigma_{m,\ell}) \qquad (9.30)$$

or

$$\Theta_2(\sigma_{m,\ell}) = \Theta_1(\sigma_{m,\ell}) + 2n\pi \qquad (9.31)$$

so that

$$\boldsymbol{f}_1(\sigma_{m,\ell}) = e^{i\Theta(\sigma_{m,\ell})}\,\psi_1(m), \qquad (9.32)$$

$$\boldsymbol{f}_2(\sigma_{m,\ell}) = e^{i\Theta(\sigma_{m,\ell})}\,\psi_2(m), \qquad (9.33)$$

$$(\boldsymbol{f}_1 + \boldsymbol{f}_2)(\sigma_{m,\ell}) = e^{i\Theta(\sigma_{m,\ell})}\left(\psi_1(m) + \psi_2(m)\right), \qquad (9.34)$$

or

$$\boldsymbol{f}(\sigma_{m,\ell}) = e^{i\Theta(\sigma_{m,\ell})}\,\psi(m), \quad \psi(m) = \left(\psi_1(m) + \psi_2(m)\right). \qquad (9.35)$$

So, for a set of homotopic functions which are single-valued up to a phase as functions on $\Pi(I\!\!E_h^2)$ to form a vector space we do require the functions to share a common path dependent phase.[14]

---
[14] Recall that a phase is a function $\Theta(\sigma_{m,\ell})$ on the path space.

The desire to form a vector space is motivated by the need for a superposition principle in quantum mechanics. In other words the state space of an orthodox quantum system must be a vector space so that we can form linear combinations of different states. It follows that if we are to use functions on path space to represent the states of a given quantum system, we must choose one definite universal path dependent phase Θ to be shared by all its states. A new universal path dependent phase would correspond to a new vector space whose vectors would describe the states of a different physical system and *vice versa*. This will have important consequences on the properties of quantum theory formulated on path space functions.

The vector space of functions introduced so far are still very general. Our aim is to establish a simple and workable model in order to investigate any possible new features of a quantum theory based on path space wave functions. So, we can afford to place further restrictions on these phase space functions to set up an explicit model theory. We can achieve considerable simplification by imposing two further conditions on the phase.

C4 *Globally homotopic functions* A homotopic function f on $\Pi(I\!\!E^2)$ is said to be *global* if the path dependent phase is dependent on the winding numbers ℓ only and is independent of $m \in I\!\!E^2$. This means that the path dependent phase is of the form $\Theta(\sigma_{m,\ell}) = \Theta(\ell)$. We can rewrite Eqs. (9.24) and (9.25) in the form of Eqs. (9.22) and (9.23), i.e.,

$$f(\sigma_{m,\ell}) = e^{i\Theta(\ell)}\psi(m). \tag{9.36}$$

To arrive at the value $\Theta(\ell)$ we can imagine adding to the value of the phase as the path $\sigma_{m,\ell}$ wraps round the hole ℓ times, with each loop round resulting in an additional value.

In orthodox quantum theory a wave function is defined as a function on the physical space, and any wave function can be given a global phase factor without affecting the state since a global phase is a constant. The situation is fundamentally different for functions on path space since a global phase is still affected by the winding number. The dependence of a global phase on ℓ could be complicated. However, for our purposes we would choose a simple dependence stated in C5 below.

C5 *Additivity of the path dependent phase* The path dependent phase of a globally homotopic function is said to be additive if it increases by the same amount as we increase the winding number by 1, i.e., $\Theta(\ell)$ in Eq. (9.36) is of the form[15]

$$\Theta(\ell) = \ell\gamma \quad \text{for some real parameter } \gamma \in [0, 2\pi). \tag{9.37}$$

---

[15] We can also choose to have $\gamma \in (-\pi, \pi]$.

9.4. QUANTUM MECHANICS ON PATH SPACE

So, each time we loop the hole we simply pick up an additional phase γ. We call γ a *universal phase constant*.

We conclude that associated with each chosen *universal phase constant* γ there is a vector space $\mathcal{F}_\gamma(\mathbf{\Pi}(I\!E_h^2))$ of functions on $\mathbf{\Pi}(I\!E_h^2)$ satisfying the five properties listed above. This vector space endowed with a scalar product defined by Eq. (9.3) can be completed to form a Hilbert space which can then form the basis for a quantum theory for a particle moving in the physical space $I\!E_h^2$. In the following section we shall give an explicit demonstration of this.

9.4 Quantum Mechanics on Path Space

A traditional quantum theory on a circle $\mathcal{C}$ of radius r has been discussed in §6.7. We have also studied a quantum theory on $\mathcal{C}_c$, a circle with a cut in §6.8. Let us consider setting up a quantum theory on the path space of a circle. We take as our coordinate on $\mathcal{C}$ the angle variable $\theta \in [0, 2\pi]$ in Eq. (1.235).[16] This variable measures the angular distance from the fixed point m_f on $\mathcal{C}$ corresponding to coordinate $\theta = 0$ to any other point m corresponding to coordinate θ.

Topologically the circle $\mathcal{C}$ is similar to $I\!E_h^2$, the plane with a hole. For example, not all paths on a circle are homotopic and we can classify loops and paths in the same way as in $I\!E_h^2$. Our previous constructs on $I\!E_h^2$ apply with little alteration. In particular a path starting from the fixed point $\theta = 0$ and ending at an arbitrary point θ can be divided into two sections:

1. The arc linking $\theta = 0$ directly to θ going in an anti-clockwise direction.

2. The number of loops $\mathcal{L}_\ell$ the path takes before ending up at the point θ.

We shall denote such a path by $\sigma_{m,\ell}$. Equation (9.11) applies here.

9.4.1 Hilbert spaces $\mathcal{H}_\gamma(\mathbf{\Pi}(\mathcal{C}))$ on path space $\mathbf{\Pi}(\mathcal{C})$

First we need to construct a Hilbert space out of functions defined on the path space $\mathbf{\Pi}(\mathcal{C})$ of the circle. We start with the set $\mathcal{F}_\gamma(\mathbf{\Pi}(\mathcal{C}))$ of functions on $\mathbf{\Pi}(\mathcal{C})$ satisfying the five properties listed in §9.3 with a chosen universal phase constant γ. A function $f_\gamma \in \mathcal{F}_\gamma(\mathbf{\Pi}(\mathcal{C}))$ will then be of the form

$$f_\gamma(\sigma_{m,\ell}) = e^{i\ell\gamma}\psi(m), \quad m \in \mathcal{C}, \tag{9.38}$$

---

[16] $\theta = 0$ and $\theta = 2\pi$ refer to the same point on $\mathcal{C}$. So, we often restrict the range of θ to the interval $[0, 2\pi)$ to achieve a one-to-one correspondence between the values of θ and points of the circle.

where ψ is a (single-valued) function on $\mathcal{C}$. Taking $\psi(m)$ as a function of the angle variable θ we can express functions in $\mathcal{F}_\gamma(\mathbf{\Pi}(\mathcal{C}))$ as functions of θ and the winding number ℓ of the path $\boldsymbol{\sigma}_{m,\ell}$:

$$f_\gamma(\boldsymbol{\sigma}_{m,\ell}) = F_\gamma(\theta, \ell) = e^{i\ell\gamma}\psi(\theta), \quad \theta \in [0, 2\pi). \tag{9.39}$$

Sharing the same universal phase constant γ, these functions can be added to form a vector space, i.e.,

$$f_{1\gamma}(\boldsymbol{\sigma}_{m,\ell}) = e^{i\ell\gamma}\psi_1(\theta), \qquad f_{2\gamma}(\boldsymbol{\sigma}_{m,\ell}) = e^{i\ell\gamma}\psi_2(\theta) \tag{9.40}$$

$$\Rightarrow \quad (f_{1\gamma} + f_{2\gamma})(\boldsymbol{\sigma}_{m,\ell}) = e^{i\ell\gamma}\psi_1(\theta) + e^{i\ell\gamma}\psi_2(\theta) \tag{9.41}$$

$$= e^{i\ell\gamma}\left(\psi_1(\theta) + \psi_2(\theta)\right) \tag{9.42}$$

is again a member of $\mathcal{F}_\gamma(\mathbf{\Pi}(\mathcal{C}))$. A natural scalar product can be defined between $f_{1\gamma}(\boldsymbol{\sigma}_{m,\ell})$ and $f_{2\gamma}(\boldsymbol{\sigma}_{m,\ell})$ by

$$\langle f_{1\gamma} \mid f_{2\gamma} \rangle_\gamma = \int_0^{2\pi} F_{1\gamma}^*(\theta, \ell) F_{2\gamma}(\theta, \ell) \, r d\theta \tag{9.43}$$

$$= \int_0^{2\pi} \psi_1^*(\theta)\psi_2(\theta) \, r d\theta. \tag{9.44}$$

This scalar product space can be completed in the usual way to form a Hilbert space $\mathcal{H}_\gamma(\mathbf{\Pi}(\mathcal{C}))$. We have in effect a family of Hilbert spaces $\mathcal{H}_\gamma = \mathcal{H}_\gamma(\mathbf{\Pi}(\mathcal{C}))$ parameterized by the universal phase constant γ. Each $\mathcal{H}_\gamma$ contains functions whose phase changes by the same amount γ as the winding number increases by 1. It should be stressed that adding functions from different Hilbert spaces $\mathcal{H}_{\gamma_1}$ and $\mathcal{H}_{\gamma_2}$ does not generally lead to a new function satisfying properties C1 to C5 stated in §9.3. If we formally add

$$F_{1\gamma_1}(\theta, \ell) = e^{i\ell\gamma_1}\psi_1(\theta) \in \mathcal{H}_{\gamma_1} \quad \text{and} \quad F_{2\gamma_2}(\theta, \ell) = e^{i\ell\gamma_2}\psi_2(\theta) \in \mathcal{H}_{\gamma_2} \tag{9.45}$$

we get

$$F_{1\gamma_1}(\theta, \ell) + F_{2\gamma_2}(\theta, \ell) = e^{i\ell\gamma_1}\psi_1(\theta) + e^{i\ell\gamma_2}\psi_2(\theta) \tag{9.46}$$

which belongs to neither $\mathcal{H}_{\gamma_1}(\mathbf{\Pi}(\mathcal{C}))$ nor $\mathcal{H}_{\gamma_2}(\mathbf{\Pi}(\mathcal{C}))$.

A path space function $F_\gamma(\theta, \ell)$ is a multi-valued function on the physical space $\mathcal{C}$, i.e., it is a multi-valued function of $\theta \in [0, 2\pi)$. A useful way of looking at the multi-valuedness of such functions is to introduce an extended coordinate variable θ_{ex} which varies from $-\infty$ to ∞ with the following interpretation:

1. The first loop round anti-clockwise corresponds to θ_{ex} taking values in the range $[0, 2\pi)$, i.e., we have $\theta_{ex} = \theta$.

9.4. QUANTUM MECHANICS ON PATH SPACE

2. The second loop round anti-clockwise corresponds to θ_{ex} taking values in the range $[2\pi, 4\pi)$ and so on.

3. The first loop round clockwise corresponds to θ_{ex} taking values in the range $(-2\pi, 0]$ and so on.

This enables us to imagine functions in $\mathcal{H}_\gamma(\mathbf{\Pi}(\mathcal{C}))$ to be defined for $\theta_{ex} \in \mathbb{R} = (-\infty, \infty)$. It follows that functions in $\mathcal{H}_\gamma(\mathbf{\Pi}(\mathcal{C}))$ can be treated explicitly in three different ways as

1. single-valued functions $\boldsymbol{f}_\gamma(\boldsymbol{\sigma}_{m,\ell})$ on the path space $\mathbf{\Pi}(\mathcal{C})$, or as

2. multi-valued functions $F_\gamma(\theta, \ell)$ on the physical space $\mathcal{C}$, or finally as

3. single-valued functions $\Psi_\gamma(\theta_{ex})$ on the real line $\mathbb{R}$.

These different representations are related by

$$\boldsymbol{f}_\gamma(\boldsymbol{\sigma}_{m,\ell}) = F_\gamma(\theta, \ell) = e^{i\ell\gamma} \psi(\theta) = \Psi_\gamma(\theta_{ex}), \tag{9.47}$$

where

$$\theta \in [0, 2\pi), \quad \theta_{ex} \in (-\infty, \infty), \quad \text{and} \quad \theta_{ex} = \theta + 2\ell\pi. \tag{9.48}$$

In what follows we shall adhere to this relation between θ_{ex} and θ. Although $\Psi_\gamma(\theta_{ex})$ are formally defined for the entire range of values of $\theta_{ex} \in (-\infty, \infty)$, the scalar product involves an integral only over the range $[0, 2\pi)$, e.g.,

$$\langle \Psi_\gamma \mid \Psi_\gamma \rangle_\gamma = \int_0^{2\pi} \Psi_\gamma^*(\theta_{ex}) \, \Psi_\gamma(\theta_{ex}) \, d\theta_{ex}. \tag{9.49}$$

The reasons for introducing $\Psi_\gamma(\theta_{ex})$ are two-fold:

1. We are more used to dealing with single-valued functions on $\mathbb{R}$ than multi-valued functions on $\mathcal{C}$ or functions on path space.

2. More importantly the variable θ_{ex} embodies a notion of continuity of motion as we go round and round the circle, a notion very useful in appreciating some important mathematical models of physical situations such as a current flowing round a conducting ring.

9.4.2 Comparing $\mathcal{H}_\gamma(\mathbf{\Pi}(\mathcal{C}))$ and $L^2(\mathcal{C}_c)$

Let us now compare the Hilbert spaces $\mathcal{H}_\gamma(\mathbf{\Pi}(\mathcal{C}))$ with the space $L^2(\mathcal{C}_c)$ introduced in §6.8. First we map the Sobolev space $S_{2,1}(\mathcal{C}_c)$ in $L^2(\mathcal{C}_c)$ to $\mathcal{H}_\gamma(\mathbf{\Pi}(\mathcal{C}))$.[17] Since $S_{2,1}(\mathcal{C}_c)$ is dense in $L^2(\mathcal{C}_c)$ this can be extended to map

---
[17]The space $S_{2,1}(\mathcal{C}_c)$ first appears in Eq. (6.323). Recall that $L^2(\mathcal{C}_c)$ is coordinated by an angle variable ϑ whose values are restricted to the open interval $(0, 2\pi)$.

$L^2(\mathcal{C}_c)$ to $\mathcal{H}_\gamma(\mathbf{\Pi}(\mathcal{C}))$. Functions $\psi(\vartheta)$ in $S_{2,1}(\mathcal{C}_c)$ possess limiting values as ϑ tends to 0.[18] We shall denote these limitng values by $\psi(0)$. We can now define a mapping

$$\widehat{U}_\gamma : L^2(\mathcal{C}_c) \mapsto \mathcal{H}_\gamma(\mathbf{\Pi}(\mathcal{C})) \qquad (9.50)$$

by mapping each $\psi(\vartheta) \in S_{2,1}(\mathcal{C}_c)$ to a corresponding element $\Psi_\gamma(\theta_{ex})$ in $\mathcal{H}_\gamma(\mathbf{\Pi}(\mathcal{C}))$ by:[19]

$$\begin{align}
\Psi_\gamma(0) &= \psi(0), & \theta_{ex} &= 0 & (9.51)\\
\Psi_\gamma(\theta_{ex}) &= \psi(\vartheta), & \theta_{ex} &= \vartheta \in (0, 2\pi) & (9.52)\\
\Psi_\gamma(2\pi) &= e^{i\gamma}\psi(0), & \theta_{ex} &= 2\pi & (9.53)\\
\Psi_\gamma(\theta_{ex}) &= e^{i\gamma}\psi(\vartheta), & \theta_{ex} &= \vartheta + 2\pi & (9.54)\\
&\vdots & & & (9.55)\\
\Psi_\gamma(2\ell\pi) &= e^{i\ell\gamma}\psi(0), & \theta_{ex} &= 2\ell\pi,\ \ell > 0 & (9.56)\\
\Psi_\gamma(\theta_{ex}) &= e^{i\ell\gamma}\psi(\vartheta), & \theta_{ex} &= 2\ell\pi + \vartheta. & (9.57)
\end{align}$$

The mapping $\widehat{U}_\gamma$ is clearly unitary. Except for the case $\gamma = 0$ functions $\Psi_\gamma(\theta_{ex})$ in $\mathcal{H}_\gamma(\mathbf{\Pi}(\mathcal{C}))$ are generally quasi-periodic with a discontinuity at $\theta_{ex} = 2\ell\pi$ even if the corresponding function $\psi(\vartheta) \in S_{2,1}(\mathcal{C}_c)$ is periodic. As an example, consider the function $\eta_{\lambda,n}(\vartheta)$ in Eq. (6.325), i.e.,

$$\eta_{\lambda,n}(\vartheta) = e^{i(n-\frac{\lambda}{2\pi})\vartheta} \in L^2(\mathcal{C}_c). \qquad (9.58)$$

This function is mapped to

$$\begin{align}
\boldsymbol{\eta}_{\gamma;\lambda,n}(\theta_{ex}) &= e^{i\ell\gamma}\eta_{\lambda,n}(\vartheta)\\
&= e^{i\ell\gamma}e^{i(n-\lambda/2\pi)\vartheta} \in \mathcal{H}_\gamma(\mathbf{\Pi}(\mathcal{C})). \qquad (9.59)
\end{align}$$

Note that γ is fixed, being the universal phase constant of the chosen Hilbert space $\mathcal{H}_\gamma(\mathbf{\Pi}(\mathcal{C}))$. The function $\eta_{\lambda,n}$, and the constant λ are arbitrarily chosen as an element of $L^2(\mathcal{C}_c)$, e.g., λ is not related to γ. There are two special cases of interest:

Case 1 with $\lambda = 0$ An element with $\lambda = 0$, i.e., $\eta_{0,n}$, is mapped to $\boldsymbol{\eta}_{\gamma;0,n}$ in $\mathcal{H}_\gamma(\mathbf{\Pi}(\mathcal{C}))$. While $\eta_{0,n}$ is periodic $\boldsymbol{\eta}_{\gamma;0,n}$ is still quasi-periodic.

Case 2 with $\lambda = \gamma$ Consider the set of elements $\eta_{\lambda,n}$, $n = 0, \pm 1, \pm 2, \ldots$ in $L^2(\mathcal{C}_c)$. When $-\lambda$ coincides with the universal phase constant γ of

---

[18]Weidmann (1980) Theorem 6.27 on p. 159.
[19]The angular position variable ϑ for $\mathcal{C}_c$ has a range $(0, 2\pi)$; ϑ is related to the angular position variable θ for $\mathcal{C}$ in accordance with Eq. (1.236).

9.4. QUANTUM MECHANICS ON PATH SPACE

$\mathcal{H}_\gamma(\Pi(\mathcal{C}))$ this set is mapped to $\boldsymbol{\eta}_{\gamma;-\gamma,n}$ in $\mathcal{H}_\gamma(\Pi(\mathcal{C}))$. We get

$$\eta_{\lambda,n} = \eta_{-\gamma,n} \mapsto \boldsymbol{\eta}_{\gamma;-\gamma,n}(\theta_{ex}) = e^{i\ell\gamma} e^{i(n+\gamma/2\pi)\vartheta} \quad (9.60)$$
$$= e^{i(n+\gamma/2\pi)\theta_{ex}}. \quad (9.61)$$

These functions are continuous in θ_{ex}, in contrast to $\boldsymbol{\eta}_{\gamma;0,n}$ in Case 1. Intuitively functions defined on the path space $\Pi(\mathcal{C})$ can be perceived to be defined on a helix as shown in Fig. 9.4.3. The extended angle variable θ_{ex} can be visualized as varying along the helix and a differentiation with respect to θ_{ex} as a differentiation along the helix. The values of $\Psi_\gamma(\theta_{ex})$ at corresponding points on successive loops, for example the three points shown as bullet points at $\theta_{ex} = 0, 2\pi$ and 4π in Fig. 9.4.3, differ by a global phase factor $e^{i\gamma}$. We can see this explicitly from the expression for $\boldsymbol{\eta}_{\gamma;-\gamma,n}$ in Eq. (9.61):

$$\boldsymbol{\eta}_{\gamma;-\gamma,n}(2\pi) = e^{i\gamma} \boldsymbol{\eta}_{\gamma;-\gamma,n}(0), \quad (9.62)$$
$$\boldsymbol{\eta}_{\gamma;-\gamma,n}(4\pi) = e^{i\gamma} \boldsymbol{\eta}_{\gamma;-\gamma,n}(2\pi), \quad (9.63)$$

and so on. While $\boldsymbol{\eta}_{\gamma;-\gamma,n}(\theta_{ex})$ is a continuous function of θ_{ex} it represents a discontinous function on $\mathcal{C}$ in the sense that $\boldsymbol{\eta}_{\gamma;-\gamma,n}(2\pi)$ and $\boldsymbol{\eta}_{\gamma;-\gamma,n}(4\pi)$ are different from $\boldsymbol{\eta}_{\gamma;-\gamma,n}(0)$ although $\theta_{ex} = 4\pi, 2\pi$ and 0 correspond to the same point on the circle. This is depicted in Fig. 9.4.3.

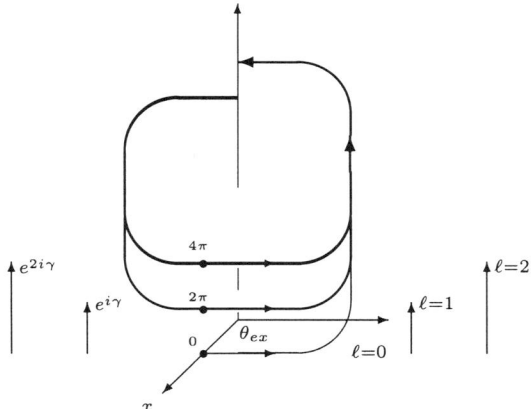

Fig. 9.4.3 Path space on $\mathcal{C}$ as a helix.

We can say that $\boldsymbol{\eta}_{\gamma;-\gamma,n}(\theta_{ex})$ is continuous along the helix but discontinuous on the circle.[20]

---
[20]Generally we should not confuse the increase due to the global phase factor and the increase of the function itself over the circle.

9.4.3 Position operators in $\mathcal{H}_\gamma(\Pi(\mathcal{C}))$

We can define an angular position operator on the Hilbert space $\mathcal{H}_\gamma(\Pi(\mathcal{C}))$ in terms of the position operator ϑ in $L^2(\mathcal{C}_c)$. Consider the following unitary transformations from $L^2(\mathcal{C}_c)$ to $\mathcal{H}_\gamma(\Pi(\mathcal{C}))$

$$\Psi_\gamma(\theta_{ex}) = \widehat{U}_\gamma \psi, \quad \widehat{\theta}_{ex,\gamma} = \widehat{U}_\gamma \vartheta \, \widehat{U}_\gamma^{-1}, \tag{9.64}$$

where $\widehat{U}_\gamma$ is given by Eq. (9.50). We have

$$\left(\widehat{\theta}_{ex,\gamma} \Psi_\gamma\right)(\theta_{ex}) = \widehat{U}_\gamma\left(\vartheta\,\psi(\vartheta)\right) = \theta_{ex} \Psi_\gamma(\theta_{ex}), \tag{9.65}$$

with the expectation value given by

$$\langle \Psi_\gamma \mid \widehat{\theta}_{ex,\gamma} \Psi_\gamma \rangle_\gamma = \int_0^{2\pi} \Psi_\gamma^*(\theta_{ex})\, \theta_{ex}\, \Psi_\gamma(\theta_{ex})\, d\theta_{ex}. \tag{9.66}$$

Note that although θ_{ex} as a variable has a formal range of $(-\infty, \infty)$ the operator expression $\widehat{\theta}_{ex}$ has only finite expectation values.

9.4.4 Momentum operators in $\mathcal{H}_\gamma(\Pi(\mathcal{C}))$

In order to establish a momentum operator in $\mathcal{H}_\gamma(\Pi(\mathcal{C}))$ we should first look back at momentum operators in $L^2(\mathcal{C}_c)$. Equation (6.322) in §6.8.1 defines a family of momentum operators $\widehat{P}_\lambda(\mathcal{C}_c)$ acting on quasi-periodic functions given by Eq. (6.323). These operators admit $\eta_{\gamma,n}(\vartheta)$ in Eq. (9.58) as eigenfunctions. Within $L^2(\mathcal{C}_c)$ there is no *a priori* reason to select a particular operator of this family, i.e., a particular value of λ, as the preferred momentum operator. In a given Hilbert space $\mathcal{H}_\gamma(\Pi(\mathcal{C}))$ consisting of functions on the path space $\Pi(\mathcal{C})$ sharing a universal phase constant γ we do naturally have a preferred momentum operator, i.e., the operator $\widehat{P}_\gamma(\Pi(\mathcal{C}))$ obtained by a unitary transformation from the momentum operator $\widehat{P}_\lambda(\mathcal{C}_c)$ in $L^2(\mathcal{C}_c)$ with $-\lambda = \gamma$. So, let

$$\widehat{P}_\gamma(\Pi(\mathcal{C})) = \widehat{U}_\gamma \, \widehat{P}_{-\gamma}(\mathcal{C}_c)\, \widehat{U}_\gamma^{-1}, \tag{9.67}$$

then

$$\widehat{P}_\gamma(\Pi(\mathcal{C}))\Psi_\gamma(\theta_{ex}) = \widehat{U}_\gamma \widehat{P}_{-\gamma}(\mathcal{C}_c)\psi(\vartheta) \tag{9.68}$$

$$= -i\hbar\, \widehat{U}_\gamma \left(\frac{d\psi(\vartheta)}{d\vartheta}\right). \tag{9.69}$$

9.5. JOSEPHSON EFFECT AND SUPERSELECTION RULES

This operator admits $\eta_{\gamma;-\gamma,n}$ in Eq. (9.61) as eigenfunctions corresponding to eigenvalues

$$p_{\gamma,n} = \frac{\hbar}{r}\left(n + \frac{\gamma}{2\pi}\right), \quad n = 0, \pm 1, \pm 2, \ldots, \qquad (9.70)$$

since[21]

$$\widehat{P}_\gamma(\mathbf{\Pi}(\mathcal{C}))\,\eta_{\gamma;-\gamma,n}(\theta_{ex}) = -\frac{i\hbar}{r}\,\widehat{U}_\gamma\left(\frac{d\eta_{-\gamma,n}(\vartheta)}{d\vartheta}\right) \qquad (9.71)$$

$$= \frac{\hbar}{r}\left(n + \frac{\gamma}{2\pi}\right)\widehat{U}_\gamma \eta_{-\gamma,n}(\vartheta) \qquad (9.72)$$

$$= p_{\gamma,n}\,\eta_{\gamma;-\gamma,n}(\theta_{ex}). \qquad (9.73)$$

We shall naturally identify $\widehat{P}_\gamma(\mathbf{\Pi}(\mathcal{C}))$ as the preferred momentum operator in $\mathcal{H}_\gamma(\mathbf{\Pi}(\mathcal{C}))$. A formal operator expression for $\widehat{P}_\gamma(\mathbf{\Pi}(\mathcal{C}))$ is

$$\widehat{P}_\gamma(\mathbf{\Pi}(\mathcal{C})) = -\frac{i\hbar}{r}\frac{\partial}{\partial\theta_{ex}}. \qquad (9.74)$$

Note that $\widehat{P}_\gamma(\mathbf{\Pi}(\mathcal{C}))$ and $\widehat{P}_{-\gamma}(\mathcal{C}_c)$ possess the same discrete spectrum.

9.5 Josephson Effect and Superselection Rules

We want to see how our present theory can be applied to model Josephson effect in a superconducting ring with a Josephson junction. Our previous treatment in §7.4 utilizes the Hilbert space $L^2(\mathcal{C}_c)$ and momentum operators $\widehat{P}_\lambda(\mathcal{C}_c)$ acting on quasi-periodic functions. In our present Hilbert space $\mathcal{H}_\gamma(\mathbf{\Pi}(\mathcal{C}))$ we have a preferred momentum operator $\widehat{P}_\gamma(\mathbf{\Pi}(\mathcal{C}))$. In $L^2(\mathcal{C}_c)$ there is a class of selfadjoint Hamiltonian operators $\widehat{H}^{(12+)}_{a,b}(\mathcal{C}_c)$ acting on functions satisfying boundary conditions (BCC12) in §6.9.12, i.e.,[22]

$$\psi'(2\pi) = a\,\psi(2\pi) + b\,\psi(0), \qquad (9.75)$$
$$\psi'(0) = -a\,\psi(0) - b\,\psi(2\pi). \qquad (9.76)$$

The corresponding Hamiltonian operators $\widehat{H}^{(12+)}_{\gamma;a;b}(\mathbf{\Pi}(\mathcal{C}))$ in $\mathcal{H}_\gamma(\mathbf{\Pi}(\mathcal{C}))$ are obtained by a unitary transformation, i.e.,

$$\widehat{H}^{(12+)}_{\gamma;a;b}(\mathbf{\Pi}(\mathcal{C})) = \widehat{U}_\gamma\,\widehat{H}^{(12+)}_{a,b}(\mathcal{C}_c)\,\widehat{U}_\gamma^{-1}. \qquad (9.77)$$

---

[21] Note that $\eta_{\lambda,n}(\vartheta)$ is given by Eq. (9.58) and is related to $\eta_{\gamma;-\gamma,n}(\theta_{ex})$ by Eq. (9.61).
[22] As remarked after Eq. (6.333) a dash represents a derivative with respective to $r\vartheta$.

The corresponding boundary conditions for the functions in the domain of $\widehat{H}_{\gamma;a,b}^{(12+)}(\mathbf{\Pi}(\mathcal{C}))$ in $\mathcal{H}_\gamma(\mathbf{\Pi}(\mathcal{C}))$ would then be:[23]

$$\Psi'_\gamma(2\pi_-) = a\,\Psi_\gamma(2\pi_-) + b\,\Psi_\gamma(0_+), \qquad (9.78)$$

$$\Psi'_\gamma(0_+) = -a\,\Psi_\gamma(0_+) - b\Psi_\gamma(2\pi_-). \qquad (9.79)$$

As in physical assumptions (PAS1) and (PAS2) stated in §7.3.2, we shall assume a superconducting state to be an eigenstate of the momentum and the Hamiltonian operator. In §7.4.3 we discuss how this requirement leads to the Josephson equation for the supercurrent. We can repeat the argument here. A superconducting state should be represented by $\eta_{\gamma;-\gamma,n}$ given by Eq. (9.61) and the Hamiltonian is chosen so that Eqs. (9.78) and (9.79) are satisfied by $\eta_{\gamma;-\gamma,n}$. A substitution of $\eta_{\gamma;-\gamma,n}$ into the above boundary conditions leads to the following equations:[24]

$$a = -b\cos\gamma, \qquad (9.80)$$

$$\frac{1}{r}\left(n + \frac{\gamma}{2\pi}\right) = -b\sin\gamma. \qquad (9.81)$$

These equations agree with Eqs. (7.37) and (7.38) in §7.4.3. We can similarly deduce the Josephson equation Eq. (7.41) for the supercurrent flow round the superconducting ring.

The physical interpretation is transparent now. The wave function $\eta_{\gamma;-\gamma,n}$ describing the system can be regarded as a continuous single-valued function of the extended position variable θ_{ex}. When the quasi-particle passes the junction in completing a loop round $\mathcal{C}$, i.e., θ_{ex} increases by 2π, the phase of the wave function increases smoothly by γ. This phase is the all important quantity here since it determines the magnitude of the supercurrent. The traditional description in terms of functions of θ will have to consider multi-valued functions on $\mathcal{C}$ where the wave function changes abruptly at the junction.

In §7.4.4 we have argued for the existence of a superselection rule forbidding a superposition of different dc superconducting currents going round in a circle. In the context of our present discussion this means no superposition of $\eta_{\gamma_1;-\gamma_1,n}$ and $\eta_{\gamma_2;-\gamma_2,n}$. In other words there should be no superposition of states of different universal phase constants γ_1 and γ_2. This result, which amounts to a superselection rule, actually emerges naturally from the fundamental structure of our present theory. Those functions defined on the path space automatically generate a whole family of Hilbert spaces $\mathcal{H}_\gamma(\mathbf{\Pi}(\mathcal{C}))$ parameterized by the phase constant γ in such a way that functions from different spaces $\mathcal{H}_{\gamma_1}(\mathbf{\Pi}(\mathcal{C}))$ and $\mathcal{H}_{\gamma_2}(\mathbf{\Pi}(\mathcal{C}))$ cannot be added together as shown

---

[23] For $\vartheta \in (0, 2\pi)$ we have $\Psi_\gamma = \psi$.

[24] The compatibility consideration in §7.4.3 should also be taken into account here, e.g., a, b, n and γ must be chosen to satisfy the two equations relating them.

in the argument following Eq. (9.46). In other words, the theory has a built in superselection rule forbidding any superposition of functions from different spaces $\mathcal{H}_\gamma(\Pi(\mathcal{C}))$ of the family. This would then explain why, physically, there is no superposition of currents corresponding to states $\eta_{\gamma_1;-\gamma_1,n}$ and $\eta_{\gamma_2;-\gamma_2,n}$ of different universal phase constants γ_1 and γ_2.

9.6 Concluding Remarks

We have seen that the path space can form the basis of a formulation of quantum mechanics which can naturally lead to new physical insights into the behaviour of certain physical systems. The application of path space formulation to closed circuit configurations is particularly interesting in view of its easy accommodation of physical quantities like the momentum operator together with its eigenfunctions which are continuous and singled-valued with respect to the extended angle variable θ_{ex}, and the current flow round the circle. This contrasts sharply with the need for discontinuous and possibly multi-valued functions in the more traditional formulation in the Hilbert space $L^2(\mathcal{C}_c)$. One can proceed further to examine point interactions, e.g., a δ-interaction discussed in §6.9.2, in a circle in our present path space formulation of quantum mechanics.[25] Superselection rules can emerge naturally on account of the inherent existence of a family of Hilbert spaces parameterized by a family of universal phase constants.

References

1. Balachandran, A. P. (1989). 'Classical Topology and Quantum Phases' in S. De Filippo, M. Marinaro, G. Marmo and G. Vilasi, eds., *Geometric Aspects of Nonlinear Field Theory*, North Holland, Amsterdam.

2. Dirac, P. A. M. (1931). *Proceedings of Royal Society A* **133** 60.

3. Horvathy, P. A., Morandi. G. and Sudarshan, E. C. G. (1989). *Nuovo Cimento D* **11** , 201.

4. Lipschutz, S. (1965). *General Topoloogy*, Schaum's Outline Series, McGraw-Hill, New York.

5. Morandi, G. (1993). *The Role of Topology in Classical and Quantum Physics*, Springer-Verlag, Berlin.

---

[25]Trueman (1999), Trueman and Wan (2000).

6. Nash, C. and Sen, S. (1983). *Topology and Geometry for Physicists*, Academic Press, London.

7. Trueman, C. (1999). *Superselection Rules, Quantization by Parts and Point Interactions*, St Andrews University PhD Thesis.

8. Trueman, C. and Wan, K. K. (2000). *J. Math. Phys.* **41** 1.

9. Weidmann, J. (1980). *Linear Operators in Hilbert Spaces*, Springer-Verlag, New York.

10. Zaccoria, F., Sudarshan, E. C. G., Nilsson, J. S., Mukunda, N., Marmo, G. and Balachandran, A. P. (1983). *Phys. Rev. D* **27**, 2327.

Bibliography

Abraham, R. and Marsden, J. E. (1978). *Foundations of Mechanics*, Benjamin/Cummings, Reading, Mass.

Aczel, A. (2002). *Entanglement: The Greatest Mystery in Physics*, Wiley, New York.

Aharonov, Y. and Bohm, D. (1959). *Phys. Rev.* **115** 485.

Aharonov, Y. and Bohm, D. (1961). *Phys. Rev.* **122** 1649.

Aharonov, Y. and Safko, L. (1975). *Ann. Phys.* **91** 279.

Akhiezer, N. I. and Glazman, I. M. (1961). *Theory of Linear Operators in Hilbert Space Vol. 1*, Frederick Ungar, New York.

Akhiezer, N. I. and Glazman, I. M. (1963). *Theory of Linear Operators in Hilbert Space Vol. 2*, Frederick Ungar, New York.

Alanson, T. (1992). *Phys. Lett. A* **63** 41.

Albeverio, S., Gesztesy, F., Hoegh-Krohn, R. and Holden, H. (1988). *Solvable Models in Quantum Mechanics*, Springer-Verlag, New York Inc.

Albeverio, S., Gesztesy, F. and Holden, H. (1992). *J. Phys. A: Math. Gen.* **26** 3903.

Ali, S. T. and Emch, G. G. (1974). *J. Math. Phys.* **15** 176.

Allcock, G. R. (1969). *Ann. Phys.* **53** 311.

Amann, A. (1988). 'Chirality as a Classical Observable in Algebraic Quantum Mechanics' in A. Amann *et al.*, ed., *Fractals, Quasicrystals, Chaos, Knots and Algebraic Quantum Mechanics*, Kluwer, Dordrecht.

Amrein, W. O. (1981). *Non-Relativistic Quantum Dynamics*, Reidel, Dordrecht.

Amrein, W. O., Jauch, J. M. and Sinha, K. B. (1977). *Scattering Theory in Quantum Mechanics*, Benjamin, Reading, Mass.

Anatopoulos, C. (2002). *Int. J. Theor. Phys.* **41** 1573.

Anderson, A. (1995). *Phys. Rev. Lett.* **74** 621.

Anderson, B. P., Dholakia, K. and Wright, E. M. (2003). *Phys. Rev. A* **67** 033601.

Anosov, D. V. and Arnold, V. I. (1988). *Dynamical Systems I*, Springer-Verlag, Berlin.

Araki, H. (1980). *Prog. Theor. Phys.* **66** 719.

Araki, H. (1986). 'A Continuous Superselection Rule as a Model of Classical Measuring Apparatus in Quantum Mechanics' in A. I. Miller, ed., *Fundamental Aspects of Quantum Theory*, a NATO ASI Series B Vol. 226, Plenum, New York.

Arndt, M., Nairz, O.,Vos-Andreae, J., Keller, C., van der Zouw, G. and Zeilinger, A. (1999). *Nature* **401** 680.

Arnold, V. I. (1978). *Mathematical Methods of Classical Mechanics*, Springer-Verlag, New York.

Atmanspacher, H. and Amann, A. (1998). *Int. J. Theor. Phys.* **37** 629.

Auletta, G. (2000). *Foundations and Interpretation of Quantum Mechanics*, World Scientific, Singapore. There are extensive bibliographic references.

Backhaus, S., Pereversev, S., Simmonds, R. W., Loshak, A., Davies, J. C. and Packard, R. E. (1998). *Nature* **392** 687.

Bacry, H. (1988). *Localizability and Space in Quantum Physics* (*Lecture Notes in Physics*), Springer-Verlag, Berlin.

Bader, R. and Essen, H. (1984). 'The Mechanics of and an Equation for the Electron Charge Density' in J. Dahl and A. Avery, eds., *Local Density Approximations in Quantum Chemistry and Solid State Physics*, Plenum, New York.

Balachandran, A. P. (1989). *Classical Topology and Quantum Phases* in S. De Filippo, M. Marinaro, G. Marmo and G. Vilasi, eds. *Geometric Aspects of Nonlinear Field Theory*, North Holland, Amsterdam.

Balian, R. (1991). *From Microphysics to Macrophysics*, Springer-Verlag, Berlin.

Ballentine, L. E. (1990). *Rev. Mod. Phys.* **42** 358.

Ballentine, L. E. (1990). *Quantum Mechanics*, Prentice-Hall, Eaglewood Cliffs, New Jersey.

Bardeen, J., Cooper, L. N. and Schreiffer, J. R. (1957). *Phys. Rev.* **108** 1175

Barone, A. and Paternò, G. (1982). *Physics and Applications of the Josephson Effect*, Wiley, New York.

Bauer, M. (1983). *Ann. Phys.* **150** 1.

Baym, G. (1969). *Lectures on Quantum Mechanics*, Benjamin, New York.

Bell, J. (1987). 'Are there quantum Jumps?' in J. Bell, *Speakable and Unspeakable in Quantum Mechanics*, Cambridge University Press, Cambridge.

Bell, J. (1990). 'Against "Measurement"' in A. I. Miller, ed., *Twenty-Six Years of Uncertainty*, a NATO ASI Series B Vol. 226, Plenum, New York.

Beltrametti, E. G. and Cassinelli, G. (1981). *The Logic of Quantum Mechanics*, Addison-Wesley, Reading Mass.

Benn, I. M. and Tucker, R. W. (1987). *An Introduction to Spinors and Geometry*, Adam Hilger, Bristol.

Berkley, A., Xu, H., Ramos, R., Gubrud, M., Strouch, F., Johnson, P., Anderson, J., Dragt, A., Lobb, C. and Wellstood, F. (2003). *Science* **300** June 1548.

Bishop, R. L. and Goldberg, S. I. (1968). *Tensor Analysis on Manifolds*, Macmillan, New York.

Blank, J., Exner, P. and Havlíček, M. (1994). *Hilbert Space Operators in Quantum Physics*, American Institute of Physics Press, New York.

Bleaney, B. I. and Bleaney, B. (1957). *Electricity and Magnetism*, Clarendon Press, Oxford.

Blum, K. (1981). *Density Matrix Theory and Applications*, Plenum, New York.

Bogoliubov, N. N. (1970). *Lectures on Quantum Statistics Vol. 2*, MacDonald Technical and Scientific, London.

BIBLIOGRAPHY 661

Bogoliubov, N. N. and Bogoliubov N. N., Jr. (1992). *An Introduction to Quantum Statistical Mechanics*, Gordon and Breach, Lausanne, Switzerland.

Bogolubov, N., Logunov, A. and Todorov, I. (1975). *Introduction to Axiomatic Quantum Field Theory*, Benjamin, Reading, Mass.

Böhm, A. (1978). *The Rigged Hilbert Space and Quantum Mechanics (Lecture Notes in Physics 78)*, Springer-Verlag, Berlin.

Böhm, A. (1979). *Quantum Mechanics*, Springer-Verlag, New York.

Bohm, D. (1952). *Phys. Rev.* **85** 166.

Bonse, U. and Rauch, H. (1979). *Neutron Interferometry*, Clarendon Press, Oxford.

Born, M. and Wolf, E. (1999). *Principles of Optics*, Cambridge University Press, Cambridge.

Bradshaw, J. E. (1995). *Superselection Rules, Quasi-Particles and Macroscopic Quantum Systems*, St Andrews University PhD Thesis.

Braginsky, V. B. and Khalili, F. Y. (1992). *Quantum Measurement*, Cambridge University Press, Cambridge.

Bransden, B. H. and Joachain, C. J. (1992). *Introduction to Quantum Mechanics*, Wiley, New York.

Bratteli, O. and Robinson, D. W. (1979). *Operator Algebras and Quantum Statistical Mechanics I*, Springer-Verlag, New York.

Bratteli, O. and Robinson, D. W. (1981). *Operator Algebras and Quantum Statistical Mechanics II*, Springer-Verlag, New York.

Brickell, F. and Clark, R. S. (1970). *Differentiable Manifolds*, Van Nostrand Reinhold, London.

Brune, M., Hagley, E., Dreyer, J., Maitre, X., Maali, A., Wunderlich, C., Raimond, J. M. and Horoche, S. (1996). *Phys. Rev. Lett.* **77** 4887.

Bub, J. (1988). *Found. Phys.* **18** 701.

Bub, J. (1997). *Interpreting the Quantum World*, Cambridge University Press, Cambridge.

Burrill, G. W. (1972). *Measure, Integration and Probability*, McGraw-Hill, New York.

Busch, P. (1985). *Int. J. Theor. Phys.* **24** 63.

Busch, P. (1990). *Found. Phys.* **20** 1.

Busch, P., Grabowski, M. and Lahti, P. J. (1994). *Phys. Lett. A* **191** 357.

Busch, P., Grabowski, M. and Lahti, P. J. (1997). *Operational Quantum Physics*, Springer-Verlag, Berlin.

Busch, P and Lahti, P. J. (1984). *Phys. Rev. D* **29** 1634.

Busch, P., Lahti, P. and Mittelstaedt, P. (1996). *The Quantum Theory of Measurement*, 2nd edition, Springer-Verlag, Berlin.

Busch, P. and Schroeck, F. E., Jr. (1989). *Found. Phys.* **19** 807.

Byron, F. W. and Fuller, R. W. (1969). *Mathematics of Classical and Quantum Physics Vol. 1*, Addison-Wesley, Reading, Mass.

Capri, A. Z. (1985). *Nonrelativistic Quantm Mechanics*, Benjamin/Cummings, Menlo Park, California.

Castellani, L. (1978). *Nuovo Cimento A* **48** 359.

Choquet-Bruhat, Y., de Witt-Morette, C., with Dilland-Bleick, M. (1982). *Analysis, Manifolds and Physics Part I: Basics*, North-Holland, Amsterdam.

Choquet-Bruhat, Y., de Witt-Morette, C. (1989). *Analysis, Manifolds and Physics Part II: 92 Applications*, North-Holland, Amsterdam.

Cini, M. and Levy-Leblond, J. M. (1990). *Quantum Theory without Reduction*, Adam Hilger, Bristol.

Cisneros, C., Martines-y-Romera, R. P., Núněz-Yépez, H. N. and Salas-Brito, A. L. (1998). *Euro. J. Phys.* **19** 237.

Clark, T. D. (1987). 'Macroscopic Quantum Objects' in B. J. Hiley and E. D. Peat, eds., *Quantum Implications, Essays in honour of David Bohm*, Routledge & Kegan, London.

Clark, T. D., Prance, H., Prance, R. J. and Spiller, T. P. eds. (1991). *Macroscopic Quantum Phenomena*, World Scientific, Singapore.

Cobden, D. H. (1999). *Nature* **397** 648.

Cohen, D. W. (1989). *An Introduction to Hilbert Space and Quantum Logic*, Springer-Verlag, New York.

Cohen, N. (1966). *J. Math. Phys.* **7** 781.

Comi, M. (1980). *IL Nuovo Cimento A* **56** 299.

Cornwell, J. F. (1984). *Group Theory in Phyiscs Vol. 1*, Academic Press, London.

Coulson, C. A. (1965). *Waves*, Oliver and Boyd, Edinburgh.

Coutinho, F., Nogami, Y. and Perez, J. (1997). *J. Phys. A* **30** 3937.

Cram, J. M. and Cram, D. J. (1978). *The Essence of Organic Chemistry*, Addison-Wesley, Reading, Mass.

Crawford, F. S., Jr. (1968). *Waves - Berkeley Physics Course Vol. 3*, MaGraw-Hill, New York.

Cycon, H. L., Froese, R. G., Kirsch, W. and Simon, B. (1987). *Schrödinger Operators*, Springer-Verlag, Berlin.

Dahl, J. P. and Avery, J. (1984). *Local Density Approximations in Quantum Chemistry and Solid State Physics*, Plenum, New York.

Dalfovo, F. and Giorgini, S. (1999). *Rev. Mod. Phys.* **71** 463.

Daneri, A., Loinger, A. and Prosperi, G. M. (1962). *Nucl. Phys.* **33** 297.

Darling, R. W. R. (1994). *Differential Forms and Connections*, Cambridge University Press, Cambridge.

Davies, E. B. (1976). *Quantum Theory of Open Systems*, Academic Press, London.

Davis, J. C. and Packard, R. E. (2002). *Rev. Mod. Phys.* **74** 741.

Davisovish, L., Brune, M., Raimond, J. M. and Horoche, S. (1996). *Phys. Rev. A* **53** 1295.

Deaver, B., Jr. and Fairbank, W. (1961). *Phys. Rev. Lett.* **7** 43.

Deb, B. M. (1984). 'Some Aspects of the Role of Single-Particle Density in Chemistry' in J. Dahl and A. Avery, eds., *Local Density Approximations in Quantum Chemistry and Solid Sate Physics*, Plenum, New York.

de Broglie, L. V. (1959). *J. Phys. Radium* **20** 936.

de Bruyn Ouboter, R. (1977). 'Macroscopic Quantum Phenomena in Superconductors' in B. B. Schwartz and S. Foner, eds., *Superconductor Applications: SQUIDs and Machines*, NATO ASI Series B Vol. 21, Plenum, New York.

de Gennes, P. G. (1963). *Phys. Lett.* **5**, 22.

de Gennes, P. G. (1964). *Rev. Mod. Phys.* **36** 225.

de Lange, O. L. and Raab, R. E. (1991). *Operator Methods in Quantum Mechanics*, Oxford University Press, Oxford.

D'Espagnat, B. (1989). *Conceptual Foundations of Quantum Mechanics*, Addison-Wesley, Reading, Mass.

de Vries, P. (1998). *Rev. Mod. Phys.* **70** 447.

Dicke, R. H. and Wittke, J. P. (1963). *Introduction to Quantum Mechanics*, Addison-Wesley, Reading, Mass., World Student Series Edition.

Dirac, P. A. M. (1931). *Proceedings of Royal Society A* **133** 60.

Dixmier, J. (1977). C^*-*Algebras*, North-Holland, Amsterdam.

Dixmier, J. (1981). *Von Neumann Algebras*, North-Holland, Amsterdam.

Doll, R. and Nabauer, M. (1961). *Phys. Rev. Lett.* **7** 51.

Eckern, U. (1986). *Nature* **319** 726.

Einstein, A., Podolsky, B. and Rosen, N. (1935). *Phys. Rev.* **47** 777.

Emch, G. G. (1972). *Algebraic Methods in Statistical Mechanics and Quantum Field Theory*, Wiley, New York.

Enss, V. (1983). *Com. Math. Phys.* **89** 245.

Everett, H. (1957). *Rev. Mod. Phys.* **29** 454.

Exner, P. and Seba, P. (1987). *J. Math. Phys.* **28** 386.

Exner, P. and Seba, P. (1989a). *Rep. Math. Phys.* **28** 7.

Exner, P. and Seba, P. (1989b). 'Quantum Junctions and the Self-Adjoint Extensions Theory' in P. Exner and P. Seba., eds., *Applications of Self-Adjoint Extensions in Quantum Physics*, Springer-Verlag, Berlin.

Exner, P., Seba, P. and Stovicek, P. (1989). 'Quantum Waveguides' in P. Exner and P. Seba. eds., *Applications of Self-Adjoint Extensions in Quantum Physics*, Springer-Verlag, Berlin.

Fano, G. (1971). *Mathematical Methods of Quantum Mechanics*, McGraw-Hill, New York.

Feynman, R. P. (1972). *Statistical Mechanics*, Benjamin, Reading, Mass.

Feynman, R. P., Leighton, R. B. and Sands, M. (1965). *The Feynman Lectures on Physics Vol. 3*, Addison-Wesley, Reading, Mass.

Flügge, S. (1974). *Practical Quantum Mechanics*, Springer-Verlag, New York.

Fountain, R. H. (1995). *Observables, Maximal Symmetric Operators, POV Measures and their Applications in Quantum Mechanics*, St Andrews University PhD Thesis.

Frankel, T. (1997). *The Geometry of Physics*, Cambridge University Press, Cambridge.

Freyberger, M., Bardroff, P., Leichtle, C., Schrade, G. and Schleich, W. (1997). *Physics World* **10** 42.

Friedman, B. (1956). *Principles and Techniques of Applied Mathematics*, Wiley, New York.

Friedman, J. R., Patel, V., Chen, W., Tolpygo, S. K. and Lukens, J. E. (2000). *Nature* **406** 43.

Fuches, C. A. (2001). 'Quantum Foundations in the Light of Quantum Information', arXiv: quant-ph/0106166.

Fujita, S. and Godoy, S. (1996). *Quantum Statistical Theory of Superconductivity*, Plenum, New York.

Gallop, J. C. (1991). *SQUIDS, The Josephson Effects and Superconducting Electronics*, Adam Hilger, Bristol.

Garden, R. W. (1984). *Modern Logic and Quantum Mechanics*, Adam Hilger, Bristol.

Gasiorowicz, S. (2003). *Quantum Physics*, Wiley International Edition.

Gerry, C. C. and Knight, P. L. (1997). *Am. J. Phys.* **65** 964.

Gesztesy, F. and Holden, H. (1987). *J. Phys. A* **20** 5157.

Ghirardi, G. C. and Rimini, A. (1990). 'Old and New Ideas in the Theory of Quantum Measurement' in A. I. Miller, ed., *Sixty-Two Years of Uncertainty and Physical Enquiries into the Foundations of Quantum Mechanics*, a NATO ASI Series B Vol. 226, Plenum, New York.

Ghirardi, G. C., Rimini, A. and Weber, T. (1976). *Nuovo Cimento B* **36** 97.

Ghirardi, G. C., Rimini, A. and Weber, T. (1986). *Phys. Rev.* **D 34** 470.

Ghirardi, G. C., Rimini, A. and Weber, T. (1988). *Nuovo Cimento B* **102** 383.

Giamarchi, T. (2004) *Quantum Physics in One Dimension*, Clarendon Press, Oxford.

Giancoli, D. C. (1988). *Physics for Scientists and Engineers*, Prentice-Hall, Englewood Cliffs, New Jersey. 2nd edition

Giannitrapani, R. (1997). *Int. J. Theor. Phys.* **36** 1575.

Gisin, N. and Percival, I. (1992). *J. Phys. A: Math. Gen.* **25** 5677.

Giulini, D., Joos, E., Kiefer, C., Kupsch, J., Stamatescu, I. and Zeh, H. D. (1996). *Decoherence and the Appearance of a Classical World in Quantum Theory*, Springer-Verlag, Berlin.

Giulini, D., Kiefer, C. and Zeh, H. D. (1995). *Phys. Lett. A* **199** 291.

Goldstein, H. (1950). *Classical Mechanics*, Addison-Wesley, Reading, Mass.

Gottfried, K. (1966). *Quantum Mechanics*, Benjamin, New York.

Grabert, H. and Devoret, M. H. (1992). *Single Charge Tunneling: Coulomb Blackage Phenomena in Nanostructures*, Plenum, New York.

Grabowsi, M. (1989). *Found. Phys.* **19** 923.

Greenberger, D. M. ed. (1986). *New Techniques and Ideas in Quantum Measurement Theory*, Annals of New York Academy of Sciences, New York.

Greenstein, G. and Zajonc, A. G. (1997). *The Quantum Challenge*, Jones and Bartlett Publishers, Sudbury, Mass.

Greiner, W. (1989). *Quantum Mechanics Vol. 1 - An Introduction*, Springer-Verlag, Berlin.

Greiner, W. and Müller, B. (1989). *Quantum Mechanics Vol. 2 - Symmetries*, Springer-Verlag, Berlin.

Gribbin, J. (1984). *In Search of Schrödinger's Cat*, Corgo Books, Reading.

Grigolini, P. (1993). *Quantum Mechanical Irreversibility and Measurement*, World Scientific, Singapore.

Gross, E. P. (1961). *Nuovo Cimento* **20** 454.

Gross, E. P. (1963). *J. Math. Phys.* **4** 195.

Gudder, S. P. (1979). *Stochastic Methods in Quantum Mechanics*, North-Holland, New York.

Guenin, M. (1966). 'Algebraic Methods in Quantum Field Theory' in W. E. Witten, A. O. Barut and M. Guenin, eds., *Boulder Lectures in Theoretical Physics Vol. IXA: Mathematical Methods of Theoretical Physics*, Gordan and Breach, New York.

Gupta, N. N. D. and Ghosh, S. K. (1946). *Rev. Mod. Phys.* **18** 225.

Gustafson, S. J. and Sigal, I. M. (2003). *Mathematical Concepts of Quantum Mechanics*, Springer-Verlag, Berlin.

Haag, R., (1973). 'Infinite Quantum Systems' in W. E. Witten, ed., *Boulder Lectures in Theoretical Physics Vol. XIVB: Mathematical Methods of Theoretical Physics*, Gordan and Breach, New York.

Haag, R. (1992). *Local Quantum Physics*, Springer-Verlag, Berlin.

Hagg, R. and Kastler, D. (1964). *J. Math. Phys.* **5** 848.

Hamilton, W. A., Klein, A. G. and Opat, G. I. (1983). *Phys. Rev.* A **28** 3149.

Harrison, F. E. (1993). *Superselection rules, Quantum Measurement Problems and Macroscopic Quantum Systems*, St Andrews University PhD Thesis.

Harrison, F. E. and Wan, K. K. (1997). *J. Phys. A: Math. Gen.* **30** 4731.

Hegstrom, R. A. and Kondepudi, D. K. (1990). *Scientific American* **263** 98.

Hegstrom, R. A. and Sols, F. (1995). *Found. Phys.* **25** 681.

Hellmuth, T., Zajonc, A. G. and Walther, H. (1985). 'Realization of a "Delayed-Choice" Mach-Zehnder Interferometer' in P. Lahti and P. Mittelstaedt, eds., *Symposium in the Foundations of Modern Physics*, World Scientific, Singapore.

Hepp, K. (1972). *Helv. Phys. Acta.* **45** 237.

Herzberg, H. (1945). *Molecular Spectra and Molecular Structure II: Infrared and Raman Spectra of Polyatomic Molecules*, Van Nostrand Reinhold, New York.

Herzog, T. J., Kwiat, P. G., Weinfurter, H. and Zeilinger, A. (1995). *Phys. Rev. Lett.* **75** 3034.

Hey, T. and Walters, P. (1987). *The Quantum Universe*, Cambridge University Press, Cambridge.

Hiley, B. and Peat, F. D. eds. (1987). *Quantum Implications: Essays in Honours of David Bohm*, Routledge & Kegan Paul, London.

Hillery, M., O'Connell, R., Scully, M. and Wigner, E. (1984). *Physics Reports* **106** 121.

Holevo, A. S. (1982). *Probabilistic and Statistical Aspects of Quantum Theory*, North-Holland, Amsterdam.

Holland, P. R. (1993). *The Quantum Theory of Motion*, Cambridge University Press, Cambridge.

Home, D. and Whitaker, A. (1992) *Physics Reports*, **210(4)** 223.

Horvathy, P. A., Morandi. G. and Sudarshan, E. C. G. (1989). *Nuovo Cim.* D **11** 201.

Howard, S. and Roy, S. K. (1987). *Am. J. Phys.* **55** 1109.

Hudson, V. and Pym, S. J. (1980). *Applications of Functional Analysis and Operator Theory*, Academic Press, London.

Hughes, R. I. G. (1989). *The Structure and Interpretation of Quantum Mechanics*, Harvard University Press, Cambridge, Mass.

Hurt, N. E. (1983). *Geometric Quantization in Action*, Reidel, Dordrecht.

Isham, C. J. (1989). *Modern Differential Geometry for Physicists*, World Scientific, Singapore.

Isham, C. J. 1995). *Lectures on Quantum Theory*, Imperial College Press, London.

Jackson, J. D. (1999). *Classical Electrodynamics*, Wiley, New York.

Jackson, T. D. (1985). *Quantum Mechanics, Locality and Asymptotic Separability*, St Andrews University PhD Thesis.

Janoschek, R. (1991). 'Chirality: From Weak Bosons to the α-Helix' in R. Janoschek, ed., *Theories on the Origin of Biomelecular Homochirality*, Springer-Verlag, Heiselberg,

Jauch, J. M. (1964). *Helv. Phys. Acta* **37** 293.

Jauch, J. M. (1968). *Foundations of Quantum Mechanics*, Addison-Wesley, Reading, Mass.

Johnson, P. R., Strauch, F. W., Gragt, A. J., Ramos, R. C., Lobb, C. J., Anderson, J. R. and Wellstood, F. C. (2003). *Phys. Rev. B* **67** 020509-1.

Jordan, T. F. (1969). *Linear Operators for Quantum Mechanics*, Wiley, Now York.

Jordan, T. F. (1978). *J. Math. Phys.* **19** 1382.

Josephson, B. D. (1962). *Phys. Lett.* **1** 251.

Kadison, R. V. and Ringrose, J. R. (1983). *Fundamentals of the Theory of Operator Algebras Vol. 1 Elementary Theory*, Academic Press, New York.

Kaiser, H., Werner, S. A. and George, E. A. (1983). *Phys. Rev. Lett.* **50** 560.

Kastler, D. ed. (1990). *The Algebraic Theory of Superselection Sectors*, World Scientific, Singapore.

Kato, T. (1996). *Perturbation Theory for Linear Operators*, Springer-Verlag, Berlin.

Kim, Y. S. and Noz, M. E. (1991). *Phase Space Picture of Quantum Mechanics: Group Theoretical Approach*, World Scientific, Singapore.

Kittel, C. (1996). *Introduction to Solid State Physics*, Wiley, New York. 7th edition.

Klauder, J. R. and Skagarstam, B. eds. (1985). *Coherent States - Applications in Physics and Mathematics*, World Scientific, Singapore.

Klein, A. G., Opat, G. I. and Hamilton, W. A. (1983). *Phys. Rev. Lett.* **50** 563.

Koopman, B. O. (1931). *Proc. Nat. Sci.* **17** 315.

Kowalski, K. (1994). *Methods of Hilbert Spaces in the Theory of Nonlinear Dynamical Systems*, World Scientific, Singapore.

Kraus, K. (1977). 'Position Observables of the Photon' in W. C. Price and S. S. Chissick, eds., *The Uncertainty Principle and the Foundations of Quantum Mechanics*, Wiley, New York.

Kraus, K. (1984). *States, Effects and Operations*, Springer-Verlag, Berlin.

Kreyszig, E. (1978). *Introductory Functional Analysis with Applications*, Wiley, New York.

BIBLIOGRAPHY

Kuper, C. G. (1968). *An Introduction to the Theory of Superconductivity*, Clarendon Press, Oxford.

Landau, L. D. and E. M. Lifshitz, E. M. (2002). *Quantum Mechanics*, 3rd edition, Butterworth-Heinemann, Oxford.

Landsman, N. P. (1991). *Int. J. Mod. Phys. A* **6** 5349.

Landsman, N. P. (1998). *Mathematical Topics between Classical and Quantum Mechanics*, Springer-Verlag, New York.

Leech, J. W. (1965). *Classical Mechanics*, Methuen & Co Ltd, London.

Leggett, A. J. (1980). *Prog. Theor. Phys.* Supplement **69** 80.

Leggett, A. J. (1986). 'Quantum Mechanics at the Macroscopic Level' in J. de Boer, E. Dal and O. Ulfbeck, eds., *The Lesson of Quantum Theory*, Elsevier Science Publishers, Amsterdam.

Leggett, A. J. (1987). 'Quantum Mechanics at the Macroscopic Level' in J Souletie and J Vannimenus, eds., *Chance and Matter*, Elsevier, Amsterdam.

Leggett, A. J. (2000). 'New Life for Schrödinger's Cat' in *Physics World* **13** No 8, 23.

Leonhardt, U. (1997). *Measuring the Quantum State of Light*, Cambridge University Press, Cambridge.

Lévy-Leblond, J. (1990). *Quantics*, North-Holland, Amsterdam.

Liboff, R.L. (1980). *Introductory Quantum Mechanics*, Addison-Wesley, Reading, Mass.

Lipschutz, S. (1964). *Set Theory*, Schaum's Outline Series, McGraw-Hill, New York.

Lipschutz, S. (1965). *General Topoloogy*, Schaum's Outline Series, McGraw-Hill, New York.

Lipschutz, S. (1974). *Theory and Problems of Probability*, Schaum's Outline Series, McGraw-Hill, New York.

Lockhart, C. M. and Misra, B. (1986). *Physica A* **136** 47.

London, F. (1961). *Superfluids , Vol. 1 and Vol. 2*, Dover, New York.

Longhurst, R. S. (1963). *Geometric and Physical Optics*, Longmans, London.

Loomis, L. H. and Sternberg, S. (1968). *Advanced Calculus*, Addison-Wesley, Reading, Mass.

Ludwig, G. (1985). *Foundations of Quantum Mechanics Vol. 2*, Springer-Verlag, New York.

Machida, S. and Namiki, M. (1980). *Prog. Theor. Phys.* **63** 1457, **63** 1833.

MacKey, G. W. (1963). *The Mathematical Foundations of Quantum Mechanics*, Benjamin, Reading, Mass.

Mandelstam L. and Tamm, I. (1945). *J. Phys.* (USSR) **9** 249-254.

Mandl, F. (1992). *Quantum Mechanics*, Wiley, Chichester.

Margenau, H. and Hill, R. N. (1961). *Prog. Theor. Phys.* **26** 722.

Martin, D. (1991). *Manifold Theory*, Ellis Harwood, New York.

Martin, J. L. (1981). *Basic Quantum Mechanics*, Clarendon Press, Oxford.

Matsushima, Y. (1972). *Differentiable Manifolds*, Marcel Dekker, New York.

McLean, R. G. D. (1983). *An Algebraic Formulation of Asymptotically Separable Quantum Mechanics*, St Andrews University PhD Thesis.

Mees, C. R. K. (1963). *The Theory of Photographic Process*, Macmillan, New York.

Mensky, M. B. (1993). *Continuous Quantum Measurements and Path Integrals*, Institute of Physics Publishing, Bristol.

Merzbacher, E. (1998). *Quantum Mechanics*, 3rd edition, Wiley, New York.

Messiah, A. (l967). *Quantum Mechanics Vol. 1*, North-Holland, Amsterdam.

Mintmire, J. W., Dunlap, B. I. and White, C. T. (1992). *Phys. Rev. Lett.* **68** 631.

Misra, B. (1978). *Proc. Natl. Acad. Sci. USA* **75** 1627.

Misra, B. (1979). *J. Stat. Phys.* **48** 1925.

Misra, B. (1995). *Found. Phys.* **25** 1087.

Misra, B., Prigogine, I. and Courbage, M. (1979). *Physica A* **98** 1

Monroe, C., Meekhof, D. M., King, B. E. and Wineland, D. J. (1996). *Science* **272** 1131.

Morandi. G. (1993). *The Role of Topology in Classical and Quantum Physics*, Springer-Verlag, Berlin.

Müler-Herold, U. (1985). *J. Chem. Education* **62** 379.

Myatt, C., King, B., Turchette, Q., Sackett, C., Kielpinski, D, Itano, W., Monroe, C. and Wineland, D. (2000). *Nature* **403** 269.

Naimark, M. A. (1968). *Linear Differential Operators Part II*, Harrap, London. Translated by E. R. Dawson.

Namiki, M. (1988). *Found. Phys.* **18** 29.

Namiki, M. (1992). *Stachastic Quantization*, Springer-Verlag, Berlin.

Namiki, M. and Pascazio, S. (1993). *Phys. Rep.* **232** 301.

Nash, C. and Sen, S. (1983). *Topology and Geometry for Physicists*, Academic Press, London.

Nelson, N. (1985). *Quantum Fluctuations*, Princeton University Press, Princeton.

Núněz-Yépez, H. N., Vargas, C. A. and Salas-Brito, A. L. (1988). *J. Phys. A: Math. Gen.* **21** L651.

Núněz-Yépez, H. N., Vargas, C. A. and Salas-Brito, A. L. (1998). *Phys. Rev. A* **39** 4306.

Ohanian, H. C. (1995). *Modern Physics*, 2nd edition, Prentice-Hall, Englewood Cilffe, N.J.

Omnès, R. (1994). *The Interpretation of Quantum Mechanics*, Princeton University Press, Princeton.

Papoulis, A. (1962). *The Fourier Integral and Its Applications*, McGraw-Hill, New York,

Park, J. L. and Margenau, H. (1968). *Int. J. Theor. Phys.* **1** 211.

Pauli, W. (1973). *Pauli Lectures on Physics Vol. 5 - Wave Mechanics*, the MIT Press, Cambridge, Mass., edited by C. P. Enz and translated by H. R. Lewis and S. Margulies.

Pearson, D. B. (1988). *Quantum Scattering and Spectral Theory*, Academic Press, London.

Peierls, R. (1935). *Nature* **136** 395.

Penrose, O. (1970). *Foundations of Statistical Mechanics*, Pergamon Press, Oxford.

Percival, I. and Richards, D. (1982). *Introduction to Dynamics*, Cambridge University Press, Cambridge.

BIBLIOGRAPHY

Percival, I. (1991). 'Quantum Measurement Theory and Experiment' in T. D. Clark, H. Prance, R. J. Prance, and T. P. Spiller, eds., *Macroscopic Quantum Phenomena*, World Scientific, Singapore.

Peres, A. (1995). *Quantum Theory: Concepts and Methods*, Kluwer Academic Publishers, Dordrecht.

Perko, L. (1991). *Differential Equations and Dynamical Systems*, Springer-Verlag, New York.

Peshkin, H. and Tonomura, A. (1989). *The Aharonov-Bohm Effect*, Springer-Verlag, Berlin.

Pfeifer, P. (1980). *Chiral Molecules - A Superselection Rule Induced by the Radiation field*, Dissertation ETH Zürich.

Pfeifer, P. (1980). *Helv. Phys. Acta.* **53** 410.

Phillips, A. C. (2003). *Introduction to Quantum Mechanics*, Wiley, New York.

Piron, C. (1976). *Foundations of Quantum Physics*, Benjamin, Reading, Mass.

Pitaevskii, L. P. (1961). *Sov. Phys. JETP* **13** 451.

Pitowski, I. (1989). *Quantum Probability - Quantum Logic*, Springer-Verlag, Berlin.

Pitt, H. R. (1963). *Integration, Measure and Probability*, Oliver & Boyd, Edinburgh.

Price, W. C. and Chissick, S. S. eds. (1977). *The Uncertainty Principle and the Foundations of Quantum Mechanics*, Wiley, New York.

Primas, H. (1983). *Chemistry, Quantum Mechanics and Reductionism*, Springer-Verlag, Berlin.

Primas, H. and Müller-Herold, U. (1978). *Advances in Chemical Physics* **38** Chapters 1-6.

Prugovečki, E. (1981). *Quantum Mechanics in Hilbert Space*, 2nd edition, Academic Press, New York.

Prugovečki, E. (1984). *Stochastic Quantum Mechanics and Quantum Spacetime*, Reidel, Dordrecht.

Rae, A. I. M. (1981). *Quantum Mechanics*, MaGraw-Hill, London.

Rae, A. I. M. (1986). *Quantum Physics: Illusion or Reality?*, Cambridge Univeristy Press, Cambridge.

Rarity, J. (1996). *Physics World* **9** 19.

Rayski, J. and Rayski, J. M., Jr. (1977). 'On the Meaning of the Time-Energy Uncertainty Relation' in W. C. Price and S. S. Chissick, eds., *The Uncertainty Principle and the Foundations of Quantum Mechanics*, Wiley, New York,

Razavy, M. (1997). *Phys. Rev.* A **55** 4102.

Recami, E. (1977). 'A Time Operator and the Time-Energy Uncertainty Relation' in W. C. Price and S. S. Chissick, eds., *The Uncertainty Principle and the Foundations of Quantum Mechanics*, Wiley, New York,

Rédei, M. and Stöltzer, M. (2001). *John von Neumann and the Foundations of Quantum Physics*, Kluwer, Dorchecht.

Redhead, M. (1987). *Incompleteness, Nonlocality and Realism*, Clarendon Press, Oxford.

Reed, M. and Simon, B. (1972). *Methods of Modern Mathematical Physics Vol. 1 Functional Analysis*, Academic Press, New York.

Reed, M. and Simon, B. (1975). *Methods of Modern Mathematical Physics Vol. 2 Fourier Analysis, Selfadjointness*, Academic Press, New York.

Reed, M. and Simon, B. (1978). *Analysis of Operators Vol. 4 Analysis of Operators*, Academic Press, New York.

Richtmyer, R. D. (1978). *Principles of Advanced Mathematical Physics Vol. 1*, Springer-Verlag, New York.

Roberts, J. E. and Roepstorff, G. (1969). *Commun. Math. Phys.* **11** 321.

Roman, P. (1965). *Advanced Qauntum Theory*, Addison-Wesley, Reading, Mass.

Roman, P. (1975). *Some Modern Mathematics for Physicists and Other Outsiders Vol. 1*, Pergamon, New York.

Roman, P. (1975). *Some Modern Mathematics for Physicists and Other Outsiders Vol. 2*, Pergamon, New York.

Rose-Innes, A. C. and Rhoderick, E. H. (1980). *Introduction to Superconductivity*, Pergamon Press, Oxford.

Saglam, M. and Boyacioglu, B. (2002a). *Int. J. Mod. Phys. B* **16** 607.

Saglam, M. and Boyacioglu, B. (2002b). *Phys. Stat. Sol. Phys. B* **230** 133.

Schiff, L. I. (1955). *Quantum Mechanics*, McGraw-Hill, New York.

Schrödinger, E. (1935). *Naturwissenschaften*, **23** 807, 823, 844. English translations in J. A. Wheeler and W. Zurek. (1983). eds., *Quantum Theory of Measurement*, Princeton University Press, Princeton, N. J.

Schroeck, F. E., Jr. (1996). *Quantum Mechanics on Phase Space*, Kluwer, Dordrecht.

Schutz, B. (1980). *Geometrical Methods of Mathematical Physics*, Cambridge University Press, Cambridge.

Scully, M. O., Lamb W. E., Jr. and Barut, A. (1989). *Found. Phys.* **17** 575.

Schwab, K. C. and Roukes, M. L. (2005). *Physics Today* **58** 36.

Schwartz, B. B. and Foner, S. *Superconductor Applications: SQUIDs and Machines*, Plenum, New York.

Seba, P. (1986). *Rep. Math. Phys.* **24** 111.

Seba, P. (1987). *Ann. Phys.* **44** 323.

Selleri, F. (1990). *Quantum Paradoxes and Physical Reality*, A. van der Merwe, ed., Kluwer Academic Publishers, Dorchecht.

Selleri, F. and Tarozzi, G. (1981). *La Rivita del Nouvo Cimento*, **4** 1.

Sewell, G. L. (1986). *Quantum Theory of Collective Phenomena*, Clarendon Press, Oxford.

Sewell, G. L. (2002). *Quantum Mechanics and its Emergent Macrophysics*, Princeton University Press, Princeton, N. J.

Simmons, G. F. (1963). *Introduction to Topology and Modern Analysis*, McGraw-Hill International Edition, Tokyo.

Sivakumar, S. (1998). *Phys. Rev. A* **58** 717.

Smirnov, V. I. (1964). *A Course of Higher Mathematics Vol. 5*, Pergamon, London, translated by D E Brown and edited by I. N. Sneddon.

Śniatycki, J. (1980). *Geometric Quantization and Quantum Mechanics*, Springer-Verlag, New York.

Spiegel, M. R. (1974). *Advanced Calculus*, a Schaum's Outline Series, McGraw-Hill, New-York.

Spiller, T. P., Clark, T. D., Prance, R. J. and Widom, A. (1992). 'Quantum Phenomena in Circuits at Low Temperature' in E. D. Brewer, ed., *Progress in Low Temperature Physics* Vol. XIII, North-Holland, Amsterdam.

Srivastava, Y. and Widom, A. (1987). *Physics Reports* **148** 1-65.

Sudbery, A. (1986). *Quantum Mechanics and the Particles of Nature, An Outline for Mathematicians*, Cambridge University Press, Cambridge.

Sumner, P. (1988). *Spatial and Generalized Phase Space Distributions of Observable Values in Quantum Mechanics*, St Andrews University PhD Thesis.

Sutherland, W. A. (1995). *Introduction to Metric and Topological Spaces*, Clarendon Press, Oxford.

Synge, J. L. and Schild, A. (1966). *Tensor Calculus*, University of Toronto Press, Toronto.

Temple, G. (1935). *Nature* **135** 957 (1935), **136** 179.

Thirring, W. (1979). *Quantum Mechanics of Atoms amd Molecules*, Springer-Verlag, New York.

Tilley, D. R. and Tilley, J. (1990). *Superfluidity and Superconductivity*, Adam Hilger, Bristol.

Timson, D. R. E. (1986). *Locality in Non-Relativistic and Relativistic Quantum Mechanics*, St Andrerws University PhD Thesis.

Timson, D. R. E. and Wan, K. K. (l988). 'Localizing Isometries, Local Comoving Evolution Operators and Observables in Quantum Mechanics' in L. Kostro, ed., *Problems in Quantum Physics; Gdansk 87*, World Scientific, Singapore.

Tonomura, A., Endo, J., Matsuda, T. and Kawasaki, T. (1989). *Am. J. Phys.* **57** 117.

Torozzi, G. and van der Merwe, A. eds. (1980). *Open Questions in Quantum Physics*, Reidel, Dordrecht.

Trueman, C. (1999). *Superselection Rules, Quantization by Parts and Point Interactions*, St Andrews University PhD Thesis.

Trueman, C. and Wan, K. K. (2000). *J. Math. Phys.* **41** 1.

Uffink, J. (1994). *Int. J. Theor. Phys.* **33** 199.

van der Wal, C. H., ter Haar, A. C. J., Wilhelm, F. K., Schouten, R. N., Harmans, C. J. P. M, Orlando, T. P., Lloyd, S. and Mooij, J. E. (2000). *Science* **290** 773.

van Fraassen, B. C. (1991). *Quantum Mechanics, An Empiricist View*, Clarendon Press, Oxford.

von Neumann, J. (1955). *Mathematical Foundations of Quantum Mechanics*, Princeton University Press, Princeton, translated from the original German book published in 1932 by R. T. Beyer.

Wan, K. K. (1980). *Can. J. Phys.* **58** 976.

Wan, K. K. (1986). 'Chronological Disordering and the Absence of Correlations between Infinitely Separated States' in V. Gorini and A. Grigerio, eds., *Fundamental Aspects of Quantum Theory*, a NATO ASI Series B Vol. 144, Plenum, New York.

Wan, K. K. (1988). *Found. Phys.* **18** 887.

Wan, K. K., Bradshaw, J., Trueman, C. and Harrison, F. E. (1998). *Found. Phys.* **28** 1739.

Wan, K. K. and Fountain, R. H. (1996). *Found. Phys.* **26** 1165.

Wan, K. K. and Fountain, R. H. (1998). *Int. J. Theor. Phys.* **37** 2153.

Wan, K. K., Fountain R. H. and Tao, Z. Y. (1995). *J. Phys. A: Math. Gen.* **28** 2379.

Wan, K. K., Green, R. and Trueman, C. (2000). *J. Opt. B.* **2** 165.

Wan, K. K. and Harrison, F. E. (1993). *Phys. Lett. A* **174** 1.

Wan, K. K. and Harrison, F. E. (1994). *Found. Phys.* **24** 831.

Wan, K. K., and Harrison, F. E. (1997). *J. Phys. A: Math. Gen.* **30** 4731.

Wan, K. K. and Jackson, T.D. (1984). *Phys. Lett. A* **106** 219.

Wan, K. K. and Jackson, T. D. (1985). *Phys. Lett. A* **111** 223.

Wan, K. K., Jackson, T. D. and McKenna, I. H. (1984). *Nuovo Cimento B* **81** 165.

Wan, K. K. and McFarlane, K. (1983). *Int. J. Theor. Phys* **22** 55.

Wan, K. K. and McKenna, I. H. (1984). *Algebras, Groups and Geometries* **1** 154.

Wan, K. K., McKenna, I. H. and Pinto, J. (1984a). *Algebras, Groups and Geometries* **1** 344.

Wan, K. K., McKenna, I. H. and Pinto, J. (1984b). *Algebras, Groups and Geometries* **1** 372.

Wan, K. K. and McLean, R. G. D. (1983a). *Phys. Lett. A* **94** 198

Wan, K. K. and McLean, R. G. D. (1983b). *Phys. Lett. A* **95A** 76.

Wan, K. K. and McLean, R. G. D. (1984a). *J. Phys. A: Math. Gen.* **17** 825. Corrigenda **17** 2363 (1984).

Wan, K. K. and McLean, R. G. D. (1984b). *J. Phys. A: Math. Gen.* **17** 837. Corrigenda **17** 2363 (1984).

Wan, K. K. and McLean, R. G. D. (1985). *J. Math. Phys.* **26** 2540.

Wan, K. K. and McLean, R. G. D. (1991). *J. Phys. A: Math. Gen.* **24** L425.

Wan, K. K. and McLean, R. G. D. (1994). *Found. Phys.* **24** 715.

Wan, K. K. and Saglam, M. (2005). *Intrinsic Magnetic Flux of the Electron's Orbital and Spin Motion*, preprint.

Wan, K. K. and Sumner, P. (1988). *Phys. Lett. A* **128** 458.

Wan, K. K. and Sumner, P. (1991). *Nuovo Cimento B* **106** 593.

Wan. K. K. and Timson, D. (1985). *Phys. Lett. A* **111** 165.

Wan, K. K. and Viazminsky, C. (1977). *Prog. Theor. Phys.* **58** 1030.

Wan, K. K. and Viazminsky, C. (1979). *J. Phys. A: Math. Gen.* **12** 643.

Weidmann, J. (1980). *Linear Operators in Hilbert Spaces*, Springer-Verlag, New York.

Wheeler, J. A. (1957). *Rev. Mod. Phys.* **29** 463.

Wheeler, J. A. and Zurek, W. H. eds. (1983). *Quantum Theory of Measurement*, Princeton University Press, Princeton, N. J.

Wheeler, J. A. (1983). 'Law without Law' in J. A. Wheeler and W. H. Zurek, eds., *Quantum Theory of Measurement*, Princeton University Press, Princeton, N. J.

Whitaker, A. (1996). *Einstein, Bohr and the Quantum Dilemma*, Cambridge University Press, Cambridge.

BIBLIOGRAPHY

Wick, G. C., Wightman, A. S. and Wigner, E. P. (1952). *Phys. Rev.* **88** 101.

Widom, A. (1979). *J. Low. Temp. Phys.* **37** 449.

Widom, A. (1991). 'Implications of Superconducting Circuits for Relativistic Quantum Electrodynamics' in T. D. Clark, H. Prance, R. J. Prance and T. P. Spiller, eds., *Macroscopic Quantum Phenomena*, World Scientific, Singapore.

Wightman, A. S. and Glance, N. (1989). *Nucl. Phys. B (Proc Suppl)* **6** 202.

Wigner, E. (1932). *Phys. Rev.* **40** 749.

Williams, H. J. (1982). *Introduction to Organic Chemistry*, Wiley, New York.

Williams, J. E. C., (1970). *Superconductivity and its Applications*, Pion Limited, London.

Williamson, J. H. (1962). *Lebesgue Integration*, Holt, Rinehart and Winston, New York.

Wilson, J. L. (1951). *The Principles of Cloud-Chamber Technique*, Cambridge University Press, Cambridge.

Wollman, D. A. et al., (1993). *Phys. Rev. Lett.* **71** 2134.

Woodhouse, N. (1986). *Geometric Quantization*, Clarendon Press, Oxford.

Yuan, L. C. L. and Wu, C. S. (1961). *Methods of Experimental Physics Vol. 5A: Nuclear Physics*, Academic Press, New York.

Yurke, B. and Stoler, D. (1986). *Phys. Rev. Lett.* **57** 13.

Zaccoria, F., Sudarshan, E. C. G., Nilsson, J. S., Mukunda, N., Marmo, G. and Balachandran, A. P. (1983). *Phys. Rev. D* **27** , 2327.

Zeh, H. D. (1992). *The Physical Basis of the Direction of Time*, Springer-Verlag, Berlin.

Zettili, N. (2001). *Quantum Mechanics*, Wiley, Chichester.

Zhao, B-H. (1992). *J. Phys. A: Math. Gen.* **25** L617.

Zhu, C. and Klauder, J. R. (1993a). *Am. J. Phys.* **61** 605.

Zhu, C. and Klauder, J. R. (1993b). *Found. Phys.* **23** 617.

Zurek, W H. (1981). *Phys. Rev. D* **24** 1516.

Zurek, W H. (1982). *Phys. Rev. D* **26** 1862.

Zurek, W H. (1991). *Physics Today* April 81-90.

Index

absolutely continuous functions $AC(I\!E)$, 115-116, 134
    with respect to $\widehat{X}$, 117
action-at-a-distance, 627
algebra of operators, 188-190
    C*-algebra, 189-191, 349, 620-622
    Lie algebra, 34, 66
    local algebra, 188-190
        local observable, 188, 621
    non-commutative, 189
    quasi-local algebra, 190
    *-algebra, 189
        quasi-local observable, 191
algebraic structure, 189, 256
almost everywhere, 117, 133-134
ammonia molecule, 599, 604
approximate eigenvector, 160, 179
approximate observable, 397-398, 407
approximation
    bounded, 249, 317
    discretized, 249
    measurement assumption, 250
assumption
    experiment and measurement, 250
    superconductivity (PAS1), (PAS2), 519
asymptotic
    decoherence, 345, 607, 620-621
    disjointness, 349
    localization, 292, 294, 299, 318-319
    separation, 292, 294, 299, 317
    state preparation, 307
    vanishing correlations, 302-303
average value, 131, 142
BCS theory, 558-561
    BCS ground state, 560
bicontinuous, 10
bilinear functional, 48

Borel
    family of Hilbert spaces, 200
    family of operators, 213
    family of subspaces, 208
    function, 133
    measurable, 133
    sets, 132, 139
bound state, 295
boundary conditions, 283, 284, 439, 442
    Dirichlet, 284, 442, 463, 498
    elastic, 285, 464, 498
    Neumann, 284, 285, 441, 465, 499
    periodic, 120, 480
    quasi-periodic, 120, 284
    (BC1) step potential, 450
    (BC2) δ-interaction, 451, 452, 454
    (BC3) δ'-interaction, 455, 457
    (BC4) perfect reflector, 463
    (BC5) elastic reflector, 464
    (BC6) open end, 464
    (BC7) ideal π-phase shifter, 465
    (BC8) high-pass π-phase shifter, 467, 468
    (BC9) low-pass π-phase shifter, 469, 470
    (BC10) ideal mid-pass $\frac{1}{2}\pi$-phase shifter, 471, 472
    (BC11) partial mid-pass filter, 473
    (BC12) ideal tunable phase shifter, 476
    (BCC1) free motion, 495
    (BCC2) δ-interaction, 495
    (BCC3) δ'-interaction, 497
    (BCC4 perfect reflector, 498
    (BCC5) elastic reflector, 498
    (BCC6) open end, 499
    (BCC7) ideal dynamic π-phase shifter, 500
    (BCC8) static π-phase shifter, 500
    (BCC9) gradient π-phase shifter, 501
    (BCC10) ideal $\frac{1}{2}\pi$-phase shifter, 502

(BCC11) static junction correlator, 503
(BCC12) ideal tunable phase shifter, 504
branch, 56, 443, 509
  branch point, 56, 443, 509
canonical coordinates, 63
  canonically conjugate, 64, 74
  non-uniqueness, 72
canonical junction observables, 612
canonical transformation, 64, 74
cat,
  classical, 592, 607
  quantum, 593
  Schrödinger's, 594
  superconducting, 618-619
Cauchy
  vectors
    convergence criterion, 82
    sequence, 82
  operator
    strong Cauchy sequence, 94
    strong sequence, 93-94
    uniform Cauchy sequence, 94
    uniform sequence, 93-94
    weak sequence, 93-94
center of localization, 192
characteristic function, 90, 103
  as local position observable, 312
  as position spectral measure, 152
  as projector, 103
  in momentum space, 154
  of selfadjoint operator, 156
characteristic wave number, 451
charge operator, 614-615, 617
chiral molecule, 605-607
  optical isomer, 350, 607
chirality, 607
  handedness, 607
chiral states, 350, 606
chronological disordering, 629-632
circle C, 57-60
  extended polar angle θ_{ex}, 653
  polar angle θ, 57
circle with a cut C_c, 57-60
  polar angle ϑ, 58
circle with a cut C_π, 57-60
  polar angle ϑ_π, 58
circuit, 443
  classical LC, 72, 377, 574
  quantized LC, 574
  quantum, 509
  superconducting, 509
  Y-shape, 542-552, 562-566

classical Hamiltonian system, 70
  in Hilbert space, 373
classical mixture, 237, 245, 342
classical observables
  for classical system, 63, 254
  for mixed quantum system, 346
classical systems, 67, 379-383
  equations of motion, 67
  first-order, 68
    in Hilbert space, 368
  second-order Hamiltonian, 69-70
    Hamilton's equations of motion, 70
    Hamiltonian as energy, 70
    Hamiltonian generator, 70
    in Hilbert space, 373
    velocity field, 70
closed linear subset of a Hilbert space, 84
closed sets, 4, 6
closely related family of observables, 408
closure in Hilbert space, 84
  operator, 89
  subset, 84
coherent, 344
  subspace, 346
  superposition, 247, 344-346
coherent state, 591
combination
  convex, 220, 340
  linear, 84
commutator, 102
commute, 102, 151
compact sets, 9
  compact support, 9
complete set, 85
  orthonormal, 85
  orthonormal basis, 85, 86
    unitary transform, 110
    in direct sum space, 197
    in tensor product space, 95
    corresponding set of projectors, 105
complete vector field, 28, 32
  sum, 32
composite
  Hamiltonian, 445
  Hilbert space, 445
  momentum, 445
  observables, 445
compound system, 384, 387
condensate, 510, 559
  in a pure or mixed state, 549-550, 552
  global nature (footnote), 550
confidence function, 396

INDEX 677

connected, 11
conserved, 290
constant family of Hilbert spaces, 200
constrained system, 282-286
continuous, 89-90, 181
continuous spectrum, 157-159, 178-179
contraction, 19, 20, 53, 55
contravariant tensor, 17-18, 47
    anti-symmetric and symmetric, 18
    intrinsic definition, 26
contravariant vector, 16, 25-26, 42
    displacement, 17, 41, 42
    intrinsic definition, 47, 77
convergence
    Cauchy criterion, 82
    strong, 93
    tangent vector, 26
    uniform or norm, 93, 190
    weak, 93
convex combination, 220
Cooper pair, 510, 559
coordinate, 13, 63
    canonical, 63
    global, 57
    local charts, 15, 57
    phase space, 373
    rectangular Cartesian, 14
    representation space, 109
    system, 13
coordinate space, 64, 73
    physical space, 73, 638
coordinate transformations, 13-14
    homogeneous orthogonal, 14
    linear, 14
correlation, 247
    asymptotically vanishing, 302
    between two states, 247
    term, 247
    two-particle, 630
        conservation law, 630
correlative quantization, 443
cotangent vector, 41-45
    coordinate representation, 41-45
    cotangent space, 41
cotangent vector field, 45
    one-form, 45
count rate, 313
countable orthonormal basis, 85, 86
counter
    Geiger, 311, 313
    size, 311
covariant tensor, 18, 47
    anti-symmetric and symmetric, 18, 48, 51
covariant vector, 16, 40, 43
    cotangent vector, 44, 40-41, 43-45
    momentum, 17
critical current, 524
critical point, 26
current density,
    electric, 513, 515
        ac, 514, 555
        dc, 514
    probability, 447
    probability current density operator, 481
curve, 27
    integral, 28
    maximal integral, 28
    path, 639
    smooth, 27
    tangent vector, 28
Darboux's theorem, 63, 64, 70
de Broglie paradox, 587
decoherence process, 348, 607
    asymptotic, 345, 607, 620-621
    dynamical, 607, 620-621
decomposition of the identity, 145
deficiency indices, 113-114
    direct sum operators, 210
    maximal symmetric extension, 115
    negative, 113-114
    positive, 113-114
    selfadjointness, 114
degeneracy, 56, 99
delta function, 136, 163
delayed choice experiment, 550, 566
delocalized, 522
dense, 6, 84
densely defined, 88
density operator, 218-221
derivation at a point on $\mathcal{C}^\infty(I\!E^n)$, 24
derivation on $\mathcal{C}^\infty(I\!E^n)$, 24
detector, 311
    photographic plate, 321
        latent image, 321
diagonalizable, 214
differential forms, 45
    closed, 55
    differential, 41-45
    exact, 55
    one-form, 45
    two-form, 51
        non-degenerate, 55
    zero-form, 51
diffeomorphism, 35

differential operator, 21-26
  in $L^2(I\!\!E)$, 115
direct integral
  constant family of Hilbert spaces, 200
  continuous $\mathcal{H}_c$, 204
  decomposable, 202
    discrete, 202
    continuous, 203, 204
  decomposition in coordinate space
    $\mathcal{H}_{I\!\!E}^{\oplus}(\mathbb{C}, dx)$, 206-207
  decomposition in momentum space
    $\mathcal{H}_{I\!\!E}^{\oplus}(\mathbb{C}, dp)$, 207
  discrete $\mathcal{H}_d$, 203
  Hilbert spaces, 201
  operators, 212-217
    decomposable, 214-217, 340
    bounded decomposable $\mathcal{B}^{\oplus}(\mathcal{H}^{\oplus})$, 217
    decomposable condition, 216, 217
    diagonalizable, 214, 217
    projector, 216
    selfadjoint, 216
    unitary, 216
  projection, 201
  subspace, 208
    non-singular and singular, 208
  tensor product, 224-225
  trivial, 207
  vector
    mixed, 202
      singular, 204
    pure, 202, 203
      singular, 204
    unit $u(\tau)$, 205, 206
direct sum
  Hilbert spaces, 196, 202-203
    external, 197
    internal, 197
  operators, 209-212, 471, 480
    decomposable, 214, 216, 340
    deficiency indices, 210
    diagonalizable, 214, 216
    maximal symmetric, 210
    selfadjoint, 210
  spectrum, 212
disjointness, 7, 346, 347, 584
  asymptotic, 349
  disjoint FAPP, 584
div (divergence), 39-40
double-well potential, 601
  finite double-well, 603
  infinite double-well, 602
dual space, 40

dynamics, see also evolution
  classical, 67, 68, 71
  non-preserving, 356, 365
    environment, 367
  preserving, 352
  orthodox quantum, 287
eigensubspace, 99, 158
eigenvalues, 99, 158
  degeneracy, 99
  generalized, 126, 159, 161-162
eigenvectors, 99
  approximate, 160-162
  generalized, 126, 159, 161
elastic potential well, 285
electro-weak nuclear force, 350
entangled states, 624
entanglement, 622-626
environment, 348, 367, 520
EPR paradox, 624-626
equilibrium system, 383, 562
  state, 566-567
  superconducting ring, 520
equivalence for all practical purposes, 583
equivalent FAPP, 252, 305
  observables, 584
  states, 584
essential supremum, 213
essentially bounded, 213
Euclidean spaces $I\!\!E^n$, 12
  $I\!\!E^-$, 33
  $I\!\!E^+$, 33
evolution
  classical, 67, 68, 71
  in Hilbert space, 368-377
  orthodox quantum, 287
  unified quantum, 351, 352
    non-preserving, 356, 365
      environment, 367
    preserving, 352
evolution operators, 287
  free, 290
exclusion principle, 626-628
expectation value $\mathcal{E}(\mathcal{F})$ in probability theory 141
expectation value in operator theory, 90, 91, 182, 183
experiment, 234, 250-252
  electron double-slit, 564
  general assumption, 250
  photon interferometer, 565
  photon beam splitter, 562-563
  statistical, 130

INDEX

yes-no, 243
Y-shape circuit, 542-552
extension of linear functional, 361
extension of operator, 89
    Friedrichs, 176, 281, 285, 544
    maximal symmetric, 113-115, 119-128
    selfadjoint, 113-115, 119-128, 129-130
        direct sum, 212
        $\widehat{K}(I\!E)$, 122-123
        $\widehat{K}_\lambda(I\!E^+)$, 128
        $\widehat{K}_\infty(I\!E^+)$, 128
        $\widehat{K}_\lambda(\Lambda)$, 121
        $\widehat{K}^\infty(\Lambda)$, 120, 121
        $\widehat{p}_\lambda(\Lambda)$, 119, 130
        $\widehat{p}(I\!E)$, 122
        $\widehat{p}(I\!E^-), \widehat{p}^-$, 127
        $\widehat{p}(I\!E^+), \widehat{p}^+$, 126
    symmetric, 113-115, 391
    von Neumann's formula, 128-130
exterior
    differential operator, 51-52
    product, 49
external direct sum, 197
family of probability functions, 400
    maximal, 401-402
    maximal symmetric operators, 402
FAPP, 252, 305, 584, 585, 589, 594, 598, 606
fixed points, 68
flow, 38
    phase, 68
flux representation, 616
for all practical purposes, 252
Fourier transform, 108, 111
    transform operator, 108
fullerene atom, 511
function, 8, 14-15
    absolutely continuous, 115-116
        with respect to $\widehat{X}$, 117
    bicontinuous, 10
    class C^ℓ, 8-9
    $C^\infty(I\!R^n)$, 9
    $C_0^\infty(I\!R^n)$, 9
    $C^\infty(I\!E^n)$, 15
    $C_0^\infty(I\!E^n)$, 15, 86
    $C^\infty(\mathcal{D})$, 9, 15
    $C^\infty(\alpha)$, 9, 15
    compact support, 9
    continuous, 10
    infinitely differentiable, 9, 14
    local, 8

    of selfadjoint operator, 154-156
        characteristic, 155
        exponential, 156
    open, 10
    smooth, 9
    staircase, 136, 139
    support, 8
function on path space, 644
    homotopic, 645
    globally homotopic, 648
    path dependent phase, 648
        additivity, 648
        phase factor, 236, 638
        universal, 646
        universal phase constant, 649
    restriction to $I\!E_h^2$, 644
    single-valuedness on $I\!E_h^2$, 645
generalized
    eigenfunctions, 126, 161-162
    eigenvalue, 126, 159, 161-162
    eigenvector, 126, 159, 161
    phase space, 419
    resolutions of the identity, 168
    spectral function, 168-169
        maximal symmetric operator, 173
        radial momentum operator, 174-175
        symmetric operator, 170-172
    spectral measure, 168-169
    spectral operators, 169
generator
    Hamiltonian, 70-71, 287
        free, 290
    one-parameter unitary group, 111-113
        selfadjoint operator, 112
    time translations, 70
global
    coordinates, 57
    operators, 582
    phase, 638
    position observable, 312
    quantity, 409, 413
globalization, 566, 567
    non-locality, 566
globally homotopic functions, 648
gradient, 39
Hamiltonian
    energy, 70
    free $\widehat{H}^{(0)}$, 290
Hamiltonian generator, 70-81, 371, 376
    velocity field, 70
    for a beam, 484, 517
    free $\widehat{H}_g^{(0)}$, 290

non-observable, 356, 367, 382
observable, 353
zero, 372, 376, 379
quantization, 282
Hamiltonian manifold, 63-64
structure, 63-64
system, 70
Hamiltonian vector field, 66, 70-71, 74, 76-77
projection on coordinate space, 74, 76-77
Hamilton's equations, 70-71
handedness, 549
Hausdorff space, 7, 56
Heisenberg picture, 352-353
Hilbert space, 83
dimension, 86
$L^2(I\!E^n)$, 84
$L^2(I\!E^-)$, 127
$L^2(I\!E^+)$, 125
$L^2(I\!M^n, d\mu)$, 84
separable, 86
homeomorphism, 10
homotopic
classes, 643
winding number, 643
equivalent, 639
function, 643
single-valuedness, 645
globally homotopic function, 648
ibar i, 108
idempotence, 90
ignorance interpretation, 246
image and inverse image, 10
impulsive interaction, 349, 387, 389
indistinguishable FAPP, 252, 305
inequalities
Schwarz, 83
Triangle, 83
infinite square potential well, 120, 275
integral curve, 28
maximal, 28
interference term, 247
interior product, 20, 53
invariant, 19
invariant subspace, 105
ionization process,
completion, 321, 322
Hamiltonian generator, 326
initiation, 321, 322
Josephson effect
ac effect, 555
dc effect, 523
quantum mechanics on path space, 655

Josephson equation, 524, 538
jump point, 157
junction, 443, 509
capacitive, 609, 613
Josephson, 522, 537-541
$\frac{1}{2}\pi$-junction, 530, 536
π-junction, 530, 536
point contact, 573
Y-shape, 542-552
kinetic energy
operator, 162
probability amplitude function, 241
quantization, 275-286
representation space, 164, 382-384
Laplacian, 39-40
latent image, 321
Lie algebra, 34, 66
Lie bracket, 34, 66
Poisson bracket, 66, 72
Lie derivative, 38
limit
strong, 93
weak, 76, 93
uniform, 93
limit vector, 84
limiting family of δ-function potentials, 458
asymmetric quadrapole, 458, 461, 466
linear functional, 40, 358-364
bilinear, 48
extension, 361
normalized positive, 359
restriction, 361
state, 361
statistically normalized positive, 359
linear operator, 21
at a point, 21
on $C_0^\infty(I\!E^n)$, 21
linear subset, 84
Liouville's theorem, 70, 339
phase volume, 71
local
coordinate chart, 15, 57
Hamiltonian, 195
momentum, 195
observables, 188
operator, 185
semi-local, 186, 375
position measurement, 320-324
position observable, 312
radial momentum operators, 274
local operator algebra, 188-190
local theory, 582

local values
    phase space approach, 413
        generalized phase space, 418-419
        value, 419
        value density, 419
    coordinate space approach, 414-415
        semi-local operator, 417
        value, 417
        value density, 416
localization
    asymptotic, 292, 294, 299, 318
    bounded operator, 187
    center, 192
    selfadjoint, 187, 193-194
    unbounded operator, 192-195
localizing function, 192-193
localizing operator, 191-192
loop, 643
macroscopic quantum system, 234, 509
macroscopic wave function, 512
macroscopic wave function hypothesis, 512
    single-particle representation, 512
magnetic flux operator, 516, 524
magnetic flux quantum
    Cooper pair, 489, 514
    electron orbital, 489
manifold
    differentiable, 56
    metric, 60
    pseudo-Riemannian, 60
    Riemannian, 60
    volume element, 62-63
mapping
    into, 9
    onto, 9
maser, 605
measure, 131-132
    Lebesgue, 133
    Lebesgue-Stieltjes, 134, 201, 203
    probability, 137-138
    set of measure zero, 133
measured value, 232
measurement, 234, 236, 250
    finite resolution, 396
    local position, 312-316, 320-324
        ionization process, 320-327
        reduction to local postion, 313-319
    spin, 319-320
    superselection rule approach, 384-390
    yes-no, 243

measuring device, 301, 384
    classical, 385-387
    finite resolution, 396
    mixed quantum, 388-390
    SQUID magnetometer, 554
Meissner effect, 513
metric, 12
    pseudo-Riemannian, 60
    Riemannian, 60
    space, 13
mixed
    quantum system, 378-381, 383, 388-390, 521, 528, 539
    singular vector, 204
    state, 341-342
    tensor, 18, 47
    vector, 202
    vector and mixed state, 344
mixed tensor, 18, 47
    intrinsic definition, 47
mixture
    classical, 237, 245, 342
    quantum, 246-247
momentum observable, 73-74
    angular, 75-76
    complete, 74, 235
        generator of transformation group, 75
        quantization, 260-275
        axiom, 270
    compressible, 262, 264
    in path space formulation, 654
    incomplete, 74, 272
        quantization, 272-275
    incompressible, 260, 262
    linear in, 269-270
        quantization, 269-272
    linear momentum, 75
    local, 195, 268
    operator, 115, 116, 122
    probability amplitude function, 181
    p_θ, 17, 247
    radial, 17
        operator in $I\!E^3$, 174, 247
        operator in $I\!E^2$, 574
momentum representation space $\widetilde{L}^2(\widetilde{I\!E})$, 109
momentum space $\widetilde{I\!E}$, 64, 151
Naimark's theorem, 168
neighbourhood, 6
non-commutative algebras, 189
non-degenerate, 56
non-equilibrium system, 383, 554-558

non-locality, 562
　globalization, 566
normalized
　positive linear functional, 358
　positive operator-valued set function, 167
　projector-valued function, 143
　projector-valued set function, 142
NPLF, 359
observables
　approximate, 397, 398, 407
　bounded in superconductivity, 517-518
　classical, 346
　concept and description, 400
　equivalent FAPP, 584
　extrinsic, 404-406
　　unsharp, 404-406
　intrinsic, 404-406
　　sharp, 404-405
　　unsharp, 404-406
　mathematical representation, 403
　measured value, 232
　one-particle and two-particle, 625
　pointer, 348, 390
　possessed value, 232, 328
　related family, 408
　unbounded, 248
　　bounded approximation, 249, 312, 313
　　discretized approximation, 249, 312, 313
open
　cube, 7
　　bounded, 9
　rectangle, 7
　set, 4, 6-7
　sphere, 13
one-parameter group
　generator, 75, 113
　transformations, 36
　translations, 75
　unitary operators, 113
operators in Hilbert space
　adjoint, 88
　bounded, 88
　　set $\mathcal{B}^{\oplus}(\mathcal{H}^{\oplus})$, 216
　　set $\mathcal{B}(\mathcal{H})$, 189, 358
　closable, 89, 92
　closed, 89
　closure, 89, 92
　commuting, 102, 151
　continuous, 89
　decomposable, 214, 216
　densely defined, 88

density, 218-221
diagonalizable, 214, 216, 217
domain, 87
essentially selfadjoint, 97
essentially strictly maximal symmetric, 100, 115
global, 185-186, 528
idempotent, 90
invertible, 98
isometric and isometry, 107
local, 186
localization, 187, 191-195
maximal symmetric, 97, 113, 402-406
norm, 88
parity, 437, 491, 592
position, 91, 152, 153, 622
positive, 97, 103
projectors, 97
　dimension of a projector, 103
　Dirac notation, 103, 110
　generated by a vector, 103
　one-dimensional, 103
　orthogonal, 104
　partial order, 104
　probability, 104
　spectral function, 143
　spectral measure, 142
range, 87
selfadjoint, 97
　function, 152
　　exponential, 154
semi-local, 186
statistical, 221-223
　trace, 222, 224
　blended, 223
　regular, 223
　singular, 223-224
strictly maximal symmetric, 100, 115
square, 175
symmetric, 97
time, 409-412, 423-424
trace, 218
　statistical operator, 221-224
translation, 113, 587
unitary, 97, 98, 107, 110
　in Dirac notation, 110
operators on a manifold
　at a point in $I\!\!E^n$, 21
　differential, 22
　linear, 21
　on $\mathcal{C}^\infty(I\!\!E^n)$, 22
　smooth differential, 22

optical isomer, 350, 607
orthodox quantum mechanics, 236
orthogonal
  complement, 85
  projectors, 104
  sequence, 85
  subspaces, 85
  to a subspace, 84
  vectors, 82
outer product, 19
paradox
  de Broglie, 587
  EPR, 624, 625-626
  Schrödinger's cat, 589
  Temple's, 256
parity operator, 437, 491, 592
partial
  Hamiltonians, 444
  Hilbert spaces, 444
  momentum operators, 444
  observables, 444
  quantities, 444
particle source
  ideal, 303, 307
  random, 305, 310
path, 639
  dependent phase, 645
  dependent phase factor, 644
  space, 639
    function on, 644-649
Pauli exclusion principle, 626-628
phase, 638
  area (volume), 71
  factor, 236, 638, 644
  flow, 38, 68
  function, 638
  global, 638
  local, 639
  objects, 465
  retardation, 464, 472
  retarders, 465
  shift, 464
  shifters, 465
  universal path dependent, 646
phase representation, 613
phase space, 64, 67
  state space, 67
  coordinates, 73
  generalized, 419
phenomenological theory, 511
photon, 623
  size, 623

physical assumptions
  (PAS1), (PAS2), 519
  experiment, measurement, 250
physical coordinate space, 73
physical space, 73, 362
physical systems classification, 379-381
  quantum/classical divide, 381, 568
plane wave, 126, 447-449
  transfer matrix, 449
point interactions in $\mathcal{C}$ general, 489-495
point interactions in $\mathcal{C}$ specific
  δ-interaction, 495
  low pass δ'-interaction, 497
  reflector
    elastic, 498
    perfect, 498
  gradient phase shifter, 501
  ideal dynamic $\pi/2$-phase shifter, 502
  ideal dynamic π-phase shifter, 500
  ideal tunable phase shifter, 504
  static junction correlator, 503
  static π-phase shifter, 500
  open end, 499
point interactions in $I\!E$ general, 446-476
  asymmetric quadrapole, 458, 461, 466
  characteristic wave number, 451
  limiting family of δ potentials, 458
point interactions in $I\!E$ specific
  filters
    high pass δ-interaction, 451
      attractive, 451
      repulsive, 454
    low pass δ'-interaction, 455
      attractive, 455
      repulsive, 457
    mid-pass
      ideal partial, 471
      partial, 473
  open end, 464
  phase shifters
    high-pass π-phase shifter, 467
    ideal mid-pass $\frac{1}{2}\pi$-phase shifter, 471
    ideal π-phase shifter, 465
    ideal tunable phase shifter, 476
    low-pass π-phase shifter, 469
  reflector
    perfect reflector, 463
    elastic, 464
point of continuous growth, 157
point of discontinuous growth, 157
point (discrete) spectrum, 157-159, 179
pointer observable, 348, 390

Poisson bracket, 66, 256, 271
position observable
　angular, 480
　approximate, 397
　discretized, 312
　global, 312, 620
　in path space formulation, 654
　local, 312
　probability amplitude function, 181
　operator, 91, 654
position variable
　linear, 480
　not as observable, 518, 522
positive
　deficiency index, 113
　deficiency subspace, 114
　operator-valued measure, 168
　operator-valued set function, 167
possessed value, 232, 328
postulates
　dynamics
　　OQD, 287
　　UQD(N), 365
　　UQD(P), 352
　statics
　　GQS, 403
　　OQS, 238
　　UQS, 339
　　UQS(F), 360
　projection, 300
　superconducting ring, 521
POV function and measure, 168-170
POVM, 168
pre-quantization, 276
preserving evolution, 352
probability
　event, 131, 139
　frequency interpretation, 131
　ignorance interpretation, 246
　selfadjoint operator, 180-183
　strictly maximal symmetric operator, 183, 184, 185
probability amplitude function
　momentum, 181
　position, 181
probability density function, 138, 140
　blended, 140
　regular, 140
　singular, 140
probability (distribution) function, 138
　absolutely continuous, 139
　blended, 140

　measure, 137
　purely discrete, 139, 181
　regular, 139
　singular, 139
probability functions on Hilbert space, 400
　maximal family, 401, 402, 420-423
probability functions, probability measures and operators
　Hamiltonian, 242
　kinetic energy, 241
　momentum, 181, 241
　position, 181, 241
　proposition, 242
　selfadjoint operator, 180
　strictly maximal symmetric operator, 183, 184, 185
projection in direct integral space, 201
projection operator, 97
projection postulate, 300, 328, 624
projector-valued function and measure, 142, 143
propagator, 287
　free, 290
proposition, 238, 242, 243
pseudo-Riemannian manifold, 60
pure
　singular state, 341
　singular vector, 204
　state, 239-240
　vector, 202
　vector and pure state, 344
purely
　continuous spectrum, 157-159, 178-179
　discrete (point) spectrum, 158, 178
PV function and measure, 143, 152-154
PVM, 143
quadrapole (asymmetric), 458, 461
quasi-local algebra, 190-191, 622
　quasi-local observables, 191
quasi-particle, 512
quasi-periodic boundary condition, 120, 284
quantization
　aim and fallacy, 478
　Born-Jordan rule, 259
　canonical conjugation rule, 257
　canonical scheme, 271, 275
　constrained system, 282-286
　coordinate representation, 254
　function preserving rule, 256
　geometric, 77-78, 211, 235, 249, 250
　Hamiltonian, 282
　identity rule, 257

INDEX 685

initial, 255
kinetic energy, 275-282
Lie algebra rule, 257
linearity rule, 256, 270-271
map, 254, 255
non-selfadjointness, 272-274, 409, 545
non-uniqueness, 272-274, 409, 490, 545
pre-quantization, 276
squaring rule, 256, 282
symmetrization rule, 259
Weyl rule, 259
quantization by parts, 443-446
orthodox quantum system, 569-573
quantum circuit
composite quantization, 444, 535-536
correlative quantization, 446, 546
partial quantization, 444, 535-536
quantized time of flight operator, 412, 423, 424-425
quantum beat, 604
quantum circuits, 509
quantum/classical divide, 379-383, 568
quantum measuring device, 384
quantum mechanics on path space, 649-657
position operator, 654
momentum operator, 654
superselection rule, 655-657
Josephson effect, 655-657
quantum potential, 414
quantum wire, 573
radial momentum, 17, 76
operator, 174-175, 274, 574
real line $I\!R$, 3
$I\!R^-$, 5
$I\!R^+$, 5
extended reals $I\!R_{ex}$, 132
reducing subspace and reduction of operator 105-107, 210
direct integral, 212-215
to singular subspace, 215
direct sum, 210, 214
residual spectrum, 178-179
resolution of the identity, 145
generalized, 168
resolvent set, 178
restriction
linear functional, 361
operator, 89
scattering state, 294-295, 307, 586-587
scattering system, 297
de Broglie paradox, 586-587
simple, 297

Schrödinger's cat states, 589-601
classical-like, 589
classically distinguishable, 589
macroscopically distinguishable, 589
paradox, 589
periodic, 599
strong sense, 594
superconducting, 618-619
weak senses, 598
Schrödinger picture, 287, 352
Schrödinger equation, 288, 289
selfadjoint extension, 99
direct sum operators, 212
selfadjoint localization, 194
self-inductance, 483
semi-local operator, 186
separable
asymptotically, 292, 294, 299
Hilbert space, 86
quantum mechanics, 620-622
spatially, 292, 294, 299
set of measure zero, 133-134
sharp observables, 404
singular
state, 341
statistical operator, 221
subspace, 209
vector, 204
SNPLF, 359
Sobolev space $S_{2,\ell}(I\!E)$, 117-118
$S_{2,1}(I\!E)$, 118
$S_{2,2}(I\!E)$, 118
$S_{2,1}(\Lambda)$, 118
$S_{2,2}(\Lambda)$, 118
spectral decomposition
identity, 149
selfadjoint operator, 148-149
continuous, 153
discrete, 158, 159
symmetric operator, 170-171
unitary operator, 156
spectral function and spectral measure, 142
continuous, 147
discrete (piece constant), 146
generalized, 167-168
strictly maximal symmetric operator, 173-175
symmetric operator, 171-172
probability measure and function, 167
kinetic energy operator, 163-166
momentum operator, 153

686   INDEX

of function of selfadjoint operator, 155, 156
orthogonality property, 145
position operator, 153
probability function, 143-144, 151, 168
projector, 152
radial momentum operator, 174-175
selfadjoint operator, 148, 154
  as characteristic function, 156
tensor product, 154
spectral projectors, 150
spectral representation, 162
  kinetic energy, 163-167
  momentum, 162-163
  radial momentum operator, 174-175
  selfadjoint operator, 162
  space
    kinetic energy, 162, 164, 165, 198
    momentum, 163, 198
  theorem, 162
spectral separation, 314, 316-318
  spin-$\frac{1}{2}$ particle, 319-320
  spectral component, 314, 316-318
spectral set, 178
spectral theorem, 148
  selfadjoint operator, 148
  strictly maximal symmetric operator, 173
spectrum $sp(\widehat{A})$
  direct sum operator, 212
  selfadjoint operator, 157
    purely continuous, 157-159
    purely discrete (point), 157-159
      eigenvalue, 157-160
  symmetric operator, 178
    continuous, 178-179
    discrete (point), 178-179
    residual, 178-179
square double-well potential, 602
  finite, 603
  infinite, 602
square-integrable, 84
  δ-function, 92
SQUID, 552-554
  magnetometer, 554
staircase function, 136, 203
states
  asymptotically localizable, 292, 294
  asymptotically separable, 292, 294
  bound, 295
  classical-like, 589, 591
  classically distinguishable, 589, 598, 605, 619
  coherent, 595
  disjoint, 346, 347
  disjoint FAPP, 584-585
  equivalent FAPP, 583
  infinitely separated states, 621
  linear functional, 358-367
  macroscopically distinguishable, 589, 598
  mixed, 239, 243, 341-342
  preparation, 300-307
    asymptotic procedure, 307
    waiting stategy, 307
  pure, 239-243, 341
  regular, 341, 343
  scattering, 295-296
  singular, 341, 343
  spatial, 307
  spin, 307-310
  stationary, 68, 70, 296, 355
stationary state, 68, 70, 296, 340, 355
statistical experiments, 130
statistical operator, 221-223
  blended, 222
  regular, 222
  singular, 222, 223
Stieltjes integrals, 134-137
  Dirac delta function, 136
Stone's theorem, 112
  unitary group, 112
strictly maximal symmetric operator, 97
  spectral theorem, 173
  square, 175-177
subsets of a Hilbert space
  closed, 84
  closure, 84
  linear, 84
  dense, 84
  orthogonal complement, 84
subspaces, 84
  Borel family, 199, 208
  coherent, 346
  invariant, 105
  orthogonal complement, 85
  reducing, 106
  singular, 209
sufficiently large set of states, 395, 396
superconducting
  circuit, 509
    assumptions (PAS1), (PAS2), 519
    dc superconductivity, 514
  ring, 514-533
    postulate, 521
    with capacitive junction, 613-617

INDEX

wire, 533-541
  with capacitive junction, 609-613
  Y-shape circuit, 542-552
superconducting quantum interference device (SQUID), 552-554
supercurrent, 509
  critical, 524, 538, 542
supercurrent operator, 516, 524, 540, 543, 552
superposition, 247, 344-345
  indistinguishable from mixture, 252
superposition principle, 247, 344, 345, 585
  breakdown, 518
  coherent superposition, 247, 344-345
  indistinguishable from mixture, 252
superselection rule, 345-346, 522, 529
  breakdown, 608, 613
  origins, 347-350, 561
  quantum mechanics on path space, 655
  supersector, 346
  superselection operator, 346
superselection rule approach to quantum measurement, 356-358, 384, 388-390
  classical measuring device, 385-388
  mixed quantum measuring device, 388-390
symmetrized product, 101, 123
symmetry breaking, 350
tangent
  space, 26
  to a curve, 27-28
  vector, 26
    coordinate representation, 26
  vector field, 26
    integral curve, 27-28
Temple's paradox, 256
tensor field of second order, 17-18, 47
  covariant
    anti-symmetric, 50-51
    symmetric, 50-51
tensor product, 47, 94
  direct integral, 224-225
  Hilbert spaces, 95, 224
  operators, 96
    direct integral, 225
  subsets, 95
time of flight, 412, 423-424
time operator, 409-412, 423-424
topological equivalence, 10
topological space, 6
  connected and disconnected, 11
topological structure, 6

topology, 6
  standard, 6, 8, 9
  relative, 13
trace, 218
  statistical operator, 222-223
transfer matrices, 448-450
  multiplicative property, 450
transform
  Fourier, 108
  Fourier transform of operators, 111
  Fourier transform operator, 108
  unitary, 107, 110-112
    between any two unit vectors, 110
    orthonormal basis, 110
transformation group, 35
  flow, 38
  vector fields, 35, 36, 37
transition
  amplitude, 248
  probability, 248
translation operator, 113, 587
trivial decomposition, 207
TSCR, 513
uncertainty in probability theory, 141
uncertainty in operator theory, 182-184
  relation, 410-411
    position-momentum, 329
    energy-time, 410-411
unitarily equivalent, 108
unitary transform, 107, 110-112
  orthonormal basis, 110
  spectral function and measure, 145, 147, 151-152
universal path dependent phase, 646
universal phase constant, 649
universality of orthodox quantum mechanics, 390-391
universality of superposition principle, 347
unsharp observables, 315, 404
  extrinsic, 404
  intrinsic, 404
unsharp measurement, 315, 406
unsharpness, 405-406
variance in probability theory, 141
variance in operator theory, 182-183
vector
  contravariant, 16
  cotangent, 41-45
  covariant, 16
  displacement, 17
  momentum, 17

tangent, 26
vector field, 20, 26-27
  compact support, 31
  complete, 28, 32-33
    Lie bracket, 34
    sum, 32-33
    transformation group, 35, 36, 37
  critical point, 26, 32
  incomplete, 28, 31
  integral curve, 28
    maximal, 28
  velocity, 29, 38, 68
vectors in Hilbert space
  mixed, 202, 342
    mixed state, 344
    mixed singular, 204
  pure, 202
    pure state, 344
    pure singular, 204
volume element, 62
von Neumann formula, 128-129
waiting strategy, 307
wave function, 235, 254
wave operators, 296
  complete, 297
weak link, 523
wedge product, 49
Wigner function, 414
winding number, 643
yes-no experiment, 243, 311, 313